AF214146

Bioresource Technology: Concept, Tools and Experiences

Tanveer Bilal Pirzadah
Chandigarh University
Punjab, India

Bisma Malik
Chandigarh University
Punjab, India

Rouf Ahmad Bhat
Government of Jammu and Kashmir
India

Khalid Rehman Hakeem
King Abdulaziz University
Jeddah, Saudi Arabia

WILEY

This edition first published 2022
© 2022 John Wiley & Sons Ltd

All rights reserved. No part of this publication may be reproduced, stored in a retrieval system, or transmitted, in any form or by any means, electronic, mechanical, photocopying, recording or otherwise, except as permitted by law. Advice on how to obtain permission to reuse material from this title is available at http://www.wiley.com/go/permissions.

The right of Tanveer Bilal Pirzadah, Bisma Malik, Rouf Ahmad Bhat and Khalid Rehman Hakeem to be identified as the author(s) of this work has been asserted in accordance with law.

Registered Office(s)
John Wiley & Sons, Inc., 111 River Street, Hoboken, NJ 07030, USA
John Wiley & Sons Ltd, The Atrium, Southern Gate, Chichester, West Sussex, PO19 8SQ, UK

Editorial Office
The Atrium, Southern Gate, Chichester, West Sussex, PO19 8SQ, UK

For details of our global editorial offices, customer services, and more information about Wiley products visit us at www.wiley.com.

Wiley also publishes its books in a variety of electronic formats and by print-on-demand. Some content that appears in standard print versions of this book may not be available in other formats.

Limit of Liability/Disclaimer of Warranty
The contents of this work are intended to further general scientific research, understanding, and discussion only and are not intended and should not be relied upon as recommending or promoting scientific method, diagnosis, or treatment by physicians for any particular patient. In view of ongoing research, equipment modifications, changes in governmental regulations, and the constant flow of information relating to the use of medicines, equipment, and devices, the reader is urged to review and evaluate the information provided in the package insert or instructions for each medicine, equipment, or device for, among other things, any changes in the instructions or indication of usage and for added warnings and precautions. While the publisher and authors have used their best efforts in preparing this work, they make no representations or warranties with respect to the accuracy or completeness of the contents of this work and specifically disclaim all warranties, including without limitation any implied warranties of merchantability or fitness for a particular purpose. No warranty may be created or extended by sales representatives, written sales materials or promotional statements for this work. The fact that an organization, website, or product is referred to in this work as a citation and/or potential source of further information does not mean that the publisher and authors endorse the information or services the organization, website, or product may provide or recommendations it may make. This work is sold with the understanding that the publisher is not engaged in rendering professional services. The advice and strategies contained herein may not be suitable for your situation. You should consult with a specialist where appropriate. Further, readers should be aware that websites listed in this work may have changed or disappeared between when this work was written and when it is read. Neither the publisher nor authors shall be liable for any loss of profit or any other commercial damages, including but not limited to special, incidental, consequential, or other damages.

Library of Congress Cataloging-in-Publication Data
Names: Pirzadah, Tanveer Bilal, author. | Malik, Bisma, author. | Bhat, Rouf Ahmad, 1981- author. | Hakeem, Khalid Rehman, author.
Title: Bioresource technology : concept, tools and experiences / Tanveer Bilal Pirzadah, Chandigarh University, Punjab, India, Bisma Malik, Chandigarh University, Punjab, India, Rouf Ahmad Bhat, Sri Pratap College Campus, Cluster University Srinagar, India, Khalid Rehman Hakeem, King Abdulaziz University, Jeddah, Saudi Arabia.
Description: First edition. | Hoboken, NJ : John Wiley & Sons, 2022. | Includes bibliographical references and index.
Identifiers: LCCN 2021054587 (print) | LCCN 2021054588 (ebook) | ISBN 9781119789383 (hardback) | ISBN 9781119789420 (pdf) | ISBN 9781119789437 (epub) | ISBN 9781119789444 (ebook)
Subjects: LCSH: Biochemical engineering. | Biological products. | Agricultural resources.
Classification: LCC TP248.3 .P575 2022 (print) | LCC TP248.3 (ebook) | DDC 660.6/3--dc23/eng/20211230
LC record available at https://lccn.loc.gov/2021054587
LC ebook record available at https://lccn.loc.gov/2021054588

Cover image: © Ralf Geithe/Shutterstock
Cover design by Wiley

Set in 9.5/12.5pt STIXTwoText by Integra Software Services Pvt. Ltd, Pondicherry, India
Printed and bound by CPI Group (UK) Ltd, Croydon, CR0 4YY

C9781119789383_030322

Contents

About the Editors

Dr Tanveer Bilal Pirzadah is Assistant Professor at University Centre for Research and Development (UCRD), Chandigarh University, Punjab, India. After completing his doctorate at the University of Kashmir, Srinagar, India, he worked as an Assistant Professor at the Cluster University Srinagar, India. Dr Tanveer has about eight years of research and teaching experience in bioresources management, biofuels, plant stress physiology and biotechnology. He is the recipient of the Young Scientist Award (RA-VD00921). He is also a fellow of the Plantae group of the American Society of Plant Biologists and a member of the World Academy of Sciences. Currently, Dr Tanveer is engaged in studying the agronomical, physiological and biochemical and nutritional aspects in underutilized crops including buckwheat. He has to his credit 32 research articles published in peer-reviewed international journals, 30 book chapters in edited volumes and four books with international publishers.

Dr Bisma Malik, PhD is Assistant Professor at University Centre for Research and Development (UCRD); Chandigarh University, Punjab, India. She is the recipient of Women Scientist Fellowship (WOS-A) from Department of Science and Technology, Government of India. She is actively engaged in research and has 9 years of research experience in plant stress physiology, molecular biology, and environmental studies. She has worked on stress biology and genetic diversity of chicory – an untapped bioresource from Himalayas in Kashmir. She has to her credit 30 publications in international journals and 30 book chapters of international repute. At present, Dr Bisma is engaged in studying plant endophytes as a promising source of lead compounds that have immense potential in the pharmaceutical sector.

Dr Rouf Ahmad Bhat (PhD) has pursued his doctorate at Sher-e-Kashmir University of Agricultural Sciences and Technology Kashmir (Division of Environmental Science) and presently working in the Department of School Education, Government of Jammu and Kashmir. Dr Bhat has been teaching graduate and postgraduate students of Environmental Sciences for the past three years. He is an author of more than 50 research articles (h-index 14; i-index 20; total citation > 650) and 40 book chapters, and has published more than 28 books with international publishers (Springer, Elsevier, CRC Press Taylor and Francis, Apple Academic Press, John Wiley, IGI Global). He has his specialization in Limnology, Toxicology, Phytochemistry and Phytoremediation. Dr Bhat has presented and participated in numerous State, National and International conferences, seminars, workshops and symposium. Also he has worked as an associate Environmental Expert in World Bank-funded Flood Recovery Project and also the Environmental Support Staff in the Asian Development Bank (ADB) funded development projects. He has received many awards, appreciations and recognition for his services to the science of water testing, air and noise analysis. He has served as editorial board member and reviewer of reputed international journals. Dr Bhat is still writing and experimenting with diverse capacities of plants for use in aquatic pollution remediation.

Dr Khalid Rehman Hakeem is a Professor at King Abdulaziz University, Jeddah, Saudi Arabia. He has completed his PhD (Botany) from Jamia Hamdard, New Delhi, India in 2011. Dr Hakeem has worked as Post Doctorate Fellow in 2012 and Fellow Researcher (Associate Prof) from 2013 to 2016 at Universiti Putra Malaysia, Selangor, Malaysia. His specialty is Plant Eco-Physiology, Molecular biology, Plant-Microbe-soil interactions, Medicinal plant research and Environmental Sciences and so far has edited and authored more than 25 books with Springer International, Academic Press, etc. He has also to his credit more than 110 research publications in peer reviewed international journals, including 40 book chapters in edited volumes with international publishers. Dr Hakeem is the recipient of many national and international awards and fellowships.

About the Book

The book provides up-to-date information on three significant aspects viz., food, energy and environment that are important and challenging issues that need to be addressed. The book consists of three sections about the role of bioresource technology in food, energy and the environment. The first section covers the recent advances in technology in the functional food industry to overcome future food challenges. The second section provides the latest research and development in the green energy sector to cater to the people's energy demand, reduce greenhouse gases (GHG), and the reliance on fossil fuels. The book's third section provides an in-depth understanding of the recent advances in bioresource technologies and management for a sustainable environment.

Foreword

Bioresource technology incorporates techniques using living organisms, parts of organisms, enzymes, proteins, and few others, which are either naturally occurring or are derived from such living systems. These techniques can be used to create or modify products, improve plant or animal productivity or develop microorganisms for special use. Emerging bioresource technology uses recombinant DNA, cell fusion and embryo manipulation among others. This technology has the potential to transform the living conditions of people through its impact on agriculture, animal husbandry, health, environmental protection, material transformation and other areas. From our increasingly deeper understanding of the intricate biochemical interactions at cellular and molecular levels, new paradigms in healthcare have emerged. We have moved from "preventive" (vaccines) and "curative" (antibiotics) medicines to "predictive and corrective" ones. The credit goes to the unraveling of the mystery of the human genome.

In the agricultural sector, new bioresource technology promises to find a solution to the problem of producing more from less, for example, more food from less arable land, use of less water per capita, and implementing less polluting energy sources. This could be attained by creating high-yield, disease- and drought-resistant crops. Due to the advent of the state-of-the art technologies in the field of bioresource technology, much progress has been achieved during the last decade. High throughput technologies are being extensively used to address various issues pertaining to sustainability. Recent advances in these technologies have revolutionized the understanding of plant responses to biotic/abiotic stress, and strategic application of this revolutionary technology will eventually lead toward attaining better environmental sustainability. Indeed, it will lead us toward a truly bio-future during the twentieth-first century.

The book entitled "Bioresource Technology: Concept, Tools and Experiences," edited by a group of four promising young editorial board, is aimed toward this direction. The chapters in this book have been contributed by eminent researchers in the broader area of sustainable bioresource management. It comprises information on the current status of the application of bioresource technology in the food, energy and environmental sectors, together with several other important topics related to nanobiotechnology and its sustainable utilization.

Looking at the topics covered this edited volume carries great weight in this field. I think it will have wider acceptance and open up avenues for scientists to engage in

consistent dialogue with the policy-makers at different levels – state, local, national and global. It can also lead to a new alliance between scientific, governmental and commercial forces and find effective ways for lobbying, advocacy and innovative actions in sustainable bioresource management.

The primary audience who will benefit from this book will be the research and teaching professionals, practitioners, policy-makers, farmers and farmer representatives like NGO professionals, who work in the area of sustainable agricultural and rural development. This kind of publication will be highly useful for the planners and practitioners in the area of bioresource for climate adaptation as well. The contributors in the fields of environment, food and industrial sectors are highly promising.

In the context of overall importance of bioresource technology in the current era, the editors have attempted to provide the state of knowledge on technological innovations and their potential for commercial exploitation and creating value from bioresources.

Prof Dr Münir Öztürk
Vice President of the Islamic World Academy of Sciences

List of Contributors

Shabbar Abbas
Department of Biosciences, COMSATS University Islamabad (CUI), Islamabad, Pakistan.

Oyebanki Joseph Adewumi
Department of Mechanical Engineering, Bells University of Technology, Ota, Nigeria.

Javid Ahmed
Proteomics and Bioinformatics Lab, Department of Biotechnology, Jamia Millia Islamia, Delhi, India.

Jaffer Ali
Department of Biosciences, COMSATS University Islamabad (CUI), Islamabad, Pakistan.

Niaz Ali
Department of Botany, Hazara University, Manshera, Khyber Pukhtunkhwa, Pakistan.

Zeeshan Ali
Nationl Agricultural Research Centre, Islamabad, Pakistan.

Pradipta Banerjee
Department of Biochemistry and Plant Physiology, Centurion University of Technology and Management, Paralakhemundi, Odisha, India.

Stephen Rathinaraj Benjamin
Department of Biomedical Engineering, Federal University of Pernambuco, Recif-PE, Brazil.

Ameer Bibi
Departmentof Plant Breeding and Genetics, University of Agriculture Faisalabad, Faisalabad, Pakistan.

Derya Boyaci
School of Engineering, Lincoln University, Lincoln, UK.

Candy Carranza-Álvarez
Faculty of Professional Studies Huasteca Zone, Autonomous University of San Luis Potosí, Mexico.

Luis Jesús Castillo-Pérez
Multidisciplinary Graduate Program in Environmental Sciences, Autonomous University of San Luis Potosí, San Luis Potosí, Mexico.

Sumit S. Chourasiya
IOL Chemicals and Pharmaceuticals Ltd, Barnala, Punjab, India.

Helon Guimarães Cordeiro
Department of Biochemistry and Tissue Biology, University of Campinas, Campinas, Brazil.

Antony De Paula Barbosa
Department of Pharmaceutical Products, Fsaculty of Pharmacy, Federal University of Minas Gerais, Antônio Carlos, Belo Horizonte, Minas Gerais-MG, Brazil.

Ian Dundas
School of Agriculture, Food and Wine, Waite Campus, The University of Adelaide, Australia.

Israel Sunday Dunmade
Earth and Environmental Sciences Department at the Mount Royal University, Calgary, Canada.

Zahra Edraki
Sharif University of Technology, Tehran, Iran.

Sreenivasan Ettammal
Research and Development, The Western Plywoods Ltd, Baliapatam, Kannur, Kerala, India.

Danilo Roberto Carvalho Ferreira
Department of Materials, Federal University of Minas Geraus, Belo Horizonte, Brazil.

Gülden Göksen
Department of Food Engineering, Vocational School of Technical Sciences,Tarsus University, Akdeniz, Mersin, Turkey.

Shabana Gulzar
Plant Reproductive Biology, Genetic Diversity and Phytochemistry Research Laboratory, Department of Botany, University of Kashmir, Srinagar, Jammu and Kashmir, India,

Afrozah Hassan
Plant Reproductive Biology, Genetic Diversity and Phytochemistry Research Laboratory, Department of Botany, University of Kashmir, Srinagar, Jammu and Kashmir, India.

Setareh Heidari
Faculty of Chemical Engineering, Sahand University of Technlogy, Tabriz, Iran.

Ahmad Fauzi Ismail
Advanced Membrane Techology Research Center, Universiti Teknologi Malaysia, Skudai, Johor Bahru, Johor, Malaysia.

Mohammad Jamil
University of Sargodha, Sargodha, Pakistan.

Hanan Javid
Plant Reproductive Biology, Genetic Diversity and Phytochemistry Research Laboratory, Department of Botany, University of Kashmir, Srinagar, Jammu and Kashmir, India.

Jun Kang
School of Life Sciences, Tianjin University, Tianjin, China.

Deepika Kathuria
University Center for Researchg and Development, Chandigarh University, Gharuan, Punjab, India.

Rumana Keyani
COMSATS, Islamabad, Pakistan.

Fatima Khalid
Quaid-e-Azam University, Islamabad, Pakistan.

Hanif Khan
Indian Institute of Wheat and Barley Research, ICAR, Karnal, Haryana, India.

Oluwaseun Kilanko
Department of Mechanical Engineering, Covenant University, Ota, Nigeria.

Philip Larkin
CSIRO, Agriculture and Food, CSIRO, Canberra, Australia.

Moline Severino Lemos
Department of Cellular Biology, Institute of Biological Science, Federal University of Minas Gerais, Belo Horizonte, Brazil.

Sagar Maitra
Department of Agronomy and Agroforestry, Centurion University of Technology and Management, Paralakhemundi, Odisha, India.

Mukhtar Ahmad Malik
Department of Environmental Science, University of Kashmir, Srinagar, J&K, India.

Juan José Maldonado Miranda
Faculty of Professional Studies Huasteca Zone, Autonomous University of San Luis Potosí, Mexico.

Alexey Morghonov
Omsk State Agrarian University, Omsk, Russia.

Abdul Mujeeb-Kazi
College Station, Texas A&M, Texas, USA.

Ghulam Mustafa
Department of Economics and Business Administration, Division of Arts and Social Sciences, University of Education, Lahore, Pakistan.

Irshad Ahmad Nawchoo
Plant Reproductive Biology, Genetic Diversity and Phytochemistry Research Laboratory, Department of Botany, University of Kashmir, Srinagar, Jammu and Kashmir, India.

Abdul Aziz Napar
Sindh University, Jamshoro, Sindh, Pakistan.

Francis Ogbonnaya
GRDC, Australia.

Oyedepo Sunday Olayinka
Department of Mechanical Engineering, Covenant University, Ota, Nigeria.

Sandipan Pine
Department of CSE, Centurion University of Technology and Management, Paralakhemundi, Odisha, India.

Tanveer Bilal Pirzadah
University Center for Research and Development, Chandigarh University, Punjab, India.

Biswajit Pramanick
Department of Agronomy, Dr Rajendra Prasad Central Agricultural University, Pusa, Samatipur, Bihar, India.

Zadia Qamar
Proteomics and Bioinformatics Lab, Department of Biotechnology, Jamia Millia Islamia, Delhi, India.

M. Urfan Qureshi
Proteomics and Bioinformatics Lab, Department of Biotechnology, Jamia Millia Islamia, Delhi, India.

Birendra K. Rajan
Environmental Science & Engineering Department, IIT, Bombay, India.

Sanjaya Rajaram
Resource Seeds International, Mexico.

Eli José Miranda Ribeiro Jnr
Faculty of Inhumas (FacMais), Avenida Monte Alegre, Inhumas, Goiás-GO, Brazil.

Erfan Sadatshojaei,
Department of Chemical Engineering, Shiraz University, Shiraz, Spain.

Mohammad Sohail Saddiq
Department of Agronomy, Ghazi University, Dera Ghazi Khan, Pakistan.

Nasir Saeed
National Institute of Biotechnology and Genetic Engineering, Faisalabed, Pakistan.

Sheikh Tanveerr Salam
Department of Zoology, Government Degree College Sopore, Baramulla, J&K, India.

Tanmoy Shankar
Department of Agronomy, Dr Rajendra Prasad Central Agricultural University, Pusa, Samatipur, Bihar, India.

Amna Syeda
Proteomics and Bioinformatics Lab, Department of Biotechnology, Jamia Millia Islamia, Delhi, India.

Ayesha Tahir
Department of Biosciences, COMSATS University Islamabad (CUI), Islamabad, Pakistan.

Mahjabeen Tariq
Arid University, Rawalpindi, Pakistan.

Venilavan Thirumavalavan
Department of Agriculture Engineering, Saveetha Nagar, Sriperumbudur Taluk, Kanchipuram, Chennai, Tamilnadu, India.

Fernanda Maria Policarpo Tonelli
Department of Cellular Biology, Institute of Biological Science, Federal University of Minas Gerais, Belo Horizonte, Brazil.

Flávia Cristina Policarpo Tonelli
Department of Biochemistry, Federal University of São João del Rei, Divinópolis, Brazil.

Nick Tucker
School of Engineering, Lincoln University, Lincoln, UK.

Richard R.C. Wang
IUSDA/ARS, Logan, Utah, USA.

Shabir Wani
Sher-e-Kashmir University of Agriculgtural Sciences and Technology of Kashmir, Jammu and Kashmir, India.

David A. Wood
DWA Energy Limited, Bassingham, Lincoln, UK.

Preface

Bioresources are important components for progress and economic activities of a nation. It is considered as the largest source of potential wealth, which remains grossly underexplored. One of the reasons for the under-utilization is the dearth of trained manpower. The current generation of biologists is largely divided into field-oriented taxonomists and ecologists on the one hand and the lab-oriented functional and molecular biologists on the other. This divide has become a limiting factor in the study of bioresources. The present book intends to bridge the gap by inculcating excellence in field and laboratory biology simultaneously. This capacity building exercise will help to generate wealth through a prudent and sustainable use of a country's bioresources. Bioresources management and utilization for human welfare is very important for the optimum utilization of bioresources. Awareness of the importance and implications of bioresources, among common people as well as elite educated citizens, for safeguarding and protecting the optimum and balanced way of using those bioresources needs critical studies to focus their wealth for the benefit of not only the present generation but for future generations for their better, healthy and peaceful living on the earth. The problem faced at present is the over exploitation of bioresources which not only has a negative impact on the environment but also sometimes totally destroys and erodes the important bioresources which are available at local, regional and national levels. Therefore, handling bioresources in a proper manner and in an appropriate way is important for optimum use without over exploitation of our bioresources wealth.

Sustainable management of the ecosystems and the rich life within them remains one of the key natural resource management challenges. The conservation and sustainable use of bioresource are of critical importance for meeting the needs of food, fodder, fiber, health, water and other needs of the growing world population, for which purpose of, access to, and sharing of both genetic resources and technologies are essential. Here we are trying to document together the various aspects of bioresources with a view to make it available for the judicious utilization by mankind. Gradual emergence of new technologies for large-scale conversions of renewable raw materials of biological origin to various industrial and energy markets has further widened the scope of bioresource technology.

Eighteen chapters are covered in this book, describing the application of bioresource technology in three sections – food, energy and environmental sectors. The first section, pertaining to the application of bioresource technology in the functional food sector comprises of seven chapters, the second section, pertaining to bioresource and future energy security comprises three chapters and the third section, pertaining to bioresource technology: solution to sustainable environment and management policies comprises eight chapters.

The editors and contributing authors hope that one result of publishing this book will be to provide a wide range of useful experimental data derived from global applications of bioresource technology. Hopefully, this book can also provide new insights about the emerging applications and sustainable management of the bioresource technology.

This book is unique and would be an ideal source of scientific information to post-graduate students, research scholars, faculty and scientists involved in agriculture, plant sciences, environmental sciences, bioenergy and biofuels, molecular biology, microbiology, biochemistry, biotechnology and food technology.

We are highly grateful to all our contributors for readily accepting our invitation for not only sharing their knowledge and research, but for venerably integrating their expertise in dispersed information from diverse fields in composing the chapters and enduring editorial suggestions to finally produce this venture.

We extend our appreciation to John Wiley & Sons for their exceptional kind support, which made our efforts successful.

Authors:
Dr. Tanveer Bilal Pirzadah
Dr. Bisma Malik
Dr. Rouf Ahmad Bhat
Dr. Khalid Rehman Hakeem

Part I

The Application of Bioresource Technology in the Functional Food Sector

1

Millets: Robust Entrants to Functional Food Sector

Sagar Maitra[1], Sandipan Pine[2], Pradipta Banerjee[3], Biswajit Pramanick[4] and Tanmoy Shankar[5]

[1] *Professor, Department of Agronomy and Agroforestry, Centurion University of Technology and Management, Paralakhemundi–761211, Odisha, India*
[2] *Associate Professor, Department of CSE, Centurion University of Technology and Management, Paralakhemundi–761211, Odisha, India*
[3] *Assistant Professor, Department of Biochemistry and Plant Physiology, Centurion University of Technology and Management, Paralakhemundi–761211, Odisha, India*
[4] *Assistant Professor, Department of Agronomy, Dr. Rajendra Prasad Central Agricultural University, Pusa–848125, Samastipur, Bihar, India*
[5] *Associate Professor, Department of Agronomy and Agroforestry, Centurion University of Technology and Management, Paralakhemundi–761211, Odisha, India*

CHAPTER MENU

Bioresource Technology: Concept, Tools and Experiences, First Edition. Tanveer Bilal Pirzadah, Bisma Malik, Rouf Ahmad Bhat, and Khalid Rehman Hakeem.
© 2022 John Wiley & Sons Ltd. Published 2022 by John Wiley & Sons Ltd.

1.1 Introduction

The word "millet" derives from the "mil" or "thousand," which refers to the huge number of grains that can be produced from a single seed. Millets are coarse cereals with a heritage of consumption since early human civilizations from the Neolithic age. Presently, these crops are cultivated in arid and semi-arid regions of Asia and Africa as staple food and animal feed (Brahmachari et al. 2018; Maitra 2020). Millets are considered as miracle crops, because cultivation of millet has multiple benefits (Figure 1.1). The grains are used as food and the stover is used as fodder. Millets are diversified cereals; hence their cultivation enriches biodiversity in the agro-ecosystem. In erosion prone steep and arid regions, short statured millets provide grassy cover, thus restricting soil and nutrient loss. Soil organic carbon depletion is a major problem in drylands and cultivation of millet with less water assures carbon sequestration (Srinivasarao et al. 2014). Millets are less nutrient demanding, are cultivated with low chemical input and emit less greenhouse gas emissions; thus, millets reduce the carbon footprint in agriculture. Under extreme weather conditions, including higher temperatures, C_4 millet plants assure CO_2 abatement opportunities and, in the future, with elevated CO_2 levels these crop will perform better than other major cereals. Food and nutritional security is a major concern,

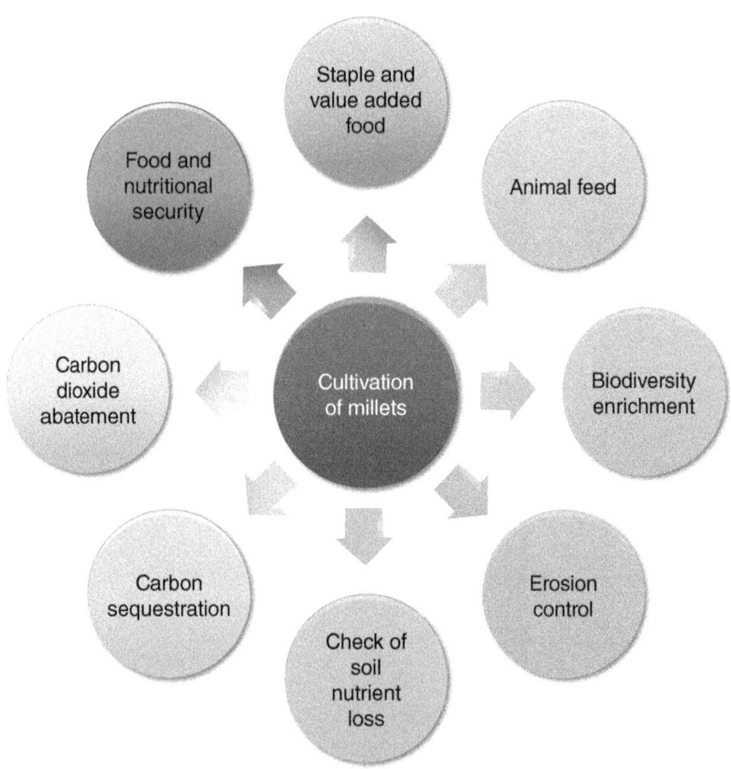

Figure 1.1 Multifaceted benefits of millet cultivation.

even in today's world, and nutrient-rich millets can fulfill this requirement. Moreover, millets are non-gluten foods richer in dietary fiber and essential macro- and micro-nutrients compared to other cereals. The world population will be 9.7 billion by 2050 and 11.2 billion by 2100 and agriculture will play a crucial role in feeding these huge populations with deteriorating and shrinking natural resources. Industrialization and urbanization have already changed the food consumption pattern and in future the transformation to value-added and energy-rich food will further intensify, leading to tremendous challenges to agriculture. The world cereal equivalent food demand is projected to be around 14,886 million tons in 2050 (Islam and Karim 2019). Food and nutritional security will add another dimension to the global food supply system. Considering all the above issues, including climate change, there is an urgent need to shift from ongoing crop production to ecologically sound and nutrient-rich crops. In this regard, millets can fulfill future requirements, because they have the desired qualities to cope with future climatic conditions. But, obviously, there is still the scope for adoption of smart technologies of precision agriculture (PA) to create a hunger-free world with sustainable agricultural production.

1.2 Nomenclature and Use

"Millet" is a generic term comprised of different coarse cereals and forage crops (Weber 1998). The millets can be placed as per prehistoric classification into mainly nine genera, namely, *Brachiaria, Digitaria, Echinochloa, Eleusine, Panicum, Paspalum, Pennisetum, Setaria* and *Sorghum*. Additionally, *Coix* and *Amaranthus*, and *Eragrotis* and *Fagopyrum* genera are also included as millets or coarse cereals. Among millets, the most prominent belong to the genera *Sorghum* and *Pennisetum*, which account for major world production, but the contribution of millets to the world's food basket is still negligible and these are essential food grains for mainly warm environments of the rainfed arid and semi-arid regions (Maitra 2020; Saxena et al. 2018). The classification of presently cultivated millets is essential to realize their uses and importance in agriculture in harsh agro-ecological regions (Table 1.1). Though the acreage of millets is less, these are a vital source of energy and staple food for sizeable populations dwelling in the dry regions of Asia and Africa. Out of total millet production in the world, about 50% of sorghum and 80% of millet are used as human food (Venkatesh Bhat et al. 2018). Grains of millets are common ingredients of animal feed and these are grown as green forage. The stover, or straw, of millets is used in animal feed, brewing, manufacturing of alcohol and for other industrial purposes. However, the use of millets varies in different countries; for example, in West African countries, a fermented thick porridge is made from pearl millet and sorghum. Finger millet and pearl millet are used for production of beer in Eastern Africa. However, these are also ingredients in Eastern European fermented drinks. Also breads are a very common food item prepared from millet in different countries like China, India, Mexico, Scotland, the United States and Ethiopia. In China, use of foxtail millet as an ingredient in soup is often observed. In cold regions of Japan, Japanese barnyard millet is used as a food. Flour from different millets is blended and used in the preparation of various value-added products such as pudding, breads, biscuits, sweets, cakes, chips,

Table 1.1 Nomenclature of millets.

S. No.	Common name	Scientific name
Major millets under cultivation		
1	Sorghum/Giant millet	*Sorghum bicolor* L.
2	Pearl millet	*Pennisetum glaucum* L.
Small millets under cultivation		
3	Finger millet	*Eleusine coracana* L. Gaertn.
4	Barnyard millet	*Echinochloa frumentacea* L.
5	Proso millet	*Panicum miliaceum* L.
6	Foxtail millet	*Setaria italic* L.
7	Kodo millet	*Paspalum scrobiculatum* L.
8	Little millet	*Panicum sumatrense* L.
Lesser known millets		
9	Brown top millet	*Brachiaria ramosa* L.
10	Crap grass	*Digitaria cruciata*
11	Tef	*Eragrotis tef*
12	White fonio	*Digitaria exilis*
13	Black fonio	*Digitaria iburua*
Extinct millet		
14	Job's tear millet	*Coix lacryma* (L.)
Pseudo millet		
15	Purple amaranththus	*Amaranthus cruentus*
16	Buckwheat	*Fagopyrum esculentum*; *F. tataricum*

Source: GOI 2020

rolls and noodles. Moreover, due to their gluten-free nature and non-allergenic properties, these millet products have gained a rapid momentum, especially in the developing countries. Millets are mainly grown in the rainy season (June–July sown), but during recent times improved varieties and hybrids have been developed which can be cultivated in winter (October–November sown) and even in summer (February sown) under irrigated conditions. These hardy crops require less water and can more easily combat extreme weather conditions, including higher temperatures. As C_4 plants, millets can utilize more CO_2 and thus assure environmental benefits (Brahmachari et at. 2018). The agriculture in India faces many constraints, like the threat and variation of the monsoon season, but millets are ecologically sound crops which can withstand different weather aberrations with greater storability (Maitra 2020). However, millets have not received much attention from researchers and policy-makers compared to other major cereals like rice, wheat, maize and barley.

1.3 Description of Important Millets

Millets exhibit diversity in grain size, structure and threshing characteristics. The edible grains of sorghum, pearl millet and teff are the naked kernels, but in finger, foxtail and proso millet, the loose husk is removed easily to reveal the edible portion. Millets are diversified crops with heterogeneous botanical characteristics. Taxonomic description of some important millets is included in Section 1.3.1 below.

1.3.1 Sorghum

Sorghum (*Sorghum bicolor* L. Moench) is a member of the subfamily *Andropogoneae*, originated in Africa, diploid in nature with 10 chromosomes (2n = 20), comprised of several subspecies and the cultivated one is *bicolor* (Oblana 2004). Sorghum is cultivated widely throughout the world, mainly for four different purposes, namely, grain crop, forage, sweet sorghum and bioenergy crop (Mullet et al. 2014). Basically, sorghum is self-pollinating with the possibility of cross-pollination to some extent (Rooney and Wayne-Smith 2000; Taylor 2019). The plant height varies from 0.5–4.0 m and has a fibrous root system and a solid stem with nodes and internodes. The leaves have parallel veins with a midrib and are 30–135 cm long. The inflorescence is known as the panicle or head, 7.5–50 cm in length and 4–20 cm in width. The grains are rounded, pointed at the base with a slight depression near the end, and color is white, pink, yellow or brownish yellow. The crop is hardy in nature, tolerant to drought, mainly cultivated as a rainfed crop during the monsoon season, requiring 25–30 cm water. Presently, a hybrid sorghum is widely cultivated and is of complex genetic origin with a high production potential. Sorghum is cultivated over an area of 42.14 million ha with a total production of 59.34 million tonnes. The United States ranks top as a sorghum producing country.

1.3.2 Pearl Millet

Pearl millet (*Pennisetum glaucum*) belongs to the family *Panicoideae* and earlier was called *Pennisetum typhoides*. Pearl millet originated in Africa and is a diploid plant (2n = 14), a self-pollinating crop, but outcrosses may occur and cause heterozygosity (Poncet et al. 2000). There are two types of pearl millet, forage and grain types (Blümmel et al. 2003). Pearl millet is a tillering annual plant with a height of 1–3 m and above, having a fibrous root system, solid stem with nodes and internodes, and long and fairly broad leaves held erect by thick midribs. The inflorescence comprises of an almost cylindrical spike packed with spikelets containing 800 to 3000 grains in each spike. The seed is caryopsis, oval in shape, with one of its ends tapering, 3–4 mm long and 2.0–2.5 mm wide. The crop is mainly grown under rainfed conditions and requires 25–30 cm water to produce a satisfactory yield. The cultivated area under pearl millet is about 31 million ha, including Africa and Asia. India is the largest producer of pearl millet (8.3 million tonnes from 9.3 million ha).

1.3.3 Finger Millet

Finger millet (*Eleusine coracana* L. Gaertn) originated in the East African highlands and entered into India during the Bronze Age (Fuller et al. 2011). The genus *Eleusine* has about 10 species including annuals and perennials. The widely cultivated finger millet is self-pollinated and tetraploid (2n = 36), an annual cereal of the *Poaceae* family, mostly grown as food in dry areas of different African countries and India. The plant is erect, growing to a height of 60 to 130 cm, with a fibrous root system and profuse tillering habit and compressed stem. The leaves are linear with a distinct midrib, ligule with a fringe of hairs. The effective tillers hold earheads comprised of spikes, named as fingers (two to eight in number) in which spikelets are arranged. The seeds are small, whitish, pale yellow or reddish brown in color. Finger millet is an ecologically sound and drought tolerant crop. It has been reported that finger millet requires less water for its cultivation and thus can withstand water stress (Harika et al. 2019).

1.3.4 Foxtail Millet

Foxtail millet (*Setaria italica* L.) was domesticated in central China during ancient periods as food grain and forage (Miller et al. 2016). It is an annual grass of the *Poaceae* family, member of the subfamily *Panicoideae*, and a diploid plant (2n = 18) grown in dry regions. The stem is slender and erect, which can grow up to a height of 80 to 150 cm. The leaves are narrow, flat and about 30 to 45 cm long. The inflorescence is cylindrical with bristling panicles often arching toward the tip. Generally each spikelet contains 1 to 4 bristles looking like the tail of a fox. The seeds are small, convex and enclosed in glume, creamy white, orange red and even purple in color. Foxtail millet is mainly cultivated in dry regions as a rainfed crop, as it shows great tolerance to drought conditions.

1.3.5 Proso Millet

Proso millet (*Panicum miliaceum* L.) is an ancient cereal first domesticated in Northern China and presently cultivated mainly in India, China, Afghanistan, Turkey and Romania, in soils with marginal and poor fertility (Hunt et al. 2008). It is an annual grass crop of the *Poaceae* family, with 2n = 36, mainly self-pollinating, but cross pollination may also take place for about 10% or more. The plant is an erect herb with profuse tillers and a plant height of 45 to 100 cm. The roots are fibrous and shallow and the stems are slender. The leaves are linear and the inflorescence is branched with spikelets at the tip of branches. The glume and palea are firmly attached to the grain and grains are white, yellow, reddish or black in colour. The crop is grown in warm regions of the world, is highly drought resistant and can easily be grown in areas with scanty rainfall.

1.3.6 Barnyard Millet

Japan is considered as the centre of origin of barnyard millet (*Echinochloa frumentacea* L.) and evidence shows that it was domesticated in China around 10,000 years ago (Sood et al. 2015). Presently, the cultivation of barnyard millet is concentrated on the Indian subcontinent and China. Barnyard millet is an annual grass of the *Panicoideae* subfamily of the family

Poaceae (Clayton et al. 2006), hexaploid (2n = 54) (Yabuno 1962; 1966), and can be cultivated up to an altitude of 2000 m above msl and possesses tolerance to water shortage.

1.3.7 Little Millet

It is believed that the origin of little millet (*Panicum sumatrense* Roth.) is peninsular India and archaeological evidence favors an Indian origin (Venkatesh Bhat et al. 2018; De Wet et al. 1983; Fuller 2011; Maitra and Shankar 2019). At present, it is cultivated mainly on the Indian subcontinent, China, Malaysia and the Philippines. Little millet is a crop of the family *Poaceae*, subfamily *Panicoideae*, tribe *Paniceae* with 2n = 36. It has two subspecies, subsp. *sumatrense* (cultivated little millet) and subsp. *psilopodium* (wild progenitor). Little millet is cultivated in the tropics and subtropics, even at an altitude of 2000 m above msl. It is a short duration crop and can withstand both drought and water stagnation for some periods of time. Like other millets, it is an annual grass that grows straight. The leaves are linear with straight and folded blades, 30 cm to 1 m long. The panicles are awned and grains are round and smooth, and brown in colour.

1.3.8 Kodo Millet

Kodo millet (*Paspalum scrobiculatum* L.) is a tetraploid species (2n = 40) and was domesticated in India around 1000 BC, belonging to the subfamily *Panicoideae*, tribe *Paniceae* (Jarret et al. 1995). Kodo millet has three races, namely, *regularis, irregularis* and *variabilis*, of which the race *regularis* is the most common. The plant is an erect herb, growing up to a height of 45–90 cm or more. The leaves are narrow, stiff and thick with hairy ligules and possess purple pigmentation. The inflorescence is raceme and spikelets are one-flowered. The crop is self-pollinated and the caryopsis is enclosed by lemma and palea and brown in color. Kodo millet has the capability to tolerate stress and is widely cultivated in drylands.

1.3.9 Brown-Top Millet

Brown-top millet (*Brachiaria ramosa* L.) is a lesser known warm-season annual or perennial grass of the *Poaceae* family originating in India (Maitra 2020; Kingwell-Banham and Fuller 2014), and cultivated mainly in the southern states of India. Besides India, brown-top millet is also grown in Africa, western Asia, Arabia, Australia and China. The plants are erect with a prostrate culm 90 cm long, with minutely hairy nodes, leaf blades of 2 to 25 cm length and 4 to 14 mm width (Clatton et al. 2006), indeterminate and stalked white flowers with open and spreading inflorescence of 1–8 cm long, and with colored ellipsoid seeds. The crop matures at 60 to 75 days after sowing (Maitra 2020).

1.4 Millets: The Ancient Crops

Millets are among the oldest foods known to humans, but due to more preference being given to rice and wheat during the second half of the previous century, they lost the respect that they deserved, considering their utility, nutritional value, compatibility in cropping systems and environmental benefits. The majority of the millets are renowned in the Old World in different countries of Asia and Africa (Weber and Fuller 2007; Jaison 2018). In China, during the Neolithic age, millets were domesticated and cultivated, as is evidenced from samples recovered from the Cishan site showing their existence at around 10,000 years ago (Lu et al. 2009). There are arguments on origin of millet in China and it is also believed that millets originated in tropical western Africa during ancient times (FAO 1995). Archaeo-botanical research from Neolithic south Indian sites indicate that brown-top millet originated in the Deccan region of south India (Maitra 2020; Kingwell-Banham and Fuller 2014). Sorghum is a native of Africa and domestication took place around 5000–8000 years ago. The Indian subcontinent is considered as the secondary center of origin of sorghum, as evidence shows cultivation at about 4500 years ago. The domestication of pearl millet was noted in north-central Sahelian Africa during 4500 BC. Finger millet is of African origin, but entered into Asia during the prehistoric period (Venkatesh Bhat et al. 2018). Other millets mostly originated in Asia and Africa (Table 1.2) and later spread and were adapted to other regions of the world. In the Neolithic age of China, proso and foxtail millets were considered as vital food crops. Foxtail millet also is an ancient cereal of Eurasia (Hunt et al. 2008; De Wet et al. 1982; Jiaju and Yuzhi 1994). The farming of foxtail millet is found mainly in China and neighboring countries as food grains, but also used as grain, bird seed, feed and forage in the United States and Australia (Sheahan 2014). The origin of proso millet is considered as Manchurian (House et al. 1995), but it has also been found in Western Asia and Europe during ancient periods (Hunt et al. 2008), presently known as an important crop of China, the central and southern states of India, Kazakhstan, Afghanistan, Turkey, Romania, the USA and Australia.

1.5 Current Scenario of Millets Production

Millets are the staple food of a large number of populations throughout the world, especially in arid and semi-arid tropicals regions, mostly located in Africa and Asia. India is the largest producer of millets in the world. India, Nigeria and China together account for 55% of global millet production. The global production of millets is about 31 million tonnes and India shares 37.5% (Table 1.3). African countries, namely, Nigeria, Mali, Burkina Faso and Sudan, are the world's leading consumers. Considerable vast acreage under millet is found in Burkina Faso, Chad, Mali, Nigeria and Sudan among the African countries (Table 1.4). About a decade back, the estimated demand for millet in 2025 was fixed at 30 million tonnes (NAAS 2013). But during recent times, millets are recognized as nutri-cereals and increasing demand is noted among the health-conscious urbanites in developing as well as developed countries. which focuses further on enhancement of production. As per the market dynamics, demand pushes production and considering the environmental issues, it may be stated that future agriculture of arid and semi-arid regions will thrive on boosting millet production.

Table 1.2 Regions of origin of different species of millets.

Scientific name	Common name	Origin
Brachiaria ramosa (L.) Stapf. (syn. *Urochloa ramosa* (L.) R.D. Webster)	Browntop millet	Deccan region, South India
Coix lacrhyma-jobi L.	Job's tears millet	Northeast India, Southeast Asia, Southern China
Digitaria exilis (Kippist) Stapf.	White fonio	West Africa
Digitaria iburua Stapf.	Black fonio, hungry rice	West Africa
Echinochloa colona (L.) Link ssp. *frumentacea* (Link) (= *E. frumentacea* Link)	Sawa millet	Peninsular India
Echinochloa crus-galli (L.) P. Beauv. (syn. E. *esculenta* (A. Braun)	Barnyard millet	Japan
Eleusine coracana (L.) Gaertn.	Finger millet	East African highlands
Eragrostis tef (Zucc.) Trotter	Tef	Ethiopian highlands
Panicum miliaceum L. ssp. *miliaceum*	Proso millet	China
Panicum sumatrense Roth. ex Roem. & Schult. Subsp. *sumatrense* (syn. P. miliare auct. pl.)	Little millet	Peninsular India
Paspalum scrobiculatum L.	Kodo millet	India
Pennisetum glaucum (L.) R. Br (= P. *americium* (L))	Pearl millet	West African Savannah
Setaria italica (L.) P. Beauv ssp. *italica*	Foxtail millet	China
S. verticillata (L.) P. Beauv	Bristley foxtail millet	South India
Sorghum bicolor (L.) Moench ssp. *bicolor*	Sorghum	African Savannahs

Source: Weber and Fuller 2007; Venkatesh Bhat et al. 2018

Table 1.3 Millet production statistics.

Sl. No.	Region/Country	Area (ha)	Yield (kg ha^{-1})	Productions (tonnes)
1	Africa	22,099,632	7180	15,867,785
2	Asia	10,971,060	13,108	14,381,318
3	India	9,107,000	12,781	11,640,000
4	World	33,560,087	9243	31,019,370

Source: FAOSTAT 2020

1.6 Nutritional Importance of Millets

Millet contains all essential nutrients like carbohydrates, dietary fibres, proteins, fats, vitamins and minerals and is comparatively nutritious to fine cereals like wheat, barley, rice and maize (Table 1.5). Interestingly, all millets are superior to fine cereals in terms of

Table 1.4 Area, yield and production of millets (except Fonio) in leading millet growing countries (2018).

Sl. No.	Country	Area (ha)	Yield (kg ha^{-1})	Production (tonnes)
1	Angola	241,059	2154	51,915
2	Australia	35,161	10,216	35,922
3	Bangladesh	32,651	3894	12,713
4	Benin	28,856	9060	26,143
5	Burkina Faso	1,393,878	8531	1,189,079
6	Burandi	8837	11,281	9969
7	Cameroon	8837	11,281	9969
8	Central African Republic	9132	11,101	10,137
9	Chad	1,221,948	6192	756,616
10	China	617,407	25,370	1,566,363
11	Congo	16,109	8832	14,227
12	Cote d'ivoire	77,673	8261	64,169
13	Democratic People's Republic of Korea	73,584	10,666	78,486
14	Democratic Republic of the Congo	60,897	6701	40,807
15	Eritrea	69,544	3440	23,921
16	Ethiopia	515,613	19,064	982,958
17	France	10,948	37,250	40,914
18	Gambia	114,568	8632	98,894
19	Ghana	179,332	10,124	181,564
20	Guinea	167,404	12,828	214,747
21	Guinea-Bissau	12,577	11,758	14,788
22	India	9,107,000	21,781	11,640,000
23	Kazakhstan	43,420	9267	40,239
24	Kenya	115,806	7215	83,556
25	Malawi	54,230	5774	31,315
26	Mali	2,158,250	8527	1,840,321
27	Mauritania	11,906	2766	3239
28	Mozambique	37,665	4521	16,995
29	Myanmar	244,113	10,155	247,895
30	Namibia	244,113	10,155	247,895
31	Nepal	263,497	11,916	313,987
32	Niger	7,033,751	5483	3,856,344
33	Nigeria	2,795,829	8015	2,240,744
34	Pakistan	455,818	7679	350,010
35	Poland	13,721	15,989	21,938

(Continued)

Table 1.4 (Continued)

Sl. No.	Country	Area (ha)	Yield (kg ha^{-1})	Production (tonnes)
36	Russian Federation	187,144	11,606	217,200
37	Rwanda	11,370	4500	5116
38	Senegal	817,901	7018	574,000
39	Sierra Leone	41,769	10,687	44,639
40	South Africa	10,225	4916	5027
41	Sudan	3,753,000	7053	2,647,000
42	Uganda	147,625	14,224	209,979
43	United Republic of Tanzania	332,637	9845	327,467
44	United States of America	163,090	20,010	326,340
45	Yemen	97,596	5926	57,836
46	Zambia	41,463	7785	32,278
47	Zimbabwe	222,032	1733	38,473

Source: FAOSTAT, 2020

mineral content and other essential constituents. Furthermore, millets are considered as functional food and consumption of millets among health-conscious people has recently increased.

1.6.1 Millets as Functional Food

Millets provide an alternative source of food and nutrition, particularly in developing countries like India and Africa, where they are now used for human consumption but were primarily restricted to animal feed in developing countries. Millets are generally gluten-free, thus considered appropriate for people who suffer from gluten allergy and celiac disease. In general, millets are rich in dietary fiber, resistant starch, oligosaccharides, lipids, antioxidants, minerals and B-complex vitamins (Banerjee and Maitra 2020). However, the anti-nutritional factors like phytates and tannins in millets affect the mineral bioavailability (Rao et al. 2017). Millets are enriched with important phytochemicals such as polyphenols, lignans, phytosterols, phyto-oestrogens and phytocyanins, which have potential health promoting benefits. Millet can be considered as a functional food, as it acts as antioxidants, immune modulators, detoxifying agents, etc. and thus protects human beings from degenerative diseases like cardiovascular diseases, hyperglycaemia, tumors, respiratory health, Parkinson's diseases, etc. (Rao and Qaim 2011; Chandrasekara et al. 2012). Thus, millets meet the necessary criteria of nutraceuticals, which are defined as any food that contains bioactive ingredients and exerts beneficial effects on human physiology to combat chronic diseases, and so can be termed as functional foods or nutraceuticals (Banerjee and Ray 2019). The vitamins, minerals and essential fatty acids that are obtained from millets are non-acid forming, easy to digest and non-allergenic in nature (Saleh et al. 2013).

Table 1.5 Nutritional value of millets and fine cereals (per 100 g of edible portion).

Crop	Carbohydrate (g)	Crude fibre (g)	Protein (g)	Fat (g)	Mineral (g)
A. Fine cereals					
Barley	69.6	3.9	11.5	1.3	1.2
Maize	66.2	2.7	11.5	3.6	1.5
Rice	78.2	0.2	6.8	0.5	0.6
Wheat	71.2	1.2	11.8	1.5	1.5
B. Millets					
Barnyard millet	74.3	14.7	11.6	5.8	4.7
Brown-top millet	71.3	–	9.0	1.9	3.9
Finger millet	72.0	3.6	7.3	1.3	2.7
Foxtail millet	60.0	8.0	12.3	4.3	3.3
Kodo millet	65.9	9.0	8.3	1.4	2.6
Little millet	75.7	8.6	8.7	5.3	1.7
Pearl millet	67.5	1.2	11.6	5.0	2.3
Proso millet	70.4	2.2	12.5	1.1	1.9
Sorghum	72.6	1.6	10.4	1.9	1.6

Source: GOI 2020; Banerjee and Maitra 2020

1.6.2 Anti-Oxidant and Anti-Aging Properties

Several human diseases are linked to oxidative stress by the action of reactive oxygen and nitrogen species. Phytochemicals which act as antioxidants can protect from oxidative damage to maintain proper physiological function in humans. Plant extracts like polyphenols can reduce the risk of neurodegenerative diseases, cancer, aging, diabetes, infections, etc. (Kaur and Kapoor 2001; Scalbert et al. 2005, Tsao 2010). The seed coats of most millets contain various phenolic compounds which exhibit antioxidant properties (Hegde et al. 2005; Chandrasekara and Shahidi 2010). The antioxidants are found mainly in free or bound form in millets. Rao and Muralikrishna (2007) found protocatechuic acid as the key free phenolic acid in finger millets. Most of the total phenolic compounds in millets are derivatives of benzoic acid (gallic acid, protocatechuic acid, *p*-hydroxybenzoic acid, vanillic acid, syringic acid), while others are either flavonoids (quercetin, proanthocyanidins like condensed tannins) (Chethan and Malleshi 2007) or derivatives of cinnamic acid (ferulic acid, *trans*-cinnamic acid, *p*-coumaric acid, Caffeic acid, sinapic acid). In bound phenolic acids, ferulic and *p*-coumaric acid are the main components (Devi et al. 2014). An *in-vivo* study in rats showed that millets in a regular diet can enhance the activities of antioxidant enzymes such as catalase, glutathione peroxidase, glutathione reductase (Hegde et al. 2005) and block cross-linking of collagen (Hegde et al. 2002), resulting in decreased stiffness of elastic tissues in tendons, skin and blood vessels. The reduction of stiffness is one of the key mechanisms of anti-aging. The processing of millets reduces the bioavailability of polyphenols. Thermal or hydrothermal treatments, germination, decortication or fermentation generally reduce the polyphenol levels (Rao and Muralikrishna 2007; Shobano and Malleshi 2007). There is an ample scope of research to increase the bioavailability of natural antioxidant in millet and to present it as a functional food ingredient.

1.6.3 Protection Against Cancer

Healthy diet and life style can prevent several health hazards, including cancer. There are several underlying causes of cancer, including genetic mutation, inheritance, etc. It has been reported that the bioactive constituents in millet like lignans possess the ability to prevent cancer. Furthermore, the frequency of spontaneous or induced tumors is diminished by consumption of anti-carcinogenic foods on a regular basis. Phytochemicals, especially antioxidants and polyphenols, are potential anti-carcinogenic agents which act as terminators of singlet oxygen species and free radicals (Banerjee and Maitra 2020). *In-vivo* experiments performed on rats using ferulic acid were found to be effective against induced carcinogenesis in tongue and colon (Kawabata et al. 2000) as well as in breast cancer cell lines (Choi and Park 2015), indicating that ferulic acid can be used as a chemotherapeutic agent against tongue, colon and breast cancer. Some millet contains nitriloside (Vitamin B_{17}), which specifically targets the cancer cell without having any detrimental side effects on human health. Previously, it has been reported that risk of esophageal cancer can be reduced by consumption of sorghum and millets. because the initiation and progression steps of cancer in tissues can be lowered by phenolic components (tannins and phytate) found in millets as reported by various authors (Chandrasekara and Shahidi 2011a; 2011b; Kumar et al. 2016). Singh et al. (2015) reported anti-carcinogenic properties of finger and pearl millet via a cyto-toxicity assay on HepG2 hepatic cancer cell lines. A good number of studies demonstrate that the dietary fibres can be used as functional food to combat breast cancer (Cade et al. 2007).

1.6.4 Anti-Diabetic Properties

Diabetes mellitus is a chronic metabolic disorder caused by a lack of sufficient insulin secretion (Type-1) or due to mutual endurance to insulin production and action (Type-2). Phenolic compounds found in the seed coats of millets (finger millet) were found effective in preventing diabetes due to their antioxidant properties and it has been revealed that the antioxidants help to lower the blood glucose levels (Shobana et al. 2009; American Diabetes Assocation 2010; Kim et al. 2006). The anti-nutrients such as tannins, phenolic and phytates found in millets decrease the glycemic response due to reduced starch digestibility and absorption.

1.6.5 Protection Against Gastro-Intestinal Disorders

Millets are a rich source of soluble and insoluble dietary fiber, or roughage, that are not degraded during digestion, but rather used as food by probiotic bacteria which in turn prevents gastro-intestinal disorders (Anderson et al. 2009). The high percentage of insoluble fibers found in millets results in stimulation of bowel mobility, prevents constipation and promotes peristalsis movement. The soluble fibers help lubrication of an inflamed alimentary tract. Moreover, the polyphenols are known to decrease peptic inflammation and exhibit anti-ulcerative properties (Chethan and Malleshi 2007). Another important fact about millets is that they contain low amounts of fat which is beneficial against obesity.

1.6.7 Protection Against Osteoporosis

Osteoporosis is caused by calcium deficiency and the costly mineral supplements available on the market for treatment possess considerable side effects (Straub 2007). Dietary intake of high amounts of naturally occurring calcium prevents osteoporosis. Millets are a rich source of calcium, containing five to ten times higher calcium content than compared to other cereal crops (Sanwalka et al. 2011; Kumar et al. 2013). It has been reported that millets have more calcium content when compared to cow's milk (Wijesinha-Bettoni and Burlingame 2013). Cow's milk is unsuitable for people who suffer from lactose intolerance and so millets prove to be a safe substitute. The value-added products developed from millet are part of an essential diet for children and older individuals for the development of their bones.

1.7 Changes in Food Consumption Pattern and Future Demand

Human civilization has witnessed massive advancements during the last century due to technological development, urbanization and improved production systems. However, there is a limitation in the supply of nutritious foods to the world population, particularly those who have comparatively less access to a healthy diet. The problem is more crucial in the developing countries of Africa and Asia, where majority of malnourished people reside (FAO 2017). The estimates show that the world population will be 9.7 billion by 2050 and 11.2 billion by 2100 and agriculture will face a tremendous challenge in the future with declining and shrinking natural resources. Interestingly, urbanization is flourishing on a rapid scale and at present 54% of the global population is urban dwelling. In 2050, about two-thirds of populations will be urbanites (UN 2015) and the rate of urbanization is more in the developing world. Earlier, in the developing countries, the majority of the population lived in rural areas and were mostly dependent, either directly or indirectly, on agricultural activities. But over the years, agriculture became less profitable due to climatic uncertainties, volatility of the market and improper progress in rural infrastructure in the agriculture-based countries. On the other hand, some technological development on the agriculture front, particularly with the use of more efficient farm machinery, reduced the use of the human workforce in agriculture, which also triggered urbanization that in turn lead to the changes in the food consumption patterns. The population of the higher income group prefers processed, value-added and animal-sourced foods, which require more energy and other inputs. Undoubtedly, the flourish in consumption of these energy diets changed the food production system where processing, value-addition, marketing and logistics started to provide employment for more people, with a subsequent decrease in the population being involved in farming activities (FAO 2017). In urban and peri-urban areas, supermarkets and value chains have increasingly shown a vertical growth during the last decade. But there is still the presence of a diversified distribution system in developing countries (The Nielsen Company 2015), with some ecological and ethical presence. Both forms of market dynamics notably realize the importance of millets. For rural populations of drylands in developing countries, consumption of millets is of local significance of their heritage, and for urbanites these are healthy and nutritious foods. Smallholders are also gaining benefits as producers of the nutritious foods and so they become part of the value chain (Rao and Qaim 2011).

Therefore, the demand for food is driven not only by population growth, but also by income growth (Valina et al. 2014). To meet future demand, farm productivity should be increased with sustainable intensification (Garnett et al. 2013), because earlier agricultural output was increased by increasing chemical inputs and non-judicious use of water at a cost to the environment. By 2050, agricultural productivity will need to be increased by about 50% of the present time (World Bank Group 2016); another estimate suggests global crop production could increase by 45 to 70% (Mueller et al. 2012). Furthermore, considering the food waste and adverse impacts of climate change with its associated uncertainties, there will be a need to produce more food to meet the food and nutritional security of the people (FAO 2018). The world cereal equivalent food demand is projected to be around 14,886 million tons by 2050 (Islam and Karim 2019), in which India and China will acquire the major share. In the present context of climate change and dwindling natural resources, it may be stated that growing populations, trending globalization and blooming urbanization have made sustainable agriculture and food, as well as nutritional security, an extraordinary challenge. The sustainable targets are to be considered at international and national level to reach the sustainable developmental goal, considering the socio-economic and ecological landscapes (Davis et al. 2016; Calicioglu et al. 2019).

1.8 Food and Nutritional Security

Worldwide, different nations and leading international organization have adopted initiatives to eliminate hunger, but still it persists in the developing world (Fraser and Rimas 2010). Besides, due to the increasing population, urbanization, mal-agricultural practices, shrinking of agriculture land and over-exploitation of resources, it is of prime importance to adopt a proper action plan for the sustainable future (Garnett 2014; Gladek et al. 2016). The issues related to climate change have an added hindrance in reaching the goal of food security. During the last five decades, the concept of food security was limited to availability and accessibility of food to all; but nowadays, food security denotes sufficient access to nutritious food to lead to a healthy life (Committee on World Food Security 2012). The recent data indicates that there were 815 million people of Africa and Asia who suffered from food insecurity (El Bilali et al. 2019), primarily due to natural calamity, worsened by climate change. Adoption of sustainable agricultural practices can only assure sustainable food system and nutritional security (El Balali 2018; Maitra and Ray 2019). The current food production system comprises of a number of limitations like deterioration of land and water resources, increased emission of greenhouse gasses, environmental degradation and erosion of genetic diversity, resulting in food insecurity (FAO 2017). Food security as well as changeover from the present system to sustainable food system is the main topic of consideration in the field of food production. Creation of attractive opportunities for enhancing farm output, reduction of food insecurity and transformation of present-day agriculture in developing countries to sustainable food production, need to be considered in a holistic manner, targeting economic prosperity of the smallholder farmers. The majority of the smallholders in dry areas of the world produce millets by subsistence farming. Cultivation, processing and value addition and consumption of millets can accomplish the food and nutritional security to the malnourished populations dwelling in the developing world.

1.9 Climate Change and Associated Threat to Agriculture

The term "climate" denotes the long-term characteristics of weather conditions of a geographical area. The concerns relating to the Earth's climate of today are not of natural origin, but rather manmade and may be defined as the change that can be accredited directly or indirectly to human interference that modifies the atmospheric composition other than natural variation over a comparable time (UNFCCC 2017). In the Earth Summit (1992), held in Rio de Janeiro, agriculture was revealed through a reference of food security and the impact of climate change on agriculture was realized for the first time (Muldowney et al. 2013; Loboguerrero et al. 2019). Over the years, it became clear that agriculture and the food value chain is one of the major contributors to different greenhouse gases (GHGs), with an estimated emission of 5.1 to 6.1 Gt CO_2-eq year^{-1} (Smith et al. 2007). Climate change is already in progress and global warming and irregular rainfall patterns have already shown negative impact on farm output, with fewer yields of maize decreasing by 3.8% and wheat by 5.5% (Lobell et al. 2011; IPCC 2013). Due to climate change the concentration of ozone (O_3) increases and negatively influences crop growth. At the global scale, 2.5 billion smallholders have been severely affected by the ill effects of climate change (Niles et al. 2017). In developing countries, about two-thirds of the population is directly or indirectly dependent on agriculture and will be adversely affected by the negative impacts of climate change (Yadav et al. 2019). Increasing temperature may cause harm to several crops if it reaches beyond the threshold level (Yadav et al. 2019). There will be a 550 ppm concentration of CO_2 in the atmosphere by 2050, which will reduce yields of C_3 plants like rice, wheat and barley. On the other hand, maize and millets (C_4 plants) can produce more photosynthates under elevated CO_2 conditions. However, protein concentrations will be reduced in some cereals and forages with more CO_2 concentration in the atmosphere (IPCC 2001). Increase in O_3 concentration also adversely affects crop yield (Avnery et al. 2011). Furthermore, water scarcity will reduce crop yield unless more drought tolerant crops or varieties are grown (Maitra and Pine 2020). Considering the future demand for food security, climate smart agriculture should be adopted with sustainable farm productivity, building resilience to climate change and reduction of GHGs emission (FAO 2019). The negative impact of climate change in terms of food insecurity will be more prominent in the ecologically

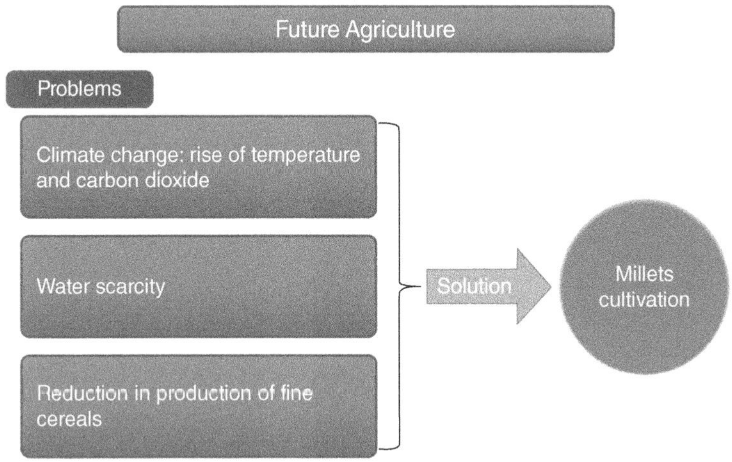

Figure 1.2 Problems of future agriculture and millets cultivation as a probable solution.

vulnerable areas, where incidence of hunger and malnutrition is common and cultivation of ecologically sound and nutritious crops may be a suitable alternative (Figure 1.2).

1.10 Millets: As Climate Smart Crops

Future agriculture will face some common environmental changes like enhanced temperature, uncertainties in rainfall, elevated CO_2 and GHGs levels, and more frequency in natural calamities (mainly drought and flood). Under these conditions, climate resilient agriculture should be adopted in which cultivation of climate smart crops will play a pivotal role. There is no doubt that millets are the climate smart crops which can simultaneously mitigate the ill effects of climate change and adapt to the changed and wider agro-climatic conditions (Bandyopadhyay et al. 2017). Millets have some efficient morphological, physiological, molecular and biochemical traits which can withstand abiotic stresses. As millets are short-duration crops, thus completing a cycle within a short time span, they can escape the possibility of environmental stress conditions under early or late sowing conditions. Besides, millets possess less leaf area (except pearl millet and sorghum), a thickened cell wall and a dense fibrous root system which facilitates their capability to tolerate abiotic stress (Li and Brutnell 2011). Being C_4 plants, millets can utilize more atmospheric CO_2 and by the process of photosynthesis can produce more assimilates, even under elevated CO_2 levels into the atmosphere (Aubry et al., 2001)). The nitrogen use efficiency of C_4 plants are higher compared to C_3 plants (like rice and wheat). The C_4 pathway of photosynthesis allows millets to perform better under hot weather conditions, including enhanced flexible allocation patterns of biomass and reduced hydraulic conductivity per unit leaf area (Sage and Zhu 2011). Furthermore, water use efficiency (WUE) of millets are higher than major cereals and in the future, under the crucial situation of water deficit in a major portion of the world, millets will automatically be chosen to combat water shortage. For example, foxtail millet needs only 257 g of water to produce 1 g of dry biomass and the water requirement is half that of maize (470 g) and wheat (510 g) to produce the same quantity of dry biomass (Bandhyophyay et al. 2017). Wheat and rice have a global warming potential of around 4 tons CO_2 eq/ha and 3.4 tons CO_2 eq/ha, respectively. Wheat, rice and maize also have high carbon equivalent emissions of 1000, 956 and 935 kg C/ha for wheat, rice and maize, respectively (Jain et al. 2016). The carbon footprint of millets is comparatively less than major cereals and therefore cultivation of millets can reduce the carbon footprint (Saxena et al. 2018). Furthermore, in the process of chemical nitrogen fertilizer production, CO_2 is generated and as per an estimate to fulfill the present requirement of chemical nitrogen fertilizers in the world, annually 300 teragram ($Tg = 10^{12}g$) of CO_2 is released into the atmosphere (Jensen et al. 2012). Millets are less nutrient demanding crops and promotion of millet cultivation will indirectly save the environment. Erosion of genetic diversity is another important issue which is the result of industrialized agriculture, and agricultural sustainability can only be achieved by assuring greater ecosystem services in which genetic diversity plays a crucial role. In this way, the rich diversity of millets will lead future farming toward sustainability.

1.11 Future Agriculture: Smart Technologies in Millet Farming

With the advancement of science and technology, agriculture has also become smart and precise. Precision agriculture (PA) is a new concept where crop management has been made easier with the latest technologies and, in future, smart technologies will enhance input use efficiency in agriculture and more production to feed the growing population. PA is actually information and technology-enabled farming practices to identify, analyze and manage the variability within the field and it aims to register optimum productivity and profitability, protection of natural resources and sustainability (Saiz-Rubio and Rovira Más 2020). In PA, new information technologies (ITs) are used to make better decisions for managing the variability in several aspects of crop production and management. Use of remote sensing, Internet of Things (IoT), different sensors, artificial intelligence, machine learning, different apps and drones for crop management will make farming more precise for a target yield output. In the smart farms of the future, there will be sensing technology, hardware and software applications with IoT-based management, data analytics and smart crop management. Presently, millet cultivation is practiced by the small and marginal farmers in dry regions of the world and in these areas investment in agriculture is less. Millets are undernourished crops managed by smallholders in developing countries. But the situation will be changed over time with proper realization of the importance of millets in food and nutritional security under changed climatic conditions. There is an urgent need of research initiatives like other major cereals in the areas of site-specific crop management with the help of PA technologies to face and fulfill the future food demand. In the advent of technological flourish it may be expected that over a period of a few decades, the orthodox millet farming may be shifted into the concept of Agriculture 5.0 that implies unmanned operations and decision support systems in which robots, drones and other forms of artificial intelligence will be used (Saiz-Rubio and Rovira Más 2020; Zambon et al. 2019).

1.12 Conclusions

At present, agriculture is facing great challenges due to the adverse effects of climate change, which have further resulted in slow progress in production enhancement with narrowed profit. The ongoing inefficient agricultural practices as well as over reliance on major cereals, urbanization and consumption of value-added processed foods and industrialization are the main causes of the increased carbon footprint triggering global warming. On the other hand, millets have the qualities to combat the ill effects of climate change and can be adapted to extreme weather conditions. Furthermore, millets are nutritionally superior and can be treated as a functional food. Considering these advantages, ancient millets can now be nutured as future food. In the near future, taking advantage of latest technological developments, millet cultivation can be intensified with smart technologies with proper transformation into Agriculture 5.0. Millets, the ancient and forgotten food, may regain its pride as an important staple of future food.

References

American Diabetes Association (2010). Diagnosis and classification of diabetes mellitus. *Diabetes Care* 33: S62–S69. https://doi.org/10.2337/dc10-S062.

Anderson, J.W., Baird, P., Davis, Jr R.H. et al.(2009). Health benefits of dietary fiber. *Nutrition Review* 67 (4): 188–205. doi:10.1111/j.1753-4887.2009/00189.x

Aubry, S., Brown, N.J., and Hibberd, J.M. (2011). The role of proteins in C_3 plants prior to their recruitment into the C_4 pathway. *Journal of Experimental Botany* 62: 3049–3059. https://doi.org/10.1093/jxb/err012.

Avnery, S., Mauzerall, D.L., Liu, J. et al. (2011). Global crop yield reductions due to surface ozone exposure. 1: Year 2000 crop production losses and economic damage. *Atmospheric Environment* 45: 2284–2296.

Bandyopadhyay, T., Muthamilarasan, M., and Prasad, M. (2017). Millets for next generation climate-smart agriculture. *Frontiers in Plant Science* 8: 1266. https://doi.org/10.3389/fpls.2017.01266.

Banerjee, P. and Ray, D.P. (2019). Functional food: A brief overview. *IJBS* 6: 57–60. https://doi.org/10.30954/2347-9655.02.2019.2.

Banerjee, P. and Maitra, S. (2020). The role of small millets as functional food to combat malnutrition in developing countries 10 (60): 20412–20417.

Blümmel, M., Zerbini, E., Reddy, B.V.S. et al.(2003). Improving the production and utilization of sorghum and pearl millet as livestock feed: Progress towards dual-purpose genotypes. *Field Crops Research* 84: 143–158

Brahmachari, K., Sarkar, S., Santra, D.K. et al. (2018). Millet for food and nutritional security in drought prone and red laterite region of eastern India. *International Journal of Plant & Soil Science* 26 (6): 1–7.

Cade, J.E., Burley, V.J., and Greenwood, D.C. (2007). Dietary fibre and risk of breast cancer in the UK Women's Cohort Study. *International Journal of Epidemiology* 36: 431–438. https://doi.org/10.1093/ije/dyl295.

Calicioglu, O., Flammini, A., Bracco, S. et al. (2019). The future challenges of food and agriculture: An integrated analysis of trends and solutions. *Sustainability* 11: 222. https://doi.org/10.3390/su11010222.

Chandrasekara, A. and Shahidi, F. (2010). Content of insoluble bound phenolics in millets and their contribution to antioxidant capacity. *Journal of Agricultural and Food Chemistry* 58: 6706–6714. https://doi.org/10.1021/jf100868b.

Chandrasekara, A. and Shahidi, F. (2011a). Anti-proliferative potential and DNA scission inhibitory activity of phenolics from whole millet grains. *Journal of Functional Foods* 3: 159–170. https://doi.org/10.1016/j.jff.2011.03.008.

Chandrasekara, A. and Shahidi, F. (2011b). Inhibitory activities of soluble and bound millet seed phenolics on free radicals and reactive oxygen species. *Journal of Agricultural and Food Chemistry* 59: 428–436. https://doi.org/10.1021/jf103896z.

Chandrasekara, A., Naczk, M., and Shahidi, F. (2012). Effect of processing on the antioxidant activity of millet grains. *Food Chemistry* 133 (1): 1–90. http://doi.org/doi:10.1016/j.foodchem.2011.09.043

Chethan, S. and Malleshi, N.G. (2007). Finger millet polyphenols optimization of extraction and the effect of pH on their stability. *Food Chemistry* 105: 862–870. https://doi.org/10.1016/j.foodchem.2007.02.012.

Choi, Y.E. and Park, E. (2015). Ferulic acid in combination with PARP inhibitor sensitizes breast cancer cells as chemotherapeutic strategy. *Biochemical and Biophysical Research Communications* 458: 520–524. https://doi.org/10.1016/j.bbrc.2015.01.147.

Clayton, W.D., Vorontsova, M.S., Harman, K.T. et al. (2006) Grass base: The online world grass flora. Available online: http://www.kew.org/data/grasses-db.html. (accessed 12 June 2020)

Committee on World Food Security (2012) Coming to terms with terminology: Food security, nutrition security, food security and nutrition, food and nutrition security. Retrieved from http://www.fao.org/fsnforum/sites/default/files/file/Terminology/MD776CFS_Coming_to_terms_with_Terminology.pdf (accessed 10 June 2020)

Davis, K.F., Gephart, J.A., Emery, K.A. et al. (2016). Meeting future food demand with current agricultural resources. *Global Environmental Change* 39: 125–132.

Devi, P.B., Vijayabharathi, R., Sathyabama, S. et al. (2014). Health benefits of finger millet (*Eleusine coracana* L.) polyphenols and dietary fiber: A review. *Journal of Food Science and Technology* 51: 1021–1040. https://doi.org/10.1007/s13197-011-0584-9.

De Wet, J.M.J., Prasada Rao, K.E. et al. (1983). Systematics and domestication of *Panicum sumatrense* (*Gramineae*). *J Agric Trad Bot Appl* 30: 159–168.

El Bilali, H. (2018). Relation between innovation and sustainability in the agro-food system. *Italian Journal of Food Science* 30: 200–225. https://doi.org/10.14674/ijfs-1096.

El Bilali, H., Callenius, C., Strassner, C. et al. (2019). Food and nutrition security and sustainability transitions in food systems. *Food and Energy Security* 8: e00154. https://doi.org/10.1002/fes3.154.

FAO (1995) Sorghum and millets in human nutrition. Available online: http://www.fao.org/docrep/t0818e/T0818E00.htm#Contents (accessed 16 May 2020)

FAO (2017) HLPE reports. High level panel of experts on food security and nutrition (HLPE). Rome, Italy. Available online: http://www.fao.org/cfs/cfs-hlpe/reports/en (accessed 12 June 2020)

FAO (2018) The future of food and agriculture, alternative pathways to 2050. Rome, Italy. Available online: http://www.fao.org/3/CA1553EN/ca1553en.pdf (accessed 12 June 2020).

FAO (2019). *Agriculture and Climate Change – Challenges and Opportunities at the Global and Local Level – Collaboration on Climate-smart Agriculture*, 52 pp. Rome, Italy.

FAOSTAT (2020) Crop Statistics. Available online: http://www.fao.org/faostat/en/#data/QC (accessed 15 June 2020)

Fraser, E.D.G. and Rimas, A. (2010). *Empires of Food: Feast, Famine and the Rise and Fall of Civilizations*. London, UK: Random House.

Fuller, D.Q. (2011). Finding plant domestication in the Indian subcontinent. *Current Anthropology* 52 (S4): S347–S362.

Fuller, D.Q., Boivin, N., Hoogervorst, T. et al. (2011). Across the Indian Ocean: The prehistoric movement of plants and animals. *Antiquity* 85: 544–558.

Garnett, T. (2014). Three perspectives on sustainable food security: Efficiency, demand restraint, food system transformation. What role for life cycle assessment. *Journal of Cleaner Production* 73: 10–18. https://doi.org/10.1016/j.jclepro.2013.07.045.

Garnett, T., Appleby, M.C., Balmford, A. et al. (2013). Sustainable Intensification in agriculture: Premises and policies. *Science* 341: 33–34. https://doi.org/10.1126/science.1234485.

Gladek, E., Fraser, M., Roemers, G. et al. (2016) The global food system: An analysis. Amsterdam. Available online: https://www.metabolic.nl/publications/global-food-system-analysis (accessed 11 June 2020)

GOI (2020) Government of India, Directorate of Millet Research. Importance of coarse cereals. Available online: http://millets.dacfw.nic.in/Importance_CoarseCereals.html (accessed 14 June 2020)

Harika, J.V., Maitra, S., Shankar, T. et al. (2019). Effect of integrated nutrient management on productivity, nutrient uptake and economics of finger millet (*Eleusine coracana* L. Gaertn). *International Journal of Agriculture Environment and Biotechnology(IJAEB)* 12 (3): 273–279.

Hegde, P.S., Chandrakasan, G., and Chandra, T.S. (2002). Inhibition of collagen glycation and cross linking *in vitro* by methanolic extracts of finger millet (*Eleusine coracana*) and kodomillet (*Paspalum scrobiculatum*). *The Journal of Nutritional Biochemistry* 13: 517–521. https://doi.org/10.1016/S0955-2863(02)00171-7.

Hegde, P.S., Rajasekaran, N.S., and Chandra, T.S. (2005). Effects of the antioxidant properties of millet species on oxidative stress and glycemic status in alloxan-induced rats. *Nutrition Research* 25: 1109–1120. https://doi.org/10.1016/j.nutres.2005.09.020.

House, L.R., Osmanzai, M., Gomez, M.I. et al. (1995). Agronomic principles. In: *Sorghum and Millets: Chemistry and Technology* (ed. D.A.V. Dendy), 27–67. St Paul, MN: American Association for Cereal Chemist.

Hunt, H.V., Vander Linden, M., Liu, X. et al. (2008). Millets across Eurasia: Chronology and context of early records of the genera *Panicum* and *Setaria* from archaeological sites in the Old World. *Vegetation History and Archaeobotany* 17: 5–18.

IPCC (2001). *Intergovernmental Panel on Climate Change, Climate Change: Impacts, Adaptation and Vulnerability, Contribution of Working Group II to the Third Assessment Report of the IPCC*. Cambridge, UK: Cambridge University Press.

IPCC (2013). *Intergovernmental Panel on Climate Change, Climate Change: The Physical Science Basis*. Contribution of Working Group I to the Fifth Assessment Report of the IPCC. Cambridge, UK: Cambridge University Press.

Islam, S.M.F. and Karim, Z. (2019). World's demand for food and water: The consequences of climate change. In: *Desalination – Challenges and Opportunities*. https://doi.org/10.5772/intechopen.85919, Online First, Available online: https://www.intechopen.com (accessed 16 June 2020)

Jain, N., Arora, P., Tomer, R. et al. (2016). Greenhouse gases emission from soils under major crops in northwest India. *The Science of the Total Environment* 542: 551–561. https://doi.org/10.1016/j.scitotenv.2015.10.073.

Jaison, G. (2018). A study on millets based on history and potential health benefits. *Asia Pacific Journal of Research* 1: 149–156.

Jarret, R.L., Ozias-Akins, P., Phatak, S. et al. (1995). DNA contents in *Paspalum* spp. determined by flow cytometry. *Genetic Resources and Crop Evolution* 42: 237–242.

Jensen, E.S., Peoples, M.B., Boddey, R.M. et al. (2012). Legumes for mitigation of climate change and the provision of feedstock for biofuels and biorefineries. A review. *Agronomy for Sustainable Development* 32: 329–364.

Jiaju, C. and Yuzhi, Q. (1994). Recent developments in foxtail millet cultivation and research in China. In: *Advances in Small Millets* (ed. R. Kw, G. Sc, A. Seetharam, et al.), 101–108. New York: International Science Publisher.

Kaur, C. and Kapoor, H.C. (2001). Antioxidants in fruits and vegetables: The millennium's health. *International Journal of Food Science & Technology* 36: 703–725. https://doi.org/10.1111/j.1365-2621.2001.00513.x.

Kawabata, K., Yamamoto, T., Hara, A. et al. (2000). Modifying effects of ferulic acid on azoxy methane-induced colon carcinogenesis in F344rats. *Cancer Letters* 157: 15–21. https://doi.org/10.1016/S0304-3835(00)00461-4.

Kim, S.H., Hyun, S.H., and Choung, S.Y. (2006). Anti-diabetic effect of cinnamon extract on blood glucose in db/db mice. *Journal of Ethnopharmacology* 104 (1–2): 119–123. https://doi.org/10.1016/j.jep.2005.08.059.

Kingwell-Banham, E. and Fuller, D.Q. (2014). Brown top millet: Origins and development. In: *Encyclopedia of Global Archaeology* (ed. C. Smith), 1021–1024. New York: Springer

Kumar, A., Goel, B.K., Karthikeyan, S. et al. (2013). Protein and calcium rich malted health drink power. *Journal of Food Processing & Technology* 4: 214. https://doi.org/10.4172/2157-7110.1000214.

Kumar, A., Metwal, M., Kaur, S. et al. (2016). Nutraceutical value of finger millet (*Eleusine coracana* (L.) Gaertn.), and their improvement using omics approaches. *Frontiers in Plant Science* 7: 934. https://doi.org/10.3389/fpls.2016.00934.

Li, P. and Brutnell, T.P. (2011). *Setaria viridis* and *Setaria italica* model genetic systems for the panicoid grasses. *Journal of Experimental Botany* 62: 3031–3037. https://doi.org/10.1093/jxb/err096.

Lobell, D.B., Schlenker, W., and Costa-Roberts, J. (2011). Climate trends and global crop production since 1980. *Science* 333: 616–620.

Loboguerrero, A.M., Campbell, B.M., Cooper, P.J.M. et al. (2019). Food and earth systems: Priorities for climate change adaptation and mitigation for agriculture and food systems. *Sustain* 11: 1372. https://doi.org/10.3390/su11051372.

Lu, H., Zhang, J., Liu, K. et al. (2009). Earliest domestication of common millet (*Panicum miliaceum*) in East Asia extended to 10,000 years ago. *Proceedings of the National Academy of Sciences of the United States of America* 106 (18): 7367–7372.

Maitra, S. (2020). Potential horizon of brown-top millet cultivation in drylands: A review. *Crop Research* 55 (1 & 2): 57–63. https://doi.org/10.31830/2454-1761.2020.012.

Maitra, S. and Pine, S. (2020). Smart irrigation for food security and agricultural sustainability. *Indian Journal of Natural Sciences* 10: 20435–20439.

Maitra, S. and Ray, D.P. (2019). Enrichment of biodiversity, influence in microbial population dynamics of soil and nutrient utilization in cereal-legume intercropping systems: A review. *International Journal of Biological Research* 6: 11–19.

Maitra, S. and Shankar, T. (2019). Agronomic management in little millet (*Panicum sumatrense* L.) for enhancement of productivity and sustainability. *International Journal of Bioresource Science (IJBS)* 6: 91–96.

Miller, N.F., Spengler, R.N., and Frachetti, M. (2016). Millet cultivation across Eurasia: Origins, spread, and the influence of seasonal climate. *The Holocene* 29 (10): 1566–1575. https://doi.org/10.1177/0959683616641742.

Mueller, N.D., Gerber, J.S., Johnston, M. et al. (2012). Closing yield gaps through nutrient and water management. *Nature* 490: 254–257.

Muldowney, J., Mounsey, J., and Kinsella, L. (2013). Agriculture in the climate change negotiations; ensuring that food production is not threatened. *International Journal of Animal Biosciences* 7 (Suppl. 2): 206–211.

Mullet, J., Morishige, D., McCormick, R. et al. (2014). Energy sorghumda genetic model for the design of C_4 grass bioenergy crops. *Journal of Experimental Botany* 65: 3479–3489.

. NAAS (2013) Role of millets in nutritional security of India. Policy Paper No. 66, National Academy of Agricultural Sciences, New Delhi, India, pp. 16.

The Nielsen Company (2015) The future of grocery. E-commerce, digital technology and changing shopping preferences around the world. New York, USA.

Niles, M.T., Ahuja, R., Esquivel, J. et al. (2017) Climate change and food systems: Assessing impacts and opportunities. Meridian Institute, 2017. Available online: http://bit.ly/2oFucpe (accessed 4 December 2019)

Obilana, A.B. (2004). Sorghum: Breeding and agronomy. In: *Encyclopedia of Grain Science*, Vol. 3 (ed. C. Wrigley, H. Corke, and C.E. Walker), 108–119. Oxford: Elsevier.

Poncet, V., Lamy, F., Devos, K.M. et al. (2000). Genetic control of domestication traits in pearl millet (*Pennisetum glaucum* L., *Poaceae*). *TAG. Theoretical and Applied Genetics. Theoretische Und Angewandte Genetik* 100: 147–159.

Rao, B.D., Bhaskarachary, K., Arlene Christina, G.D. et al. (2017). *Nutritional and Health Benefits of Millets*, 1–112. Rajendranagar, Hyderabad, India: ICAR-Indian Institute of Millets Research (IIMR).

Rao, E.J.O. and Qaim, M. (2011). Supermarkets, farm household income, and poverty: Insights from Kenya. *World Development* 39 (5): 784–796.

Rao, M.V.S.S.T.S. and Muralikrishna, G. (2002). Evaluation of the antioxidant properties of free and bound phenolic acids from native and malted finger millet (ragi, *Eleusine coracana* Indaf-15). *Journal of Agricultural and Food Chemistry* 50: 889–892. https://doi.org/10.1021/jf011210d.

Rao, R.S.P. and Muralikrishna, G. (2007). Structural characteristics of water soluble ferulyl arabinoxylans from rice (*Oryza sativa*) and ragi (finger millet, *Eleusine corcana*): Variations upon malting. *Food Chemistry* 104: 1160–1170. https://doi.org/10.1016/j.foodchem.2007.01.015.

Rooney, W.L. and Wayne Smith, C. (2000). Techniques for developing new cultivars. In: *Sorghum: Origin, History, Technology and Production* (ed. S.C. Wayne and R.A. Frederiksen), 329–347. New York: John Wiley & Sons.

Sage, R.F. and Zhu, X.-G. (2011). Exploiting the engine of C_4 photosynthesis. *Journal of Experimental Botany* 62: 2989–3000. https://doi.org/10.1093/jxb/err179.

Saiz-Rubio, V. and Rovira-Más, F. (2020). From smart farming towards Agriculture 5.0: A review on crop data management. *Agronomy* 10: 207. https://doi.org/10.3390/agronomy10020207.

Saleh, A.S.M., Zhang, Q., Chen, J. et al. (2013). Millet grains: Nutritional quality, processing, and potential health benefits. *Comprehensive Review in Food Science and Food Safety* 12: 281–295. https://doi.org/10.1111/1541-4337.12012.

Sanwalka, N.J., Khadilkar, A.V., and Chiplonkar, S.A. (2011). Development of non-dairy, calcium-rich vegetarian food products to improve calcium intake in vegetarian youth. *Current Science* 101: 657–663.

Saxena, R., Vanga, S.K., Wang, J. et al. (2018). Millets for food security in the context of climate change: A review. *Sustainability* 10: 2228. https://doi.org/10.3390/su10072228.

Scalbert, A., Manach, C., Morand, C. et al. (2005). Dietary polyphenols and prevention of diseases. *Critical Reviews in Food Science and Nutrition* 45: 287–306. https://doi.org/10.1080/1040869059096.

Sheahan, C.M. (2014). *Plant Guide for Foxtail Millet (Setaria italica).* Cape May, NJ: USDA-Natural Resources Conservation Service.

Shobana, S. and Malleshi, N.G. (2007). Preparation and functional properties of decorticated finger millet (*Eleusine coracana*). *Journal of Food Engineering* 79: 529–538. https://doi.org/10.1016/j.jfoodeng.2006.01.076.

Shobana, S., Sreerama, Y.N., and Malleshi, N.G. (2009). Composition and enzyme inhibitory properties of finger-millet (*Eleusine coracana L.*) seed coat phenolics: Mode of inhibition of a-glucosidase and pancreatic amylase. *Food Chemistry* 115: 1268–1273.

Singh, N., Meenu, G., Sekhar, A. et al. (2015). Evaluation of anti-microbial and anti-cancer properties of finger millet (*Eleusine coracana*) and Pearl millet (*Pennisetum glaucum*) extracts. *Journal of Pharmaceutical Innovation (JPI)* 3: 82–86.

Smith, P., Martino, D., Cai, Z. et al. (2007). Agriculture. In: *Climate Change 2007: Mitigation.* Contribution of Working Group III to the Fourth Assessment Report of the Intergovernmental Panel on Climate Change (ed. B. Metz, O.R. Davidson, P.R. Bosch et al.). Cambridge, UK and New York, USA: Cambridge University Press.

Sood, S., Khulbe, R.K., Gupta, A.K. et al. (2015). Barnyard millet A potential food and feed crop of future. *Plant Breeding* 134: 135–147.

Srinivasarao, C., Lal, R., Kundu, S. et al. et al. et al.(2014). Soil carbon sequestration in rainfed production systems in the semiarid tropics of India. *The Science of the Total Environment* 487: 587–603. https://doi.org/10.1016/j.scitotenv.2013.10.006

Straub, D.A. (2007). Calcium supplementation in clinical practice: A review of forms, doses, and indications. *Nutrition in Clinical Practice : Official Publication of the American Society for Parenteral and Enteral Nutrition* 22: 286–296. https://doi.org/10.1177/0115426507022003286.

Taylor, J.R.N. (2019). *Sorghum and Millets: Taxonomy, History, Distribution, and Production. Sorghum and Millets*, Second Edition, 01–21. Published by Elsevier Inc. in cooperation with AACC International.

Tsao, R. (2010). Chemistry and biochemistry of dietary polyphenols. *Nutrients* 2: 1231–1246. https://doi.org/10.3390/nu2121231.

UN (United Nations) (2015) World Population Prospects: The 2015 Revision. Available online at: https://esa.un.org/unpd/wpp (accessed 16 June 2020)

UNFCCC (2017) United Nations Framework Convention on Climate Change, Koronivia Joint Work on Agriculture. Decision/CP.23. 2017. Available online: https://unfc (accessed 12 June 2020)

Valina, H., Sands, R.D., Van Der Mensbrugghe, D. et al. (2014). The future of food demand: Understanding differences in global economic models. *Agricultural Economics* 45: 51–67.

Venkatesh Bhat, B., Dayakar Rao, B., and Tonapi, V.A. et al.(2018). *The Story of Millets*, 110. Bengaluru, India: Karnataka State Department of Agriculture. with ICAR-Indian Institute of Millets Research, Hyderabad, India.

Weber, S.A. (1998). Out of Africa: The Initial impact of millets in South Asia. *Current Anthropology* 39 (2): 267–274.

Weber, S.A. and Fuller, D.Q. (2007). Millets and Their Role in Early Agriculture. Paper presented in the International Seminar on the First Farmers in Global Perspective, Lucknow, India. *Pragdhara* 18: 69–90.

Wijesinha-Bettoni, R. and Burlingame, B. (2013). Milk and dairy product composition. In: *Milk and Dairy Products in Human Nutrition* (ed. E. Muehlhoff, A. Bennett and D. McMahon), 41–90. Rome, Italy: FAO.

World Bank Group (2016). *Global Monitoring Report 2015/16: Development Goals in an Era of Demographic Change*. Washington DC: World Bank.

Yabuno, T. (1962). Cytotaxonomic studies on the two cultivated species and the wild relatives in the genus *Echinochloa*. *Cytologia* 27: 296–305.

Yabuno, T. (1966). Biosystematic study of the genus Echinochloa. *Journal of Japanese Botany* 19: 277–323.

Yadav, S.S., Hegde, V.S., Habibi, A.B. et al. (2019). Climate change, agriculture and food security. In: *Food Security and Climate Change* (ed. S.S. Yadav, R.J. Redden, J.L. Hatfield, et al.), 1–24. John Wiley & Sons Ltd.

Zambon, I., Cecchini, M., Egidi, G. et al. (2019). Revolution 4.0: Industry vs. Agriculture in a Future Development for SMEs. *Processes* 7: 36. https://doi.org/10.3390/pr7010036.

2

The Art and Science of Growing Microgreens

Sreenivasan Ettammal

Research and Development, The Western India Plywoods Ltd, Mill Road, Baliapatam, Kannur, Kerala-670010, India

CHAPTER MENU

2.1 Introduction

With the rising global population, the current rate of progress in agriculture production will not be enough to eradicate widespread hunger and malnutrition by 2030. Hence, transformation of food systems is urgently needed, not only to provide sufficient food, but also to improve the nutritional status of the diet both in the developed and developing countries (FAO 2017; Weber 2017). One way to address this increasing demand for nutrient-rich food is by producing fresh and pesticide residue-free vegetables with lots of potential to cure various vitamin and mineral deficiencies. The use of affordable and sustainable cultivation practices, such as the soil-less culture, microgreens, etc., offers a promising approach to increase the production of vegetables without causing damage to the supporting ecosystems. Microgreens, the popularly known "new Superfoods." are immature edible vegetables. Their normal size is between a sprout and a well-grown seedling and they are harvested anywhere from a week to two weeks after germination with two fully developed cotyledons and a pair of partially developed true leaves (Xiao et

Bioresource Technology: Concept, Tools and Experiences, First Edition. Tanveer Bilal Pirzadah, Bisma Malik, Rouf Ahmad Bhat, and Khalid Rehman Hakeem.
© 2022 John Wiley & Sons Ltd. Published 2022 by John Wiley & Sons Ltd.

al. 2016). Microgreens are valued for their flavor and are widely acclaimed as an essential component to enrich diets by adding texture and color to our dishes. The visually appealing greens can also be eaten in salads, in sandwiches or wraps, in pizzas, stir-fries or added to soups (Treadwell et al. 2010). As it is an emerging field, scientific studies on microgreens are limited and its immense potential is being explored by various research teams to generate new insights. The present chapter reviews the art and science of growing microgreens and also examines the prospects and challenges of future projects on microgreens.

2.2 Historical Background

The marvel of microgreens is known to have existed for about only 90 years, with the early records of cultivation of wheat grass for medicinal purposes in the 1930s and winter greens like sunflower during the 1960s and home grown "grasses" in the 1970s for their health benefits (Kumar et al. 2018). Later, in the 1980s, chefs in San Francisco, California started growing "cresses" and "seedlings" for garnishing their menus (Treadwell et al. 2010). Regarding the large-scale ventures, at present the US, at present the microgreen industry consists of a variety of seed companies and growers. In recent years, microgreens have entered the Indian market also, beginning with the early initiatives by a few chefs and food entrepreneurs in growing and blending various microgreens in the restaurant menu. Their relentless efforts have created interest in growing these miniature plants and put them in the spotlight among many people about their importance in healthy living.

2.3 Health Benefits of Microgreens

2.3.1 Source of Functional Food Components

Microgreens are well known for their health benefits and they provide almost all the nutrients required for our healthy living (Xiao et al. 2012). The phytonutrients are beneficial to human health and also known to lower the risk of certain cancers and heart diseases. For example, the most common nutrients found in *Brassica* microgreens are β-carotene and other carotenoids and vitamins K, C and E (Herr and Büchler 2010). A recent study on microgreens of radish, French basil, roselle, etc. also revealed that they are excellent sources of essential vitamins such as ascorbic acid, vitamin E and also β-carotene, a precursor to vitamin A (Ghoora et al. 2020). These tiny plants can be used to brighten up a wide variety of main dishes with their vivid color ranges. Microgreens are added to various dishes whole or chopped to release their unique flavors and aroma. Usually, microgreens are not heated to avoid deterioration and so are generously used raw in a variety of cold dishes. However, at present, many are interested in cooked microgreens as side dishes for vegetarian meals and prefer the tasty recipes with the addition of other common leafy vegetables.

2.3.2 Component of Space Life Support Systems

Microgreens are grown as "space crops" to provide fresh, nutritious food to astronauts as they can be harvested directly ensuring freshness and high quality (Turner et al. 2020). Important factors impacting space plant growth and consumption are water delivery to the root zone in microgravity, sanitation methods for microbiological safety, plant responses to the light quality/spectrum and identifying optimal edible plants suitable for growth on the International Space Station (ISS). Availability of most suitable substrates and efficient lighting for growing crops in microgravity hydroponic systems are some of the major constraints of food production on space flights. In the case of substrates, synthetic fibrous media are found to be suitable for the microgreens Phytochemical Factory (MPF) in the space root module (Nyenhuis and Drelich 2015). Perchonok et al. (2012) reported the ability of fibrous media in alleviating the problems associated with crop performance in low oxygen and nutrient solubility conditions.

2.3.3 Component of Nutritional Diet of Troops and Residents of High Altitude Regions

The majority of residents at high altitudes are facing various degrees of food scarcity and under-nutrition and providing conventional vegetables and fruits in such locations is found to be extremely difficult during all seasons. In order to solve this problem, the Defense Institute of High Altitude Research, Leh Ladakh (DIHAR-DRDO) has carried out many investigations and been able to develop cultivation techniques for producing microgreen vegetables for high altitudes to supplement the food of their army personnel despatched to such regions, as well as for the local residents (Singh et al. 2019).

2.4 Cultivation Practices

2.4.1 Species Selection

Table 2.1 shows the major microgreen crops that are suitable for growing at home and for commercial production units. An exclusive list of plant species and the most common families used at present as microgreens is presented by Ebert (2012). The list is currently being expanded based on trials on a wide range of species, other than those of the *Solanaceae* or nightshade family which are reported as unfit as microgreens as they are known to contain toxins in their stems.

2.4.2 Growing Media and Propagation Felts

To set up a modest-scale or home unit, any low-cost or easily available nursery trays or shallow containers can be used for successful cultivation. Size and number of containers can be selected based on the availability of space for cultivation and also the convenience in handling and transporting the fresh produce. For commercial units, it is good to choose any containers with provision for proper drainage of water from the substrates. Selection of substrates is also very important and usually includes the list of conventional growing media.

Table 2.1 Common microgreen species suitable for growing at home and for commercial production units.

Species for production at home	Species production at commercial production units
Green gram, chickpeas, fenugreek, fennel, mustard, mung bean, green peas, cucumber, wheatgrass, garlic, onion, coriander and amaranth	Spinach, amaranth, cauliflower, table beet, mustard, buckwheat, arugula, red cabbage, white cabbage, broccoli, radish, lettuce, carrot, cress, basil, garlic, onion, leek, chive, fennel, lemongrass, popcorn, sweet pea, fennel, radish, melon, cucumber, lentils, sunflower, rice, oats, wheat, corn, barley and celery

Microgreens are usually not grown in water, but in soil or soil substitutes such as a peat-based soil-less growing mix or other fibrous materials (cellulose pulp) filled into nursery flats or plug trays (Ebert 2012). It is also very important to consider the cost of substrates and conventional growers use a variety of locally available inert materials such as coir fiber, coir pith, jute fiber, vermiculite, etc. Natural substrates are preferred over synthetic ones, where the performance and phytochemical composition of the microgreens are also considered. For instance, the nitrate and overall macro-mineral concentrations of coriander, kohlrabi and *pak choi* microgreens grown on natural fibee substrates (agave fiber, coconut fiber and peat moss) are reported to be higher when compared to those on synthetic alternatives (capillary mat and cellulose sponge) (Kyriacou et al. 2020). Some growers prefer to sell microgreens as a "living product" so that the customer harvests and washes them as they are needed to serve the freshest quality. Hydroponic pads and soil-less substrates tend to be favored for this practice for ease of transport and perception of cleanliness in a kitchen environment (Turner et al. 2020; Renna et al 2017). Though the wood fiber-based substrates such as Fibralur^{R} and softboard were reported to be useful for hydroponic cultivation systems (Muro et al. 2005; Sreenivasan 2018), efforts on establishing their potential for the "clean-harvest" of microgreens are rather limited. The common brands of biofiber-based propagation mats available in the international market are listed in Table 2.2. However, in India, the production technology of microgreens recently gained momentum due to its immense potential in the functional food sector. Furthermore, the growing media like organic potting mixes based on peat, pine bark, sawdust, rice hulls and coir pith are used in other conventional farming systems. It is worth mentioning here that most of the products available on the market at present either do not readily break down in the compost pile or are synthetic petroleum-based materials such as polymeric foams, plastic beads or are inorganic mineral-based, viz., perlite and stone wool. Since there are no commercial microgreen units in India today, the new low-cost propagation felts (e.g., wood and coir fiber-based mats), which are under various stages of development, are targeted mainly at the modest-scale growers in metros who sell high-quality fresh produce to the restaurant industry and also to those who are planning to invest in this upcoming agri-business field in the near future.

2.4.3 Growing Process

Microgreens are grown either indoors or outdoors depending on the scale of the operation. The different stages involved in the process of growing microgreens are depicted

Table 2.2 Growing media and propagation felts available in the market.

Product	Mat content	Company
BioStrate™	Bio-based textile	AmHydro, Arcata, California
Sucohans 12	Hemp fibers	Amazing-USA
Terrafibre™	Hemp fibers	BiocompositeGroup,Canada
PureGrown™	Hemp fibers	Grow-Tech, USA
Green Ease	Jute fibers	EnRoot Products LLC,USA
Coir fibere Mat	Coir fibers	Damrongslip Coconut, Thailand
Microgreen Coir Grow Mat	Coir fibers	Ascar Products, Erode, India
Biodegradable Micro-Mat™ Pads	Wood fibers	Handy Pantry, USA
Microgreens Hemp mat, Jute Microgreen Grow mat, Coir fiber Roll and Coco mats	Hemp, Jute and Coir fibers	Biodegradable Mats, Malaysia

Figure 2.1 Stages in growing green gram microgreens: (A) Sprouting stage; (B) First true stage; and (C) Harvesting stage.

in Figure 2.1. To begin with this cultivation, we need to partially fill the trays with a standard, sterile, loose, soil-less germinating media or other fibrous materials as seeding beds. The fibrous mats are often used in large-scale production systems. The burlap is commonly regarded as an excellent choice for certain crops (Kyriacou et al. 2016). Seed treatments through matrix priming or by using hydrogen peroxide may be useful to some seeds for their speedy germination or for improving the germination percentage, as observed in beet and chard by a research team led by Lee et al. (2004). Though it is very common to apply seeds as thickly as possible onto the substrates to maximize production, many units do not follow this practice to avoid the unwanted elongation of shoots. However, some growers have noticed the possibility of a "second harvest" of microgreens, just after the majority of tall-grown microgreens are carefully harvested from the mat having a very high density of seeds, leaving a certain percentage of

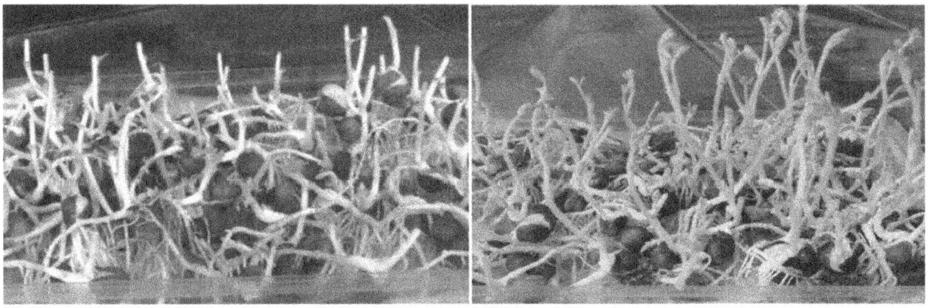

Figure 2.2 Chickpea microgreens showing regrowth of shoots after the first harvest.

ungerminated seeds and sprouts to develop further. In some rare cases such as chickpeas, "multiple harvests" were obtained from one set of seeds by the regrowth of shoots after successive cuts (Sreenivasan 2020) (Figure 2.2). Fertilizers are not required for growth of microgreens in general but some species benefit from the application of liquid fertilizers in low concentrations to the bottom of the trays. For producing premium-quality microgreens a adequate nutrients may also be provided to the growing media (Kyriacou et al. 2016). Attention may also be given to the strength of the chemical contents of the nutrients used for fertilizing (Palmitessa et al. 2020). Unlike the fast-growing species, fertilizers are not essential for the large-seeded, slow growing crops such as the chickpeas because of their self-contained nutrients which are not exhausted within a few days of germination.

2.5 Quality and Shelf Life

Microgreens are perishable crops and the short shelf life is attributed to the post-harvest degradation due to the rapid respiration of the microgreens after the removal of roots. Pre-harvest calcium chloride spray, chitosan powder coating, *Aloe vera* gel coating and organic acid treatments are some of the methods developed for improving the shelf life of *Brassica oleracea*, *Raphanus sativus*, *Hibiscus sabdarifa* and *Helianthus annus* microgreens, based on the standard techniques for storage of fruits and vegetables. Further investigations are required to explore the potential of these methods for extending to other microgreens (Kou et al 2015; Lu et al. 2018; Ghoora and Srividya 2020; Dalal et al. 2020). Microgreens are also prone to pre- and post-emergence damping-off caused by soil-borne pathogenic fungi. The disease can be successfully controlled in beet microgreens by a combination of *Trichoderma harzianum* and *T. virens* strains (Pill et al. 2011). There are several research groups working on the occurrence and growth of bacteria, yeast and mold on microgreens and how to develop an effective protocol to address such problems (Bergspica et al. 2020; Chandra et al. 2012). Misra and Gibson (2020) reported that survival of a bacterial pathogen is dependent on the type of soil-less media and recommend selecting a suitable matrix for reducing the risk of contamination in microgreens. Microgreens are grown on conventional and artificial soil substitutes and eating of unclean raw greens is known to cause common bacterial infections from *Escherichia coli, Salmonella, Listeria monocytogenes* and also from Norovirus (Bergspica et al. 2020; Li and Uyttendaele 2018).

2.6 Market Trends

India is emerging as a good market and is expected to grow further with the boost in consumer economy driven by elite and affluent income segments, hospitality and the travel sectors. Recently, microgreens are gaining increasing popularity among the general public in India, as they have realized that this indoor farming activity is relatively easy and inexpensive to undertake during their "stay at home" in the lockdown period declared by the Central government for the safety of the citizens during the Covid-19 pandemic. Due to the awareness created by a few agri-news reporters through the press and social media, interest in the cultivation and use of microgreens has received great acceptability within a short period, as predicted by Weber (2017) in a recent research article on broccoli microgreens. Also, the benefits of using microgreens is good for attracting new entrants to the business of propagation and marketing and the product is expected to witness an increase in demand year after year.

2.7 Future Outlook

During the last two decades, in addition to ongoing efforts to establish production units, research work is also underway to select new species of vegetables and other edible plants having potential for commercialization. Similarly, the factors affecting the growth and development, such as light and temperature, on growth and nutritive value of microgreens are also being explored. Other research priorities under various stages of experimentations in India and elsewhere include: i) development of novel low-cost, biodegradable propagation felts; ii) discovering new treatment methods for maintaining both quality and safety of microgreens that includes studies to optimize nutrient content of microgreens; iii) optimize the washing and drying techniques for delicate greens; iv) standardize new techniques to store and transport microgreen products; and v) strategies to be adopted to reduce the generation of wastes and/or to manage the wastes being generated from microgreens production operations.

2.8 Conclusions

According to the Food and Agriculture Organization (FAO), in order to solve the problem of feeding our growing population in both developed and developing countries, transformation of food systems is urgently needed not only to provide sufficient food, but also to improve the nutritional status of the diet (FAO 2017). Over the years, the use of modern sustainable cultivation practices, such as hydroponics, aquaphonics and microgreens, are considered as promising approaches to increase the nutrient food production in order to combat food and hidden hunger crisis. Microgreens possess the potential to be included in the list of our ongoing attempts to eliminate the problems of malnourishment from the developing countries.

References

Bergspica, I., Ozola, A., Miltiņa, E. et al. (2020). Occurrence of pathogenic and potentially pathogenic bacteria in microgreens, sprouts, and sprouted seeds on retail market in Riga, Latvia. *Foodborne Pathogens and Disease* 17: 420–428. doi:10.1089/fpd.2019.2733.

Chandra, D., Kim, J.G., and Kim, Y.P. (2012). Changes in microbial population and quality of microgreens treated with different sanitizers and packaging films. *Horticulture, Environment, and Biotechnology* 53: 32–40.

Dalal, N., Siddiqui, S., and Neeraj, (2020). Sensory attributes of sunflower microgreens with storage under ethanol vapour and organic acid treatments. *International Journal of Current Microbiology and Applied Sciences* 9: 208–214. https://doi.org/10.20546/ijcmas.2020.902.906.

Ebert, A.W. (2012). Sprouts, microgreens, and edible flowers: The potential for high value specialty produce in Asia. In: *Proceedings of the Regional Symposium on High Value Vegetables in Southeast Asia: Production, Supply and Demand (SEAVEG2012)*. Publication no. 12-758 (ed. R. Holmer, G. Linwattana, P. Nath, et al.), 216–227. Taiwan: AVRDC-The World Vegetable Center.

FAO. (2017). *The Future of Food and Agriculture Trends and Challenges*, 180. Rome: Food and Agriculture Organization of the United Nations.

Ghoora, M.D. and Srividya, N. (2020). Effect of packaging and coating technique on post-harvest quality and shelf life of *Raphanus sativus* L. and *Hibiscus sabdarifa* L. Microgreens. *Foods* 9: 653. doi:10.3390/foods9050653.

Ghoora, M.D., Babu, D.R., and Srividya, N. (2020). Nutrient composition, oxalate content and nutritional ranking of ten culinary microgreens. *Journal of Food Composition and Analysis* 91: 103495. doi:10.16/j.jfca.2020.103495.

Herr, I. and Büchler, M.W. (2010). Dietary constituents of broccoli and other cruciferous vegetables: Implications for prevention and therapy of cancer. *Cancer Treatment Reviews* 36: 377–383. doi:10.1016/j.ctrv.2010.01.002.

Kou, L., Yang, T., Liu, X. et al. (2015). Effects of pre- and post-harvest calcium treatment on shelf life and postharvest quality of broccoli microgreens. *HortScience* 50: 1801–1808.

Kumar, S., Jasmin, B.L., and Saravaiya, S.N. (2018). Microgreens: A new beginning towards nutrition and livelihood in urban-periurban and rural continuum. In: *Technologies and Sustainability of Protected Cultivation for Hi-valued Vegetable Crops* (ed. S. Kumar, N.B. Patel, S.N. Saravaiya, et al.), 246–261. Navsari, Gujarat, India: Navsari Agricultural University.

Kyriacou, M.C., Rouphael, Y., Di Gioia, F. et al. (2016). Micro-scale vegetable production and the rise of microgreens. *Trends in Food Science and Technology* 57A: 103–115. doi:10.1016/j.tifs.2016.09.005.

Kyriacou, M.C., El-Nakhel, C., Panico, A. et al. (2020). Phenolic constitution, phytochemical and macronutrient content in three species of microgreens as modulated by natural fiber and synthetic substrates. *Antioxidants* 9: 252. doi:10.3390/antiox9030252.

Lee, J.S., Pill, W.G., Cobb, B.B. et al. (2004). Seed treatments to advance greenhouse establishment of beet and chard microgreens. *The Journal of Horticultural Science and Biotechnology* 79: 565–570.

Li, D. and Uyttendaele, M. (2018). Potential of human norovirus surrogates and *Salmonella enterica* contamination of pre-harvest Basil (*Ocimum basilicum)* via leaf surface and plant substrate. *Frontiers in Microbiology* 9: 1728. https://doi.org/10.3389/fmicb.2018.01728.

Lu, Y., Dong, W., Alcazar, J. et al. (2018). Effect of preharvest $CaCl_2$ spray and post-harvest UV-B radiation on storage quality of broccoli microgreens, a richer source of glucosinolates. *Journal of Food Composition and Analysis* 67: 55–62.

Misra, G. and Gibson, K.E. (2020). Survival of *Salmonella enteric* subsp. *enterica* serovar Javiana and *Listeria monocytogenes* is dependent on type of soil-free microgreen cultivation matrices. *Journal of Applied Microbiology* 129: 1720–1732. doi:10.1111/jam.14696.

Muro, J., Ignacio, I., Samitier, P. et al. (2005). Wood fiber as growing medium in hydroponic crop. *Acta Horticulturae* 697: 179–185. doi:10.17660/ActaHortic.2005.697.22.

Nyenhuis, J. and Drelich, J.W. (2015). Essential micronutrient biofortification of sprouts grown on mineral fortified fiber mats. *International Journal of Bioengineering and Life Sciences* 9: 1029–1032.

Palmitessa, O.D., Renna, M., Crupi, P. et al. (2020). Yield and quality characteristics of *Brassica* microgreens as affected by the NH_4:NO_3 molar ratio and strength of the nutrient solution. *Foods* 9: 677. doi:10.3390/foods9050677.

Perchonok, M.H., Cooper, M.R., and Catauro, P.M. (2012). Mission to Mars: Food production and processing for the final frontier. *Annual Review of Food Science and Technology* 3: 311–330. doi:10.1146/annurev-food-022811-101222.

Pill, W.G., Collins, C.M., Gregory, N. et al. (2011). Application method and rate of *Trichoderma* species as a biological control against *Pythium aphanidermatum* (Edson) Fitzp. in the production of microgreen table beets (*Beta vulgaris* L.). *Scientia Horticulturae* 129: 914–918.

Renna, M., Di Gioia, F., Leoni, B. et al. (2017). Culinary assessment of self-produced microgreens as basic ingredients in sweet and savory dishes. *Journal of Culinary Science and Technology* 15: 126–142.

Singh, N., Aditika, S., Rani, S. et al. (2019). Vegetable microgreens farming in high-altitude region of trans-Himalayas to maintain nutritional diet of Indian troops. *Proceedings of the National Academy of Sciences, India Section B: Biological Sciences* The National Academy of Sciences, India. doi:10.1007/s40011-019-01147-0.

Sreenivasan, E. (2018). Softboard as a novel wood fibre-based substrate for hydroponic systems: A preliminary study. *International Journal of Research and Scientific Innovation* 5: 129–131.

Sreenivasan, E. (2020). Preliminary report on multiple harvests of microgreens from Chickpea (*Cicer arietinum)* seeds. *International Journal of Trend in Scientific Research and Development* 4: 306–308.

Treadwell, D., Hochmuth, R., Landrum, L. et al.(2010). Microgreens: a new specialty crop (Gainesville, FL: University of Floridla. IFAS Extension HS1164)

Turner, E.R., Luo, Y., and Buchanan, R.L. (2020). Microgreen nutrition, food safety, and shelf life: A review. *Journal of Food Science* 85: 870–882.

Weber, F.C. (2017). Broccoli microgreens: A mineral-rich crop that can diversify food systems. *Frontiers in Nutrition* 4: 1–9. doi:10.3389/fnut.2017.00007.

Xiao, Z., Lester, G.E., Luo, Y. et al. (2012). Assessment of vitamin and carotenoid concentrations of emerging food products: Edible microgreens. *Journal of Agricultural and Food Chemistry* 60: 7644–7651.

Xiao, Z., Codling, E.E., Luo, Y. et al. (2016). Microgreens of Brassicaceae: Mineral composition and content of 30 varieties. *Journal of Food Composition and Analysis* 49: 87–93. doi:10.1016/j.jfca.2016.04.006.

3

Novel Nutraceuticals From Marine Resources

Zadia Qamar, Amna Syeda, Javed Ahmed and M. Irfan Qureshi

Proteomics and Bioinformatics Lab, Department of Biotechnology, Jamia Millia Islamia, Delhi, India

CHAPTER MENU

3.1 Introduction

According to WHO: "Health is not only the absence of infirmity or disease but it is a state of total mental social and physical well-being" (Callahan 1973). There is substantial data supporting that diet has a major impact on health, and diet is a vital determinant for the prevention of disease. Current health can only be affected via eating habits (Panico et al. 2014). The term "nutraceutical" includes dietary supplements, fortified food or raw food comprised of bioactive molecules (Palthurk et al. 2010), that not only provide basic nutrition but also are good for health (Liu 2003). In 1989, S. DeFelice put forward the idea of "nutraceutical," which includes the terms "pharmaceutical" and "nutrition." These bioactive compounds known as "bioceuticals" have remarkable potential to be used as active pharmaceutical ingredients and

Bioresource Technology: Concept, Tools and Experiences, First Edition. Tanveer Bilal Pirzadah, Bisma Malik, Rouf Ahmad Bhat, and Khalid Rehman Hakeem.
© 2022 John Wiley & Sons Ltd. Published 2022 by John Wiley & Sons Ltd.

can also be used for designing several nutraceuticals in food supplements (Dewapriya and Se-kwon 2014). Nutraceuticals are usually extracted from marine natural sources by a non-denaturing process in order to retain their original properties. Microflora of marine resources have overwhelming biodiversity and thus there is a crucial need to exploit and explore their potential as nutraceutical and bio-therapeutic agents. These bioactive molecules contain vitamins, peptides, fatty acids, polysaccharides and phytochemicals that are present naturally in the foods and can be further formulated into dietary supplements (Thorpe et al. 2000; Ruiz-Torres et al. 2017). This group also contains bacteria, macroalgae, some fishes, microalgae, crustaceans and cyanobacteria that under hostile environment produce bioactive secondary metabolites. Microflora of marine resources have overwhelming biodiversity and thus there is a crucial requirement to exploit and explore their capability as nutraceutical and biotherapeutic molecules. It has been well estimated that in the marine environment the diversity is greater than in the tropical forests. The aquatic environment is known for its huge biodiversity which also makes it a great source of new bioactive natural compounds. The microbiota of this environment are soft bodied that normally have a sedentary type of life-style. The toxic components from marine sources help the organisms in sensing predators and they can even be used to paralyze their competitors.

The organisms from marine resources are a great reserve of bioactive natural compounds, which are be used for the treatment of disorders/diseases such as AIDS, cancer and heart problems. It was in 2004, when FDA approved zinconotide obtained from the cone snail, that it became evident that the bioactive compounds from marine resources could be a potential way to gain novel entities of great therapeutic value. Marine organisms generate bioactive compounds to defend themselves from the harmful effects of high concentrations of oxygen and light. In the marine environment, diverse types of bacteria, fungi, microalgae, actinomycetes, microalgae, vertebrates and invertebrates are present that generate bioactive molecules which have pronounced pharmacological properties. The natural bioactive products from marine resources include various secondary metabolites such as peptides, sterols, carotenoids, sulfated polysaccharides, derivatives of chito-oligosaccarides, chitosan and chitan (Ruis-Torres et al. 2017). Some important metabolites such as phlorotannins are produced by marine brown algae which possess anti-diabetic, antimicrobial, anti-inflammatory, anti-hypersensitive and radioprotective activities (Erdmann et al. 2008). Marine algae also produce sulfated polysaccharides that have an important role in activity against AIDS and cancer (Agarwal and Rao 2000). In the marine environment, the most simply organized and primitive type of algae are the microalgae. They are a great source of several food nutrients such as polyunsaturated fatty acid (PUFA), vitamins, polysaccharides and also astaxanthin (Grobbelaar 2004). The bioactive constituents of microalgae are used as dietary supplements, for fortifying infant milk and as food additives (Volkman 2003).

3.2 Marine Microorganisms as a Source of Nutraceuticals

Natural bioactive molecules produced from marine sources are highly effective as therapeutic agents for various non-infectious and infectious diseases (Figure 3.1). The group of symbiotic microorganisms has been confirmed to be an important source of natural

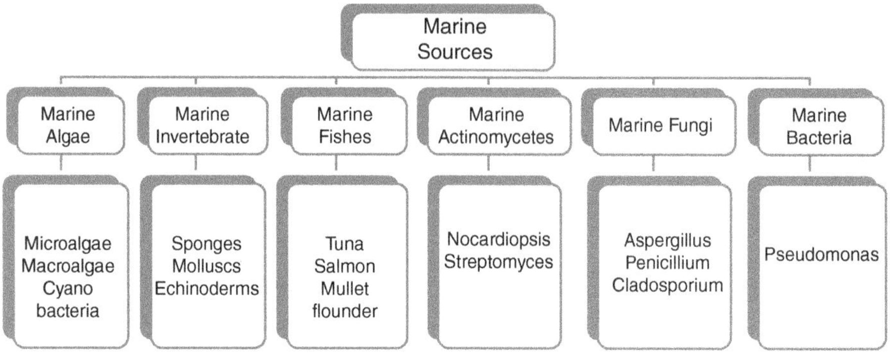

Figure 3.1 Different marine organisms as a source of nutraceuticals.

bioactive compounds having therapeutic value. Several fungi and bacteria have now been sampled from invertebrates' internal tissues and also from marine plants' surfaces and they have found to be of growing interest.

3.2.1 Marine Algae

Marine algae are the source of food ingredients as well as food in the nutraceutical industry (Tseng 2001). They are further divided into two categories, macroalgae and microalgae, and both have nutraceutical applications. Blue-green algae, also known as Cyanobacteria, is also included in this group and it is a Gram negative bacteria that produce a group of secondary bioactive metabolites having antibiotic, antiviral, antifungal and various other biological activities (Leandro et al. 2020). They also show bioactivity against invertebrates, vertebrates, microalgae, viruses, bacteria, plants and also in human cell lines. Hence they are of great nutraceutical and pharmacological value. Furthermore, Patellamide A and Ulithiacyclamide are produced by class Cyanobactins possessing antitumor, antimalarial and also multidrug reversal activities (Sivonen et al. 2010). Microalgae are simply organized and the most primary member of the marine plants and are rich sources of important food constituents like polysaccharides, astaxantin, PUFA, Vitamins A, H, D, C and B_{12}, B_6, B_2, B_1 (Grobbelaar 2004; Luiten et al. 2003; Gó,ez-Guillén et al. 2002). In the nutraceutical industries, the most common type of algae used are the macroalgae, which are also commonly known as seaweeds (Figure 3.2). In Asian countries such as China, Japan and Korea, macroalgae provide a variety of food ingredients and food (Indegaard and Minsaas 1991). In 2007, the estimated value of the production of macroalgae was found to be US$5 billion (Pickering et al. 2007). Agarose is obtained from macroalgae and they are also a source of natural products and metabolites with unique therapeutic and nutritional properties. They have also been used to improve the digestibility of protein and in the preparation of bioactive peptides. Recently, a bioactive peptide having antihypertensive properties have been isolated from them that may act as ACE (angiotensin converting enzyme) inhibitors (Kim et al. 2011). Macroalgae are also a good source of both soluble and insoluble dietary fiber, as in the gut they are not hydrolyzed by enzymes (Ruperez and Toledano 2003). Examples of storage and structural

Figure 3.2 Marine seaweeds (A) *Ascophyllum nodosum*; (B) *Osmundia pinnatifida*; and (C) *Aspargopsis armata.*

polysaccharides found in brown and red seaweeds include agar, alginates, carrageenan, fucan and laminaran. The alginates obtained from marine brown algae due to their biological activity are used as hydrocolloids (Rioux et al. 2007). Fucans obtained from brown seaweeds can be utilized in the cosmetics and food industries. Seaweeds provide a source of phycocolloids and thermopolysaccharides which have been used for centuries because of their emulsifying and gelling properties (Shalaby 2011; Stengel et al. 2011). Seaweeds are the chief source of effective molecules having numerous pharmacological activities such as anti-tumor, anti-inflammatory (Lindequist and Schweder 2001) and anti-microbial (Newman and Cragg 2004) activities, making them suitable for curing several diseases. For example, there is a significant drop in the proliferation and progression of tumor cells when treated with the extract of *Palmaria palmate* (Yuan et al. 2005). Seaweeds Halimeda species is a source of various novel phenolic bioactive natural compounds like catechuic acid, trihydobenzoic acid and dexcyanidanol (Yoshie et al. 2002). However, there still are many bioactive molecules from these algae that remain to be identified (Figure 3.3).

3.2.2 Marine Invertebrates

Marine invertebrates like crutaceans, molluscs, echinoderms and sponges are sources of alkaloids, phenols, bioactive peptides, ether, terpenoids, steroids and strigolactones (Hu et al. 2011). Compounds or extracts from the marine invertebrates have anti-cancer, antifungal, antiviral, anthelmintic, antibacterial, immuno-modulatory and antihypertensive properties (Hu et al. 2011).

3.2.2.1 Sponges

Sponges are the simplest of all multicellular organisms, consist of gelatinous material, and are sessile metazoans. They body is hollow pitcher-like that is covered by collagen fibers (spongia) and spicules (calcium structure or silica) (Uniz et al. 2003). There are several chemically different compounds isolated from the sponges such as sterols, alkaloids, peptides, terpenoids, polyphenolic compounds, macrolides and polyketides that have the potential to cure many diseases (Thoms and Schupp 2007).

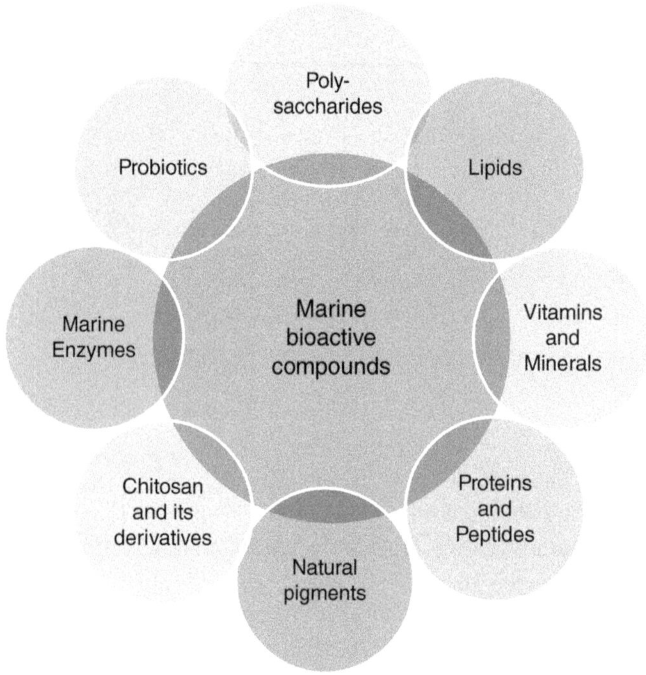

Figure 3.3 Marine environment as a reservoir of bioactive compounds.

3.2.2.2 Crustaceans, Echinoderms and Molluscs

Crustaceans constitute the most economical and largest class of the marine arthopoda in fish markets and they also play an important role in the nutraceutical and food industries (Kim 2020). Due to their health benefits and utilization, shrimps, prawns and crabs have gained huge attention (Smith et al. 2010). It was found that the composition of nutrients in crustaceans like krill and shrimp was studied and it has been shown to improve the levels of Vitamin A, omega-3 fatty acids and eicosapentaenoic and also reduce blood lipids. Molluscs have been consumed widely as marine foods and are also as known as functional foods. Fermentation of oyster sauces and blue mussel produces bioactive peptides that decrease hypertension, whereas eye disease can be treated by ground abalone and also its shells (Kim and Pallela 2012). In many Asian countries, octopus, nautiloids, squid and cuttlefish are consumed as a result of their therapeutic potential such that ear inflammation, gastrointestinal disorders and rickets can be cured (Kim and Pallela 2012). Hence due to their greater nutritional content, they are suggested to be used for human consumption and also in health food products development (Tou et al. 2007).

3.2.2.3 Marine Fishes

The uppermost position in consumption of marine animals is occupied by fish and is significant to the economy of world. In 2012, approximately 16% of the world's total protein requirements was provided by fish, with tuna, herring, cod, mullet, salmon, anchovy and flounder being the most common species used for that consumption (Kim and

Pallela 2012). Worldwide, one of the largest commercially canned products is tuna. The nutritional benefits of fish consumption is the presence of unsaturated fatty acid, proteins, Vitamins (D, A, E and B_{12}, B_6 and B_3) and minerals (iron, zinc, calcium and selenium). Research has also revealed that the peptides obtained from the fermentation of fish followed by enzymatic treatment might be valuable therapeutics for curing many chronic and acute diseases like cancer, hypertension, Alzheimer's diseases and viral infections (Wijesekara and Kim 2010; Vo and Kim 2010). The collagen of fish is used in the treatment of bone as a substitute to collagen of mammals which is referred to as immunogenica (Mos et al. 2004).

3.2.2.4 Marine Actinomycetes

Marine actinomycetes are known to be secondary metabolites producers. Many antibiotics having potential biological activities such as tetracycline, erythromycin and streptomycin are produced by them (Takahashi and Omura 2003). Many novel anti-microbial compounds (El-Gendy et al. 2008), antimalarial (Prudhomme et al. 2008) and anti-cancerous compounds (Olano et al. 2009) are produced from marine actinomycetes. Furthermore, *Nocardiopsis lucentensis* are other marine actinomycetes that produce Lucentamycins A-D peptides from their fermentation broth (Cho et al. 2007). They are known to possess cytotoxicity activity in HCT-116 colon carcinoma cells. Similarly, *Streptomyces* sp. marine derived isolates are known to produce Mansouramycin A-D (four new derivatives). These marine isolates are known to have strong cytotoxicity, having selectivity for breast cancer, non-small cell lung cancer, prostate cancer and melanoma, which suggests that these isolates can be used as anti-tumor drugs (Hawas et al. 2009). Similarly, Tartrolon-D and macrodiolide were isolated from a *Streptomyces* fermentation broth by Perez Pérez et al. (2009). The Tartrolon was found to have cytotoxic activity against colon cancer (HT29) and lung cancer (A549) human cell lines (Pérez et al. 2009). In another study, *Saccharomonospora* sp. secondary metabolite produces a new alkaloid compound Lodopyridone, which is known to possess cytotoxic activity against human colorectal cancer cells (HCT-116) (Maloney et al. 2009). Moreover, these marine actinomycetes also possess antimicrobial property besides their anti-cancerous and antitumor activity. Two new bioactive compounds were isolated (Carlson et al. 2009), namely tirandamycin D and E, dienoyl tetramic acids from marine *Streptomyces* species. These isolated compounds were extremely effective against the infection of *Enterococcus faecalis*. The trioxacarcins isolated from the marine environment also possess good antimalarial, antitumor and antimicrobial activity. The *Streptomyces* strain crude extract isolated from the deep sediments also showed strong antifouling properties (Xu et al. 2010). Loreneic acid isolated from marine *Actinomyces* had substantial inhibition activity against the enzyme phosphodiesterase (Iwata et al. 2009).

3.2.2.5 Marine Fungi

Marine fungi produce a variety of different bioactive compounds that are known to possess antibacterial, anti-inflammatory, antiviral and anti-cancer properties (Bhadury et al. 2006; Newman and Hill 2006). These unique molecules are used as main structures for protection of plants and for medicines. The complete process for their cultivation and isolation from different oceanic organisms (i.e., sponges, algae and mangrove plants) has been described by Kjer et al. (2010). The structure of the secondary metabolites formed by these fungal species are also explained. An anthraquinone derivative known as aspergiolide A was also isolated from *Aspergillus glaucus*, which is a filamentous marine fungus in Fujian China (Du et al. 2007). It was also known to show the cytotoxic effect against P388 and K562 cell lines. A marine fungi isolate known as *Exophiala* was found to show antimicrobial activity along with nematicidal activity. Another strain of marine fungi *Phoma herbarum* was found to have radical scavenging property due to the occurrence of halogenated benzoquinones (Henkep et al. 2010).

3.2.2.6 Marine Bacteria

Marine bacteria thrive in extreme oceanic environments and are known to be the manufacturers of various bioactive secondary metabolites. Bacterial strain KMM is non-pigmented and aerobic and known to produce various bioactive compounds such as quinolone, pyrroles, phenazine, phthalate, moiramides and phloroglucinol (Romanenko et al. 2008). Quinones and alkaloids secondary metabolites, which are isolated from the marine bacteria, are accountable for their anti-tumor activities (Solanki et al. 2008). Furthermore, derivatives of quinine analogs, such as mitocin, daunorubicin, driamycin and lapachol, show antitumor activity by affecting the mitochondrial pathways and development of hydroxyl radicals as lethal products in human cell lines (Solanki et al. 2008). Antimicrobial secondary metabolites are also isolated from *Anadara broughtoni* (ark shell). which associates to the heterotrophic bacteria and is found to have potent hemolytic, antimicrobial activities (Romanenko et al. 2008). Pseudo-alteromonas tunicate associates with marine isolates tambjamine to form a yellow color pigment that is known to have anti-fungal properties and also possess antiproliferative, anti-microbial and immunosuppressive activities (Bhatnagar and Kim 2010).

3.3 Classification of Different Nutraceuticals Obtained from Marine Environment

3.3.1 Polysaccharides

There are many uses of the polysaccharides obtained from marine resources in beverages and food supplements. Marine polysaccharides obtained from crustaceans, algae and various other aquatic organisms include hydrocolloids, carrageenans, glycosaminoglycans and fucanoids, etc. Seaweeds (macroalgae) have greater amounts of polysaccharides such as carrageenans, alginates and agar. Seaweeds are collectively known as hydrocolloids or phycocolloids and they also act as a food fiber. A few other polysaccharides, which are isolated from seaweeds such as carrageenan, alginates, xylofucans, fucoidan, sulfated polysaccharides, caulerpicin, Sphinganine amide and alginic acid, can also be used as nutraceuticals (Berteau and Mulloy 2003). These molecules obtained from marine resources have numerous biological functions such as proliferative, anti-inflammatory, antithrombotic, anticoagulant and antiviral activities (Berteau and Mulloy 2003). Alginates and carrageenans are both biopolymers that have been recognized as the plentiful polysaccharides found in brown algae and red algae (Kim et al.

2008). Brown algae have a sulfated matrix and have complex polysaccharides called fucoidans. The fucoidans structure obtained from various aquatic organisms differs in content of sulfur, composition of saccharides, molecular weight, saccharide residue sequence and mode of linkage (Li et al. 2008). The presence of the sulfate groups further increase the activity of fucoidans, which permits their nutraceutical use in the dairy industries (Vidanarchchi and Kurukulasuriya 2011). These secondary metabolites obtained from marine resources also have benefits in human health, which aids them to be used as nutraceuticals. The capsules of bacteria generally have polysaccharides. From the nutraceutical point of view they are significant classes of the secondary metabolites. The exopolysaccharides obtained from these marine bacteria are used as food processing agents and also in numerous pharmaceutical formulations. Various research studies have shown that they can be used as adhesive, gelling agents, thickeners, and also as stabilizers and coagulating agents, etc. (Sapute et al. 2007). These exopolysaccharides are known to possess pseudo-plasticity and a very good viscosity that help them to impart the capability to resist the extremes of salinity, pH and the temperature. Hence this property raises their capability to be used in different industrial processes (Satpute et al. 2007) (Table 3.1).

3.3.2 Marine Lipids

The lipids from the marine environment originate from either crustaceans, fish or any other aquatic organisms. Sterols, phospholipids, wax esters and triacyl glycerol and their metabolic products are the chief composition of lipids from the marine environment. Lipids, such are glycolipids, hydrocarbons, glycerol esters and sulpholipids are present in minor amounts. Lipids of marine origin are mainly derived from marine mammals' blubber, lean fish liver and the fatty flesh of fish. Marine mammal oil and fish oils are rich sources of Docosahexaenoic acid and Eicosapentaenoic acid and both have anti-inflammatory activity (Raposo et al. 2013). Crustaceans contain spingomyeline which have anti-tumor and anti-bacterial activity. Approximately 08-2.0% of total lipids is present in crabs and lobsters. Marine microalgae lipids have extensive applications and are used as anti-viral, anti-inflammatory, anti-allergic and therapeutic agents (Raposo et al. 2013). These activities are due to the presence of components such as HUFA (highly unsaturated fatty acid), PUFA (polyunsaturated fatty acids) and various other substances (Sijtsma and de Swaaf 2004).

3.3.3 Natural Pigments from Marine Sources

Marine microalgae and macroalgae, besides lipids and polysaccharides, also provide different bioactive marine natural pigments. Marine algae natural pigments also provide pigmentation and food by the process of photosynthesis. The bioactive pigments also found to show benefits for health and for this reason they are one of the significant marine nutraceuticals. For instance, bioactive compounds such as lutein, Chlorophyll-a canthaxanthin and zeaxantin have anti-tumor properties; Phycocyanin, Zeaxanthin, Pheophorbide-a, Fucoxanthin and Astaxanthin all have excellent anti-oxidant properties; and pheophytin-a possesses anti-inflammatory and neuro-protective activity (Lee et al. 2007). Alpha-carotene

Table 3.1 Categories of bioactive molecules, sources and their applications in the nutraceutical industry.

Category	Bioactive molecule	Sources	Applications	Health perspective	References
Polysacchrides	Fucanoid and fucan	Sea urchin eggs Sea cucumber	Nutraceutical supplements	Antiviral, anti-inflammatory anticoagulant, antiproliferative, antithrombotic.	(Berteau et al. 2003)
	Carrageenan	Macroalga, e.g., *B. gelatinum*	Coatings in meat Gel forming	Antibacterial, anti-HIV. Anti-tumor, anti-coagulant.	(Vlieghe et al. 2002)
	Agar Agar	Red algae *Gigartina*	Food gums and gel formation		(Freile-Pelegrin and Murano 2005)
	Alginates	Cell wall of brown algae	Food industries as additives, plasticizers	Reduces intestinal absorption. Stimulates immune system.	(Taylor et al. 2005)
	Chitosan, chitin and its derivatives	Lobster, crab, krill, shrimp, prawn	De-acidification and clarification of fruits, edible protective films.	Fungicidal, bactericidal anti-tumor, increase dietary fibers.	(Shahide et al. 2005)
Lipids	Omega-3 fatty acids	Almost all marine sources	Nutraceuticals (fish capsules and oil)	Reduce risk of cardiovascular diseases, ameliorate diseases like hypertension and arthritis.	(Sijtsma and de Swaaf 2004)
	Docosahexaenoic acid (DHA)		Nutraceutical	Brain development.	(Raposo et al. 2013)
Pigments	Carotenoids β-carotene and lutein	*Chlorella vulgaris*, *Haematococcus pluvialis*, *Dunaliella salina*	Nutraceutical agent, natural food coloring	Anti-inflammatory anti-carcinogenic, anti-oxidant Vitamin A precursor.	(Maeda et al. 2005)
	Chlorophylls	*S. platensis*	Beverage, colorant Natural food	Anti-cancer activity.	(Bhattacharya and Shivaprakash 2005)

	Fucoxanthin	Brown seaweeds		Neuroprotective, anti-obesity, anti-cancer, anti-angiogenic, anti-inflammatory.	(Pangestuti and Kim 2011)
Protein and peptides	Albumin	Crustaceans, low fat fish and mollusks	Stabilizing agent, suspending and whipping agent	Anti-oxidant, and anti-coagulant.	(Nicholson et al. 2000)
	Gelatin	Fish (cod, hake, haddock, pollock)	Texturizer, confectionaries, thickener in jam, ice-cream, yogurt, cream	Prevention and treatment of atrophic gastritis	(Gomez-Ordonez et al. 2010)
	Collagen	fish (yellow sea bream albacore tuna, rainbow trout, common horse mackerel silver-line grunt)	Used in meat industry as coating of meat (e.g., sausages)	Anti-hypertensive, anti-oxidant, anti-skin-aging activities.	(Lai et al. 2008)
Minerals and vitamins	Zinc, iodine, iron, manganese. Water soluble and fat soluble vitamins	Almost all marine sources. Seaweeds are richest source	Nutraceutical, food and pharmaceutical industries	Minerals and vitamins, perform many essential functions such as cofactors and provide transport inside cell.	(Parr et al. 2006)
Marine enzymes	Cysteine and serine, Protease	Short-finned squid, mollusks and crustaceans	Meat tenderizing, squid-fermentation Preventing unwanted color changes in food products.		(Shahidi et al. 2005)
	Trans-glutaminase	Rainbow trout, Pollock liver, Red sea bream, scallop, atka mackerel	Creates protein cross-links to improve rheological properties of gels.		(Chen et al. 2003)

(*source*: based on Satpute et al. 2007)

can also be used in food industries as a food additive (Ande et al. 2017). In brown seaweed chloroplasts, the major carotenoid present is fucoxanthin. It accounts for more than 10% of total natural production of carotenoids and thus is the most abundant (Matsuno 2001). Fucaxanthin forms a complex with chlorophyll and hence plays a significant role in photoprotection and light harvesting. The chief allenic carotenoids are neoxanthin from higher plants, perinijin from the microalgae and fucoxanthin derived from brown algae (Dembitsky and Maoka 2007).

3.3.4 Chitosan and Its Derivatives

Chitosan is derived from chitin and is a natural polymer. After cellulose, it is the other most rich polysaccharide. Its recovery is from downstream processing of discarded lobster, crab, shrimp and crayfish following demineralization followed by de-proteinization. The chitin is then deacetylated to form chitosan (Park and Kim 2010). The cell walls of fungi are also chitin rich. Chitosan and chitin exhibit a variety of applications and can be used as biomaterial for improving quality of nutrition, as a food additive in the edible film industry, recovery of solid material from food processing and water purification. Furthermore, chitosan also possesses anti-microbial, antioxidant, anti-inflammatory and hypocholesterolemic activity (Je and Kim 2012). Chitosan also ruptures the cell membrane of Gram negative bacteria and also interrupts the barrier properties of the bacterial outer membrane as a result of ionic interactions between microbial membrane anionic group and chitosan cationic groups. Chitosan is also known to possess antifungal activity by interfering with fungal growth, formation of coats that are gas-permeable, stimulation of defense systems like chitinases build-up, protein inhibitor production and synthesis of callus stimulators (Shahidi et al. 2005). Chitosan have the ability to combine with the molecules of lipids and also chelates metal ions, which attributes for its anti-oxidative activity (Shahidi et al. 2005). Since chitosan have such special properties, it can be used in the food industry, pharmaceuticals, and also in agriculture and health due to its non-toxic nature, biodegradability and biocompatibility. Moreover, it can also be utilized as a vehicle for pharmacological compounds as well as nutraceutical compounds via encapsulation (Ande et al. 2017). The derivatives of chitosan may be produced to obtain a product that is more effective in having significant applications (Ande et al. 2017).

3.3.5 Proteins and Peptides

Proteins obtained from marine sources including seaweeds, crustaceans, fish (hake, pollock, tuna, rout, cod, haddock, herring), extremophiles such as *Dunaliella*, and molluscs have distinctive properties like antioxidant, gel-forming activity, antimicrobial and anticoagulant, and foaming and film capacity. The most common proteins from marine resources are albumin, gelatin and collagen that are used in diets and are hydrolyzed by enzymes for the bioactive peptides formation, which also have the capability to be utilized in nutraceutical industries (Agyei and Denquah 2011). In the food industry, protein protamine can also be used as an anti-bacterial agent (Sibilla et al. 2015). During processing or digestion, bioactive protein fragments or peptides are generated from the parent proteins that range from size 2–20 amino acids (De-Bernardini et al. 2011). Moreover, bioactive protein fragments can also be produced from fermentation, cooking or hydrolysis (alkaline or acid). These protein fragments have anti-hypertensive, anti-microbial, anti-thrombotic and immunomodulatory bioactivites (Murray and FitzGerald 2007). Researchers have focused on the peptides from marine sources for the development of pharmaceutical bioactive compounds for antihypertensive and ACE (Angiotensin converting enzyme) inhibition function (Byun and Kim 2001). The richest sources of bioactive natural compounds marine proteins are obtained from crustaceans, molluscs and fish (Rajapakse et al. 2005). Collagen is obtained from bovine meat and is also used in various industries such as pharmaceuticals, cosmetics, biomedicine and the food industry.

3.3.6 Minerals, Vitamins and Enzymes

Minerals and vitamins also perform essential functions, such as during metabolic processes they serve as cofactors, and help in transport inside the cells. Seaweeds are the richest source of minerals and vitamins including iodine, iron, zinc and manganese (Pena-Rodriguez et al. 2011). Cod liver oil produced commercially is rich in Vitamins D and A. Microalgae like *Tetraselmis*, *Arthrospira*, *cruentum* and *I. galbana* are rich in Vitamin alpha-carotene, A, K, E and C, having excellent anti-oxidant activity (Suleria et al. 2015). Vitamin E (alpha-tocopherol and Vitamin A (alpha-carotene and provitamin A) have strong free radical scavenging and antioxidant activity that protect the cells from damage by the oxidants. From pavlova (a small golden brown flagellate that is similar to Isochrysis), Vitamin K is isolated and helps in coagulation or blood clotting (Suleria et al. 2015). Enzymes have the ability to catalyze the reaction and convert the molecule into a valuable biological active compound that can be further used in nutraceutical and food industries. Enzymes as food ingredients have the ability to influence factors like processing, storage, spoilage and safety. Furthermore, enzymes from aquatic resources are chitinolytic enzymes, polyphenol oxidase (phendase, cresolase, catechol oxidase, tyrosinase, catecholase and polyphenolase), lipase and enzymes from red algae, which are involved in the degradation pathway of starch (eg α-1, 4-glucanase) (Okada and Morrisse 2007). When compared with their terrestrial counterparts they are known to possess excellent physical, chemical or catalytic properties (Diaz-Lopez and Garcia-Carreno 2000). They are being used in the processing of food due to their diverse properties, salt tolerance properties and great activity at mild pH and specificity (Okada and Morrisse 2007).

3.3.7 Marine Probiotics and Phenolic Compounds

The marine environment microbial diversity is rich and also very useful in producing effective and safe probiotics, which may be helpful in combatting antibiotic resistance. Probiotics are the compounds that are selectively fermented and non-digestible that are known to stimulate the activity and growth of valuable gut microflora, that in turn provides health benefits to the host. They are usually oligosaccharides, for example, chitosan oligosaccharides and moreover prebiotic activity is also possessed by the algal polysaccharides (Lee et al. 2007). From marine lactic acid bacteria exoplysaccharides, bifidogenic benefits have been reported (Honypattarakere et al. 2012). Biomass of cyanobacteria *Spirulina platensis* stimulates species of *Bifidobacterium* and *Lactobacillus* and thus stimulates their prebiotic effects. Generally, phlorotannins found in brown algae are the most abundant type of polyphenols and the flavonoids in green algae contribute most to the total content of phenolic compounds. The profile of marine brown algae is generally made up of dieckol, eckol and phloroglucinol (Shibata et al. 2008). Phlorotannins are known to possess anti-oxidant properties which enables them as active ingredients in the food industry and as nutraceuticals.

3.4 Important Bioactive Metabolites and Their Biological Properties

Some significant microorganisms produce bioactive secondary metabolites with important biological activities. For instance, marine microbial metabolites such as Dolastatin-10 (antimicrotubule property and act as the synthetic analog, TZT-1027, and thus can be used as an antitumor compound); Dolastatin-15 (antimicrotubule and acts as a synthetic analog, ILX-651, thus as an antitumor compound); Curacin A (antimicrotubule); and Toyocamycin (anti-fungal) are obtained from cyanobacteria (Suleria et al. 2015). Furthermore, Norharman (Enzyme inhibitor) and Calothrixin-A (antimalarial and anti-cancerous) are obtained from algae. Also from symbiotic microbes, BE-43472B (antibacterial) and Macrolactin V (antibacterial and antilarval) are obtained (Suleria et al. 2015). Similarly, Daryamide C (antitumor) and Marinomycin A (antitumor and antibiotic) are obtained from Actinomycetes (Bhatnagar and Kim 2010).

3.5 Current Status of Nutraceuticals in Market

Nutraceuticals, which include dietary supplements and functional foods, are known to have incredible market potential. The global market of nutraceuticals include dietary supplements, beverages and functional foods, and in 2014 it was valued at approximately $250 billion and this amount will continue to increase and likely to reach approximately $385 billion by 2020 (Mordor Intelligence Global Nutraceuticals Market 2015–2020). The market potential is led by Europe and the USA, which accounts for about 85% of the nutraceutical market (Research and Markets: Nutraceutocals 2012). In the UK, the Mintel survey was carried out on mineral and vitamin supplements and it was found that about 25% of the total adult population were pleased with the effects of the products of nutraceuticals. It is strongly believed that in future the nutraceutical industry will remain in the leading market with countries such as China and India, because of more responsiveness in consumers toward their health (Suleria et al 2015). In the USA, nutraceutical consumption is relatively higher and accounts for about 40% of the adult population. In Spain, only about 9% of people aged between 35 and 80 years consume dietary supplements as the major source of minerals and vitamins. Moreover, about 72% of women that are educated aged between 35 and 49 years are more likely to choose dietary supplements and nutraceuticals (Rovira et al. 2013).

3.6 Conclusion and Future Recommendations

Nowadays, nutraceuticals from marine sources are gaining attention in the supplement and food industries. The aquatic environment is enriched with a variety of microflora that has the potential to produce numerous bioactive secondary metabolites that have both pharmaceutical and nutraceutical value and to formulate novel drugs that are proved to be highly effective against several pathogens. The current chapter précises marine bioactive natural compounds and their possible application in the food industries as well as in the supplement industries. Furthermore, collaborative and interdisciplinary research are needed among marine microbiologists, scientists, medical practitioners and biotechnologists to provide inventive research and also isolate novel therapeutic molecules having nutraceutical potential.

References

Agarwal, S. and Rao, A.V. (2000). Carotenoids and chronic diseases. *Drug Metabolism and Drug Interactions* 17: 189–210.

Agyei, D. and Danquah, K. (2011). Industrial-scale manufacturing of pharmaceutical-grade bioactive peptides. *Biotechnology Advances* 29: 272–277.

Ande, M.P., Syamala, K., Rao, P.S. et al. (2017). Marine nutraceuticals. *Aquaculture Times* 3 (2): 06–09.

Bhattacharya, S. and Shivaprakash, M.K. (2005). Evaluation of three Spirulina species grown under Similar conditions for their growth and biochemicals. *Journal of the Science of Food and Agriculture* 85: 333–336.

Berteau, O. and Mulloy, B. (2003). Sulfatedfucans, fresh perspectives: Structures, functions, and biological properties of sulfatedfucans and an overview of enzymes active toward this class of polysaccharide. *Glycobiology* 13: 29R–40R.

Bhadury, P., Mohammad, B.T., and Wright, P.C. (2006). The current status of natural products from marine fungi and their potential as anti-infective agents. *Journal of Industrial Microbiology & Biotechnology* 33: 325–337.

Bhatnagar, I. and Kim, S.-K. (2010). Immense essence of excellence: Marine microbial bioactive compounds. *Marine Drugs* 8: 2673–2701.

Byun, H.G. and Kim, S.K. (2001). Purification and characterization of angiotensin I converting enzyme (ACE) inhibitory peptides from Alaska pollack (*Theragrachalcogramma*) skin. *Process Biochemistry* 36: 1155–1162.

Callahan, D. (1973). The WHO definition of "health." *Studies – Hastings Center* 1: 77–88.

Carlson, J.C., Li, S., Burr, D.A. et al. (2009). Isolation and characterization of Tirandamycins from a marine-derived *Streptomyces* sp. *Journal of Natural Products* 72: 2076–2079.

Chen, T., Embree, H.D., Brown, E.M. et al. (2003). Enzyme-catalyzed gel formation of gelatin and chitosan: Potential for *in situ* applications. *Biomaterials* 24: 2831–2841.

Cho, J.Y., Williams, P.G., Kwon, H.C. et al. (2007). Lucentamycins A–D, cytotoxic peptides from the marine-derived actinomycete *Nocardiopsis lucentensis*. *Journal of Natural Products* 70: 1321–1328.

Dembitsky, V.M. and Maoka, T. (2007). Allenic and cumulenic lipids. *Progress in Lipid Research* 46: 328–375.
Dewapriya, P. and Se-kwon, K. (2014). Marine microorganisms: An emerging avenue in modern nutraceuticals and functional foods. *Food Research International* 56: 115–125.
Diaz-Lopez, M. and Garcia-Carreno, F.L. (2000). Applications of fish and shellfish enzymes in food and feed products. In: *Seafood Enzymes* (ed. N.F. Haard and B.K. Simpson), 571–618. New York, USA: Marcel Dekker, Inc.
Di-Bernardini, R., Harnedy, P., Bolton, D. et al. (2011). Antioxidant and antimicrobial peptidic hydrolysates from muscle protein sources and by-products. *Food Chemistry* 124: 1296–1307.
Du, L., Zhu, T., Fang, Y. et al. (2007). Aspergiolide A, a novel anthraquinone derivative with naphtho(1,2,3-de)chromene-2,7-dione skeleton isolated from a marine-derived fungus *Aspergillus glaucus*. *Tetrahedron* 63: 1085–1088.
El-Gendy, M.M., Hawas, U.W., and Jaspars, M. (2008). Novel bioactive metabolites from a marine derived bacterium *Nocardia* sp. ALAA 2000. *Journal of Antibiotics (Tokyo)* 61: 379–386.
Erdmann, K., Cheung, B.W.Y., and Schroder, H. (2008). The possible roles of food-derived bioactive peptides in reducing the risk of cardiovascular disease. *Journal of Nutritional Biochemistry* 19: 643–654.
Freile-Pelegrín, Y. and Murano, E. (2005). Agars from three species of *Gracilaria* (Rhodophyta) from Yucatán Peninsula. *Bioresource Technology* 96: 295–302.
Gómez-guillén, M.C., Turnay, J., Fernández-Díaz, M.D. et al. (2002). Structural and physical properties of gelatin extracted from different marine especies: A comparative study. *Food Hydrocolloids* 16: 25–34.
Gomez-Ordonez, E., Jimenez-Escrig, A., and Ruperez, P. (2010). Dietary fibre and physico-chemical properties of several edible seaweeds from the Northwestern Spanish coast. *Food Research International* 9: 2289–2294.
Grobbelaar, J.U. (2004). Algal biotechnology: Real opportunities for Africa. *South African Journal of Botany* 70: 140–144.
Hawas, U.W., Shaaban, M., Shaaban, K.A. et al. (2009). Cytotoxic isoquinolinequinones from a marine streptomycete. *Journal of Natural Products* 72: 2120–2124.
Honypattarakere, T., Cherntong, N., Wickienchot, S. et al. (2012). *In vitro* prebiotic evaluation of exopolysaccharides produced by marine isolated lactic acid bacteria. *Carbohydrate Polymers* 87: 846–852.
Hu, G.P., Yuan, J., Sun, L. et al. (2011). Statistical research on marine natural products based on data obtained between 1985 and 2008.
Indegaard, M. and Minsaas, J. (1991). Animal and human nutrition. In: *Seaweed Resources in Europe: Uses and Potential* (ed. M.D. Guiry and G. Blunden), 21–64. Chichester, UK: Wiley.
Iwata, F., Sato, S., Mukai, T. et al. (2009). Lorneic acids, trialkyl-substituted aromatic acids from a marine-derived actinomycete. *Journal of Natural Products* 72: 2046–2048.
Je, J.Y. and Kim, S.K. (2012). Chitosan as potential marine nutraceutical. *Advances in Food and Nutrition Research* 65: 121–135.
Kjer, J., Debbab, A., Aly, A.H. et al. (2010). Methods for isolation of marine-derived endophytic fungi and their bioactive secondary products. *Nature Protocols* 5: 479–490.

Kim, S.K. (2010). *Chitin Chitosan, Oligosaccharides and Their Derivatives: Biological Activities and Applications*, 666. Boca Raton, FL: CRC Taylor & Francis.

Kim, S. and Pallela, R. (2012). Medicinal foods from marine animals: Current status and prospects. *Advances in Food and Nutrition Research* 65: 1–9.

Kim, S.K., Ravichandran, Y.D., Khan, S.B. et al. (2008). Prospective of the cosmeceuticals derived from marine organisms. *Biotechnology and Bioprocess Engineering* 13: 511–523.

Kim, E.Y., Kim, D.G., Kim, Y.R. et al. (2011). *An improved method of protein isolation and proteome analysis with Saccharina japonica* (Laminariales) incubated under different pH conditions. *Journal of Applied Phycology* 23: 123–130.

Lai, G., Yang, L., and Guoying, L. (2008). Effect of concentration and temperature on the rheological behavior of collagen solution. *International Journal of Biological Macromolecules* 42: 285–291.

Leandro, A., Pereira, L., and Gonçalves, A.M.M. (2020). Diverse applications of marine macroalgae. *Marine Drugs* 18: 17.

Lee, J.H., Seo, Y.B., Jeon, S.Y. et al. (2007). Functional analysis of combinations in astaxanthin biosynthesis genes from *Paracoccushaeundaensis*. *Biotechnology and Bioprocess Engineering* 12: 312–317.

Li, L.Y., Sattler, I., Deng, Z.W. et al. (2008). A seco-oleane-type triterpenes from *Phomopsis* sp. (strain HK10458) isolated from the mangrove plant *Hibiscus tiliaceus*. *Phytochemistry* 69: 511–517.

Lindequist, U. and Schweder, T. (2001). Marine biotechnology. In: *Biotechnology* (ed. H.J. Rehm and G. Reed), 441–484. Weinheim: Wiley-VCH.

Liu, R.H. (2003). Health benefits of fruit and vegetables are from additive and synergistic combinations of phytochemicals. *American Journal of Clinical Nutrition* 78: 517S–520S.

Luiten, E.E., Akkerman, I., Koulman, A. et al. (2003). R.H. Realizing the promises of marine biotechnology. *Biomolecular Engineering* 20: 429–439.

Maeda, H., Sakuragi, Y., Bryant, D.A. et al. (2005). Tocopherols protect *Synechocystis* sp. strain PCC 6803 from lipid peroxidation. *Plant Physiology* 138: 1422–1435.

Maloney, K.N., Macmillan, J.B., Kauffman, C.A. et al (2009). Lodopyridone, a structurally unprecedented alkaloid from a marine actinomycete. *Organic Letters* 11: 5422–5424.

Matsuno, T. (2001). Aquatic animal carotenoids. *Fisheries Science : FS* 67: 771–783.

Mordor Intelligence Global Nutraceuticals Market: Growth, Trends and Forecasts (2015–2020). Available from: http://www.mordorintelligence.com/industryreports/global-nutraceuticals-marketindustry

Mos, L., Jack, J., Cullon, D. et al. (2004). The importance of marine foods to a near-urban first nation community in coastal British Columbia, Canada: Toward a risk-benefit assessment. *Journal of Toxicology and Environmental Health, Part A* 67: 791–808.

Murray, B.A. and FitzGerald, R.J. (2007). Angiotensin converting enzyme inhibitory peptides derived from food proteins: Biochemistry, bioactivity and production. *Current Pharmaceutical Design* 13: 773–791.

Nenkep, V.N., Yun, K., Li, Y. et al. (2010). New production of haloquinones, bromochlorogentisylquinones A and B, by a halide salt from a marine isolate of the fungus *Phoma herbarum*. *The Journal of Antibiotics* 63: 199–201.

Newman, D.J. and Cragg, G.M. (2004). Marine natural products and related compounds in clinical and advanced preclinical trials. *Journal of Natural Products* 67: 1216–1238.
Newman, D.J. and Hill, R.T. (2006). New drugs from marine microbes: The tide is turning. *Journal of Industrial Microbiology & Biotechnology* 33: 539–544.
Nicholson, J.P., Wolmarans, M.R., and Park, G.R. (2000). The role of albumin in critical illness. *British Journal of Anaesthesia* 85: 599–610.
Okada, T. and Morrisse, M.T. (2007). *Marine enzymes from seafood by-products. In: Maximising the Value of Marine Byproducts* (ed. F. Shahidi), 374–396. Boca Raton, FL: CRC Press; Woodhead Publishing Limited: Cambridge, UK.
Olano, C., Méndez, C., and Salas, J.A. (2009). Antitumor compounds from marine actinomycetes. *Marine Drugs* 7: 210–248.
Palthur, M.P., Sajala Palthur, S.S. et al. (2010). Nutraceuticals: Concept and regulatory scenario. *International Journal of Pharmacy and Pharmaceutical Sciences* 2: 14–20.
Pangestuti, R. and Kim, S.K. (2011). Biological activities and health benefit effects of natural pigments derived from marine algae. *Journal of Functional Foods* 3: 255–266.
Panico, S., Mattiello, A., Panico, C. et al. (2014). Mediterranean dietary pattern and chronic diseases. *Cancer Treatment and Research* 159: 69–81.
Park, B.K. and Kim, M.M. (2010). Applications of chitin and its derivatives in biological medicine. *International Journal of Molecular Sciences* 11 (12): 5152–5164.
Parr, R.M., Aras, N.K., and Iyengar, G.V. (2006). Dietary intakes of essential trace elements: Results from total diet studies supported by the IAEA. *Journal of Radioanalytical and Nuclear Chemistry* 270: 155–161.
Pena-Rodriguez, A., Mawhinney, T.P., Ricque-Marie, D. et al. (2011). Chemical composition of cultivated seaweed *Ulvaclathrata* (Roth) C. *Agardh. Food Chem.* 129: 491–498.
Pérez, M., Crespo, C., Schleissner, C. et al. (2009). Tartrolon D, a cytotoxic macrodiolide from the marine-derived actinomycete *Streptomyces* sp. MDG-04-17-069. *Journal of Natural Products* 72: 2192–2194.
Pickering, T.D., Posa, S., and Reuben, J.S. (2007). Intentional introductions of commercially harvested alien seaweeds. *Botanica Marina* 50: 338–350.
Prudhomme, J., McDaniel, E., Ponts, N. et al (2008). Marine actinomycetes: A new source of compounds against the human malaria parasite. *PLoS One* 3: e233587.
Rajapakse, N., Jung, W.K., Mendis, E. et al. (2005). A novel anticoagulant purified from fish protein hydrolysate inhibits factor XIIa and platelet aggregation. *Life Sciences* 76: 2607–2619.
Raposo, M.F.D., De Morais, R. et al. (2013). Health applications of bioactive compounds from marine microalgae. *Life Sciences* 93: 479–486.
Research and Markets: Nutraceuticals (2012). Global Strategic Business Report Annual Estimates and Forecasts for 2010–2018. 2012. Available from: http://www.researchandmarkets
Rioux, L., Sl, T., and Beaulieu, M. (2007). Characterization of polysaccharides extracted from brown seaweeds. *Carbohydrate Polymers* 69: 530–537.
Romanenko, L.A., Uchino, M., Kalinovskaya, N.I. et al. (2008). Isolation, phylogenetic analysis and screening of marine mollusc-associated bacteria for antimicrobial, hemolytic and surface activities. *Microbiological Research* 163: 633–644.

Rovira, M.A., Grau, M., Castañer, O. et al.and Investigators, R.E.G.I.C.O.R. (2013). Dietary supplement use and health-related behaviors in a Mediterranean population. *Journal of Nutrition Education and Behavior* 45 (5): 386–391

Ruiz-Torres, V., Encinar, J.A., Herranz-Lopez, M. et al. (2017). An updated review on marine anticancer compounds: The use of virtual screening for the discovery of small-molecule cancer drugs. *Molecules* 22: 1037.

Ruperez, P. and Toledano, G. (2003). Indigestible fraction of edible marine seaweeds. *Journal of the Science of Food and Agriculture* 12: 1267–1272.

Satpute, S.K., Banat, I.M., Dhakephalkar, P.K. et al. (2007). Biosurfactants, bioemulsifiers and exopolysaccharides from marine microorganisms. *Biotechnology Advances* 28: 436–450.

Shahidi, F. and Abuzaytoun, R. (2005). Chitin, chitosan, and co-products: Chemistry, production, applications, and health effects. *Advances in Food and Nutrition Research* 49: 93–135.

Shalaby, E. (2011). Algae as promising organisms for environment and health. *Plant Signaling & Behavior* 6: 1338–1350.

Shibata, T., Ishimaru, K., Kawaguchi, S. et al. (2008). Antioxidant activities of phlorotannins isolated from Japanese *Laminariaceae*. *Journal of Applied Phycology* 20: 705–711.

Sibilla, S., Martin, G., Sarah, B. et al. (2015). An overview of the beneficial effects of hydrolysed collagen as a nutraceutical on skin properties: Scientific background and clinical studies. *The Open Nutraceuticals Journal* 8: 29–42.

Sijtsma, L. and De Swaaf, M.E. (2004). Biotechnological production and applications of the omega-3 polyunsaturated fatty acid docosahexaenoic acid. *Applied Microbiology and Biotechnology* 64: 146–153.

Sivonen, K., Leikoski, N., Fewer, D.P. et al. (2010). Cyanobactins-ribosomal cyclic peptides produced by cyanobacteria. *Applied Microbiology and Biotechnology* 86: 1213–1225.

Smith, V.J., Desbois, A.P., and Dyrynda, E.A. (2010). Conventional and unconventional antimicrobials from fish, marine invertebrates and micro-algae. *Marine Drugs* 8: 1213–1262.

Solanki, R., Khanna, M., and Lal (2008). Bioactive compounds from marine actinomycetes. *Indian Journal of Microbiology* 48: 410–431.

Stengel, D.B., Connan, S., and Popper, Z.A. (2011). Algal chemodiversity and bioactivity: Sources of natural variability and implications for commercial application. *Biotechnology Advances* 29: 483–501.

Suleria, H.A.R., Osborne, S., Masci, P. et al. (2015). Marine-based nutraceuticals: An innovative trend in the food and supplement industries. *Marine Drugs* 13: 6336–6351.

Takahashi, Y. and Omura, S. (2003). Isolation of new actinomycete strains for the screening of new bioactive compounds. *The Journal of General and Applied Microbiology* 49: 141–154.

Taylor, C., Pearson, J.P., Draget, K.I. et al. (2005). Rheological characterization of mixed gels of mucin and alginates. *Carbohydrate Polymers* 6: 189–195.

Thoms, C. and Schupp, P.J. (2007). Chemical defense strategies in sponges: A review. In: *Porifera Research: Biodiversity, Innovation and Sustainability* (ed. M.R. Custodio, G.L. Hajdu, E. Hajdu et al.), 627–637. Rio de Janeiro: IMOS. Museu Nacional, Brazil.

Thorpe, J.P., Sole-Cava, A.M., and Watts, P.C. (2000). Exploited marine invertebrates: Genetics and fisheries. *Hydrobiologia* 420: 165–184

Tou, J.C., Jaczynski, J., and Yc, C. (2007). Krill for human consumption: Nutritional value and potential health benefits. *Nutrition Reviews* 65: 63–77.

Tseng, C.K. (2001). Algal biotechnology industries and research activities in China. *Journal of Applied Phycology* 13: 375–380.

Uriz, M., Turon, X., Becerro, M.A. et al. (2003). Siliceous spicules and skeleton frameworks in sponges: Origin, diversity, ultrastructural patterns, and biological functions. *Microscopy Research and Technique* 62: 279–299.

Vidanarachchi, J.K. and Kurukulasuriya, M. (2011). *Industrial Applications of Marine Cosmeceuticals, Marine Cosmeceuticals: Latest Trends and Prospects*. New York, USA: Taylor & Francis.

Vlieghe, P., Clerc, T., Pannecouque, C. et al. (2002). Synthesis of new covalently bound kappa-carrageenan-AZT conjugates with improved anti-HIV activities. *Journal of Medicinal Chemistry* 45: 1275–1283.

Vo, T.S. and Kim, S.K. (2010). Potential anti-HIV agents from marine resources: An overview. *Marine Drugs* 8: 2871–2892.

Volkman, J.K. (2003). Sterols in microorganisms. *Applied Microbiology and Biotechnology* 60: 495–506.

Wijesekara, I. and Kim, S.K. (2010). Angiotensin-i-converting enzyme (ACE) inhibitors from marine resources: Prospects in the pharmaceutical industry. *Marine Drugs* 8: 1080–1093.

Yoshie, Y., Wang, W., Hsieh, Y.P.et al. (2002). Compositional difference of phenolic compounds between two seaweeds *Halimeda* spp J. *Tokyo University of Fisheries* 88: 21–24.

Yuan, Y.V., Carrington, M.F., and Walsh, N.A. (2005). Extracts from dulse (*Palmaria palmata*) are effective antioxidants and inhibitors of cell proliferation *in vitro*. *Food and Chemical Toxicology* 43: 1073–1081.

Xu, Y., He, H., Schulz, S. et al (2010). Potent antifouling compounds produced by marine *Streptomyces*. *Bioresource Technology* 101: 1331–1336.

4

Bioprospecting of Bioresources: Creating Value From Bioresources

Deepika Kathuria[1] *and Sumit S. Chourasiya*[2]

[1] *University Center for Research and Development, Chandigarh University, Gharuan, Punjab, 140413, India.*
[2] *IOL Chemicals and Pharmaceuticals Ltd, Barnala, Punjab, 148101, India.*

CHAPTER MENU

4.1 Introduction

Bioprospecting (bio = biological/natural source/; prospect = search) is the process of exploring natural resources such as plants, microbes, animals, etc. in the search for materials or compounds with interesting properties which can be commercialized for the benefit of human beings. The main objectives of bioprospecting are: a) sustainable use of bioresources; b) conservation of bioresources; and c) socioeconomic and scientific development. Bioprospecting has commercial importance since it provides scope for the generation of revenue by combining traditional knowledge and genetic information (Purkayasha 2016; Toly 2004; Harvey and Gericke 2011). This process has to follow

Bioresource Technology: Concept, Tools and Experiences, First Edition. Tanveer Bilal Pirzadah, Bisma Malik, Rouf Ahmad Bhat, and Khalid Rehman Hakeem.
© 2022 John Wiley & Sons Ltd. Published 2022 by John Wiley & Sons Ltd.

official guidelines/procedures. If the researcher or organization uses bioresource or related information (traditional or indigenous knowledge) without gaining official permission for their profit, this is known as biopiracy (Balasubramanian 2017; Robinson 2010; Lee-Huang et al. 1996; Brink 2000). Table 4.1 represents some examples of biopiracy. The bioprospecting-derived resources and their developed products have application in several industries such as pharmaceutical, agriculture, cosmetic, biosensors, textile, food, etc. Among them, pharmaceutical and agribusiness make major contributions. The process of bioprospecting of pharmaceuticals involves four main stages. The first stage involves the onsite or agency collection of a sample from terrestrial or marine or both sources. It is also referred to as the "Harvesting phase." It is undertaken by the scientists in the field. The second stage is identification and isolation of the compound followed by the characterization and then culture of the compound of interest. This process is carried out in the laboratory. The third step involves the screening of the compounds for their bioactivity and toxicity/safety or any other uses which are taken care of by the commercial parties. The final step is the development of the finished product for commercialization which includes patenting, sales and marketing (Figure 4.1) (Bhatia and Chugh 2015; Leary 2007).

Animal, plant and microbes are the major sources of enzymes. The traditional method of bioprospecting of enzymes involves the collection of the industrially relevant enzyme from the environmental genome. However, this process is slow as the time required from selection of the enzyme to develop it into a commercial product is time-consuming. The alternate and faster approach is the use of novel metagenome enzyme sequencing obtained by direct sequencing, which is subjected to high-throughput expression for subsequent production and characterization. Enzymes obtained from the microbes represent an attractive option for bioprospecting as: i) they can be cultured in large quantities in a relatively short time span; ii) they are more stable than enzymes obtained from plants or animals sources; and iii) can be stored for a week under non-ideal conditions without losing their biological activity. The overall global utilization of enzymes in 1995 was $1 billion, which

Table 4.1 List of biopirated resources.

Biological source	Country	Biopirate country	Purpose
Turmeric (*Curcuma longa*)	India	US	Healing of wound
Neem (*Azadirachta indica*)	India	US	Pesticides
Guggul (*Commiphora mukul*)	India	US	cholesterol-lowering properties
Basmati Rice	India	US	Food
Bitter Melon (*Momordica charantia*)	China	US	Treatment of tumors
Llang-Ilang (*Cananga odorata*)	Philippines	France	Cosmetics, perfumeries

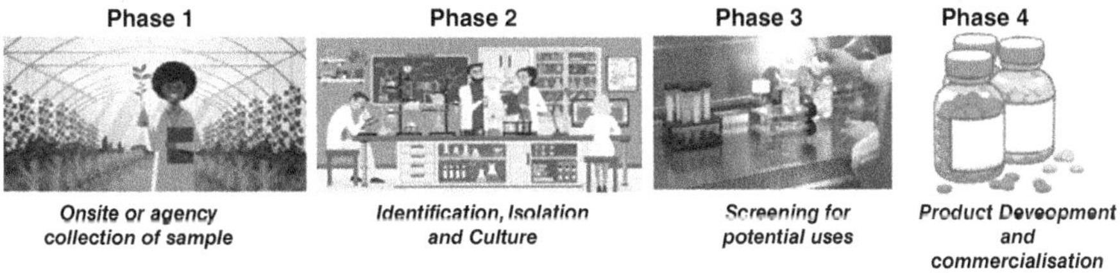

Figure 4.1 Phases of bioprospecting process taking an example of pharmaceutical product.

was increased to $1.5 billion in 2000. Environmental enzymes have been successfully commercialized in different industrial segments and will be discussed in this chapter (Schäfer and Vedel Borchert 2004; Adrio and Demain 2014; Ferrer et al. 2016).

4.2 Bioprospecting in Various Industrial Fields

4.2.1 Pharmaceutical Industries

Natural products are important sources of drugs. The natural product derived compounds have been popular and play a vital role in drug discovery. It has been estimated that approximately 50% of the approved drugs for different therapeutic activities are of natural origin. This can be seen in the case of cancer research where about 85 drugs out of 175 are natural products or their derivatives (Ahmad et al. 2006; Veeresham 2012; Newman and Cragg 2016). A few notable examples of drugs obtained from plant sources include digoxin, atropine, ephedrine, vincristine, colchicine, artemisinin, artemether, galantamine, morphine, caffeine, salicylic acid, etc.

4.2.1.1 Drugs From Plants

Morphine was the first natural product-based drug developed and commercialized for its analgesic property by Merck in 1826. The biological source of morphine is *Papaver somniferum*. The worldwide consumption of morphine and its congeners exceeds 400 tons per year (Veeresham 2012). About 30 total syntheses are reported in the literature and efforts are being made to have a more efficient and cost-effective procedure for the synthesis of morphine. Another example of a natural product-based drug is aspirin. It is known that the salicin isolated from *Salix alba* is the natural product which is transformed inside the body into salicylic acid, which has anti-inflammatory and analgesic properties (Newman and Cragg 2016; Sneader 2000). Since salicylic acid was associated with causing ulcers, its acetyl derivative was developed and synthesized by Bayer in 1899 which is devoid of this drawback. They dubbed the acetyl derivatives of salicylic acid as Aspirin (Ekinci et al. 2011). Galantamine is alkaloid, first isolated in 1947 from *Galanthus nivalis*. The interest toward this alkaloid increased only after 1960 when studies suggesting its inhibitory effect toward acetyl cholinesterase appeared in the literature. Galantamine was commercialized for the use of Alzheimer's disease in 2000. The galantamine is only available from plant sources, and until now no industrial acceptable synthesis is available (Mucke 2015).

Quinine and artemisinin are examples of natural product-derived antimalarial drugs on the market. Chemically, quinine is an alkaloid present in the bark of the Cinchona tree and in use for many years for its medicinal importance. For treating severe and complicated malaria, quinine is a drug of choice in most parts of the world. Apart from its antimalarial activity, it is also used as an anthelmintic as well as bitter tonic in soft drinks. Quinine is marketed in the form of its sulfate form (Mojab 2012). The reports suggest that quinine captured a global market of approximately \$804.98 million in 2018, which is expected to touch around \$1,184.15 million by the end of 2025 (Research and Market 2020). Currently, Cinchona trees are the only economic source of quinine to fulfill market demand. Though many total syntheses of quinine have been reported, none of them are commercially viable in comparison with its isolation from natural sources. Another example of a natural antimalarial agent is artemisinin isolated from the extract of sweet wormwood. Artemisinin is chemically a sesquiterpene lactone with endoperoxide moiety. The artemisinin was first isolated in 1970 and later discovered as a highly effective antimalarial drug (Mohammadi et al. 2020; White 1985; Phillipson and O'Neill 1987). Artemisinin-based combination therapy was recommended by WHO in 2004 to treat malaria caused by *P. falciparum*. Later, in 2006, the discontinuation of manufacturing and marketing of all artemisinin monotherapies was requested to prevent drug resistance (Majori 2004). The isolation of artemisinin from natural sources is a lengthy process involving a time span of 12–14 months, beginning from planting of *A. annua* to the production of artemisinin for commercialization. To date, plants are the sole source of artemisinin worldwide. To meet the market demand of artemisinin and to make it commercially viable, an alternative way for the supply of artemisinin is considered, which includes chemical synthesis, recombinant technologies and microbial production. All these process are only partly commercially viable and need further exploration (Hale et al. 2007).

There are anti-cancer drugs on the market which are obtained from plant sources. These include: i) vinca alkaloids; ii) epipodophyllotoxins; iii) taxanes; and iv) derivatives of camptothecin. Vincristine, vinblastine, vinorelbine, vindesine and vinflunine (the most recent) are clinically useful vinca alkaloid extracted from *Catharanthus roseus* (pink periwinkle). Chemically, they are monoterpenoid indole alkaloids. They exert their anti-cancer property by interactions with tubulin, and disruptions of microtubule function vinca alkaloids are the second-most used class of cancer drugs. Vincristine is isolated in very low yields, thus making them expensive to fulfill market demand. Whereas vinblastine is present 1000 times higher than vincristine, its cost is one-third that of vincristine. Commercially, vinca alkaloids are now produced by chemical synthesis. Paclitaxel and docetaxel are the currently most effective anti-cancer drugs from taxanes. Paclitaxel is originally obtained from the bark of *Taxus brevifolia* (a Pacific Yew) and docetaxel is a semisynthetic compound derived from 10-deacetylbaccatin III obtained from *Taxus baccata* (European Yew) leaves. Due to their poor solubility, toxicity and allergic reactions, many analogs and new formulations of these two taxanes have been approved or under clinical trial. Camptothecin, originally isolated from *Camptotheca acuminate*. Topotecan, irinotecan and belotecan are the semisynthetic derivatives belonging to the camptothecin class and had been clinically approved (Cragg and Pezzuto 2016).

Silymarin is obtained from the seeds of *Silybum marianum*. Silymarin possesses various pharmacological properties such as hepatoprotective, antioxidant, anti-inflammatory, anti-cancer and regenerative properties. It is a complex mixture of flavonolignans, consisting of silibinins, isosilibinins, silidianin and silicristin. The current statistics indicate that silymarin is one of the most widely-used natural compounds for the treatment of hepatic diseases worldwide; milk thistle extracts are available on the market (Karkanis et al. 2011; Gillessen and Schmidt 2020). The current global annual market of silymarin is $87 million, which is projected to reach $105.4 million by 2026 (Market Watch 2020).

Podophyllotoxin is a natural lignan isolated from the roots and rhizomes of *Podophyllum emodi*. Currently, podophyllotoxin is used as a precursor to semisynthetic anti-cancer drugs, etoposide, teniposide and etopophos. They are potent anti-cancer agents and show their anti-cancer activity by arresting cell growth by inhibiting DNA *topoisomerase* II, which causes double strand breaks in DNA. The supply of podophyllotoxin is thus a big challenge. The slow growth of species, scarcity due to intensive collection, and lack of cultivation are the factors which further restrict its supply. Currently, the commercial demand of podophyllotoxin is fulfilled by biotechnological approaches (Giri and Narasu 2000; Canel et al. 2000; Seegers et al. 2017).

4.2.1.2 Drugs From Bugs

4.2.1.2.1 Microbes

The exploration of microbes as a source of drugs/drug substances began seriously in 1928 with the discovery of penicillin from a fungus *Penicillium notatum* by Fleming (Ligon 2004). This area of research was greatly appreciated for its direct benefit to society. Streptomycin is another example of antibiotic against tubercular infection, for which Professor Waksman was awarded a noble prize in 1952 (Woodruff 2014). Later, major classes of antibiotics were discovered using a rational-based drug designing approach which includes tetracycline, quinoline, cephalosporine, etc. (Table 4.2). Various semi-synthetic and synthetic variants of natural antibiotics were also developed to combat resistance. *Tolipocladium infantum* is an ascomycete fungus from which cyclosporine A is produced. Cyclosporine A is an immunosuppressant mainly used to treat autoimmune disease or during organ transplantation (Table 4.3, entry 10). It also possesses antifungal properties (Svarstad et al. 2000). As compared to cancer drugs, the approval rate of antibiotics is low. As per the BCC report, the market size for antibiotics will rise from $44.4 billion (from 2020) to $65.4 billion (by 2025) (BCC Research 2021).

4.2.1.2.2 Enzymes

Enzymes have been used as therapeutic agents for the treatment of diseases that occur due to enzyme deficiencies and other medical conditions in human beings. The development of recombinant pharmaceutical protein isolated from the saliva of leeches (*Hirdu medicinalis*) having anticoagulant activity represents an example of bioprospecting of enzymes. The scarcity of hirudin led to its production through genetic engineering (Markwardt 1994).

L-asparaginase represents another example of bioprospecting and is obtained from *Erwinia chrysanthemi*. It is used to treat acute lymphoblastic leukemia (ALL), and is obtained from *E. coli* or *Erwinia chrysanthemi*. The treatment with *E. chrysanthemi*

Table 4.2 List of major classes of antibiotics.

Class	Antibiotic	Microbial source
β-lactams	Penicillin G	*Penicillium notatum*
	Cephalosporin C	*Cephalosporium acremonium*
	Monobactams	*Nocardia uniformis, Pseudomonas acidophilia*
	Thienamycin	*Streptomyces cattleya*
Tetracyclines	Chlorotetracycline	*Streptomyces aureofaciens*
	Oxytetracycline	*Streptomyces rimosus*
Aminoglycosides	Kanamycin	*Streptomyces kanamyceticus*
	Gentamicin	*Micromonospora pupurea*
	Sisomycin	*Micromonospora inyoensis*
	Streptomycin	*Streptomyces anulatus (previously S. griseus)*
	Spectinomycin	*Streptomyces spectabilis*
Macrolides	Erythromycin	*Saccharopolyspora erythrea*
Streptogramins	Virginiamycin	*Streptomyces virginiae*
Ansamycins	Rifamycin	*Amycolatopsis rifamycinica*
Lipopeptides	Daptomycin	*Streptomycin roseoporus*
Lincosamides	Lincomycin	*Streptomyces lincolnensis*

asparaginase is an alternative for patients who are hypersensitive toward an *E. coli*-derived asparaginase, as these two are immunologically distinct and thus exhibit no cross-reactivity (Egler et al. 2016). Pegylated formulations of asparaginase are available under the brand names PEG-ASNase, Oncaspa; Pegasparaginase. They have been developed for patients who are hypersensitive toward L-asparaginase. The anti-cancer mechanism of L-asparaginase is shown in Figure 4.2.

Pegademase is a pegylated modified adenosine deaminase. It is marked under the trade name of Adagen. The pegademase is used as an enzyme replacement therapy to adenosine deaminase (ADA), the deficiency which leads to the severe combined immunodeficiency (SCID) syndrome. The SCID syndrome is characterized by abnormality in cellular functions and humoral immunity. Among various types of SCIDs, a quarter are due to an adenosine deaminase (ADA) deficiency (Labrou 2019; Hershfield 1995). *Dornase alfa*, developed by genetic engineering of Chinese Hamster Ovary (CHO) cells, is a recombinant human deoxyribonuclease I (rhDNase) selectively cleaving DNA. It is produced by genetically engineered cells which contain DNA encoding for the native human protein, deoxyribonuclease I (DNase). It is marketed under the brand name of Pulmozyme. Dornase alfa is used to treat and improve lung infections in cystic fibrosis (Lazarus and Wagener 2019; https://www.rxlist.com/pulmozyme-drug.htm).

Streptokinase is a plasminogen activator (Figure 4.3) supplied under the brand name of Sreptase®. It is isolated from group C (beta)-hemolytic *streptococcus* and is indicated for health complications associated with pulmonary embolism, myocardial infarction

Table 4.3 List of common natural products as drug.

Sr. No.	Drugs	Source	Structure	Use	MOA	Marketed names
1.	Morphine	*Papaver somniferum*	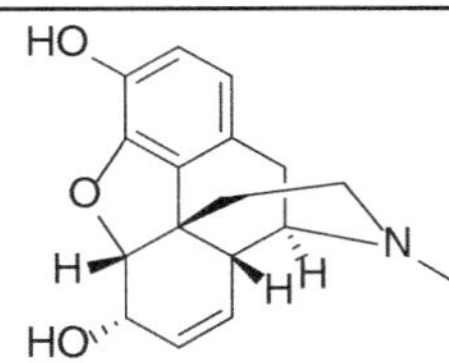	Pain	Opioid receptor agonist	Statex, MSContin, Oramorph, Sevredol
2.	Galantamine	*Galanthus nivalis*	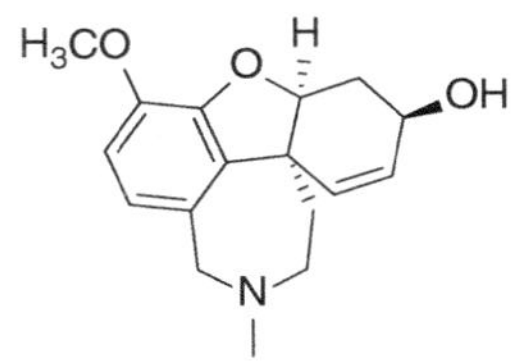	Alzheimer diseaseg	Acetylcholinesterase inhibitor	Razadyne
3.	Artemisinin	*Artemisia annua*	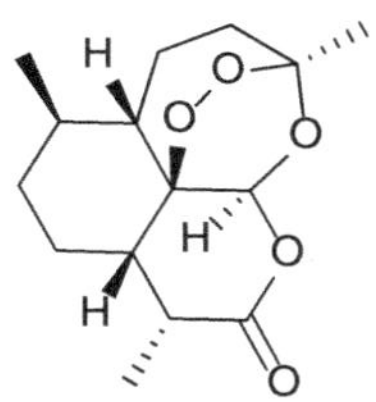	Anti-malarial	Generation of free radical by heme-mediated decomposition of endoperoxide bridge	Alaxin Armiqin Artequik Artequik

(*Continued*)

Table 4.3 (Continued)

Sr. No.	Drugs	Source	Structure	Use	MOA	Marketed names
4.	Artemether	Semi-synthetic derivative of artemisinin	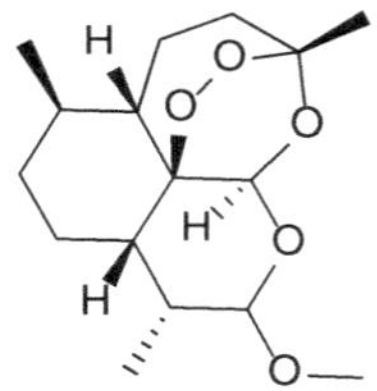	Anti-malarial	Generation of free radical by heme-mediated decomposition of endoperoxide bridge	Coartem
5.	Vincristine	*Catharanthus roseus*	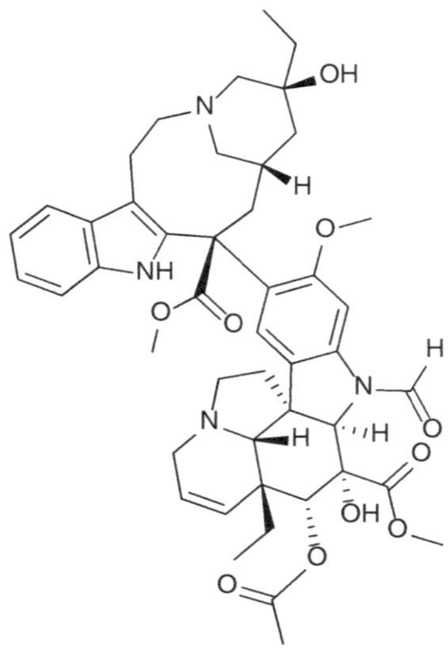	Anti-Cancer	Binds to tubulin and inhibit formation of microtubule	Oncovin

6.	Vinblastine	*Catharanthus roseus*	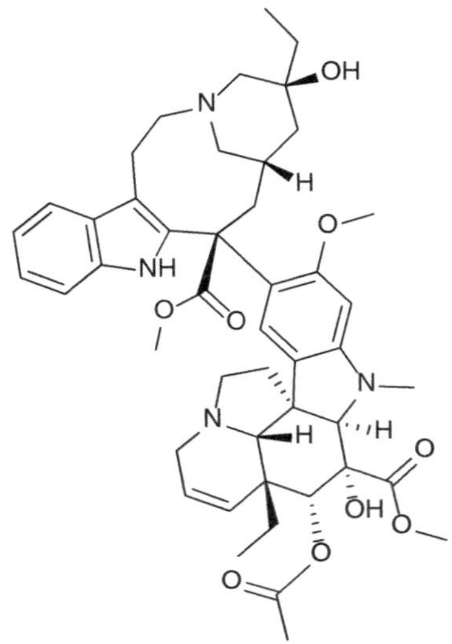	Anti-Cancer	Binds to tubulin and inhibit formation of microtubule	Velban
7.	Silymarin (Silybin)	*Silybum marianum*	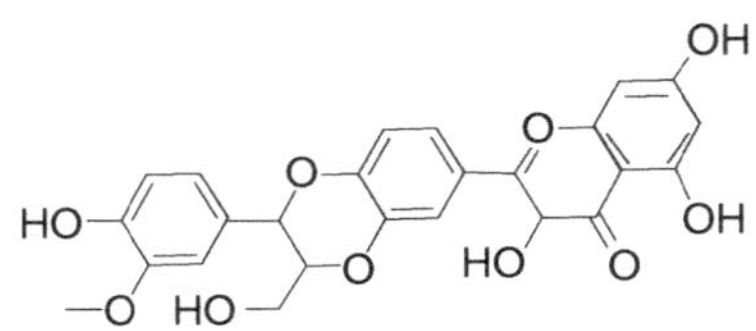	Treatment of liver diseases	antioxidant defense, RNA polymerase I activity stimulation	Eurobio, Eloliv, Healthvit, Hepacon-L

(Continued)

Table 4.3 (Continued)

Sr. No.	Drugs	Source	Structure	Use	MOA	Marketed names
8.	Podophyllotoxin	*Podophyllum peltatum* *Podophyllum emodi*	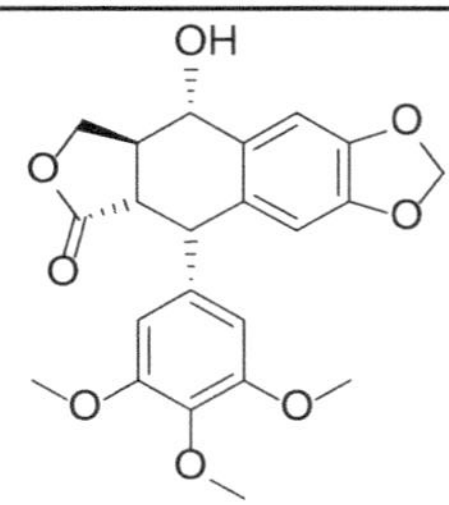	Treatment of genital warts and molluscum contagiosum	Binds to tubulin and thus destabilize microtubules	Podowart
9.	Etoposide Teoposide	Semisynthetic derivative of podophyllotoxin	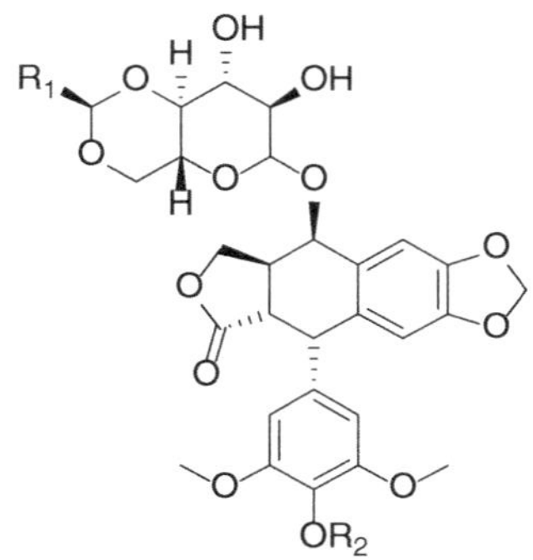 Etoposide, R_1 = Me, R_2 = H Teniposide, R_1 = 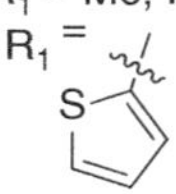 , R_2 = H	Anti-Cancer	Inhibit topoisomerase- II	Toposar VePesid Etopophos

10.	Cyclosporine A	*Tolypocladium inflatum*	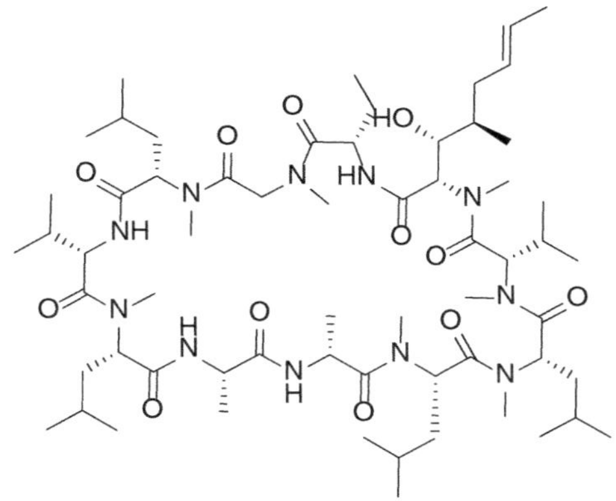	Immune Suppressant	Calcineurin inhibitor	Gengraf, Sandimmune, Neoral

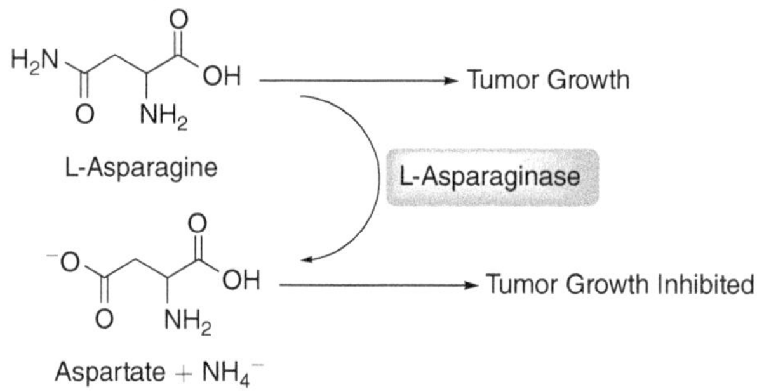

Figure 4.2 Anti-tumor action of L-asparaginase.

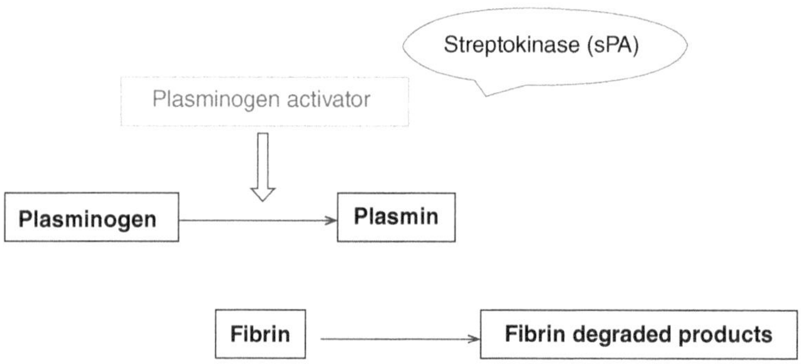

Figure 4.3 Mechanism of action of Streptokinase.

and thrombosis. The mechanism of action of streptokinase involves the conversion of plasminogen to plasmin which further dissolves fibrin and thus helps in the lysis of blood clots (Banerjee et al. 2004; Zia 2020).

Rasburicase, developed by recombinant technology, is a urate oxidase mainly involved in the conversion of uric acid (non-water soluble) in blood into allantoin (water soluble), which and indicates its use to reduce uric acid levels. This enzyme is obtained from a cDNA code of a modified *Aspergillus flavus* strain and expressed in a modified yeast strain of *Saccharomyces cerevisiae* (Ueng 2005; Cho et al. 2020). A galsidase beta (Fabrazyme®) is a recombinant human α-galactosidase A enzyme developed from CHO mammalian cells by recombinant DNA technology. It is marketed under the brand name of Fabrazyme®. It is indicated for patients with Fabry disease, categorized as a rare genetic disease caused by the deficiency of alpha galactosidase and affects many parts of the body such as skin, eye, GIT, kidney, heart and brain (Keating and Simpson 2007; Hopkin et al. 2020).

Serratiopeptidase (Silkworm enzyme) is a well-known enzyme commercialized for its use as a potent pain killer. Seratiopeptidase is an enzyme of the protease class which is produced by *Serratia marcescens* ATCC21074/Serratia sp. E-15, a non-pathogenic entero-bacterium originally isolated from the intestine of the silkworm (*Bombyx mori L.*). It shows it anti-inflammatory activity by inhibiting the release of bradykinin from inflamed tissues. It also breaks the insoluble fibrin and thus protects against artherosclerotic

Table 4.4 List of serratiopeptidase containing marketed products.

Sr. No.	Brand name	Cost/10 tablet (INR)	Marketed by
1.	Enzoflam	132–138	Alkem Laboratory Ltd
2.	Scradic P	108	Obsurge Biotech Ltd
3.	Ebility	149	Abbott
4.	Flozen-Plus	88	Mankind Pharma Ltd
5.	Muscodac	121–135	Indchemie Health Specialities Pvt. Ltd
6.	Esgipyrin SP	129	Abbott
7.	Lyser DP	112	Comed Chemical Ltd
8.	Diclotal SP	41	Blue Cross Loboratories Ltd
9.	Zinase-DP 134	134	Rapross Pharmaceutical Pvt. Ltd
10.	Diclogesic-SP	75	Torrent Pharmaceutical Ltd

plaque. Apart from its anti-inflammatory properties, it is also used as a dietary supplement for cardiovascular, anti-inflammatory or immune support. Table 4.4 lists the serratiopeptidase-containing anti-inflammatory products in combination with diclofenac and paracetamol which are currently on the market (Yamasaki et al. 1967; Jadhav et al. 2020; Bhagat et al. 2013; https://www.verywellhealth.com/serrapeptase-89513).

Nattokinase is a serine protease class of enzyme classified as a dietary supplement. It is produced during the fermentation of soyabeans by *Bacillus subtilis* to produce natto (a food item). It acts as an antithrombotic agent by hydrolyzing the fibrin and plasmin substrate (Weng et al. 2017; Chen et al. 2013).

4.2.1.3 Drugs From Aquatics

Marine sources are seen as a potent source of drugs to combat life-threatening diseases. Cytarabine, vidarabine, ziconotide and trabectedin (Figure 4.4) are examples of drugs derived from marine sources and are approved by the USFDA. Apart from these, many other compounds are at different stages of clinical trials. Cytarabine, also known as cytosine arabinoside, is a synthetic pyrimidine nucleoside. Cytarabine was patented by Pfizer Pharmaceuticals in 1960 as an anti-cancer agent. It was originally developed by synthetic means from spongothymidine (a nucleoside) isolated from the Caribbean sponge *Tethya crypta* (Alejandro et al. 2010). Pfizer obtained FDA approval in 1969 and marketed it under the brand names of Ara-C or Cytosar-U in its conventional or liposomal form. Cytarabine is used to treat various types of leukemia, and the liposomal formulation (Depocyt®) of cytarabine is indicated to treat lymphomatous meningitis (Martins et al. 2014). In 2017, a liposomal formulation (Vyxeos) containing the fixed combination of daunorubicin and cytarabine has been approved by the FDA for therapy related to acute myeloid leukemia. The export/import data suggest that the major suppliers of cytarabine globally are Italy, China, Australia, the United States and France (Krauss et al. 2019).

Vidarabine is a synthetic purine nucleoside isolated from a naturally occurring nucleoside spongouridine from *Tethya crypta* (a Caribbean sponge) and is currently obtained

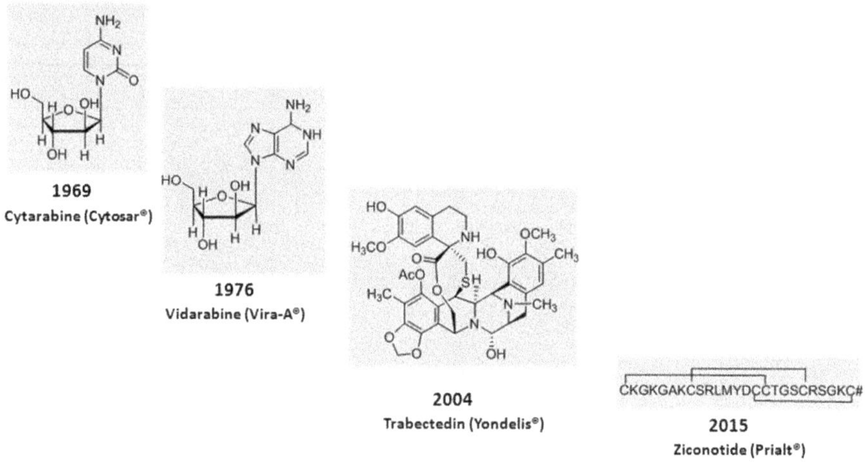

Figure 4.4 FDA approved drugs developed from marine source.

from *Streptomyces antibioticus*. It was approved by the FDA in 1976 for the treatment of various ophthalmic infections. It is marketed under the brand name VIRA-A in the form of an ointment. Marketing for vidarabine started in 1976 and ended in 2001 (Malve 2016; https://www.rtxlist.com/vira-a-drug.htm).

Ziconotide is a synthetic molecule extracted from venom of the marine snail *Conus magus*. During the early stages of clinical trials, it was known as SNX-111 and is currently supplied as an injection under the brand name Prialt®. It was approved by the FDA in 2004 as a new-generation intrathecal drug for its analgesic potential. It is the only known peptide approved by the FDA with calcium channel blocking properties (McGivern 2007; https://www.onclive.com/view/fda-approves-trabectedin-in-soft-tissue-sarcoma).

Trabectedin is another drug from marine sources with anti-cancer potential. It was approved by the European Union (EU) for the treatment of soft tissue sarcoma and in relapsed cases of ovarian cancer (https://www.onclive.com/view/fda-approves-trabectedin-in-soft-tissue-sarcoma). Trabectedine is marketed under the brand name Yondelis, ET-743. It is basically an alkaloid (tetrahydroisoquinoline class) extracted from *Ecteinascidia turbinate*, a tunicate species (Alejandro et al. 2010). In October 2015, Trabectedin was approved by the US-FDA for patients suffering from soft tissue sarcoma, mainly unresectable or metastatic liposarcoma or leiomyosarcoma. Trabectedin is exported from India to Brazil, Switzerland, the United Kingdom, France and Sri Lanka (Martins et al. 2014).

4.3 Chemical Industries

4.3.1 Biocatalysis

We have described how the chemical compounds from plants, microbes and marine organisms have been used for the treatment of various diseases. Various enzymes have also been utilized to synthesize a diverse set of chemical compounds possessing a wide variety of industrial applications. Enzymes as biocatalysts are of current interest to the chemical and pharmaceutical industries. Nitrilase and nitrile hydratase are nitrile-hydrolyzing enzymes that have found their potential as biocatalysts in organic synthesis to convert nitriles into their corresponding acid or amide. Nitrilase is an enzyme that can be obtained from nitrilase-producing enzymes including bacteria, fungi, yeast and plants. The production of (*R*)-(-)-mandelic acid from racemic mandelonitrile (Scheme 4.1a) was successfully achieved using nitrilase, which was only otherwise possible by the chemical route following tedious methods for isolation of the pure isomer. The enzymatic route for the synthesis of (*R*)-(-)-mandelic acid was found to be commercially viable on an industrial scale in comparison to the chemical route (Schmidt et al. 2001; Zhang et al. 2020; Jiang et al. 2017). Acrylic acid is another example of a chemical compound which has been synthesized on an industrial scale using nitrilase (Scheme 4.1b). This technology was patented by Ciba. The demand of acrylic acid in the market is rising owing to its industrial applications. Acrylic acid is a key starting material for the production of acrylic esters which have diverse applications, such as in targeted drug delivery systems, coating of tablets, adhesives, flocculants, etc. (Armitage et al. 1999).

Nitrile hydratase (NHase), a class of nitrile-hydrolyzing enzymes, has gained importance recently due to its industrial application as a biocatalyst. This enzyme is used to convert nitrile into amide. Acrylamide is an important intermediate which has found wide applications in the polymer industry, paper industry, pharmaceuticals, cosmetics, agriculture and in effluent treatment plants. The conventional synthesis of acrylamide involves the use of copper or palladium catalysts, which is not a commercially viable process. This organic catalyst now has been replaced by NHase for the production of acrylamide (Scheme 4.2a) on a commercial scale, with an annual production exceeding 100,000 tons (Nagasawa and Yamada 1990; Jiao et al. 2020).

DuPont, an American chemical company, used a similar strategy to convert adiponitrile to 5-cyanovaleramide, an important intermediate stage required for the synthesis of a new herbicide by regioselective hydration (Scheme 4.2b). This technology was

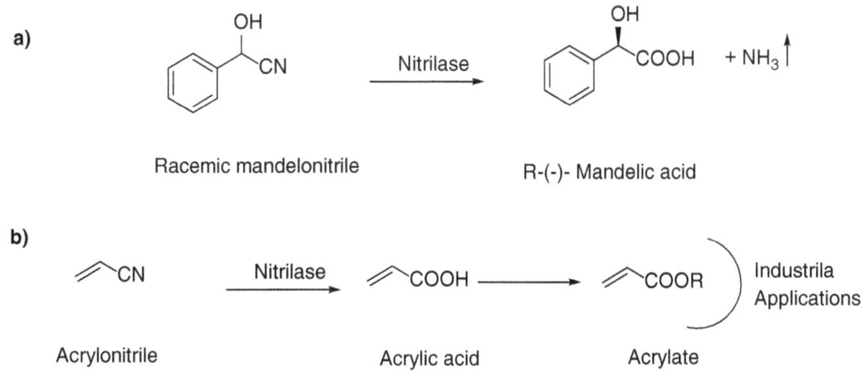

Scheme 4.1 Application of nitrilase in biocatalysis: a) conversion of racemic mandelonitrile to the desired R-(-) Mandelic acid using nitrilase; and b) formation of acrylic acid from acrylonitrile.

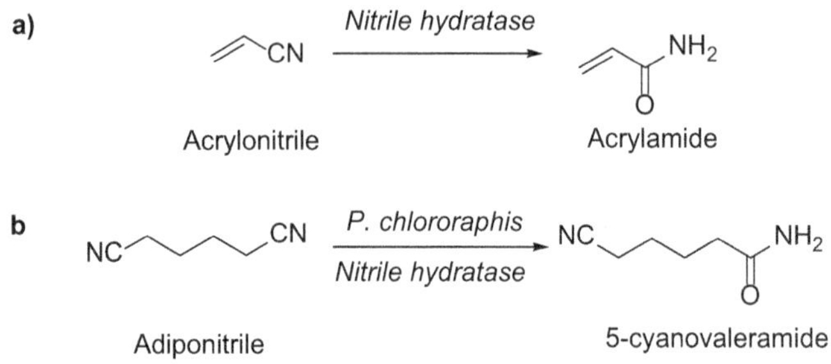

Scheme 4.2 Biocatalysis by nitrile hydratase for the production of industrially important (a) acrylamide and (b) 5-cyanovaleramide.

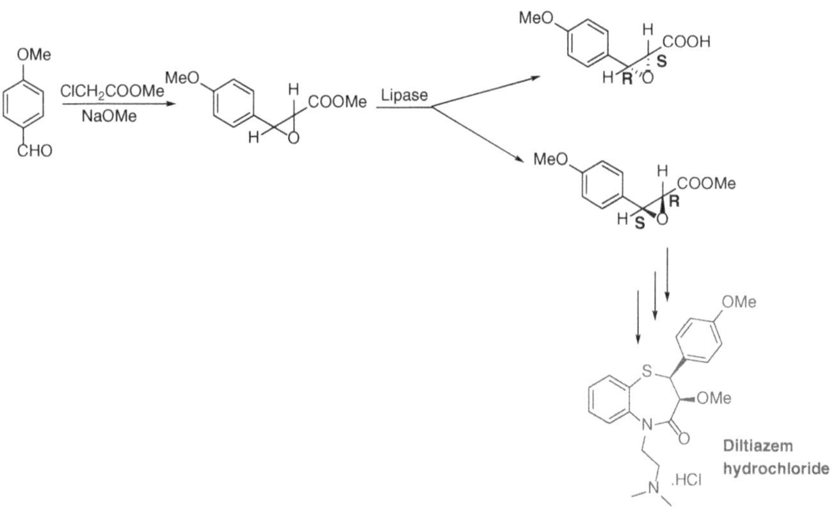

Scheme 4.3 Biocatalytical synthesis of diltiazem using lipase via kinetic resolution of trans-3-(4- methoxyphenyl) glycidic acid methyl ester (a key intermediate).

developed after realizing the drawbacks of chemical methods such as inefficiency and generation of significant amounts of by-products which are difficult to dispose of. Biocatalysis-based technology developed by DuPont has many advantages such as efficiency, is cleaner and generates fewer by-products and thus is more environmentally-friendly. Moreover, the catalyst, nitrile-hydratase, can be recycled multiple times during bulk production and thus reduces production costs (Zaks 2001; Kamble et al. 2013).

Lipases are the class of enzymes that hydrolyzes ester of triacylglycerol to give glycerol and fatty acids. They have been industrially accepted because of their chemical stability, catalytic activity, high enantio-selectivity and substrate specificity. Chiral glycidic ester is a key chemical intermediate required for the production of diltiazem hydrochloride (Scheme 4.3). Diltiazem hydrochloride is an antihypertensive and antiarrhythmic agent. The conventional synthesis of this key intermediate is non-economical due to its long time cycle and poor yield of final isolated product. An alternative to conventional

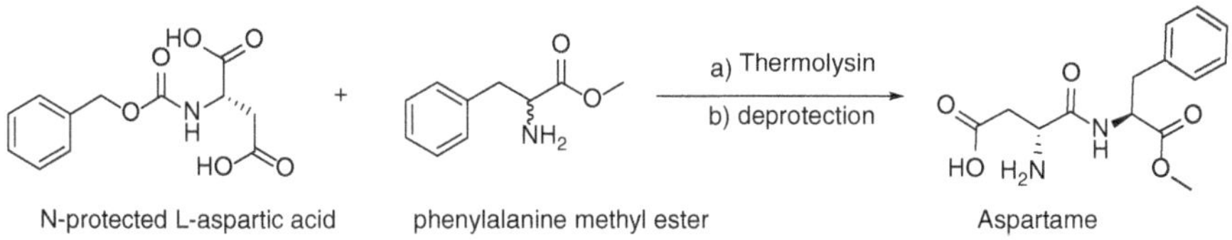

Scheme 4.4 Biocatalytic production of aspartame using thermolysin.

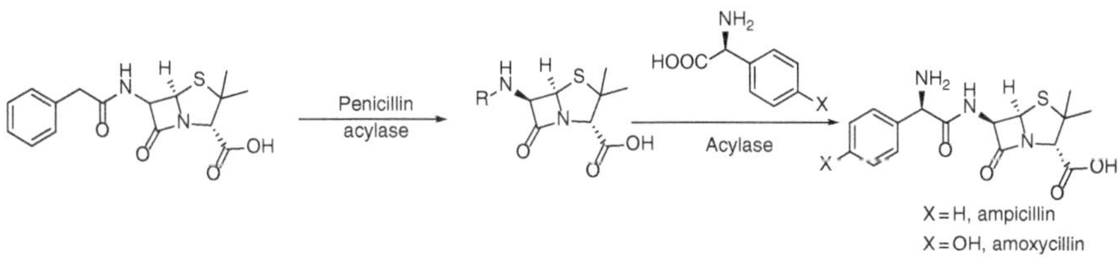

Scheme 4.5 Production of semisynthetic penicillin using acylase.

synthesis is the lipase-mediated biocatalysis process which is more economical and greener. The process involves the lipase mediated kinetic resolution of racemate intermediate trans-3-(4-methoxyphenyl) glycidic acid methyl ester to (2R, 3S)-ester and (2S, 3R)-acid. The lipase-mediated kinetic resolution has an advantage over conventional methods as it reduces time of production and generates less waste (Contesini et al. 2020; Sarmah et al. 2018; Angajala et al. 2016; Zhao et al. 2010; Matsumae et al. 1993).

Aspartame, chemically known as L-α-aspartyl-L-phenylalanine methyl ester, is a low-calorie sweetener. It is a dipeptide obtained from *N*-protected L-aspartic acid and phenylalanine methyl ester (Scheme 4.4). This dipeptide formulation on commercial scale is now achieved by thermolysin obtained from *B. subtilis*. Regioselectivity (aspartic acid carries two carboxylic acids) is another significant advantage of this biocatalysis over the conventional method, where the formation of unwanted isomer is the major concern (Rozzell 1999; Alsoufi and Aziz 2019; Birrane et al. 2014).

Penicillin acylase is used in one of the steps for the commercial production of 6-aminopenicillanic acid (Scheme 4.5). It hydrolyzes the side chains of Pen-G and Pen-V, which is then converted to 6-aminopenicillanic acid. This technology was discovered in the early 1960s and is now in multi-ton-scale production. D-p-hydroxyphenyl glycine and D-phenylglycine, the side chains of the 6-APA back-bone in the semisynthetic antibiotics amoxicillin and ampicillin, are also reported to be catalysed by acylases (Bruggink et al. 1998; Sawant et al. 2020).

4.4 Bioprospecting in Agriculture

4.4.1 Biofertilizers and Biopesticides

The replacement of chemical fertilizers with biofertilizers in agriculture is of current interest. Biofertilizers are the substances derived from microorganisms. They contain microbial cells live or in dormant state, which provide essential nutrients such as nitrogen, proteins and vitamins to the soil. They also help in the production of phytohormones. Biofertilizers possess advantages such as: i) improve yield and texture of plant; ii) help to destroy harmful substances present in soil; iii) reduce risk of plant disease; iv) protect plant from stressful conditions; v) cost-effective; and vii) environmentally-friendly. Their environmental-friendliness is a unique property that makes them a lucrative option as compared to chemical fertilizers which are harmful to the environment. Common microorganisms used as biofertilizers are *Rhizobium*, *Azospirilium* and *Azotobacter* – nitrogen fixing bacteria; *Anabaena* – a nitrogen fixing cyanobacteria; *Pseudomonas* species – a phosphate solublizing bacteria. Insect resistance can be accomplished through the expression of insecticidal proteins from soil bacteria like cholesterol oxidase from *Streptomyces s.p* and delta-endotoxin from *Bacillus thuringiensis* (Wani et al. 2020; Kumar 2018). Since 1961, *B. thuringiensis* has been registered by the US Environmental Protection Agency (EPA) for its use in pesticides (Leahy et al. 2014).

4.4.2 Bioremediation

Biomediation is a process which employs microorganisms, plants, or microbial or plant enzymes which have the ability to degrade the contaminants and thus help in the detoxification of pollutants from soil, water or the environment. *In-situ* and *ex-situ* bioremediation are the categories of bioremediation. In the *in situ* bioremediation process, the contaminated material is treated at the same place, whereas in *ex situ* bioremediation, the contaminated material is treated somewhere else. Bioventing, bioaugmentation, biostimulation and bioslurping are examples of *in-situ* bioremediation. Composting, land farming and biofiltering are examples of *ex-situ* bioremediation (Kumar et al. 2018).

4.5 Bioprospecting in Beautification/Cosmetics

Various beauty products containing one or more ingredients derived from natural products represent the best example of bioprospecting (Table 4.5). Data bridge market research suggests that the global market for cosmetic products was $435.47 billion in 2018, which is estimated to grow to $648.31 billion by 2026. Bloomberg reported that the global market for natural cosmetics is expected to reach $48.04 billion at 5.01% CAGR from 2019 to 2025. Skin care products form the largest segment in the global cosmetics market as it accounts for 36% of the global cosmetics market. The key players in natural cosmetics are Unilever, Burt's Bees, Beiersdorf Global AG, 100% Pure, Korres, Weleda, Arbonne International LLC, Bare Escentuals Beauty Inc., Coty Inc., Avon Product Inc., L'Oréal SA and AVEENO. (Data Bridge Market Research 2019; https://www.bloomberg.com/press-releases/2019-06-11/natural-cosmetics-market-worth-48-04-billion-by-2025-cagr-5-01-grand-view-research-inc). Examples of natural ingradients present in cosmetic products are now discussed.

Natural products have long been known for their medicinal value and are used for the treatment of skin ailments (Mahesh et al. 2019). Hyaluronic acid (HA) is a glycosaminoglycan (GAG) having a unique capacity to retain water and so is used in many cosmetic products such as moisturizer, sunscreen, masks and anti-aging skin treatments. It is used in up to 2% in cosmetics as skin conditioning agents. Examples of marketed formulations

Table 4.5 List of marketed cosmetics with natural ingredients.

Company	Product	Key Natural Ingredients	Uses
Burt's Bees	Goodness Glows Illuminating Face Primer	Avocado Oil	Hydrate skin
Burt's Bees	Goodness Glows Miracle Balm	Squalene	Moisturize and smooth skin and lips
Burt's Bees	Refreshing Cucumber Foaming Cleanser	Cucumber Extract, Mint	Refresh skin, Cleanset and remove excess oil
Burt's Bees	Soothing Cucumber Sheet Mask	Cucumber Extract	Moisturize skin, balance pH and improve texture of skin
Burt's Bees	Sensitive Toner With Aloe Vera	Aloe Vera extract Witch Hazel	Balance pH, remove excess oil, nourish and soothe skin
Korres	Black Pine 3D Eye-Lift Super Serum	Black Pine	A serum for eyes which helps to brighten the area around the eye
Korres	Greek Yoghurt Foaming Cream Cleanser	Flower extract of *Lonicera caprifolium* and *Lonicera japonica*, *Laminaria digitata* extract	Nourish the skin and cleanse
Korres	Apothecary Wild Rose Night-Brightening Sleeping Facial	Seed oil of *Simmondsia chinensis* (jojoba) and *Helianthus annuus* (sunflower), *Glycine soja* (soybean) oil, sterols, lecithin, *Imperata cylindrical* root extract, *Rosa canina* fruit oil, *Rosa canina* fruit extract, leaf extract of *Rosmarinus officinalis* (rosemary)	Improve tone, brighten skin, remove dullness
Weleda	Skin food	*Helianthus Annuus* (Sunflower) seed oil, *Prunus amygdalus dulcis* oil (sweet almond oil), *Rosemary officinalis* (rosemary) leaf extract, flower extract of *Chamomilla recutita* (matricaria), *Calendula officinalis*	Nourish and hydrate dry skin
Weleda	Almond Soothing Intensive Facial Cream	*Prunus amygdalus dulcis* oil (sweet almond oil), plum kernel oil	Nourish, soothe and moisturize all types of skin

(Continued)

Table 4.5 (Continued)

Company	Product	Key Natural Ingredients	Uses
Weleda	Skin Food Body Butter	*Helianthus Annuus* (sunflower) seed oil, flower extract of *Chamomilla recutita* (matricaria) and *Calendula officinalis*, *Viola tricolor* (Wild pansy) extract, *Butyrospermum parkii* (shea) butter	Moisturize and soften the skin
Arbonne	RE9 Advanced	Stem cells of sustainable Orange (*Citrus aurantium dulcis callusbculture*), Sea Buckthorn Ceramide, Brown Algae Extract	Improve radiance, tone and texture, remove wrinkles
Avon	Rose Hand & Body Lotion	Rose and Peach	Moisturize and soften skin, provide aroma
Avon	AT Antihairfall Conditioner	*Pouzolzia Pentandra* Extract, Bamboo extract, Peppermint oil	Reduce hairfall by 95%
Soul Tree	Anti-wrinkle cream	Turmeric, Amla Brahmi	Reduce wrinkles, boosts radiance, nourishes the skin
Soul Tree	Anti-Acne Kit	Neem, Nutgrass, Tulsi, Neroli Oil, and Tea Tree Oil	Remove excess oil, treats damaged skin
Soul Tree	Nourishing Cream	Saffron Sweet Almond Oil Ashwagandha Shatavari	Helps in skin brightening, anti-aging, stimulates the production of collagen
Soul Tree	Sun Protection Cream SPF 30	Aloe, Sunflower Oil, Olive Oil, Carrot Seed Oil, Coconut Oil, Almond Oil, Green Tea and Vitamin E extract	Skin texture improvement, protection from UV rays, antioxidant
Soul Tree	Cleansing Lotion	Aloe Vera, Rose water, Licorice	Hydrate and smooth skin, dirt removed by licorice
Mamaearth	Vitamin C Skincare Regimen Kit	Vitamin C, Turmeric	Cleansing, toning, and moisturizing
Mamaearth	Retinol Face Serum	Retinol, Linseed Oil Bakuchi Cotton Seed Oil	Boosts collagen production, reduces wrinkles and hyper-pigmentation, hydrates, high Vitamin E conc.

(Continued)

Table 4.5 (Continued)

Company	Product	Key Natural Ingredients	Uses
Mamaearth	Face Scrub	Turmeric Saffron Walnut Beads Carrot Oil	Removes tan, antioxidant, skin brightening
Mamaearth	Tea Tree Face wash	Neem Tea tree oil Aloe vera	Deap cleansing, controls excess oil
Mamaearth	Charcoal Face wash	Activated Charcoal Coffee	Removes impurities, remove dead skin
Patanjali	Kesh kanti Oil	Aloe Vera, Wheat Germ Oil, Sunflower oil and Bhringraj	Improve strength and shinier hair, reduce dandruff
Patanjali	Tejus Tailum	Sunflower oil, Soyabean oil, Almond oil, Walnut oil, Sesame oil, Olive oil, Mustard oil, Groundnut oil and and castor oil	Moisturize skin, nourish and strengthen the hair, reduce dandruff
Patanjali	Aloe Vera Gel	Aloe Vera	Anti-oxidants, heal and repair the skin, remove skin tanning
Patanjali	Patanjali Kesh Kanti Hair Cleanser	Reetha, Bhringraj, Baguchi, Turmeric, Neem, Amla, Tulsi, Heena, Shikakai, and Aloe Vera	Treatment of hair loss, treat dryness, remove dandruff
Himalayas	Tulsi Soap	Tulsi extract	For cleaning of hands, antimicrobial action
Himalayas	Cocoa Butter Intensive Body Lotion	Cocoa Butter, Wheat Germ Oil	Cocoa butter moisturizes and smooths the skin, Wheat germ oil provides Vitamin E
Himalayas	Lip Balm	Coconut Oil, Wheat Germ Oil, Carrot Seed Oil, Castor Oil, Sweet Indrajao	Moisturize, soften and condition the lips
Himalayas	Nourishing Body Lotion	Aloe Vera, Winter Cherry	Improves appearance of skin, moisturize the skin
Khaddi	Khadi Natural Amla & Bhringraj Hair Cleanser	Amla, Neem, Harathki, Almond oil, Reetha and Bhringraj	Makes hair healthy, prevents graying and hair loss, stimulates hair growth

are BareAirNight Serum, Sesderma C-Vit, minimalist, Charmis, etc. (Olejnik et al. 2012; Bukhari et al. 2018). Aloe vera, cucumber, rosemary, lavender, tea tree, calendula, basil, turmeric, amla, jojoba, chamomile, brahmi, etc. are examples of natural plants which are frequently used for the production of cosmetic products. A list of commercial cosmetic products with the natural ingredients is given in Table 4.5.

A wide variety of oils such as castor oil, sunflower oil, coconut oil and olive oil is used in cosmetic formulations due to their smoothing and emollient property. Jajoba oil is a complex mixture of long-chain esters obtained from the seeds of the jojoba plant. It helps in softening and conditioning of skin, improves skin tone and decreases the excessive production of sebum (a cause of acne). Jajoba oil is available in conditioners, hair oils, lotions, moisturizers, creams, etc. Castor oil has long been considered important in the commercialization of cosmetic products. It is obtained from the seeds of the castor bean, *Ricinus communis*. The key benefits of castor oil are as an emollient, protecting from harsh environments, moisturizing, and smoothing the skin. The hair smoothing and thickening properties of castor oil make them an attractive ingredient in manyformulated products for hair. Sunflower oil is expressed from the seeds of *Helianthus annuus*, mainly consisting of triglycerides of linoleic and oleic acids. It is commonly used as an edible oil. It is also used for skin disorders (i.e., psoriasis) and for treating bruises. It improves skin quality, moisturizes the skin and protects the skin from harsh environments. Additionally, it is a rich source of Vitamins A, D, E and K. Olive oil is the most common vegetable oil, obtained from the fruits of *Olea europaea*. It helps to beautify of the hair and moisturize the skin (Patil et al. 2017).

4.6 Bioprospecting in Detergent Industry

Many companies prepare laundry and dishwashing detergents using natural products. Natural or bio-based detergents have various advantages such as being environmentally-friendly, leaving the hands soft, having antimicrobial action and are non-allergic, etc. Soapnuts or soap berries have been known from thousands of years for washing clothes. They are basically the dried fruit of the tree Ritha (*Sapindus mukorossi*), also known as the Indian soapberry. *Shepherdia Canadensis*, native to North America, is also known as the soapberry. *Quillaja saponaria* (soap bark), *Saponaria officinalis* (soapwort) and *Chlorogalum* (soaproot) are common plants with detergent properties. Their detergent action is due to the presence of saponin. Some of the well-known companies manufacturing eco-friendly detergents are Allegro natura, Greennatural, Krya Consumer Products LLP, Daily Dump, Bubblenut Wash, Rustic Art, etc. (https://www.thehindu.com/life-and-style/homes-and-gardens/stains-are-great-chemicals-are-not/article23303108.ece; Kregiel et al. 2017). A few products with their natural ingredients are listed in Table 4.6. Instead of chemical compounds obtained from plants, enzymes have been widely used in the detergent industry as a functional ingredient.

In many of the formulations of detergents, enzymes are used as one of the ingredients which enhance the detergent's ability to remove tough stains in an efficient, eco-friendly and energy-saving manner. Examples of enzymes used in the detergent industry include protease, amylase, lipase, cellulase and mannose (Table 4.7). Proteases are the most

Table 4.6 Listing the few examples of marketed natural detergents.

Company	Product	Ingredients	Use
Krya	Natural Dishwashing Powder	Lemongrass, Citriodora, Soapnuts, Shikakai, Chebulic Myrobalan, Neem Leaf, Indian Gooseberry	Dishwashing
Daily Dump	108 Soapynuts	Soapynuts	Fabric washing
Almawin	Verbena Laundry Scent	Xanthan Gum, Limonene, Citral, Geraniol, Linalool, Citronellol, *Verbena officinalis* (verbena) leaf extract	Fabric washing
Allegro Natura	Organic Neutral Marseille Soap	*Olea Europaea* (Olive) Oil and *Cocos Nucifera* (Coconut) Oil	Fabric washing, can be use to Body wash
Bubble nut wash	BubbleNut Wash Laundry Detergent	Soapnuts	Fabric washing
Rustic Art	Natural Little Laundry Powder	Coconut oil soap (Sodium Cocoate), Neem (Sodium Neemate) Lavender (*Lavandula angustifolia*) essential oil	Fabric washing

Table 4.7 Uses of enzymes in detergent industry.

Enzymes	Use in detergent industry	Commercial detergents	Commercial enzymes
Proteases	To degrade proteinaceous stains such as blood, milk, egg, grass and sauces	Tide, Aerial, Persil	Alkalase, durazym, esperase, everlase, protamex
Amylases	To degrade starchy foods such as potatoes, gravies, custard, chocolate, etc. to dextrins and other oligosaccharides	Puracy – Natural laundry detergent	BAN, Termamyl, stainzyme, duramyl, Fungamyl, Maxamyl
Lipase	Helps in removal of fat-containing stains such as salad oils, fried fat, butter, fat-based sauces, lipstick, soups and human sebum	Persil	Lipolase, Lipolase ultra, Lipolase prime, Lumafast, Lipomax,
Cellulase	To impart brightness to the clothes		Celluzyme, carezyme
Mannanases	To remove gum like material	Mannaway, Effectenz	Novozymes

dominant enzyme used in the detergent industry (Hasan et al. 2010). Proteases are produced from *B. amyloliquefaciens, B. subtilinis* and *B. licheniformis*. Alkaline proteases obtained from these species are subtilisins. Subtilisins have found their potential application in all types of laundry detergents and in automatic dishwashing detergents. They show their action by proteolysis of protein stains such as blood, grass, egg, etc. Amylases are the second type of enzymes used in the formulation of enzymatic detergent. Amylases are derived from *Bacillus* or *Aspergillus*. Amylase is mainly used to remove starch-containing stains, e.g., potato, chocolate, pasta, gravy and baby food (Hamza 2017; Ito et al. 2005; Gupta et al. 2003).

The combination of fats and oil stains are difficult to remove during laundering at low temperatures due to their hydrophobic nature. In this context, lipases are widely used as one of the important ingredients in detergent formulations to remove fat-containing stains such as salad oil, fat-based sauces, fried fat, butter, soups, etc. They convert triglyceride into mono- and diglyceride, free fatty acid and glycerol, which are hydrophilic and thus can be removed easily in water (Houde et al. 2004). Cellulase not only removes stains effectively but also gives smoothness and brightness to the clothes. Examples of cellulase used in the detergent industry are derived from *Bacillus sps*. or *Trichoderma sp*. (Juturu and Wu 2014).

Alkaline mannanases are used in the detergent industry as stain removal boosters. They hydrolyze mannan-containing materials such as guar gum, glucomannas, galactomannans and glucomannans which are mainly present as stabilizers or thickeners in many household products and also in food items such as icecreams. Mannanases break the β-1,4-linkages of mannans into smaller carbohydrate fragments (Dhawan and Kaur 2007). The key players in bio-based enzyme market are DuPont Nutrition & Biosciences, Advanced Enzymes Technologies, Americos Industries, Novozymes, etc.

4.7 Bioprospecting in Textile Industry

Humans have used compounds obtained from natural resources such as plants or animals for thousands of years as dyes to impart color to clothes and food, and for artistic work. A few examples are discussed here. Indigo is a dark-blue dye, originally obtained from the leaves of the *Indigofera tinctoria* plant but now produced synthetically for commercial purposes. It is mainly used for dyeing cotton yarn and for denim jeans production. Examples of natural red dyes are Alizarin and Purpurin isolated from the *Rubia tinctorum* roots, which fall under the anthraquinone class, were used for cotton dyeing. Juglone (*Juglans nigra*) is a dye from the napthoquinine class, found to be a natural cosmetic and food dye and also imparts color to fabrics. In the past, synthetic dyes were mostly utilized for fabric dyeing, which resulted in environmental pollution and health hazards, therefore research is underway to promote the use of natural products in the textile industry (Samanta and Konar 2011).

The reduction of energy and water consumption to control environmental pollution is a big challenge for the textile industry. Enzymes are useful in the textile industry as they replace harsh chemicals, are substrate specific, work under mild conditions, and are safe and biodegradable. Among various classes of enzymes, the hydrolase class of enzymes is mostly used in the textile industry. The important enzymes in this category include cellulase, amylase, laccase, pectinase and catalase. Cellulases are used in the production of denim material attaining the attractive faded look by removing the surface-bound indigo dye. The treatment of cotton yarn with cellulases results in the removal of indigo dye from the yarn surface. Cellulase breaks the fibers on the yarn surface and thus loosens the indigo layer so that it can be easily removed by mechanical abrasion. Commercially, cellulases are a mixture of endoglucanases, cellobiohydrolases and cellobiases (Araujo et al. 2008).

Amylases are used for textile desizing and play a crucial role in the preparation of cotton fabrics. Before the discovery of amylase, desizing of fabrics was done by treating the fabric with acid, alkali or oxidizing agents at high temperature. These chemical treatments lead to various problems such as degradation of cotton fibers and incomplete removal of starch causing improper dyeing. Nowadays, this chemical treatment has been replaced by amylases which offer several advantages, such as complete removal of size without degrading the fabric, in an environmentally-friendly and efficient process. Laccases are the enzymes belonging to the multicopper oxidase family. These are mainly known for decolorization of textile effluents as they have the capability to degrade dyes of diverse structures. Laccases are also studied for the treatment of water contaminated with dyes and bleaching of cotton (Etters and Annis 1998; Abadulla et al. 2000; Hou et al. 2004).

4.8 Bioprospecting in Paper Industry

The pulp and paper industry forms a major sector to boost the world's economy. The demand is to reach 23.5 million tons by 2024/25. India is the lucrative market for the paper industry, where the annual growth rate is about 6% yearly. Approximately half of the fibers for the making of paper come from wood, as trees are a rich source of cellulose fibers (https://smeventure.com/growth-paper-industry-india/). Softwoods have mainly been used for the production of paper pulp, as their fibers possess good length and slenderness. The examples of plants in the low density softwood category are spruce, Scots pine, fir and lodgepole pine. Denser softwoods are longleaf pine, slash pine, loblolly pine and shortleaf pine. Hard woods have been utilized for manufacture of printing paper. When it comes to the paper industry, how can we forget the name of the fastest growing plant (in a day growing more than a meter), i.e., bamboo. Bamboos have been used for paper-making for hundreds of years, while bamboo canes have been utilized for producing bamboo pulp from which paper can be prepared (Asif 2009; Chen et al. 2019; Singh et al. 2019).

The major challenge for paper industries is the treatment of effluent generated during paper production. The process for the production of paper mainly consists of three steps: i) pulping; ii) bleaching; and iii) finishing. Bleaching is necessary to add brightness and whiteness to the paper. It is this stage where chemical substances are used to improve the desired appearance of paper. The generation of large quantities of waste products is an environmental hazard. As a result, paper industries are looking for an alternative to replace the chemical bleaching process with facile bio-based economic technologies. The concept of using enzymes in the pulp and paper industry was first introduced in 1986. An example of a bleaching agent of biological origin is xylanase and laccase, or in a cocktail. Xylanase has been established as a prebleaching agent. The use of xylanases have increased the bleaching speed in both totally chlorine-free (TCF) and elemental chlorine-free (ECF) printing processes (Kumar et al. 2017; Bajpai 1999; Sharma et al. 2020).

4.9 Bioprospecting in Food Industry

Food processing is carried out in the beverages and food industries to make the original foodstuffs suitable for human consumption. The history of use of enzymes go back to 6000 BC. Enzymes are used in the production of beer, wine, bread, cheese, etc. The food and beverages industries are the major sector using industrial enzymes. Most of the food enzymes are considered as food processing aids and only a few enzymes such as lysozyme and invertase are used as food additives (Adrio and Demain 2014; Freedonia Group 2011).

a) **Starch Industry**: *Glucose isomerase* is used as a catalyst for the conversion of glucose to fructose via reversible isomerization for the production of high-fructose corn syrup (HFCS). The fructose syrup is 1.3 times sweeter then glucose and is being extensively used in food and drink products as a sweetener. It also prevents crystallization and stickiness. Two variants of HFCS are available, i.e., HFCS-55 and HFCS-42. HFCS-55 contains 55% fructose and 45% glucose, whereas HFCS-42 contain 42% fructose and 58% glucose (Bhosale et al. 1996; Arbige et al. 2019).
b) **Dairy industry**: Enzymes are the key component of the dairy industry. The examples of enzymes possessing industrial applications include lipases, lactase, proteases, esterases and catalase. Rennets are famous as exogenous enzymes in diary processing. They are a mixture of pepsin and chymosin, isolated from calf stomachs (for animal rennet) or *Mucor miehei* (for microbial rennet) and are widely used for milk coagulation and cheese ripening. Rennet from microorganisms constitutes 30% of the global cheese production. The use of the rennet enzyme by cheese manufacturers represents the largest application of enzymes in dairy processing (Fox and McSweeney 1997; Ben Amira et al. 2017; Hellmuth and Van Den Brink 2013).

   Lipases are used in the food industry as a flavoring agent. Apart from this use, lipases are used to prepare customized milk products, reduce the time cycle of cheese processing, and help in the breakdown of milk fat. Lipases enhance softening of cheeses with specific flavor characteristics originated with free fatty acids. The lipases derived from *A. oryzae, M. miehei* and *A. niger* are of special importance in the dairy industry as they modify the fatty acid chain lengths to enhance the flavors of cheeses and cause lipolysis of fatty products like cream and cheese.

   Lactase is known in the dairy industry for its property to enhance solubility and to improve the sweetness of a wide variety of dairy products. Lactases are also used in milk products to reduce and remove lactose for lactose-intolerant patients. The increasing demand for lactose-free products is increasing the demand for lactase production (Ray 2012; Negi 2019; Ferreira-Lazarte et al. 2018).

c) **Baking**: The microbial amylases have been extensively used in the baking industry. Mainly α-amylases (α-1, 4-glucanohydrolase) have been used for baking, brewing, fruit juices and starch syrups. These enzymes are isolated from bacterial, fungal, cereal and genetically engineered bacterial sources. These amylases have been added to bread dough which breaks the starch into dextrins. It enhances the fermentation rate, reducing the viscosity and texture of the dough. Proteases are also used commercially to produce baked items such as bread, waffles and crackers. Proteases provide many added advantages, such as improving dough consistency, reducing blending times, improving the texture and flavor of bread and to regulate gluten strength in bread. Protease also acts on the protein material, weakening the gluten (Putseys and Schooneveld-Bergmans 2019).
d) **Soyabean products**: Soyabean products have a high demand in the market due to their richness in protein. Microbial transglutaminases (MTGase) are useful in the preparation of the famous soybean product, tofu. Tofu comprises of soya proteins that contain 11S and 7S globulins which are good substrates for MTGase. The addition of MTGase helps to maintain a smooth texture of retorted tofu for a long time. It is also used in the preparation of Yuba, an edible soyabean film (Wang et al. 2019; Duarte et al. 2020).

4.9.1 Bioprospecting in Brewing Industry

The application of bioprospecting can be seen in the brewing industry as relevantenzymes from microorganism are utilized for alcoholic fermentation. Protease, β-glucanase, α-amylase and β-amylase are the four most commonly used enzymes in the brewing industry. Barley is the basic ingredient of beer which contains starch, which has to be broken down to fermentable sugars before the yeast can make alcohol. The brewing process consists mainly of nine general steps, which are malting, milling, mashing, lautering, hopping, fermentation, conditioning, filtering and canning or bottling. During malting, the germinating seeds release mainly α-amylase, β-amylase and β-glucanase (endogenous enzymes), which help in the conversion of long-chain starch into small sugars such as glucose and maltose which promote fermentation and improve the mashing process. However, the level of endogenous enzymes may not be sufficient to produce the desired brew. The addition of commercially available exogenous enzymes (α-Amylase, β-Glucanase, xylanase, amyloglucosidase, pullulanase) helps to improve fermentation and filtration (Gomaa 2018; Bamforth 2009).

4.10 Diagnostic

Taq polymerase has been commercialized by Perkin Elmer Cetus USA under the brand name Taq. Taq pol is a thermostable DNA polymerase obtained from *Thermus aquaticus*. Taq polymerase is a crucial component in performing polymerase chain reaction (PCR), which has recently become a crucial technology for diagnostics. The discovery of the Taq polymerase offered the advantage of thermal stability (~95°C) over the usual polymerases. The commercial Taq polymerase is produced by some users of PCR having replaced

Taq polymerase by Pfu polymerase, which was isolated from *Pyrococcus furiosis* ("flaming fireball") and works best at 100°C (Zhu and Sun 2020; Gelfand 1989).

4.10.1 Application of Enzymes for the Detection of Pyrogens in Pharmaceutical Products

Bacterial endotoxin testing (BET) is a key quality control test to detect the presence of endotoxin toxic contaminants in pharmaceuticals and biomedical products below a prescribed limit. There are various methods for the detection of pyrogen, which includes limulus amebocyte lysate (LAL), rabbit pyrogen test (RPT) and monocyte activation test (MAT). The LAL test is widely used for the detection of the endotoxin as it is more sensitive and specific for pharmaceutical and biotechnological products. The LAL reagent is formulated from the aqueous extract of the blood cells of horseshoe crabs (*Limulus polyphemus*). Fuzifilm, Thermofischer and Biogenix systems are the suppliers of LAL assay test kits (Novitsky 1994; Uchida et al. 2019).

4.10.2 Bioprospecting in Biofuel Production

The transportation sector has shown potential growth in the recent past (https://www.eia.gov/outlooks/ieo/pdf/transportation.pdf). Currently, the most dominant source of transportation energy is fossil fuels. The burning of fossil fuels leads to global warming, indicating the need for alternatives which are more environmentally-friendly, renewable and sustainable resources. In this context, microbial biofuel has emerged as a reliable alternative, which can reduce the dependency on fossil energy source, reduce the emission of greenhouse gases and is more ecofriendly. Thus, the bioprospecting of microbes/microbial enzymes for biofuel is of current interest to the industry. Lipase and phospholipase have emerged as promising biocatalysts for biofuel production. Lipase converts free fatty acids (FFA) and triacylglycerol (TAG) into fatty acid methyl esters, which are a main component of biodiesel, whereas phospholipase transforms phospholipid into diacylglycerol, and a substrate for lipase. Currently, these biochemical processes are not economical and efforts are being made to make them more so (Borrelli and Trono 2015; Cesarini et al. 2014).

4.11 Conclusions and Future Perspectives

To encourage the sustainable use of biodiversity, bioprospecting has been proposed as a potential means. Despite the discovery of so many successful drugs from natural source, this approach is not taken seriously nowadays, probably due to the complexity in accessing the lead molecules (screening and isolation), less amounts affecting availailithy and benefit sharing. The development of new technologies, databases and screening softwares give hope to scientists to screen new molecules and to lead to the natural product drug discovery in a comparably easy manner. In this modern world, looks became an important parameter and so people are looking more toward cosmaceuticals products and thus lots of natural product-based cosmaceuticals have been commercialized and valued as they possess advantages such as: i) nourish the skin with no side effects; ii) environmentally-friendly; and iii) have natural fragrance, etc. Natural dyes have been used by humans from many years for dyeing fabrics, but the poor bonding of natural dyes with fabric forced the industry to switch toward synthetic dyes which is creating environmental issues. Thus, research should be promoted toward the application of natural dyes on natural fibers to produce textiles which are environmentally-friendly. Enzymes have been extensively used in various sectors of industries such as pharmaceutical, food, detergent, pulp and paper, etc. Enzymes play a very crucial role in the production of many important chemicals at accelerated rates. The available information on bioprospecting and the technology to speed up the process of bioprospecting may attract many more industries to commercialize many different products of natural origin. Also it will provide an option to an industrialist to replace hazardous chemicals with ecofriendly substances from natural sources and thus help to prevent environmental pollution.

References

Abadulla, E., Tzanov, T., Costa, S. et al. (2000). Decolorization and detoxification of textile dyes with a laccase from *Trametes hirsuta*. *Applied and Environmental Microbiology* 66: 3357–3362.

Adrio, J.L. and Demain, A.L. (2014). Microbial enzymes: Tools for biotechnological processes. *Biomolecules* 4: 117–139.

Ahmad, I., Aqil, F., and Owais, M. (2006). *Modern Phytomedicine: Turning Medicinal Plants into Drugs*. John Wiley & Sons.

Alejandro, M.S., Glaser, K.B., Cuevas, C. et al. (2010). The odyssey of marine pharmaceuticals: A current pipeline perspective. *Trends in Pharmacological Sciences* 31: 255–265.

Alsoufi, M.A. and Aziz, R.A. (2019). Production of aspartame by immobilized thermolysin. *Iraqi Journal of Science* 60: 1232–1239.

Angajala, G., Pavan, P., and Subashini, R. (2016). Lipases: An overview of its current challenges and prospectives in the revolution of biocatalysis. *Biocatalysis and Agricultural Biotechnology* 7: 257–270.

Araujo, R., Casal, M., and Cavaco-Paulo, A. (2008). Application of enzymes for textile fibres processing. *Biocatalysis and Biotransformation* 26: 332–349.

Arbige, M.V., Shetty, J.K., and Chotani, G.K. (2019). Industrial enzymology: The next chapter. *Trends in Biotechnology* 37: 1355–1366.

Armitage, Y.C., Hughes, J., and Webster, N.A., Ciba Specialty Chemicals Water Treatments Ltd, (1999). Nitrilase from *Rhodoccus rhodochrous* for converting acrylonitrile directly to acrylic acid. US Patent 5,998,180.

Asif, M. (2009). Sustainability of timber, wood and bamboo in construction. In: *Sustainability of Construction Materials* (ed. J.M. Khatib), 31–54.

Bajpai, P. (1999). Application of enzymes in the pulp and paper industry. *Biotechnology Progress* 15: 147–157.

Balasubramanian, S. (2017). *India: Traditional Knowledge and Patent Issues: An Overview of Turmeric, Basmati, Neem Cases*. Singh & Associates.

Bamforth, C.W. (2009). Current perspectives on the role of enzymes in brewing. *Journal of Cereal Science* 50: 353–357.

Banerjee, A., Chisti, Y., and Banerjee, U.C. (2004). Streptokinase – a clinically useful thrombolytic agent. *Biotechnology Advances* 22: 287–307.

BCC Research (2021) Antibiotics: Technologies and Global Markets.

Bhagat, S., Agarwal, M., and Roy, V. (2013). Serratiopeptidase: A systematic review of the existing evidence. *International Journal of Surgery (London)* 11: 209–217.

Ben Amira, A., Besbes, S., Attia, H. et al. (2017). Milk-clotting properties of plant rennets and their enzymatic, rheological, and sensory role in cheese making: A review. *International Journal of Food Properties* 20: S76–S93.

Bhatia, P. and Chugh, A. (2015). Role of marine bioprospecting contracts in developing access and benefit sharing mechanism for marine traditional knowledge holders in the pharmaceutical industry. *Global Ecology and Conservation* 3: 176–187.

Birrane, G., Bhyravbhatla, B., and Navia, M.A. (2014). Synthesis of aspartame by thermolysin: An X-ray structural study. *ACS Medicinal Chemistry Letters* 5: 706–710.

Bhosale, S.H., Rao, M.B., and Deshpande, V.V. (1996). Molecular and industrial aspects of glucose isomerase. *Microbiological Reviews* 60: 280–300.

Borrelli, G.M. and Trono, D. (2015). Recombinant lipases and phospholipases and their use as biocatalysts for industrial applications. *International Journal of Molecular Sciences* 16: 20774–20840.

Brink, W.D. (2000) Weight control product and method of treating hyperlipidemia and increasing vigor with said product. US Patent 6,113,949.

Bruggink, A., Roos, E.C., and De Vroom, E. (1998). Penicillin acylase in the industrial production of β-lactam antibiotics. *Organic Process Research and Development* 2: 128–133.

Bukhari, S.N.A., Roswandi, N.L., Waqas, M. et al. (2018). Hyaluronic acid, a promising skin rejuvenating biomedicine: A review of recent updates and pre-clinical and clinical investigations on cosmetic and nutricosmetic effects. *International Journal of Biological Macromolecules* 120: 1682–1695.

Canel, C., Moraes, R.M., Dayan, F.E. et al. (2000). Podophyllotoxin. *Phytochemistry* 54: 115–120.

Cesarini, S., Haller, R.F., Diaz, P. et al. (2014). Combining phospholipases and a liquid lipase for one-step biodiesel production using crude oils. *Biotechnology for Biofuels* 7: 29.

Chen, H., McGowan, E.M., Ren, N. et al. (2018). Nattokinase: A promising alternative in prevention and treatment of cardiovascular diseases. *Biomarker Insights* 13: 1177271918785130.

Chen, Z., Zhang, H., He, Z. et al. (2019). Bamboo as an emerging resource for worldwide pulping and papermaking. *BioResources* 14: 3–5.

Cho, M.H., Ahn, Y.H., Hee Lim, S. et al. (2020). Rasburicase improves the outcome of acute kidney injury from typical hemolytic uremic syndrome. *Pediatric Nephrology* 35: 2183–2189.

Contesini, F.J., Davanço, M.G., Borin, G.P. et al. (2020). Advances in recombinant lipases: Production, engineering, immobilization and application in the pharmaceutical industry. *Catalysts* 10: 1032.

Cragg, G.M. and Pezzuto, J.M. (2016). Natural products as a vital source for the discovery of cancer chemotherapeutic and chemopreventive agents. *Medical Principles and Practice* 25: 41–59.

Data Bridge Market Research. (2019). *Global Cosmetics*. Market – Industry Trends and Forecast to 2026.

Dhawan, S. and Kaur, J. (2007). Microbial mannanases: An overview of production and applications. *Critical Reviews in Biotechnology* 27: 197–216.

Duarte, L., Matte, C.R., Bizarro, C.V. et al. (2020). Review transglutaminases. Part II: Industrial applications in food, biotechnology, textiles and leather products. *World Journal of Microbiology and Biotechnology* 36: 1–20.

Egler, R.A., Ahuja, S.P., and Matloub, Y. (2016). L-asparaginase in the treatment of patients with acute lymphoblastic leukemia. *Journal of Pharmacology and Pharmacotherapeutics* 7: 62–71.

Ekinci, D., Şentürk, M., and Küfrevioğlu, Ö.İ. (2011). Salicylic acid derivatives: Synthesis, features and usage as therapeutic tools. *Expert Opinion on Therapeutic Patents* 21: 1831–1841.

Etters, J.N. and Annis, P.A. (1998). Textile enzyme use: A developing technology. *American Dyestuff Reporter* 87: 18–23.

Ferrer, M., Martínez-Martínez, M., Bargiela, R. et al. (2016). Estimating the success of enzyme bioprospecting through metagenomics: Current status and future trends. *Microbial Biotechnology* 9: 22–34.

Ferreira-Lazarte, A., Moreno, F.J., and Villamiel, M. (2018). Application of a commercial digestive supplement formulated with enzymes and probiotics in lactase non-persistence management. *Food & Function* 9 (9): 4642–4650.

Fox, P.F. and McSweeney, P.L.H. (1997). Rennets: Their role in milk coagulation and cheese ripening. In: *Microbiology and Biochemistry of Cheese and Fermented Milk* (ed. B.A. Law), 1–49.

Freedonia Group; Cleveland, OH: 2011. World Enzymes.

Gelfand, D.H. (1989). Taq DNA polymerase. In: *PCR Technology* (ed. H.A. Erlich), 17–22.

Gillessen, A. and Schmidt, H.H.J. (2020). Silymarin as supportive treatment in liver diseases: A narrative review. *Advances in Therapy* 37: 1279–1301.

Giri, A. and Narasu, M.L. (2000). Production of podophyllotoxin from *Podophyllum hexandrum*: A potential natural product for clinically useful anticancer drugs. *Cytotechnology* 34: 17–26.

Gomaa, A.M. (2018). Application of enzymes in brewing. *Journal of Nutrition and Food Science Forecast* 1: 1002.

Gupta, R., Gigras, P., Mohapatra, H. et al. (2003). Microbial α-amylases: A biotechnological perspective. *Process Biochemistry* 38: 1599–1616.

Hale, V., Keasling, J.D., Renninger, N et al. (2007). Microbially derived artemisinin: A biotechnology solution to the global problem of access to affordable antimalarial drugs. *American Journal of Tropical Medicine and Hygiene* 77: 198–202.

Hamza, T.A. (2017). Bacterial protease enzyme: Safe and good alternative for industrial and commercial use. *International Journal of Chemical and Biomolecular Science* 3: 1–10.

Harvey, A.L. and Gericke, N. (2011). Bioprospecting: Creating a value for biodiversity. In: *Research in Biodiversity – Models and Applications* (ed. Pavlinov, I.), 323–338.

Hasan, F., Shah, A.A., Javed, S. et al. (2010). Enzymes used in detergents: Lipases. *African Journal of Biotechnology* 9: 4836–4844.

Hellmuth, K. and Van Den Brink, J.M. (2013). Microbial production of enzymes used in food applications. In: *Microbial Production of Food Ingredients, Enzymes and Nutraceuticals* (eds. B. McNeil, D. Archer, I. Giavasis, et al.), 262–287.

Hershfield, M.S. (1995). PEG-ADA replacement therapy for adenosine deaminase deficiency: An update after 8.5 years. *Clinical Immunology and Immunopathology* 76: S228–S232.

Hopkin, R.J., Feldt-Rasmussen, U., Germain, D.P. et al. (2020). Improvement of gastrointestinal symptoms in a significant proportion of male patients with classic Fabry disease treated with agalsidase beta: A Fabry Registry analysis stratified by phenotype. *Molecular Genetics and Metabolism Reports* 25: 100670.

Hou, H., Zhou, J., Wang, J. et al. (2004). Enhancement of laccase production by *Pleurotus ostreatus* and its use for the decolorization of anthraquinone dye. *Process Biochemistry* 39: 1415–1419.

Houde, A., Kademi, A., and Leblanc, D. (2004). Lipases and their industrial applications. *Applied Biochemistry and Biotechnology* 118: 155–170.

Ito, S., Kobayashi, T., Hatada, et al. (2005). Enzymes in modern detergents. In: *Microbial Enzymes and Biotransformations* (ed. J.L. Barredo), 151–161.

Jadhav, S.B., Shah, N., Rathi, A. et al. (2020) Serratiopeptidase: Insights into the therapeutic applications. *Biotechnology Reports*, e00544.

Jiang, W., Tao, R., Yang, Y. et al. (2017). Production of (R)-(−)-mandelic acid with nitrilase immobilized on D155 resin modified by l-lysine. *Biochemical Engineering Journal* 127: 32–42.

Jiao, S., Li, F., Yu, H., and Shen, Z. (2020). Advances in acrylamide bioproduction catalyzed with *Rhodococcus* cells harboring nitrile hydratase. *Applied Microbiology and Biotechnology* 104: 1001–1012.

Juturu, V. and Wu, J.C. (2014). Microbial cellulases: Engineering, production and applications. *Renewable and Sustainable Energy Reviews* 33: 188–203.

Kamble, A.L., Banoth, L., Meena, V.S. et al. (2013). Nitrile hydratase of *Rhodococcus erythropolis*: Characterization of the enzyme and the use of whole cells for biotransformation of nitriles. *3 Biotech* 3: 319–330.

Karkanis, A., Bilalis, D., and Efthimiadou, A. (2011). Cultivation of milk thistle (*Silybum marianum* L. Gaertn.), a medicinal weed. *Industrial Crops and Products* 34: 825–830.

Keating, G.M. and Simpson, D. (2007). Agalsidase Beta. *Drugs* 67: 435–456.

Krauss, A.C., Gao, X., Li, L. et al. (2019). FDA approval summary:(daunorubicin and cytarabine) liposome for injection for the treatment of adults with high-risk acute myeloid leukemia. *Clinical Cancer Research* 25: 2685–2690.

Kregiel, D., Berlowska, J., Witonska, I. et al. (2017). Saponin-based, biological-active surfactants from plants. In: *Application and Characterization of Surfactants* (ed. R. Nazzar), 183–205.

Kumar, V.V. (2018). Biofertilizers and biopesticides in sustainable agriculture. In: *Role of Rhizospheric Microbes in Soil*, 377–398.

Kumar, D., Kumar, S.S., Kumar, J. et al. (2017). Xylanases and their industrial applications: A review. *Biochemical and Cellular Archives* 17: 353–360.

Kumar, V., Shahi, S.K., and Singh, S. (2018). Bioremediation: An eco-sustainable approach for restoration of contaminated sites. In: *Microbial Bioprospecting for Sustainable Development* (eds. J. Singh, D. Sharma, G. Kumar, et al.), 115–136.
Labrou, N. (2019). *Therapeutic Enzymes: Function and Clinical Implications*, vol. 1148. Springer Nature.
Lazarus, R.A. and Wagener, J.S. (2019). Recombinant Human Deoxyribonuclease I. *Pharmaceutical Biotechnology* 14: 471–488.
Leahy, J., Mendelsohn, M., Kough, J. et al. (2014). Biopesticide oversight and registration at the US Environmental Protection Agency. In: *Biopesticides: State of the Art and Future Opportunities* (eds. D.G. Aaron, R.C. Joel, O.D. Stephen et al.), 3–18.
Leary, D.K. (2007). *International Law and the Genetic Resources of the Deep Sea*, 157–158. Martinus Nijhoff Publishers.
Lee-Huang, S., Huang, P.L., Nara, P.L. et al. (1996) Plant protein useful for treating tumors and HIV infection. US Patent 5,484,889.
Ligon, B.L. (2004). Penicillin: Its discovery and early development. *Seminars in Pediatric Infectious Diseases* 15: 52–57.
Mahesh, S.K., Fathima, J., and Veena, V.G. (2019). Cosmetic potential of natural products: Industrial applications. In: *Natural Bio-active Compounds* (eds. J. Singh, D. Sharma, G. Kumar, et al.), 215–250.
Majori, G. (2004). Combined antimalarial therapy using artemisinin. *Parassitologia* 46: 85–87.
Malve, H. (2016). Exploring the ocean for new drug developments: Marine pharmacology. *Journal of Pharmacy and Bioallied Sciences* 8: 83–91.
Markwardt, F. (1994). The development of hirudin as an antithrombotic drug. *Thrombosis Research* 74: 1–23.
Martins, A., Vieira, H., Gaspar, H. et al. (2014). Marketed marine natural products in the pharmaceutical and cosmeceutical industries: Tips for success. *Marine Drugs* 12: 1066–1101.
Matsumae, H., Furui, M., and Shibatani, T. (1993). Lipase-catalyzed asymmetric hydrolysis of 3-phenylglycidic acid ester, the key intermediate in the synthesis of diltiazem hydrochloride. *Journal of Fermentation and Bioengineering* 75: 93–98.
McGivern, J.G. (2007). Ziconotide: A review of its pharmacology and use in the treatment of pain. *Neuropsychiatric Disease and Treatment* 3: 69–85.
Mohammadi, S., Jafari, B., Asgharian, P. et al. (2020). Medicinal plants used in the treatment of Malaria: A key emphasis to Artemisia, Cinchona, Cryptolepis, and Tabebuia genera. *Phytotherapy Research* 34: 1556–1569.
Mojab, F. (2012). Antimalarial natural products: A review. *Avicenna Journal of Phytomedicine* 2: 52–62.
Mucke, H.A. (2015). The case of galantamine: Repurposing and late blooming of a cholinergic drug. *Future Science* 1: FSO73.
Nagasawa, T. and Yamada, H. (1990). Application of nitrite converting enzymes for the production of useful compounds. *Pure and Applied Chemistry* 62: 1441–1444.
Negi, S. (2019). Lipases: A promising tool for food industry. In: *Green Bio-processes* (ed. B. Parameswaran, S. Varjani, and S. Raveendran), 181–198.

Newman, D.J. and Cragg, G.M. (2016). Natural products as sources of new drugs from 1981 to 2014. *Journal of Natural Products* 79: 629–661.

Novitsky, T.J. (1994). Limulus amebocyte lysate (LAL) detection of endotoxin in human blood. *Journal of Endotoxin Research* 1: 253–263.

Olejnik, A., Goscianska, J., and Nowak, I. (2012). Significance of hyaluronic acid in cosmetic industry and aesthetic medicine. *Chemik* 66: 129–135.

Patil, A.S., Patil, A.V., Patil, A.H. et al. (2017). A review on: Standardization of herb in new era of cosmaceuticals: Herbal cosmetics. *World Journal of Pharmaceutical Research* 6: 303–320.

Phillipson, J.D. and O'Neill, M.J. (1987). *Antimalarial and Amoebicidal Natural Products. Biologically Active Natural Products*, 49–64. Oxford: Clarendon Press.

Purkayastha, J. (2016). *Emerging Trends in Sustainable Bioprospecting of Bioresources, Bioprospecting of Indigenous Bioresources of North-East India*, 3–19. Springer.

Putseys, J.A. and Schooneveld-Bergmans, M.E.F. (2019). Enzymes used in baking. In: *Industrial Enzyme Applications* (eds. A. Andreas Vogel and O. May), 95–123.

Research and Market (2020) Global Quinine Market Review 2015–2018 and Forecast 2019–2025.

Ray, A. (2012). Application of lipase in industry. *Asian Journal of Pharmacy and Technology* 2: 33–37.

Robinson, D. (2010). *Confronting Biopiracy: Challenges, Cases and International Debates*. Routledge.

Rozzell, J.D. (1999). Commercial scale biocatalysis: Myths and realities. *Bioorganic and Medicinal Chemistry* 7: 2253–2261.

Samanta, A.K. and Konar, A. (2011). Dyeing of textiles with natural dyes. *Natural Dyes* 3: 30–56.

Sarmah, N., Revathi, D., Sheelu, G. et al. (2018). Recent advances on sources and industrial applications of lipases. *Biotechnology Progress* 34: 5–28.

Sawant, A.M., Sunder, A.V., Vamkudoth, K.R. et al. (2020). Process development for 6-Aminopenicillanic acid production using Lentikats-encapsulated *Escherichia coli* cells expressing penicillin V acylase. *ACS Omega* 5: 28972–28976.

Schäfer, T. and Vedel Borchert, T. (2004). Bioprospecting for industrial enzymes: Importance of integrated technology platforms for successful biocatalyst development. In: *Microbial Diversity and Bioprospecting* (ed. A.T. Bull), 375–390.

Schmid, A., Dordick, J.S., Hauer, B. et al. (2001). Industrial biocatalysis today and tomorrow. *Nature* 409: 258–268.

Seegers, C.L., Setroikromo, R., and Quax, W.J. (2017). Towards metabolic engineering of podophyllotoxin production. In: *Natural Products and Cancer Drug Discovery. Intech Open* (ed. F.E. Koehn), 287–306.

Sharma, D., Chaudhary, R., Kaur, J. et al. (2020). Greener approach for pulp and paper industry by Xylanase and Laccase. In: *Biocatalysis and Agricultural Biotechnology* (ed. H. Ching), 101604.

Singh, G., Kaur, S., Khatri, M. et al. (2019). Biobleaching for pulp and paper industry in India: Emerging enzyme technology. *Biocatalysis and Agricultural Biotechnology* 17: 558–565.

Sneader, W. (2000). The discovery of aspirin: A reappraisal. *British Medical Journal* 321: 1591–1594.

Svarstad, H., Bugge, H.C., and Dhillion, S.S. (2000). From Norway to Novartis: Cyclosporin from *Tolypocladium inflatum* in an open access bioprospecting regime. *Biodiversity and Conservation* 9: 1521–1541.

Toly, N.J. (2004). Globalization and the capitalization of nature: A political ecology of biodiversity in Mesoamerica. *Bulletin of Science, Technology and Society* 24: 47–54.

Uchida, T., Kaku, Y., Hayasaka, H. et al. (2019). Utility of an automatic limulus amebocyte lysate kinetic turbidimetric test for endotoxin screening of dialysate samples. *Medical Devices (Auckland, NZ)* 12: 429–433.

Ueng, S. (2005). Rasburicase (Elitek): A novel agent for tumor lysis syndrome. *BUMC Proceedings* 18: 275–279.

Veeresham, C. (2012). Natural products derived from plants as a source of drugs. *Journal of Advanced Pharmaceutical Technology and Research* 3: 200–201.

Wang, C., Li, J., Zhou, S. et al. (2019). Application of transglutaminase for quality improvement of whole soybean curd. *Journal of Food Science and Technology* 56: 233–244.

Wani, J.A., Wali, A.F., Majid, S. et al. (2020). Bio-pesticides: Application and possible mechanism of action. *Bioremediation and Biotechnology* 2: 97–119.

Watch, M. (2020) Silymarin Market 2020: Top Countries Data with Future Scope, Market Size, Definition, Opportunities with Strategic Growth and Top Key Players Analysis and Forecasts to 2026.

Weng, Y., Yao, J., Sparks, S. et al. (2017). Nattokinase: An oral antithrombotic agent for the prevention of cardiovascular disease. *International Journal of Molecular Sciences* 18: 523.

White, N.J. (1985). Clinical pharmacokinetics of antimalarial drugs. *Clinical Pharmacokinetics* 10: 187–215.

Woodruff, H.B. (2014). Selman A. Waksman, winner of the 1952 Nobel Prize for physiology or medicine. *Applied and Environmental Microbiology* 80: 2–8.

Yamasaki, H., Tsuji, H., and Saeki, K. (1967). Anti-inflammatory action of a protease, TSP, produced by Serratia. *Nihon Yakurigaku Zasshi* 63: 302–314.

Zaks, A. (2001). Industrial biocatalysis. *Current Opinion in Chemical Biology* 5: 130–136.

Zhang, X.H., Wang, C.Y., Cai, X. et al. (2020). Upscale production of (R)-mandelic acid with a stereospecific nitrilase in an aqueous system. *Bioprocess and Biosystems Engineering* 43: 1299–1307.

Zhao, L.L., Pan, J., and Xu, J.H. (2010). Efficient production of diltiazem chiral intermediate using immobilized lipase from *Serratia marcescens*. *Biotechnology and Bioprocess Engineering* 15: 199–207.

Zhu, Z. and Sun, D., Abclonal Science Inc. (2020) Mutant Taq polymerase for faster amplification. US Patent 10,865,441.

Zia, M.A. (2020). Streptokinase: An efficient enzyme in cardiac medicine. *Protein and Peptide Letters* 27: 111–119.

https://www.rxlist.com/pulmozyme-drug.htm.

https://www.verywellhealth.com/serrapeptase-89513.

https://www.rxlist.com/vira-a-drug.htm.

https://www.onclive.com/view/fda-approves-trabectedin-in-soft-tissue-sarcoma

https://www.bloomberg.com/press-releases/2019-06-11/natural-cosmetics-market-worth-48-04-billion-by-2025-cagr-5-01-grand-view-research-inc.

https://www.thehindu.com/life-and-style/homes-and-gardens/stains-are-great-chemicals-are-not/article23303108.ece.

https://smeventure.com/growth-paper-industry-india.

https://www.eia.gov/outlooks/ieo/pdf/transportation.pdf.

5

Green and Smart Packaging of Food

Gülden Gökşen[1], Derya Boyacı[2] and Nick Tucker[2]

[1] *Department of Food Engineering. Vocational School of Technical Science, Tarsus University, Akdeniz, Mersin, Turkey*
[2] *School of Engineering, Lincoln University, Lincoln, UK*

CHAPTER MENU

5.1 Introduction

The technology of food packaging is highly sophisticated and draws on a range of scientific disciplines to achieve a variety of aims that extend far beyond the simple notion of protecting the food from damage and deterioration in passage from the farm to the consumer. For example, the aesthetics of packaging are an important factor in influencing the consumer to buy the food product.

In pursuit of higher profitability, the food industry seeks to increase the range and volume of foodstuffs offered for sale to the consumer. This has in turn increased demand for longer storage periods without appreciable loss of food quality.

Bioresource Technology: Concept, Tools and Experiences, First Edition. Tanveer Bilal Pirzadah, Bisma Malik, Rouf Ahmad Bhat, and Khalid Rehman Hakeem.
© 2022 John Wiley & Sons Ltd. Published 2022 by John Wiley & Sons Ltd.

The development of better methods for the preservation of foodstuffs requires improved solutions that can quickly and accurately monitor and communicate the state of the freshness and therefore the safety of those foodstuffs. This will maximize shelf life (defined as time to manufacture, store, ship and ultimately consume) (Silvestre et al. 2011).

Packaging is the most fundamental component part of a sophisticated system that aims to provide stable and long-lasting quality food products by preserving the contents through the stages of storage, transportation, delivery and consumption. The main purpose of packaging is to protect the payload from mechanical damage and prevent or inhibit chemical, biochemical and microbiological deterioration (Ghaani et al. 2016). In addition, the packaging plays important roles in providing information about the nutritional values and caloric content of the product within, in addition to displaying the consistency, shape, volume and color of the product (Ghosal et al. 2018). It has also been demonstrated that the aesthetics of food packaging have a measurable effect on the perception of the quality of the food. In addition, matters such as choice of packaging color or use of an image on the packaging can affect the level of sales by as much as 10% (Ahmad et al. 2012) (Figure 5.1),

Food packaging is predominantly seen as a single-use article, and due to the number of materials used, the packaging is sent to landfill, as recovery and recycling is not

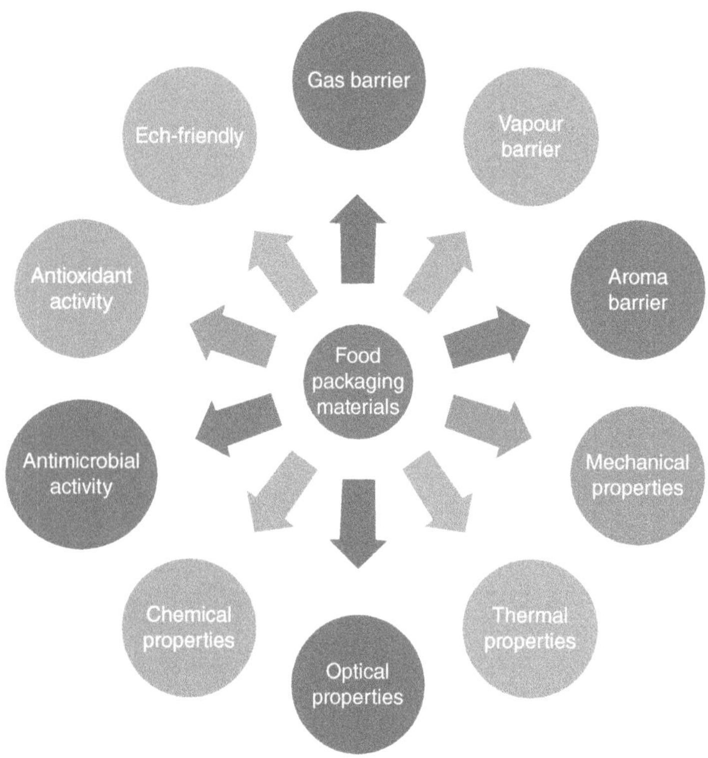

Figure 5.1 Properties of food packaging materials.

economical. Consider as an example the film used to package potato crisps. This film is made in four layers:

- First (inside) layer – metallized Biaxaily Oriented Polypropylene (BOPP)
- Second layer – LDPE
- Third layer – unmetallized BOPP
- Fourth layer – ionomer resin

BOPP has a high oxygen transmission rate, and if it were used alone, it would permit oxygen into the package. This would oxidize the fat in the potato crisps causing spoilage. This is the reason that the inner layer is metallized with a 40–50 nm layer of aluminum. This provides an effective barrier to external atmospheric oxygen, aromas and light. This prevents spoilage by oxidation of the frying fat and prevents the loss of the internal nitrogen protective atmosphere. The protective role is two-fold, first stopping oxidation, and second, providing a slight internal pressure to prevent crushing in bulk storage.

The LDPE layer is cheap and gives flexibility, moisture protection, toughness, chemical resistance and sealing properties. It cannot be used alone because of its poor gas resistance, inability to retain ink, and its strong tendency to develop a static charge that may attract dust, which can be unsightly in the retail setting.

The ionomer layer adds impact toughness, and abrasion and chemical resistance. It is also readily heat sealable. The ionomer layer is also suitable for printing of decorative designs. This complex material is technically recyclable (https://www.terracycle.com/en-GB/brigades/crisppacket, accessed 19 July 2020) but the difficulties associated with collection and separation of crisp packets mean that only about 0.01% of crisp packets in the UK are recycled (https://www.theguardian.com/environment/2018/oct/05/walkers-answers-critics-with-launch-of-uk-crisp-packet-recycling-plan, accessed 19 July 2020).

5.2 Green Packaging in Food

In recent years, petroleum-derived plastics have become perceived as a major environmental problem. To address this issue, the industries concerned are focusing on the development of biodegradable food packaging materials and novel packaging technologies (Rabnawaz et al. 2017; Lörcks 1998). Life cycles of green packaging materials are shown in Figure 5.2.

5.3 Properties of Green Packaging Materials

Food packaging materials are generally categorized as either glass, metal, paper or plastic, and of these plastic materials have been researched most intensively to improve sustainability. The concept of green packaging implies that sustainable packaging materials having a minimum impact on the environment, human health and safety. Materials with prefix "bio" (biocomposites, bioplastics, biopolymers, etc.) are of recent crop origin and thus naturally biodegradable through microbial action.

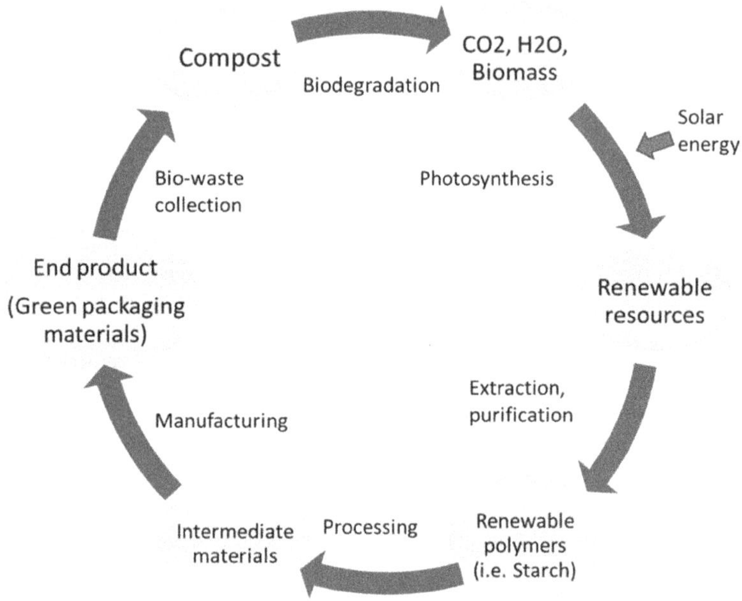

Figure 5.2 Life cycle of green packaging materials (*source*: Reprinted with permission from Lörcks (1998); © 1998 Published by Elsevier Ltd.

The biodegradable properties of "green" packaging materials are important because they offer a solution to disposal problems as they fit in with natural cycles of decay to CO_2, water and non-toxic inorganic compounds. Green materials are used in classical, active and smart food packaging applications, drug delivery, biomedical applications and in construction (Ramesh et al. 2017; Jafarzadeh et al. 2018; Nurul Fazita et al. 2016; Medina-Jaramillo et al. 2017; Talegaonkar et al. 2017; Wang et al. 2019).

Not all the biodegradable materials are renewable, and not all the renewable materials are biodegradable. Materials are either bio-based or biodegradable, or both are considered as "bioplastic" by European Bioplastics (european-bioplastics.org). If a material is produced using biomass instead of petrochemical resources, it is classified as bio-based. However, petroleum-based polymers such as PBAT (polybutylene adipate terephthalate), PBS (polybutylene succinate) and PCL (polycaprolactone) are biodegradable. There also exist non-biodegradable bioplastics, namely bio-based PET (polyethylene terephthalate), PE (polyethylene), PA (polyamide) and PTT. In this chapter, both bio-based and biodegradable polymers are discussed.

Biodegradable and renewable materials can be made from polymers directly extracted from natural sources such as carbohydrates (chitosan, cellulose, lignin, starch, pectin, agar, alginate) and proteins (whey, zein, casein, gluten, gelatin, soy protein), and may be composed of bio-derived polymers (i.e., PLA, PGA, PVA) and nanoparticles (i.e., graphene, clays, starch nanocrystals, carbon nanotubes) or obtained via microbial fermentation (PHAs) (Sadeghi and Mahsa 2015; Su et al. 2010).

Today, complete replacement of fossil-origin materials with renewable green packaging materials does not yet seem entirely possible. Poor mechanical and barrier properties, higher cost compared to petroleum-based plastics, and the lack of recycling and composting infrastructure limit the take-up of green commercial packaging applications. Nevertheless, many reports showed that a significant portion of consumers in different countries (India, North America) will consider paying more for environmentally-friendly packaging (Laroche et al. 2001; Barber 2010).

Renewable and biodegradable polymers generally possess poor barrier and mechanical properties when used alone. Use of plasticizers, cross-linking and reinforcing bio-based polymers with other particulates and polymers are the most commonly applied solutions to overcome these technical problems (Todkar and Patil 2019; Tian et al. 2017; Wang et al. 2017; Garavand et al. 2017; Farah et al. 2016).

5.4 Mechanical Properties of Green Packaging Materials

The mechanical properties of biodegradable materials depend on chemical composition, polymer architecture, processing conditions, storage and aging. Green packaging materials tend to have poor mechanical properties when biodegradable polymers are used alone. To improve the strength and durability, plasticizers, reinforcing materials and cross-linking are used.

Since mechanical properties vary with relative humidity and temperature, the materials should be tested at the targeted use conditions, i.e., refrigeration temperatures or ambient conditions. Typically, tensile strength (MPa), elongation at break (%) and Young's modulus (elastic modulus) (GPa) are determined to assess the mechanical performance. Mechanical tests may include tensile tests, puncture tests and tearing strength for films, and compression strength, burst strength and 3-point bending tests for thermoformed materials. Mechanical strength and durability are critical requirements for green packaging materials.

Natural fibers such as coir, hemp, jute, rice husk, sisal and bagasse are widely used to reinforce thermoset or thermoplastic materials, both to improve the mechanical strength and to use waste materials. They are preferred due to their low density (1.25–1.50 g/cm^3), low cost, high availability, biodegradability and non-toxic properties. Plant-origin fibers are obtained from seeds, stems, stalks, leaves and fruits. Their contribution to mechanical properties depends on their composition, their interfacial adhesion within the matrix, and the fiber diameter and length. Since natural fibers are hydrophilic, the resulting poor adhesion between fibers and hydrophobic polymers poses a difficulty. Bamboo, flax, kenaf, wood and hemp fibres may be incorporated into PLA to improve mechanical properties (Nurul Fazita et al. 2016). For stronger green materials, incorporation of nanoparticles into green polymers is another well-documented approach (Talegaonkar et al. 2017; Sun et al. 2014; López et al. 2015; Lorevice et al. 2016; Oymaci and Altinkaya 2016).

Metal and metal oxide nanoparticles such as nanoscale silver, zinc oxide or titanium oxide are inorganic examples. Organic nano-particles such as liposomes, protein clusters, micelles and dendrimers are non-toxic but are relatively sensitive to thermal or non-thermal high shear processing. Nano-particles, nano-clays and nano-emulsions have better compatibility, higher availability and lower cost compared to macro-scale reinforcing materials (Moustafa et al. 2019). It is cost-efficient to obtain desired properties using less material by using nanoscale reinforcing particles. Lignocellulosic nanomaterials such as nano-fibrillated cellulose, nano-crystalline cellulose and cellulose nano-whiskers are environmentally-friendly and abundant. Nano-crystalline cellulose has a high aspect ratio (70:100), strength (tensile strength: 7500 MPa) and stiffness to enhance the mechanical properties of a polymer matrix when used as a nanofiller (Moustafa et al. 2019; Flauzino Neto et al. 2016). Also, using lignocellulosic waste-origin nano-particles in packaging formulations reduces embodied energy as well as using otherwise waste materials.

For an integrated polymer matrix, cross-linking of polymer chains via covalent or non-covalent links is effective. Proteins and carbohydrates are particularly suitable for inter- and intra-molecular cross-linking using agents such as formaldehyde, glutaraldehyde, polyphenols, sodium benzoate, citric acid, boric acid, calcium chloride and the Maillard reaction (Garavand et al. 2017; Qin et al. 2020; Tavassoli-Kafrani et al. 2018), although it should be noted that formaldehyde is not suitable for food contact applications. Cross-linkers are matched to the polymer type. It is a commonly used cross-linking agent since it can react with functional groups of both proteins and carbohydrates, and result as improved tensile strength (Çay et al. 2014; Fan et al. 2018). However, due to its cytotoxic potential at concentrations above 8%, greener cross-linkers are preferred (Reddy et al. 2015). When toxic cross-linking agents are replaced by natural ones, such as citric acid or glucose, the reaction may need to be enhanced by catalysis (i.e., sodium hypophosphite) and/or heating (Azeredo and Waldron 2016; Siimon et al. 2015; Reddy and Yang 2010). Cross-linking is attributed as safer, cost-effective and more efficient than other reinforcement applications (Garavand et al. 2017). Generally, as flexibility of a biodegradable material increases, tensile strength decreases. Percentage elongation of brittle zein films increased from 3.34 to 142.24% by addition of catechins (phenolic compounds naturally found in tea), whereas tensile strength was reduced ten-fold (Arcan and Yemenicioğlu 2011).

5.5 Barrier Properties of Green Packaging

Barrier properties of food packaging materials are critical if a prolonged shelf life is to be maintained. Barrier requirements vary depending on the food composition, biochemical reactions occurring within the food during storage and intended use of the material in packaging applications. For food products high in lipids, good oxygen and UV light barrier properties are essential to prevent lipid oxidation.

Oxygen and water vapor permeation rates are the two main factors considered while designing a packaging targeting a certain type of food. Also, depending on the packed food, permeability of other gases (CO_2 or N_2 for modified atmosphere-packed (MAP) foods), oil resistance, flavor and UV transmission can also be important (Han 2014). Oxygen transmission is important for fresh products (i.e., fruits and vegetables, fresh meat) and lipid oxidation-sensitive products (i.e., nuts, crisps). Permeability is generally expressed as the oxygen permeability coefficient (kg mm^{-2} s^{-1} Pa^{-1}) or the oxygen transmission rate (OTR) (cc m^{-2} s^{-1}). Water vapor barrier ability is critical to avoid high moisture loss or gain resulting in deterioration in the physical and chemical properties of the food material. This ability is expressed as the water vapor permeability coefficient (WVPC) (kg mm^{-2} s^{-1} Pa^{-1}). Barrier properties vary with temperature and relative humidity. As the temperature increases, the permeant gains higher energy and the rate of transmission increases. Also, at the molecular level, polymer structural changes associated with phase transition temperatures affect the migration rate of substances. The degree of polymer crystallinity affects barrier properties. The ordered and tightly packed structure of polysaccharides provides excellent oxygen barrier properties; however, they have very poor water vapor barrier qualities due to their hydrophilicity. Starch-based materials are especially sensitive to moisture (Han et al. 2006).

One of the strategies to improve barrier properties is multi-layer film production. A two-layered soy protein isolate-poly(lactide) (SPI-PLA) film had 40-fold lower water vapor permeability than single SPI film and 26-fold lower oxygen permeability than single PLA film (Rhim et al. 2006). By producing PLA-wheat gluten-PLA multilayered film, oxygen and water vapor permeability of PLA film was lowered by 2000% and 20%, respectively (Rocca-Smith et al. 2019). As with mechanical improvement strategies, the barrier properties of packaging materials can also be adjusted by reinforcement of particles, especially nano-particles. Clay nano-platelets with a high surface area, high aspect ratio and thermal stability were distributed to this effect within a polymer matrix to improve the transmission of gas, water vapor and UV light (Duncan 2011). Effective dispersion of exfoliated nano platelets is the key factor in providing strong barrier behavior. Soy protein isolate (SPI)-based composite films were incorporated with bio-based hyperbranched polyester and phenolic cardanol. The tensile strength, elongation and UV-barrier properties of the films increased significantly due to the increased internal cavities and cross-linking (Gu et al. 2019).

5.6 Green Packaging Materials with Active Properties

Green packaging materials with active properties can extend shelf life of the foods by preventing microbial and chemical deterioration. In this way, while the rate of food-borne diseases is reduced, less food loss provides further economic and environmental advantages besides the sustainability aspects of green packaging materials. Basically, the mechanisms of action of active packaging materials are:

- releasing active agents (i.e., antimicrobial enzymes, phenolic compounds),
- scavenging undesirable components (i.e., oxygen, ethylene, moisture),
- emitting preservative substances (i.e., sulfur dioxide, carbon dioxide), or
- providing an antimicrobial contact material onto food surface (i.e., chitosan) (Prasad and Kochhar 2014; Yildirim et al. 2018) (Table 5.1).

For these active applications, scavengers and emitters are generally applied in sachet, pad or label form. Green packaging materials may be used in manufacturing of these tools; however, their major role in active packaging is as active and biodegradable films or coatings (Ahmed et al. 2017).

Antimicrobial packaging is a form of active packaging designed to suppress or inhibit microbial growth on the food surface. An antimicrobial packaging system is made using

Table 5.1 Active biodegradable packaging systems applied to foodstuff.

Category	Incorporated active agent	Improved physical properties	Food application	Source
Cassava starch film	Microalgae (*Heterochlorella luteoviridis* and *Dunaliella tertiolecta*)	Elongation, rupture	Antioxidant activity in salmon	(Carissimi et al. 2918)
Cassava starch film	Encapsulated grape pomace extract	Biodegradability, UV barrier	Antioxidant activity in extra virgin olive oil	(Stoll et al. 2017)
Chitosan/ Gelatin film	Silver nanoparticles	Light barrier, mechanical properties	Extended shelf life of red grapes	(Kumar et al. 2018)
Potato peel powder/ Bacterial celulose film	Curcumin	Tensile strength	Antioxidant activity in fresh pork	(Xie et al. 2020)
Chitosan film	Olive pomace particles	Tensile strength	Antioxidant activity in walnut	De Moraes Crizel et al. 2018)
PLA-PHB film	Cinnamaldehyde	Tensile strength, water vapor barrier	Antimicrobial and antioxidant activity in fresh salmon	(Ma et al. 2018)
Cellulose acetate film	Carotenoids (norbixin, lycopene and zeaxanthin)	UV barrier	Antioxidant activity in sunflower oil	(Assis et al. 2020)
Chitosan film	Kombucha tea	UV and water vapor barrier	Extended shelf life of minced beef	(Ashrafi et al. 2018)
Chitosan/ Pectin film	Tea polyphenols	Elongation, release rate	Prevention of color degradation in fresh beef	(Gao et al. 2019)
PHBV film	Oregano and clove essential oils	–	Antimicrobial activity in cheese	(Requena et al. 2019)

a polymer matrix containing an antimicrobial agent. Antimicrobial compounds may be incorporated into polymer films and coatings, adsorbed onto polymer film surfaces or immobilized to polymers by ion or covalent linkages (Appendini and Hotchkiss 2002). Another option is using polymer matrices made of inherently antimicrobial polymers (i.e., chitosan). Antimicrobial packaging materials are able to maintain sustained activity on the food surface by gradual release or constant contact of antimicrobial agents, compared to direct addition of the antimicrobials in the food formulation (Quintavalla and Vicini 2002). This mechanism of antimicrobial materials is critical since the food surface has the highest susceptibility to microbial spoilage.

Antioxidant packaging materials are designed to prevent lipid oxidation, a major reason of food spoilage after microbial spoilage. Release of antioxidants into packaging environment or scavenging lipid oxidation factors (oxygen, reactive oxygen species, metal ions) are two main antioxidant packaging strategies (Gómez-Estaca et al. 2015). Numerous biodegradable active packaging systems have been applied to foodstuffs (Table 5.1). Incorporation of active agents into green polymer-based packaging materials may negatively affect their mechanical and barrier properties, depending on the plasticizing effect and the compatibility between incorporated agents and the polymer matrix (Carissimi et al. 2018; Adilah et al. 2018).

In keeping with the green packaging ethos, non-toxic and natural active agents are used in green active packaging materials. Organic acids, bacteriocins, antimicrobial enzymes, essential oils, natural phenolic compounds, antimicrobial peptides and spices are examples of active agents with GRAS (generally recognized as safe) status. Safety issues regarding nanoparticle use are increasing due to the high potential of migration of these nano-particles into foodstuffs (Kim et al. 2020). Since nano-particles possess very high surface area to volume ratio, their physicochemical and biological activity is expected to be different than macro-scale particles when introduced into biological systems (Espitia et al. 2012).

5.7 Biodegradation Mechanisms of Green Packaging

The amount of plastic waste is increasing each year and the reported proportion of recycled plastic waste is less than 10% of the total used (D'ambrières 2019). A major part of the municipal solid waste includes packaging materials and food containers. Packaging waste disposal methods are mainly landfill followed by recycling, incineration and composting. Recycling of glass, paper, aluminum and plastic are efficient unless the materials are soiled with food or other biological residuals (Kale et al. 2007). Composting in industrial reactors or domestic compost heaps is a promising way to reduce the amount of solid waste attributed to packaging materials.

Composting is a natural process performed by mesophilic and thermophilic microorganisms including bacteria, fungi and algae ubiquitous in soil and water. Composted material is degraded into CO_2, water, methane and biomass through chemical and biological processes without leaving any toxic residues. Biomaterials provide the organic material for microorganisms to use as an energy source and as anabolic building blocks for cell structures. The availability of water and oxygen, and the ratio of C to N, are critical parameters for an optimum level of microbial activity. For thermophilic microorganisms, a 20:1 carbon-to-nitrogen (C:N) ratio is ideal for composting (Siddiqui et al. 2011). Composting has two stages, namely the active stage, and curing. During the active stage, the temperature rises, and microbial activity is at its highest. This active stage is followed by curing where the temperature lowers, but organic materials are still being converted into CO_2, albeit at a slower rate, continuing until no organic carbon is left in the reaction medium (Kale et al. 2007). A compostable material should degrade within a given time at a certain rate. This period is 6 months for industrial composting and 12 months for home composting. Industrial composting is applied following specified standards, mainly ASTM D6400 in the USA, EN 13432 in Europe and ISO 14855 internationally.

Biodegradation, which is basically an enzymatic reaction, is the critical step of composting. In the presence of microbial activity, aerobic degradation converts organic compounds to carbon dioxide, water and inorganic compounds. In the absence of oxygen, the process shifts to anaerobic digestion where the end products are biogas, mainly methane and CO_2, and digestate (Kale et al. 2007; Avella et al. 2001; Petersen et al. 1999). In general, polymers degrade through scission of their main chain and side chains. Biodegradation of a polymer involves both chemical and biological mechanisms. Enzymes play an important role in degradation of the polymers, nevertheless they are not responsible of the hydrolysis alone. Enzymatic reactions caused by various microorganisms are enzymatic oxidation and enzymatic hydrolysis. Enzymatic oxidation takes place in the presence of aerobic microorganisms, where enzymatic hydrolysis may occur under both aerobic and anaerobic conditions. Enzymatic hydrolysis of aliphatic polyesters takes place in two steps on the polymer surface: i) adsorption of the enzyme on the polymer surface via its active site; and ii) hydrolysis of the ester bonds.

During enzymatic hydrolysis, ester groups of polymer chain are detached randomly, and the molecular weight of polymer reduces. Functional groups, copolymer composition (if present) and chemical bonds are the critical determinants of a polymer's biodegradability. Hydrolyzing enzymes released by microorganisms can only attack the specific sites of the functional groups. The presence of hydrolyzable functional groups such as ester, amide and carbonate bonds, increase the biodegradation rate of polymers. Also, polymers conformational flexibility plays an important role in enzymatic activity. Poly(3-hydroxybutyrate) depolymerases, lipases and proteases are the commonly examined enzymes responsible for the hydrolysis. Non-enzymatic hydrolysis depends on the water diffusion into the polymer, so hydrolysis may occur on the surface as well as throughout the material's bulk. Water diffused through the amorphous regions is critical, while diffusion through the crystalline regions has no significant effect (Kale 2007).

Biodegradation does not depend on the source of the material, but rather depends on the environmental factors and the chemical structure of the subjected material. Biodegradation reaction rate depends on the temperature (ideally between 50 and 70°C), humidity, number and type of microorganisms, pH and oxygen. As more water is diffused through the polymer, faster biodegradation occurs (european-bioplastics.org).

Factors affecting the biodegradation related to the polymer can be classified as:

- Factors related to first-order structure (chemical structure, molecular weight (MW), MW distribution)
- Factors related to higher-order structure (crystallinity, glass transition temperature (T_g), melting point (T_m), elastic modulus)
- Factors related to surface properties (surface area, hydrophobicity)
- Particle size and shape, isomer percentage, residual monomer concentration and water diffusion are other significant factors in the susceptibility of polymers to biodegradation.

Rather than forming crystalline or amorphous solids, polymers have crystalline and amorphous phases. Degree of crystallinity, expressed as a percentage, is the ratio affecting biodegradability of a polymer. Crystalline parts of a polymer are more resistant to biodegradation than amorphous regions. Higher MW and higher melting point make the polymer more resistant to biodegradation (Tokiwa and Calabia 2006).

Biodegradation of natural polymers, such as starch and cellulose, occurs via extracellular enzymes. After cleavage of long polymer chains in an extracellular environment, the smaller chains are metabolized inside the microorganisms' cells. Here, the molecular weight of polymer plays an important role in the biodegradation rate. Materials dominantly made of starch, cellulose and poly(hydroxyalkanoate)s (PHAs) are the subjects of this type of biodegradation. An exceptional carbon-chain polymer, poly(vinyl alcohol), biodegrades by enzymatic oxidation of hydroxy groups (-OH), forming carbonyl groups and enzymatic hydrolysis of two carbonyl groups leading to lower molecular weight chains of which metabolizing by microorganisms is possible. Polymers having atoms other than carbon (i.e., oxygen or nitrogen) are more susceptible to hydrolysis (Van Der Harst et al. 2014). Poly(lactic acid) (PLA), poly(glycolic acid) (PGA), PHAs and poly(ε-caprolactone) (PCL) are examples to these type of heteropolymers.

Rapid mineralization in an important criterion for biodegradation. Enzymatically cleaved degradation products should be assimilated into CO_2 and H_2O at a similar rate to prevent accumulation in the environment. Natural polymers like rubber and lignin do not fit into this criterion because they degrade via an oxo-biodegradation mechanism. Oxo-biodegradation, triggered by heat and light, is slower than hydro-biodegradation at ambient temperatures. By means of peroxidation, low molar compounds such as carboxylic acid, alcohol, aldehydes and ketones are produced (Siracusa et al. 2008). Generated low MW substances are assimilated by microorganisms.

Unlike hydrophobic synthetic polymers, hydrophilic polymers like cellulose and starch have high wettability and swelling properties, therefore faster biodegradation rates. Hydro-biodegradable aliphatic polymers PLA and PHAs stand between these two points. Aliphatic polyesters (PLA, PCL and PHAs) are promising biodegradable polymers. These polymers have one or more hydrolyzable bonds of ester, ether, amide or carbonate. The hydrolyzable bonds also increase the polymer flexibility and conformation making them more suitable for enzymatic hydrolysis. Longer PLA chains are more susceptible to biodegradation than shorter ones due to susceptibility to random hydrolysis. Blending biodegradable polymers may slow down the biodegradation rate compared to unblended grades (Su et al. 2010). PVA/SPI blend films showed slower biodegradation than pure SPI films. PVA has a slow biodegradation rate. Interactions within the composite or blend materials affect the total biodegradability of the article. It was reported that plasticizers have no effect on biodegradation. Using filler particles to enhance water and air diffusion toward the inner part of the matrix has an accelerating effect in biodegradability.

Biodegradability of commercial packaging materials is tested by composting procedure. According to EN 13432, chemical characterization, biodegradation, disintegration and ecotoxicity tests are applied to biodegradable materials for compostability certification. Generally, CO_2 production is measured to determine biodegradability. Under industrial composting conditions (58°C), at least 90% organic carbon should be converted to CO_2 gas within a 6 months' period.

5.8 Main Green Food Packaging

5.8.1 Poly(lactic Acid) (PLA)

Poly(lactic Acid (PLA) is a promising green material produced by fermentation of sugar-rich feedstock (corn, potato) to produce lactic acid, and subsequent controlled polymerization of the obtained lactic acid monomers. It is a fully compostable and recyclable material with good processability, transparency and water resistance (Siracusa et al. 2008). L(+) or D(−) stereoisomer PLA semi-crystalline forms of PLA both have high tensile strength and low elongation. DL-PLA consists of both L-lactic acid and D-lactic acid and the ratio of these two enantiomers determines the degree of crystallinity of PLA. For thermoforming processes applied in food packaging applications, amorphous PLA with around 12% D-lactic acid is desirable (Tokiwa and Calabia 2006). PLA is synthesized either by condensation polymerization of lactic acid or by ring opening polymerization of lactide, the cyclic dimer of lactic acid.

PLA-based materials exhibit comparable tensile strength to plastic packaging materials with its advantageous properties such as high melting point (175°C), high transparency and ease of manufacturing. It is used as nonedible, biodegradable film and/or container in the food packaging industry. It was shown that PLA/sugar beet pulp films showed much higher tensile strength and elasticity than Styrofoam (Li et al. 2012). Also, good temperature stability and processability of PLA makes it comparable in performance to Polystyrene (PS). PLA performs better than some synthetic fossil-based materials under many conditions, i.e., oriented polystyrene (OPS) and PET, and it has lower sealing temperature than PET and PP. PLA film is an oil resistant material with good water vapor, flavor and gas barrier, and is comparable to PET. Its CO_2 and O_2 permeability coefficients are lower than PS and higher than PET; however, its water vapor permeability (WVP) is not as good as those of PS and PET. Since most of the aroma compounds are of a hydrophobic nature, PLA possesses good aroma barrier properties (Colomines et al. 2010).

PLA, due to its hydrophobic nature, is not a good carrier of hydrophilic antimicrobials. Therefore, it is mostly combined with another biodegradable hydrophilic matrix, such as pectin, to bind and release compatible active agents (Jin and Liu 2020). In case of extrusion of PLA, most natural antimicrobials are not resistant to the process temperature and lose their efficiency.

PLA is compostable under industrial composting conditions. It will biodegrade after a prior hydrolysis. Previously, reports have suggested that degradation of PLA occurs through hydrolysis with no enzyme involvement (Tokiwa et al. 2009). Despite this, there have been 39 bacterial strains isolated from soil reported for biodegradation of aliphatic polyesters (PHB, PCL and PBS), none of which were able to degrade PLA. This result suggested that microorganisms capable of biodegrading PLA do not widely exist in the natural environment. Nevertheless, recently there have been microorganisms reported for PLA biodegradation. *Amicolatopsis* sp., *Tritirachium album*, *Bacillus stearothermophilus* and *Geobacillus thermocatenulatus* are some the microorganisms degrading PLA by releasing protases, and most of the detected strains were thermophilic (Tokiwa et al. 2009). Another study showed that nano-clay incorporated PLA showed higher biodegradation rates (Fukushima et al. 2013). Reinforcing particles into the polymer matrix may accelerate or slow down biodegradation, depending on the structural change they cause. If crystallinity increases as a result of reinforcement, slower biodegradation occurs. To the contrary, if the composite-making particles allow better water diffusion and therefore microbial activity, accelerated biodegradation is obtained (Souza and Fernando 2016).

5.8.2 Polyhydroxyalkaonate (PHA)

Polyhydroxyalkaonates are water insoluble bio-based and biodegradable polymers consisting of a hydroxy acid monomer produced by bacterial fermentation utilizing sugar and lipid. They are harvested directly after bacterial fermentation. Besides homopolymers, co-polymers such as poly(3-hydroxybutyrate-co-3-hydroxyvalerate (P(3HB-co-3HV)) or poly(3-hydroxybutyrate-co-4-hydroxybutyrate) (P(3HB-co-4HB)) can be produced in the case of using mixed substrates (Możejko-Ciesielska and Kiewisz 2016). Centrifugation, filtration or flocculation-centrifugation are methods to separate biomass containing PHAs, and PHAs are recovered by methods of solvent extraction, floatation, supercritical fluid extraction and differential digestion of cellular material. During the recovery of PHAs, polymer denatures and its crystallinity increases. Around 300 bacterial strains, including *Alcaligenes eutrophus*, *Pseudomonas* sp., *Azotobacter vinelandii* and recombinant *Escherichia coli*, produce PHAs as energy storage compounds (Raza et al. 2018). Depending on the synthase enzyme type in microorganisms, short-chain-length (3–5 C atoms) or medium-chain-length (6–14 C atoms) PHAs are produced.

PHAs have a thermoplastic behavior with low melting points (40–180°C) (Raza et al. 2018). Among the PHAs, poly(3-hydroxyautyrate (P(3HB)) is the most researched polymer. It has 55 to 80% crystallinity, therefore is stiffer and more brittle than PP. Its melting temperature (175–179°C) is close to the temperature at which it decomposes (200°C), which poses a problem during processing. Medium-chain-length has lower melting temperatures (39–61°C) and 25% crystallinity. These characteristics result in softer and more flexible materials than short-chain-length PHAs (Możejko-Ciesielska and Kiewisz 2016). To improve the physical properties of PHB, biosynthesis of copolymers of PHB contain 3-hydroxyalkanoates such as PHBV (polyhydroxybutyrate/valerate). PHBV has a high elastic modulus, low tensile strength, is less stiff and provides an excellent gas barrier. Another processing strategy for improved thermal and mechanical properties is blending with plasticizers and nucleation agents to lower Tg and crystallinity, and to blend with other polymers (Bugnicourt et al. 2014).

Since the enzymes degrading PHAs have a wide specificity, they are readily biodegraded under aerobic or anaerobic conditions within approximately 5 to 7 weeks, although in anaerobic environments their degradation is faster (Zheng et al. 2005). The main enzyme responsible for degradation of PHB is PHB depolymerase. PHA-degrading bacteria, including *Bacillus*, *Pseudomonas* and *Streptomyces* species found in nature, were estimated as 0.5 to 9.6% of the total colonies (Bugnicourt et al. 2014).

PHAs are relatively expensive polymers (15–18 times more expensive than poly propylene) due to their limited large-scale production. Substrate cost is attributed as the main factor of the high cost of PHAs, which requires further research in terms of alternative carbon sources (Możejko-Ciesielska and Kiewisz 2016; Raza et al. 2018; Reddy et al. 2003).

5.8.3 Starch-based Materials

Starch is a biodegradable, safe and inexpensive polymer found abundantly in nature. Potato, corn, rice, wheat, tapioca and cassava are the common sources for starch extraction. Starch consists of 20 to 25% linear amylose chains and 75 to 80% branched amylopectin chains. Thermo-processing of starch combined with plasticizers results in an amorphous thermoplastic moldable material. Thermoplastic starch (TPS) has good oxygen barrier properties. However, its main drawback is its sensitivity to atmospheric water vapor causing loss in its physical characteristics. High water vapor permeability of starch can be improved by blending with other polymers such as PLA and PHAs. Also, nanofillers such as cellulose nanofibers are effective to obtain better mechanical and barrier properties in starch-based materials. Heat sealing resistance was improved by addition of acetylated starch into starch films (Gadhave et al. 2018).

Starch foam trays have the potential to replace expanded PS foam trays, one of the mostly used plastics in the food industry and food service chains. Starch batter prepared with plasticizers can expand into a foam product and shaped as a food container using a compression molding machine. Starch trays' water affinity can be reduced by adding lignocellulosic fibers, blending with other polymers such as PVOH and PLA, or laminating with a water-resistant biodegradable film (Engel et al. 2019). Water resistance of potato starch trays incorporated with corn fiber was improved by PVOH addition. Potato starch matrix was combined with brewer's spent grain, chitosan and glyoxal cross-linker, and the trays were coated with beeswax to obtain similar properties to expanded polystyrene. Coating TPS/PLA trays with beeswax is an effective method to improve mechanical and water vapor barrier properties (Cinelli et al. 2006; Ferreira et al. 2019; Reis et al. 2018).

The high wettability of starch helps it to biodegrade rapidly. Most starch-based material can disintegrate within 5 days in an aqueous aerobic environment, and in 45 days under composting conditions. At 25°C, around 40% of a starch-based plastic bag was degraded after 90 days. For cassava starch trays blended with 20% orange bagasse, an agricultural waste fiber, full biodegradation was observed in soil after 60 days, which was faster than control starch trays. Presence of lignocellulosic filler materials accelerates biodegradation of thermo-processed starch-based trays (Ferreira et al. 2019; Accinelli et al. 2012; Chiellini et al. 2009).

5.8.4 Cellulose-based Materials

Cellulose is the most abundant biopolymer found in nature, composed of linearly linked β(1,4)glucopyranose units. Plant origin cellulose and bacterial cellulose have the potential to be used in food packaging applications. Extraction of cellulose from plant sources requires use of harsh chemicals to remove lignin, hemicellulose and pectin. Bacterial strains such as *Acetobacter xylinum* and *A. pasteurianus* can produce cellulose with very high purity (Weber et al, 2002). The principal source of cellulose is wood pulp, consisting of 40 to 50% cellulose (Vilarinho et al. 2018). When cellulose is modified, thermoplastic cellulose acetate (CA) is obtained. CA is an ester of cellulose, used as membrane, film or fiber in packaging applications, or as insulation material. The degree of substitution determines the solubility of CA. Films are produced from cellulose triacetate, and for membrane and fiber production cellulose, 2–2.5 acetate are suitable (Fischer et al. 2008). CA films are preferred since they are biodegradable, ecofriendly, water insoluble, odorless and non-toxic (Puls et al. 2011). Hydrophilicity of CA films is low, so the films are resistant to swelling in water (Marrez et al. 2019). CA can form films at low temperatures (i.e., room temperature), therefore incorporation of heat sensitive active agents into the film matrix is possible. CA films are widely employed in active food packaging applications (Assis et al. 2020; Dairi et al. 2019; Indumathi et al. 2019; Gouvêa et al. 2015).

Degradation of CA films takes place after an initial deacetylation by chemical hydrolysis and acetylesterases, so cellulase enzyme can access the cellulose backbone for further degradation. Combined activity of esterase and cellulase helps CA to biodegrade completely. It was reported that as the degree of substitution increases, the biodegradation rate of CA reduces. In an activated *in vitro* biodegradability medium, CA with 1.7 DS, biodegraded by 91 to 98% after 4 to 5 days. In another test conducted in wastewater, high levels of biodegradation took 27 days (Buchanan et al. 1993, 1996).

5.9 Life Cycle of Green Packaging Materials

To evaluate the environmental performance of a green packaging material, life-cycle assessment (LCA) is standardized in ISO 14000 and ASTM D7075. Considering the life cycle of a packaging material, composting is the end of life for biodegradable green materials. Many studies examined the environmental impact of bio-based and biodegradable materials compared to petroleum-based counterparts. In such studies, the parameters taken into account may result in different conclusions. For example, disposable plates made from leaf had higher global warming potential due to transport and energy use, but lower acidification and esterification impact compared to paper plates (Korbelyiova 2019). In another study, biodegradable tableware causing 50% lower GHG emission compared to petroleum-based equivalents had higher land-use for production of raw materials (Fieschi and Pretato 2018). Main processes responsible for the environmental impact of a material are determined as production of the intermediate material, manufacturing of the article and its waste management. Recycling and reusing a material is proposed to be more susceptible (Van Der Harst et al. 2014). However, recycling of single-use food packaging materials is ineffective, since the containers and films are generally heavily soiled with food compounds, which require cleaning steps and cause a larger water footprint than compostable materials. Also, recycling is energy expensive. So, biodegradable green food packaging materials are the most promising solution to prevent environmental pollution sourced by food packaging. When a bio-based material is composted, the environmental benefit grows if the compost is further used in agricultural practices in substitution for peat and fertilizer (Fieschi and Pretato 2018). Further waste management infrastructures are required to avoid contamination of recycling facilities and to assure effective industrial composting.

Biopolymers and bio-based additives are highly promising components for green food packaging material production. Significant developments have already occurred in compostable green packaging materials with sufficient barrier, mechanical and thermal properties, as well as with compliance to international standards and legislations. As the political and social pressure increases toward the use of eco-friendly packaging materials, a higher driving force causes advances in manufacturing, recycling and composting phases of green packaging materials.

5.10 Smart Packaging in Food

"Smartness" in food packaging is a broad concept that includes the implementation of a variety of technologies during production to provide both useful information and beneficial properties for the consumer. Present and prospective functional "smartness" would be embodied in packages retaining integrity and preventing food contamination, actively responding to changes in the product or package environment, communicating user information, product history or condition, identifying, tracking, authenticating and finally acting as a protection against adulteration and imitation (Silvestre et al. 2011; Ghosal et al. 2018; Kuswandi et al. 2011; 2012). Table 5.2 indicates comparison between smart and traditional packaging.

There is a crucial need for new technologies to provide food safety and sustainability, developing packaging that does much more than contain and protect the contents. Smart packaging with its "intelligent/smart" approach will feed back information on the product itself, aspects of consumer behavior, affordability and safety (Vanderroost et al. 2014; Chen et al. 2020).

Intelligent packaging features may switch on and off in response to shifting environmental or internal variations, and may communicate to the consumer information on the product status. A simple smart packaging definition is that it senses and informs (Kuswandi et al. 2011; Yam et al. 2005). Intelligent packaging is characterized by effective communication and marketing functionality (Sharma et al. 2017). The notion of intelligent sensors has been known for the past three and a half decades, and is defined as a measurement system that evaluates a physical variable and employs a digital computer to perform the signal and information processing (Biji et al. 2015).

Table 5.2 Comparison between smart and traditional packaging.

Smart packaging	Traditional packaging
Designed to monitor and communicate information about food quality of the packed food during transport and storage	Designed to delay the adverse effects of the environment and handling on the food product
Digital components	No-digital components
Shelf life of food is maximized	Shelf life of food is not affected
Highly automated handling	Labor intensive handling
High initial cost, low total costs	Low initial cost, high total cost

(*source*: based on Biji et al. 2015)

Factors related to the growth of the intelligent packaging sector include customer and socio-economic factors as well as technical capabilities related to food preservation (Mihindukulasuriya and Lim 2014). Smart packaging was not a commercially feasible idea until a decade ago, because matching packaging and computer technologies were too costly and offered relatively limited functionality. There is now readily accessible, more efficient and affordable technology providing a positive environment for market expansion (Fang et al. 2017).

The rising demand for packaging information as a symbol of consumer choice requires a considerable advance in packaging technology. Increasingly, customers will want to know what components are in the product, as well as how to store and use the product. For example, an intelligent labeling system will be able to communicate directly with the consumer that provides sound and visual information, either in reply to contact, movement or some other tracking or stimulation means (Abreu et al. 2012).

The guidelines used for household food and packaging waste ideally steer the customer to efficiently dispose of post-consumption waste so that it can profitably enter the recycling and reuse plant, minimizing the need to separate packaging materials from foodstuff waste (Han et al. 2018).

Food in smart packaging may be programmed to communicate end-of-life information back to various stages in the supply chain stock systems via intermediate devices such as mobile phones or computers (Vanderroost et al. 2014; Yam et al. 2005).

Quality assurance can be assisted by using intelligent tags (Ahvenainen and Hurme 1997). Information relating to consumer safety is important, especially for perishable foods. For example, it is of legitimate concern if a chilled ready meal is safe to eat: this is currently addressed with "best by" label. However, a passive label does not take into account whether the substance was exposed unintentionally to higher temperatures during storage or transport. Microbial growth and temperature-time indicators based on physical, chemical or enzymatic activity in the food can provide a simple and reliable description of quality, security and remaining extent of shelf life (Kuswandi et al. 2012; Kerry et al. 2006; Heising et al. 2014). Furthermore, intelligent packaging could also contribute to improving Hazard Analysis and Critical Control Points (HACCP) and Critical Control Points (CCP) systems (Vanderroost et a. 2014; Heising et al. 2014). For smart materials to be implemented in packaging, they must be cheap compared to the product value, durable, accurate, replicable, eco-sustainable and safe for food interaction (Ahvenainen and Hurme 1997; Heising et al. 2014).

Three key technologies are commonly recognized as allowing intelligent packaging systems to be realized:

1) Indicators intended to provide greater convenience and/or inform consumers of the quality of the food;
2) Sensors enable the rapid and reliable quantitative analysis of food; and
3) Data carriers, such as barcodes and radio-frequency identification (RFID) tags, intended primarily for storage, delivery and traceability purposes (Ghaani et al. 2016) (Figure 5.3).

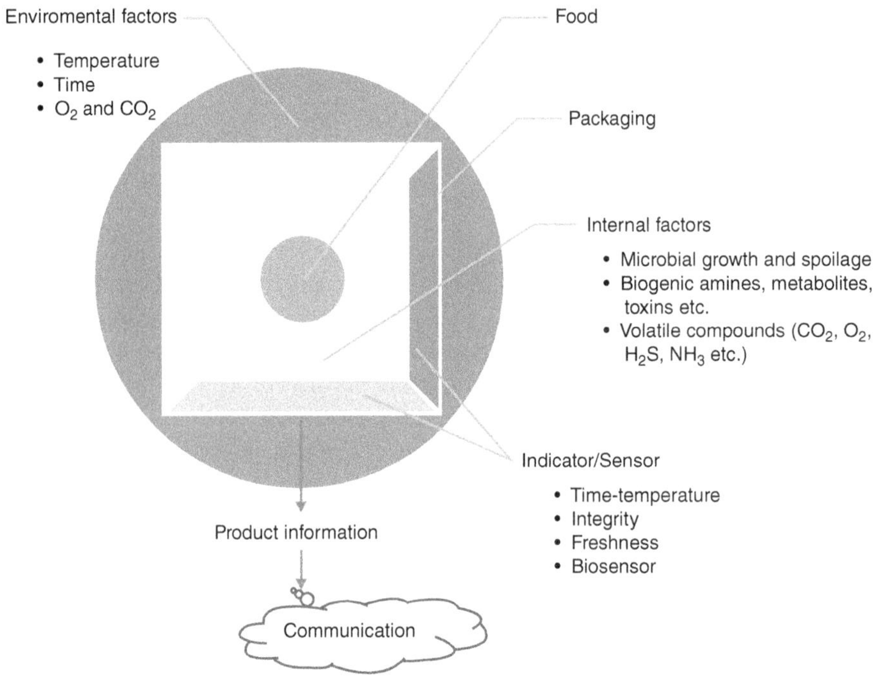

Figure 5.3 Smart packaging for food packaging materials (source: based on Weber 2017).

5.11 Indicators for Smart Packaging

Indicators can provide benefits by confirming the presence, exclusion or concentration of another substance, or the degree of reaction between two or more compounds, and signal this to the consumer through a distinctive change, particularly in color (Biji et al. 2015; Hogan and Kerry 2008). The most commonly used smart apparatus for food quality can be classified as indirect or direct indicators. While indirect indicators cannot provide direct guidance to help consumers in assessing quality of food and sensory characteristics, they can indirectly evaluate the effects of the environment on the food based on the shelf life that could lead to a concealed hazard (Han et al. 2018).

Direct indicators can directly inform about food properties. So, these devices are usually located inside the packaging so that they have direct contact with the atmosphere surrounding the food or with the food itself. Some even can monitor properties such as freshness, edibility and quality indirectly from the outside via an optical system (Venderroost et al. 2014; Kerry et al. 2006; Heising et al. 2014).

5.11.1 Time-Temperature Indicator (TTI)

Temperature is generally the most important risk known environmental factor affecting the deterioration of food. It is particularly important to give warning of temperature abuse for perishable foodstuffs such as chilled or frozen products. The Food and Drug

Administration (FDA) defines requirements for time and temperature control of food to ensure food safety and quality (Han et al. 2018). The aim of time temperature indicators is to provide information as to whether a reference temperature has been exceeded over time and/or to quantify the total amount of time a product has spent above the reference temperature (Vanderroost et al. 2013; Kerry et al. 2006).

There are three basic types of time-temperature indicators (TTI) based on their response mechanism:

1) critical temperature indicators,
2) partial history indicators, and
3) full history indicators (Han et al. 2018).

The critical temperature indicators only indicate whether food has been exposed to a temperature above or below a threshold temperature. The partial history indicators do not respond unless a temperature threshold has been surpassed and show that food has been exposed to a temperature sufficient to affect its quality. The full history indicators include continuous monitoring of the way temperature changes over time, over the history of a product, and so are the focus of current industrial research (Biji et al. 2015; Kerry et al. 2006; Hogan and Kerry 2008).

The basic operation principles of TTIs are based on the temperature-related and irreversible responses of enzymatic, chemical or microbiological reactions (Fang et al. 2017; Han et al. 2018). These changes are usually conveyed as a clear response in the manner of a deformation, formation of color or movement of color (Ghosal et al. 2018; Kruijf et al. 2002). TTIs currently available on the market work by chemical or physical response, for example by an acid–base reaction, melting, polymerization, biological responses including growth of microorganisms, fungi, yeast or enzymes based on time or temperature (Kuswandi et al. 2011). TTIs must be cost-effective, tiny, accurate, fully adaptable into a package/container by printing or coating and not give false positives due to the influence of non-temperature ambient condition changes. In addition, TTIs should also be versatile in choice of temperature range, durable, effectively express data, and should not carry any toxicological or safety potential risk (Kuswandi et al. 2011; Kerry et al. 2006; Park et al. 2015).

TTIs are used during distribution and storage to monitor exposure to inappropriate temperature changes and are an indication of security and quality for the manufacturer, as they ensure that the product arrives with the end-user in optimum circumstances (Abreu et al. 2012).

TTIs can facilitate the guarantee of fresh food for the consumer by tracking and inspection of food products. This has the potential to decrease food waste by extending shelf life through a reliable indication of the duration of post processing stages and lengthening the window of opportunity for safe consumption (Ghosal et al. 2018).

5.11.2 Freshness Indicators

Freshness indicators provide an indication of the spoilage or loss of freshness resulting from microbiological and chemical changes. This mechanism works by changes associated with microbial growth or chemical metabolites that are incorporated into indicators within the packaging, giving visual information about food quality and safety for consumption (Biji et al. 2015; Kerry et al. 2006; Müller and Schmid 2019). Furthermore, the indicator system must comply with food-contact regulations, since the indicator will be directly in contact with the foodstuff or with the headspace of the packaging Mohebi and Marquez 2015).

Freshness indicators react to changes in the concentration of volatile metabolites and the formation of organic acids resulting from aging during the storage of meat, fruits and vegetables. Carbon dioxide, hydrogen sulfide, ammonia, biogenic amines, n-butyrate, L-lactic acid and diacetyl are examples of the range of compounds that result from microbial growth (Biji et al. 2015; Realini and Marcos 2014). The freshness of meat and fish products is measured by the degree of formation of biological amines occurring as a result of the catabolic degradation of proteins to amino acids and enzymatic decarboxylation (Abrey et al. 2012; Mustafa and Andreescu 2018). A freshness indicator should be highly sensitive to such metabolite compounds (Mohebi and Marquez 2015).

Monitoring of food freshness can be achieved by using color-based pH indicators. The formation of CO_2 in meat products causes a reaction with pH dyes and the consequent change in color of the indicator. A particular example of a pH dye is bromothymol blue, with other pH dyes such as xylenol blue, bromocresol purple, bromocresol green, cresol red, phenol red, methyl red and alizarin also recommended as freshness indicators: targeting changes in pH can track and quantify the freshness of foodstuffs (Fang et al. 2017; Kalpana et al. 2019).

The number of convenient market applications of smart package indicators for detecting freshness is still very limited. Despite this, the production of freshness indicators based on proven knowledge of quality indicating metabolites has considerable potential (Kerry et al. 2006). Note that to build up sufficient detectable concentrations of vapor or gas, packaging material may need to be non-breathable or have limited gas permeability to allow effective use of an indicator incorporated into packaging. This of itself may lead to a compromise in the quality life of the product within the packaging. Hence, compromise is critical, which are may prevent developers from embracing the use of such indicators (Mohebi and Marquez 2015).

Freshness detection may be achieved using color changes based on aromatic fingerprinting coupled with a visual indicator. Naked eye imaging and interpretation provides particular advantages for perceiving food quality and safety. These types of indicator are mostly focused on easily perishable expensive foods (Poyatos-Racionero et al. 2018).

5.11.3 Packaging Integrity Indicators

Modified atmosphere packaging (MAP) provides the product with a controlled mixture of gases atmosphere to extend the quality and shelf life of fresh food products. The cost of this technology is offset by advantages for both the consumer and the manufacturer. MAP provides a convenient and non-invasive packaging system, guaranteeing the delivery to the consumer of a high-quality and additive-free product (Hempel et al. 2013). However, there may be concern about failure of the packaging due to leakage or failure of the physical integrity of the package, which can occur at any stage of the storage and distribution chain. Package failure can cause significant variation from optimum gas concentrations, and these concentration changes can have deleterious effects on food quality, particularly sensory acceptability. To eliminate these negative effects, it is desirable to use integrity indicators to protect the consumer by monitoring leakage or other damage to the packaged products (Kalpana et al. 2019). Knowledge of the integrity of packaging is an important prerequisite for ensuring the quality and security of foods in MAP. Among the available types of integrity indicators, O_2 and CO_2 indicators are popular for MAP assessment applications (Realini and Marcos 2014). MAP usually includes high levels of CO_2 (20–80%) and low levels of O_2 (0.1–1%). Therefore, indicators based on O_2 levels can quickly detect leaks inside MAP due to an increase in O_2 concentration and a consequent reduction in the CO_2 concentration (Kuswandi et al. 2012). These indicators are available commercially in the form of labels or tablets printed or affixed inside the package and so have the important advantage of detection of quality variation, without opening the package, usually by a color change (Ahvenainen and Hurme 1997; Mohebi and Marquez 2015) (Table 5.3).

Oxygen indicators are the one of the most widely used indicators used for food package integrity, as oxygen can have adverse impacts on food quality, as oxidation can cause lipid rancidity, undesirable changes in color, and microbial growth (Fang et al. 2017). Most O_2 indicators are developed from colorimetric-based indicators involving a redox dye, such as methylene blue, 2,6-dichloroindophenol or N,N,N0,N0-tetramethyl-p-phenylenediamine and a strong reducing compound, for example glucose in an alkaline media such as sodium hydroxide or potassium hydroxide. The redox dye is entrapped by oxygen; hence giving a color shift that can be readily monitored. A disadvantage is that the color shift is reversible and if the oxygen levels return to the original concentration, the indicator will no longer show a problem condition. Reversibility such as this is unacceptable for leakage control because oxygen can be consumed by microbial growth (Kuswandi et al. 2011; Realini and Marcos 2014). Several of these types of leakage indicators were tested, and reported as reliable for the example applications of MAP minced meat and pizzas (Kerry et al. 2006).

With regard to the use of CO_2 indicators, the detection of leaks does not appear to be any more accurate than equivalent O_2 indicators. In addition to CO_2 produced by respiration, gas dissolved in the food can defuse out into the packaging headspace. A significant decrease in CO_2 concentration after can be a clear indication of leakage in the packaging. The second drawback of the CO_2 indicator is directly linked to CO_2 production by aerobic microbial spoilage. This evolution of CO_2 may maintain the headspace carbon dioxide at a desirable level, so product quality is still perceived as good and safe by the consumer. To address this issue, the color change of this type of indicator should be irreversible (Ghaani et al. 2016; Kuswandi et al. 2012; Ahvenainen and Hurme 1997).

5.12 Sensor Applications for Smart Packaging

Several advanced systems for the identification of pathogenic microorganisms and contaminants have been developed. Some of these are instruments that require sample extraction to assess the presence of the target contaminant. There are technological challenges to incorporate these systems within commodity packaging so the application for detection of contaminants or pathogens is still very limited (Kuswandi et al. 2011). The ideal sensor should possess the following characteristics:

Table 5.3 List of commercially available indicators for food packaging.

Category	Products	Developer	Principles	Application
Time-Temperature	OnVu®	Freshpoint and Ciba	Based on printable, intelligent, temperature-sensitive ink technology.	Chilled or frozen products
Time-Temperature	Freshness Check®	Temptime Corp.	Diacetylene crystals polymerize via 1,4 addition polymerization resulting in an increase in color intensity related to time and temperature.	Ready-to-eat fruit and meat
Time-Temperature	Keep-it	Keep-it Technologies	A label that constantly monitors temperature over time, and therefore shows the effect of temperature on product shelf life.	Seafood and meat
Time-Temperature	FreshCode	Varcode Ltd.	Barcodes printed with fading inks. The rate of fading depends on time and temperature	Poultry products
Time-Temperature	Tempix	Tempix AB	Barcodes printed with fading inks. The rate of fading depends on time and temperature.	Frozen and refrigerated products
Time-Temperature	TopCryo	TRACEO company	A Microbiological Time Temperature Integrator (TTI) interpreting time/temperature impact on the conservation of temperature sensitive products by an irreversible and easy measurable color change from green to red.	Real-time monitoring
Time-Temperature	3 M Monitor Mark®	3 M Company	A non-reversible indicator using the time and temperature dependence of the diffusion rate of a colored fatty acid ester along a porous wick.	
Time-Temperature	Check-Point® SmartLabel™	Vitsab International	Tracks the time-temperature relationship by the colour change of a dot, starting as white changing to green on activation, and yellow to red toward the end of shelf life.	Meat, seafood and vegetable
Time-Temperature	Coors Light®	Coors Brewing Company	A two-stage color change label indicating that beverage is cold or “supercold.”	Beer

(*Continued*)

Table 5.3 (Continued)

Category	Products	Developer	Principles	Application
Time-Temperature	Thermax	Thermographic Measurements Ltd.	A tamper-proof sensor designed to show the highest temperature that a food product has been exposed to by an irreversible color change.	Food
Freshness	FreshTag®	COX Technologies	Color change in a tag is directly related to freshness of the product.	Seafood
Freshness	UWI Label	UWI Technology Limited	The label color change is initiated when the when the container is first opened indicating useful life of the contents.	Babyfood
Freshness	RipeSense	Ripesense Limited	Reacting to volatiles emitted by ripening fruits shows the degree if ripeness.	Fruits
Freshness	SensorQ™	Food Quality Sensor International Inc.	As the product spoils, the bacterial count increases and the label changes color.	Packed meat and poultry
Integrity	Ageless Eye®	Mitshubishi Gas Chemical	Inserted inside the container and changes in color from pink to blue when the oxygen concentration is above 0.5%.	Chill-stored foods
Integrity	Tell-Tab	IMPAK	Designed to monitor the oxygen consumption of meat products from the composition of packaging head space gases.	Meat
Integrity	Novas®	Insignia Technologies Ltd.	A specialized pigment for use in modified atmosphere plastic packaging which shows a clear color change when packaging has been damaged.	Meat products
Integrity	O_2Sense	FreshPoint Lab	Based on luminescence – an O_2 indicator label to detect leakages in modified atmosphere packaging.	Fruit and vegetables

- A high degree of selectivity;
- High sensitivity to changes in bacteria concentrations;
- Rapid reaction time;
- Long shelf life; and
- Small size and low-cost (Ghaani et al. 2016).

Biosensors detect biological reactions: this type of sensor consists of a bioreceptor for an analyte such as an enzyme, antigen, microorganism, hormone or nucleic acid coupled to a transducer, which may be electrochemical, optical or acoustic and is capable of converting biological signals into an observable and quantifiable response (Biji et al. 2015; Müller and Schmid 2019). The type of transducer required depends on the parameters being analyzed (Abreu et al. 2012). The specificity, sensitivity, durability, portability and usability of a biosensor are all important (Yam et al. 2005).

An example of a commercial biosensor developed to detect food pathogens uses a mechanism linking specific antibodies to a membrane which is part of the product barcode – presence of pathogens causes the barcode to be unreadable due to regionalized dark bar formation (Realini and Marcos 2014).

Other visual biosensors have been developed based on antibodies and printed on packaging material, which detects specific target foodborne pathogens such as *Salmonella*, *Campylobacter*, *E. coli* and *Listeria* sp. (Fang et al. 2017; Müller and Schmid 2019). These sensors give improved quantitative measurement of microorganism cultures to detect pathogens (Kuswandi et al. 2011). Additionally, biosensors using synthetic antimicrobial peptides as recognition agents can detect and identify pathogen strains. These peptides have advantages of high stability and low cost over immunological and DNA-based receptors that rely on inactivation of bacteria. Particularly advantageous for the milk and meat industries, *L. monocytogenes* can be detected at low levels using colorimetric biosensor strips made with peptides immobilized on a gold chip (Mustafa and Andreescu 2018). The detection of *E. coli* in meat packaging can be realized using a biosensor using nano-spheres of silica immobilizing a fluorescent dye that is technically and economically compatible with packaging (Kuswandi et al. 2011).

The ideal packaging material will provide a simple visual message to warn the customer, manufacturer or regulator when the antibodies come across the targeted pathogen. Such devices are designed to detect gross infection, as they are not yet sensitive enough to detect extremely low levels of disease-causing pathogens (Yam et al. 2005).

Selective optical sensors have been designed to detect biogenic amines in beer based on a covalent interaction between the active vinyl groups and the tryptamine molecule (Ramon-Marquez et al. 2016). Another example of a biosensor detects xanthine that is a product of adenine nucleotide degradation in animal tissue by immobilization of xanthine oxide onto the transducer electrodes. These biosensors can detect biogenic amines formed either by the decarboxylation of amino acids or by amination and transamination of aldehydes and ketones (Biji et al. 2015). A biosensor has been developed for milk based on an oxygen electrode that assesses oxygen consumption by the respiration rate of the immobilized microorganism *Arthrobacternicotianae* that degrades short-chain fatty acids that are generated by lipolysis of milk fat (Heising et al. 2014) (Table 5.4).

Table 5.4 List of recent five years' works related to indicators/sensors on the food packaging.

Ingredients	Applications	Features	References
Agar incorporated with natural dye extracted from *Arnebia euchroma* root.	Fish	pH indicator	(Huang et al. 2019)
Artemisia sphaerocephala Krasch. gum-carboxymethyl cellulose sodium film capable of anchoring the natural dye extracted from red cabbage.	–	pH and NH_3 sensor	(Liang et al. 2019)
Films (chitosan, gelatin or starch) containing bioactive compounds from aqueous hibiscus extract.	–	pH indicator	(Peralta et al. 2019)
Chitosan films including immobilized black chokeberry (*Aronia Melanocarpa*) pomace extract.	–	pH indicator	(Halász and Csóka 2018)
The filter paper including immobilized methyl red and bromo cresol purple.	Beef	On-package dual sensors for pH indicator	(Kuswandi and Nurfawaide 2017)
Microcrystalline cellulose and a eutectic mixture films contain chitosan/ curcumin.	–	pH indicator	(Pereira and Andrade 2017)
The poly(ether-block-amide) film immobilized pH dyes.	Chicken	pH indicator	(Lee et al. 2019)
Chitosan/PVA/bentonite films containing black carrot anthocyanins.	–	pH indicator	(Koosha and Hemedi 2019)
Extracts from blueberry and blackberry pomace extracts incorporated into chitosan.	–	pH indicator	(Kurek et al. 2019)
Bacterial cellulose nanofibers doped with anthocyanins extracted from red cabbage (*Brassica oleracea*).	–	pH indicator	(Pourjavaber et al. 2017)
The chitosan with anthocyanin-rich purple eggplant extract or black eggplant extract.	Milk	pH indicator	(Yong et al. 2019)
The blueberry residue incorporated into cassava starch film.	Chicken	pH indicator	(Luchese et al. 2018)
The alizarin in a starch-cellulose paper.	Fish	pH indicator	(Ezati et al. 2019a)
The gelatin film containing anthocyanins from red cabbage.	–	pH indicator	(Musso et al. 2019)
An immobilization of red grape extract on bioorganic agarose membrane.	–	pH indicator	(Abolghasemi et al. 2016)
Extracts from grape skins incorporated into tara gum/cellulose nanocrystal matrix.	Milk	pH indicator	(Ma and Wang 2016)

(*Continued*)

Table 5.4 (Continued)

Ingredients	Applications	Features	References
The starch/polyvinyl alcohol/roselle anthocyanins film.	Pork	Freshness indicator	(Zhang et al. 2019
The cassava starch incorporated with *L. ruthenicum* anthocyanins.	Pork	Freshness indicator	(Qin et al. 2019)
The poly(N,N-dimethyl acrylamide-co-methacryloyl sulfadimethoxine).	–	Freshness indicator	(Baek et al. 2017)
Curcumin incorporated into a tara gum and polyvinyl alcohol blended matrix.	Shrimp	Freshness indicator	(Ma et al. 2017)
κ-carrageenan and hydroxypropyl methyl- cellulose incorporating *Prunus maackii* juice.	Pork	Freshness indicator	(Sun et al. 2019)
κ-carrageenan incorporated with extract from *Lycium ruthenicum* Murr.	Milk and shrimp	Freshness indicator	(Liu et al. 2019)
The alizarin in cellulose-chitosan film.	Minced beef	Freshness indicator	(Ezati et al. 2019b)
The bromocresol green immobilized with ethylene vinyl acetate emulsion.	Chicken breast	Freshness indicator	(Lee et al. 2019a)
The methylene blue onto a modified polypropylene film.	Milk	Freshness indicator	(Cavallo et al. 2014
Indicator contains an inner poly(ether-block-amide) film, a color-changing layer of eight polymer-immobilized pH dyes, and an outer poly(ethylene terephthalate) film.	Chicken breast	Freshness indicator	(Lee et al. 2019b)
The laccase immobilized on the electrospun zein fiber.	–	Time–temperature indicator	(Jhuang et al. 2020)
The laccase immobilized on the electrospun chitosan/poly(vinyl alcohol) fiber.	–	Time–temperature indicator	(Jhuang et al. 2020)
The isopropyl palmitate diffusion.	Non-pasteurized angelica juice	Time–temperature indicator	(Kim et al. 2016)
A polydiacetylene/silica nanocomposite.	Wheatgrass juice	Time–temperature indicator	(Suppakul et al. 2018)
Tryptic Soy Agar enriched with glycerol, spot-inoculated with the bacterium.	Ground pork meat with *Pseudomonas* sp. and sliced, cooked and cured meat product with lactic acid bacteria (LAB)	Time–temperature indicator	(Mataragas et al. 2019)

(*Continued*)

Table 5.4 (Continued)

Ingredients	Applications	Features	References
Films produced from chitosan, PVA and anthocyanin extracted from red cabbage.	Pasteurized milk	Time-temperature Indicators	(Pereira et al. 2015)
The extruded low-density polyethylene film containing the bromophenol blue.	Fish	Volatile amine sensor	(Wells et al. 2019)
Lysine, ε-poly-l-lysine and anthocyanin extract from red cabbage	Chicken breast	Carbon dioxide sensor	(Saliu and Della Pergola 2018)
Zein nanofibers containing alizarin.	Fish	Halochromic nanosensor	(Aghaei et al. 2020)
Gellan gum capped silver nanoparticles.	Chicken breast and silver carp	H_2S sensor	(Zhai et al. 2019)

(*source*: based on Heising et al. 2014)

5.13 Data Carriers for Smart Packaging

The last category of smart packaging systems is data carriers which are used to improve information transfer efficiency and good communication between the manufacturer and the customer. Data carriers are neither sensors nor indicators but instead reflect a different mode of smart packaging using electronic information processing (Kerry et al 2006). Data carriers such as barcode labels and RFID tags are used in traceability procedures for the purpose of theft prevention, anti-counterfeiting and the production of auditable trails (Heising et al. 2014).

Data carrier systems can augment automatic detection technologies, allowing the reliable and convenient information flow concerning the food supply chain. Due to size and expense, data carriers are placed more frequently on tertiary packaging that is known as multi-box containers, shipping crates, pallets, large packages of paperboard rather than point-of-sale packaging (Ghaani et al. 2016).

If devices recording the total history of the package through the delivery cycle are available, then a market for use is justified because of the resulting guarantee of quality of the commercial product (Kalpana et al. 2019).

Since the early 1970s, the barcodes have been available for machine-readable automatic identification and data capture (AIDC) (Fang et al. 2017). These labels are the most common and cheapest AIDC method and operate using an optical concept based on a sequence of linear white and black bars (Yam et al. 2005). There are two different types of barcodes: one-dimensional (1D) and two-dimensional (2D). The choice between the two depends on data storage requirements. The simple parallel arrangement of gaps and lines in 1D barcodes gives limited data storage but the stored information can be translated through a barcode reader forming part of a long established and very widespread system (Müller and Schmid 2019). 2D barcodes are combined dots and spaces arranged in an array or a matrix. These can store a greater amount of data including packaging date and weight, lot number, nutritional details or preparation guides. The 2D system needs a more sophisticated barcode scanner that can simultaneously read both vertically and horizontally (Ghaani et al. 2016; Müller and Schmid 2019). The coded information is gathered by the optical reader that scans the bar arrangement, which can be a unique web address containing a large amount of relevant information (Kalpana et al. 2019).

The first barcode was implemented commercially in the form of Universal Commodity Code (UPC) in the 1970s, and has become omnipresent in grocery stores to enable inventory management, stock reordering and checkout, and also in use-by date (Fang et al. 2017). This type of barcode is a linear graphic pattern that consists of bars of vary thicknesses and spacing encoding 12 digits (Yam et al. 2005).

New barcode formats were developed to tackle the growing demand for greater data storage, ideally in a smaller space (Ghaani et al. 2016; Fang et al. 2017):

- Reduced Space Symbology (RSS, RSS-14 stacked omnidirectional barcode and the RSS expanded barcode)
- Portable Data File (PDF) 417 (2D barcode)
- Quick Response Codes (QR, 2D barcode)
- Composite Symbology (combining a 2D barcode such as PDF 417 with a linear barcode such as UPC)

The second type of data carrier is the radio-frequency identification tag (RFID). RFID depends on automated remote communication by magnetic field or electromagnetic waves. RFID can be used to populate a real-time database with information such as temperature, humidity and manufacturing details in real time as the product passes through the food supply chain, thereby increasing traceability and maintaining a high and certified level of food quality and safety (Han et al. 2018).

There are three component parts in an RFID tag system:

1) *The tag* – a small antenna and transponder that produces a response (a specific number or alphanumeric sequence) when it receives a radio-frequency interrogation
2) *The reader* – emits the interrogating radio signals to initiate the response from the tag
3) *Middleware* – hard- and software – controlling the real-time passage of the collected data to a database server that is in turn connected to the Internet. The data can then be assessed and appropriate response actions can be initiated (Fang et al. 2017).
4) RFID systems are classified into three types:
   i) Passive tags have a coiled antenna inside the tag and use power supplied as radio waves from the reader to generate a magnetic field is. The tag uses the energy to dispatch the data to the reader
   ii) Semi-passive tags use battery to power memory storage and power the electronics that respond to the reader
   iii) Active tags use a battery both to power a microprocessor and to transmit response signals to the reader (Abreu et al. 2012; Kalpana et al 2019).

These active tags are more expensive but have a storage capacity of up to 1 MB data and an increased reading range of 100 m. Note that the reading range given is a maximum and may be reduced by several factors including operating frequency, reader power supply and potential interference from metal items (Yam et al. 2005). Low frequency

(~125 kHz) tags are considered most suitable for packaging meat products as not only are they cheaper, but are better able to penetrate non-metallic items (Fang et al. 2017).

RFID tags may be embedded into an object, placed inside food packaging or implanted into live animal bodies. These properties make them suitable for various food applications, allowing product recognition and traceability, cold chain tracking to continuously monitor the temperature variation experienced by the product, livestock management and forecasting of shelf life (Ghaani et al. 2016; Han et al. 2018). The focus of research is mostly on sensor-augmented RFID tags for use in food supply chains.

5.14 Regulatory Aspects

There was no comprehensive legislation for smart packaging in Europe until 2004, which slowed penetration of the technology into the EU market. Regulation 1935/2004/EC and regulation 450/2009/EC laid down a new legislative framework for the application, marketing and distribution of smart packaging (Restuccia 2010). These European Food Safety Authority (EFSA) regulations defined smart packaging materials as "materials and articles monitoring the condition of packaged food or the food environment" and the legislation also covers active materials which are "intended to extend the shelf life or to maintain or improve the condition of packaged food; they are designed to deliberately incorporate components that would release or absorb substances into or from the packaged food or the environment surrounding the food." The material may communicate with the consumer, but does not interact with the foodstuffs. The purpose of the technology is to track the food and facilitate the useful exchange of information captured during storage and distribution from manufacture to end-user. These devices can be placed outside or inside packaging, depending on the data properties required. Interpretation of the data by consumer or manufacturer must be simple, comprehensible and convenient (Müller and Schmid 2019).

In common with all materials that come into contact with food stuffs, smart packaging systems are evaluated to ensure that undesirable chemical migration into the foodstuff does not occur. Migration is the transfer of chemical substances from food contact materials into food. The acceptable limits for chemical migration are governed by Commission Regulation (EU) No. 10/2011. External indicators such as time-temperature indicators are placed on the outer surface of a package. Internal indicators for say oxygen and carbon dioxide levels may be placed in the headspace of the packaging. Migration is not a problem with external indicators, as they never come into direct contact with the food, so external mechanisms usually do not need migration testing as they are outside a "functional barrier" that sufficiently decreases any possible migration. It should be noted that most smart systems installed on the outside of the packed food are to monitor environmental conditions over a relatively short time period. Internal indicators are not expected to come into direct contact with packed food either, because they are installed either in the free headspace above the product or fixed to the internal surface of the package or to the lid. If this is the case, these smart materials therefore do not require a migration assessment but are still not beyond the control Regulation 450/2009/EC (Ahvenainen and Hurme 1997). Certain non-authorized substances can be used inside the functional barrier, as long as they meet requirements that migration stays under the limit of detection and should not in any event exceed 0.01 mg per kg food (Restuccia et al. 2010).

In Regulation 1935/2004/EC, articles 3 and 4 are of particular importance. Article 3 refers to the labeling of smart components used in packaging, specifically that "the labelling, advertising and presentation of a material or article shall not mislead the consumers" and article 4 provides that "Active materials and articles shall not bring about changes in the composition or organoleptic characteristics of food, for instance by masking the spoilage of food, which could mislead consumers...and...intelligent materials and articles shall not give information about the condition of the food which could mislead consumers." Additionally, there must be adequate labeling to allow ready identification of non-edible parts and also to indicate that "the materials or articles are active and/or intelligent" (Dainelli et al 2008).

The European and United States regulatory concepts about food-contact materials differ not only in detail but in fundamental approach. The European principle is that all products should be specifically approved in legislation, and that these authorizations must be done on a toxicological assessment of the identified chemicals. The US approach provides tremendous legitimacy to the notion that "the dosage makes the poison" and that no toxicological rationale is required and thus toxicity tests are greatly reduced in scope (Restuccia et al. 2010). There has been no legal framework for smart packaging in the USA. This is why the systems are not widely adopted in the USA. Moreover, such smart packaging as there is in the the USA cannot easily be adopted in Europe because of the tighter EU rules (Müller and Schmid 2019).

5.15 Conclusion and Future Perspectives

Smart package systems have been introduced over the last few years to increase the profitability of the food industry. This is achieved by extending shelf life, and reducing the amount of food that is discarded due to suspicion of inappropriate storage or handling. There is also a benefit to consumers in that foods can be guaranteed to be presented at an optimal level of freshness and general condition. The globalization of the food market, bringing extended food supply chains and higher consumer expectations have much affected retail processes. Both these drivers have helped to establish new packaging technologies that are designed to maximize food shelf life tracking and quantifying factors that can affect food safety and quality. Smart packaging is still under technical development and there is still much scope for further commercial take-up. In other words, there is still a technology indicating a gap between potential and commercially practical applications. The purpose of smart packaging systems appeals to the food industry by having demonstrable effects to increase customer demands for a perceived higher quality of food products. This can be achieved by notifying both retailers and end users about the history of the product during distribution and storage.

There are three important applications for smart packaging:

1) Indicators, such as time temperature indicators (TTIs), effectively focus on providing to the consumer information on freshness, microbiological safety and quality, and the shelf life of foods. In addition, data systems such as barcodes and RFID tags deliver

real-time information on product verification and tracking, thus also improving supply chain performance and effectiveness.

2) Smart packaging materials demonstrably provide benefits that exceed the cost of provision, but their commercial application yet remains restricted. To promote further uptake, it is essential to narrow the gap between findings based on laboratory tests and actual commercially viable foodstuff applications.
3) Hand-in-hand with these technical and commercial developments, it is important to educate consumers about the benefits of these products – what function they serve, how they operate, and the safety and quality benefits that result from their application. In general, we can say that smart packaging helps improve food safety and quality, shelf life, validity, traceability and protection from fraud. The use of smart packaging has certainly introduced a change in the packaging industry and will give rise to further novel strategies and developments in food packaging applications in the near future.

References

Abolghasemi, M.M., Sobhi, M., and Piryaei, M. (2016). Preparation of a novel green optical pH sensor based on immobilization of red grape extract on bioorganic agarose membrane. *Sensors and Actuators, B: Chemical* 224: 391–395.

Abreu, D.A.P., De Cruz, J.M. et al. (2012). Active and intelligent packaging for the food industry. *Food Reviews International* 28: 146–187.

Accinelli, C., Saccà, M.L., Mencarelli, M. et al. (2012). Deterioration of bioplastic carrier bags in the environment and assessment of a new recycling alternative. *Chemosphere* 89: 136–143.

Adilah, A.N., Jamilah, B., Noranizan, M.A. et al. (2018). Utilization of mango peel extracts on the biodegradable films for active packaging. *Food Packaging and Shelf Life* 16: 1–7.

Aghaei, Z., Ghorani, B., Emadzadeh, B. et al. (2020). Protein-based halochromic electrospun nanosensor for monitoring trout fish freshness. *Food Control* 111: 107065.

Ahmad, N., Billoo, M., and Lakhan, A. (2012). Effect of product packaging on consumer buying decision. *Journal of Business Strategies* 6: 1–10.

Ahmed, I., Lin, H., Zou, L. et al. (2017). A comprehensive review on the application of active packaging technologies to muscle foods. *Food Control* 82: 163–178.

Accinelli, C., Saccà, M.L., Mencarelli, M. et al. (2012). Deterioration of bioplastic carrier bags in the environment and assessment of a new recycling alternative. *Chemosphere* 89: 136–143.

Ahvenainen, R. and Hurme, E. (1997). Active and smart packaging for meeting consumer demands for quality and safety. *Food Additivies and Contaminants* 14: 753–763.

Appendini, P. and Hotchkiss, J.H. (2002). Review of antimicrobial food packaging. *Innovative Food Science & Emerging Technologies* 3: 113–126.

Arcan, I. and Yemenicioğlu, A. (2011). Incorporating phenolic compounds opens a new perspective to use zein films as flexible bioactive packaging materials. *Food Research International* 44: 550–556.

Ashrafi, A., Jokar, M., and Mohammadi Nafchi, A. (2018). Preparation and characterization of biocomposite film based on chitosan and kombucha tea as active food packaging. *International Journal of Biological Macromolecules* 108: 444–454.

Assis, R.Q., Rios, P.D.A., Rios, A.D. et al. (2020). Biodegradable packaging of cellulose acetate incorporated with norbixin, lycopene or zeaxanthin. *Industrial Crops and Products* 147: 112212.

Avella, M., Bonadies, E., Martuscelli, E. et al. (2001). European current standardization for plastic packaging recoverable through composting and biodegradation. *Polymer Testing* 20: 517–521.

Azeredo, H.M.C. and Waldron, K.W. (2016). Crosslinking in polysaccharide and protein films and coatings for food contact: A review. *Trends in Food Science and Technology* 52: 109–122.

Baek, S., Kim, D., Jeon, S.L. et al. (2017). Preparation and characterization of pH-responsive poly(N,N-dimethyl acrylamide-co-methacryloyl sulfadimethoxine) hydrogels for application as food freshness indicators. *Reactive and Functional Polymers* 120: 57–65.

Barber, N. (2010). "Green" wine packaging: Targeting environmental consumers. *International Journal of Wine Business Research* 22: 423–444.

Biji, K.B., Ravishankar, C.N., and Mohan, C.O. (2015). Smart packaging systems for food applications: A review. *Journal of Food Science and Technology* 52: 6125–6135.

Buchanan, C.M., Gardner, R.M., and Komarek, R.J. (1993). Aerobic biodegradation of cellulose acetate. *Journal of Applied Polymer Science* 47: 1709–1719.

Buchanan, C.M., Dorschel, D., Gardner, R.M. et al. (1996). The influence of degree of substitution on blend miscibility and biodegradation of cellulose acetate blends. *Journal of Environmental Polymer Degradation* 4: 179–195.

Bugnicourt, E., Cinelli, P., Lazzeri, A. et al. (2014). Polyhydroxyalkanoate (PHA): Review of synthesis, characteristics, processing and potential applications in packaging. *Express Polymer Letters* 8: 791–808.

Carissimi, M., Flôres, S.H., and Rech, R. (2018). Effect of microalgae addition on active biodegradable starch film. *Algal Research* 32: 201–209.

Cavallo, J.A., Strumia, M.C., and Gomez, C.G. (2014). Preparation of a milk spoilage indicator adsorbed to a modified polypropylene film as an attempt to build a smart packaging. *Journal of Food Engineering* 136: 48–55.

Çay, A., Miraftab, M., and Kumbasar, E.P.A. (2014). Characterization and swelling performance of physically stabilized electrospun poly(vinyl alcohol)/chitosan nanofibres. *European Polymer Journal* 61: 253–262.

Chen, S., Brahma, S., Mackay, J. et al. (2020). The role of smart packaging system in food supply chain. *Journal of Food Science* 85: 517–525.

Chiellini, E., Cinelli, P., Ilieva, V.I. et al. (2009). Environmentally-compatible foamed articles based on potato starch, corn fiber, and poly(vinyl alcohol). *Journal of Cellular Plastics* 45: 17–32.

Cinelli, P., Chiellini, E., Lawton, J.W. et al. (2006). Foamed articles based on potato starch, corn fibers and poly(vinyl alcohol). *Polymer Degradation and Stability* 91: 1147–1155.

Colomines, G., Ducruet, V., Courgneau, Ć. et al. (2010). Barrier properties of poly(lactic acid) and its morphological changes induced by aroma compound sorption. *Polymer International* 59: 818–826.

Dainelli, D., Gontard, N., Spyropoulos, D. et al. (2008). Active and intelligent food packaging: Legal aspects and safety concerns. *Trends in Food Science and Technology* 19: S103–S112.

Dairi, N., Ferfera-Harrar, H., Ramos, M. et al. (2019). Cellulose acetate/AgNPs-organoclay and/or thymol nano-biocomposite films with combined antimicrobial/antioxidant properties for active food packaging use. *International Journal of Biological Macromolecules* 121: 508–523.

D'ambrières, W. (2019) Plastics recycling worldwide: Current overview and desirable changes. *Field Actions Science Report*, (Special Issue 19): 12–21.

De Moraes Crizel, T., De Oliveira Rios, A., Alves, V. et al. (2018). Active food packaging prepared with chitosan and olive pomace. *Food Hydrocolloids* 74: 139–150.

Duncan, T.V. (2011). Applications of nanotechnology in food packaging and food safety: Barrier materials, antimicrobials and sensors. *Journal of Colloid and Interface Science* 363: 1–24.

Engel, J.B., Ambrosi, A., and Tessaro, I.C. (2019). Development of a Cassava starch-based foam incorporated with grape stalks using an experimental design. *Journal of Polymers and the Environment* 27: 2853–2866.

Espitia, P.J.P., Soares, N.D.F.F., Coimbra, J.S.D.R. et al. (2012). Zinc oxide nanoparticles: synthesis, antimicrobial activity and food packaging applications. *Food and Bioprocess Technology* 5: 1447–1464.

Ezati, P., Tajik, H., Moradi, M. et al. (2019a). Intelligent pH-sensitive indicator based on starch-cellulose and alizarin dye to track freshness of rainbow trout fillet. *International Journal of Biological Macromolecules* 132: 157–165.

Ezati, P., Tajik, H., and Moradi, M. (2019b). Fabrication and characterization of alizarin colorimetric indicator based on cellulose-chitosan to monitor the freshness of minced beef. *Sensors and Actuators, B: Chemical* 285: 519–528.

Fan, H.Y., Duquette, D., Dumont, M.J., and Simpson, B.K. (2018). Salmon skin gelatin-corn zein composite films produced via crosslinking with glutaraldehyde: Optimization using response surface methodology and characterization. *International Journal of Biological Macromolecules* 120: 263–273.

Fang, Z., Zhao, Y., Warner, R.D. et al. (2017). Active and intelligent packaging in meat industry. *Trends in Food Science and Technology* 61: 60–71.

Farah, S., Anderson, D.G., and Langer, R. (2016). Physical and mechanical properties of PLA, and their functions in widespread applications: A comprehensive review. *Advanced Drug Delivery Reviews* 107: 367–392.

Ferreira, A.M., Martins, J., Carvalho, L.H. et al. (2019). Biosourced disposable trays made of brewer's spent grain and potato starch. *Polymers* 11: 923.

Fieschi, M. and Pretato, U. (2018). Role of compostable tableware in food service and waste management. A life cycle assessment study. *Waste Management* 73: 14–25.

Fischer, S., Thümmler, K., Volkert, B. et al. (2008). Properties and applications of cellulose acetate. *Macromolecular Symposia* 262: 89–96.

Flauzino Neto, W.P., Mariano, M., Da Silva, I.S.V. et al. (2016). Mechanical properties of natural rubber nanocomposites reinforced with high aspect ratio cellulose nanocrystals isolated from soy hulls. *Carbohydrate Polymers* 153: 143–152.

Fukushima, K., Tabuani, D., Arena, M. et al. (2013). Effect of clay type and loading on thermal, mechanical properties and biodegradation of poly(lactic acid) nanocomposites. *Reactive and Functional Polymers* 73: 540–549.

Gadhave, R.V., Das, A., Mahanwar, P.A. et al. (2018). Starch based bio-plastics: The future of sustainable packaging. *Open Journal of Polymer Chemistry* 8: 21–33.

Gao, H.X., He, Z., Sun, Q. et al. (2019). A functional polysaccharide film forming by pectin, chitosan, and tea polyphenols. *Carbohydrate Polymers* 215: 1–7.

Garavand, F., Rouhi, M., Razavi, S.H. et al. (2017). Improving the integrity of natural biopolymer films used in food packaging by crosslinking approach: A review. *International Journal of Biological Macromolecules* 104: 687–707.

Ghaani, M., Cozzolino, C.A., Castelli, G. et al. (2016). An overview of the intelligent packaging technologies in the food sector. *Trends in Food Science and Technology* 51: 1–11.

Ghosal, K., Chandra, A., Praveen, G. et al. (2018). Electrospinning over solvent casting: Tuning of mechanical properties of membranes. *Scientific Reports* 8: 5058.

Gómez-Estaca, J., Gavara, R., and Hernández-Muñoz, P. (2015). Encapsulation of curcumin in electrosprayed gelatin microspheres enhances its bioaccessibility and widens its uses in food applications. *Innovative Food Science and Emerging Technologies* 29: 302–307.

Gouvêa, D.M., Mendonça, R.C.S., Soto, M.L. et al. (2015). Acetate cellulose film with bacteriophages for potential antimicrobial use in food packaging. *LWT – Food Science and Technology* 63: 85–91.

Gu, W., Liu, X., Li, F. et al. (2019). Tough, strong, and biodegradable composite film with excellent UV barrier performance comprising soy protein isolate, hyperbranched polyester, and cardanol derivative. *Green Chemistry* 21: 3651–3665.

Halász, K. and Csóka, L. (2018). Black chokeberry (*Aronia melanocarpa*) pomace extract immobilized in chitosan for colorimetric pH indicator film application. *Food Packaging and Shelf Life* 16: 185–193.

Han, J.H. (2014). A review of food packaging technologies and innovations. In: *Innovations in Food Packaging* Chapter 1: 3–12.

Han, J.H., Seo, G.H., Park, I.M. et al. (2006). Physical and mechanical properties of pea starch edible films containing beeswax emulsions. *Journal of Food Science* 71: E290–296.

Han, J.W., Ruiz-Garcia, L., Qian, J.P. et al. (2018). Food packaging: A comprehensive review and future trends. *Comprehensive Reviews in Food Science and Food Safety* 17: 860–877.

Heising, J.K., Dekker, M., Bartels, P.V. et al. (2014). Monitoring the quality of perishable foods: Opportunities for intelligent packaging. *Critical Reviews in Food Science and Nutrition* 54: 645–654.

Hempel, A.W., O'Sullivan, M.G., Papkovsky, D.B. et al. (2013). Use of smart packaging technologies for monitoring and extending the shelf life quality of modified atmosphere packaged (MAP) bread: Application of intelligent oxygen sensors and active ethanol emitters. *European Food Research and Technology* 237: 117–124.

Hogan, S.A.and Kerry. K.P. (2008). Smart packaging technologies for fast moving consumer goods. In: *Smart Packaging of Meat and Poultry Products*. Chapter 3, pp. 33–54.

Huang, S., Xiong, Y., Zou, Y. et al. (2019). A novel colorimetric indicator based on agar incorporated with *Arnebia euchroma* root extracts for monitoring fish freshness. *Food Hydrocolloids* 90: 198–205.

Indumathi, M.P., Saral Sarojini, K., and Rajarajeswari, G.R. (2019). Antimicrobial and biodegradable chitosan/cellulose acetate phthalate/ZnO nano composite films with

optimal oxygen permeability and hydrophobicity for extending the shelf life of black grape fruits. *International Journal of Biological Macromolecules* 132: 1112–1120.
Jafarzadeh, S., Rhim, J.W., Alias, A.K. et al. (2019). Application of antimicrobial active packaging film made of semolina flour, nano zinc oxide and nano-kaolin to maintain the quality of low-moisture mozzarella cheese during low-temperature storage. *Journal of the Science of Food and Agriculture* 99: 2716–2725.
Jhuang, J.R., Lou, S.N., Lin, S.B. et al. (2020). Immobilizing laccase on electrospun chitosan fiber to prepare time-temperature indicator for food quality monitoring. *Innovative Food Science and Emerging Technologies* 63: 102370.
Jin, T.Z. and Liu, L. (2020) Roles of green polymer materials in active packaging. In: *Innovative Uses of Agricultural Products and Byproducts*. Chapter 5, pp. 83–107. American Chemical Society.
Kale, G., Kijchavengkul, T., Auras, R. et al. (2007). Compostability of bioplastic packaging materials: An overview. *Macromolecular Bioscience* 7: 255–277.
Kalpana, S., Priyadarshini, S.R., Maria Leena, M. et al. (2019). Intelligent packaging: Trends and applications in food systems. *Trends in Food Science and Technology* 93: 145–157.
Kerry, J.P., O'Grady, M.N., and Hogan, S.A. (2006). Past, current and potential utilisation of active and intelligent packaging systems for meat and muscle-based products: A review. *Meat Science* 74: 113–130.
Kim, J.U., Ghafoor, K., Ahn, J. et al. (2016). Kinetic modeling and characterization of a diffusion-based time-temperature indicator (TTI) for monitoring microbial quality of non-pasteurized angelica juice. *LWT – Food Science and Technology* 67: 143–150.
Kim, I., Viswanathan, K., Kasi, G. et al. (2020). ZnO nanostructures in active antibacterial food packaging: preparation methods, antimicrobial mechanisms, safety issues, future prospects, and challenges. *Food Reviews International* 1–29.
Koosha, M. and Hamedi, S. (2019). Intelligent Chitosan/PVA nanocomposite films containing black carrot anthocyanin and bentonite nanoclays with improved mechanical, thermal and antibacterial properties. *Progress in Organic Coatings* 127: 338–347.
Korbelyiova, L. (2019). Paper vs. Leaf: Comparative Life Cycle Assessment of Single-use Plates Made of Renewable Materials. Master Thesis (Dissertation. Retrieved from: http://urn.kb.se/resolve?urn=urn:nbn:se:uu:diva-386367
Kruijf, N.D., Beest, M.V., Rijk, R. et al. (2002). Active and intelligent packaging: Applications and regulatory aspects. *Food Additivies and Contaminants* 19: 144–162.
Kumar, S., Shukla, A., Baul, P.P. et al. (2018). Biodegradable hybrid nanocomposites of chitosan/gelatin and silver nanoparticles for active food packaging applications. *Food Packaging and Shelf Life* 16: 178–184.
Kurek, M., Garofulić, I.E., Bakić, M.T. et al. (2018). Development and evaluation of a novel antioxidant and pH indicator film based on chitosan and food waste sources of antioxidants. *Food Hydrocolloids* 84: 238–246.
Kuswandi, B. and Nurfawaidi, A. (2017). On-package dual sensors label based on pH indicators for real-time monitoring of beef freshness. *Food Control* 82: 91–100.
Kuswandi, B., Wicaksono, Y., and Ahmad, M. (2011). Smart packaging : Sensors for monitoring of food quality and safety. *Sensory & Instrumental Food Quality* 5: 137–146.
Kuswandi, B., Jayus, Restyana, A. et al. (2012). A novel colorimetric food package label for fish spoilage based on polyaniline film. *Food Control* 25: 184–189.

Laroche, M., Bergeron, J., and Barbaro-Forleo, G. (2001). Targetting consumers who are likely to pay more for ethical products. *Journal of Consumer Marketing* 18: 503–520.

Lee, K., Park, H., Baek, S. et al. (2019). Colorimetric array freshness indicator and digital color processing for monitoring the freshness of packaged chicken breast. *Food Packaging and Shelf Life* 22: 100408.

Lee, K., Baek, S., Kim, D. et al. (2019). A freshness indicator for monitoring chicken-breast spoilage using a Tyvek® sheet and RGB color analysis. *Food Packaging and Shelf Life* 19: 40–46.

Li, W., Coffin, D.R., Jin, T.Z. et al. (2012). Biodegradable composites from polyester and sugar beet pulp with antimicrobial coating for food packaging. *Journal of Applied Polymer Science* 116: 2658–2667.

Liang, T., Sun, G., Cao, L. et al. (2019). A pH and NH_3 sensing intelligent film based on *Artemisia sphaerocephala* Krasch. gum and red cabbage anthocyanins anchored by carboxymethyl cellulose sodium added as a host complex. *Food Hydrocolloids* 87: 858–868.

Liu, J., Wang, H., Guo, M. et al. (2019). Extract from *Lycium ruthenicum* Murr. Incorporating κ-carrageenan colorimetric film with a wide pH–sensing range for food freshness monitoring. *Food Hydrocolloids* 94: 1–10.

López, O.V., Castillo, L.A., García, M.A. et al. (2015). Food packaging bags based on thermoplastic corn starch reinforced with talc nanoparticles. *Food Hydrocolloids* 43: 18–24.

Lörcks, J. (1998). Properties and applications of compostable starch-based plastic material. *Polymer Degradation and Stability* 59: 245–249.

Luchese, C.L., Abdalla, V.F., Spada, J.C. et al. (2018). Evaluation of blueberry residue incorporated cassava starch film as pH indicator in different simulants and foodstuffs. *Food Hydrocolloids* 82: 209–218.

Ma, Q. and Wang, L. (2016). Preparation of a visual pH-sensing film based on tara gum incorporating cellulose and extracts from grape skins. *Sensors and Actuators, B: Chemical* 235: 401–407.

Ma, Q., Du, L., and Wang, L. (2017). Tara gum/polyvinyl alcohol-based colorimetric NH3 indicator films incorporating curcumin for intelligent packaging. *Sensors and Actuators, B: Chemical* 244: 759–766.

Ma, Y., Li, L., and Wang, Y. (2018). Development of PLA-PHB-based biodegradable active packaging and its application to salmon. *Packaging Technology and Science* 31: 739–746.

Marrez, D.A., Abdelhamid, A.E., and Darwesh, O.M. (2019). Eco-friendly cellulose acetate green synthesized silver nano-composite as antibacterial packaging system for food safety. *Food Packaging and Shelf Life* 20: 100302.

Mataragas, M., Bikouli, V.C., Korre, M. et al. (2019). Development of a microbial Time Temperature Indicator for monitoring the shelf life of meat. *Innovative Food Science and Emerging Technologies* 52: 89–99.

Lorevice, M.V., Otoni, C.G., De Moura, M.R. et al. (2016). Chitosan nanoparticles on the improvement of thermal, barrier, and mechanical properties of high- and low-methyl pectin films. *Food Hydrocolloids* 52: 732–740.

Ma, Q. and Wang, L. (2016). Preparation of a visual pH-sensing film based on tara gum incorporating cellulose and extracts from grape skins. *Sensors and Actuators, B: Chemical* 235: 401–407.

Medina-Jaramillo, C., Ochoa-Yepes, O., Bernal, C. et al. (2017). Active and smart biodegradable packaging based on starch and natural extracts. *Carbohydrate Polymers* 176: 187–194.

Mihindukulasuriya, S.D.F. and Lim, L.T. (2014). Nanotechnology development in food packaging: A review. *Trends in Food Science and Technology* 40: 149–167.

Mohebi, E. and Marquez, L. (2015). Intelligent packaging in meat industry: An overview of existing solutions. *Journal of Food Science and Technology* 52: 3947–3964.

Moustafa, H., Youssef, A.M., Darwish, N.A. et al. (2019). Eco-friendly polymer composites for green packaging: Future vision and challenges. *Composites Part B: Engineering* 172: 16–25.

Możejko-Ciesielska, J. and Kiewisz, R. (2016). Bacterial polyhydroxyalkanoates: Still fabulous? *Microbiological Research* 192: 271–282.

Müller, P. and Schmid, M. (2019). Intelligent packaging in the food sector: A brief overview. *Foods* 8: 16.

Musso, Y.S., Salgado, P.R., and Mauri, A.N. (2019). Smart gelatin films prepared using red cabbage (*Brassica oleracea* L.) extracts as solvent. *Food Hydrocolloids* 89: 674–681.

Mustafa, F. and Andreescu, S. (2018). Chemical and biological sensors for food-quality monitoring and smart packaging. *Foods* 7: 1–20.

Nurul Fazita, M.R., Jayaraman, K., Bhattacharyya, D. et al. (2016). Green composites made of bamboo fabric and poly (lactic) acid for packaging applications: A review. *Materials* 9: 435.

Oymaci, P. and Altinkaya, S.A. (2016). Improvement of barrier and mechanical properties of whey protein isolate based food packaging films by incorporation of zein nanoparticles as a novel bionanocomposite. *Food Hydrocolloids* 54: 1–9.

Park, Y.W., Kim, S.M., Lee, J.Y. et al. (2015). Application of biosensors in smart packaging. *Molecular and Cellular Toxicology* 11: 277–285.

Peralta, J., Bitencourt-Cervi, C.M., Maciel, V.B.V. et al. (2019). Aqueous hibiscus extract as a potential natural pH indicator incorporated in natural polymeric films. *Food Packaging and Shelf Life* 19: 47–55.

Pereira, P.F. and Andrade, C.T. (2017). Optimized pH-responsive film based on a eutectic mixture-plasticized chitosan. *Carbohydrate Polymers* 165: 238–246.

Pereira, V.A., De Arruda, I.N.Q., and Stefani, R. (2015). Active chitosan/PVA films with anthocyanins from *Brassica oleraceae* (red cabbage) as time-temperature indicators for application in intelligent food packaging. *Food Hydrocolloids* 43: 180–188.

Petersen, K., Væggemose Nielsen, P., Bertelsen, G. et al. (1999). Potential of biobased materials for food packaging. *Trends in Food Science and Technology* 10: 52–68.

Pourjavaher, S., Almasi, H., Meshkini, S. et al. (2017). Development of a colorimetric pH indicator based on bacterial cellulose nanofibers and red cabbage (*Brassica oleraceae*) extract. *Carbohydrate Polymers* 156: 193–201.

Poyatos-Racionero, E., Ros-Lis, J.V., Vivancos, J.L. et al. (2018). Recent advances on intelligent packaging as tools to reduce food waste. *Journal of Cleaner Production* 172: 3398–3409.

Prasad, P. and Kochhar, A. (2014). Active packaging in food industry: A review. *IOSR Journal of Environmental Science, Toxicology and Food Technology* 8: 01–07.

Puls, J., Wilson, S.A., and Hölter, D. (2011). Degradation of cellulose acetate-based materials: A review. *Journal of Polymers and the Environment* 19: 152–165.

Qin, Y., Liu, Y., Yong, H. et al. (2019). Preparation and characterization of active and intelligent packaging films based on cassava starch and anthocyanins from *Lycium ruthenicum* Murr. *International Journal of Biological Macromolecules* 134: 80–90.

Qin, Z., Jia, X., Liu, Q. et al. (2020). Enhancing physical properties of chitosan/pullulan electrospinning nanofibers via green crosslinking strategies. *Carbohydrate Polymers* 247: 116734.

Quintavalla, S. and Vicini, L. (2002). Antimicrobial food packaging in meat industry. *Meat Science* 62: 373–380.

Rabnawaz, M., Wyman, I., Auras, R. et al. (2017). A roadmap towards green packaging: The current status and future outlook for polyesters in the packaging industry. *Green Chemistry* 19: 4737–4753.

Ramesh, M., Palanikumar, K., and Reddy, K.H. (2017). Plant fibre based bio-composites: Sustainable and renewable green materials. *Renewable and Sustainable Energy Reviews* 79: 558–584.

Ramon-Marquez, T., Medina-Castillo, A.L. et al. (2016). Novel optical sensing film based on a functional nonwoven nanofibre mat for an easy, fast and highly selective and sensitive detection of tryptamine in beer. *Biosensors and Bioelectronics* 79: 600–607.

Raza, Z.A., Abid, S., and Banat, I.M. (2018). Polyhydroxyalkanoates: Characteristics, production, recent developments and applications. *International Biodeterioration and Biodegradation* 126: 45–56.

Realini, C.E. and Marcos, B. (2014). Active and intelligent packaging systems for a modern society. *Meat Science* 98: 404–419.

Reddy, N. and Yang, Y. (2010). Citric acid cross-linking of starch films. *Food Chemistry* 118: 702–711.

Reddy, C.S.K., Ghai, R., and Kalia, V. (2003). Polyhydroxyalkanoates: An overview. *Bioresource Technology* 87: 137–146.

Reddy, N., Reddy, R., and Jiang, Q. (2015). Crosslinking biopolymers for biomedical applications. *Trends in Biotechnology* 33: 362–369.

Reis, M.O., Olivato, J.B., Bilck, A.P. et al. (2018). Biodegradable trays of thermoplastic starch/poly (lactic acid) coated with beeswax. *Industrial Crops and Products* 112: 481–487.

Requena, R., Vargas, M., and Chiralt, A. (2019). Eugenol and carvacrol migration from PHBV films and antibacterial action in different food matrices. *Food Chemistry* 277: 38–45.

Restuccia, D., Spizzirri, U.G., Parisi, O.I. et al. (2010). New EU regulation aspects and global market of active and intelligent packaging for food industry applications. *Food Control* 21: 1425–1435.

Rhim, J.W., Mohanty, K.A., Singh, S.P. et al. (2006). Preparation and properties of biodegradable multilayer films based on soy protein isolate and poly(lactide). *Industrial and Engineering Chemistry Research* 45: 3059–3066.

Rocca-Smith, J.R., Pasquarelli, R., Lagorce-Tachon, A. et al. (2019). Toward sustainable PLA-based multilayer complexes with improved barrier properties. *ACS Sustainable Chemistry and Engineering* 7: 3759–3771.

Sadeghi, G.M.M. and Mahsa, S. (2015). Compostable polymers and nanocomposites: A big chance for planet earth. In: *Recycling Materials Based on Environmentally Friendly Techniques*, Chapter 4, p. 63.

Saliu, F. and Della Pergola, R. (2018). Carbon dioxide colorimetric indicators for food packaging application: Applicability of anthocyanin and poly-lysine mixtures. *Sensors and Actuators, B: Chemical* 258: 1117–1124.

Sharma, C., Dhiman, R., Rokana, N. et al. (2017). Nanotechnology: An untapped resource for food packaging. *Frontiers in Microbiology* 8: 1735.

Siddiqui, Z., Horan, N.J., and Anaman, K. (2011). Optimisation of C:N ratio for co-digested processed industrial food waste and sewage sludge using the BMP test. *International Journal of Chemical Reactor Engineering* 9 (1).

Silvestre, C., Duraccio, D., and Cimmino, S. (2011). Food packaging based on polymer nanomaterials. *Progress in Polymer Science* 36: 1766–1782.

Siimon, K., Siimon, H., and Järvekülg, M. (2015). Mechanical characterization of electrospun gelatin scaffolds cross-linked by glucose. *Journal of Materials Science. Materials in Medicine* 26: 1–9.

Siracusa, V., Rocculi, P., Romani, S. et al. (2008). Biodegradable polymers for food packaging: A review. *Trends in Food Science and Technology* 19: 634–643.

Stoll, L., Silva, A.M., Da, I. et al. (2017). Active biodegradable film with encapsulated anthocyanins: Effect on the quality attributes of extra-virgin olive oil during storage. *Journal of Food Processing and Preservation* 41: 1–8.

Su, J.F., Yuan, X.Y., Huang, Z. et al. (2010). Properties stability and biodegradation behaviors of soy protein isolate/poly (vinyl alcohol) blend films. *Polymer Degradation and Stability* 95: 1226–1237.

Souza, V.G.L. and Fernando, A.L. (2016). Nanoparticles in food packaging: Biodegradability and potential migration to food. A review. *Food Packaging and Shelf Life* 8: 63–70.

Sun, Q., Xi, T., Li, Y. et al. (2014). Characterization of corn starch films reinforced with $CaCO_3$ nanoparticles. *PLoS ONE* 9: 1–6.

Sun, G., Chi, W., Zhang, C. et al. (2019). Developing a green film with pH-sensitivity and antioxidant activity based on κ-carrageenan and hydroxypropyl methylcellulose incorporating *Prunus maackii* juice. *Food Hydrocolloids* 94: 345–353.

Suppakul, P., Kim, D.Y., Yang, J.H. et al. (2018). Practical design of a diffusion-type time-temperature indicator with intrinsic low temperature dependency. *Journal of Food Engineering* 223: 22–31.

Talegaonkar, S., Sharma, H., Pandey, S. et al. (2017) Bionanocomposites: Smart biodegradable packaging material for food preservation. In: *Food Packaging*. Chapter 3, pp. 79–110. Academic Press.

Tavassoli-Kafrani, E., Goli, S.A.H., and Fathi, M. (2018). Encapsulation of orange essential oil using cross-linked electrospun gelatin nanofibers. *Food and Bioprocess Technology* 11: 427–434.

Tian, H., Yan, J., Rajulu, A.V. et al. (2017). Fabrication and properties of polyvinyl alcohol/starch blend films: Effect of composition and humidity. *International Journal of Biological Macromolecules* 96: 518–523.

Todkar, S.S. and Patil, S.A. (2019). Review on mechanical properties evaluation of pineapple leaf fibre (PALF) reinforced polymer composites. *Composites Part B: Engineering* 174: 106927.

Tokiwa, Y. and Calabia, B.P. (2006). Biodegradability and biodegradation of poly(lactide). *Applied Microbiology and Biotechnology* 72: 244–251.

Tokiwa, Y., Calabia, B.P., Ugwu, C.U. et al. (2009). Biodegradability of plastics. *International Journal of Molecular Sciences* 10: 3722–3742.

Van Der Harst, E., Potting, J., and Kroeze, C. (2014). Multiple data sets and modelling choices in a comparative LCA of disposable beverage cups. *Science of the Total Environment* 494: 129–143.

Vanderroost, M., Ragaert, P., Devlieghere, F. et al. (2014). Intelligent food packaging: The next generation. *Trends in Food Science and Technology* 39: 47–62.

Vilarinho, F., Sanches Silva, A., Vaz, M.F. et al. (2018). Nanocellulose in green food packaging. *Critical Reviews in Food Science and Nutrition* 58: 1526–1537.

Wang, Y.T., Zhao, H.B., Degracia, K. et al. (2017). Green approach to improving the strength and flame retardancy of poly(vinyl alcohol)/Clay Aerogels: Incorporating biobased gelatin. *ACS Applied Materials and Interfaces* 9: 42258–42265.

Wang, Y.H., Zhao, M., Barker, S.A. et al. (2019). A spectroscopic and thermal investigation into the relationship between composition, secondary structure and physical characteristics of electrospun zein nanofibers. *Materials Science and Engineering: C* 98: 409–418.

Weber, C.J., Haugaard, V., Festersen, R. et al. (2002). Production and applications of biobased packaging materials for the food industry. *Food Additives and Contaminants* 19: 172–177.

Wells, N., Yusufu, D., and Mills, A. (2019). Colourimetric plastic film indicator for the detection of the volatile basic nitrogen compounds associated with fish spoilage. *Talanta* 194: 830–836.

Xie, Y., Niu, X., Yang, J. et al. (2020). Active biodegradable films based on the whole potato peel incorporated with bacterial cellulose and curcumin. *International Journal of Biological Macromolecules* 150: 480–491.

Yam, K.L., Takhistov, P.T., and Miltz, J. (2005). Intelligent packaging: Concepts and applications. *Journal of Food Science* 70: 1–10.

Yildirim, S., Röcker, B., Pettersen, M.K. et al. (2018). Active packaging applications for food. *Comprehensive Reviews in Food Science and Food Safety* 17: 165–199.

Yong, H., Wang, X., Zhang, X. et al. (2019). Effects of anthocyanin-rich purple and black eggplant extracts on the physical, antioxidant and pH-sensitive properties of chitosan film. *Food Hydrocolloids* 94: 93–104.

Zhai, X., Li, Z., Shi, J. et al. (2019). A colorimetric hydrogen sulfide sensor based on gellan gum-silver nanoparticles bionanocomposite for monitoring of meat spoilage in intelligent packaging. *Food Chemistry* 290: 135–143.

Zhang, J., Zou, X., Zhai, X. et al. (2019). Preparation of an intelligent pH film based on biodegradable polymers and roselle anthocyanins for monitoring pork freshness. *Food Chemistry* 272: 306–312.

Zheng, Y., Yanful, E.K., and Bassi, A.S. (2005). A review of plastic waste biodegradation. *Critical Reviews in Biotechnology* 25: 243–250.

6

Nanosensors: Diagnostic Tools in the Food Industry

Stephen Rathinaraj Benjamin[1], *Eli José Miranda Ribeiro Junior*[2], *Vennilavan Thirumavalavan*[3] *and Antony De Paula Barbosa*[4]

[1] *Department of Biomedical Engineering, Federal University of Pernambuco (UFPE), Recife-PE, Brazil*
[2] *Faculty of Inhumas (FacMais), Avenida Monte Alegre, Inhumas, Goiás-GO, Brazil*
[3] *Department of Agriculture Engineering, Saveetha Nagar, Sriperumbudur Taluk, Kanchipuram, Chennai, Tamilnadu, India.*
[4] *Department of Pharmaceutical Products, Faculty of Pharmacy, Federal University of Minas Gerais (UFMG), Belo Horizonte, Minas Gerais-MG, Brazil*

CHAPTER MENU

6.1 Introduction

Nanotechnology is a method of research of both the implementation and advancement of material properties on a nanometer scale. Over recent years, the food and agricultural industries have made major developments in the area of nanotechnology to boost the quality of the product and to reduce total costs. It improves food processing, the preparation of useful and bioactive materials, the construction of advanced and creative methods and the creation of technologies to increase the quality of food, nutritional value, packaging, transport and traceability while securing food safety. Estimation of the international economic effect of nanotechnology was at least $3 trillion in 2020, which

Bioresource Technology: Concept, Tools and Experiences, First Edition. Tanveer Bilal Pirzadah, Bisma Malik, Rouf Ahmad Bhat, and Khalid Rehman Hakeem.
© 2022 John Wiley & Sons Ltd. Published 2022 by John Wiley & Sons Ltd.

employed 6 million people in the world's rising nanotechnology industries (He et al. 2019). It is very promising and has attracted many food companies involved in the production and marketing of new nanomaterial materials and improved the quality of processing, food attributes, flavor and security. Nanomaterials would be used primarily in food packaging and formulations. Furthermore, the beneficial properties of nanoparticles (NPs) in food production including safety concerns have also recently been reported by researchers. Foods with nanomaterial additives predominantly provide more enhanced, homogeneous and stable structure, good taste, ideal presence, and safer nanomaterial components than conventionally formulated products. New developments are focused on the production of cheaper, simpler and also more gainful nanosensors toward molecular observation and analysis. In this sense, nanomaterials provide a substantial advance in the construction of various nanoprobes for molecular identification in a broad range of materials. The molecules can indeed be measured at lower concentrations due to the use of nanomaterials like nanosensors. A nanosensor is a biosensitive substrate that either incorporates a molecule of biological recognition or has the ability to attach a molecule of biological recognition to it. The high area-to-volume ratio of NPs encourages more bioreceptors to be immobilized, consequently expanding the performance of the sensor. Nanosensors are developing important tools for agriculture and food production applications. Compared to traditional chemical and biological approaches, they offer substantial enhancements in specificity, speed and susceptibility. Nanosensors can be used to assess the bacteria, toxins, pathogens and freshness of the food. The nanosensors used during food evaluation incorporate biological, chemical and nanotechnological information and may also be designated as nano-biosensors. The immobilization strategy plays a major role in designing a sensitive nanosensor using the unique optical attributes of NPs and therefore needs to be optimized for specific usage. Gold nanoparticles, quantum dots and carbon-based materials are commonly used for bio-(immuno) sensing among the nanomaterials and will, therefore, be described in this chapter.

6.2 Identification of Foodborne Pathogens and Toxins

A growing number of researchers in recent years have concentrated on the diagnosis of foodborne pathogens utilizing nano biosensors techniques. Hence, we describe recent approaches in nanosensors that always had to identify various food-borne pathogens in this study (Table 6.1). Pathogenic bacteria have been detected throughout human civilization. Some bacterial infections are dangerous, and bacterial infectious diseases caused by pathogenic bacteria can result in the deaths of millions and costly treatments are required every year owing to extreme bacterial infections. These infections are today one of the top causes of mortality on Earth (Hall-Stoodley et al. 2004). WHO report that approximately 600 million, i.e., nearly 1 in 10 people globally, become infected after consuming contaminated food and 420,000 people die annually, resulting in the loss of 33 million healthy decades of existence. Bacteria is the most prevalent foodborne pathogen and accounts for 91% of total foodborne disease outbreaks in the United States (Beran Ames et al. 1991).

Table 6.1 List of different nanosensors applications in food industry

Bacteria analyte	Detection technique	Linear range (CFU ml^{-1})	Detection limit (CFU ml^{-1})	Food analysis	References
Salmonella enteritidis	Label-free strip sensor	10^2 to 10^8	10^2	Cabbage salad, watermelon juice	(Z. Wang et al. 2019)
	Immunochromatographic strips	10^4 to 10^7	5×10^2	Beef, pork, milk	(Q. Wang et al. 2020)
E. coli O157:H7	Immunosensing	3.5×10^2 to 3.5×10^8	1.83×10^2	Milk	(Ye et al. 2018)
	Impedimetric immunosensor	3×10^3 to 3×10^6	3	Milk	(Malvano et al. 2018)
Vibrio parahaemolyticus	Electrochemiluminescence immunosensor	10 to 10^8	5	Seafood and water	(Sha et al. 2016)
	Electrochemical Immunosensor	10^2 to 10^8	33	Seafood	(Song et al. 2017)
Listeria monocytogenes	Amperometric biosensors	10^1 to 10^1	2 log	Food	(Davis et al. 2013)
	Immuno impedimetric biosensors	3.0×10^1 to 3.0×10^4	30	Spiked lettuce	(Q. Chen et al. 2015)
Campylobacter Jejuni	Colorimetric aptasensor	10^5 to 10^8	7.2×10^5	chicken carcass	(Kim et al. 2018)
	Fluorescence nanobiosensor	10 to 300 pM	4.4 pM	Milk	(Dehghani et al. 2019)
Brucella melitensis	Label-free Impedance immunosensor	4×10^4 to 4×10^6	1×10^4, 4×10^5	Milk	(Haiyun Wu et al. 2013)
	Quartz crystal microbalance (QCM) aptasensor	1.0^2 to 1.0^7	103 cells	Milk products	(Bayramoglu et al. 2019)

(*Continued*)

Table 6.1 (Continued)

Bacteria analyte	Detection technique	Linear range (CFU ml^{-1})	Detection limit (CFU ml^{-1})	Food analysis	References
Salmonella typhimurium	Aptasensor	10^1 to 10^6	10^1	Raw chicken	(Dinshaw et al. 2017)
	Electrochemical impedance	10^2 to 10^6	80	Chicken	(L.Wang et al. 2020)
Yersinia enterocolitica	Electrochemical biosensor	10^0 to 10^8	10^4	Kimchi	(Sobhan et al. 2019)
Bacillus cereus,	Electrochemical DNA biosensor	10^0 to 10^7	10^0	Raw milk	(Izadi et al. 2016)
Clostridium perfringens	Electrochemical DNA biosensor	1.0×10^{-14} to 1.0×10^{-7} mol L^{-1}	7.06×10^{-15} mol L^{-1}	Milk	(Qian et al. 2018)
Staphylococcus aureus	Paper-based immunosensor	10 to 10^7	13	Milk	(Bhardwaj et al. 2017)

Escherichia coli (*E. coli*) is a coliform bacterium in the shape of a Gram-negative rod. In fact, *E. coli* O157: H7 can be transmitted as one of the extremely pathogenic bacteria by milk, drinking water, etc., which exhibits not only significant risk to human health but also major economic damage. Recently, the process of colony counting and colony methods has been used as the traditional approach for defining *E. coli*. It takes time-consuming laborious procedures, given its strong resilience down to one colon. The existing detection methods (including polymerase chain reaction-PCR and enzyme-linked immune-sorbent assay – ELISA) are sufficiently sensitive and selective complex steps in the sample processing and longer exposure periods (Berg et al. 2015). The combined effect of magnetic and fluorescent approaches helps to detect and measure contamination by the pathogen in diverse concentration ranges with notable efficiency in different stages of identification of the contaminating microorganisms. Multiparametric magneto fluorescent nanosensors (MFnS) are considered as substitute methods for detection and identification of bacteria. In this technique, the intensity of the MFnS fluorescent signal increases, allowing for the rapid detection of microbial contamination. This offers a dual-modal detection principle able to detect low and high concentrations (i.e., 1–20 and 1–100 CFU ml^{-1}) and low detection values (i.e., 1 CFU ml^{-1}) of bacterial contaminants present in milk samples (Shelby et al. 2017). Furthermore, Y. Li et al. (2017) developed an electrochemical biosensor allowing ultrasensitive simultaneous identification of the *E. coli* and *Vibrio cholerae* O1 using multifunctional nanoconjugates including nanoparticles (CdS and PbS), which contain capture probes via C60@AuNPs as nanocarriers and HCR (hybridization chain reaction) amplification. They demonstrated the effective implementation of monitoring of certain bacterial pathogens that would allow early identification of a number of diseases with the use of electrochemical biosensors.

Nanocomposite components were utilized extensively for the enhancement of electrochemical biosensors due to their biocompatibility and their configurable conduction properties and extensive efficient surface area for desired adsorption biomolecules. The advancement of metal oxide thin film-focused amperometric biosensors has recently become significantly increased due to their remarkable stability. For the rapid identification of the target pathogenic *E. coli* O157: H7, a label-free amperometric sensing platform based on NiO thin film was successfully developed. The biosensor exhibits the premise of standard antigen-antibody interactions which are covalently linked to the nanostructured NiO matrix surface. The studies of cyclic voltammetry (CV) for milk samples were exploited to analyze the amperometric response of the bioelectrode developed against the concentration of *E. coli* O157: H7 ranging from 10^1 to 107 cells ml^{-1} and a relatively low limit value of 10^0, i.e., 1 cell ml^{-1}, and including high sensitivity, selectivity and specificity to certain bacterial organisms (Dhull et al. 2019). Electrochemical immunosensors are more suitable for different biosensors than other sensors, as the electrochemical transducers utilized to provide good sensitivity, including the usage of immunochemical interaction, allows specific analysis. The enzyme-free electrochemical immunosensors and immunoassays could also be produced by utilizing nanostructures doped with metal ions or quantum dots as labels. For example, a sandwich amperometric immunosensor designed to identify *E. Coli* employed cadmium sulfide quantum dots (CdS QDs) zeolitic imidazolate framework-8 (ZIF-8) as labels, as the anti-*E. coli* O157: H7 immobilized antibody maintains high catalytic activity on a biocompatible substrate. The immunosensor succeeded in detecting *E. Coli* in measured milk (Zhong et al. 2019).

Salmonella is among the most dangerous foodborne pathogens worldwide and responsible for the majority of warm-blooded animal diseases. *Salmonella* belongs to the Gram-negative bacterium (Enterobacteriaceae family). The main sources of human infection with *Salmonella* are contaminated foods or water. *Salmonella*'s most common clinical symtoms are recognized as diarrhea, vomiting, gastroenteritis, severe dehydration, and other stages of typhoid fever (Bula-Rudus et al. 2015). C. Xiang et al. (2015) have shown that the efficiency of the ultrasensitive electrochemical immunosensor can be enhanced by employing gold nanoparticles on chitosan composite. Owing to the biopolymer's electroactive and effective film forming, an oxidized film of chitosan could be used to conjugate an antibody capture. A sandwich electrochemical immunosensor was assembled following incubation in the *Salmonella* suspension and horseradish peroxidase (HRP) covalently linked secondary antibody solution. The film was grafted with the antibody *Salmonella* to detect Gram-negative *Salmonella choleraesuis* in dairy samples. The sensor showed a good linear concentration range 10^4 to 10^9 CFU ml^{-1} and the biosensor reached a very good limit of detection (LOD) of 3.0×10^3 CFU ml^{-1}. The electrochemical immunosensor had already been provisionally developed for the detection of bacteria in past years, owing to its large selectivity for immune responses and good sensitivity for electrochemical analysis for the detection of analytes such as biological and chemical pathogens, and contaminants. In this research, a sandwich novel electrochemical immunosensor was established for the detection of *S. pullorum* and *S. gallinarum* in different food samples such as chicken, egg and meat. In addition, a novel electrochemical immunosensor has been developed by electrodeposition of gold nanoparticles with the incorporation of modified different materials (ionic liquid, sodium alginate, β-cyclodextrin and chitosan) dropped onto the SPCE. Nevertheless, the immunosensor involves the electrodeposition of gold nanoparticles used for antibody capture and signal enhancement. A secondary antibody (HRP) was added to form the sandwich after exposure to the analytes. The biosensor demonstrates a considerable linear range (10^4–10^9 CFU ml^{-1}) low detection limit (3.0×10^3 CFU ml^{-1}), with the high specificity signals evaluated using CV and electrical impedance spectroscopy (EIS) techniques (Fei et al. 2015). Recent research conducted by Hongsu Wang et al. (2019) proposed an electrochemical immunosensor for the detection of *Staphylococcus aureus*. According to this report, this immunosensor improved bacteria sensing based on the immobilization of antibodies on hierarchical mesoporous silica. The immobilized conjugate resulted in the signals being amplified. In this electrochemical immunosensor, the loading of Ab-HMS bio-conjugate on glass carbon electrode detection (20 mts) of *S. aureus* used DPV and EIS. The LOD and linear range were respectively 12 CFU ml^{-1} and 10 and 2×10^3 CFU ml^{-1}. Additionally, Bhardwaj et al. (2017) used a low-cost paper-based biosensor to identify the *S, aureus* in spiked dairy products within just 30 minutes. Through covalent bonding, single-walled carbon nanotubes were integrated with antibodies and deposited on a carbon paste electrode onto a printed paper sensor that remained relatively stable for more than a month. This device is particularly suitable for analysis of *S. aureus* concentration with an LOD value of 13 CFU ml^{-1} and bacterial concentration ranges from 10 to 10^7 CFU ml^{-1} with high specificity and sensitivity.

Listeria monocytogenes (LM), one of the most dangerous pathogens of foodborne diseases, is a rod-shaped Gram-positive bacterium. Despite the relatively small number of people infected with *L. monocytogenes*, this can be found not only in unprocessed animal-derived foods. The determination of *L. monocytogenes* is on screen-printed interdigitated electrodes for quantitative measurement of the impedance change of the catalysate (D. Wang et al. 2017). The monoclonal anti-*Listeria* antibodies were covalently linked to the MNPs (magnetic nanoparticles) surface based on the streptavidin-biotin binding. The optimized biosensor performed the identification of *L. monocytogenes* in lettuce samples using the competitive effect between gold nanoparticles modified with polyclonal antibodies (PAbs) and urease. The LOD of this biosensor was 1.6×10^3 CFU ml^{-1} with a linear relationship that was found to be 1.9×10^3 to 1.9×10^6 CFU ml^{-1}. The research group consequently reported the same methodology in other electrochemical biosensors detection for *L. monocytogenes*. The developed impedance biosensor was fabricated by covalent immobilization for MNP-MAb-Listeria-PAb-AuNP-urease sandwich complexes to show rapid screening of foodborne pathogens in spiked lettuce samples (Q. Chen et al. 2015). In a similar way, Lu et al. (2016) prepared a specific and sensitive electrochemical immunosensor for *L. monocytogenes*, which was constructed by depositing MWCNTs fibers using immobilized labelled antibody HRP onto the surface. By both physical and chemical immobilization processes, HRP crystallizations on the immunoelectrode surface have been identified. Moreover, the developed immunosensor exhibited the linear range from 10^2 to 10^5 CFU ml^{-1} and an LOD value 1.07×10^2 CFU ml^{-1} with excellent stability and reproducibility in storage (RSD < 6.5%) in milk samples.

Campylobacter jejuni is a Gram-negative bacterium with helical shape. *C. jejuni* has been one of the major reasons for foodborne human degenerative infections. Livestock and poultry items with this pathogenic bacterium were identified as the primary pathways of human transmission. Different methodologies, such as lateral flow test strip assay, have recently reported new analytical evaluations (Poonlapdecha et al 2018). Hong Wang et al. (2018) prepared a sensitive immunosensor established to detect *C. Jejuni* using a quartz crystal microbalance (QCM) with magnetic nanobeads (MNBs) in poultry products to identify pathogenic bacteria and gold nanoparticles for amplification measurement. The results showed that the QCM immunosensor was capable of detecting a pathogen in poultry products with a 20 to 30 CFU ml^{-1} low detection value and a maximum detection period of time lower than 30 minutes. In a similar manner, Masdor et al. (2016) have described the distinctive detection of *C. jejuni* in food samples developed using the quartz crystal microbalance (QCM) sensor system. Furthermore, Kim et al. (2018) developed a two-stage label-free aptamer functionalized by expressly binding to *C. jejuni* and *C. coli* using AuNPs conjugated with aptamers. The aptamer showed high sensitivity and distinct detection of food pathogens from chicken samples, which requires less than 30 minutes.

Clostridium perfringens is a microbe-forming, Gram-positive bacterium and among the species is the most bacterial pathogen in the *clostridium* genus found widely across different foods such as dairy products and meats. *C. perfringens* is among the most widespread foodborne pathogens. In this work, the authors explored fabricating novel electrochemical DNA sensors developed for the detection of *C. perfringens* DNA sequences by the differential pulse voltammetry technique (DPV). Although used for constructing the biosensors, cerium dioxide nanoparticles and chitosan composite allow an efficient method to amplify label-free electrochemical detection of DNA hybridization by enhancing charge transfer between surface-anchored DNA sequences and nanocomposite (Qian et al. 2918). The proposed biosensor complex linear range for detecting the *C. perfringens* oligonucleotide sequence ranged from 1.0×10^{-14} to 1.0×10^{-7} mol L^{-1} and the LOD was found to be 7.06×10^{-15} mol L^{-1} in dairy products. Over recent years, more focus has been on self-assembled monolayers (SAMs) (Vaisocherová-Lisalová et al. 2016), electrochemical biosensors (Freitas et al. 2014), light scattering sensors (Abdelgaseib et al. 2019), and the uses of food contaminant studies. Recently, Zheng et al. (2019) fabricated the microcantilever array biosensor modified with gold nanoparticles (Au NPs) suitable for evaluating the ultralow concentration of pathogenic microorganisms, such as *V. parahaemolyticus*, *E. coli* O157: H7, *Salmonella*, *L. monocytogenes*, *Shigella*, *S. aureus*, etc., employing milk samples. In this method, the linear range and LOD was 3–1.2 × 105 cells ml^{-1} and 1–9 cells ml^{-1}, respectively.

Accordingly, hybrid nanomaterials use of the development of amperometric immunosensors of sandwich-type is key to accelerate the accurate determination of the microorganism. For example, Freitas et al. (2014) demonstrated the use of anti-*Salmonella* antibodies for the rapid detection of *S. Typhimurium* obtained from milk samples. In this technique, the electrochemical labels proposed the use of SPE with iron oxide/gold core/shell nanomagnetic particles coated with antibodies and CdS nanocrystals. Besides, the modified sensor used in this case to analyze samples of milk spiked with *Salmonella* 10^3 and 10^4 CFU ml^{-1} achieved sufficient recovery.

6.3 Pesticides and Carcinogenic Detection

Residue detection of pesticides has become a major issue for food safety specialists, as well as a rapid method for detection of residues of pesticides that is needed to prevent affecting humans. Due to their low persistence compared to other pesticides, organophosphorus pesticides (OPs) are among the most commonly used pesticides. Gold nanoparticles were used to detect organophosphorus and malathion pesticides as colorimetric techniques and non-electrochemical sensors (D. Li et al. 2019). Nanocomposite graphene oxide containing 3D porous chalcopyrite ($CuFeS_2$) and PEDOT:PSS modified electrodes were used for the sensing of paraoxon pesticides in vegetables (Hryniewicz et al. 2018; Rajaji et al. 2019). Nickel oxide nanoplatelets (NPs) modified SPE and gold nanorod (Au NR)-based (SERS) platforms have been used for the detection of parathion (Khairy et al. 2018; Huixiang Wu et al. 2019). Bisphenol A is extensively used throughout the manufacturing of polycarbonates while making containers for food storage and also it belongs to the category of Endocrine Disruptive chemicals (EDCs). 4-mercaptobenzoic acid (4-MBA) and chitosan (CS) capped Au@Ag core-shell NPs have been used as an extremely sensitive detection of BPA using the SERS technique. Additionally, a ZnO/CNTs/ionic liquid composite has been used to detect both Sudan I and bisphenol A in food samples (Najafi et al. 2014; C.-Y. Wang et al. 2018). Similarly, MWCNTs/DCHP (Dicyclohexylphthalate) composites were proposed by D. Chen et al. (2017) for the immobilization of acetylcholinesterase (AChE). The pesticide residue detection was based on pesticide inhibition of the activity of AChE, chlorpyrifos used as model compounds, reaching an LOD of 0.05 μg/L with considerable selectivity. A great performance was also achieved during examining real vegetable samples like cabbage, rape and lettuce. However, electrochemical microanalytical device-based sensors have been successfully employed to detect chlorpyrifos in food samples (Islam et al. 2019; Nagabooshanam et al. 2019).

6.3.1 Nitrites-Carcinogenic Detection

Over many decades, sulfite compounds including sodium metabisulfite, sodium sulfite and sodium bisulfite have been legal food products familiar in global widespread use as food additives as well as in beverages that contribute to food preservation by preventing the browning reaction, intracellular oxidative stress and microbial spoilage. Among these compounds, sodium hydroxy methane sulfinate (HMS) is a commercial chemical substance, used as an outstanding decoloring reagent for certain organic materials, as well as for caramel and sugar juice, etc. Wu et al. (2018) fabricated the electrochemical nonenzymatic sensor modified with CTAB-Chit functionalized CNT composite to detect illegally added nonfood substances (HMS) in real samples (yuba, wheat flour and corn starch). The presence of components have a synergetic effect and great electrocatalytic occurrence toward using HMS oxidation. This proposed sensor showed a linear range of 30 to 800 μM with an LOD value of 9.6 μM. For centuries, nitrite and nitrate were used in the ingestion and preservation of meat and fish, as well as in the manufacturing process of some cheeses. The average nitrite levels in cured meat products are in the range of 10 to 40 mg kg^{-1} (San and Baddoo 1997). Prussian Blue Analog (PBA) is a suitable material for producing electrochemical sensors adhered to a composite of multi-walled carbon nanotubes with metallic nanoparticles like Cu and Co for nitrite detection (Liu et al. 2019). In this process, the modified electrochemical sensor was employed to determine nitrite compounds in food samples (ham, sausage and mustard) with an acceptable linear range (10–400, 400–2100 μM) and LOD value (0.5 μM). Furthermore, gold nanoparticles are also showing a good effect in the detection of nitrite in sausage samples with adequate results (Üzer et al. 2016).

6.3.2 Mycotoxin Detection

Mycotoxins are potentially toxic metabolic byproducts in numerous fungi that have a substantial effect on global safety and quality, from disclosure to contaminants and from economic crop losses. There are various types of mycotoxins that have been described to date, including aflatoxins (AFs), citrinine (CIT), ochratoxins (OTs), fumonisins, trichothecenes, patulin (PTL), deoxynivalenol (DON) and zearalenone (ZEA), all possessing prospective impacts on human health as well as from the agronomic view. Recent research has shown promise that nanotechnology will set the stage for the growth of nanosensors that will be beneficial to detect these mycotoxins. Ochratoxin A is a secondary fungal metabolite, mostly released by the species Aspergillus and Penicillium. These mycotoxins occur in a variety of agricultural food products like grape juice, beer, coffee, wine, soy milk and lager. Xiang et al. (2018) reported that the simple voltammetric modification of nanosensors was fabricated using gold nanoparticles with functionalized BP (black phosphorene) nanosheets. The sensor was used for voltammetric assessment of OTA in food samples, including beer and grape juice, utilizing the DPV technique. Detection of Aflatoxin M1 (AFM1) in milk was described by Shadjou et al. (2018). Silver nanoparticles were electrodeposited on a green and conductive nanocomposite comprising of α-cyclodextrin as a conductive matrix, as well as graphene quantum dots as an amplifier showing the development electrochemical assay for the detection of AFM1 in milk. Thus, the developed method was found to be very sensitive, having a detection limit of 2 μM with a linear range of 0.015 to 25 mm. Additionally, screen-printed electrodes, a commonly used platform for the manufacture of disposable and accessible electrochemical sensors that are modified as a direct sensor by a hybrid among gold nanoparticles and graphene quantum dots for a potential substitute to the immunosensors, are widely used in malted barley samples for determining AFB1 (Gevaerd et al. 2020). In another report, El Hassani et al. (2020) constructed an electrochemical imprinted sensor by combining two types of nanocomposites of polypyrrole with chitosan (CS) (ZnO-NPs/CS/PPy/AFB2) and zinc oxide nanocomposite (ZnO-NPs/CS/AFB2). Using the hybrid functionalized nanocomposite, the electrochemical sensing of AFB2 was carried out using signal amplification with low LOD values (0.6 and 2.0 fg ml^{-1} in ZnO-NPs/CS/AFB2 and 0.2 and 0.7 fg ml^{-1} in ZnO-NPs/CS/PPy/AFB2). This promising technique constitutes one of the most sensitive imprinted sensors for AFB2 detection in milk samples. Moradi et al. (2020) have fabricated an electrospun sensor for detection of zearalenone mycotoxin in foods, especially in dairy products, including fruit juices and other liquid foods. A nanofiber-modified pencil-graphite electrode was fabricated utilizing electro-spinning technique. This modified electrode had good sensitivity for zearalenone (LOD of 0.25 ng ml^{-1}) with linear ranges of 5 to 230 nm and 60 to 100 nm employed for quantification based on the square wave voltammetry (SWV) technique.

In recent years, the growing number of GMO (genetically-modified organisms) based products has prompted several countries to involve legislation and labeling of grains and foodstuffs containing GMOs and monitoring of mislabeled products to enable informed choice for consumers. S. Wang et al. (2016) developed an electrochemical DNA biosensor based on the EIS technique to monitor GM soybean screening by detection of 35S from the cauliflower mosaic virus (CaMV 35S). In this sensor, multiwalled carbon nanotube-reduced graphene oxide nanoribbon (MWCNT-rGONR) and the AuNPs/MWCNT-rGONR composites could effectively improve the sensitivity of target DNA detection. The proposed DNA sensor exhibited a linear range from 1×10^{-16} M to 5×10^{-10} M, with a detection limit of 3.3×10^{-17} M in soya bean samples.

6.3.3 Food Packaging

Recently, using food nanotechnology in industrial applications has increased from intelligent packaging through immersive food creation, and smart packaging is commonly used in the packaging of foodstuffs, in which the condition of packaged food products is altered to enhance the sensory quality and the security of the food, thus sustaining the quality of foodstuffs. Developed packaged food systems and nanomaterial-based systems eliminate loss of food ascribed to various contaminations by the microbes. However, various nanoparticles, including magnesium oxide, titanium dioxide, silver nanoparticles, zinc oxide, carbon nanotube derivatives, zerovalent iron and fullerene, have exhibited a great effect as the antimicrobial agent. Additionally, the application of distinct nanomaterials has recently been developed to identify/minimize hazardous chemicals/pathogens in food products. An integrated electronic tongue in developed food packaging includes a compilation of nanosensors that are especially penetrating to gases emitted from waste food. This device provides a simple and visible indication signifying if the food is fresh or not, by using sensor strips which change color. Nanosensors identify toxic compounds and contaminants found in various food products by potential identification, which is beneficial for food security. Nanotechnology applications in the food industry endorse advances in various measurements of food products such as texture, taste, sensory characteristics, product color change and shelf-life/stability.

6.3.4 Food Freshness Detection

In food packaging, various types of sensors based on nanoparticles are being used. In contrast to the ideas used for indices of integrity, freshness markers offer additional specific information on the quality of product caused by changes in microbiological growth, pH and spoilage-induced volatile metabolites. Although seafood freshness indicates consumer orientation and product safety, speedy evaluation of deterioration in seafood is deemed essential (Lee et al. 2018). It produces the xanthine (XA) level throughout the storage of fish so it can be regarded as a great tool for assessing the freshness of the fish. Several nanomaterials widely used during xanthine detection include gold (Dervisevic et al. 2017), ferric oxide (Dervisevic et al. 2015) and multiwalled carbon nanotubes (Dalkiran et al. 2014). Yazdanparast et al. (2019) developed an enzyme-based electrochemical biosensor for xanthine detection using a poly(L-aspartic acid)/MWCNT bionano composite for meat freshness detection. In this study, the authors prepared a simple nanocomposite and immobilized on xanthine oxidase (XO) in the glassy carbon electrode. The established biosensor was used to determine the xanthine level with linear ranges (0.001–0.004 μM and 0.005–50.0 μM) with an LOD value (3.5 μM) and good selectivity at different storage times. Multicolor colorimetric biosensor detection of hypoxanthine was developed by using gold nanorods with peroxidase-like activity. The H_2O_2 produced by XO from hypoxanthine oxidation undergoes a Fenton reaction in the presence of Fe^{2+} to produce hydroxyl radicals. The sensor showed a hypoxanthine in the linear range of 0 to 1.13 mm The sensor demonstrated the potential of the naked eye to semi-quantitatively determine hypoxanthine concentrations in fish extract (Z. Chen et al. 2017). Biogenic amines (BA) are low molecular weight organic nitrogen compounds that have formed naturally by pathogenic decarboxylation of correlating amino acids throughout food and which participate in various physiological activities in biological systems. Biogenic amines are substances widely found in foods and drinks. El-Nou et al. (2017) developed the fast screening optical sensor with sensitivity to the existence of biogenic amines in samples of poultry meat. Gold nanoparticles allow for the rapid detection of BAs with normal fluoremetric and spectrophotometric techniques. The modified biosensor exhibited linear values from 0.6 to 12 μM and an LOD of 0.6 μM.

Additionally, when contaminated with bacteria or toxins, some foodstuffs are fungible and susceptible to deterioration. Some foodstuffs perish due to time and temperature change in storage. Subsequently, changes in color, size and spectral properties of nanoparticles were developed in order to establish time-temperature indicators. Thus, time-temperature indicators based on nanoparticles were developed to resolve these requirements. The freshness of a processed food product is its susceptibility to spoilage by exposure to oxygen. However, oxygen enhances the oxidation of antioxidants, as well as inhibiting the growth of bacteria. Electrospun protein nanofibers, developed by the authors, Aghaei et al. (2020), were used in the production of halochromic nanosensors and applied to the quality assessment of rainbow trout fillets in real-time. This nanosensor can evaluate the fish freshness with variations of color in real-time. The proposed optical gas nanosensor (polyetherimide (PEI) and silver (Ag) nanoparticles) have the potential to monitor the freshness of meat and can be implemented into smart packaged foods (Ryspayeva et al. 2018).

6.4 Chemicals and Food Additives Detection

6.4.1 Preservatives

The use of preservatives is becoming an increasingly important food technology due to the increasing demand for processed food. A number of organic acids have been used as preservatives and their salts, which include lactates, Na and K salts, Benzoic acid and benzoates, both prevent spoilage by inhibiting mold and yeast. Melamine has been reported as a potential endocrine and neurotoxicant disruptor. Owing to the presence of its rich nitrogen structure, melamine is prohibited from being added to milk products to artificially enhance the content of total nitrogen and has led to a number of foodborne diseases. Barium sulfate-coated $CsPbBr_3$ perovskite nanocrystals ($CsPbBr_3$ NCs@$BaSO_4$) were proposed by Q. Li et al. (2020) for rapid melamine discrimination. This novel turn-on fluorescent nanosensor is an effective procedure for melamine detection with a very high sensitivity limit of 0.42 nmol L^{-1} and in the range of 5.0 to 500.0 nmol L^{-1}. Additionally, Dong et al. (2016) developed an aptamer-modified SERS nanosensor and oligonucleotide chip to quantitatively detect melamine in milk samples with an LOD of 1.0 pg ml^{-1}. Zhu et al. (2016) developed a simple nanosensor containing cross-linked polyelectrolyte(CPE)-stabilized silver nanoparticles (P1-AgNPs), which favors the fluorescence and colorimetric determination of melamine with an LOD of 0.1 and 0.45 nM respectively.

6.4.2 Dyes

Food additives are always incorporated into the food for various determinations, particularly preservation, coloring and increased levels of nutrients. The commonly used synthesized colorants in the food industry consist of indigotine, Brilliant blue, allura red, erythrosine, sunset yellow (SY) and tartrazine, etc. Many nanomaterial-based sensors, including exfoliated graphene oxide (eGO) (Tran et al. 2019), GO with polyaniline nanocomposite (Alqami et al. 2018) and ZnO nanoparticles doped with Pd (L. Li et al. 2019) were established for the electrochemical recognition of SY. Several nanosensors-based materials, such as polypyrrole-functionalized SWCNT/GCE (Meiling et al. 2014),

1-allyl-3-methylimidazolium chloride-graphene oxide functionalized AuNPs (Meiling Wang and Zhao 2015) and graphene oxide with modified MWCNTs (Qiu et al. 2016), were exploited for the determination of tartrazine and SY dyes in food and drink samples. Recently, Sivasankaran et al. (2017) determined Brilliant blue FCF in sports drink samples and showed the low nanomolar LOD to subnanomolar levels, high sensitivity and effective robustness conditions. Another example was reported using an electrochemical nanosensor technique for direct simultaneous estimation of metanil yellow and fast green that are toxic food dyes (Shan 2020). This sensor was modified with gold nanoparticles and calixarene to detect toxic food dyes like metanil yellow and fast green simultaneously. The sensor showed an LOD of 9.8 and 19.7 nm, and a linearity of 0.05 to 45 μM to target food contaminants in fruit juice and water samples.

6.4.3 Sweeteners

Sweeteners are chemical compounds, with the use of sucrose as a sweet substitute. Throughout the food industry, saccharose is used as a reference sugar, and as well as attributed to saccharose, every sweet substance has its comparative flavor, quantified by human sensory analysis. Nanosensors for determining intense sweeteners (saccharin) in beverage samples have recently been explored with the use of SWCNT-modified GCE (Yang et al. 2014), and polypyrrole-doped selective membranes (Álvarez-Romero et al. 2010). Those sensors have low micromolar to submicromolar levels of detection values. Samphao et al. (2018) fabricated the bienzymatic sensor determination of glucose and ethanol in wine samples. The fabricated sensor platform is based on core-shell iron oxide with gold (Fe_3O_4@Au) nanoparticles. In addition to the high sensitivity, excellent reproducibility, repeatability, wide linear range (glucose and ethanol from 0.3 to 7.8 mm and 0.1 to 0.7 M) and LOD values (glucose (0.1 mm) and ethanol (60 mm)), have been achieved using the developed system. Amani et al. (2018) determined the fructose from apple and orange juice by using copper oxide nanoparticles with a carbon paste electrode (CuO-CPE). This sensor displayed a linear range of 0.06 to 3000 mm with an LOD of 0.07 μM.

6.4.4 Antioxidants

Antioxidants are active substance classes commonly found in foodstuffs. Antioxidants can also be used to stabilize polymeric items, petroleum, foods, cosmetics and pharmaceuticals. Standard methods are essential for the identification and evaluation, as well as for the quantitative study of antioxidants present in beverages, foods and biological fluids. The José Palacios group has developed a rapid and simple electrochemical sensor manufactured with a sonogel glass carbon electrode for ascorbic acid detection from commercial baby apple juice (Abdelrahim et al. 2013). The electrochemical sensor presented a low LOD value limit of 1.59 and 2.93 μM for ascorbic acid.

6.4.5 Food Allergens

Food allergy is among the major health concerns affecting people worldwide and over the past few decades the food industry has made considerable endeavors as well as acquisitions to provide good-quality food to vulnerable consumers. Food allergy correlated to gluten consumption impacts 0.2 to 0.5% of the populous, as well as gluten sensitivity impacting almost 10% of the US community (Vasagar and Leonard 2014). In recent times, Alves et al. (2015) fabricated the screen-printed immunosensor covered with gold nanoparticles to detect major allergen in peanut (Ara h) 1 in foodstuffs. In a sandwich-type immunoassay, two monoclonal antibodies have been used, and antibody-antigen activity has been observed electrochemically by stripping technique in silver ions placed enzymatically. The established immunosensor permitted Ara h 1 is quantified between 12.6 and 2000 ng ml^{-1}, with a detection limit of 3.8 ng ml^{-1}. Additionally, Manfredi et al. (2016), developing the fast determination of celiotoxic prolamins, established the first stable disposable amperometric immunosensor premised on gliadin-functionalized carbon/nanogold SPE (screen-printed electrode). This proposed immunosensor demonstrates an LOD value of 8 ng ml^{-1} in ethanolic extracts of gliadin and could be a valuable resource for assessing the security of raw resources used throughout the manufacture of dietary products for celiac patients.

6.5 Nano-based Sensors for Smart Packaging

Despite many studies to evaluate biological and chemical contaminants in food products, in packaging technology, the use of nanosensors essentially involves the detection of gas. The construction of sensors are designed to detect the impact of carbon dioxide, oxygen, moisture, ethylene, amines, ethanol and acetone in various real samples. The "smart" and "active" nano-based food packaging has numerous advantages, including barrier coating technology for products that capture ethylene gas to slow maturation and thinner and stronger packaging and so less waste. Intelligent packaging uses colorimetric nanosensors on food packages to warn consumers whenever an item is unsafe to eat (Duncan 2011).

Photo-activated ink based on nano-sized titanium dioxide (TiO_2) may cause changes in oxygen, which impact the color of meat and gaseous amines specifying bacterial growth of meat and fish (Ramachandraiah et al. 2014). Nanosensors can detect changes in temperature, moisture and oxygen levels and identify contamination by microbes or toxic products, including heavy metals or residues of pesticides. Smart packaging is grouped under developed packaging which integrates nanosensors that interact within the packaging to the physical or chemical change in food samples to delay bacterial growth or contamination. In food packaging, various types of sensors based on nanoparticles are being used. Nanosensor packaging can also be used to establish food product conditions in the food system. For plastic packaging, nanosensors can diagnose food gases through contamination, as well as the packaging itself changing color to alert for consumer. Additionally, spot-based nanoparticles are sensitive and easily labelable on individual packages (Jiang et al. 2015). From a microbial perspective, nanosensors are mainly designed to reduce pathogen identification time (Fuertes et al. 2016a). Additionally, different types of nanomaterials, including silver shell gold nanorods (Zhang et al. 2013), polydiacetylene–silica (Nopwinyuwong et al. 2013) and iron oxide–titanium dioxide (Lou et al. 2013) are used to develop a time-temperature indicator for foodstuffs application.

6.5.1 Nanobarcodes

The use of nanobarcodes are becomes extensive for tracking and safety. The uses of smart sensors consequently benefit consumers in terms of long-term quality identification, and manufacturers in order to quickly distribute and authenticate foodstuffs. Recently, specific nanoparticular encoding processes and nanodisk codes were investigated (Miao Wang et al. 2015). Du et al. (2018) developed a new small molecule of bio-bar amplifying code with a competitive immunoassay for tracking pesticides in a microwell plate with higher stability and sensitivity in food samples like apple and cabbage. In a similar strategy, gold and nickel nanodisks comprising chromophores work by observing the improvement of biological labels for DNA detection and food adulteration. Hereafter, nanotechnology will be used to develop an effective nanobarcode system to ensure the quality and efficiency of the food. In order to improve food life/safety, food additives have been incorporated to implement quality properties. Nevertheless, undesirably high food additive concentrations and industrial pollutions of food and water supplies of veterinary product contaminants and pollutants were becoming major public health problems.

6.5.2 e-NOSE and e-TONGUE

The Electronic nose (e-NOSE) as well as the Electronic tongue (e-TONGUE) are also becoming alternative options and replace skilled human sensory and consumer panel set-ups for performance evaluation and product quality throughout production methods to satisfy consumers to the maximum extent. It is also possible to use nanosensors in the type of techniques to detect chemicals produced throughout food spoilage by inserting them in close contact with packing materials to ensure a better performance. By using the most suitable device of this type of e-NOSE and e-TONGUE, sensors showed sensitivity, redundancy, selectivity, reliability and accuracy for sensing applications of the response in the food industry. The release of volatile organic compounds (VOCs) from the food elements makes e-NOSE and e-TONGUE more relevant in the field of food science foe testing of unprocessed and manufactured product cooking as well as the fermenting system, food packaging and long-term storage in quality analysis. The e-NOSE is made up of a quartz crystal microbalance coated with modified metal porphyrins, and associated substances can also be used to understand the extent of contamination in red wine and pulped tomatoes (Baietto and Wilson 2015). However, these experiments showed that acceptable LOD values for an e-NOSE are significantly lower than that of sensory perception in humans. In this respect, a bioelectronic nose may be suitable for applications in different fields such as food quality control, environmental monitoring and even clinical diagnostics. Son et al. (2016) established a bioelectronic nose based on the synthetic peptide and carbon nanotube field-effect transistor (CNT-FET) derived from *Drosophila* odorant-binding protein (OBP) for identification of *Salmonella* contamination in ham samples with low detection limit value 1 fM and good sensitivity.

6.5.3 Oxygen Sensors

With regard to oxygen sensors, notable effects have been obtained with TiO_2 and SnO_2 nanomaterials which act as a photoactivated color change indicators in response to low

O_2 concentrations in the presence of a redox-active dye (methylene blue) (Ollis et al. 2016). OxyDot® is the marketed nanosensor used to directly measure the dissolved oxygen in the food packaging product as well as a sealed drink. The OxyDot is a powerful, non-destructive, sensitive and fast (<0.1 s) method for detecting oxygen which can measure oxygen concentrations in real time. This nanosensor is used in horticulture to distinguish the diffusivity of oxygen in food (Bhunia et al. 2016), to avoid anaerobic pathogenic microbes from evolving in pasteurized chilled products (Muñoz et al. 2017).

6.5.4 Humidity Sensors

In order to sense the storage conditions, nanosensors are incorporated into the packaging by assessing the physiological function, including humidity, temperature, oxygen content, pH, toxins, freshness and pathogens, by evaluating the fermented byproduct under the food preservation conditions. Specific nanoparticles have been developed with polymer composite and nanostructures usually made up of nanoparticles, including ZnO, TiO_2, Al_2O_3, antimony (Sb), tin oxide (SnO_2), potassium niobate ($KNbO_3$) and reduced graphene oxide (rGO) (Y. Wu et al. 2019). The nanosensor depending on 3D hierarchical SnO_2 dodecahedral nanocrystals (DNCs) displayed unique characteristics for sensing humidity. The presence of 3D open nanostructures can encourage water molecules to penetrate and diffuse, and the exposed facets can improve water molecules' adsorption capability (Feng et al. 2017). Recently, Ganeshkumar et al. (2017) developed $KNbO_3$ perovskite nanofibers to be outstanding components for extremely sensitive remote sensing and humidity sensing applications. Y. Wu et al. (2017) developed a flexible humidity sensor with an easy fabrication method from carbon material along with electrodes and functional layers (rGO). The authors found the thickness of GO film achieves 100 nm, and the sensitivity demonstrates the best performance and the best time of adsorption and desorption can be as fast as 0.9 s and 4.5 s.

6.5.5 Carbon Dioxide (CO_2) Sensor

In general, the presence of elevated levels of CO_2 in packaged foods is also the primary indication of spoilage (Puligundla et al. 2012), and it is important to maintain it at optimal levels to reduce contamination in food products packaged under modified-atmosphere packaging (MAP) requirements. Carbon dioxide sensors can be used to detect food product freshness and safety, with a focus on qualitative signals, including their transformations (Meng et al. 2014). Trimethylamine oxide (TMAO) is a further marker that can be used to evaluate the degradation of fish. TMAO is an essential component of the odor of stored fish that gives the usual fishy odor. Normally, the colorimetric model for the identification of TMAQ was established utilizing pH indicator dyes immobilized on microparticles of cellulose to indicate whether the food had been spoiled, changing from green to red (Schaude et al. 2017). In addition, the total volatile basic nitrogen (TVB-N) was analyzed through organic semiconductor gas sensors with a nanostructured surface (polystyrene (PS) nanospheres) porous top metal electrode sensor to detect fish freshness (Chang et al. 2017). RFID (radio frequency identification) technology comprises a small electronic gadget consisting of a tiny chip and a transmitter that can be used to collect real-time package data to track and control counterfeiting, as well as to enhance tracking. RFID tagging is also a powerful tool in the field of the intelligent packaging system. In this sense, different types of promising areas of monitoring by which the proliferation of nano-scale barcodes, magnetic nanoparticles, as well as quantum dots, will contribute to prospective nanosensors, are being developed (Fuertes et al. 2016b). There have been many advances in this field, like temperature development and gas concentration incorporated into a radio frequency transmitter with the batteryless smart sensor, for *in situ* control of processes of degradation of fishery products (Chung et al. 2017). There is an RFID tag to control the freshness of meat (Eom et al. 2014) and fish (D. Wu et al. 2021), an RFID tag premised on PDMS (silanol-terminated polydimethylsiloxane) for sensing humidity in food-fresh applications (Belsey et al. 2017), and an RFID developed and implemented in conjunction with a compact and long-range batteryless sensor tag to monitor the freshness of packed vegetables (Le et al. 2016). The nanosensors have a constrained measurement field and the advancement of wireless nanosensor networks (WNSNs) is indeed important for the intelligent packaging industry. Recently, Ma et al. (2018) developed the nanostructured conductive polymer (PAni) doped *p*-toluene sulfonate hexahydrate (PTS)-based gas sensor to detect food spoilage in meat samples with a smartphone. In addition, this technique incorporated with near-field communication (NFC) in wireless sensor showed high sensitivity.

6.6 Challenges

The major focus of nanosensor experts is to design economically viable sensors for its field implementation from both the qualitative assessment of nanosensors and classically oriented technologies. Robust distinctions with conventional developments also would help in addressing a connected test including new techniques, assuring major advantages. Nevertheless, the wide adoption of nanosensor technological innovation in new space nano-sensing is a relatively novel area through several inspiring opportunities, both for food manufacturing and regulatory agencies. Certainly, there are many challenges to overcome in order to commercialize this technique effectively. Indeed, ongoing research and development as well as, in particular, a new emphasis on validation in contradiction of the further recognized detection standards, will benefit to strengthen the key role of nanosensing in building a healthier and potentially more desirable food supply worldwide. In addition, with the advancement of the lab-on-a-chip methodology, the incorporation of analytes into a microfluidic chip to establish a biotic micro-electromechanical scheme provides a modern scientific dimension mostly in the biosensors field, since these technologies necessitate negligible sample quantities and provide strong sensitivity of analyte detection. The major drawback of immobilization is indeed the existing resistance to diffusion induced by entraped material, which would result in loss of efficacy. While the world has become severely aware of the impact of food on public health as well as biowar security, the previously mentioned opportunities will become a major advancement in biosensor targets. Moreover, there are also several barriers to be resolved that reduce the advancement of technology development and marketing, connected mostly to the challenges of integrating all of the elements into a single portable system. However, there is a lot of work ahead for full adaptation of this emerging field.

6.7 Conclusions and Future Perspectives

Nanosensors ensure that microorganisms, pathogens and adulterants are detected rapidly and easily, compared to existing types of sensors. However, nanoparticles are beneficial for detecting degradable food products, such as vitamins and antioxidants. Certain other fields of nanoparticles are used as individual package quality determinants and intelligent, robust packaging materials. The use of nanoparticles of the electronic tongue as well as the electronic nose have already prompted the human-like effectiveness of artificial detecting of taste and smell. Remarkably, nanoparticles are of enormous importance in the food industry. These types of nanosensors will enhance food and feed quality, and ultimately minimize health hazards due to the consumption of contaminated food. Developing new sensors by using nanomaterials has been restricted mostly to the analysis. A wide number of applications and nanosensor adaptability can create research choices. Furthermore, based on the merits of nanosensors, comprehensive research is needed to improve the existing nanosensors due to their use and relevance limitations. Additionally, the development of nanosensors requires additional attempts to analyze the food industry at this time. All of these restrictions should always be taken into account while improving established methods. Nanomaterial enhancement of biosensors could allow them to identify foodborne pathogens and toxins at quite low concentrations. The use of nanomaterials for the immobilization of biomolecules as well as the acknowledgment of microorganisms and mycotoxins permits more investigation into increased stability and robustness throughout the development of nanosensors.

References

1 Abdelhaseib, M.U., Singh, A.K., and Bhunia, A.K. (2019). Simultaneous detection of *Salmonella enterica, Escherichia coli* and *Listeria monocytogenes* in food using a light scattering sensor. *Journal of Applied Microbiology* 126 (5): 1496–1507. https://doi.org/10.1111/jam.14225.

2 Abdelrahim, M.Y.M., Benjamin, S.R., Cubillana-Aguilera, L.M. et al. (2013). Study of the electrocatalytic activity of cerium oxide and gold-studded cerium oxide nanoparticles using a Sonogel-Carbon material as supporting electrode: Electroanalytical study in apple juice for babies. *Sensors (Basel, Switzerland)* 13 (4): 4979–5007. https://doi.org/10.3390/s130404979.

3 Aghaei, Z., Ghorani, B., Emadzadeh, B. et al. (2020). Protein-based halochromic electrospun nanosensor for monitoring trout fish freshness. *Food Control* 111: 107065. https://doi.org/10.1016/j.foodcont.2019.107065.

4 Alqarni, S.A., Hussein, M.A., and Ganash, A.A. (2018). Highly sensitive and selective electrochemical determination of Sunset Yellow in food products based on AuNPs/PANI-co-PoAN-co-PoT/GO/Au electrode. *ChemistrySelect* 3 (46): 13167–13177. https://doi.org/10.1002/slct.201802528.

5 Álvarez-Romero, G.A., Lozada-Ascencio, S.M., Rodríguez-Ávila, J.A. et al. (2010). Potentiometric quantification of saccharin by using a selective membrane formed by pyrrole electropolymerization. *Food Chemistry* 120 (4): 1250–1254. https://doi.org/10.1016/j.foodchem.2009.11.072.

6 Alves, R.C., Pimentel, F.B., Nouws, H.P.A. et al. (2015). Detection of Ara h 1 (a major peanut allergen) in food using an electrochemical gold nanoparticle-coated screen-printed immunosensor. *Biosensors & Bioelectronics* 64: 19–24. https://doi.org/10.1016/j.bios.2014.08.026.

7 Amani-Beni, Z. and Nezamzadeh-Ejhieh, A. (2018). Construction of a sensitive non-enzymatic fructose carbon paste electrode – CuO nanoflower: Designing the experiments by response surface methodology. *New Journal of Chemistry* 42 (2): 1021–1030. https://doi.org/10.1039/C7NJ03124F.

8 Baietto, M. and Wilson, A. (2015). Electronic-nose applications for fruit identification, ripeness and quality grading. *Sensors* 15 (1): 899–931. https://doi.org/10.3390/s150100899.

9 Bayramoglu, G., Ozalp, V.C., Oztekin, M. et al. (2019). Rapid and label-free detection of *Brucella melitensis* in milk and milk products using an aptasensor. *Talanta* 200: 263–271. https://doi.org/10.1016/j.talanta.2019.03.048.

10 Belsey, K.E., Parry, A.V.S., Rumens, C.V. et al. (2017). Switchable disposable passive RFID vapour sensors from inkjet printed electronic components integrated with PDMS as a stimulus responsive material. *Journal of Materials Chemistry C* 5 (12): 3167–3175. https://doi.org/10.1039/C6TC05509E.

11 Beran Ames, G.W., Iowa, S.U., Shoeman, H.P. et al. (1991). Food safety: An overview of problems. *Dairy, Food and Environmental Sanitation (USA)* 11: 189–194.

12 Berg, B., Cortazar, B., Tseng, D. et al. (2015). Cellphone-based hand-held microplate reader for point-of-care testing of enzyme-linked immunosorbent assays. *ACS Nano* 9 (8): 7857–7866. https://doi.org/10.1021/acsnano.5b03203.

13 Bhardwaj, J., Devarakonda, S., Kumar, S. et al. (2017). Development of a paper-based electrochemical immunosensor using an antibody-single walled carbon nanotubes bio-conjugate modified electrode for label-free detection of foodborne pathogens. *Sensors and Actuators. B, Chemical* 253: 115–123. https://doi.org/10.1016/j.snb.2017.06.108.

14 Bhunia, K., Sablani, S.S., Tang, J. et al. (2016). Non-invasive measurement of oxygen diffusion in model foods. *Food Research International* 89: 161–168. https://doi.org/10.1016/j.foodres.2016.07.015.

15 Bula-Rudas, F.J., Rathore, M.H., and Maraqa, N.F. (2015). Salmonella infections in childhood. *Advances in Pediatrics* 62 (1): 29–58. https://doi.org/10.1016/j.yapd.2015.04.005.

16 Chang, L.-Y., Chuang, M.-Y., Zan, H.-W. et al. (2017). One-minute fish freshness evaluation by testing the volatile amine gas with an ultrasensitive porous-electrode-capped organic gas sensor system. *ACS Sensors* 2 (4): 531–539. https://doi.org/10.1021/acssensors.6b00829.

17 Chen, D., Liu, Z., Fu, J. et al. (2017). Electrochemical acetylcholinesterase biosensor based on multi-walled carbon nanotubes/dicyclohexyl phthalate modified screen-printed electrode for detection of chlorpyrifos. *Journal of Electroanalytical Chemistry* 801: 185–191. https://doi.org/10.1016/j.jelechem.2017.06.032.

18 Chen, Q., Lin, J., Gan, C. et al. (2015). A sensitive impedance biosensor based on immunomagnetic separation and urease catalysis for rapid detection of *Listeria* *monocytogenes* using an immobilization-free interdigitated array microelectrode. *Biosensors & Bioelectronics* 74: 504–511. https://doi.org/10.1016/j.bios.2015.06.007.

19 Chen, Z., Lin, Y., Ma, X. et al. (2017). Multicolor biosensor for fish freshness assessment with the naked eye. *Sensors and Actuators. B, Chemical* 252: 201–208. https://doi.org/10.1016/j.snb.2017.06.007.

20 Chung, W.-Y., Le, G.T., Tran, T.V. et al. (2017). Novel proximal fish freshness monitoring using batteryless smart sensor tag. *Sensors and Actuators. B, Chemical* 248: 910–916. https://doi.org/10.1016/j.snb.2017.01.134.

21 Dalkiran, B., Kaçar, C., Erden, P.E. et al. (2014). Amperometric xanthine biosensors based on chitosan-CO_3O_4-multiwall carbon nanotube modified glassy carbon electrode. *Sensors and Actuators. B, Chemical* 200: 83–91. https://doi.org/10.1016/j.snb.2014.04.025.

22 Davis, D., Guo, X., Musavi, L. et al. (2013). Gold nanoparticle-modified carbon electrode biosensor for the detection of listeria monocytogenes. *Industrial Biotechnology* 9 (1): 31–36. https://doi.org/10.1089/ind.2012.0033.

23 Dehghani, Z., Hosseini, M., and Mohammadnejad, J. (2019). A fluorescence nanobiosensor for detection of *Campylobacter jejuni* DNA in milk based on Au/Ag bimetallic nanoclusters. *Journal of Food Measurement and Characterization* 13 (3): 1797–1804. https://doi.org/10.1007/s11694-019-00098-4.

24 Dervisevic, M., Custiuc, E., Çevik, E. et al. (2015). Electrochemical biosensor based on REGO/Fe_3O_4 bionanocomposite interface for xanthine detection in fish sample. *Food Control* 57: 402–410. https://doi.org/10.1016/j.foodcont.2015.05.001.

25 Dervisevic, M., Dervisevic, E., Çevik, E. et al. (2017). Novel electrochemical xanthine biosensor based on chitosan–polypyrrole–gold nanoparticles hybrid bio-nanocomposite platform. *Journal of Food and Drug Analysis* 25 (3): 510–519. https://doi.org/10.1016/j.jfda.2016.12.005.

26 Dhull, N., Kaur, G., Jain, P. et al. (2019). Label-free amperometric biosensor for *Escherichia coli* O157: H7detection. *Applied Surface Science* 495: 143548. https://doi.org/10.1016/j.apsusc.2019.143548.

27 Dinshaw, I.J., Muniandy, S., Teh, S.J. et al. (2017). Development of an aptasensor using reduced graphene oxide chitosan complex to detect Salmonella. *Journal of Electroanalytical Chemistry* 806: 88–96. https://doi.org/10.1016/j.jelechem.2017.10.054.

28 Dong, N., Hu, Y., Yang, K. et al. (2016). Development of aptamer-modified SERS nanosensor and oligonucleotide chip to quantitatively detect melamine in milk with high sensitivity. *Sensors and Actuators. B, Chemical* 228: 85–93. https://doi.org/10.1016/j.snb.2015.12.089.

29 Du, P., Jin, M., Zhang, C. et al. (2018). Highly sensitive detection of triazophos pesticide using a novel bio-bar-code amplification competitive immunoassay in a micro well plate-based platform. *Sensors and Actuators. B, Chemical* 256: 457–464. https://doi.org/10.1016/j.snb.2017.10.075.

30 Duncan, T.V. (2011). Applications of nanotechnology in food packaging and food safety: Barrier materials, antimicrobials and sensors. *Journal of Colloid and Interface Science* 363 (1): 1–24. https://doi.org/10.1016/j.jcis.2011.07.017.

31 El-Nour, K.M.A., Salam, E.T.A., Soliman, H.M. et al. (2017). Gold nanoparticles as a direct and rapid sensor for sensitive analytical detection of biogenic amines. *Nanoscale Research Letters* 12 (1): 231. https://doi.org/10.1186/s11671-017-2014-z.

32 El Hassani, N.E.A., Bouchikhi, B., and El Bari, N. (2020). Recent development of an electrochemical imprinted sensor for the detection of trace-level of unmetabolized aflatoxin B2 in dairy milk. *Journal of Electroanalytical Chemistry* 114123. https://doi.org/10.1016/j.jelechem.2020.114123.

33 Eom, K.-H., Hyun, K.-H., Lin, S. et al. (2014). The meat freshness monitoring system using the smart RFID tag. *International Journal of Distributed Sensor Networks* 10 (7): 591812. https://doi.org/10.1155/2014/591812.

34 Fei, J., Dou, W., and Zhao, G. (2015). A sandwich electrochemical immunosensor for *Salmonella pullorum and Salmonella gallinarum* based on a screen-printed carbon electrode modified with an ionic liquid and electrodeposited gold nanoparticles. *Microchimica Acta* 182 (13–14): 2267–2275. https://doi.org/10.1007/s00604-015-1573-x.

35 Feng, H., Li, C., Li, T. et al. (2017). Three-dimensional hierarchical SnO_2 dodecahedral nanocrystals with enhanced humidity sensing properties. *Sensors and Actuators. B, Chemical* 243: 704–714. https://doi.org/10.1016/j.snb.2016.12.043.

36 Freitas, M., Viswanathan, S., Nouws, H.P.A. et al. (2014). Iron oxide/gold core/shell nanomagnetic probes and CdS biolabels for amplified electrochemical immunosensing of *Salmonella typhimurium*. *Biosensors & Bioelectronics* 51: 195–200. https://doi.org/10.1016/j.bios.2013.07.048.

37 Fuertes, G., Soto, I., Carrasco, R. et al. (2016a). Intelligent packaging systems: Sensors and nanosensors to monitor food quality and safety. *Journal of Sensors* 2016: 1–8. https://doi.org/10.1155/2016/4046061.

38 Fuertes, G., Soto, I., Vargas, M. et al. (2016b). Nanosensors for a monitoring system in intelligent and active packaging. *Journal of Sensors*. https://doi.org/10.1155/2016/7980476.

39 Ganeshkumar, R., Cheah, C.W., and Zhao, R. (2017). Annealing temperature and bias voltage dependency of humidity nanosensors based on electrospun $KNbO_3$ nanofibers. *Surfaces and Interfaces* 8: 60–64. https://doi.org/10.1016/j.surfin.2017.04.010.

40 Gevaerd, A., Banks, C.E., Bergamini, M.F. et al. (2020). Nanomodified screen-printed electrode for direct determination of Aflatoxin B1 in malted barley samples. *Sensors and Actuators. B, Chemical* 307: 127547. https://doi.org/10.1016/j.snb.2019.127547.

41 Hall-Stoodley, L., Costerton, J.W., and Stoodley, P. (2004). Bacterial biofilms: From the natural environment to infectious diseases. *Nature Reviews. Microbiology* 2: 95–108. https://doi.org/10.1038/nrmicro821.

42 He, X., Deng, H., and Hwang, H. (2019). The current application of nanotechnology in food and agriculture. *Journal of Food and Drug Analysis* 27 (1): 1–21. https://doi.org/10.1016/j.jfda.2018.12.002.

43 Hryniewicz, B.M., Orth, E.S., and Vidotti, M. (2018). Enzymeless PEDOT-based electrochemical sensor for the detection of nitrophenols and organophosphates. *Sensors and Actuators. B, Chemical* 257: 570–578. https://doi.org/10.1016/j.snb.2017.10.162.

44 Islam, S., Shukla, S., Bajpai, V.K. et al. (2019). Microfluidic-based graphene field effect transistor for femtomolar detection of chlorpyrifos. *Scientific Reports* 9 (1): 276. https://doi.org/10.1038/s41598-018-36746-w.

45 Izadi, Z., Sheikh-Zeinoddin, M., Ensafi, A.A. et al. (2016). Fabrication of an electrochemical DNA-based biosensor for *Bacillus cereus* detection in milk and infant formula. *Biosensors & Bioelectronics* 80: 582–589. https://doi.org/10.1016/j.bios.2016.02.032.

46 Jiang, X., Valdeperez, D., Nazarenus, M. et al. (2015). Future perspectives towards the use of nanomaterials for smart food packaging and quality control. *Particle & Particle Systems Characterization* 32 (4): 408–416. https://doi.org/10.1002/ppsc.201400192.

47 Khairy, M., Ayoub, H.A., and Banks, C.E. (2018). Non-enzymatic electrochemical platform for parathion pesticide sensing based on nanometer-sized nickel oxide modified screen-printed electrodes. *Food Chemistry* 255: 104–111. https://doi.org/10.1016/j.foodchem.2018.02.004.

48 Kim, Y.-J., Kim, H.-S., Chon, J.-W. et al. (2018). New colorimetric aptasensor for rapid on-site detection of *Campylobacter jejuni* and *Campylobacter coli* in chicken carcass samples. *Analytica Chimica Acta* 1029: 78–85. https://doi.org/10.1016/j.aca.2018.04.059.

49 Le, G.T., Tran, T.V., Lee, H.-S. et al. (2016). Long-range batteryless RF sensor for monitoring the freshness of packaged vegetables. *Sensors and Actuators. A, Physical* 237: 20–28. https://doi.org/10.1016/j.sna.2015.11.013.

50 Lee, K.M., Son, M., Kang, J.H. et al. (2018). A triangle study of human, instrument and bioelectronic nose for non-destructive sensing of seafood freshness. *Scientific Reports* 8 (1): 547. https://doi.org/10.1038/s41598-017-19033-y.

51 Li, D., Wang, S., Wang, L. et al. (2019). A simple colorimetric probe based on anti-aggregation of AuNPs for rapid and sensitive detection of malathion in environmental samples. *Analytical and Bioanalytical Chemistry* 411 (12): 2645–2652. https://doi.org/10.1007/s00216-019-01703-7.

52 Li, L., Zheng, H., Guo, L., Qu, L. et al. (2019). Construction of novel electrochemical sensors based on bimetallic nanoparticle functionalized graphene for determination of sunset yellow in soft drink. *Journal of Electroanalytical Chemistry* 833: 393–400. https://doi.org/10.1016/j.jelechem.2018.11.059.

53 Li, Q., Wang, H., Yue, X. et al. (2020). Perovskite nanocrystals fluorescence nanosensor for ultrasensitive detection of trace melamine in dairy products by the manipulation of inner filter effect of gold nanoparticles. *Talanta* 211: 120705. https://doi.org/10.1016/j.talanta.2019.120705.

54 Li, Y., Xiong, Y., Fang, L. et al. (2017). An electrochemical strategy using multifunctional nanoconjugates for efficient simultaneous detection of *Escherichia coli* O157: H7 and *Vibrio cholerae* O1. *Theranostics* 7 (4): 935–944. https://doi.org/10.7150/thno.17544.

55 Liu, H.-Y., Wen, -J.-J., Huang, Z.-H. et al. (2019). Prussian blue analogue of copper-cobalt decorated with multi-walled carbon nanotubes based electrochemical sensor for sensitive determination of nitrite in food samples. *Chinese Journal of Analytical Chemistry* 47 (6): e19066–e19072. https://doi.org/10.1016/S1872-2040(19)61168-0.

56 Lou, Z., Li, F., Deng, J., Wang, L. et al. (2013). Branch-like hierarchical heterostructure (α-Fe_2O_3 /TiO_2): A novel sensing material for trimethylamine gas sensor. *ACS Applied Materials & Interfaces* 5 (23): 12310–12316. https://doi.org/10.1021/am402532v.

57 Lu, Y., Liu, Y., Zhao, Y. et al. (2016). A novel and disposable enzyme-labeled amperometric immunosensor based on MWCNT fibers for listeria monocytogenes detection. *Journal of Nanomaterials* 2016: 1–8. https://doi.org/10.1155/2016/3895920.

58 Ma, Z., Chen, P., Cheng, W. et al. (2018). Highly sensitive, printable nanostructured conductive polymer wireless sensor for food spoilage detection. *Nano Letters* 18 (7): 4570–4575. https://doi.org/10.1021/acs.nanolett.8b01825.

59 Malvano, F., Pilloton, R., and Albanese, D. (2018). Sensitive detection of *Escherichia coli* O157: H7 in food products by impedimetric immunosensors. *Sensors* 18 (7): 2168. https://doi.org/10.3390/s18072168.

60 Manfredi, A., Giannetto, M., Mattarozzi, M. et al. (2016). Competitive immunosensor based on gliadin immobilization on disposable carbon-nanogold screen-printed electrodes for rapid determination of celiotoxic prolamins. *Analytical and Bioanalytical Chemistry* 408 (26): 7289–7298. https://doi.org/10.1007/s00216-016-9494-z.

61 Masdor, N.A., Altintas, Z., and Tothill, I.E. (2016). Sensitive detection of *Campylobacter jejuni* using nanoparticles enhanced QCM sensor. *Biosensors & Bioelectronics* 78: 328–336. https://doi.org/10.1016/j.bios.2015.11.033.

62 Meng, X., Kim, S., Puligundla, P. et al. (2014). Carbon dioxide and oxygen gas sensors. Possible application for monitoring quality, freshness, and safety of agricultural and food products with emphasis on importance of analytical signals and their transformation. *Journal of the Korean Society for Applied Biological Chemistry* 57 (6): 723–733. https://doi.org/10.1007/s13765-014-4180-3.

63 Moradi, S., Azizi-Lalabadi, M., Bagheri, V. et al. (2020). Fabrication of electrospun sensor based on a synthesized component doped into PAN (polyacrylonitrile) nanofibers for electrochemical detection of zearalenone mycotoxin in foods simulant. *Sensing and Bio-Sensing Research* 28: 100321. https://doi.org/10.1016/j.sbsr.2020.100321.

64 Muñoz, N., Bhunia, K., Zhang, H. et al. (2017). Headspace oxygen as a hurdle to improve the safety of in-pack pasteurized chilled food during storage at different temperatures. *International Journal of Food Microbiology* 253: 29–35. https://doi.org/10.1016/j.ijfoodmicro.2017.04.018.

65 Nagabooshanam, S., Roy, S., Mathur, A. et al. (2019). Electrochemical micro analytical device interfaced with portable potentiostat for rapid detection of chlorpyrifos using acetylcholinesterase conjugated metal organic framework using Internet of Things. *Scientific Reports* 9 (1): 19862. https://doi.org/10.1038/s41598-019-56510-y.

66 Najafi, M., Khalilzadeh, M.A., and Karimi-Maleh, H. (2014). A new strategy for determination of bisphenol A in the presence of Sudan I using a ZnO/CNTs/ionic liquid paste electrode in food samples. *Food Chemistry* 158: 125–131. https://doi.org/10.1016/j.foodchem.2014.02.082.

67 Nopwinyuwong, A., Kaisone, T., Hanthanon, P. et al. (2014). Effects of nanoparticle concentration and plasticizer type on colorimetric behavior of polydiacetylene/silica nanocomposite as time-temperature indicator. *Energy Procedia* 56: 423–430. https://doi.org/10.1016/j.egypro.2014.07.175.

68 Ollis, D., Mills, A., and Lawrie, K. (2016). Kinetics of methylene blue (MB) photocatalyzed reduction and dark regeneration in a colorimetric oxygen sensor. *Applied Catalysis. B, Environmental* 184: 201–207. https://doi.org/10.1016/j.apcatb.2015.11.011.

69 Poonlapdecha, W., Seetang-Nun, Y., Wonglumsom, W. et al. (2018). Antibody-conjugated ferromagnetic nanoparticles with lateral flow test strip assay for rapid detection of *Campylobacter jejuni* in poultry samples. *International Journal of Food Microbiology* 286: 6–14. https://doi.org/10.1016/j.ijfoodmicro.2018.07.009.

70 Puligundla, P., Jung, J., and Ko, S. (2012). Carbon dioxide sensors for intelligent food packaging applications. *Food Control* 25 (1): 328–333. https://doi.org/10.1016/j.foodcont.2011.10.043.

71 Qian, X., Qu, Q., Li, L., Ran, X. et al. (2018). Ultrasensitive electrochemical detection of *Clostridium perfringens* DNA based morphology-dependent DNA adsorption properties of CeO_2 nanorods in dairy products. *Sensors* 18 (6): 1878. https://doi.org/10.3390/s18061878.

72 Qiu, X., Lu, L., Leng, J. et al. (2016). An enhanced electrochemical platform based on graphene oxide and multi-walled carbon nanotubes nanocomposite for sensitive determination of Sunset Yellow and Tartrazine. *Food Chemistry* 190: 889–895. https://doi.org/10.1016/j.foodchem.2015.06.045.

73 Rajaji, U., Murugan, K., Chen, S.-M. et al. (2019). Graphene oxide encapsulated 3D porous chalcopyrite ($CuFeS_2$) nanocomposite as an emerging electrocatalyst for agro-hazardous (methyl paraoxon) detection in vegetables. *Composites Part B: Engineering* 160: 268–276. https://doi.org/10.1016/j.compositesb.2018.10.042.

74 Ramachandraiah, K., Han, S.G., and Chin, K.B. (2014). Nanotechnology in meat processing and packaging – Potential applications: A review. *Asian-Australasian Journal of Animal Sciences* 28 (2): 290–302. https://doi.org/10.5713/ajas.14.0607.

75 Ryspayeva, A.A., Jones, T.D., Hughes, P.A. et al. (2018). PEI/Ag as an optical gas nano-sensor for intelligent food packaging. In *2018 IEEE 18th International Conference on Nanotechnology (IEEE-NANO)* (pp. 1–4). IEEE. https://doi.org/10.1109/NANO.2018.8626269.

76 Samphao, A., Butmee, P., Saejueng, P. et al. (2018). Monitoring of glucose and ethanol during wine fermentation by bienzymatic biosensor. *Journal of Electroanalytical Chemistry* 816: 179–188. https://doi.org/10.1016/j.jelechem.2018.03.052.

77 Schaude, C., Meindl, C., Fröhlich, E. et al. (2017). Developing a sensor layer for the optical detection of amines during food spoilage. *Talanta* 170: 481–487. https://doi.org/10.1016/j.talanta.2017.04.029.

78 Sen, N.P. and Baddoo, P.A. (1997). Trends in the levels of residual nitrite in Canadian cured meat products over the past 25 years. *Journal of Agricultural and Food Chemistry* 45 (12): 4714–4718. https://doi.org/10.1021/jf9706641.

79 Sha, Y., Zhang, X., Li, W. et al. (2016). A label-free multi-functionalized graphene oxide based electrochemiluminscence immunosensor for ultrasensitive and rapid detection of *Vibrio parahaemolyticus* in seawater and seafood. *Talanta* 147: 220–225. https://doi.org/10.1016/j.talanta.2015.09.058.

80 Shadjou, R., Hasanzadeh, M., Heidar-Poor, M. et al. (2018). Electrochemical monitoring of aflatoxin M1 in milk samples using silver nanoparticles dispersed on α-cyclodextrin-GQDs nanocomposite. *Journal of Molecular Recognition* 31 (6): e2699. https://doi.org/10.1002/jmr.2699.

81 Shah, A. (2020). A novel electrochemical nanosensor for the simultaneous sensing of two toxic food dyes. *ACS Omega* 5 (11): 6187–6193. https://doi.org/10.1021/acsomega.0c00354.

82 Shelby, T., Sulthana, S., McAfee, J. et al. (2017). Foodborne pathogen screening using magneto-fluorescent nanosensor: Rapid detection of *E. Coli* O157: H7. *Journal of Visualized Experiments* (127): 1–7. https://doi.org/10.3791/55821.

83 Sivasankaran, U., Cyriac, S.T., Menon, S. et al. (2017). Fluorescence turn off sensor for brilliant blue FCF: An approach based on inner filter effect. *Journal of Fluorescence* 27 (1): 69–77. https://doi.org/10.1007/s10895-016-1935-8.

84 Sobhan, A., Lee, J., Park, M.-K. et al. (2019). Rapid detection of *Yersinia enterocolitica* using a single–walled carbon nanotube-based biosensor for Kimchi product. *LWT* 108: 48–54. https://doi.org/10.1016/j.lwt.2019.03.037.

85 Son, M., Kim, D., Kang, J. et al. (2016). Bioelectronic nose using odorant binding protein-derived peptide and carbon nanotube field-effect transistor for the assessment of salmonella contamination in food. *Analytical Chemistry* 88 (23): 11283–11287. https://doi.org/10.1021/acs.analchem.6b03284.

86 Song, X., Xv, L., Wu, L. et al. (2017). Electrochemical immunosensor for detection of *Vibrio parahaemolyticus* based on faradaycage-type anodic stripping voltammetry. *Journal of the Electrochemical Society* 164 (13): B704–B709. https://doi.org/10.1149/2.1961713jes.

87 Tran, Q.T., Phung, T.T., Nguyen, Q.T. et al. (2019). Highly sensitive and rapid determination of sunset yellow in drinks using a low-cost carbon material-based electrochemical sensor. *Analytical and Bioanalytical Chemistry* 411 (28): 7539–7549. https://doi.org/10.1007/s00216-019-02155-9.

88 Üzer, A., Sağlam, Ş., Can, Z. et al. (2016). Electrochemical determination of food preservative nitrite with gold nanoparticles/p-Aminothiophenol-modified gold electrode. *International Journal of Molecular Sciences* 17 (8): 1253. https://doi.org/10.3390/ijms17081253.

89 Vaisocherová-Lísalová, H., Víšová, I., Ermini, M.L. et al. (2016). Low-fouling surface plasmon resonance biosensor for multi-step detection of foodborne bacterial pathogens in complex food samples. *Biosensors & Bioelectronics* 80: 84–90. https://doi.org/10.1016/j.bios.2016.01.040.

90 Vasagar, B. and Leonard, M. (2014). US perspective on gluten-related diseases. *Clinical and Experimental Gastroenterology* 25. https://doi.org/10.2147/CEG.S54567.

91 Wang, C.-Y., Zeng, Y., Shen, A.-G. et al. (2018). A highly sensitive SERS probe for bisphenol A detection based on functionalized Au@Ag nanoparticles. *Analytical Methods* 10 (47): 5622–5628. https://doi.org/10.1039/C8AY01966E.

92 Wang, D., Chen, Q., Huo, H. et al. (2017). Efficient separation and quantitative detection of *Listeria monocytogenes* based on screen-printed interdigitated electrode, urease and magnetic nanoparticles. *Food Control* 73: 555–561. https://doi.org/10.1016/j.foodcont.2016.09.003.

93 Wang, H.O.N.G., Wang, L., Hu, Q. et al. (2018). Rapid and sensitive detection of *Campylobacter jejuni* in poultry products using a nanoparticle-based piezoelectric immunosensor integrated with magnetic immunoseparation. *Journal of Food Protection* 81 (8): 1321–1330. https://doi.org/10.4315/0362-028X.JFP-17-381.

94 Wang, H., Xiu, Y., Chen, Y. et al. (2019). Electrochemical immunosensor based on an antibody-hierarchical mesoporous SiO_2 for the detection of *Staphylococcus aureus*. *RSC Advances* 9 (28): 16278–16287. https://doi.org/10.1039/C9RA00907H.

95 Wang, L., Huo, X., Qi, W. et al. (2020). Rapid and sensitive detection of *Salmonella typhimurium* using nickel nanowire bridge for electrochemical impedance amplification. *Talanta* 211: 120715. https://doi.org/10.1016/j.talanta.2020.120715.

96 Wang, M., Gao, Y., Sun, Q. et al. (2014). Sensitively simultaneous determination of sunset yellow and Tartrazine in foods based on polypyrrole modified oxidized single-walled carbon nanotubes. *Journal of the Electrochemical Society* 161 (14): B297–B304. https://doi.org/10.1149/2.0081414jes.

97 Wang, M. and Zhao, J. (2015). Facile synthesis of Au supported on ionic liquid functionalized reduced graphene oxide for simultaneous determination of Sunset yellow and Tartrazine in drinks. *Sensors and Actuators. B, Chemical* 216: 578–585. https://doi.org/10.1016/j.snb.2015.04.053.

98 Wang, M., Duong, B., Fenniri, H. et al. (2015). Nanomaterial-based barcodes. *Nanoscale* 7 (26): 11240–11247. https://doi.org/10.1039/C5NR01948F.

99 Wang, Q., Long, M., Lv, C. et al. (2020). Lanthanide-labeled fluorescent-nanoparticle immunochromatographic strips enable rapid and quantitative detection of *Escherichia coli* O157: H7in food samples. *Food Control* 109: 106894. https://doi.org/10.1016/j.foodcont.2019.106894.

100 Wang, S., Liu, Q., Li, H. et al. (2016). Fabrication of label-free electrochemical impedimetric DNA biosensor for detection of genetically modified soybean by recognizing CaMV 35S promoter. *Journal of Electroanalytical Chemistry* 782: 19–25. https://doi.org/10.1016/j.jelechem.2016.09.052.

101 Wang, Z., Yao, X., Wang, R. et al. (2019). Label-free strip sensor based on surface positively charged nitrogen-rich carbon nanoparticles for rapid detection of *Salmonella enteritidis*. *Biosensors & Bioelectronics* 132: 360–367. https://doi.org/10.1016/j.bios.2019.02.061.

102 Wu, D., Zhang, M., Chen, H. et al. (2021). Freshness monitoring technology of fish products in intelligent packaging. *Critical Reviews in Food Science and Nutrition* 61: 1279–1292. https://doi.org/10.1080/10408398.2020.1757615.

103 Wu, H., Zuo, Y., Cui, C. et al. (2013). Rapid quantitative detection of *Brucella melitensis* by a label-free impedance immunosensor based on a gold nanoparticle-modified screen-printed carbon electrode. *Sensors* 13 (7): 8551–8563. https://doi.org/10.3390/s130708551.

104 Wu, H., Luo, Y., Hou, C. et al. (2019). Rapid and fingerprinted monitoring of pesticide methyl parathion on the surface of fruits/leaves as well as in surface water enabled by gold nanorods based casting-and-sensing SERS platform. *Talanta* 200: 84–90. https://doi.org/10.1016/j.talanta.2019.03.004.

105 Wu, Y., Huang, Q., Nie, J. et al. (2019). All-carbon based flexible humidity sensor. *Journal of Nanoscience and Nanotechnology* 19 (8): 5310–5316. https://doi.org/10.1166/jnn.2019.16821.

106 Wu, Z., Guo, F., Huang, L. et al. (2018). Electrochemical nonenzymatic sensor based on cetyltrimethylammonium bromide and chitosan functionalized carbon nanotube modified glassy carbon electrode for the determination of hydroxymethanesulfinate in the presence of sulfite in foods. *Food Chemistry* 259: 213–218. https://doi.org/10.1016/j.foodchem.2018.03.080.

107 Xiang, C., Li, R., Adhikari, B., She, Z. et al. (2015). Sensitive electrochemical detection of *Salmonella* with chitosan–gold nanoparticles composite film. *Talanta* 140: 122–127. https://doi.org/10.1016/j.talanta.2015.03.033.

108 Xiang, Y., Camarada, M.B., Wen, Y. et al. (2018). Simple voltammetric analyses of ochratoxin A in food samples using highly-stable and anti-fouling black phosphorene nanosensor. *Electrochimica Acta* 282: 490–498. https://doi.org/10.1016/j.electacta.2018.06.055.

109 Yang, L., Wang, L., Li, K. et al. (2014). Sensitive voltammetric determination of neohesperidin dihydrochalcone based on SWNTs modified glassy carbon electrode. *Analytical Methods : Advancing Methods and Applications* 6 (23): 9410–9418. https://doi.org/10.1039/C4AY02004A.

110 Yazdanparast, S., Benvidi, A., Abbasi, S. et al. (2019). Enzyme-based ultrasensitive electrochemical biosensor using poly(l-aspartic acid)/MWCNT bio-nanocomposite for xanthine detection: A meat freshness marker. *Microchemical Journal* 149: 104000. https://doi.org/10.1016/j.microc.2019.104000.

111 Ye, L., Zhao, G., and Dou, W. (2018). An electrochemical immunoassay for *Escherichia coli* O157: H7 using double functionalized Au@Pt/SiO_2 nanocomposites and immune magnetic nanoparticles. *Talanta* 182: 354–362. https://doi.org/10.1016/j.talanta.2018.01.095.

112 Zhang, Z., Lin, M., Zhang, S. et al. (2013). Detection of aflatoxin M1 in milk by dynamic light scattering coupled with superparamagnetic beads and gold nanoprobes. *Journal of Agricultural and Food Chemistry* 61 (19): 4520–4525. https://doi.org/10.1021/jf400043z.

113 Zheng, F., Wang, P., Du, Q. et al. (2019). Simultaneous and ultrasensitive detection of foodborne bacteria by gold nanoparticles-amplified microcantilever array biosensor. *Frontiers in Chemistry* 7: 232. https://doi.org/10.3389/fchem.2019.00232.

114 Zhong, M., Yang, L., Yang, H. et al. (2019). An electrochemical immunobiosensor for ultrasensitive detection of *Escherichia coli* O157: H7 using CdS quantum dots-encapsulated metal-organic frameworks as signal-amplifying tags. *Biosensors & Bioelectronics* 126: 493–500. https://doi.org/10.1016/j.bios.2018.11.001.

115 Zhu, X., Xiao, Y., Jiang, X. et al. (2016). A ratiometric nanosensor based on conjugated polyelectrolyte-stabilized AgNPs for ultrasensitive fluorescent and colorimetric sensing of melamine. *Talanta* 151: 68–74. https://doi.org/10.1016/j.talanta.2016.01.012.

7

Harnessing Genetic Diversity for Addressing Wheat-based Time Bound Food Security Projections

A Selective Comprehensive Practical Overview

Abdul Mujeeb-Kazi[1], Niaz Ali[2], Ian Dundas[3], Philip Larkin[4], Alexey Morghonov[5], Richard R-C Wang[6], Francis Ogbonnaya[7], Hanif Khan[8], Nasir Saeed[9], Shabir Wani[10], Mohammad Sohail Saddiq[11], Mohammad Jamil[12], Abdul Aziz Napar[13], Fatima Khalid[14], Mahjabeen Tariq[15], Rumana Keyani[16], Zeeshan Ali[17] and Sanjaya Rajaram[18]

[1] *College Station, Texas A&M, Texas, USA*
[2] *Department of Botany, Hazara University (HU), Manshera-21300, Khyber Pukhtunkhwa, Pakistan*
[3] *School of Agriculture, Food and Wine, Waite Campus, The University of Adelaide, Australia*
[4] *CSIRO, Agriculture and Food, CSIRO, Canberra, Australia*
[5] *Omsk State Agrarian University, Omsk 644008, Russia*
[6] *USDA/ARS, Logan, Utah, USA*
[7] *GRDC, Australia*
[8] *Indian Institute of Wheat and Barley Research, ICAR, Karnal 13200, Haryana, India*
[9] *National Institute of Biotechnology and Genetic Engineering (NIBGE), Faisalabad, Pakistan*
[10] *Sher-e-Kashmir University of Agricultural Sciences and Technology of Kashmir. Jammu and Kashmir, India*
[11] *Department of Agronomy, Ghazi University, Dera Ghazi Khan, Pakistan*
[12] *University of Sargodha, Sargodha, Pakistan*
[13] *Sindh University, Jamshoro, Sindh, Pakistan*
[14] *Quaid-e-Azam University (QAU), Islamabad, PAkistan*
[15] *Arid University, Rawalpindi, Pakistan*
[16] *COMSATS, Islamabad, Pakistan*
[17] *National Agricultural Research Centre, Islamabad, Pakistan*
[18] *Resource Seeds International, Mexico DF, Mexico*

CHAPTER MENU

Bioresource Technology: Concept, Tools and Experiences, First Edition. Tanveer Bilal Pirzadah, Bisma Malik, Rouf Ahmad Bhat, and Khalid Rehman Hakeem.
© 2022 John Wiley & Sons Ltd. Published 2022 by John Wiley & Sons Ltd.

7.1 The Global Wheat Scenario

Global wheat production increased from 220 million tons in 1961 to 764.5 million tons in 2018 at 3.52 mt ha^{-1}. There is still a gap between the prevalent 1.0% annual genetic gain and the demand of 1.7%, a target challenged by the unknown magnitude of possible negative effects that are off-shoots of climate change. Crop improvement exploits genetic diversity and has been voiced in many circles that since wheat possesses a narrow genetic base, we may have reached the maximum limit from which the way forward will require a widening of this genetic base and then identifying the practicality of the new variability for increased yield potential. The main question would be as to whether wheat yield per se can be increased sufficiently to meet the expected 1.7% genetic gain demand. a gain indicative that the 1 billion tons wheat production 2050 target will be achieved. Expectations are that the new diversity harnessed possesses allelic richness that not only improves productivity but could also counter constraints emerging from abiotic and biotic stresses, be micronutrient rich and cover minor aspects that influence wheat production as well as being compatible with wheat cultivation across all global mega- and micro-environments. These are aspects that will attract global researchers to exploit the new germplasm if it can provide a complete novel and diverse varietal component. Practical outputs of harnessing new genetic diversity would have to meet the challenges of global wheat production that currently embraces 764.493 million tons at 3.52 mt ha^{-1} for 7.8 billion people (USDA) and for 2050 to meet the genetic gain demand of 1.7% per annum (UN 2019). This review and our experimental findings are inclined to convey an optimistic picture, a way forward that is very doable, and to emphasize the fact that research creativity and operational integration focused on productivity is paramount for bringing a change that will deliver the products necessary to combat wheat production constraints posed by climate change. Hence, new germplasm and breakthrough technologies will be essential tools with applied time bound goals. Historically, gene pool widening efforts have been in the scientific arena since the mid-1870s, rigorously investigated for practically a century thereafter, generating outputs versus time that allow us today to share positive perceptions where we can well integrate these diverse outputs. Creative science in conjunction with novel diversity, innovative technologies, allelic richness and distant hybridization, etc. are all within the framework of wide crosses that started in 1876 and have generated unique germplasms and have multiplied our arsenal for combating wheat production constraints. The measure of the wide crosses research lies in the fact that the elucidated diversity of the outputs is pivotal for current wheat production over a cultivated area of 217 million hectares, influencing the lives of 7.8 billion people that consume wheat in any form. This comprehensive review is indicative of how the new diversity will be able to influence achieving high yields, providing broad adaptation, possess multifaceted trait variation and embody varietal output resilience. Our aspiration is that a smart package of novel diversity be on hand to impact all current wheat production constraints, offer futuristic protection to optimistically provide an "evergreen" revolution well covering the suggested 4D's of wheat, i.e., domestication, diversity, discovery and delivery (BBC NEWS). These aspects are all embedded in the broader version of wheat-wide crosses and forms the theme of this contribution which is heavily slanted toward practical output requirements at the field level. Nevertheless, we must stay cognizant of all conventional strategies that are given a brief coverage here, the base on which novel contributions would be additive and identifying stringent smart protocols that offer rapid outputs to be employed, thus minimizing efforts that dominate academic basic research endeavors rather than the practical gains that are measured by tons/hectare in fields output performance.

7.2 Food Security: The Challenge of Feeding Over 9 Billion by 2050

Annual wheat production has spread from 592 million metric tons in 1990/91 to the current 765.41 million metric tons in 2019/20. Global wheat yield in 2015/16 was 735.9 million metric tons, with the figures indicative that annual increases are not proportional. A drastic drop in 2018/19 to 731.45 million metric tons from 762.88 in 2017/18 is disconcerting and suggests that a strategy is required that can be more inclined toward increasing yields by incorporation of any additive multidisciplinary improvement protocol/s. The surge that uplifted wheat production in the last century through the efforts of Dr Borlaug and his global partners was founded on the smart exploitation of the dwarfing gene Norin 10 that in nearly two decades resulted in the generation of semi-dwarf wheats by 1967 in Pakistan and India, known as the Green Revolution, earning him the 1970 Nobel peace prize. In essence, it was novel genetic diversity exploited conventionally, tested on-farm performance and put to superb use. This contribution subsequently led to incorporation of novel diversity of a winter wheat resource in the mid-1970s, also in CIMMYT, Mexico when Dr Rajaram engineered the creative winter/spring wheat cross in an attempt to widen the spring wheat gene pool conventionally, where the major limitation was getting the two growth habits of the wheat initial parents to niche. The final parentage involved four wheats: KAVKAZ (with a wheat/alien chromosome translocation; T1BL.1RS), BUHO, KALYANSONA and BLUEBIRD. Derivatives from these pedigree combinations significantly enhanced wheat productivity globally, supporting the fact that genetic diversity was paramount for crop improvement. These efforts were under normal field-based hybridization and were conventionally managed. It was, however, the success of the Green Revolution that triggered curiosity to harness other forms of diversity, leading to vibrant distant hybridization undertakings. The strategy involved various levels of complexities, and opened up investigative areas during the late 1960s, and until the early 1980s was mainly an academic exercise. Little attention was given to linkage drag and the attributes that the alien parent would contribute, as to obtain a hybrid was itself a huge hurdle. Thus, the entire spectrum of diversity across genomes and polyploidy levels within the Triticeae members was open for utilization. Until the early 1980s, limited success ensued, where the most popular combinations involved durum wheat, bread wheat, rye, oats and barley around prominent wide cross teams located in Denmark, Australia, Canada, the USA and Mexico. The major objective of their ventures was primarily to obtain well-documented hybrids with the prime obstacle being to overcome cross-incompatibility. This basic bottleneck was significantly overcome by the early t1980s and swiftly led to the infusion of strategic goals that allowed researchers to envision practical food productivity gains and link these with time-bound futuristic goals.

Food security targets for 2050 have been projected and food supply for approximately 9.7 billion inhabitants on Earth has to be ascertained. The current annual wheat production estimates are slightly less than 780 million tons, where an increase of 1.7% per annum to deliver 1 billion tons by 2050 is targeted, an annual increase that has fluctuated over the past few years. The yield enhancement modes encompass various means that embrace research, management and inputs of multiple disciplinary facets to overcome prevalent yield gaps across many countries. Can the stagnancy bottlenecks be countered, and how? It first behoves us to determine if increase within the ideal experimental spring wheat plot yields of approximately ten tons per hectare is conventionally possible, and if not, rigorous exploitation strategies of novel genetic diversity has the potential for enhancing yield coupled by harnessing other influencing variables capable of contributing toward varietal development via their integration for augmenting holistic farm-based returns.

The latter option now has high priority as conventional strategies are not enough. Novel genetic resources are allele rich and we shall essentially focus on their priority usage in an attempt to elucidate a way forward for reaching the 2050 target of food security. Wheat is the conduit, unequivocally recognized globally, and it is fortuitous that the Triticeae family has a huge spectrum of genera and accessional resource that offer immense diversity for enriching its genetic complement. The supporting protocols for research-based yield increases, including first conventional and then aided by practical innovative aspects. This chapter provides a comprehensive coverage of vital contributions from the late eighteenth century. The early efforts shed light on the complexities of generating quality products that are well integrated in a varietal package. Hence, our focus shoulders stringently on the practicality of the challenge of food security that is threatened by various climate change constraints and ensure that well integrated outputs are made over a wide spectrum of characteristics. Though basic research efforts are important for generating scientific contributions, emphasis is justified toward demonstrating the way forward, i.e., practical output gains that translate innovative research involvements within a tight time frame by enriching new and resilient wheat varieties with unique alleles. We hope that our message well projects that food is in dire demand globally and the luxury of basic science providing promises of futuristic benefits/advantages stand secondary. It is prudent to express that current experimentations must exude timely practical returns measured as tons per hectare in fields, a challenge that is time bound requiring regular annual increments to meet the projected demands of 2050 food security. We thus proceed with futuristic demands as the conventional route that leads into innovative creative dimensions. When both these operate in unison and are well integrated, we are very likely to deliver practically useful products linked with sustainable food security for the burgeoning populace.

7.3 Conventional Wheat Improvement Strategies

7.3.1 Breeding Methods

Mass selection in wheat has been practiced by farmers since cereal domestication, without scientific information on hand to guide on how to select and maintain progeny lines. In the pre-Mendelian era, many wheat cultivars were developed by identifying and selecting superior genotypes from mixed cultivars or landraces. Breeding efforts in this era were limited by the insufficient knowledge and understanding of wheat biology and the genetic basis of key agronomic traits. The practice of hybridization followed by selection as a wheat improvement strategy was initiated by Vilmorin in France around 1874 (Vilmorin 1880), followed in 1875 by Wilhelm Rimpau in Germany (Bonjeau and Angus 2001). The understanding that selection was effective only upon heritable variation led early plant breeders to use the nascent understanding of sexuality of plants to cross different wheat varieties. Wheat breeding was put on a scientific foundation following the rediscovery of Mendel's laws of heredity in 1900. Johannsen's (1903) work with seed yield in beans led to the development of the pure line theory of selection. Geneticists and breeders then looked at the genetic basis of important traits. Demonstration of the simple Mendelian inheritance of resistance for stripe rust in rivet wheat was studied (Biffen 1905), enabling wheat breeders to select homogeneous resistance in cultivars, and thus started a human-guided management system for host resistance genes.

The basic breeding procedures applied in conventional wheat improvement include: i) introduction and collection of germplasm; ii) pure line selection; iii) hybridization; iv) mutation breeding; and v) hybrid cultivar development. Backcrossing is often employed to add desirable genes to established cultivars. The major goals in wheat breeding are: i) improvement of productivity in unit area and time; ii) stabilizing yield levels by minimizing losses due to abiotic and biotic stresses; and iii) providing grains of acceptable end-use quality. Wheat being an autogamous crop, conventional breeding involves searching for a superior, homozygous genotype or a group of homozygous genotypes. The chances for transgressive improvements in crosses within related, already highly adapted material are high due to the polyploid nature of its genome. The conventional methods can be classified as the pedigree or line method and bulk method. All other methods involve principles of these two approaches. These breeding methods aim toward: i) recombination of genes from different parents; and ii) fixation of genotypes with or without selection/s.

Recombination breeding consists of controlled crossing between parents of choice followed by pedigree or mass-pedigree selections in the segregating generations for targeted trait(s), which is the widely employed approach in wheat breeding. The rapid fixation of wheat genotypes can be achieved by: i) shuttle breeding; ii) speed breeding; and iii) doubled haploid breeding.

Initial wheat cultivar development and release was undertaken via private ventures primarily in Europe. In India, the systematic wheat improvement work was started by Howard and Howard in 1905 with the establishment of the Imperial Agricultural Research Institute at Pusa in Bihar (Howard and Howard, 1910). The breeding priority has changed over decades from purification of landraces placing emphasis on semi-dwarf wheats, resistance to rust diseases, lodging, drought and end-use quality. Wheat cultivars released in the latter half of the twentieth century and in the twenty-first century are primarily pure lines, with relatively few hybrids and multiline cultivars (Baenziger and DePauw 2009).

Genetic variability is the feedstock for improvement in the desired direction of crop plants, no matter what breeding approach is resorted to. Over the decades, a long history of wheat breeding, and natural and induced variability, have been effectively exploited for crop improvement. There are many different ways in which wheat breeders manage material flow in their breeding programs, each with pros and cons. Preference of method depends on many factors, including cost of cultivar development, which makes the shortening of the breeding cycle a high priority, resources accessible to the breeding program, and the availability of efficient selection methodologies for the target traits.

Section 7.3.2 outlines mechanisms to generate heritable genetic variation and subsequently examines methods for its assessment and utilization in wheat breeding. Genetic variation originates via genetic recombination (through sexual hybridization), manipulation in chromosome composition or number, and mutations. With advances in the understanding of biological mechanisms and technology (e.g., recombinant DNA, gene transfer, somaclonal variation, gene-editing), new mechanisms have been developed to generate and expand genetic variability, that is, variation that does not exist within the species and its wild relatives.

7.3.2 Recombination Breeding

A key step in wheat cultivar development is to generate heritable genetic variability. By far the most common way of creating heritable variation is to cross two or more wheat genotypes that have complementary traits. Recombination breeding consisting of controlled crossing between selected parents followed by selections employing pedigree or bulk method of selection or a combination of bulk and pedigree method in the segregating generations. Unlike pure line selection, which is limited to identification of the best in the naturally available variability, recombination breeding is a tool to generate variability and recombine desired traits for any given condition. Recombination breeding is employed in different forms depending on the breeding objectives and genetics of the trait(s) to be recombined. For simply inherited traits like resistance to rust or powdery mildew disease, plant height, waxiness, etc. depending upon the agronomic potential of the parents, either single cross or backcross followed by pedigree selection is preferred. For quantitatively inherited traits in wheat like yield, tolerance to complex abiotic stresses like drought, heat, lodging, etc., single or multi-parent crosses followed by selection through pedigree or bulk method is the usual practice. If the breeding objective is to combine a set of traits that make a variety ideal to a given situation or wider adaptability, a "convergent breeding" approach is followed. This involves stepwise accumulation of component traits. Variety with resistance to three rusts (stripe, leaf and stem) of wheat is a good example of products of convergent breeding.

One approach to parent selection and/or progeny is to follow an ideotype concept. Donald (1968) proposed the ideotype approach of breeding for wheat and several other crops, in addition to two existing breeding philosophies of "defect elimination" and "selection for yield." Donald aimed at wheat ideotypes capable of high grain yield when grown as a crop community and under high sowing density. Subsequently, several other researchers emphasized developing new ideotypes for wheat that will provide a basis for a better understanding of crop ecology and physiology. In order to develop high-yielding wheat ideotypes, several traits have been identified as being crucial for improvement of wheat yield potential. These traits include improved photosynthetic capacity and efficiency, lodging resistance, optimal developmental pattern to maximize spike fertility, optimal partitioning to grain, improved grain filling, potential grain size and many others.

Wheat cultivars developed by hybridization often have complex pedigrees due to the practice of crossing superior selections from previous crosses. The repetitive sequence of crossing, selection and then intercrossing superior selections as parents in a new generation of crosses constitutes a form of recurrent selection in which a selection cycle may take 5 to 8 years. In a way, the ongoing worldwide effort, including CIMMYT's wheat breeding program to breed varieties for the major wheat ecologies or "Mega Environments," requires stepwise accumulation of the desired characteristics that would combine high and stable yield, adaptability and acceptable quality also as a convergent approach. The principal philosophy in the wheat breeding programs is a broad-based recurrent selection system where the best genotypes are cycled back into the crossing program based on performance at many locations (Dubin and Rajaram 1996). This system retains lines resistant or tolerant to important diseases.

7.3.3 Pedigree or Line Breeding

The two main methods of conducting selections in segregating populations of wheat were developed in Europe more than 100 years ago. The first of them, the genealogical, involves selection simultaneously with generation advance. It was developed at the Swedish Seed Association in Svalof, Sweden, by Hjalmar Nilson at the end of the nineteenth century. William Farrer of Australia developed several important wheat cultivars during the late nineteenth century, using a precursor of the pedigree method that featured hybridization to combine desirable characters of two landraces (Hayes et al. 1955).

The pedigree method is derived from two basic concepts: hybridization and pureline selection of Johannsen (1903). The important features of the pedigree method are the detailed study of individuals during the segregating generations and the records kept so that parent-progeny relations can be followed. In this method, the F_2's are space planted so that the record of individual plant can be kept. Some characters with high heritability, such as rust resistance, leaf angle, growth habit, plant height, flowering, maturity and seed size can be selected. During the segregating generations, disease and insect epidemics can be created artificially to select resistant lines and plants. F_2 or F_3 generations can be grown in areas with abiotic stresses, such as drought, terminal heat and salinity/alkalinity. The progeny from each F_2 selected plant are sown as F_3 rows. Generally, F_3 head-rows from each of the selected F_2 plants are planted at a row-to-row spacing of 25–30 cm. The size of each F_3 is decided by the resources available. Selection from F_3 onwards is based on line performance. Two to ten plants per line are retained from best performing F_3 lines and grown for the F_4 generation. In the F_4 and F_5, evaluation and selection is made on the family basis. Any line that is phenotypically similar can be bulk harvested and the progeny used for yield testing. If necessary, the F_6 generation can be handled in a similar way to F_4 and/or F_5.

As wheat genotypes respond differently to changes in plant density, the ones selected for their superior performance under low density might not give similar performance when tested under high density of commercial production. The problems caused by selection and testing in differential planting densities are, thus, more pronounced for yield. A dismal 81% failure of newly bulked lines against checks was observed in Indian wheat material evolved by the pedigree method (Kulshrestha 1989).

7.3.4 Bulk Method

Bulk selection, a concept developed by the Swedish wheat breeder Nilsson-Ehle in 1908, as described by Newman (1912), compared with the pedigree method by Love (1927), is akin to mass selection in that plants are chosen which express individual advantages, and a sample of the seed aggregate is propagated in the next inbreeding cycle. The bulk method of breeding depends primarily on natural selection during segregating generations to increase the frequency of superior genotypes. The segregating generations are handled as bulks and not as single plants, up to about F_6 generation. Selection of plants or spikes in the F_6 or F_7 generation will provide fairly homozygous lines. Although this procedure requires less labor and is cost-effective, it is crucial to carry on with a large quantity of seed to give a satisfactory representation of genotypes, since both superior and inferior genotypes will be included. Genetic studies on bulk selection in wheat have shown that natural selection exerts a dynamic influence on the population composition at each generation, resulting in changes in gene frequencies as the hybrid moves toward homozygosity (Tee and Qualset 1975).

7.3.5 Single Seed Descent (SSD) Method

Golden (1939) suggested the process which Brim (1966) called Single Seed Descent (SSD). SSD is a method to achieve homozygosity, while often practicing minimal selection. In the SSD method, a single seed is taken from each of the F_2 plants and the seeds are grown to produce the next generation. In F_2 and the subsequent generation, selection can be done for the traits that have high heritability. The objective is to reduce the amount of material carried from one generation to the next and speed up generation advancement. Crosses are advanced for 5–6 generations using SSD. At each generation, some plants are lost due to various reasons. Care should be taken to select a sufficient number of plants in F_2 to provide a reasonable number of SSD-derived lines. In the final generation, plants are space planted to produce sufficient seed to begin testing of the homozygous lines. Evidence from several studies shows that SSD is a useful breeding method, particularly when heritability is low and early generation selection is ineffective. Because SSD samples only one seed per plant, the probability of selecting individuals with the maximum number of desirable alleles is greatly reduced. There are many alternative schemes like mechanical mass selection, spike selection, simplified pedigree selection of bulking at F_3 stage, etc. Knott and Kumar (1975) compared SSD with early-generation yield testing in wheat and found that, in general, early-generation yield testing identified higher yielding lines on average. In addition to being used in applied breeding wheat programs, SSD is often used to create mapping populations (Marza et al. 2006; Thépot et al. 2015), where a random sample of lines needs to be developed which best represents the genetic segregation in the F_2. In this case, SSD is done strictly without selection.

7.3.6 Backcross Breeding

Backcrossing is a method of recurrent hybridization by which a desirable allele for a specific trait is substituted for the alternative or undesirable allele. This basic element in recombination strategies, of shifting population means toward one parent, was first described by Harlan and Pope (1922). Despite the advantages and successes, backcross breeding has not gained wide acceptance. A strategy combining single backcrossing with selected bulk breeding has been successfully used and recommended in wheat improvement at CIMMYT to introgress rust resistant genes from donor parents to elite adapted cultivars (Wang et al. 2009). This breeding strategy has advantages in retaining or improving the adaptation of the recurrent parents, and at the same time transferring most of the desired donor genes in a wide range of situations. With the advent of DNA markers, marker assisted backcrossing (MABC) has been successfully deployed in many wheat breeding programs. It includes marker-assisted foreground selection along with phenotypic selection to identify plants carrying desirable genes in segregating populations and background selection, utilizing the polymorphic markers to ensure the satisfactory recurrent parent genome (RPG) recovery.

7.3.7 Modified Pedigree Bulk

IUYUO;POPedigree selection was primarily used in wheat breeding from 1945 to 1985 (Wang et al. 2003). From 1985 until the second half of the 1990s, the main selection method was a modified pedigree-bulk method (MODPED) (Van Ginkel et al. 2002). In MODPED, the crosses may be single (A × B), three-way (A × B × C) or limited backcross

(A × B × A). Only the best F_1's are promoted to F_2. In this method, the F_2 populations are space planted. Within the better F_2 populations, the best plants are selected by the breeders and experienced research assistants during 3–5 rounds of selection based on good agronomic type, "durable" disease resistance, synchronous tillering, desired spike type, good fertility, appropriate height and maturity. F_3 to F_5 lines are identified for phenotypic and disease resistance superiority. Then 10 heads are harvested and bulked. From F_5, the material is treated as the pedigree method. In F_5, 10 single heads are selected but planted individually as F_6 lines. Bulk selections are made in F_6 or F_7 and those with appropriate characteristics are put into yield experiments. This modification of the bulk method provides an opportunity for the breeder to exercise skill and judgment in selecting superior plants in early generations, but avoids the time and labor required in the pedigree method.

7.3.8 Selected Bulk

In an attempt to improve resource-use efficiency, CIMMYT replaced modified pedigree-bulk method for wheat breeding by the selected bulk method (SELBLK) (Van Ginkel et al. 2002) in the late 1990s. In this method, spikes of selected F_2 plants within one cross are harvested in bulk and threshed together, resulting in one F_3 seed lot per cross. This selected bulk selection is also used from F_3 to F_5, while pedigree selection is used only in the F_6. Assuming that plant density is similar, SELBLK used two-thirds of the land allocated to the MODPED and produces smaller numbers of families. Therefore, when SELBLK is applied, fewer seed bulks need to be handled at both harvest and sowing, resulting in significant saving in labor, time and cost.

7.3.9 Multiline Breeding

Wheat varieties developed via the pure-line breeding method quickly succumbed to the new races of the rust pathogen. The use of planned heterogeneity in self-pollinated crop varieties to provide stability of protection to diseases via host resistance was proposed by Rosen (1949) and refined by Jensen (1952). The concept of using multiline varieties in oat breeding by intra-varietal diversification was conceived as a method for reducing the risk of losses in crop productivity from attacks by unknown pathogens (Jensen 1952). By definition, a multiline is a mixture of near isogenic lines which are agronomically uniform but carry diversified genes of resistance against disease-causing organisms. Each component line possesses a different gene for vertical resistance; hence, are resistant to a large proportion of the pathogen population. Multiline cultivar can be developed by selecting a high yielding, well adapted cultivar and by incorporating the resistance genes from various genetically different sources through backcrossing. Borlaug (1958) initiated a multiline breeding program for bread wheat in Mexico as a strategy for providing consistent and persistent protection of this crop to rust diseases. He proposed to blend of 8 to 16 isogenic lines (each with a different stem-rust resistance gene) into a phenotypically uniform multiline variety of wheat. Borlaug (1958) also proposed the concept of growing mixtures of different cultivars carrying diverse resistance genes. The idea behind the "cultivar mixture" strategy was similar to that of multilines, except that the multilines are more laborious to construct and manage. Although multiline wheat varieties were not released in Mexico, "Miramar 63" and "Miramar 65" varieties that were heterogeneous for resistance to stem and stripe rusts, were released in Colombia (Rockefeller Foundation 1965). During the 1970s, a number of wheat multilines were developed in the USA and India. In India, three multilines of the popular wheat "Kalyansona," namely KSML-3, MLKS-11 and KML 7406 (Bithoor) were developed through cooperative international effort between the CIMMYT and the Indian programs in 1978 (Gill et al. 1981) and were released for use in some wheat-growing areas in the northern plains of India. Sonalika multiline SKML-1 was developed for late-sown conditions of cultivation (Gill et al. 1982). Results from multilines in cereals showed that they yield higher than pure line cultivars in the face of biotic stresses (Barrett 1978). Multiline KSML-3 significantly out-yielded the recurrent parent Kalyansona for four consecutive years from 1975 to 1978. However, wheat breeders have since realized that although multiline breeding can manage rust diseases, it is a very conservative and slow in approach with regard to yield, because new cultivars may rapidly surpass the recurrent parent. Beside this, problems also arose in the multiplication of each component line and the replacement of components that became susceptible. For these reasons, the multiline strategy in wheat did not succeed. Rajaram and Torres (1986) suggested modified strategies of breeding multiline cultivars whereby, instead of undertaking a full-fledged backcross breeding program, a limited number of backcrosses would be made to recover essential phenotypic characters of the recurrent parent.

7.3.10 Shuttle Breeding

Shuttle breeding refers to a method where breeding generations from the same crosses are alternately selected in environmentally-contrasting locations as a way of combining desirable characters. The main elements of the wheat improvement program at CIMMYT have been shuttle breeding at two contrasting locations in Mexico (Ortiz et al. 2007). The shuttle breeding scheme between Mexican field sites (Ciudad Obregon in North-Western Mexico during winter and Toluca or El Batan in the highlands near Mexico City during summer) was initiated by Dr N.E. Borlaug in the mid-1940s is well known for its impact on developing widely adapted photo-insensitive and rust resistant spring bread wheat cultivars that triggered the Green Revolution in the 1960s. While shuttle breeding, wide adaptation, durable rust resistance, and international multi-location testing improved wheat yield stability, the right use of genetic variation improved yield gains of subsequently produced genotypes (Rajaram 1999). More recently, CIMMYT wheat breeders have decentralized their shuttle breeding program by collaboration with NARS, each one concentrating on a different disease or pest or unique crop ecology. Such summer and winter season, shuttle breeding is dominantly possible in spring bread wheat and has been adopted in wheat breeding programs of many countries having suitable locations for a summer wheat nursery. Winter wheat generally requires vernalization and therefore it is grown during the winter cycle under field condition. Further examples of shuttle breeding operations were also in place earlier.

Aluminum (Al^{3+}) toxicity is a significant factor limiting growth of wheat cultivated in the acid soils of Brazil, the Andean region of South America, Central Africa and the

Himalayas of the Indian subcontinent. In 1974, CIMMYT launched a shuttle breeding program primarily to combine Al toxicity tolerance of Brazilian wheats with high yield potential and wide adaptability of Mexican semi-dwarf wheats (Rajaram et al. 1991). The breeding and laboratory screening methodologies were combined with shuttling of alternate segregating populations between locations in Mexico and Brazil. Promising segregating materials identified in the laboratory were selected and grown under different acid soil conditions in Brazil, and at the two non-acid sites at Toluca and Ciudad Obregon, Mexico. The CIMMYT shuttle wheat breeding program thus resulted in the development of a large number of high-yielding, semi-dwarf, broadly adapted wheat cultivars with high levels of tolerance to soluble aluminum.

The Yangtze river basin of China, with about 7 million ha, has traditionally been known to be highly prone to the Fusarium head blight (FHB) or scab epidemics. A shuttle wheat breeding and germplasm exchange program was established between CIMMYT and China in the mid-1980s, which significantly assisted the incorporation of Chinese germplasm scab resistance into high yield, semi-dwarf and rust resistant CIMMYT wheats (He et al. 2013). As a result, many Chinese derivatives were included in the CIMMYT international nurseries distributed globally by CIMMYT, created and assembled via a series of FHB Screening Nurseries (FHBSN).

7.3.11 Doubled Haploid

The production of haploid plants from hybrids, followed by chromosome doubling, provides the wheat breeder with a means of accelerating the process of true breeding line development (Henry and De Buyser 1990; Liu et al. 2002). The doubled haploid (DH) system allows completely homozygous lines to be developed from heterozygous parents in a single generation. The two systems of another culture and wheat × maize cross are the most commonly used induction methods in wheat, the latter being preferred as it is genotype non-specific. Details are elaborated on in Section 7.4, which explains wide crossing. The DH system can be used to increase the efficiency of wheat cultivar development programs (Zheng et al. 2002). DHs are treated in the same way as the conventional breeding program's F_6-derived F_7 advanced lines. The best DHs were evaluated in yield trials along with the best advanced lines through conventional breeding from the other crosses. In the last few decades, several wheat cultivars developed through DH breeding strategies have been released in many countries such as China, France, Hungary, Romania, India and Canada. Besides varietal improvement, wheat DH populations have also been used in the creation of molecular marker maps and identification of new alleles/QTLs and unveiling of their action mechanisms (Dwivedi et al. 2015; Forster and Thomas 2010). The DH breeding strategy in wheat may have an advantage when genetic gains per unit of time is considered, and only one generation is grown per year. Li et al. (2013) compared conventional wheat breeding (SELBLK) with two DH strategies: DH lines directly derived from F_1 hybrids (F_1-DH), and DH lines derived from F_3 individuals that are retained following selection for agronomic traits in the F_2 generation (F_3-DH). They estimated that genetic gains per cycle, per year and per dollar spent were consistently higher for conventional breeding than for DH breeding strategies, especially gains per dollar. Though the F_1-DH strategy saved 1 year in completing a breeding cycle, genetic gains per year for the adaptation trait from F_1-DH were much lower than those from shuttle breeding, where two growing seasons are used per year. When Inagaki et al. (1998) compared genotypes developed from pedigree, SSD and DH methods, no significant differences in grain yield were found among any populations when the parental varieties were closely related in their pedigrees.

7.3.12 Mutation Breeding

Applied since the 1930s to accelerate the process of developing and selecting new valuable agronomic traits, mutation breeding uses a plant's own genetic make-up, mimicking the natural process of spontaneous mutation. The mutation process generates random genetic variations, resulting in mutant plants with new and useful traits. According to the joint FAO-IAEA Mutation Breeding Program (https://www.iaea.org/topics/mutation-breeding), 264 wheat cultivars originating from mutation events have been released globally. Jointly with the FAO, the IAEA assists in developing and implementing technologies that, through the use of gamma irradiation, X-rays and chemicals, can induce mutations and thereby considerably speed up the breeding process. There are several recent examples of developing wheat mutants with valuable traits. In Pakistan, chemical mutagenesis was applied (0.8% EMS) to improve stripe and leaf rust resistance in variety NN-Gandum-1 (Hussain et al. 2018). A total of 3634 M_2 fertile plants were produced and out of these, 33 (0.91%) and 20 (0.55%) plants showed absolute resistance to leaf and yellow rust, respectively. In the M_4 generation, a total of nine absolute resistant lines were selected. The SNP identified in chimeric allele (*Lr21*) of a resistant mutant (N1-252) was located in an NBS domain of chr.1BS. Through computational analysis, it was demonstrated that this identified SNP causes a substitution of glutamic acid with alanine, resulting in a predicted altered protein structure. This mutation, therefore, is a candidate for contributing to the resistance phenotype in the mutant line. In Ukraine, six bread winter wheat lines have been identified as high-grain productivity mutants which out-yielded national standard variety for the agriculture zone by summarizing three years' results. Gamma-ray dosage of 100 Gy as a mutagen factor was the most successful in induction of productive mutations. Medium dosage of gamma rays (100 Gy) and concentrations of nitrosoalkylureas (0.01–0.0125%) are recommended for winter wheat mutation breeding on grain productivity. In Romania, the mutant and mutant/recombinant wheat DH lines were obtained at NARDI Fundulea by using a specific mutagenic protocol including two modern wheat genotypes, applications of two irradiation cycles, hybridization and DH technology (Barbu et al. 2018). More than 20 lines were selected with stable and high genetic potential for productivity and quality traits, significantly superior to the two parents.

The application of mutagenesis in wheat breeding and research from the modern genomic perspective has been reviewed by Uauy et al. (2017). For wheat, increase in the ploidy level, from diploid to hexaploid, led to a progressive decrease in the number of visible mutants. This buffering effect of polyploidy limits identification of mutants. Chemical mutagenesis agents such as ethyl methanesulfonate (EMS) became commonplace in wheat research in the 1960s. EMS was shown to produce point mutations in wheat (G to A and C to T) contrary to radiation, which produced large deletions and chromosome breaks. EMS was adopted both for developing new alleles for breeding and for functional studies. An important breakthrough came with the development of the TILLING (targeting induced local lesions in genome) approach, which provides a relatively simple strategy to identify mutants (or lesions) in a target sequence independently of its phenotypic effect. TILLING approaches require a population of typically EMS-induced mutants and a screening method to identify individuals with mutations in the target gene. Mutations are then prioritized based on their potential to disrupt protein function. Truncated proteins resulting from a premature termination codon or a change to a canonical splice site provide a high probability of functional disruption. However, non-synonymous mutations resulting in amino acids with different chemical properties and located in evolutionary conserved positions of the protein also have a good probability of altering protein function. The higher the number of mutant alleles recovered, the higher the probability of identifying an allele leading to a non-functional protein. Sequenced mutant populations and genome editing have changed the paradigm of what is possible in functional genetic analysis of wheat.

There are several examples where application of mutagenesis and modern genomic tools have improved highly important traits. Resistance to *Fusarium graminearum*, a destructive wheat disease, is controlled by *Fhb1*, a deletion mutant locus discovered in a Chinese germplasm (Su et al. 2019a). TaHRC, a gene that encodes a putative histidine-rich calcium-binding protein, is the key determinant of Fhb1-mediated resistance. TaHRC encodes a nuclear protein conferring FHB susceptibility and a deletion spanning the start codon of this gene results in FHB resistance. Identical sequences of the TaHRC-R allele in diverse accessions indicate that Fhb1 had a single origin. This discovery opens a new avenue to improve FHB resistance in wheat, and possibly in other cereal crops, by manipulating the TaHRC sequence through bioengineering approaches (Su et al. 2019b). Five novel wheat mutant lines with variable seed color were developed using gamma irradiation of a hexaploid wheat inbred line K4191 (purple seed color) (Hong et al. 2019). The total anthocyanin contents of three mutant lines (L47, L167 and L925) were significantly higher than those of wild-type lines, including K4191 and "Keumkang" (white seed color). L925 showed the highest total anthocyanin content, and cyanidin-3-glucoside was presented as the most predominant anthocyanin. Compared with Keumkang, the expression of anthocyanin biosynthesis genes was significantly up-regulated in purple seed mutant lines. The expression of a few antioxidant-related genes and total anthocyanin content were positively correlated with antioxidant capacity. The objective of the study in Kenya was to determine adult plant resistance of wheat mutant lines to stem rust across three different locations during the 2015/2016 cropping season (Ogutu and Charimbu 2020). Sixty-three mutant lines and six checks (NJBWII, Duma and Kwale, Kingbird, Robin and Cacuke) were evaluated under field conditions. The mean for the area under the disease progress curve and coefficient of infection revealed that Duma200gry (1026) and Duma200gry (1124) were best disease performers. The calculated variance (Si) distinguished stable genotypes in terms of disease and yield, which included Duma100gry (995) and Kwale100gry (1483), respectively. There was positive effect of dosage 400gry on the mutant lines in terms of disease, yield and 1000 kernel weight, mostly with the Duma mutant lines. The mean grain yield for the genotypes ranged from 5.5 to 14.1 tons per hectare. Overall, it appears that mutation breeding tools have great potential in combination with the modern genomic tools.

7.3.13 Hybrid Wheat

Hybrid wheat research has been going on ever since Wilson and Ross (1962) reported that the interaction of *T. timopheevi* cytoplasm and *T. aestivum* nucleus resulted in complete and stable male sterility. Commercial hybrid wheat kept on eluding wheat breeders for almost half a century, after its noticeable initial promise in the early 1960s (Gupta et al. 2019a). The early programs of hybrid wheat were largely based on cytoplasmic male sterility (CMS), which could not be exploited for production of commercial hybrid wheat for a variety of reasons. Fertility restoration in lines carrying the timopheevii cytoplasm has often been found inadequate. In 1996, with support from Monsanto, hybrid wheat exploration at CIMMYT started again, this time using a chemical hybridizing agent (CHAs) (Çukadar et al. 1997). Commercial hybrid breeding and seed production in wheat is still limited to a niche sector in comparison to other cereals like maize or rice (Longin et al. 2012; Whitford et al. 2013). In South Africa, SENSAKO started the first hybrid wheat breeding programme in 1970, and a range of hybrid wheat cultivars that differ in adaptation according to factors including planting dates and production potential, have been specifically developed for South African conditions (Smit et al. 2010; Van Niekerk 2001). In China and India, hybrid wheat is produced based on the CMS systems or photoperiodic sensitivity sterility systems (Longin et al. 2012).

Hybrid wheat cultivars based on CHAs are commercially produced for the European market. Starting in the 1990s, more than 60 hybrid wheat cultivars have been released in Europe and USA, with the majority of varieties commercialized in Europe. According to some reports, the area under hybrid wheat in Europe increased from approximately 100,000 ha in 2002 to 560,000 ha in the year 2017/2018 (Gupta et al. 2019a). Currently, commercial hybrid wheat production is largely based on the in-field application of CHA by preventing the formation of viable pollen on the female parent (Cisar and Cooper 2002). Croisor®100 (Sintofen, France) is currently the only CHA for wheat registered in Europe for commercial production. Some CHA-based wheat hybrids have been reported to give 10–20% yield advantage over the best available pure lines (Gowda et al. 2012; Longin 2016). Although modern CHAs are in principle functional for a broad spectrum of wheat genotypes and display relatively low phytotoxicity on wheat, they still have limitations such as compromised seed set on treated plants (Adugna et al. 2004) or variation in field efficiency depending on the weather conditions at the time of application.

In the last two decades, four new alien male-sterile cytoplasms (*Ae. kotschyi*, *Ae. uniaristata*, *Ae. mutica* and *H. chilense*) were found promising (Hohn and Lukaszewski 2016; Lukaszewski 2017; Martin et al. 2008, 2010; Tsunewaki 2015). At present, more than 70 different cytoplasms are known, which induce cytoplasmic male sterility (CMS) in wheat. Among these CMS systems, *T. timopheevii* (T-type) and *Hordeum chilense* (msH) systems are currently at the center of attention. The CMS system is based on the application of three parent systems involving the A, B and R parents, where A and B are identical except that B is not cytoplasmic sterile and thus it is used to maintain the A parent which is cytoplasmic sterile. Pyramiding of two or more Rf genes in the same male parent has been achieved through the use of associated molecular markers. This system is used in the USA, China, India and Australia (Tadesse et al. 2019). Currently, Syngenta, Bayer and Dow DuPont are working to develop their respective hybrid wheat varieties for Europe, the USA and Australian markets (https://www.futurefarming.com). While CMS is efficient in other important cereals such as rice and maize, it has turned out to be difficult to develop, tricky to maintain and partially reliable for wheat. Exploitation of transgene technologies was reported to be promising for the establishment of hybrid wheat production systems (Kempe et al. 2014).

Hybrid wheats are expected to adapt more to marginal areas and produce more stable and higher yields as compared to the conventional varieties. Economically viable large-scale implementation of hybrid wheat will need the combination of a practical low-cost hybrid seed production system, high heterosis in traits of interest such as grain yield and yield stability and an efficient breeding scheme for further improvement (Longin et al. 2012). Moreover, the modification of the naturally closed inflorescence structure of wheat adapted to selfing to a more open structure allowing more efficient open-pollination should be an important breeding objective (Whitford et al. 2013).

Hybrid-enabled line profiling (HELP), described by Van Ginkel and Ortiz (2018) for self-pollinated crops, can help to exploit maximum heterosis and reduce breeding cycles in wheat. Heterosis in self-pollinated crops is often driven by additive and additive × additive gene action. Additive forms can be relatively easily fixed in homozygous lines. The newly integrated HELP strategy shows how hybrids can help identify the best performing crosses with mostly additive and additive × additive gene action, which predicts high progeny performance. The HELP strategy enables the performance of F_1 hybrids to guide the profiling and development of F_1-derived lines or DHs. Such lines or DHs are the direct source for a new cultivar emanating from HELP. Hybridizing diverse, complementary "selfing" parents to create the desired trait or allele line profile requires strict male sterility of the female, which can now be achieved relatively easily through current and emerging chemical, environmental or genetic techniques (Van Ginkel and Ortiz 2018). Practically small amounts of hybrid seed are needed, with no need to scale up seed production, as the hybrid is not to be produced commercially. After multilocation testing, homozygous lines from only the best hybrids, driven mainly by additive effects and additive × additive gene action, are fast derived using techniques such as DHs. Thereafter, multilocation evaluation and molecular validation of target line profiles is done to identify superior lines for release to farmers. This strategy can result in significant increases in selection efficiency and can overcome the yield plateauing as seen in major selfing food crops.

Established in the mid-1960s, Hybrid Wheats Australia (HWA) produced the world's first commercially released hybrid wheat. HWA was bought by Cargill in 1983. Hybrid wheat varieties were released by Cargill in the USA and Australia in the 1980s. Cargill dropped hybrid wheat in the USA, but continued in Australia and Argentina (Pickett and Galwey 1997). In 1999, it became part of Sunprime seeds, which in turn was acquired by AGT in 2005.

7.3.14 The XYZ System

A three-line system for hybrid wheat seed production, based on the utilization of a recessive male sterile mutant (ms), was proposed by Driscoll (1972). This system uses an addition chromosome carrying a gene for pollen fertility and a gene for a phenotype marker –

blue-colored aleurone layer. Three lines, X, Y and Z, are homozygous for the male-sterility factor and contain 2, 1 and 0 doses of the rye alien addition chromosome, respectively. This three-line system of hybrid seed production was modified by Driscoll (1985). In theory, it allows for efficient hybrid seed production. However, a drawback of the system is the poor coloration of grains carrying one dose of the blue aleurone gene, and thus the insufficiently effective sorting of grains. It has not reached common practice for this reason (Dowell et al. 2009).

Researchers at the University of Adelaide, in collaboration with DuPont Pioneer, announced the discovery of Ms_1, a naturally occurring wheat gene that prevents self-pollination when turned off but enables cross-pollination (Tucker et al. 2017).

7.4 Innovative Technologies for Accessing Novel Genetic Diversity

To complement the existent genetic diversity prevalent in conventional wheat germplasm, harnessing the variation that resides in approximately 325 annual and perennial Triticeae genera and their abundant accessional strength is a valid option. Given the breadth since 1876 until the mid-1980s, efforts of various researchers have resulted in diverse outputs that involved efforts to combine Triticeae genera via variable success levels. Over this time, the term "wide crosses" surfaced and became in common usage. The primary interest was on intergeneric hybridization that led to the interspecific area and, with the latter practicality, emerged intraspecific hybridization. The commonality of all three facets was on harnessing new forms of genetic diversity by setting sights on exploiting distant relatives that had been under-utilized or un-utilized and required complex handling from initial hybridization to full practical utilization. Incorporation of this distant variation was a great challenge and of prime interest to provide additive genetic diversification, for which the Triticeae members formed the germplasm strength for global experimentation. This unique diversity was present either in a user-friendly access form or was available in a form that required complex protocols to utilize creatively. Nevertheless, from both routes outputs were generated by researchers.

Genetic diversity is now globally present within research groups but in the form of perennial hybrids is available only in the CIMMYT program in Mexico where, over a hundred unique intergeneric hybrid combinations were produced and maintained as clones in the form of a living herbarium. They have been maintained by physical cloning twice annually and serve for continuous production of additional amphiploids and different wheat parent backcross 1 derivatives (Mujeeb-Kazi et al. 2017). All annual genera combined were represented as amphiploids. These resources are arsenals of novel alleles with the potential to improve wheat at a swift pace, circumventing the effort needed to generate new hybrids, a task that is cumbersome. Access to such user-friendly perenniality is not available anywhere else globally

Exploiting genetic stocks next from those conserved in gene banks is another potential option that precludes one from having to remake the hybrid combination that, apart from being a complex process, is a low frequency event beset with constraints like hybrid necrosis, hybrid bunching/clumping, etc. where to overcome one requires tedious additive protocols. The exploitation of amphiploids, self-fertile backcross 1's, disomic addition lines, etc. is exciting, where these can be manipulated by the *ph* gene yielding alien introgressions in the form of homoeologous or the less desirable non-homoeologous translocations.

This also opens up the usage of the pairing inhibitor gene Ph^{I} reported by Chen et al. (1994) and subsequently used to demonstrate its additional effectiveness by Aghaee et al. (2000) and Wang et al. (2003), utilizing it for agricultural practicality as the substitute for the Chinese Spring *phph* genetic stock.

In order to have a stable wheat productivity system, it is thus essential that a vigilant balance is maintained that well covers the dynamic system of pathogen (biotic stresses) aggressiveness complimented by the strength of the supportive static (abiotic stresses) system. Within this framework, conventional field-based programs have been prospered globally since their inception but now, with progressive science and progressive climatic shifts, the pace has slowed down. Emphasis on harnessing under- or un-utilized genetic diversity has opened up new horizons that since the 1980s have shown promise. Complexity of handling this new diversity has been successfully tackled and the mystique of gains from "wide crossing" diminished significantly (Fedak 1977, 1979, 1980; Islam et al. 1978, 1981; Jiang et al. 1995; Mujeeb-Kazi and Asiedu 1990; Mujeeb-Kazi and Bernard 1985; Mujeeb-Kazi and Kimber 1985; Mujeeb-Kazi and Rodriguez, 1980, 1981; Mujeeb-Kazi et al. 1987, 1989; Sharma and Gill 1983a, 1983b, 1983c; Wang 1989) prompting Sharma (1995) to express "How wide can a wide cross be." Our perception is that the basic barriers relative to hybridization hae been well overcome and it is quite simplistic to combine members within the Triticeae readily. This output success has led to the phase that selectivity of parents can be prioritized, generation of genetic stocks predicted, their screening for stress attributes feasible, and defining targets required to be met on the practical end of the spectrum of crop improvement translated into YIELD outputs. The shift toward practicality saw the basic investigation areas cross into a strategic phase and pre-breeding blossomed where derivatives emerged as being agronomically more refined for plant type and more user-friendly for breeders to work further with ease.

We argue that from several decades of experimentation globally the outputs from intergeneric hybridization are not that abundant and final varietal products are time-consuming to attain. However, gains all around from close progenitor sources hold a distinct advantage, as they have profound genetic homology, and above all exhibit the potential to produce swift practical gains. Since the preferred two (A and D genome) progenitor are diploids, residing in the primary wheat gene pool, their preference is natural to our perception. The major emphasis over the last three decades has been on the D genome diploid *Aegilops tauschii* ($2n = 2x = 14$; DD). Genetic stocks derived from this resource are in the form of synthetic hexaploid wheats (SHWs), also now referred to in the literature as "primary synthetics." The other form is as pre-bred derivatives, where the SH wheats have been crossed onto the conventional bread wheat and advanced for selecting stress tolerances and appropriate phenotypes that would suit breeders to utilize in their additional breeding efforts or even release directly as cultivars. These derivatives have later become known to as synthetic derived lines referred by the acronym SDL's (Rosyara et al. 2019). Due to the vast variation that has been found in the D genome synthetics for production traits, plus the global distribution of the pure synthetic stocks and

SDL's, experimental curiosity has been aroused and reports of Li et al. (2018a) support the value of using SHW or SDL wheats in varietal development. Close to 70 varieties have been released in 14 countries so far in the direct manner of release by acquisition of a SDL or as a donor (SH) in breeding efforts globally. Contribution of the D genome molecularly is validated by Rosyara et al. (2019), where up to 17% of the advanced materials from CIMMYT have sequences from the D genome and about 27% of the international distribution of seed nurseries from CIMMYT, Mexico, possess SH blood in them. The above information substantiates the potential that distant diversity stocks have shown toward aiding productivity and counter-productivity output constraints. Hence, this paves the way forward to stringently focus on adopting a wheat improvement course that gives positive returns across a wide spectrum of attributes and possesses multiplicity swiftly, essentially because the wheat/alien resources are perfectly homologous and genetic recombination exudes minimal to no complexity. These are the attributes that suggest the D genome progenitor as being a choice donor of variability that can well address the essential wheat production constraints simplistically covering an array of multiple stress factors as well within a single accession, thus endorsing our alien parental preference for harnessing novel alleles for wheat improvement.

7.5 Major Global Locations of Wheat Genetic Diversity

Genetic resources of wheat, where species fall in the primary, secondary and tertiary gene pools of the Triticeae, are excellent sources of novel genes and have received worldwide attention for improving productivity (Mujeeb-Kazi et al. 2013; Ogbonnaya et al. 2013). During the last two decades, survival of species has been of increasing concern as a consequence of climate change and human activities (i.e., urbanization and pollution), thus environmental stresses are posing critical challenges (McCouch et al. 2013). Fortunately, seed banks around the world have secured wild relatives of wheat collected from different global eco-geographical locations, which so far remain largely untapped. Globally, there are more than 80 wheat collections, holding more than 800,000 accessions. Some of the important wild relative collections of wheat are housed in the CIMMYT seed bank at El Batan, Mexico; National Small Grains Collection, Aberdeen, Idaho, USA; Institute of Crop Germplasm Resources, Chinese Academy of Agricultural Sciences, China; National Bureau of Plant Genetic Resources (NBPGR), New Delhi, India; ICARDA gene bank, Morocco; National Institute of Agrobiological Sciences, Japan; NI Vavilov Research Institute of Plant Genetic Resources (VIR), St Petersburg, Russia; Istituto di Genetica Vegetale (IGV) Italy; and IPK, Gatersleben, Germany. Our efforts have enabled us to structure Table 7.1 below.

7.6 Utilization of Genetic Diversity

The gene banks and global research facilities are also households of genetic diversity that are allelic treasures of utilization for wheat improvement. These germplasms or genetic stocks have been generated over the past several years and comprise of amphiploids,

Table 7.1 Global wheat gene bank locations and their germplasm holding strength.

			Accession type percentage (%)				
Genus	**Gene bank**	**Accession numbers**	**WS**	**LR**	**BL**	**AC**	**OT**
Triticum	CIMMYT, Mexico	110,281	6	31	5	7	6
Triticum	NSGC, USA	57,348	4	57	24	14	<1
Triticum	ICGRCAAS, China	43,039	5				95
Triticum	NBPGR, India	35,889	4	2	9	1	84
Triticum	ICARDA, Morocco	34,951	5	75		<1	27
Triticum	NIAS, Japan	34,652	3	4	31		61
Triticum	VIR, Russia	34 253	1	43	20	35	<1
Triticum	IGV, Italy	32,751	2	98			
Triticum	IPK, Germany	26,842	4	49	12	32	
Triticum	TAMAWC, Australia	23,811		3	50	32	16
Triticum	NPGBI-SPII, Iran	18,442					100
Triticum	RIA, Kazakhistan	8000					100
Triticum	IBC, Ethiopia	13,421		100			<1
Triticum	IPGR, Czech Republic	12,539					
Triticum	IHAR, Poland	11,586		3	88	7	3
Triticum	INRA-CLERMON, France	10,715					100
Triticum	PGRC, Canada	10,514	19	14	35	28	3
Triticum	RICP, Czech Republic	10,419	2	7	27	64	<1
Triticum	IPSR, UK	9462		11	28	25	36
Triticum	INIA QUIL, Chile	9333					100
Triticum	UzRIPI, Uzbekitan	9277					100
Triticum	RCA, Hungary	8569		2	<1	12	86
Triticum	ARI, Cyprus	7696		1	99		
Triticum	RAC, Switzerland	7266					100

(*Continued*)

Table 7.1 (Continued)

Genus	Gene bank	Accession numbers	Accession type percentage (%)				
			WS	LR	BL	AC	OT
Triticum	IR, Ukraine	7220		4	42	53	1
Triticum	UNALM, Peru	7000					100
Triticum	OTHERS (202)	237,428	5	14	15	22	42
Triticum	TOTAL	856,168	4	24	20	13	39
Aegilops	ICCI-TELAVUN, Israel	9146	100				<1
	ICARDA, Morocco	3847	100				<1
	NPGBI-SPII, Iran	2653	99				1
	NIAS, Japan	2433	5				95
	VIR, Russia	2248					100
	NSGC, USA	2207	100				
	LPGPB, Armenia	1827	100		<1		
	IPK, Germany	1526	100				<1
	CIMMYT, Mexico	1326	99		<1		<1
	INRA-RENNES, France	1070					100
	Others (52)	12,643	81	3	2		14
	Total	40,926	80	1	1		18
Triticosecale	CIMMYT, Mexico	17,394	<1		97	3	<1
	VIR, Russia	2030					100
	NSGC, USA	2009		1	83	16	
	SCRDC-AAFC, Canada	2000					100
	IR, Ukraine	1748			86	13	1
	LUBLIN, Poland	1748			63	33	3
	IPK, Germany	1577		2	81	17	<1
	Others (62)	8934	4	<1	36	11	49
	Total	37,440	1	<1	68	8	23

Source (Modified from FAO 2010); WS: wild species; LR: landraces/old cultivars; BL: research materials/breeding lines; OT: (others) the type is unknown or a mixture of two or more types; FAO (2010). The second report on the state of the Worlds Plant Genetic Resources for Food and Agriculture Rome. ISBN 978-92-5-106534-1, pp. 243–249 partial amphiploids, backcross 1 self-fertile combinations, specific translocations, as well as alien introgressions, etc., for which stress descriptive data is available or can be generated anew. However, many of these stocks are available in wheat backgrounds that are agronomically inferior and do not fit the standard field cultivation crop cycle, thus limiting their exploitation in wheat breeding. Chinese Spring wheat is a common wheat that has been widely exploited as it is highly crossable with distant genera of which these stocks are on hand as a consequence of three of the four crossability *kr* genes it possesses. Researchers have advocated to transfer these stocks in well adapted wheat backgrounds by recurrent backcrossing, an aspect not yet completed. Breeding protocol variations will show later how the Chinese Spring background can be circumvented and practical returns be realized (Mujeeb-Kazi and Asiedu 1990).

Utilization of genetic diversity has varied but the ideal form is to first stress screen the resources for the various traits, identify those that are positive for the stress and then evolve them for improving wheats that are susceptible to stresses. There is always the possibility that the ensuing hybrid from such a well charted cross combination may not possess the trait of interest, as such genetic suppression does occur when targeted crosses are made. This has been reported for hybrids between goat grass (*Aegilops tauschii*) and wheat, where the rust resistant grass accession when combined with wheat had its rust resistance suppressed (Ma et al. 1995). To counter this, the wide cross program in CIMMYT, Mexico indiscriminately hybridized *Aegilops tauschii* accessions with durum wheats screened for their trait susceptiblity to produce a huge number of D genome synthetics and identified those synthetics that were resistant/tolerant in which, as the durum parent was susceptible, gave the inference that the stress positivity observed was contributed by the goat grass accession. Similarly, where the grass accession was to be hybridized to bread wheat as in a direct cross (Gill and Raupp 1987), knowing that the wheat cultivar was susceptible upon breeding if segregates showed resistance or tolerance, the inference favored the grass resource as being the contributor to the D genome.

Wheat breeding has encompassed a multitude of aspects that have importance for its productivity as to where it is cultivated. This output could be across mega-environments or restricted to micro-environments. In essence. the constraints are grouped within stresses that embrace biotic, abiotic, quality components, mineral stresses, nutrient levels, varietal aspects, insect and viral influences, etc. Ideally, in a variety that is desired to have multiple stress resistance/tolerance plus other attributes, thus the choice of a contributing parent having many desirable descriptors stands as an advantage. Biotic stresses span from all three rust resistances, head scab, spot blotch, leaf spot, tan spot, mildews, karnal bunt, rice blast, *Alternaria*, etc. Abiotic stresses cover drought, salinity, waterlogging, frost/cold and heat tolerance. The other significant constraints associated with wheat productivity include BYDV, aphids, greenbug, saw fly, hessian fly, nematodes, phytic acid, boron toxicity, aluminum tolerance, grain component quality, micronutrient content (Fe and Zn), etc. Additionally and also of significance is to have a variety that covers all aspects of physiological parameters, e.g., pre-harvest sprouting, plant habit, major yield components, leaf pubescence, root growth, coleoptile emergence, early growth habit, waxiness, stay green character, awn plus or minus, etc. These traits alone or singularly have the potential to offset yield returns and varietal development that encompasses multiplicity, having an advantage associated with resistances that impart durability by genetic means or by exploiting major gene pyramiding protocols leading to sustainable outputs. The process of wheat improvement is continuous and distributed across two main scenarios, the dynamic and the static. The dynamic phase resides in a living entity capable of mutating due to pathogen presence that renders varieties vulnerable as resistance can be overcome versus time. The static facet involves lack of a pathogen and tolerance once introgressed has the potential to be long lasting, as a mutable base is lacking.

The above aspects now allow us to continue with functional modes of harnessing alien genetic diversity. We do need to be cognizant of the fact that chromosome variations (aneuploidy) are rampant in almost all genetic stocks. Figures 7.1 and 7.2 show this for an amphiploid and a backcross 1 derivative. Usually, a 56 chromosome stock tends to be more stable than one of higher ploidy (Mujeeb-Kazi et al. unpublished)

7.6.1 Wide Crosses: The Historical Build-up

Wheat/alien crosses were done sporadically toward the end of the eighteenth century, and to a very limited extent, until the 1970s (Ciferri 1955). Foremost were the exciting

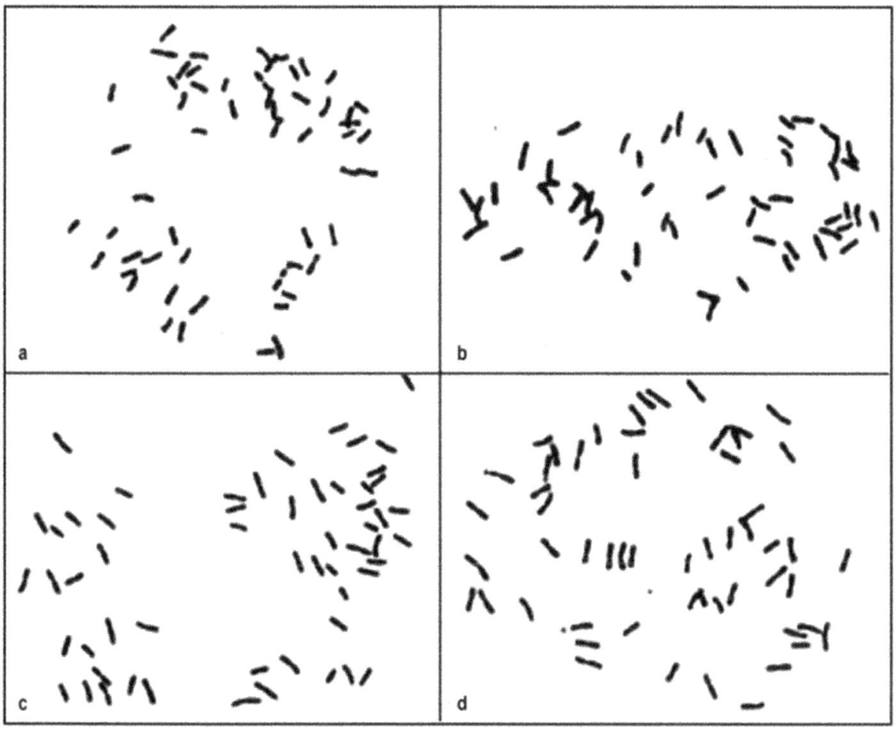

Figure 7.1 Aneuploidy in amphiploid advance during maintenance of a *Triticum aestivum/ Aegilops variabilis* combination shows various somatic cells of individual plants having: a) 55 chromosomes; b) 56 chromosomes + 1 telocentric; c) 57 chromosomes; and d) 56 chromosomes. The normal somatic count in such amphiploids is 2n = 8x = 56, AABBDD UUSS. This is an initial indicator to expect such aberrations, even in a 56 chromosome derivative where structural changes are possible.

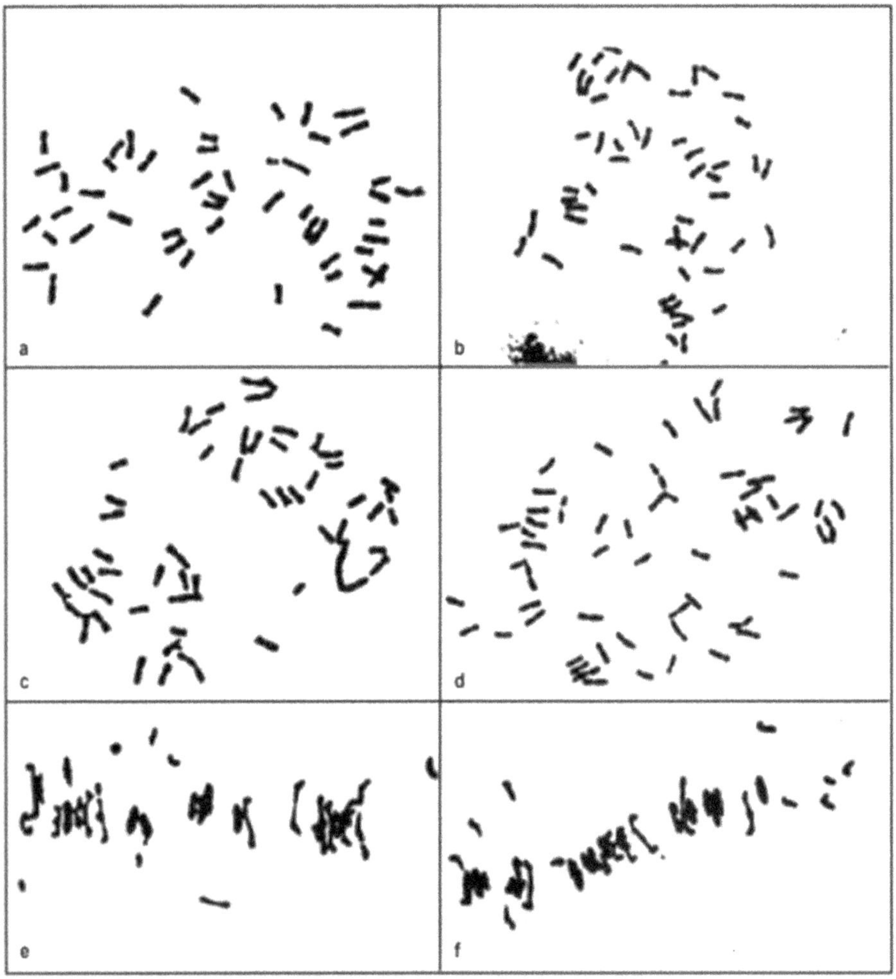

Figure 7.2 Mitotic and meiotic chromosomal details of some backcross 1 derivatives of somatic cells and meiocytes are shown as follows: a) 56 chromosomes plus a telocentric; b) 53 chromosomes; c) 56 chromosomes plus a telocentric; d) normal count of 56 chromosomes and two meiocytes showing in e) 5 univalents + 25 rod and ring bivalents and in f) 6 univalents + 11 rod bivalents + and 14 ring bivalents. These basic cytological details are indicative that genetic stock maintenance is prone to structural changes and aberrations, requiring stringent assessment for maximum purity to be achieved.

publications of Kruse (1967, 1969, 1973, 1974) that inspired young professionals to investigate the area further.

A major effort with wide crosses at the start of the 1970s involved Kansas State University (KSU) and CIMMYT. It was essentially based upon what Dr Norman Borlaug, after winning the Nobel Prize, expressed that researchers should be prepared to tackle surprising wheat production constraints in order to combat stresses that may require new allelic diversity should this not be accessible in conventional germplasms. The question asked was would we start a search for such sources when the problem arose or should we invest in futuristic research. This sparked a young researcher, Lynn Bates at KSU, to embark on addressing the need for futuristic research and thus dawned his popular paper in *Economic Botany* (Bates and Deyoe 1973), where he proposed that use of immuno-suppressants to treat the female parent and then pollinating with distant sources would overcome cross incompatibility barriers and allow distant cross combinations to be produced, thus combining diverse genera. Success was negligible until 1978, which led CIMMYT to undertake within their Mexico location an independent wide crossing program under the leadership of the first author of this manuscript.

The new program gave birth in CIMMYT from late 1979, and within the first two years its consolidation took place with the structure partitioned into basic, strategic and applied facets where the basic and strategic areas dove-tailed and led to the practical final applied undertakings. Research progression stages involved were intergeneric, interspecifically linked with intraspecific hybridization and these areas charted the progress of wide crosses.

The initiation of the program effort under this new leadership at CIMMYT set the work mandate to define pre-breeding as being a laboratory/greenhouse/screen house/field venture, separating distant crosses from other diversity resources that could be 100% field doable. Hence, pre-breeding was characterized by the following three major criteria:

1) An intergeneric hybrid had to be embryo rescued, was a low frequency cross, required a special growing environment, preferably exhibited a co-dominant phenotype indicative of alien parents genetic expressivity and was cytologically validated by presence of both the parental chromosomal compositions.
2) The resultant hybrid be self-sterile but female-fertile.
3) The pre-breeding protocols involve controlled environmental conditions to be functional, such as a laboratory, growth room/chamber, greenhouse, screenhouse, etc., after which the derivatives be moved to field conditions and conventionally handled.

The above framework thus separated CIMMYT's way forward from the efforts from Kansas State that were more interested in pursuing the role of immunosuppressants. These criteria also defined the program structure to be adopted that focused largely on applied cytogenetics and thus this chapter has extensive outputs covering its intricacies on a very large scale of applying wide cross outputs for practical wheat productive gains.

With a cursory initial coverage done here at the start of the conventional crop improvement facets we shall branch off to elucidate the intricacies of salient complex wide hybridization by defining the major basic areas, introducing strategic advances, and linking their outputs with the major interest on applied outputs vital for impacting food security targets.

7.7 Distribution of Genetic Diversity: Gene Pools, Their Potential Impact and Research Integration for Practicality

The literature is stacked with information on genetic resources that can be utilized for wheat improvement and more recently several options have surfaced. Publications have elucidated details and there is no dearth of sources to exploit. The advance on wheat/

alien introgressions with a major emphasis on the potency of *Aegilops* species diversity has recently surfaced and has come to the forefront. We feel that it is essential to set usage priorities for the future direction in exploiting genetic diversity. This shift will then target on swiftly yielding applied outputs of agricultural practicality measured by crop performance in the farmer's field and sustainability. The importance of basic research efforts and strategic perceptions are to be recognized, but the immediate need is to combat food security projections that are paramount for providing for an enlarging global populace estimated at 9.7 billion by 2050.

The 325 annual and perennial members of the Triticeae (Dewey 1984) are distributed into the primary, secondary and tertiary gene pools (Friebe et al. 1996; Jiang et al. 1995) and abbreviated as GP-1, GP-2 and GP-3, with enormous accessional strength harboring allelic diversity suitable for wheat improvement. This stronghold of genetic resources is housed in global gene banks and is freely available for research endeavors. More often, the collections are duplicated and given their own bank record numbers; hence, experimental outputs are duplicated. With the advent of new molecular tools now available, this aspect can be overcome and pre-breeding/breeding become well directed. Similarly, wild relatives may also benefit from "accessional purity," but this is a cumbersome exercise that may be an aspect for gene ban curators to consider.

7.7.1 The Gene Pool Structure

7.7.1.1 Primary Gene Pool Species

The primary gene pool (GP1) includes hexaploid bread wheat *T. aestivum* ($2n = 6x = 42$, AABBDD), hexaploid landraces, *T. carthlicoides, T. spelta*, $2n = 4x = 28$, AABB tetraploids *T. turgidum, T. dicoccum, T. dicoccoides, T. carthlicum, T. polonicum*, etc. and two diploid donors, $2n = 2x = 14$, AA and DD genomes. The members have homologous chromosome associations with the genomes of bread and durum wheat and this facilitates ease of recombination upon hybridization via normal integration with standard protocols of conventional breeding. Some cross combinations require embryo rescue, but no cytogenetic manipulation procedures are necessary.

Accessions of *A. tauschii*, the donor of D genome, are an excellent source of stress resistances/tolerances plus other attributes essential for wheat varietal performance (Mujeeb-Kazi and Hettel 1995; Ogbonnaya et al. 2013). Its numerous accessions are widely distributed and it is currently the most explored species of the primary gene pool because of its tolerance to all major biotic and abiotic stresses (Ferrahi et al. 2017; Mujeeb-Kazi et al. 2013). Also present in this pool and gaining importance in breeding circles are landraces that possess immense diversity importance and are addressed in depth here (Gill et al. 2014; Karagoz 2014; Lodhi et al. 2020; Lopes et al. 2015; Morgounov et al. 2016). The Mediterranean durum landraces are also rich diversity sources (Kyzeridis et al. 1995). Genetic transfers from the two progenitor genomes A and D occur as a consequence of interspecific hybridization and homologous recombination and their exploitation mode for efficient wheat improvement is elucidated. Its usage is defined by adopting the bridge crossing and direct crossing avenues that will be discussed later in this chapter.

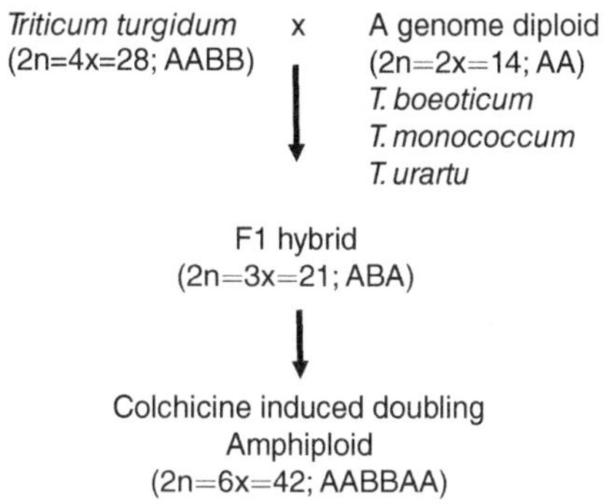

Figure 7.3 Schematic showing the production of A genome genetic stocks as a consequence of hybridizing durum cultivars with A genome accessions.

7.7.1.2 The A Genome (*Triticum Boeoticum, T. Monococcum, T. Urartu*; 2n = 2x = 14, AA)

One avenue of using the A-genome accessional diversity is via bridge-crossing of the AABBAA amphiploids (Figure 7.3). In general, the durum parents × A-genome accession are high frequency simple crosses, exhibit meiosis in the F_1's with up to six bivalents metaphase I chromosome associations per meiocyte indicative of genomic exchange among the A genomes, and are meiotically stable in their AABBAA, 2n = 6x = 42 amphiploids, suggesting maintenance ease of these stocks and good fertility. The durum cultivars, as female parents in these amphiploids, are susceptible to the stresses being addressed Thus, upon stress screening, a resistant amphiploid implies that the particular A-genome accession involved has contributed to the expressed resistance and has been so reported (Delgado et al. 2001).

7.7.1.3 The D Genome (*Aegilops Tauschii* = Goat Grass; 2n = 2x = 14, DD)

The D genome demonstrates genetic diversity for multiple stresses as observed in their AABBDD synthetic (Mujeeb-Kazi, 2001a). Their production protocol is shown in Figure 7.4. Crossing the resistant synthetic hexaploids with elite but susceptible bread wheat (BW) cultivars has yielded resistant BW/SH derivatives (Mujeeb-Kazi et al. 2001b, 2001a).

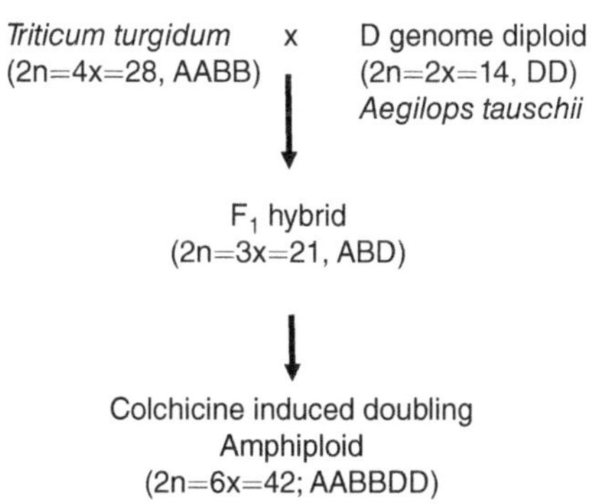

Figure 7.4 Schematic showing the production of D genome genetic stocks as a consequence of hybridizing durum cultivars with D genome accessions.

Currently, over a thousand SH wheats of spring and winter habit have been produced, and a wide array of these wheats are being globally utilized for wheat improvement, either at the SH (syn. primary SH) or at the BW/SH advanced derivative (syn. SDL) level.

Incorporation of the above D genome synthetics allows enrichment of all three bread genomes i.e., A, B and D and is known as "bridge crossing," a route preferred by breeders. For better definition of a transfer from the D genome into bread wheat. "direct crosses" between the wheat and the diploid accession is ideal so the transfer is restricted to the D genome and is well defined (Gill and Raupp 1987)

7.7.1.4 Secondary Gene Pool Species

The secondary gene pool is formed by the polyploid *Triticum* plus *Aegilops* species, which share at least one genome with wheat. The diploid species of the Sitopsis section are included in this pool that reportedly have affinity with the B genome of wheat but demonstrate reduced chromosome pairing that permits homologous exchange between the related genomes. Embryo rescue aids hybrid production.

7.7.1.5 Selected Secondary Gene Pool Species Utilization Example

This gene pool comprises of approximately 35 polyploid *Triticum* and *Aegilops* species having one genome with wheat, thus offering a distinct possibility of gene transfer from these species to cultivated bread and durum wheats. The relationship of the B genome of wheat is within this pools Sitopsis section of five diploid species, but the chromosomal association is not as perfect as exists between wheat and its A and D genome chromosomes. where 6 to 7 bivalents are the norm. The B genome is designated as B^S. Special emphasis is currently given to the use of *Ae. speltoides* accessions ($2n = 2x = 14; B^SB^S$) via the hexaploid amphiploid bridge-crossing route ($2n = 6x = 42$, $AABBB^SB^S$) (Figure 7.5). These newly-produced amphiploids have shown initial promise for a few biotic stress resistances and are an additional diversity resource for wheat improvement in the future.

7.7.1.6 Tertiary Gene Pool Species

The tertiary gene pool is comprised of species of variable ploidy levels. Their genomes are non-homologous but genomic homoeology of these species does permit the transfer of

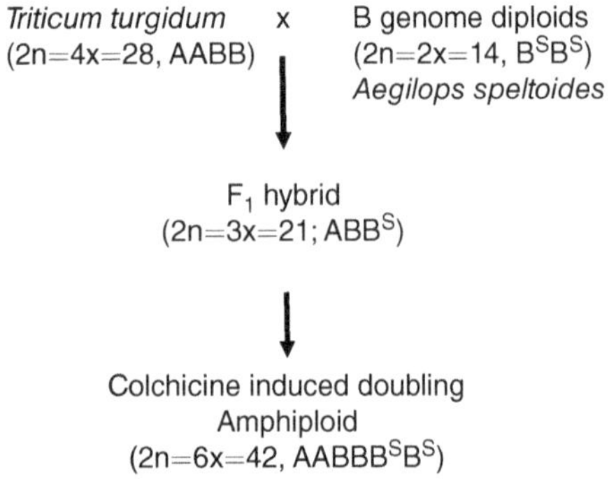

Figure 7.5 Schematic showing the production of B genome genetic stocks as a consequence of hybridizing durum cultivars with *Aegilops speltoides* accessions of the Sitopsis section.

genes by somewhat complex protocols for effecting homoeologous exchanges, facilitated by cytogenetic manipulation, irradiation or callus culture mediated translocation induction. The allelic variation within the members of this pool has great significance (Ali 2012; Ali et al. 2016). Parental selection for introgressing traits of agricultural importance holds astute decisiveness and has importance. Preference of selectivity for breeding has greater preference for a diploid rather than a higher ploidy member. Exemplifying this for salt tolerance preference are diploids like *Thinopyrum. bessarabicum* or *Lophopyrum elongatum* instead of the tetraploid or hexaploid *elongatum* or the decaploid *ponticum*. Despite the complexity of achieving practical outputs, the potential use in terms of additive diversity in wheat improvement is very high but influenced by slow returns. The traits in GP3 may often be inaccessible for introgression due to cyclical translocations, fertility genes or the presence of pairing genes (Jiang et al. 1995). Therefore, the genetic diversity of the tertiary gene pool has been largely un- or under-utilized in hexaploid wheat improvement. However, some of the noticeable species having a direct impact on cultivated wheat improvement are *Secale cereale* L, *Th. curvifolium* Lange, *Th. intermedium, Th. elongatum* and *Th. distichum* Thunb, to mention a few (Graybosch et al. 2009; Mujeeb-Kazi et al. 2013).

To recap, we introduce the use of the *ph* gene genetic stock. Of the approximately 325 species in the tribe Triticeae, about 250 are perennials (Dewey 1984) with the rest being annuals. The bulk of these species have been difficult to exploit in cereal improvement, primarily because their genomes are non-homologous to those of wheat, and genetic transfers cannot be made by homologous recombination. Genomic homoeology is prevalent, which enables gene transferring via an array of complex and tedious cytogenetic protocols of which the *ph* gene stock (Sears 1977) is of preferred utility. The use of the *ph* genetic stock has many forms and a modified manipulation strategy.where an amphiploid is shown in Figure 7.6 with the *PhPh* composition that is utilized. This strategy has applications for other such genetic stocks available in gene banks, thus overriding the need to produce from scratch new amphiploids, etc.

Hybridization with wheat per se is a low frequency event but is no longer a major obstacle within this pool of members. Versus time hybrid production has been highly simplified and the range of hybrids dramatically increased (Sharma 1995). However, achieving swifter practical outputs from genera within this pool remains a definite constraint.

The schematic (Figure 7.6) adopted to engineer translocation events is as follows:

7.7.1.7 The Gene Pool Potential Recap

The justification to embark on this diverse route was a prompt from Dr Borlaug, after his Nobel Prize, that well identified novel diversity on hand and that was extremely useful for crop improvement. It ensured and allowed incorporation of unique genetic resources that were not in mainstream usage for combatting constraints that penalized productivity. In addition to this, in the early 1980s, critics had started to voice that wheat had a very narrow genetic base and this threat could influence global wheat productivity in the future. Both these factors (excluding very limited pre-1970 efforts), apart or in unison, were instrumental to encourage some researchers to embark rigorously in utilizing distant resources that required innovative ways and means to provide

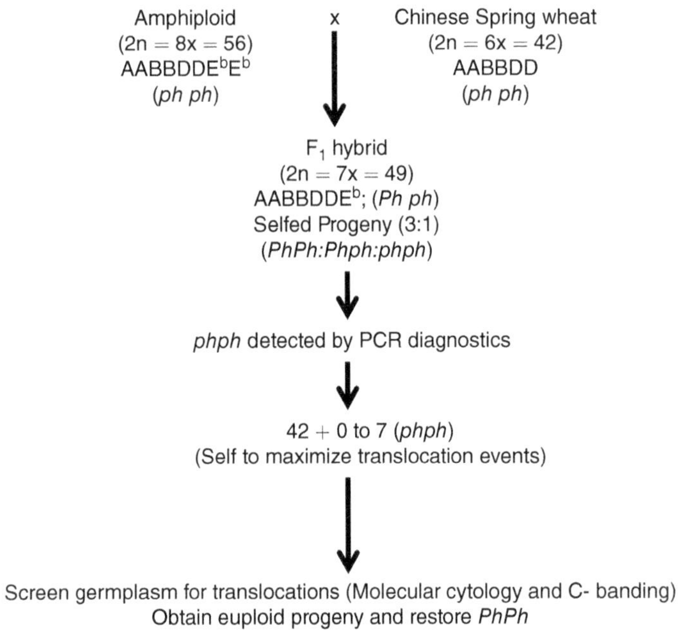

Figure 7.6 Schematic elucidating the recurrent backcrossing steps used to transfer the existent wheat/alien chromosome translocations into agronomically superior wheat cultivars (*Source*: Modified from Mujeeb-Kazi 2006) by exploiting the *ph1b ph1b* genetic stock.

successful returns initially in the form of genuine cytologically well-documented hybrids. Selection of distant sources were associated with the stresses that influenced wheat varietal development comprising of major biotic, abiotic traits complimented by supportive multidisciplinary aspects that altogether had to culminate in giving high yielding returns. A major factor also was to be cognizant of the structure and usage priority of wheat's genetic resources, as this was crucial to achieving practical success from the diversity that had not been a major concern of the initial basic academic research pursuits. These resources are distributed in three gene pools known as the primary, secondary and tertiary, including genera with close or distant relationships with the genetic structure of bread and/or durum wheats. The affinity of these pools resides upon genetic homology that is established by how the wheats A, B and D genome chromosomes pair with those of the related genera chromosomes. Members within the primary pool have maximum closeness with the A, B and D genomes and thus have the highest priority to be utilized, since perfect homology would prevail and genetic information could be readily incorporated. These members have varied ploidy levels and some can be easily utilized using conventional breeding protocols, while others require special conditions like embryo rescue, etc. The two diploid donors to wheat are in this gene pool. These are the A and D genome diploid donors that each have enormous accessional strength with phenomenal allelic richness that we see as being the ideal top priority for utilization in wheat improvement. The balance is tilted toward the D genome, as historically when bread wheat originated around 10,000 year ago, only a few accessions of the D genome were involved, which has been the explanation commonly expressed of the narrow diversity within this genome in wheat. There is no dearth of accessing variability from other Triticeae members where these two genomes are present but considering the task of having on hand a system that provides perfect (D genome with 7 bivalent chromosomal associations and maximized recombining opportunity) to near perfect (6 bivalents) homology, the presence of other genomes with little to no homology will be extraneous and difficult to dispense during the breeding facets. The B or B^S genome diploids of the Sitopsis section are an additive third resource. All these three genome members have over the past few decades enriched available genetic stocks via generation of synthetic hexaploids ($2n = 6x = 42$), of which so far the most in use have been the D genome stocks that have shown great promise in farmers' fields (Ogbonnaya et al. 2013). So far, the best applied example of wheat/alien varietal practical outputs are seen on a highly significant global scale via the famous spontaneous wheat/rye T1BL.1 RS Robertsonian translocation that was integrated via winter/spring crossing into the spring habit Veery wheats of CIMMYT (Rajaram et al. 1983). There are a few other translocations (Mujeeb-Kazi et al. 2013) that have contributed as well, but to a lesser extent. We perceive that a major impact on future wheat production is within the *Ae. tauschii*'s D genome as evident in Sichuan, China where the variety Chaunmai 42 gave a 22.7% yield increase over a large planting area. With over 1000 synthetics available in winter and spring habits, possessing multiple disciplinary values, the way forward for timely success to aid food security targets are doable and an in-depth coverage of benefits accruing from alien exchanges is covered in Section 7.10.8 later in the chapter.

7.7.1.8 Conclusion: Transfer Prerequisites Across Gene Pools

Some prerequisites for achieving gene transfers from the annual and perennial Triticeae across all three gene pools are related with varying degrees of complexity to hybrid production, embryo rescue, plant regeneration, hybrid validation through multiple diagnostic protocols, breeding methodology, biotic and abiotic stress screening and stability of the wheat/alien derivatives achieved by the use of sexual homozygosity inducing protocols. In essence, these were the initial challenges within the basic research category to connect with the strategic phase but have been successfully overcome (Mujeeb-Kazi 2003), as the applied aspects took priority with the task of delivering output gains suited for the 2050 food security targets.

7.8 Underexplored Areas

7.8.1 Land Races: Definitions, General Characteristics and Practicality Potential

Wheat landraces represent an important and highly diverse genetic resource. The simple fact that they have been cultivated by farmers for centuries until the modern breeding and improvement started some 100 years ago, confirms their uniqueness and importance for agriculture and humanity. More recently, there is growing interest in wheat research and production communities in landraces and the Web of Science database is indicative of this interest. For example, search for "wheat landraces" resulted in: 2001–2005 – 29; 2006–2010 – 83; 2011–2015 – 117; 2016–2018 – 112 publications, respectively. There are scientific journal issues that are entirely devoted to wheat landraces. Additionally, there are scientific conferences and workshops that have focused on wheat or cereal landraces and not to mention numerous projects on landraces that were supported by diverse donors. This renewed interest and priority given to wheat landraces on one hand is driven by the general concern related to degradation or under-utilization of plant genetic resources. While, on the other hand, it is also been driven by their uniqueness in terms of resistance to abiotic stresses, diseases, pests and importantly by their end-use plus nutritional quality.

The definition of landrace as such is important to agree upon. There are numerous names used to describe this genetic resource: heritage or heirloom varieties, ancient grains, local, traditional or primitive varieties or cultivars, etc. The first comprehensive review on the subject was published by Zeven (1998), who presented various definitions of a landrace, with intentions on how these definitions have evolved by the end of the nineteenth century. However, his conclusion was that the landraces were undefinable, since there were so many landrace properties affected by so many factors. The concluding definition was: "The landrace grown for a long period in the farming system concerned. As the environment changes annually and as the landrace becomes 'contaminated' – purposely or not – with few genotypes of other landrace(s), or cultivar(s) it will continuously adapt itself." When the European Union and the UK started projects on national inventories of the landraces, this definition proved too weak to be used to classify this genetic resource (Villa et al. 2006). The authors listed several defining characteristics associated with landraces: historical origin, recognizable identity, lack of formal genetic improvement, high genetic diversity, local genetic adaptation and association with traditional farming systems. Their definition of a landrace was "a dynamic population(s) of a cultivated plant that has historical origin, distinct identity and lacks formal crop improvement, as well as often being genetically diverse, locally adapted and associated with traditional farming systems." The meaning of this definition is used throughout this chapter.

The landraces Gene Bank collections and on-farm preservation increased interest to wheat landraces stimulated their inventories in national and international Gene Banks as well as *in-situ* in farmers' fields. A comprehensive review by Carvalho et al. (2012) identified 859,472 accessions of wheat in Gene Banks globally using World Information and Early Warning System on Plant Genetic Resources for Food and Agriculture (WIEWS). The landraces accounted for 26% or 225,120 accessions. The top six wheat germplasm collections included: CIMMYT, Mexico, which maintains a total of 111,396 accessions, 31% of which are identified as landraces; the National Small Grains Germplasm Research Facility (NSGC) at the USDA, ARS, USA, with 57,788 accessions, 57% of which are classified as landraces; the Institute of Crop Germplasm Resources, Chinese Academy of Agricultural Sciences (ICGR-CAAS), P.R. China, maintaining a total of 41,030 accessions (no accessions identified as landraces); the NBPGR, India, with a total of 35,889 accessions, of which only 2% are classified as landraces; ICARDA, Syria, maintaining a total of 34,983, 75% of which are landraces; and the NIAS, Japan, maintaining a total of 34,652 with only 4.4% classified as landraces. The European Wheat Database (EWDB) currently (EWDB 2012) consists of 160,262 records of which 38,865 are identified as landraces and may be accessed at http://genbank.vurv.cz/ewdb/default.htm.

Wheat landraces are still grown by farmers in many countries, primarily in remote, mountainous or harsh dry regions. They are big in numbers but small in area. The reports of wheat landraces continuously grown by farmers from time unknown until the present have been published for Afghanistan (Buerkert et al. 2006), Ethiopia (Eticha et al. 2006; Tsegaye and Berg 2007), Georgia (Akhalkats et al. 2010), India-Himalaya (Tripathi et al. 2017), Iran (Abbasabad Zarei et al. 2017), Italy (Porfiri et al. 2002), Jordan (Rawashdeh et al. 2007), Mexico (Delacy et al. 2000), Morocco (Chentoufi et al. 2014), Pakistan-Balochistan (Khan et al. 2010), Sahara desert (Zaharieva et al. 2014), South-Eastern Europe (Bosnia and Herzegovina, Bulgaria, Macedonia, Moldova, Montenegro, Romania and Slovenia) (Simeonovska et al. 2013), Syria (Bishaw et al. 2014), Tajikistan (Husenov et al. 2015), Tunisia (Robbana et al. 2019), Turkey (Morgounov et al. 2016) and Uzbekistan (Baboev et al. 2017). These landraces are still grown by farmers and represent a unique genetic resource being adapted to the changing climates and serving the rural communities despite the immense pressure from the modern high yielding varieties.

7.8.2 Wheat Landraces: An Additive Diversity Source

Wheat landraces have been attracting the attention of researchers for their diversity in morphology, adaptation pattern, grain quality and reflected in their genetic viability. Their diversity as such is considered useful for a positive character or a valuable trait required to be introduced or be maintained in wheat cultivars. From a practical breeding perspective, phenotypic and genetic diversity may have very limited values unless it contributes toward superior agronomic performance, tolerance to stresses or quality traits and overall practical outputs. On the other hand, DNA polymorphism holds importance to groups due to the germplasm based on its geographical origin, environmental conditions, use by farmers, parental relationship or other factors. Several studies using molecular markers documented the loss of genetic diversity of wheat landraces compared to modern varieties in worldwide durum wheat collections (Ren et al. 2013), Swiss bread wheat (Müller et al. 2018), Pakistan (Rasheed et al. 2016b) and France (Bonnin et al. 2014).

Ghimire et al. (2006), analyzing the diversity of 324 East Asian wheat landraces, was able to trace historical movement of wheat from the West to East along two "silk way" corridors and the third group of landraces originating from Nepal-Tibet movement to Western China. The study of SNP diversity of 172 Mediterranean durum wheat landraces from 21 countries identified four distinct groups: east Mediterranean, east Balkan and Turkey, west Balkan and Egypt and west Mediterranean (Soriano et al. 2018). The frequency of alleles positively affecting grain productivity was higher in western landraces compared to other groups. An international group of scientists (Cavanagh et al. 2013) developed a high-throughput array to interrogate 9000 gene-associated single-nucleotide polymorphisms (SNP) in a worldwide sample of 2994 accessions of hexaploid wheat, including landraces and modern cultivars. The impact of crop improvement on genomic and geographic patterns of genetic diversity was characterized. Extensive use of ancestral variation was often traceable to founders of cultivars from diverse geographic regions. Selective sweeps were found around genes involved in the regulation of flowering time and phenology. Breeder's selection likely acts on distinct targets or multiple alleles in different portions of the geographic range of wheat. The majority of selected alleles were present at low frequency in local populations, suggesting either weak selection pressure or temporal variation in the targets of directional selection during breeding and is probably associated with changing agricultural practices or environmental conditions over time.

7.8.3 Important Collections of Wheat Landraces

Wheat landraces, being diverse from the modern cultivars, attracted attention of breeding programs to incorporate them into crossing programs. However, undesirable genes that negatively affect crop performance are often co-introduced and this most often limits their use in modern breeding programs (Müller et al. 2018). Detailed genetic characterization is an important prerequisite to solve this problem and to make landraces more accessible to breeding. There are several important wheat landrace collections, which have been selected, described and thoroughly evaluated. In the 1930s, A.E. Watkins acquired landrace cultivars of bread wheat from official channels of the board of Trade in London, many of which originated from the local markets of 32 countries (Wingen et al. 2014). The geographic distribution of 826 landraces (called the Watkins collection) covers many Asian, European and some African countries. The collection was genotyped with 41 microsatellite markers and high genetic diversity was found, more significant than in a collection of modern European winter bread wheat varieties developed during 1945 to 2000. A core set, which captures the majority of the genetic diversity, was established. Further research on the Watkins collection identified a substantial amount of novel genetic diversity that either had not been captured in current breeding programs or which was lost through previous selection pressures (Winfield et al. 2018). A number of accessions carrying novel alleles and interesting chromosomal rearrangements were selected. Furthermore, using nested association mapping, panel 60 segregating biparental populations were developed from the Watkins collection (Wingen et al. 2017). A consensus map, named as the landrace consensus map (LRC), was constructed that contained 2498 genetic loci. In general, marker order was highly correlated, which provides support for strong synteny between bread wheat accessions. However, many exceptional cases of incongruent linkage groups and increased marker distances as well as translocations were found. Loci involved in recombination rate were identified by QTL analyses using the crossover counts as a trait. Genome-wide association studies (GWAS) on the Watkins collection identified SNPs contributing resistance to three wheat rusts (Pasam et al. 2017), single gene of resistance to stem base disease Eyespot (Burt et al. 2014) and allelic variation at the grain texture-controlling linked loci *Pina-D1* and *Pinb-D1* (Qamar et al. 2014).

The Vavilov Research Institute of Plant Genetic Resources (VIR) in St Petersburg, Russia holds one of the richest collection of wheat landraces tracing back to the early 1900s. The VIR panel, comprising of 295 accessions, includes landraces, breeding lines and cultivars that were genotyped using the genotyping-by-sequencing Diversity Arrays Technology platform (Riaz et al. 2017a). Cluster analysis identified two distinct groups, one comprising mostly of breeding lines and cultivars, and the other comprising of landraces. Diversity was benchmarked in comparison to a set of standards, which revealed a high degree of genetic similarity among breeding materials from Australia and CIMMYT. Furthermore, 11,025 markers were polymorphic in the diversity panel only, thus representing allelic diversity potentially not present in Australian or CIMMYT germplasm. The VIR panel was also screened for resistance to leaf rust (Riaz et al. 2017b). Based on DNA marker evaluation, 83 of the 300 lines carried known adult plant resistance (APR) genes, namely *Lr34*, *Lr46* and *Lr67*. Interestingly, lines carrying *Lr67* were mostly landraces from India and Pakistan, reconfirming the likely origin of this gene. Field testing identified a subset of 13 lines that consistently displayed high levels of APR across years and pathotypes. The Vavilov panel was also subjected to GWAS under inoculation by *Septoria nodorum* (Phan et al. 2017) and several QTLs contributing resistance were detected.

Wheat landraces introduced into Mexico from Europe, also known as Creole wheats, were adapted to a wide range of climatic regimes and represent a unique genetic resource (Vikram et al. 2016). Genotyping-by-sequencing technology characterized 8416 wheat landraces representing all dimensions of Mexico. Results revealed sub-groups adapted to specific environments of Mexico. A large percentage of landrace accessions were genetically very close, although belonged to different regions most likely due to the recent (nearly five centuries before) introduction of wheat into Mexico. Some of the groups adapted to extreme environments and accumulated a high number of rare alleles. Core reference sets were assembled simultaneously using multiple variables, capturing 89% of the rare alleles present in the complete set. Creole wheats were grouped into seven botanical varieties and demonstrated wide variations for HMW-glutenin subunit composition (Caballero et al. 2010). Similarly, Crossa et al. (2016) examined genomic prediction within the 8416 Mexican landrace accessions and 2403 Iranian landraces. These were evaluated under optimum environment as well as under drought and heat stress environments. Genomic prediction models included genotype × environment interaction (G × E). Population structure, when accounted for, decreased prediction accuracy by 15–20% as compared to prediction accuracy obtained when population structure was not accounted for. Prediction accuracy varied from 0.412 to 0.654 for Mexican landraces, and from 0.182 to 0.647 for Iranian landraces. Results showed promising prediction accuracies for potential use in germplasm enhancement and rapid introgression of exotic germplasm into elite materials. This experiment is an excellent demonstration of how large number of landraces can be evaluated and selected for breeding purposes.

Japanese scientific expeditions to Afghanistan in 1955, 1967 and 1979 collected wheat landraces from 16 provinces to establish the so-called Kihara collection (Sohail et al. 2015). The landraces and varieties were genotyped using 30,000 diversity array technology (DArT) genetic markers, of which 15,817 were polymorphic. The landraces were grouped into 15 sub-populations based on population structure and phylogenetic studies. The Afghan landraces were highly diverse compared with lines from other origins, possessing unique genes that might allow enrichment of the global wheat gene pool. Evaluation of the Kihara collection under rust and Fusarium Head Blight infection resulted in selection of a number of resistant accessions and their genetic characterization (Manickavelu et al. 2014). Afghan landraces mainly originated from dry areas and demonstrated superior drought tolerance and long roots (Winfield et al. 2016b). A genome-wide association study on 220 un-utilized Afghan wheat landraces against stripe rust resistance in eight locations resulted in 36 MTAs clustered in 9 QTL (Manickavelu et al. 2016). The study revealed that stripe rust epidemics in the geographical origin of this germplasm through time have permitted evolution of novel resistance loci.

There are extensive studies of wheat landraces collections in China, Ethiopia, Iran, Spain, Turkey and other countries. However, the Watkins, Vavilov and Kihara collections and Creole wheats demonstrate that application of modern phenotyping and genomic tools on a unique set of landraces produces spectacular results. We now digress to address the food security futuristic projections of 2050 and what the landraces have to offer.

To enhance the efficiency of breeding, climate change resilient wheat genotypes with wider genetic bases are of paramount significance (Trethowan and Mujeeb-Kazi 2008). This diversity is distributed within the three gene pools of Triticeae, of which the landraces occupy the primary gene pool and are thus easy to exploit as crossing onto wheat is characterized by perfect homologous chromosome associations and en-bloc genetic information transfer.

Wheat landraces have been grown for centuries on marginal lands, as well as under severe weather conditions. Due to this practice, significant adaptation traits, particularly those enhancing duration of photosynthesis, have been preserved in landraces (Lopes et al. 2015). Wheat breeding programs can be enriched with utilization of these preserved landraces after they are characterized (Robbana et al. 2019; Royo et al. 2014; Soriano et al. 2018). Wheat improvement programs for adaptation to global climate change, yield potential and wheat processing quality can be accelerated with holistic efforts among plant breeders, plant pathologists, plant physiologists, molecular biologists, wheat quality scientists and by data and germplasm sharing among the multidisciplinary team of scientists (Warburton et al. 2006; Chen et al. 2019, Velu et al. 2019). Landraces of wheat have been utilized to study various traits that are crucial for varietal development and performance. Genes have been identified that can result in development of wheats capable to combat climate change embracing multiple production constraints as exemplified below (Table 7.2).

The renewed interests of the last decade toward wheat landraces have mainly focused on addressing the narrow genetic basis of modern wheat cultivars. More recently, Lodhi et al. (2020) have elucidated various aspects that relate to the utilization and exploitation of landraces integrated with other aspects of wheat improvement. Evidence that wheat landraces, being very diverse and an important genetic resource for a number of traits which require improvement in modern wheat, is overwhelming. Modern phenotyping and genomic tools have made it possible to dissect the genetic mechanism of many traits, whereby wheat landraces may provide avenues for their

Table 7.2 Landraces studied with the potential to contribute to wheat varietal development and combat climate change factors.

Attribute Supporting References
Characterization: Royo et al. 2014; Soriano et al. 2018; Robbana et al. 2019; Warburton et al. 2006; Chen et al. 2019; Velu et al. 2019
d: Elings 1993; Royo et al. 2006; Sukumaran et al. 2015, Lopes et al., 2015; Soriano et al. 2018
Photosynthesis: Lopes et al. 2015
Early maturity: Sukumaran et al. 2016,
Biotic Stress Resistance
Rust: Van Dijk et al. 1988; Kertho et al. 2015; Lodhi et al. 2018; Wang et al. 2019; Chen et al. 2019; Gessese et al. 2019; Newcomb et al. 2013; Mengistu et al. 2014; Bux et al. 2012; Kolmer et al. 2018
Scab: Li et al. 2011
Septoria: Kidane et al. 2017
Abiotic Stress Resistance
Drought tolerance: Blum et al. 1989; Dodig et al. 2012; Reynolds et al. 2007; Sukumaran et al. 2018; Abu-Zaiton et al., 2018; Chaichi et al. 2019; Sehgal et al. 2016; Bektas et al. 2016; Ruiz et al. 2018; Winfield et al. 2016a; Zhang et al. 2018
Salt tolerance: Munns et al. 2003; Rames et al. 2008; Shamaya et al. 2017; Ayed-Slama et al., 2018; Mujeeb-Kazi et al. 2019
Heat stress tolerance: Hede et al., 1999; Ortiz et al. 2008; Lopes et al. 2015
Cold tolerance: Cao et al. 2012; Mohammadi et al. 2014; Kandel et al. 2018
Insect Resistance and Other Aspects
Aphid resistance: Du Toit 1989; Liu et al. 2018
Saw fly: Varella et al. 2017
Nematode: Li et al. 2011; Thompson et al. 2016
Iron and zinc: Xu et al. 2011; Guzmán et al. 2014; Velu et al. 2019; Velu et al. 2018; Goel et al. 2018
Quality: Peng et al. 2015; Guo et al. 2011; Lee et al. 2018; Ayala et al. 2016; Shewry 2018; Caballero et al. 2010
Molecular/Diversity: Bonman et al. 2015; Kidane et al. 2017; Li et al. 2011; Liu et al. 2018; Varella et al. 2017; Riaz et al. 2017b; Sohail et al. 2015; Phan et al. 2017
Durum landraces: Mohammadi et al. 2014
For drought, yield tability, cold, and adaptation: Ruiz et al. 2018

utilization in breeding assisted by molecular approaches. However, there are limited reports on actual use of wheat landraces in the development of new varieties. There is certainly a lag time between the discovery of the traits and associated genes and their translation into breeding. On the brighter side, the tremendous knowledge generated on wheat landraces in the last 5–10 years will be utilized in breeding new cultivars with a wider genetic base and will contribute to global food security. In essence, it is evident that landraces are rich reservoirs of additive diversity and as they are components of the primary gene pool upon crossing with wheat due to homologous associations, gene transfers are the norm. A more elaborate conclusion will be made toward the end of this chapter, when the overall practicality of genetic resource exploitation is covered around some priority *modus operandi* aspects.

7.9 Perennial Wheat

7.9.1 The Concept of a More Sustainable Perennial Wheat-Like Cereal. Is It Feasible?

7.9.1.1 What Benefit/s Would Come?

There have been much interest in developing perennial grain crops. Van Tassel et al. (2010) pointed out that despite 10,000 years of plant domestication, no single herbaceous perennial species for grain production has been domesticated in the human niche. The authors speculated this to be a consequence of the conditions prevailed during crop domestication. Farmers developed an unusual environment with regular soil disturbance, artificial dispersing of seeds and suppression of interplant competition; hence, annuals were best placed to thrive and adapt to this type of environment. Perennials generally allocate less resources to sexual reproductive effort than annuals; however, examples such as apple, olive and oil palm trees illustrate that selection can deliver perennials which allocate greater than 50% of mass or energy to reproductive tissues.

A perennial grain crop could contribute substantially to a sustainable food production system, with greater biomass and penetration compared to annual systems, and manage water and nutrients more efficiently. The development of perennial grain crops would be anticipated to provide more options for farmers under diverse and generally more marginal production conditions (Bell et al. 2010; Glover et al. 2010a, 2010b). Annual grain crops can lose 5 times as much water and 35 times as much nitrate as perennial crops. On the other hand, perennial crops require fewer passes of farm equipment, and less fuel, fertilizer and herbicide (Glover et al. 2010b). They offer soil stabilization through the seasons against erosion. We would predict improved nutrient balance against soil acidification, rising water-tables and salinity, and in general, enhanced ecosystem services. Such agro-ecological systems should provide farmers with greater flexibility, diversity and income stability in their enterprises.

Perennial cereals develop their canopies faster from established root systems and crowns and consequently have a longer photosynthetic season giving greater potential plant productivity. Bell et al. (2008, 2010) used bio-economic modeling to show that forage from a perennial wheat could be used to replace annual wheat stubble over summer and grain supplement at break of season, and permit increased farm stock numbers. The additional value added by grazing reduced the relative yield required for perennial wheat to be profitable. This analysis suggested that dual-purpose perennial wheat would be a profitable option for mixed crop–livestock farmers, and suggested that a perennial cereal yielding 40–60% of annual wheat could be economically attractive in southern Australian farming systems, especially if it was to provide additional forage for livestock.

7.9.1.2 Potential Pitfalls

While conceding potential benefits, some have considered the concept of perennial cereals as an unacceptable disease threat to annual cereal crops. The thought here is that a perennial crop would become a persisting accumulator of disease load, especially in the surviving root system, without the clean-out that comes with the death of annual plants and land use rotation. The type of perennial cereal being developed does die back to the crown over summer; however, the rapid emergence from the crown through leaf residue does represent the potential for persistent disease pressure. Furthermore, the survival of root systems also means greater potential for amplified root disease burden.

However, improvement of soil health and shift to beneficial soil microbiome populations has been documented in landscapes under perennial grass stands, where soil nutrients and organic matter composition is improved (Chan et al. 2001; Ghini and Morandi 2006; Weller et al. 2002). Such soils reduce the ability of pathogens to infect their hosts (Cox et al. 2005). Furthermore, the types of perennial cereal crops in some of the recent studies (see below) are hybrids of annual cereals and perennial relatives and have been shown to inherit the many disease resistances associated with the perennial parent species. Precisely because the disease pressures on perennials are greater, they have evolved strong genetic resistance mechanisms. The transfer of resistances to the perennial hybrids has been demonstrated in a number of studies (Cox et al. 2005, 2002, 2010; Hayes et al. 2012).

Despite these mitigating factors, it remains imperative that breeders of perennial cereals should be vigilant in producing germplasm with resistance to a wide range of foliar and root diseases, as well as developing and advocating cultural practices for the perennial crop farming system which will meet the challenges of sustainable agriculture and increased adaptation.

The grain yield of wild perennial grasses is much lower than intensively bred annual cereals. However, studies of related annual and perennial species across 28 genera found no relationship between seed production and lifespan, including wild annual grass species and wild progenitors to modern annual cereals (Cox et al. 2006; Vico et al. 2016), suggesting no intrinsic genetic link that might prevent the development of higher-yielding perennial grain crops through intense selection.

7.9.1.3 What Approaches Can Be Conceived?

Two broad options may be considered (Cox et al. 2006; Larkin et al. 2014): improve the grain yield and grain characteristics of existing perennial species (the domestication route); or find a way to add perennial growth habit to one of our cereal crops without substantially compromising grain yield or grain qualities (perennialize an existing cereal).

Domesticating a species that is already persistently perennial in the target environment so that it becomes a profitable grain crop requires attention to a number of desirable traits, including: non-shattering seed heads, so that seed is retained until harvest; free threshing seed, so that the grain can be readily separated from the husk; large seed size, so that the mechanical separation from the husk can be achieved efficiently; synchronous maturity, so that immature seed is not mixed with mature seed at harvest; stems strong and short enough to resist lodging; nutrient rich and palatable seed; flour functionality suited to intended food use; high seed yield; and forage yield and quality to enable use as a dual-purpose crop. For the purposes of this chapter, the question may be reduced to: Can an existing perennial grass be bred to become a substitute for wheat in the human diet with the nutrition and special food functionality associated with wheat? The genetic dissection of many of these traits has progressed greatly in the last decade, making it feasible to explore modifying a grass genome to achieve some of these objectives. For example, the *Q* gene and mutation in wheat is known, which is responsible for free-threshing grain and with influence on a number of other domestication-related traits including glume shape and tenacity, rachis fragility, spike compactness, plant height and spike emergence time (Simons et al. 2006). Likewise, gene families are known to control flowering (*Vrn* and *Ppd*) or plant height (*Rht* genes). It is feasible to look for mutations in gene homologs in perennial grasses, which might mimic these desirable phenotypes. And while the expanding power of gene editing offers some potentially non-GM prospects, currently some of the desirable traits, such as wheat-like gluten proteins, are likely achievable only through inter-specific gene transfer.

The other general approach is to introduce a perennial habit to an existing annual cereal species. Just as there is no single domestication gene, there appears to be no single perennation gene. Indeed, it also depends on which type of perennial habit is the target. The most attractive form of perenniation for our purposes is likely to be a plant that forms a complex crown composed of a number of ramets, some or all of which survive after harvest (the death of the reproductive tillers). Depending on the target environment, some ramets must survive either through the lack of moisture and heat of summer or the extreme cold of winter, and this is best achieved by some form of dormancy and subsequent appropriate response to seasonal signals in returning to a vegetative growth phase. Curwen-Mcadams and Jones (2017) give a good description of the biology at work. These considerations and experiences (see below) confirm that transfer of perennation will involve many genes and the most feasible approach will be to transfer whole genomes of perennial grasses to taxonomically-related annual cereals.

7.9.1.4 What Progress?

Some progress has been made with both of these strategies. The most advanced domesticated perennial grass for these purposes is intermediate wheatgrass, *Thinopyrum intermedium*. Efforts to domesticate *Th. intermedium* as a grain crop have been ongoing for over 20 years (DeHaan et al. 2005; Wagoner 1990). The resulting selections have now been given the common name *kernza* by The Land Institute, Salina, Kansas. Culman et al. (2013) compared kernza to annual wheat over two years under three management systems. In the second year, kernza reduced lower depth moisture and total nitrate leaching by more than 86% compared to wheat. The grain yield was very small in the first year but increased to 33% of the wheat in the second year and with a harvest index of 0.1. The progress of selection in kernza is described in DeHaan et al. (2016). Two generations of recurrent selection increased seed mass by 23%, and yield per land area in solid seeded plots by 77%. This was followed by eight generations of selection for individual seed weight achieving an increase of 0.52 mg per seed per generation. If progress continues in a linear fashion for another 44 years, 30 mg seed may be achieved. Progress has also been made in breeding a perennial habit into wheat. Cui et al. (2018) summarize the wheat × wheatgrass hybrid work in China from the early 1950s. Hybridization between wheat and *Th. intermedium* was initiated by Shancheng Sun at Northeast Agricultural University in 1953 and subsequently a large number of perennial wheat lines were selected from the progeny of backcrosses between the octoploid and hexaploid wheat–wheatgrass hybrids and wheat. Hybrids between diploid *Th. elongatum* and common wheat exhibited only weak regrowth in dry and cold conditions of the Shanxi Province.

Emphasis therefore remained on *Th. intermedium* as the perennial donor. The best lines had 50–56 chromosomes, including 8–14 *Th. intermedium* chromosomes, and were able to survive in the field for more than two years. The wheat–*Th. intermedium* partial amphiploids, especially of the "Zhong" series (Qi et al. 1979), possess multiple disease resistances as well as a moderate perennial habit.

Hayes et al. (2012) evaluated a diverse array of wide cross wheat × wheatgrass derivatives. including germplasm from Washington State University, The Land Institute (Kansas), the Russian State Agrarian University (Moscow), the Main Botanical Gardens (Moscow), the Chinese Academy of Sciences (Beijing), CIMMYT, and assorted lines from other germplasm collections. This initial evaluation of more than 150 wheat × wheatgrass derivatives assessed capacity to regrow post-harvest and yield grain over successive years. Several entries persisted to produce grain over three successive years. Although there was genomic instability within many of the accessions, regrowth was consistently associated with the presence of at least one whole-genome equivalent (14 chromosomes) from the perennial donor species. This research established that developing a perennial wheat might be feasible, although even the best of the tested germplasm was not sufficiently developed to be deployed commercially. Furthermore, the persistence and yield in subsequent years was too low. Larkin et al. (2014) extended these studies and with evaluation of new accessions. Some lines were able to perennate in one or both of the field environments in which they were grown, and a few were able to regrow through four seasons. Again, the only wheat-derived lines that could survive multiple seasons contained seven or more pairs of chromosomes from the perennial parent. Individual Triticeae genomes have seven pairs of chromosomes and, therefore, the chromosome constitutions of stable allopolyploids, both naturally occurring and synthetic amphiploids, stabilize at multiples of 14. Many fertile and stable wheat-derived amphiploids have been formed at the octoploid level ($2n = 8x = 56$; Mujeeb-Kazi and Riera-Lizarazu 1996) and this seems to be a benchmark of stability. Successful donor perennial parents in the production of perennial amphiploids have included *Th. ponticum* ($2n = 10x = 70$, decaploid), *Th. intermedium* ($2n = 6x = 42$, hexaploid) and *Th. elongatum* ($2n = 2x = 14$, diploid). From these experiences, the best near-term prospect for a perennial, wheat-derived grain crop is likely to be a full or partial amphiploid containing the tetraploid (AABB) or hexaploid (AABBDD) wheat chromosomes plus one genome equivalent (XX) from the donor. If the perennial donor is a polyploid, the extra genome is usually a synthetic genome, consisting of a mixture of chromosomes from the polyploid parent genomes, but where each of the seven homoeologous groups are represented in the synthetic genome. This creates a significant difficulty in that each time a primary partial amphiploid is produced, the synthetic genome may consist of a different mix of donor chromosomes (Banks et al. 1993; Han et al. 2004; Jones et al. 1999); intercrossing of such primary partial amphiploids may result in significant infertility and loss of the donor chromosomes.

7.9.1.5 What Lessons?

Curwen-Mcadams and Jones (2017) remind us that there is a range of types of perenniality in grasses, including caespitose (storing energy in the crown), rhizomatous (with thickened roots traveling underground) and stoloniferous (with stem-like spreading organs). The perennial wheat breeder also needs to consider if dormancy is required to survive harsh winters or hot and dry summers. A major recent study (Hayes et al. 2018) across four continents, involving a large diversity of wheat × wheatgrass hybrids, came to a number of conclusions: longevity was generally greater at sites with milder summers and winters; no genotype was superior in all environments; derivatives of *Th. intermedium* generally performed better at higher latitudes; derivatives of *Th. ponticum* or *Th. elongatum* generally performed better at lower latitudes; derivatives using a tetraploid wheat showed greater longevity than those using hexaploid wheat; and yield in year 1 was a very poor predictor of yield in year 2.

Any successful new crop will need ongoing breeding to bring continuous improvement and responsiveness to new disease threats, as well as new and changing environments. In practice, for the wheat-wheatgrass hybrid approach, it means new genetically diverse hybrids need to be generated, which are fully compatible with the existing perennial hybrids. Mixed synthetic genomes in partial amphiploids would make such breeding difficult. The mismatch of chromosomes in independently derived, partial amphiploids becomes problematic in establishing a breeding program. Robust perennial habit might be achieved in partial amphiploids from a polyploid perennial donor. However, ongoing breeding will need the ability to inter-cross compatible primary amphiploids in order to generate genetic segregation on which to impose heavy selection.

7.9.1.6 Suggested Way Forward?

The early successes of *Th. intermedium* (kernza) herald ongoing progress with the domestication option. Genomic selection may further accelerate grain yield and adaptation progress in kernza. Also, with expanding knowledge of target genes, it might be possible to accelerate genetic variability and progress through targeted mutagenesis via TILLING and gene editing. There may be other perennial grasses which offer opportunities for domestication, such as *Microlaena stipoides* (Henry 2012) and *Secale montanum* (Jaikumar et al. 2012; Kotvics et al. 2001; Reimann-Philipp 1995).

For wheat-wheatgrass hybrids, we suggest the use of a diploid perennial donor so that primary amphiploids are chromosomally compatible with each other, and consequently a breeding program can be readily achieved. Of the wheatgrasses that have been used in wheat × perennial breeding attempts, only *Th. elongatum* (EE) is a diploid and sometimes there has been confusion in nomenclature, such that the decaploid *Th. ponticum* historically has been called *Agropyron elongatum* and it is not always obvious which species was actually employed. However, as noted above from the Chinese experience, amphiploids with *Th. elongatum* (AABBDDEE or AABBEE) have so far proved only weakly perennial. *Elymus scaber* is an Australian native diploid perennial species of Triticeae, and adapted to a wide range of climatic zones and soil types in south-eastern Australia. Attempts were made to produce amphiploids following crosses to wheat (Newell et al. 2015). While fertilization and seed formation was readily achieved, two methods of embryo rescue failed to recover hybrid plants. Furthermore, the essential summer dormancy of various accessions of *E. scaber* proved to be facultative, and the authors conclude that *E. scaber* is not a priority candidate to be used as a perennial donor species in perennial grain, at least for south-eastern Australia. To date, the most promising perennial hybrids have involved polyploid donors, *Th. ponticum* or *Th. intermedium*. Otrastayuschaya 38 is a 56 chromosome grain and fodder perennial partial amphiploid between wheat and *Th. intermedium*, bred by VF Lybimova and NV Tsitsin, released in the former Soviet Union in 1978 (Berezhnoi 1987), but it had little impact.

It has been suggested that the task ahead for introducing the perennial habit to wheat through wide hybridization is analogous to the task of breeding the first commercially viable Triticales (Curwen-Mcadams and Jones 2017; Larkin et al. 2014). Delayed DNA replication at the heterochromatic telomeres appears to have been a problem for coordinating the behavior of wheat and rye chromosomes in meiosis in early triticales. Fully fertile triticales have lost much of the telomeric heterochromatin on the rye chromosomes. Similar issues may be problematic in the current wheat-wheatgrass amphiploids and close attention to the history of triticale breeding may offer a targeted way forward. Feldman and Levy (2012) describe the dramatic and rapid genetic changes that occur following allopolyploidization, including rapid loss of non-coding DNA, and both low and high copy DNA. A further lesson from the history of triticale is that hexaploid lines (AABBRR) have been generally more successful than octoploids (AABBDDRR). So too for perennial wheat, it may prove preferable to use a tetraploid wheat parent, thereby also achieving a greater proportion of the amphiploid genome being from the perennial parent. This was the approach in the development of Montana-2 (MT-2), the first commercially released North American perennial wheat in 1987, being a cross between durum wheat (*Triticum turgidum* L. var. *durum*) and *Th. intermedium* at Montana State University (Schulz-Schaeffer and Haller 1987). Jones et al. (1999), showed that the chromosome composition was still variable in MT-2, averaging 26.2 wheat chromosomes, 9.4 St chromosomes, 18.8 E chromosomes and 1.5 St/E translocations, giving an average total of 55.9. Cui et al. (2018) remind us that the interaction between nuclear genomes and cytoplasmic genomes can also be a source of instability and incompatibility. This is an area totally unexplored in the context of wheat-wheatgrass perennial hybrids.

Experience of chromosome instability in the early generations and loss of the ability to regrow, suggests early generation selection should emphasize regrowth capacity. While it is most practical to do this selection in the greenhouse at F_2 and F_3, ideally subsequent selections should be done in the field under targeted environmental conditions. Other traits for early generation selection should include seed size, plant height, self-fertility and chromosome constitution. Later generation selection should emphasize traits such as grain yield and quality, chromosome stability, forage biomass, determinant flowering, disease resistance, appropriate crown dormancy, robust spring regrowth and second harvest yield. Finally, the environmental services anticipated should be verified experimentally.

Funding support for the ongoing development of perennial wheat will be aided by finding a commercial niche for early generation product. This might be a graze and grain, dual-purpose perennial for mixed farming enterprises, in which the biomass and forage value will offset the expected reduced grain yield. Other possibilities are the use for stabilizing marginal parcels of land on otherwise profitable farms, or creating niche food or beverage markets such as are emerging for kernza.

7.9.2 Genetic Engineering for Wheat Improvement Focused on a Few Major Food Security Aspects

Wheat is the most important food crop throughout the world. To achieve yield potential and produce more food to feed growing populations, concerted efforts are needed for genetic improvement of wheat through breeding and biotechnology. Biotechnology is the genetic modification through insertion of foreign DNA into the wheat genome, called genetically-modified (GM or Biotech wheat). Traits for genetic improvement of wheat may include grain yield, diseases and herbicides resistance, abiotic stresses (drought, salinity, heat, oxidative stress, etc.), nutrient use efficiency (nitrogen, phosphorus, etc.), nutritional quality (iron, zinc, vitamins, etc.) and others depending upon the crops problem and demand. Several crop problems may be addressed with precise genetic changes at the molecular level in the genome for a specific trait that is regulated by a specific gene/pathway. Contrary to conventional breeding and hybridization, wheat developed through biotechnology tools needs approval from the regulatory agencies (National Biosafety Commissions) for lab manipulations, limited field trials and commercialization. A large number of genes for various traits have been identified in crop plants and other organisms. These genes were transformed into wheat and tested in labs and limited field trials with economically useful results, but so far, unfortunately, no GM Biotech wheat has been approved and released for commercial cultivation globally. The major bottleneck is the approval from regulatory biosafety committees and reluctance from the seed councils and variety approval bodies. Wheat biotech programs are going on at many public sector institutions with huge investments. Some multinational biotech companies have discontinued their efforts. Some of the breakthroughs, but not limited to, in wheat biotech research have been addressed here. With the addition of new genome modification/editing tools coupled with doubled haploids and speed breeding approaches, in the future it may be possible to produce transgene free wheat with minimum regulatory hurdles in the shortest possible time. Salient functional facets are elucidated in this coverage.

7.9.2.1 Tissue Culture and Transformation of Wheat

Unlike many other dicot model plants, wheat is a monocot with specific *in-vitro* culture requirements for growth of callus and regeneration of plants. Selection of explants is the basic requirement to grow wheat tissue on artificial media in controlled environments. Only a few wheat tissues have the capacity to divide and regenerate. Immature embryos with scutellum layers are the best explants. These explants are taken from immature embryos 14 days post anthesis from greenhouse or field grown plants (Habib et al. 2014). Alternatively, callus raised from mature embryos may be used for transformations. Unfortunately, other tissues do not have the capacity to grow and make complete plants. Before the start of transformation experiments, availability of genes, vectors and transformation methods are decided. Both Biolistic (gene gun) and *Agrobacterium* methods work well for wheat transformation with some modifications. At present, *Agrobacterium* is the preferred method of transformation and widely used globally (Ishida et al. 2015; Sparks and Jones 2016; Hayta et al. 2019; NIAB and Rothamsted, UK, NIBGE, Pakistan and many other laboratories). Some labs have also tried direct insertion of genes via biolistics (Hamada et al. 2017) and *Agrobacterium* infiltration into the meristem region of mature wheat embryos. In nature, *Agrobacterium* infects dicots only, but this problem has been solved by the use of *Acetosyringone* in the culture medium. The *Agrobacterium* method is less expensive and capable of inserting low copy numbers into the wheat genome and is more convenient for testing the transgenic material in segregating generations and observe gene expression. However, the very large and complex polyploid wheat genome (2n = 6x = 42; AABBDD, 17,000 MB) hinders PCR selection of transformants in subsequent generations.

7.9.2.2 Production of Genetically-Modified Wheat

Initially, wheat tissue culture was started on the cultivar Bobwhite as a model system. With the advancement in media combinations and skills, it is possible to regenerate and transform any wheat variety with variable transformation efficiency (Jones et al. 2005). It is also possible to transform locally adapted wheat varieties with any desirable characters or traits. The wheat genome has already been sequenced and a wealth of information is publicly available. Since the era of the Green Revolution in the late 1960s, wheat yield has been stagnant and annual genetic gains need augmentation annually to meet the food security targets. The genetic gain in grain yield potential for wheat has mainly been achieved by increasing the grain number per unit of land (Griffiths et al. 2015). In some countries (UK, Germany, Denmark, France), yield of winter wheat is quite high (8 t h^{-1}) due to a long growing season and favorable weather conditions. In the rest of the world, spring wheat is cultivated and the average yield is very low (2.0 to 2.5 t h^{-1}), due to a short growing window and prevailing growth conditions. The biotech success stories in other crops (cotton, maize, etc.), mostly centered on herbicide and insect resistance genes, which have been taken from sources other than plants. The origins of most of these genes was from microorganism and have nothing to do with the physiology of the plants. Most of the important plant traits are polygenic and there are only a limited number of plant genes which have a profound impact on the overall yield increment of the plant. In wheat, a remarkable increase in research publications and academic reports has been noted, but the real practical breakthroughs are very limited commercially. There are only success stories of insect and herbicide tolerant traits in four crops. Efforts are underway to translate this knowledge, transform and test genes for drought, salinity, heat, nitrogen use efficiency, quality traits, etc., with economic impact providing optimism that transgenics may offer (Ray et al. 2011).

7.9.2.3 CRISPR/Cas9 Genome Editing in Wheat

Precise genome engineering (genome surgery) by the CRIPSR/cas9 (clustered regularly interspaced short palindromic repeats) system has recently emerged as a tool for targeted alteration of DNA and crop improvement. This system is versatile, enabling editing of multiple genes simultaneously based on the guidance of small RNAs. This system works well in diploid species with small genomes but its application remains a challenge for plants with polyploidy and complex genomes like wheat. Shan et al. (2014) demonstrated the application of CRISPR/Cas9 in wheat protoplasts for *TaMLO* powdery mildew resistance gene editing. Out of 72 lines, 4 lines were found to be edited for the restriction enzyme sites (Wang et al. 2014a). Kim et al. (2018) reported genome editing for two abiotic stress related genes (*TaDREB2*) and *TaEFR3*). The technology is advancing rapidly with modifications in Cas protein and transfection of RNPs (ribonucleo-protein complexes) directly into wheat germ-line cells via a biolistic gun in a transgene free manner (Liang et al. 2017; Zhang et al. 2016). Some specific defects of crop or addition/deletion of single nucleotides in important genes (like grain size, grain weight, disease resistance) may be possible (Wang et al. 2018).

7.9.2.4 Potential Traits for Genetic Improvement of Wheat Through Biotechnology

There is an enormous arsenal of genetic diversity that permits allelic variation to be incorporated into bread or durum wheat and setting priorities of what resource to utilize needs extraordinary vigilance. There are conventional sources of genetic variability, close relatives or distant relatives of wheat providing various options and described in previous sections. These routes have the advantage that variation from these sources can be ideally incorporated into homoeologous recipient chromosomes, thus over-riding misplaced effects of extraneous genes. Because of the time-bound target of food security, it is paramount that novel areas be researched and with the spectrum of traits having significant advantage to give crop resilience and combat the hazards of climate change, such research should be the subject of such novel digressions. The arena of improvement covers all biotic/abiotic stresses, yield components, insect and pest resistance, micronutrient needs, grain quality, mineral toxicities, etc. The most important linkage with climate change are environmental stresses, on which we focus here, are drought, salinity, heat and yield per se. These are polygenically controlled traits and complex to handle but complementary inputs are suitable. For varietal development, a holistic approach is required that permits en-bloc improvement to realize any yield advantage, thus the task of a single gene to bestow such a product is extremely challenging and we must be cognizant of this limitation. Some selected major traits significant toward food security contributions are touched upon and the potential of their success elucidated.

7.9.2.5 Yield Potential

The development of biotechnology tools allows us to characterize the role of key genes in plant productivity and crop yield. The large and complex wheat genome offers a large reservoir of genes that can be targeted through breeding, transgenics and genome editing (Nadolska-Orczyk et al. 2017). These genes are classified in four functional groups:

1) Transcription factors that regulate spike development which mainly affects grain number
2) Genes involved in metabolism or signaling of growth regulators-cytokinins, gibberellins and brassinosteroids, which control plant architecture, stem hardiness and grain yield
3) Genes determining cell division and proliferation, mainly impacting grain size
4) Floral regulators influencing inflorescence architecture and seed number and impact on plant architecture and grain yield.

Many of these genes (e.g., *TaCKX6-D1, TaTEF-7A, TaGW2, TaTGW6, TaGS5, TaGS-D1, TaSus1, NFYAs, NFYBs, NFYCs, TaNAC2-5A* and *TaCW1*) are members of a multi-gene family. To break the yield barrier and yield stagnation, these and many others related genes are under testing and are viewed with optimism.

7.9.2.6 Climate Change

Crop yields are projected to decrease under future climatic conditions (Ray et al. 2019). This has likely led to approximately 1% average reduction (-3.5×10^{13} kcal year^{-1}) in consumable food calories in ten crops. Asseng et al. (2018) have studied effects of CO_2, water, nitrogen (N) and temperature on wheat grain protein concentration in a changing climate for the world's main wheat-producing regions. This necessitates the need for molecular genetic analysis of genes responsible for these changes in the wheat genome. The abiotic stresses of highest priority are covered under drought, salinity and heat below.

7.9.2.7 Drought

Climate change creates a need to find genes responsible for drought tolerance and also to mitigate the effects of drought. Following the discoveries in plant genes isolation and transformation methods (Jones et al. 2005), several laboratories started working on wheat improvement. Four drought-related genes (*DREB1A, HVA1, AtNCED3* and *WXP1*) were transformed into three wheat varieties and these wheats gave significantly higher yield under severe drought stress conditions (Malik et al. 2013; Saeed et al. 2011). Transgenic wheat with *DREB1A* transcription factor gene under rd29 promoter gave 19.49% higher grain yield than the best drought tolerant control variety (*Dharabi*) in a desert-like climate where no irrigation was applied for 40 days after sowing (Saeed et al. 2011). Since drought is a complex polygenic trait, it is expected that many genes are contributors to yield. Major genes include (but not limited to) protein kinases (*MAPKs, CIPK, CDPK*, other kinases), transcription factors (*AP2/ERF, bZIP, NAC*, Zinc finger), protein degradation, protein modification, other nuclear protein, metabolism of abscisic acid, metabolism of other hormones, osmotic adjustments, dehydrin/*LEA*, other transporters, reactive oxygen species and amino acid metabolism. Genes also controlling root architecture and stomatal development play an important role in soil moisture extraction and its retention. A few key genes have been identified to be positive or negative for genetic modifications (Kulkarni et al. 2017). When *HaHB4* TF was transformed into wheat and tested across a range of growing conditions, it exhibited adaptation to drought-prone environments and gave 20% more yield (Gonzelez et al. 2018). Validya et al. (2019) have developed opabactin (OP), an ABA receptor antagonist that mimics the effect of abiotic stress tolerance and water use. This OP enhances drought tolerance and hyperactivate ABA transcriptional responses in wheat. NAC TF *TaSNAC8-6A* is tightly associated with drought and improves water use efficiency (Mao et al. 2020). Many studies explain the potential role of genes in controlling drought stress, which is an essential productivity constraint resulting from climate change.

7.9.2.8 Salinity

Wheat is a moderately salt tolerant crop, so will require improvement for salt tolerance. For wheat, the ECe threshold level is 5.0 dS/m at 25°C, while 50% reduction in crop yield is achieved at an ECe of 9.9 dS/m. Salinity problems are more common in arid/semiarid regions and canal irrigated plains of many countries. There are many adaptive mechanisms at both cell and organ level (Basil and Blumwald 2014; Mujeeb-Kazi et al. 2019; Munns and Gilliham, 2015; Munns and Tester 2008; Rauf et al. 2016). The *AtNHX1* has been transformed into wheat with significant improvement in salinity tolerance. Moreover, homologs of this gene have further been isolated from wheat, barley, *Leptocloa fusca* and many other plants and tested in wheat under constitutive and stress inducible promoters alone or in combination with the *AVP1* gene that is a proton (H^+) generating pump at the tonoplast membrane. These genes may include *HVA1* (Habib et al. 2014), *LfNHX1* (Rauf et al. 2014), HKT1 (Shahzad et al. 2014), HVP1 (Haq et al. 2019), *Nax1*, *Nax2*, SOS pathway, *AtCIPK16*, *DERB1A*, *AtJUB1*, *HVGSNOR1*, *AtPLC3*, *AtPLC5*, *AsKa*, *PQL1*, *ROS/SOD* scavenging mechanisms, heat shock proteins, transcription factors, miroRNAs, etc. Novel transformation and genome modification methods are being used in major crops to improve their performance and to deliver positive gains in yield under saline conditions.

7.9.2.9 Heat

Heat has emerged as another major climate change hazard that threatens wheat productivity. For each degree Celsius rise in temperature, losses occur (Ullah et al. 2019); even up to 6% globally. Since, this is also a polygenic trait, breakthroughs in developing heat stress tolerance has been limited. Some physiological parameters, like canopy temperature depression and membrane thermo-stability, appear to be effective indicators of heat tolerance. Through functional genomics and transgenic studies, some genes and heat shock proteins (*TamiR159, TaHsfA6f, TaMBF1c, TaFER-5B, TaOEP16-2-5B, TaGASR1, TaB2, TaHsfC2a, TaWRKY33, TaNAC2l, TaLTP3, TaHsfA2d, TaHSF3* and *HSP26*) and their regulators have been identified (Ni et al. 2018; Singh et al. 2019). Similarly, Su et al. (2019a) found that transcriptional regulation of zeatin, brassinosteroid and flavonoid biosynthesis pathways may play an important role in wheat heat tolerance.

7.10 Historical Non-Conventional Trends for Exploiting Wheat's Genetic Resources

We trace the progressive build-up initiated in the late 1960s that attempted to access novel genetic diversity for wheat research. This effort was minimal and took over a century to gain volatile interest in order to become a routine phenomenon as a viable component of wheat improvement. The stages of progress were very distinctive, with the 1970s showing a surge in widespread activity. These are elucidated across distinctive time spans that now enable us to consider alien integrated alleles as routine in pre-breeding and breeding activities since various hurdles were overcome. Interspersed within these time spans are pertinent digressions that have a crucial role in facilitating progress.

7.10.1 Pre-1900

We trace the progressive build-up initiated in the late 1960s that attempted to access novel genetic diversity for wheat research. This effort was minimal and took over a century to gain volatile interest in order to become a routine phenomenon as a viable component of wheat improvement. The stages of progress were very distinctive, with the 1970s showing a surge in widespread activity. These are elucidated across distinctive time spans and enable us to consider alien integrated alleles as routine in pre-breeding and breeding activities since various hurdles were overcome. Interspersed within these time spans are pertinent digressions that have a crucial role in facilitating progress.

7.10.2 1901–1920

The early twentieth century marked the possibility of using the findings of successes of wide crosses of wheat with its relatives for the benefit of agricultural productivity, although one of the major objective was elucidating the origin of wheat. It also coincided with the rediscovery of Mendel's Laws of Inheritance by three scientists, Dutch biologist Hugo de Vries, German geneticist Carl Correns and Austrian plant breeder Erich von Tschermak-Seysenegg (Eucarpia 2020; Tschermak-Seysenegg 1951). Of these three, Tschermak-Seysenegg was actively involved in crossing of *T. dicoccoides* with *Ae. ovata* and *Ae. cylindrica* and *T. monococcum* with rye (Percival 1921) and even coined the term "Agrotricum" (Schmidt et al. 1953). According to McFadden and Sears (1947), M.A. Carleton suggested as early as 1901 that rust resistance and certain other characteristics could be transferred from the tetraploid wheats (emmers and durums) to common wheat. According to Percival (1921) and Reitz et al. (1945), Hillman gave an account of a hybrid made in 1903 between wheat and couch grass (*T. vulgare* × *Agropyrum repens*), but both papers indicated the absence of supportive evidence for the hybrid. (Note: early European workers refer to the genus as *Agropyrum*, while North American workers refer to the genus as *Agropyron*). Vavilov (1914) described resistance to *Puccinia triticina* (leaf rust) and *Erysiphe* (powdery mildew) in *T. durum*, *T. polonicum* and *T. turgidum*, and "perfect immunity" to leaf rust in *T. monococcum*. *T. dicoccum* had both susceptible and immune types to both diseases (Vavilov 1914).

Hayes et al. (1920) presented a summary of the history of wide crosses involving *Triticum* spp. In their work, they focused on stem rust (*P. graminis*) resistance in crosses of susceptible *T. vulgare* with resistant *T. durum* and with immune varieties of *T. dicoccum*. Hayes et al. (1920) acknowledged that the breeding of spring wheat for rust resistance in Minnesota, USA, began as early as 1908 by Dr E.M. Freeman and E.C. Johnson. Around this time, McFadden (1917) suggested the possibility of obtaining greater winter hardiness from rye, *S. cereale*. Edgar McFadden (in McFadden and Sears 1947) briefly reported on his crossbreeding experiments involving *Agropyron* dating back to 1914, and called attention to numerous characteristics in that genus that would be of value in the cultivated wheats. McFadden (1930) also reported that in 1916 he had initiated crosses between Yaroslav emmer (*T. dicoccum*) and wheat "in part as a practical breeding project and in part to test...a theory...that...wheat originated from intercrossing between certain wild wheat-like grasses, such as...wild emmer (*T. dicoccoides*), wild einkorn (*T. monococcum*) and one or more various species of *Aegilop*." This work by McFadden is one of the most spectacularly successful in the history of wheat breeding, as the stem rust resistance gene transferred to wheat from Yaroslav emmer (named *Sr2*) became an important part of the Green Revolution wheats of the twentieth century, and is still used in hundreds of wheat cultivars around the world.

7.10.3 1921–1930

Solving the mystery of the actual origin of wheat still intrigued scientists well into the twentieth century. Classification of wheat into three distinct groups (Einkorn, emmer and vulgare) had been earlier based on crossability, but Sax (1922, 1923) also studied chromosome pairing and segregation in crosses between these three *Triticum* groups and concluded, "many of the desirable characteristic properties of the emmer and vulgare wheats could be combined." After studying crosses between *T. vulgare* and *Ae. cylindrica*, they came to the conclusion that "suggests the possible origin of the *vulgare* wheats from a cross of *Aegilops* with a wheat species of the Emmer series." Sando (1935) reported crosses dating back to 1922 between *Aegilops*, *Secale*, *Triticum* and *Haynaldia*, with most success between *Aegilops* and *Haynaldia* with emmer wheats. According to Reitz et al. (1945), N.B. Zizine (also spelled as Tzitzin), made the first successful cross between wheat and *Agropyron* (later identified as *Ag. intermedium* (Verushkine and Shechurdine 1933)), although Reitz et al. (1945) states that Sando had also made crosses between wheat and *Agropyron*, but fertile derivatives were not obtained until 1935. Other researchers reporting wide crosses during this period were Thompson (1926) and Leighty and Sando (1927). McFadden (1930), in McFadden and Sears (1947), discussed obtaining desirable characteristics from *Aegilops*.

7.10.4 1931–1950

Russian workers were actively involved in wide hybridization during the 1930s, notably Tzitzin (1936, 1937) in Omsk, Siberia, Verushkine and Shechurdine (1933) in Saratov (all investigating wheat – *Agropyron*/*Agropyrum* crosses), Udol'skaja (1937) (drought tolerance), Blinkova (1937) (winter hardiness), Ljubimova (1936) (*Secale-Agropyrum* hybrids) and Weruschkin (1937) (*Agropyron-Triticum*, *Aegilops* and *Secale* hybrids). Tzitzin (1937) mentioned that crosses between *Agropyron* and wheat were immune to rusts, smut, bunt and showed lower winter frost damage and drought tolerance and inherited luxuriant root systems and tolerance to saline soil characteristics (Tzitzin 1936). The Russian workers were possibly inspired to develop more productive wheat cultivars following the great famine of the early 1930s and attempting "daring and resolute methods of reconstruction" of agriculture under the Socialist system (Verushkine and Shechurdine 1933).

Armstrong (1936) also attributes the first successful wheat–*Agropyron* cross to N.V. Tzitzin, who apparently aimed to develop perennial wheat. Armstrong (1936) crossed *Agropyron* with both tetraploid and hexaploid wheats, with the goal of preventing soil drift on the Canadian prairies. The potential value of *Agropyron* for wheat improvement continued with the papers of Novitskii (1941 in Schmidt et al. 1953) (*Helminthosporum*, *Fusarium* and stem saw fly resistance), Reitz et al. (1945) (aphid resistance), McKinney and Sando (1951) (yellow streak mosaic virus) and Jones et al. (1952, in Schmidt et al., 1953) (wheat jointworm resistance). Smith (1942) reported extensive attempts to hybridize wheat with 15 species of *Agropyron*, 7 of *Elymus*, 2 of *Festuca* and one each of *Lolium* and *Secale*. Problems were encountered in this era with failure to produce hybrid seeds, sterility and failure to breed true in subsequent generations (Schlehuber and Sebesta 1958).

7.10.5 The Post-1950 Era: Preamble

Before embarking into the post-1950s development, it is crucial that the complexities of affecting alien gene transfers is addressed. These are more pronounced when there is a lack of homologous chromosome pairing in wheat/alien hybrids, particularly involving members of the secondary and tertiary gene pools. We thus discuss the era that emerged post-1950s, involving transferring small segments of alien variation into wheat and their shortening before an in-depth elucidation of utilization of genetic resources for wheat improvement that gained momentum over time. Similarly, the evolution of strategies exuded greater simplicity of agricultural practicality, thus warranting here a detailed exposure to present an adequate story of advance. Hence, there followed a progressive build-up from a complex scenario prevalent until the 1970s, leading subsequently to somewhat simplistic useful practical output options keeping food security targets in perspective.

The impinges on alien allelic introgression into wheat marked the new era of transfer of shortened alien chromosome segments to wheat in the post-1950s and was considered a pre-requisite for operational research functionality. Details are addressed here to avoid prevalent gaps in the progression facets of this innovative area by maintaining a continuum.

Apart from genes introduced to wheat from other *Triticum* species (e.g., *Sr2*), it is unclear if any commercial cultivars of wheat were released prior to 1950 carrying characteristics from the more distantly related species. This may be a consequence of failure of those chromosomes from non-*Triticum* species to pair with wheat chromosomes, and the presence of too much alien chromatin carrying detrimental characters from the wild species. Addition lines carrying non-pairing alien chromosomes are generally unsuitable for agriculture because of loss of the introduced chromosomes over generations. Even Godron (1858) noted that "hybrids...tend to return after a certain number of generations to one of the two types which have given them birth," which today we attribute to chromosome loss at meiotic divisions. Alien chromosome addition lines were first developed by O'Mara (1940) to determine the effect of individual rye chromosomes in wheat. He backcrossed a wheat-rye amphiploid with normal wheat followed by a selection for the rye chromosomes but none of these addition lines were released as commercial cultivars.

Chromosome translocations are a significant advance toward solving the problem of gene transfer between wheat and alien chromosomes that do not normally pair at meiosis; translocations can be induced or spontaneous. The first reported use of irradiation for transfer of a disease resistance gene was that of Sears (1956), who translocated a segment of an *Ae. umbellulata* chromosome carrying the leaf rust resistance gene *Lr9* to bread wheat. Later, Knott (1961) used this method to transfer the stem rust resistance gene *Sr26* from *Ag. elongatum* chromosome 6Ae#1 to wheat chromosome arm 6AL. The history of *Sr26* may be traced back to 1939, when crosses were made at the University of Saskatchewan, Canada, between decaploid *Th. ponticum* and hexaploid *T. aestivum* cultivar Chinese Spring (Shebeski and Wu 1952). The transfer of *Sr26* has proven to be one of the most successful examples of utilization of an alien resistance gene. This gene is still effective against all known races of stem rust (including the entire Ug99 lineage) and derived shortened recombinants (Dundas et al. 2015b) have been distributed to wheat breeding programs worldwide. Irradiation as a means of alien chromosome transfer is still used. Recently, Chen et al. (2013) used ^{60}Co γ-rays to induce chromosome segment transfer of 6VS chromosome segments from *H. villosa* to the wheat chromosomes.

The most common type of spontaneous translocation is the centric fusion or Robertsonian translocation. Sears (1972) demonstrated that in wheat plants that are monosomic for two chromosomes; both monosomes can mis-divide through their centromeres at meiosis and re-join to give chromosomes carrying arms from each original chromosome. The most successful of these types of translocations in agriculture has been the 1BL/1RS translocation in wheat (Zeller 1973), involving a segment of rye chromosome 1R carrying the disease resistance genes leaf rust (*Lr26*), stripe rust (*Yr9*), powdery mildew (*Pm8*) and stem rust (*Sr31*) fused to the long arm of wheat chromosome 1B. According to Schlegel and Korzun (1997), in the last century, two research stations in Germany worked with lines carrying the 1R chromosome; Geog Riebesel in Salzmunde used Petkus rye while G. Katterman in Weihenstephan used triticale. Riebesel was apparently inspired by the publications of Erich von Tschermak-Seysenegg to make his first crosses in 1924 (Rabinovich 1998). This translocation chromosome was incorporated into many European cultivars (Bartos 1993), including the Russian cultivars Aurora and Kavkaz, and subsequently into many lines released by CIMMYT in Mexico (Rajaram et al. 1983). The resistance provided by *Lr26*, *Yr9*, *Pm8* (Bartos 1993) and *Sr31* (Pretorius et al. 2000) has been overcome by virulent pathotypes of leaf rust, stripe (yellow) rust, powdery mildew and stem rust (respectively), particularly the Ug99 lineage of stem rust pathotypes for the latter gene.

7.10.6 Homoeologous Pairing

The donor species in many cases are wild grasses, which carry many genes enabling survival in uncultivated environments but may not necessarily be beneficial to agricultural production. Translocation chromosomes involving alien segments have two disadvantages that limit their application in commercial agriculture:

Firstly, the inserted alien chromosome segment may not fully compensate for the loss of indispensable genes on the deleted wheat chromosome. Secondly, the size of the introduced alien segments may be excessively large, carrying deleterious genes reducing yield and/or quality (Ali et al. 2016). These problems might be addressed through induced homoeologous chromosome pairing (i.e., chromosome engineering). Chromosome pairing behavior between homoeologous chromosomes of the three genomes of wheat (A, B and D) is normally prevented by the action of several genes, the principal one being *Ph1* on chromosome arm 5BL (Okamoto 1957; Sears and Okamoto 1958; Riley and Chapman 1958). Wheat lacking chromosome 5B shows homoeologous pairing between wheat chromosomes of different genomes and also with other species in the tribe Triticeae. Sears (1973; 1978a, 1978b) produced a number of chromosome 3Ae#1 transfers from *Ag. elongatum* to wheat 3D by exploiting homoeologous chromosome pairing in the absence of the *Ph1* gene. The wheat line carrying the 3Ae#1 chromosome (TAP 67) came from Oklahoma, but it is unclear if the Agrotricum ancestor was developed in Kansas (Bakshi and Schlehuber 1957) or by L.H.J. Shebeski in Canada (Schmidt et al. 1953). Several white seeded forms of these recombinants were selected by workers at the University of

Sydney and have since provided the basis of stem rust resistance (*Sr24*) in many Australian wheat cultivars (Friebe et al. 1996; McIntosh 1991).

The most common method of inducing homoeologous recombination is the *ph1b* mutant (Sears 1977), which is a deletion of an interstitial section of the 5BL arm carrying the *Ph1* locus. Sears (1983) first described the production of wheat–3Ae#1 recombinants in a *ph1bph1b* background. Transfer of disease resistance into wheat using the *ph1b* mutation has been reported from *Ae. triuncialis* (Fan et al. 1995), *T. longissimum* (Ceoloni 1984; Ceoloni et al. 1989, 1988, 1992), *Agropyron elongatum* (Kibirige-Sebunya and Knott 1983), *Secale cereale* (Anugrahwati et al. 2008; Koebner and Shepherd 1986, 1987, 1988), *T. cylindricum* (Bai et al. 1995), *T. timopheevii* (Bhowal et al. 1993), and *Ae. speltoides* (Mago et al. 2013, 2009). Anugrahwati et al. (2008) described the development of a wheat line carrying the *SrR* (*Sr50*) gene on an interstitial 1 RS chromosome segment by intercrossing overlapping primary recombinants and then selecting a secondary recombinant followed by inducing the production of tertiary recombinants. Use of wheat nullisomic for the entire 5B chromosome carrying the *Ph1* gene for transfer of disease resistance has been from *Ag. elongatum* (Kibirige-Sebunya and Knott 1983), *Ae. caudata* (Dyck et al. 1990), *S. cereale* (Joshi and Singh 1978; Koebner and Shepherd 1986, 1987).

An alternative to the *ph1b* mutant in wheat is the characteristic of some *Ae. speltoides* accessions of inducing homoeologous recombination. For example, Riley et al. (1968) was able to transfer the yellow (stripe) rust resistance gene *Yr8* from the *Aegilops comosa* chromosome 2 M to wheat by genetically-induced homoeologous recombination using the *Ph1* suppressor factor in *Ae. speltoides* and produced the line Compair. Chen et al. (1994) transferred two duplicate and independent Ph^1 genes from *Ae. speltoides* and later Chen et al. (2003) developed wheat–*H. villosa* translocations using this tool. Similarly, Cai et al. (2008) used the Ph^1 inhibitor to transfer Fusarium Head Blight resistance from related species and also later by Aghaee et al. (2000) to show enhancement of homoeologous pairing and by Wang et al. (2003) in outputs of salt tolerant wheats.

Mujeeb-Kazi (2003) proposed another strategy of using the *ph1b* genetic stock, modified the schematic in 2006, shown here in Figure 7.6 under gene pools and the tertiary gene pool. This allows all genetic stocks stored in gene banks and available with cytogeneticists to be utilized for effecting transfers, with the flexibility to also define breeding efforts more efficiently after screening these stocks, and if available their derived alien disomic addition lines, etc., thus establishing what to utilize for agriculture via homoeologous incorporation protocols.

7.10.7 Isolation of Homoeologous Recombinants

The questions arise as to:

(i) how a researcher can determine if a recombinant line between homoeologous chromosomes has been produced;
(ii) how much alien chromatin is involved in the transfer; and
(iii) which chromosome of the cultivated species is involved in the recombination event?

There are three basic methods to accomplish this: i) chromosome pairing analysis; ii) molecular marker studies; and iii) *in situ* hybridization of chromosome spreads. Sears

(1978a, 1978b) selected recombinants between *Ag. elongatum* 3Ag (3Ae#1) and 7Ag chromosomes and wheat 3D and 7D chromosomes. Sears then test-crossed critical plants with wheat carrying telocentric chromosomes of both wheat and *Agropyron* chromosomes and examined meiotic cells for evidence of pairing between the telocentric chromosomes and the suspected wheat–*Agropyron* recombinant chromosomes. Later, Zhang and Dvorák (1990), using an *Agropyron*-specific DNA probe, was able to largely confirm the relative sizes of the *Agropyron* segments, which had been predicted earlier by Sears (1978a, 1978b).

The post-1950 era of harnessing alien genetic resources until the mid-1980s was one where the *modus operandi* of intercrossing wheat and alien species overcame all major hybrid production complexities. There were many subtle often overlooked details that upon overcoming helped resolve the programs major progress constraints. These are well defined and addressed in this review. Hybrid production we perceive now as a rather simple exercise but, despite this, still requires state-of-the-art experimentation modes to be in place, and reiteration that accessing alien diversity does not remain an over-complicated subject currently.

During the 1960s, Kruse pioneered wide crossing of wheat with other genera and published his findings (Kruse 1967, 1969, 1973, 1974), but satisfactory cytological verification remained elusive. He did elude to using hormones to aid these crossing attempts. The early 1970s generated huge interest in the use of some chemicals (E. caproic acid, salicyclic acid, chloramphenicol, etc.) for treating wheat spikes to be used in wide crossing for overcoming crossability barriers by acting as immuno-suppressants (Bates and Deyoe 1973); an effort where outputs could not be cytologically validated. But later in Mexico (Mujeeb-Kazi and Rodríguez 1980, 1981), the United States (Sharma and Gill, 1983a, 1983b, 1983c), Canada (Fedak 1979, 1980; Fedak. 1977) and Australia (Islam et al. 1978, 1981) found success with adequate cytological validation. Additive supporting information on this subject came via publications (Mujeeb-Kazi and Bernard 1985; Mujeeb-Kazi and Kimber 1985; Wang 1989; Mujeeb-Kazi et al. 1987; 1989; Mujeeb-Kazi and Asiedu 1990; Mujeeb-Kazi and Hettel 1995; Sharma 1995; Jiang et al. 1995; Fedak 1998, etc.), that consolidated many missing links prevalent in wheat wide crossing with diverse alien resources. Initial interest was devoted to wheat/barley or barley/wheat crosses plus use of rye and *Agropyron*, using its broad commonly used terminology then in place. The alien resources were indiscriminately hybridized with both bread and durum wheat and the salient output was a wealth of complex hybrids across a wide array of Triticeae genera produced and/or reported upon by the above researchers. Significant to attain these outputs and make dramatic advances were a few operational protocols initially related exclusively to intergeneric hybridization. The interspecific aspects shall follow after this first step:

7.10.8 Intergeneric Hybridization Steps for Wheat/Alien Crossing

a) Crossing protocol: pollen load, time of pollination, hormone treatment pre- and post pollination
b) role of wheat around its superior crossability promoting genes
c) embryo rescue, pre-conditioning to overcome dormancy and its differentiation

d) integrating a pre-breeding protocol to overcome the poor agronomic plant type of the wheat parent used in the initial hybridization
e) development of genetic stocks like amphiploids, backcross-1 self-fertiles, disomic chromosome addition
f) developing a living herbarium of all self-sterile F_1 hybrids for continued production of amphiploid production and a wide array of backcross 1 self-fertile derivatives, thus precluding the need for global researchers to produce new hybrids if so desired.

These details have been previously been published in a scattered manner. Thus covering them in-depth may be informative and are an off-shoot of the protocols adopted and elucidated by Mujeeb-Kazi and Hettel (1995). All figure schematics have been modified for use in this chapter and originate from various communications of A. Mujeeb-Kazi during 1985 until 2017.

A step-wise manner is the recommended way forward in the area of wide crossing, where in the future researchers may be interested in producing more unique hybrids. The steps reviewed are a consensus of the major program researchers have mentioned earlier, adopted in wide crossing and shared information in conferences and symposia reporting, to some extent in the *International Wheat Genetics Symposia*, the Annual Agronomy meetings between 1980 until early 2000, the *Wheat Newsletter* between 1995 to over a decade where the basic, strategic and applied significance of wide crosses was elucidated on for agricultural gains, with all the researcher outputs complimented by the unpublished file archives with information from the CIMMYT wide cross program. The latter program focused mainly on generating practical output gains to aid food security targets, and thus dispensed the cutting edge scientific efforts to other international centers of scientific excellence. The *modus operandi* in CIMMYT was to develop genetic stocks of practical productivity value, distribute them internationally from the program or via CIMMYT'S international nursery section and thus aid global wheat pre-breeding/breeding programs that were a reflection of what priorities were formulated for the way forward for wheat improvement using alien genetic resources. The wide crossing steps of general interest were common to all wide cross professionals, had subtle variations, never entirely used as a full package, and are as follows:

a) In the crossing protocol from 1875 until 1965, all hybrids were produced by adopting standard crossing protocols and fortuitously did give limited success. It was only in the early 1980s that the merit of using field grown range grass nurseries was utilized for wheat/alien intergeneric hybrid production, where the wheat cultivars were also field grown in the same location. An initial attempt was made by the wide cross group of Dr B.S. Gill at Kansas State University in the grass nursery at Evans Farm in Logan, Utah. The following year, in late 1981, Dr Douglas Dewey of USDA/ARS at Utah State was instrumental in further extending the Logan farm facilitative links to the CIMMYT group and thus in 1982 and 1983 this led to the astounding production of just over 1000 intergeneric hybrids across 117 hybrid combinations, of which several were unique. He also made available his special planting plot of grasses at the Paradise location close to Evans farm. The wheat parents were both bread and durum cultivars planted at the start of spring in 1982 and 1983 at the Evans farm, which was utilized for crossing during the respective summers by the members of the CIMMYT team.

Field work was done over six weeks followed by embryo culturing in Logan by CIMMYT's Mujeeb-Kazi and Silverio Roldan, then brought down to Mexico where subsequent studies were conducted under controlled conditions prior to the fertile derivatives being taken for field studies in Mexico and derivatives also provided to internal breeders, CIMMYT'S outreach colleagues and international/national program collaborators that formed the foundation of the applied global practicality dimension of wide crosses.

Crucial to achieving this hybrid production, successes were from the following experimental steps:

Early pollination or bud pollination before stigma appeared receptive or even late pollination was a variable step from the standard, as these two facets circumvented fertilization incompatibility present under conventional schedules. In addition to this was the time of pollination over the working day, where pollen shed occurred very early just after dawn, particularly for the small anthered Elymus species. Some were restricted to shed pollen from noon and 2:00 p.m. and a very few even into the evening under the long daylight conditions of Logan, where natural light permitted emasculation up to 8:30 p.m. In essence the hybridization efforts were a consistent full long day exercise from 6:15 a.m. until 9:15 p.m. over near two months. No planned crosses were missed. The anthers appeared yellowish and partially extruded, in which case pollen had questionable viability and did not deliver hybrids. This prompted new combinations to be given immediate astute attention of their anther extrusion and utilized immediately. Some variations that aided hybrid production were:

1) to use the difficult alien species as female parents and obtain success as occurred for *Ag. ciliare, Ag. fibrosum, Ag. trachycaulum, Ag. yezoense, E. canadensis*, etc.,
2) an additive variation that ensured success chances was to make double pollinations, recognizing that this research effort was a costly daily financial venture where successful outputs were mandatory, and
3) additional variation was in the novel use of the alien diploid grass parent where hybrid production was elusive, even when more than 2000 florets were pollinated by it.

Our subsequently fortuitously locating of the grasses late flowering autotetraploid from permitted hybridization success is shown for *Psathyrostachys juncea* in Figure 7.7. This led to our then suggesting that using alien autotetraploids had a distinct advantage, as the very first backcross with wheat onto the self-sterile F_1 hybrid would and did instantly produce a fertile amphiploid.

Pre- and post-pollination hormone treatment, where GA3 and 2,4-D were the two main chemicals, aided success. Usage of these two had been mentioned earlier in literature and its mandatory involvement took off widely after the reports of Kruse (1973, 1974). Crossability genes in wheat is governed by three genes *kr1*, *kr2* and *kr3*. Chinese Spring wheat possesses these genes and thus intergeneric hybrids have been produced with considerable ease. The highly successful role of Chinese Spring has been attributed to these genes that this cultivar possesses (Falk and Kasha 1981; Fedak and Jui 1982) earlier reported for its crossability with rye (Riley and Chapman 1967). The loci are located on chromosomes of homoeologous group 5 in *kr*1 (5B), *kr*2 (5A) and *kr*3 (5D) respectively, acting in a complementary manner, with the *kr*1 *kr*1 loci contributing most

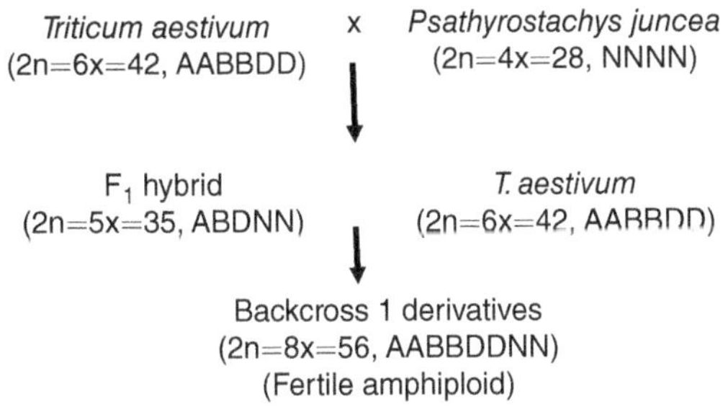

Figure 7.7 Schematic showing production of F_1 hybrids from crossing bread wheat with an auttetraploid *Psathyrostachys juncea* source and its backcrossing to bread wheat to produce an amphiploid.

to the crossability frequencies. The *kr4* locus on chromosome 1A in a Chinese Sichuan wheat is now added (Youliang et al. 1992), as are some other crossable cultivars (Asakazikomugi and Fukuhokomugi), with both possessing a more superior agronomic type than Chinese Spring (Fedak 1998; Jauhar 1995). Their role has been widely reported. Other improved agronomic type wheats with improved crossability have been since identified, of which one is TH3929 (McIntosh, 2013). Chinese Spring wheat nevertheless, when wide crosses was gaining global interest, was still a favorite with all cytogeneticists. Numerous alien exchanges were being made in it and the genetic stocks produced by Dr Sears. Chinese Spring also figured in wheat/alien translocation stocks. The stunning practical gains take-off by the T1BL.1 RS wheats in mid-1970s, as later reported by Rajaram et al. (1983), catalyzed researchers to use other translocations. Very few translocations gave promising practical outputs. Hence, efforts got into place in 2009 at an international forum in Mexico to transfer these stocks into widely adapted wheats using a simple backcrossing process, a concept that has never effectively materialized. The main drawback with the Chinese Spring wheat background is that its agronomic phenotype was not adapted to field conditions. as it was late, tall, had susceptibility to rusts, and screening for other agronomic traits remained non-informative. Thus, breeders in general shunned its use, including intergeneric hybrids produced with this wheat, though having tremendous academic appeal, were of little interest for applied wheat productivity targets. It was only later that a pre-breeding protocol was innovatively introduced in 1984 that derivatives from F_1 hybrids with Chinese Spring were integrated with the main frame breeding and became widely reported by Mujeeb-Kazi and Asiedu (1990), supported by field validation that advanced derivatives of hybrids produced with Chinese Spring could be phenotypically significantly improved for plant type and fit the crops cultivation cycle.

7.10.8.1 Embryo Extraction and Handling

Treating the crossed spikes with GA3 supported longevity of the spikes. GA3 was considered to be instrumental in cell elongation and maintenance of substantial moisture that allowed the pollinated florets to prolong their turgor and give a few extra days for the embryo to gain size and then culture between 15 to 18 days. Early culturing has been done after 12 days but the embryos were so minute that planting properly and having the scutellum placed correctly was tedious. Getting a few extra days was a huge help. A media for use to aid small embryos has been little used, but shows positivity (Tiara and Larter 1978).

After embryo rescue, keeping them in the dark and introduction of refrigeration for 15 days was helpful in overcoming any incipient dormancy that is common in mature harvested seed, and this step was automatically applied without any experimentation to validate the claim. After refrigeration, the tubes were placed in the dark under air-conditioning at 18 C and observed for seedling initiation when they were kept in a 14 l–10 d (light/dark) hours set growth chamber. These steps were critical and aided in giving very high success, i.e., up to 90% embryo differentiation. Transplanting of the grown seedling was routine into jiffy pots, transferred to the greenhouse and after tillering, each hybrid was physically cloned to increase its number. A batch was used for inducing amphiploidy, another batch used for generating backcross progeny for use in breeding, and a few clones maintained as pure hybrids for a living herbariums bi-annual sub-cloning and increase.

7.10.8.2 Pre-Breeding Protocol

In order to make wide cross products breeder user-friendly, it was essential to generate progeny derivatives that would adapt well to the regular wheat crop growth cycle conditions, since having a Chinese Spring wheat phenotype was definitely a huge disadvantage for exploiting wide cross outputs. There exist three major options of charting the pre-breeding options:

1) Hybrid between two parents, i.e., A × B, where the AB hybrid is advanced by backcrossing it to the same wheat parent, i.e., A shown as: A × B = AB × A = backcross ABA, where the A wheat parent is the Chinese Spring cultivar. This mode is for the purists who like to develop cytogenetic stocks within a pure Chinese Spring background.
2) Hybrid between two parents, i.e., A × B, where the AB hybrid is advanced by backcrossing it to a different well-adapted wheat parent, i.e., C shown as: A × B = AB × C = backcross 1 ABC, where the A wheat parent is the Chinese Spring cultivar and C is a well-adapted wheat cultivar. Just one backcross by an elite wheat cultivar improves the BC_1 phenotype dramatically. Since the derivative will carry the alien chromosomes, it is prudent to have all cytogenetic manipulation aspects addressed before the plant type improvement falls in place, i.e., any involvement of the *ph* stock where in its homozygous state it promotes wheat/alien exchanges, as shown in Figure 7.6, or make the initial hybrid with a *ph* recessive stock. The process is lengthy and accordingly subsequent efforts led to use of interspecific hybridization as a priority procedure to harness alien variation, as described in the interspecific section that follows.
3) Hybrid between two parents, i.e., A × B, where the AB hybrid is advanced by backcrossing it to two different well-adapted wheat parents, i.e., C and D shown as: A × B = AB × C = BC_1 derivative/s ABC, where the A wheat parent is Chinese Spring cultivar and C is a well-adapted wheat cultivar followed by the second backcross with another elite cultivar D to give BC_2 derivatives AB × C = ABC × D = ABCD

combination. This is akin to what is known as the F_1 top crossing procedure in routine operations. The presence of the alien chromosomes are a constraint that will warrant taking care of, as expressed above.

7.10.8.3 Development of Genetic Stocks

By using a combination of the above pre-breeding options, genetic stock development has resulted in which the major outputs have been for fertile backcross 1 derivatives, either in a pure Chinese Spring background or a mix with one elite bread wheat cultivar. The consistent use of clones from the self-sterile intergeneric hybrids has been the source to progressively generate fertile wheat/alien fertile amphiploids. Both these genetic stocks are of great value in utilizing them for subsequent alien introgression programs and to conserve genetic diversity in a user-friendly background.

7.10.8.4 Establishing a Living Herbarium

All F_1 self-sterile hybrids that were perennials were an excellent source for being maintained in the form of a living herbarium, as this allowed researchers to produce new backcross 1 derivatives upon their need to have their desired wheat parents involvement, with the first backcross done in the herbarium location in CIMMYT, Mexico. The F_1 perennial hybrids have been so maintained for over 100 intergeneric combinations (Mujeeb-Kazi et al. 2017), by growing them under a screen house or greenhouse conditions, where twice a year each hybrid was physically re-divided into clones. The clones form the hybrid population for producing additional amphiploids, backcross 1's and further pure maintenance.

7.10.9 Interspecific Hybridization

In view of the lengthy process in generating practical on-field wheat productivity goals, using alien resources that have high homologous potential between their wheat chromosomes led to add on to the wide crosses aspects the interspecific option where wide crosses was defined by the criteria mentioned earlier, thus excluding all those genera that could be easily hybridized under a total field operation. This emphasis shift went on to tap into the genomic diversity that involved the wheat progenitor diploids of the primary gene pool as a priority and then the B^s genome sources of the secondary gene pool. Of these three forms, the order of interest set its priority for the D genome diploids, next came the A genome and lastly the B^s genome diploids. Of these, the latter two were a futuristic investment, as to have their genetic stocks produced and conserved in gene banks for subsequent use should additive variation be required or a durum improvement program so desired for new under- or un-utilized variation.

7.10.10 Additive Durum Wheat Improvement

An optional source of durum improvement is an additional use of trait-rich *Ae. tauschi* accessions identified from screening of their hexaploid synthetic stocks. Durum wheat cv. Cappelli has a *ph1c* genetic stock, the use of which in direct crosses between the

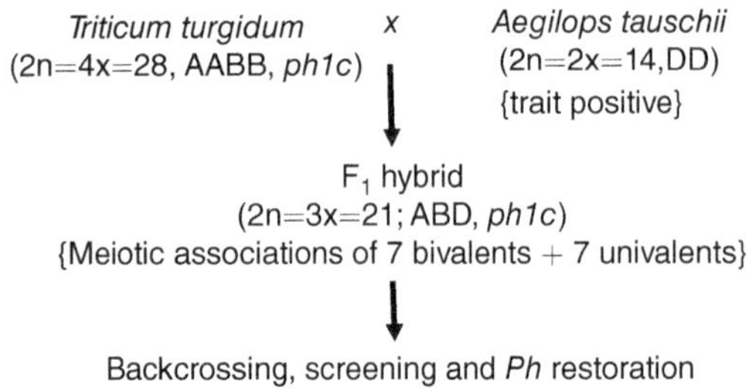

Figure 7.8 Durum wheat with the *ph1c* genetic stock permitting homoeologous association of up to 7 bivalents between the A genome of durum and the D genome of *Aegilops tauschii*; modified from Mujeeb-Kazi (2006). The bivalents are due to the homoeology between the A/D genomes, promote transfers from the D genome into the A and the F_1, then used in pre-breeding further by backcrossing, etc. with non-*ph1c* durum cultivars.

cultivar with the desired *Ae. tauschii* accession of practical significance facilitates preferential pairing of the A and D genome chromosomes up to 7 bivalents to obtain the D genome transfer initiated at the F_1 stage, following which the F_1 is utilized further in pre-breeding by crossing it with the normal non-*ph1c* cultivars (Figure 7.8).

The production of these interspecific stocks utilized to some extent the major criteria mentioned above for intergeneric hybrids but it was rare that the progenitor diploids were difficult to hybridize with durum wheat. The alien species were all vernalized and field grown, so pollen quality was excellent, and quantity abundant. The wheat crossability genes had little effect when the D genome was involved with direct crosses into bread wheat cultivars (Gill and Raupp 1987, for procedure). The schematic of the direct hybridization route is a potent means of improving wheat and is shown Figure 7.9. Where it was desired to hybridize other Aegilops resources in experiments and hybridization, success was low, and the earlier mentioned hormonal manipulation and pollination timings were beneficial. An observation was that where Aegilops was used as the female parent and pollinated with durum wheat to produce the D genome synthetics, the embryo recovery touched 80%. However, regeneration frequency was extremely poor, suggesting that media formulations like that for ill-defined embryos, etc. (Tiara and Larter 1978) may improve the situation, but has not been taken up in practice.

The D genome diploid has shown to aid wheat improvement by either direct crossing of the diploid into bread wheat or by its pollinating into durum wheat to produce the synthetic hexaploid. Direct crossing ensures that the transfers are restricted to the D genome of bread wheat and in the latter case transfers happen across all three wheat genomes; hence, using susceptible bread wheats are required to discern the contribution of the positivity of the trait within the diploid parental accession. However, the latter "Bridge Crossing" route has been a favorite for applied breeding, as intraspecific hybridization is automatically included, over-riding the durum/bread wheat pentaploid breeding option.

An additive strategy to maximize practical gains vs. time demand targets of food security, which we perceive are to deliver timely outputs that combat food security constraints, require prioritizing the *modus operandi* across at least three major facets:

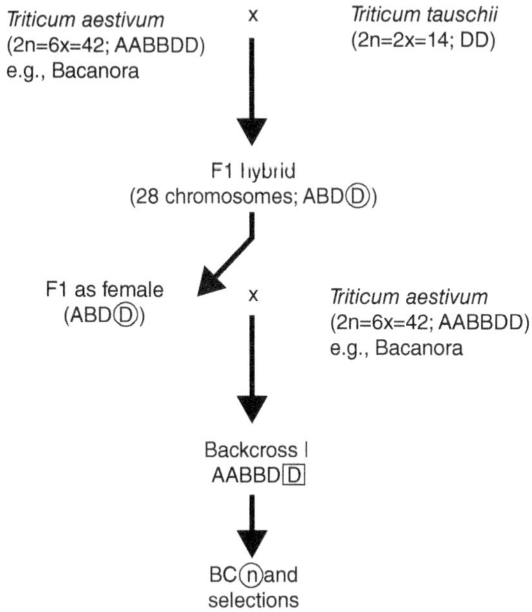

Figure 7.9 Schematic demonstrating alien transfers from *Aegilops tauschii* to elite *Triticum aestivum* cultivars requiring specific trait additions via direct crossing and backcrossing (*Source*: Mujeeb-Kazi and Hettel 1995).

1) parental choice to tap on allelic diversity
2) incorporate a process that will generate swift outputs
3) integrate cutting edge molecular technology/ies.

The first two are discussed here and the latter is left for a special section at the end.

7.10.10.1 The Parental Choice

We take this step cognizant of the fact that plant scientists have innumerable options to generate valued wheat germplasm and way forward strategies also vary enormously. However, as the projected timed targets due to food security tasks is set for 2050, it is prudent that some crisp pointers surface and preferences are established. There is ample evidence from the publications between the early 1990s over a decade of support focusing on exploiting the D genome synthetic hexaploids, as there are close to 1500 synthetics available from all global sources in the public domain but they are underutilized. Furthermore, the A and B^s genome synthetics are also a back-up option but so far completely un-utilized for harnessing their breeding potential. The D genome synthetics are demonstrated resources for all breeding characters so far evaluated and thus stand as a topmost priority. The synthetics possess multiple assets across multiple disciplines and also many synthetics have multiple stress resistances. The DNA polymorphism within these stocks provides an easy infusion of using molecular tools for enhancing program efficiency.

7.10.10.2 Shortening the Breeding Cycle by Inducing Homozygosity in Desired Early Breeding Generations

Anther culture had been an early choice for inducing homozygosity, despite its genotypic specificity and inability to integrate with wheat breeding programs. The exciting finding by Barclay (1975), too being genotype specific, could not be integrated with wheat breeding. It was only in the mid-1980s, when the lead publication of Laurie and Bennett (1986) demonstrated the effectiveness of a wheat/maize haploid production system that led to the refinement made by workers with the usage of hormones (Suenaga and Nakajima 1989), use of other sources like pearl millet (Ahmed and Comeau 1990), sorghum (Laurie and Bennett 1988; Ohkawa et al. 1992), Teosinte (Ushiyama et al. 1991), Tripsacum (Riera-Lizarazu and Mujeeb-Kazi 1993) and protocol modifications using detached tillers from field grown plants (Riera-Lizarazu and Mujeeb-Kazi 1990) in breeding populations that permitted handling of large populations under controlled conditions more simplistically. These steps have been reviewed during the last two decades and progressive details are described by Patial et al. (2017) that are very informative. They cover the role of *Imperata cylindrica* as being a very effective option for wheat haploid production. However, the majority preference so far remains with the wheat/maize system. Within a decade following Laurie and Bennet's 1986 initial report and the need for a post-pollination hormone application by Suenaga and Nakajima (1989, Mujeeb-Kazi and Riera-Lizarazu (1996) concluded that embryo formation frequencies from wheat/maize combinations came to about 25%, with 90 to 95% regeneration of these embryos and upon colchicine treatment gave 100% doubled haploids for all bread wheat cultivars, thus the integration with wheat breeding was perfect. The detached tiller procedure was perfectly adapted to breeding material in any filial generation where tillers cut from breeder selected plants became the source for producing the doubled haploids on those specific plants under controlled growing conditions. This symbiotic unison of field and laboratory inputs was initiated as early as 1998 by the wide cross group in Mexico when a MONSANTO/CIMMYT hybrid testing program to evaluate the MONSANTO chemical hybridizing agent was in progress. Breeder selected F_1 plants were so treated. This concept has been extended for the HELP (hybrid enabling line profiling) program discussed by Van Ginkel and Ortiz (2018). In general, the overall value of applying the haploid induction protocol as described above has been defined in detail by Mujeeb-Kazi et al. (2006), where its advantage, other than for use in breeding swiftly and delivering homozygous progeny, has been addressed for several other plant research areas. The protocol is adapted to durum cultivars also. This single additional step can augment a program's progress significantly and is worthy of widespread research facet integration, as it has been taken up by many private companies who have made a major shift in their breeding procedures with doubled haploidy being pivotal. There is a preference in some wheat breeding circles to delay the doubled haploid production a for few generations allowing a selection pressure to take place as the promising DH success frequency is enhanced which favors cost-effectiveness instead of producing DH's on an F_1 population, unless the objective is to capture F_1 heterosis.

There does exist the need for involving cytological checking of the haploids, both for bread and durum wheats as aneuploidy (both hyper- and hypo-ploidy plus chromosome fragmentation) does occur (Figures 7.10 and 7.11).

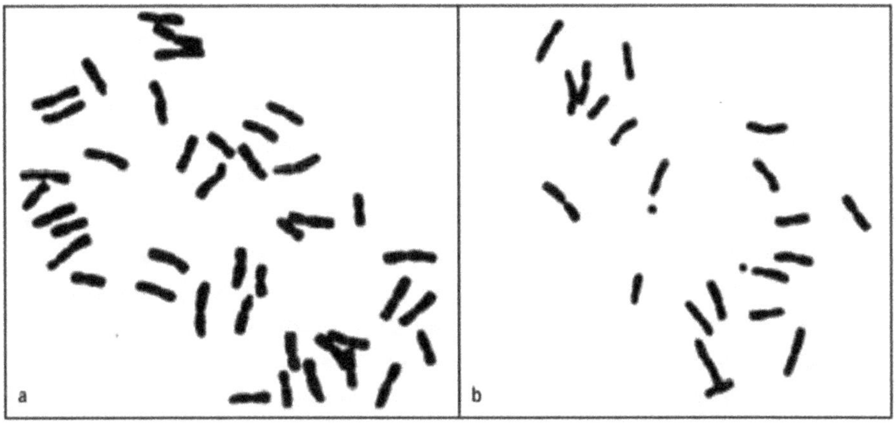

Figure 7.10 a) Mitotic metaphase spread showing its 42 chromosome complement; and b) its n = 3x = 21 polyhaploid with 21 chromosomes.

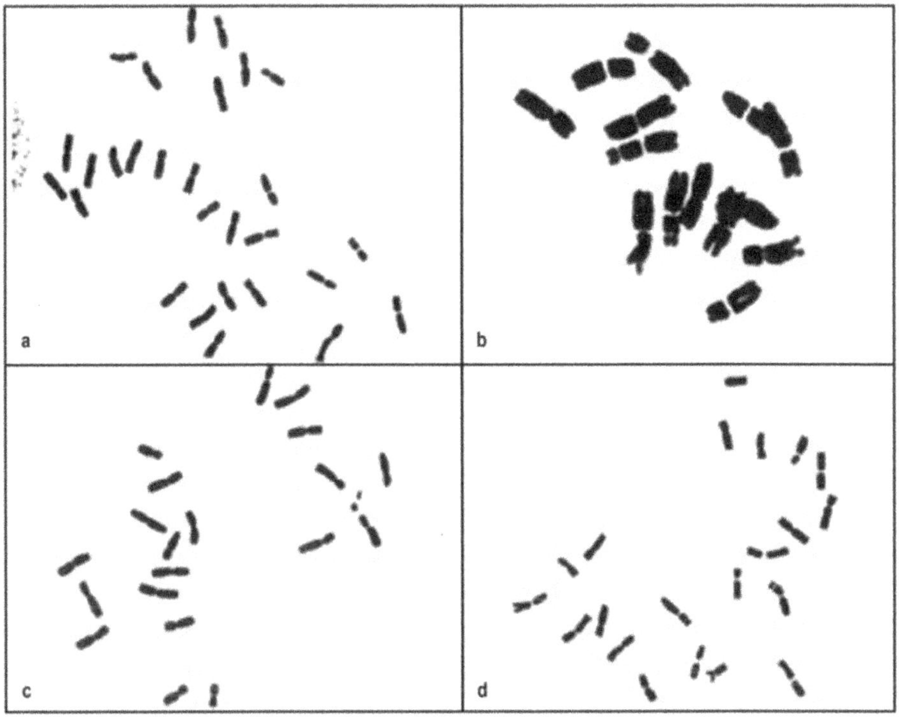

Figure 7.11 a) Mitotic metaphase spread of *Triticum turgidum* with its 28 chromosome complement; b) a polyhaploid with n = 2x = 14, AB chromosomes; c) and d) cells with aneuploid somatic counts in the polyhaploids of *T. aestivum* with 20 and 20 + 1 telocentric.

7.10.10.3 The Integration of Molecular Development Options for Efficiency and Precision

This is discussed in the following section on the genomics overview presented.

7.11 Alleviating Wheat Productivity Constraints via New Genetic Variation

Wheat productivity around the world is heavily dependent on commercial cultivars carrying genes derived from related species. Almost all wheat cultivars would carry at least one important character of this type. We have much to owe to the foresight and pioneering efforts of researchers of the past who not only developed germplasm (amphiploids, alien chromosome addition, substitution or translocation lines), but also the tools to induce recombination (*ph1b* mutant plus *ph1c* in durum) and the markers to select lines with introgressed and shortened alien chromosome segments

Use of genetic resources in related species and genera will continue to expand further, mostly by non-GM methods. The good news for wheat breeders and those concerned with increasing wheat production in response to increasing global population is that there seems to be no shortage of genetic diversity to keep exploiting. Working with wild relatives of wheat is rarely straightforward and researchers routinely encounter obstacles from non-crossability, sterility or lack of recombination (Kishii 2019). The keys to successfully using this germplasm are two-fold: i) initiate the program with as many sources as are available; and ii) produce as many introgression derivatives as possible with resources available.

We have hardly begun to exploit the vast genetic resources available to us. Barkworth (2020) has listed 36 genera in the Triticeae tribe. Of these, only a few (e.g., Triticum, Aegilops, Thinopyrum and Dasypyrum) have contributed agriculturally useful characters that have been released in commercial cultivars. By far the majority of alien-derived genes confer resistance to diseases such as stem, leaf and stripe rusts. Nevertheless, the potential value of wild relatives of wheat for many other characters has been recognized since the beginning of last century but is yet to be deployed widespread in cultivars.

Many of the perceptions mentioned above have been addressed and advances have rendered use of alien resources to be rather routine; but the practicality of on-farm performances remain limiting, for which the need is to set stringent priorities instead of delving into basic scientific findings. The need is to combat the food demands projected for 2050 that can only be achieved by focusing on applied research endeavors.

The arsenal of beneficial genetic variation on hand provides optimism that these resources need step-wise exploitation to help in delivering what the food demands are, and are thus complimentary to other strategies that have gained preference in various research circles that we too are stressing here. Presented are these other resources of variation that also hold wheat improvement promise for the following few addressed here.

7.11.1 Biotic Constraints

Biotic organisms of greatest concern are those which: a) cause massive damage to wheat crops; b) mutate into new virulent pathotypes; and c) have the ability to spread widely and quickly on wind and clothing. Without doubt, stem rust (caused by *Puccinia graminis* f. sp. *tritici*) is the most dangerous disease affecting wheat worldwide. The fact that resistance to this disease (and leaf rust) among wild relatives of wheat was specifically mentioned by researchers early last century (Percival 1921; Vavilov 1914) indicates its impact on wheat production. Historically, the rusts have been a scourge of cereal crops for innumerable years (Zadoks 1985). The recent response by wheat researchers around the world to the appearance of the Ug99 stem rust races (BGRI, 2020) still shows the importance of resistance genes derived from wild relatives for protection of wheat productivity (RustTracker 2020a). There is no realistic alternative to respond to a major crisis potentially affecting wheat production. The Ug99 lineage of stem rust races (numbering 13 as at March 2020) (RustTracker 2020b) together overcome more than 30 resistance genes. Kishii (2019) listed 34 named genes resistant to Ug99 stem rust, and of these, 30 originate from relatives of wheat. It is in the nature of the rusts (stem, leaf and stripe) to mutate and overcome existing resistance genes and undoubtedly new even more dangerous pathotypes than the Ug99 races will probably appear and so the search for new effective resistance will continue.

7.11.2 Insect Resistance

The potential value of *Agropyron* as a sources of insect resistance for wheat improvement was noted last century by Novitskii (1941 in Schmidt et al. 1953) (stem saw fly resistance), Reitz et al. (1945) (aphid resistance) and Jones et al. (1952, in Schmidt et al., 1953) (wheat jointworm resistance). Table 7.3 lists the more recent papers reporting insect resistance in related species.

Table 7.3 Recent literature reporting on insect resistance in wheat related species.

Insect pest	Resistance source	Reference
Hessian fly	*Triticum araraticum*	Gill et al. (1983)
(*Mayetiola destructor*)	*T. tauschii*	Hatchett and Gill (1983)
	T. boeoticum	Raupp et al. (1988)
	rye chromosome 1 R and *Ae. speltoides* chromosome 7S	Crespo-Herrera et al. (2019}
Greenbug	*T. boeoticum*	Gill et al. (1983)
(*Schizaphis graminum*)	*Ae. squarrosa*	Raupp et al. (1988)
	Ae. speltoides 7S chromosome	Crespo-Herrera et al. (2013)
Wheat curl mite (*Eriophyes tulipae*)	Chromosome 6Ag from *Agropyron elongatum*	Whelan and Thomas (1988)
Russian wheat aphid (*Diuraphis noxia*)	rye and *Ae. speltoides* 7S chromosome	Crespo-Herrera et al. (2019)
Wheat stem saw fly (*Cephus pygmeaus*)	rye and *Ae. speltoides*	Crespo-Herrera et al. (2019)
Bird cherry-oat aphid (*Rhopalosiphum padi*)	rye chromosome 1 R	Crespo-Herrera et al. (2013)
English grain aphid (*Sitobion avenae*)	rye chromosome 1 R	Crespo-Herrera et al. (2013)

7.11.3 Root Diseases

The Cereal Cyst Nematode (CCN) (*Heterodera avenae* and *H. filipjevi*) is another important disease of wheat reported across Europe, Asia, Australia, North America and Africa (Yuan et al. 2011). Asiedu et al. (1990) reported the discovery of a line of triticale T701-4-6 highly resistant to the Australian strain of CCN and development of a substitution 6 R(6D) line. Yuan et al. (2011) screened 75 wheat lines for resistance to *H. filipjevi* in China and found that 6 R(6D) was the most resistant of all lines tested. Li et al. (2020) named this CCN resistance gene *Cre10* and described the transfer of the 6 RL arm carrying *Cre10* as well as *Yr83* to a wheat chromosome as a centric fusion. This line still is the best source of resistance to the cereal cyst nematode available. So far, there have been no commercial cultivars released carrying the *Cre10* resistance gene. Dundas et al. (2001) described the attempt to transfer the gene by homoeologous recombination but no recombinants were produced. Translocations of 6 RL segments to wheat have been produced (Li et al. 2020) and attempts are underway at the University of Adelaide, Australia, to produce a compensating translocation carrying the *Cre10* gene along with *Yr83* (stripe rust resistance) and powdery mildew resistance.

The Root Lesion Nematode (*Pratylenchus neglectus*) is another internationally distributed pest of wheat. Fanning et al. (2018) screened 62 cereal varieties and found that triticale cultivars Fusion and Tahara were highly resistant. There is no evidence yet if the resistance in these triticales against *Pratylenchus* is located on a rye chromosome or in the wheat background. Resistance to Common Root Rot has been reported to be associated with J–J^{S} translocation chromosomes derived from *Th. ponticum*. Literature references on root rot are listed in Table 7.4.

7.11.4 Abiotic Stresses

Salt tolerance has been reported in 14 species of *Agropyron* (Dewey 1960), and also in *Thinopyrum bessarabicum* (Forster et al. 1988; King et al. 1997; Mujeeb-Kazi and Bernard 1985; Tohidinejad et al. 2008), *Th. distichum* (Marais and Marais 2003; Mujeeb-Kazi and Asiedu 1990), *Th. ponticum* (Chen et al. 2004) or triticale (Koebner and Martin 1996).

Table 7.4 Literature reporting on root rot resistance in wheat related species.

Root disease character	Source of resistance	Reference
Heterodera avenae, H. filipjevi (Cereal Cyst Nematode)	*XTriticosecale* (Rye chromosome 6 R)	Asiedu et al. (1990); Yuan et al. (2011); Li et al. (2020)
	Aegilops triuncialis	Dhaliwal et al. (1998; Martin-Sanchez et al. 1998)
	Ae. ventricose	Delibes et al. (1993)
Pratylenchus neglectus (Root lesion nematode)	*XTriticale cv.* Fusion, Tahara	Fanning et al. (2018)
Cochliobolus sativus (Common Root rot)	*Thinopyrum ponticum*	Chen et al. (2003a)

Except for *Th. distichum* as a source of commercial releases in Pakistan for salinity and drought with Pasban 90 and Rohtas 90, not many commercial cultivars have been released with these salt tolerance characters Other unexploited species known to carry salt tolerance are *Elytrigia scirpea* (*Th. scirpeum*), *E. diae* (*Th. runemarkii*) (McGuire and Dvořák 1981) and *Hordeum marinum* (Colmer et al. 2006; Islam et al. 2007).

Aluminum toxicity limits growth for wheat cultivated in the acid soils of South America, Central Africa and the Indian Subcontinent (Rajaram et al. 1991). Tolerance to aluminum has been described in *Ae. uniaristata* (Berzonsky and Kimber 1986; Miller et al. 1997), and more specifically on chromosome 3 N from that species (Miller et al. 1995), rye chromosome 5 R (Manyowa et al. 1988), *Leymus racemosus* (Mohammed et al. 2013), plus in rye and triticale (Osborne and Rengel 2002). No commercial cultivars have been released carrying aluminum tolerance from these sources but very recently it has also been been identified in D genome synthetics (Emebiri et al. 2020).

Drought tolerance has been reported in *Triticum dicoccoides* (Krugman et al. 2008; Saranga et al. 2003) and on rye chromosomes in triticale (Nachit and Tahir 1983) and the 2RL arm from rye (Lahsaiezadeh et al. 1983). Other abiotic tolerances of potential value to wheat improvement have been reported by Osborne and Rengel (2002) (phosphorus uptake, and tolerance to low and high pH in triticales and rye), Sasanuma et al. (2008) (water tolerance in *Elymus humidus*), adaptation to phosphorus insufficiency (Li et al. 2013a), copper efficiency on rye chromosome 5RL (Graham et al. 1987; Schlegel et al. 1993), high Zn efficiency (Cakmak et al. 1997) and high light stress in wheat–*Th. ponticum* derivatives (Li et al. 2013b).

7.11.5 Grain Yield

Extensive wide crossing of bread wheat with distantly related species over the past century has produced a vast number of introgression lines carrying alien-derived chromosome segments, but relatively few lines of this material have been released in commercial cultivars. Introduction of alien chromosome segments to wheat is generally deleterious to yield and/or quality. The most notable contribution to yield improvement by an alien-derived chromosome is the T1BL.1 RS translocation carrying a segment of Petkus rye. This translocation carrying an entire rye chromosome arm is strangely contradictory to the established philosophy that the smaller the segment of alien chromatin the better. Zeller and Hsam (1983) and Mujeeb-Kazi et al. (2013) have reviewed its history. Effects on increased yield have been associated with this translocation (Rajaram et al. 1983) and it was found that the rye chromosome arm 1 RS carries a locus for increased root biomass in wheat (Ehdaie et al. 2003; Waines and Ehdaie 2005; 2007). The yield improving effect was present in only some environments (Villareal et al. 1997, 1998) and in spring rather than winter wheats (Lukaszewski 2015). However, despite this translocation being successfully deployed in dozens of wheat cultivars worldwide (Rabinovich 1998), many of the cultivars carrying these translocations showed deleterious effects on dough quality characteristics such as poor mixing tolerance, inadequate dough rheology and stickiness (Dhaliwal et al. 1987; Graybosch et al. 1990; Martin and Stewart 1990; Pena et al. 1990). This inferior bread-making quality has severely limited the use of these breeding materials containing the 1 RS segment in some countries.

A second example of yield enhancement associated with alien chromosome introgression is the 7elL arm from *Thinopyrum ponticum* (*Agropyron elongatum*) carrying the *Lr19* gene (Reynolds et al. 2001). Ceoloni et al. (2015) described structural-functional dissection of the 7A-7Ag chromosome and mapping yield-contributing traits. Yield improvement as a consequence of alien chromosome introgression is rare but occasionally yield component characters such as enlarged grain size (Figure 7.1) or increased spikelet number per head (Figure 7.2) have been observed in wheat–*Aegilops* derivatives at the University of Adelaide. However, these potentially yield-enhancing characters are often associated with low grain number per spike.

7.11.6 Bio-Fortification

In many parts of the developing world, micronutrient deficiency in people's diets may lead to serious health issues (Welch and Graham 1999). Development of wheat lines carrying higher levels of Zn has been promoted as a means of alleviating deficiency of that micronutrient in India and Pakistan (HarvestPlus 2020). Chhuneja et al. (2006) screened many accessions of *Aegilops kotschyi* and *Ae. tauschii* and found several with higher levels of grain Fe and Zn. Tiwari et al. (2008) extended the screening of wild relatives and also reported accessions of *Ae. peregrina*, *Ae. longissima*, *Ae. geniculata* and *Ae. ventricosa* with two- to three-fold higher grain Fe and Zn. Rawat et al. (2009) described the initial stages of hybridizations between wheat and *Aegilops* species for enhancing Fe and Zn content using homoeologous recombination. Tiwari et al. (2010) found that substitutions of chromosomes 2S and 7 U from *Ae. kotschyi* were associated with high grain Fe and Zn. Verma et al. (2016) induced introgressions of *Ae. kotschyi* 2S chromosome segments by irradiation and produced lines with high Fe and Zn. High Zn cultivars of wheat, developed through crosses with wild species and landraces, are now being grown across India with great success (CIMMYT 2020). The d genome synthetics are also identified as excellent sources that possess high levels of Fe and Zn.

7.11.7 Future Directions and Strategies

The time taken from initial hybridizations to development of recombinant chromosomes carrying introgressed alien chromatin may take decades. However, recent successes have indicated that existing wheat–alien chromosome recombinants can be used for introduction of new alien-derived genes which saves many years of effort. For example, resistance genes *Lr26* and *Yr9* on the 1 RS chromosome from Petkus rye are ineffective against new leaf and stripe rust pathotypes. Szakács et al. (2017) crossed wheat carrying the T1BL.1RS translocation (+Lr26 and Yr9) with a leaf and stripe rust resistant *Secale cereanum* line. Backcrossed derivatives confirmed resistance to leaf and stripe rusts and to carry a different 1RS chromosome segment, most likely the result of a rye–rye recombination event. Similarly, Mago et al. (2019) demonstrated that a new stem rust resistance gene *SrB* (now *Sr61*) on an entire *Th. ponticum* 6Ae#3 chromosome was transferred onto an existing 6A-6Ae#1 recombinant (named WA-5; Dundas et al. 2015b) by homologous recombination of *Thinopyrum* segments. The WA-5 recombinant carries the *Sr26* stem rust resistance gene from *Th. ponticum* on a short 6Ae#1 segment. Mago et al.

(2019) presented evidence that the newly derived T6A.6Ae#1-6Ae#3 recombinant, developed at the University of Adelaide, carried both *Sr26* and *Sr61*. Both of these resistance genes are effective against all known stem rust pathotypes.

The success of incorporation of a new alien-derived gene into wheat by use of existing homoeologous recombinants has presented a time-saving approach to gene transfer. Recombinant chromosome sets exist for rye chromosome 1R (Anugrahwati et al. 2008), *Ae. speltoides* chromosomes 2S#1 (Mago et al. 2013), 2S#2 (Mago et al. 2009), 2S#3 (Dundas et al. 2015a), *Th. ponticum* chromosome 6Ae#1 (Dundas et al. 2015b) and *Th. intermedium* chromosome 7Ai#1 (Khan 2000). King et al. (2017) have published details of whole genome introgressions of *Amblyopyrum muticum* chromosome segments in wheat using a high throughput SNP array. Future endeavors at interspecific and intergeneric gene transfer will utilize sets of homoeologous recombinants.

Despite the successes mentioned in overcoming crossing barriers, it should be mentioned that there are still some prevalent reservations. Only a few genera in the Triticeae tribe have so far contributed to the development of commercial wheat cultivars. Fedak (1998) outlined some of the procedures and constraints involved with transferring agronomic traits from alien species to wheat. The first problem a researcher encounters is the low success rate with hybridization. The more distantly related a species from wheat, the harder it is to obtain hybrids. Subsequently, if a hybrid is obtained, there is a high likelihood it will be sterile. Use of wheat varieties like Chinese Spring and TH3929, which possess a more suitable set of crossability (Kr) genes (McIntosh et al. 2013), increase the chances of successful hybridization. A second difficulty may be the low level of recombination between the introduced alien chromosome and wheat. Dundas et al. (2001) screened over 3000 wheat plants in a *ph1bph1b* background for recombination between rye chromosome 6 R and wheat and recovered no recombinants. Similarly, Leach et al. (2006) found no wheat-5RL recombinants after screening over 600 seedlings in a *ph1b-ph1b* genotype. These reports are very supportive of our contention that for achieving wheat productivity gains of significance that aid food security targets, the use of priority packages of strategies is paramount.

So, how do research programs in the future make use of genetic resources in species unable to be crossed with wheat? Some parts of the world willingly accept wheat developed using GM techniques. Recent advances in high throughput sequencing techniques has enabled identification and cloning of genes in a number of species, particularly genes identified through mutation techniques (MutRenSeq; Steuernagel et al. 2016), (MutChromSeq; Sánchez-Martín et al. 2016) and (AgRenSeq; Arora et al. 2019). Potentially valuable genes in species uncrossable with wheat may be accessible in the future by using sequencing and cloning and transformation procedures.

Most success with alien-derived genes in commercial wheats has been achieved with disease resistance (Kishii 2019; McIntosh et al. 2013). Disease resistance genes are usually single, dominant genes and can be inserted on shortened alien chromosome segments (Anugrahwati et al. 2008). However, if a character that is controlled by several loci is on different chromosomes, it will be more difficult to incorporate into wheat. A possible solution to this problem could be in the "Genetically Modified Cassette" approach (Wulff and Moscou 2014), where a set of cloned genes are artificially stacked and then incorporated into wheat by transformation techniques. However, there is no certainty that a character observed in a wild species will be expressed in a wheat genetic background nor will the public accept the resulting wheat line owing to its "GM" status. Recognizing these constraints and the urgency to deliver climate resilient wheat varieties, utilizing genetic resources amenable to interspecific techniques hold greater preference and their values have been well demonstrated throughout this chapter and will be further emphasized as the way forward conclusion.

7.12 Accruing Potental Practical Benefits

Table 7.5 consists of 170 genes in the *Catalogue of Gene Symbols for Wheat*, which are classed as resistance against biotic factors. Included are all transfers from any *Triticum/Aegilops/Thinopyrum/Leymus/Elymus/Secale* species to either bread or durum wheats. At the end of the attachments is a summary (Table 7.6). This only includes "named" genes, or in a few cases, where the gene is un-named but used in agriculture (e.g., *YrCn17, PmCn17*). Included is the resistance against wheat blast that is causing an issue in Bangladesh and is thought to be on the VPM segment from *Ae. ventricosa*, which may already be in use across the world in cultivars. Searching through all updates of the Catalogue (up to and including 2019), each gene was manually inserted into the database search facility at http://www.wheatpedigree.net. Scanning through the germplasm listing for each alien gene gave an indication of the commercial releases for each gene. Genes like Sr31/Lr26/Yr9/Pm8 (1RS) had hundreds of breeding lines/cultivars listed. Another website of use was the http://www.rye-gene-map.de/rye-introgression by Rolf Schlegel for rye transfers. The Texas line with greenbug resistance Gb2; TAM-107 has the T1AL.1RS translocation from Amigo. There has not been much commercial development of the Gb2 gene.

Table 7.5 Resistance genes against biotic factors transferred to bread or durum wheat by interspecific hybridization. Only named genes are listed, except where an un-named gene has been used in agriculture.

Character[1]	Gene	Source	Use in agriculture[2]
Reaction to barley Yellow Dwarf Virus	*Bdv2*	*Thinopyrum intermedium*	Australia, USA
Reaction to *Blumeria graminis* (powdery mildew)	*Pm2b*	*Agropyron cristatum*	
	Pm4b	*Triticum persicum/ T. carthlicum*	>240 breeding lines/ cultivars. Europe, USA, Brazil, China
	Pm5a	*T. dicoccum*	>100 breeding lines/ cultivars. Europe, USA, New Zealand, Australia
	Pm6	*T. timopheevii*	>350 breeding lines/ cultivars worldwide
	Pm7	*Secale cereal*	>100 breeding lines/ cultivars. USA, Australia, Netherlands, Germany

(*Continued*)

Table 7.5 (Continued)

Character[1]	Gene	Source	Use in agriculture[2]
	Pm8	*Secale cereal*	>1000 breeding lines/ cultivars. Many European, Asian, North/ South American, African countries
	Pm12	*Aegilops speltoides*	Great Britain
	Pm13	*Ae. longissimi*	
	Pm16	*T. dicoccoides*	
	Pm17	*Secale cereal*	>100 breeding lines/ cultivars. Azerbaijan, Ukraine, USA
	Pm19	*Ae. tauschii*	
	Pm20	*Secale cereal*	USA
	Pm21	*Dasypyrum villosum*	China
	Pm26	*T. dicoccoides*	
	Pm27	*T. timopheevii*	
	Pm29	*Ae. ovata*	
	Pm32	*Ae. speltoides*	
	Pm33	*T. carthlicum*	
	Pm34, Pm35	*Ae. tauschii*	
	Pm36	*T. dicoccoides*	
	Pm37	*T. timopheevii*	
	Pm40	*Th. intermedium*	
	Pm41, Pm42	*T. dicoccoides*	
	Pm43	*Th. intermedium*	
	Pm49, Pm50	*T. dicoccum*	
	Pm51	*Th. ponticum*	
	Pm53	*Ae. speltoides*	
	Pm55	*Dasypyrum villosum*	
	Pm56	*Secale cereal*	
	Pm57	*Ae. searsii*	
	Pm58	*Ae. tauschii*	
	Pm66	*Ae. longissimi*	
	PmCn17	*Secale cereal*	China
Reaction to *Diuraphis noxia* (Russian wheat aphid)	*Dn3*	*Ae. tauschii*	
	Dn7	*Secale cereal*	South Africa, Malawi

(Continued)

Table 7.5 (Continued)

Character[1]	**Gene**	**Source**	**Use in agriculture**[2]
Reaction to *Fusarium* spp. (head scab)	*Fhb3*	*Leymus racemosus*	
	Fhb6	*Elymus tsukushiensis*	
	Fhb7	*Th. ponticum*	
Reaction to *Heterodera avenae* Woll. (cereal cyst nematode)	*Cre2*	*Ae. ventricosa*	
	Cre3	*Ae. tauschii*	India
	Cre4	*Ae. tauschii*	
	Cre5	*Ae. ventricosa*	India, Mexico, China, Thailand, Bulgaria, Australia
	Cre6	*Ae. ventricosa*	
	Cre7	*Ae. triuncialis*	Spain
	Cre10	*Secale cereal*	
Reaction to Pyricularia oryzae (Magnaporthe oryzae/M. grisea) wheat/rice blast	*Rmg7*	*T. dicoccum*	
	Un-named gene on 2NS	*Ae. ventricosa*	
Reaction to *Mayetiola destructor* (Say) (Hessian fly)	*H13*	*Ae. tauschii*	USA
	H21	*Secale cereal*	
	H23, H26	*Ae. tauschii*	
	H25	*Secale cereal*	USA
	H27	*Ae. ventricosa*	Italy
	H30	*Ae. triuncialis*	
	Hdic	*T. dicoccum*	USA
Reaction to *Meloidogyne* spp. (root knot nematode)	*Rkn1*	*Ae. tauschii*	USA
	Rkn2	*Ae. peregrina*	
	Rkn3	*Ae. ventricosa*	USA
Reaction to *Mycosphaerella graminicola* (*Septoria tritici* blotch)	*Stb5*	*Ae. tauschii*	
	Stb19	*Ae. tauschii?*	Australia
Reaction to *Phaeosphaeria nodorum* (*Septoria nodorum* blotch)	*SnbTM*	*T. timopheevii*	USA
Reaction to *Puccinia graminis* (stem rust)	*Sr2*	*T. turgidum* var. *dicoccum*	>1700 breeding lines/ cultivars worldwide
	Sr9d	*T. dicoccum*	Australia, Europe, Nth America, India,

(*Continued*)

Table 7.5 (Continued)

Character[1]	Gene	Source	Use in agriculture[2]
	Sr9e	*T. dicoccum*	Canada, Ethiopia, Australia, USA, South Africa,
	Sr9g	*T. durum*	>100 breeding lines/ cultivars worldwide
	Sr11	*T. durum*	>300 breeding lines/ cultivars worldwide
	Sr12	*T. durum*	>150 breeding lines/ cultivars worldwide
	Sr13	*T. dicoccum*	Australia, India, Canada, South Africa, New Zealand, Poland
	Sr14	*T. dicoccum*	India, Australia, Canada, USA
	Sr17	*T. dicoccum*	>400 breeding lines/ cultivars worldwide
	Sr21	*T. monococcum*	Ethiopia
	Sr22	*T. monococcum*	Australia, Mexico, USA
	Sr24	*Th. ponticum*	>900 breeding lines worldwide. Many Australian cultivars
	Sr25	*Th. ponticum*	South Africa, Australia, Canada, India, USA, Egypt, Russia, Pakistan, Mexico
	Sr26	*Th. ponticum*	Australia, India, Finland
	Sr27	*Secale cereal*	Australia, Italy, Russia, USA
	Sr31	*Secale cereal*	>1000 breeding lines/ cultivars worldwide
	Sr32	*Ae. speltoides*	Great Britain
	Sr33	*Ae. tauschii*	Australia
	Sr34	*Ae. comosa*	Great Britain
	Sr35	*T. monococcum*	
	Sr36	*T. timopheevii*	USA, Australia, Europe, South Africa, India
	Sr37	*T. timopheevii*	Ethiopia
	Sr38	*Ae. ventricosa*	Europe, USA, Canada, Australia, South Africa, India
	Sr39	*Ae. speltoides*	
	Sr40	*T. timopheevii*	Great Britain
	Sr43	*Th. ponticum*	
	Sr44	*Th. intermedium*	USA, Russia
	Sr45	*Ae. tauschii*	Australia
	Sr46	*Ae. tauschii*	
	Sr47	*Ae. speltoides*	

(Continued)

Table 7.5 (Continued)

Character[1]	Gene	Source	Use in agriculture[2]
	Sr50	*Secale cereal*	
	Sr51	*Ae. searsii*	
	Sr52	*Dasypyrum villosum*	
	Sr53	*Ae. geniculate*	
	Sr59	*Secale cereal*	
	Sr60	*T. monococcum*	
	Sr61	*Th. ponticum*	
	Sr1RS^{Amigo}	*Secale cereal*	USA
Reaction to *Puccinia striiformis* (yellow rust, stripe rust)	*Yr5*	*T. spelta*	Iran, Slovakia, China, Germany, USA
	Yr7	*T. durum*	>200 breeding lines/ cultivars Europe, Australia, China, USA, New Zealand, Canada, Africa
	Yr8	*Ae. comosa*	Great Britain, China, India, Pakistan, Mexico
	Yr9	*Secale cereal*	>1000 breeding lines/ cultivars worldwide
	Yr15	*T. dicoccoides*	India, China, USA, Australia, Denmark, Germany
	Yr17	*Ae. ventricosa*	>300 breeding lines/ cultivars worldwide
	Yr28	*Ae. tauschii*	
	Yr35	*T. dicoccoides*	
	Yr37	*Ae. kotschyi*	
	Yr38	*Ae. sharonensis*	
	Yr40	*Ae. geniculate*	
	Yr42	*Ae. neglecta*	
	Yr50	*Th. intermedium*	
	Yr69	*Th. ponticum*	
	Yr70	*Ae. umbellulata*	
	Yr83	*Secale cereal*	
	YrCn17	*Secale cereal*	China
Reaction to *Puccinia triticina* (leaf rust)	*Lr9*	*Ae. umbellulata*	>250 breeding lines/ cultivars. USA, Europe, India, Africa, Sth America
	Lr18	*T. timopheevii*	>200 breeding lines/ cultivars. Nth America, Africa, Europe, Asia, Australia
	Lr19	*Th. ponticum*	>200 breeding lines/cultivars Many worldwide

(*Continued*)

Table 7.5 (Continued)

Character[1]	Gene	Source	Use in agriculture[2]
	Lr21	*Ae. tauschii*	Canada, USA, Europe, Egypt, India, Ethiopia
	Lr23	*T. turgidum* var. *durum*	>550 breeding lines/ cultivars worldwide. USA, Mexico, Iraq, Australia, Brazil, Chile, India, Ukraine, China
	Lr24	*Th. ponticum*	>850 breeding lines/ cultivars. Australia, Canada, USA, India, Egypt, South Africa
	Lr25	*Secale cereal*	Russia, Australia, Canada, USA
	Lr26	*Secale cereal*	>1000 breeding lines/ cultivars Many worldwide
	Lr28	*Ae. speltoides*	Canada, USA, Argentina, Europe, China, South Africa, Great Britain, India
	Lr29	*Th. ponticum*	Egypt, Slovakia
	Lr32	*Ae. tauschii*	China
	Lr35	*Ae. speltoides*	Egypt
	Lr36	*Ae. speltoides*	Egypt, Australia, China
	Lr37	*Ae. venticosa*	>450 breeding lines/ cultivars. Argentina, Europe, Australia, USA, South Africa, India, Uruguay, Kazakhstan, Switzerland, China
	Lr38	*Th. intermedium*	
	Lr39	*Ae. tauschii* var *meyeri*	USA, Egypt
	Lr41	*Ae. tauschii*	USA, Egypt
	Lr42	*Ae. tauschii*	China, USA
	Lr44	*T. spelta*	Mexico, Ethiopia, USA, Uruguay
	Lr45	*Secale cereal*	
	Lr47	*Ae. speltoides*	Argentina
	Lr50	*T. armeniacum*	Russia
	Lr51	*Ae. speltoides*	
	Lr53	*T. dicoccoides*	
	Lr54	*Ae. kotschyi*	
	Lr55	*Elymus trachycaulus*	
	Lr56	*Ae. sharonensis*	
	Lr57	*Ae. geniculate*	
	Lr58	*Ae. triuncialis*	
	Lr59	*Ae. peregrina*	

(*Continued*)

Table 7.5 (Continued)

Character[1]	Gene	Source	Use in agriculture[2]
	Lr62	*Ae. neglecta*	
	Lr63	*T. monococcum*	
	Lr64	*T. dicoccoides*	
	Lr65	*T. spelta*	
	Lr66	*Ae. speltoides*	
	Lr76	*Ae. umbellulata*	
	LrTr	*Ae. triuncialis*	Russia, Brazil
Reaction to *Pyrenophora tritici-repentis* (tan spot)	*Tsr1*	*T. dicoccoides*	
	Tsr3	*Ae. tauschii*	
Reaction to *Schizaphis graminum* (greenbug)	*Gb2*	*Secale cereal*	USA
	Gb5	*Ae. speltoides*	
	Gb6	*Secale cereal*	
	Gb7	*Ae. tauschii*	
Soil-borne Cereal Mosaic Virus	*SBWMV*	*Ae. tauschii*	
Reaction to *Tapesia yallundae* (eyespot, strawbreaker footrot)	*Pch1*	*Ae. ventricosa*	Many European countries, USA
Wheat Spindle Streak Mosaic Bymovirus	*Wss1*	*Dasypyrum villosa*	
Wheat Streak Mosaic Virus	*Wsm1*	*Th. intermedium*	USA
	Wsm3	*Th. intermedium*	
Colonization by *Eriophyes tulipae*- Wheat Curl Mite	*Cmc1*	*Ae. tauschii*	Canada
	Cmc2	*Thinopyrum elongatum*	
	Cmc3	*Secale cereal*	USA
	Cmc4	*Ae. tauschii*	

[1]McIntosh et al. (2013); GrainGenes (2020)
[2]http://www.wheatpedigree.net (accessed 5 July 2020).

7.13 Summary of the Practical Potential Benefits

A survey was conducted of characteristics conferring resistance/low reaction to a number of biotic factors listed in the *Catalogue of Gene Symbols* for wheat (McIntosh et al. 2013 and supplements 2013/14 to 2019 in GrainGenes 2020), which had been transferred across species into either bread or durum wheat (Table 7.6). Not included in this survey were reports of the transfer of un-named genes. About 170 genes were listed which were

Table 7.6 Summary of alien-derived genes transferred to bread or durum wheat. Named genes only have been included (*Catalogue of Gene Symbols for Wheat* (GrainGenes 2020; McIntosh et al. 2013)].

Category	Number of genes	Species origin or type of resistance gene
Total Number of alien-derived resistance genes against biotic factors	170	14 *Aegilops* spp.
		8 *Triticum* spp.
		3 *Thinopyrum/Agropyron* spp.
		2 *Elymus* spp.
		1 *Leymus* spp.
		1 *Dasypyrum* spp.
		1 *Secale* sp.
Released in cultivars	82	30 genes from *Aegilops*
		27 genes from *Triticum*
		15 genes from *Secale*
		9 genes from *Thinopyrum/ Agropyron*
		1 gene from *Dasypyrum*
Genes used in > 100 breeding lines/ cultivars	23	6 *Pm* genes
		7 *Sr* genes
		7 *Lr* genes
		3 *Yr* genes

derived from 7 genera. Of these, 82 genes from 5 genera have been released in commercial cultivars, with the majority derived from *Aegilops* spp. (30 genes), *Triticum* spp. (27 genes) and *Secale* (15 genes) (http://www.wheatpedigree.net).

After assessing the characteristics deployed in more than 100 breeding lines/cultivars, by far the most widely-used alien-derived genes in cultivars were for resistances to stem and leaf rusts (7 genes each), powdery mildew (6 genes) and stripe/yellow rust (3 genes). By far the most widely-used alien-introgressions have been the *Sr2* gene from *Yaroslav emmer* (*T. dicoccum*) (>1700 breeding lines/cultivars) and the T1BL.1RS chromosome carrying *Pm8, Sr31, Lr26* and *Yr9*. Schlegel (2020) listed about 90 breeding lines/cultivars worldwide carrying the T1AL.1RS translocation derived from Amigo and the CIMMYT database listed more than 200 breeding lines or cultivars carrying the 7Ag chromosome with *Lr26* from *Thinopyrum ponticum* (http://www.wheatpedigree.net).

7.14 The Role of Genomics Information Including Molecular Markers in Wheat

The elucidation of the double helix structure of deoxyribonucleic acid (DNA) in the mid-twentieth century (Watson and Crick 1953) and others major milestones such as the advent of recombinant DNA technology (Cohen et al. 1973), included the landmark paper by

Botstein et al. (1980) on fragment length polymorphisms (RFLPs). This heralded the development of the first "molecular" maps (Burr et al. 1988) using small DNA fragments (probes) that detected restriction RFLP, resulting in the development and coverage of the molecular genetic maps to a few hundred markers across the genome. The watershed moment occurred with the invention of polymerase chain reaction, which radically transformed biological science from the time it was first discovered (Mullis 1990) and made possible the generation of molecular markers routine. Over the decades, various types of molecular markers were developed for plant species: random amplified polymorphic DNA (RAPD) markers (Lanza et al. 1997); amplified fragment length polymorphism (AFLP) markers (Castiglioni et al. 1999); and simple sequence repeat (SSR) to markers and single nucleotide polymorphism (SNP). The development of RFLP markers in wheat was generally attributed to Sharp et al. (1989) and Chao et al. (1989). However, it is little known that Sharp and Soltes-Rak reported on RFLP markers of wheat group 2 chromosome arms at the *7th International Wheat Genetics Symposium* held at Cambridge, UK, 13–19 July 1988. Similarly, the use of RAPD was attributed to only Williams et al. (1990) by most authors. However, the publisher of *Nucleic Acids Research* had made a statement that the paper by Welsh and McClelland (1990) must be cited along with Williams et al. (1990) as simultaneous publications (see Table 7.7 for details on the advent of various molecular marker technologies). Over the past 15 years, there have been quantum leaps in scientific advances which have resulted in the generation, sequencing and annotation of plant genome facilitated by the so-called next generation sequencing (NGS). The first human full-genome sequencing projects took years to complete with costs as high as $3 billion (human genome project) while today, sequencing efforts are measured in weeks and in thousands of dollars. In plants, molecular markers are generally developed first for species with small genome. As such, the first plant genome, thale cress (*Arabidopsis thaliana*) was sequenced in 2000 (The Arabidopsis Genome Initiative 2000), while the first food crop rice (*Oryza sativa*) was sequenced in 2002 (Goff et al. 2002; Yu et al. 2002).

Bread wheat (*T. aestivum*), which is composed of three genomes (A, B and D), has a large and complex genome with a large amount of repetitive sequences. As such, in Triticeae, molecular tools were developed first in diploid species, then tetraploids, and then hexaploids. This is exemplified by the whole-genome sequencing of D genome in diploid *Ae. tauschii* (Jia et al. 2013; Luo et al. 2017), A genome of *T. urartu* (Ling et al. 2018; 2013), AB genome of *T. turgidum* ssp. *dicoccoides* (Avni et al. 2017) and recently ABD genome of *T. aestivum* (Appels et al. 2018). The release of an annotated reference whole genome for bread wheat provides the opportunity to enhance the use of genetic resources for wheat improvement by facilitating the identification and selection of valuable genes and alleles associated with abiotic, biotic stress tolerances and improved yield and end-use qualities. The perennial Triticeae species have also been sequenced including the rye genome (Bauer et al. 2017), H genome (Mascher et al. 2017) and the E genome of *Th. elongatum* (Wang et al. 2020b). SNPs are the most abundant type of sequence variations in plant genomes that reflect allelic variations in functional genes. Therefore, they are suitable for various genetic studies and useful in plant improvement. The newer marker technologies based on PCR, including GBS, SLAF, STARP and IT, require reference genome sequences for analysis to identify SNPs or to design primers (Table 7.7). The availability of reference genome sequences in rice, barley and wheat (cited above), and the synteny/collinearity among

Table 7.7 Advent and progress of molecular marker technologies.

Molecular marker	Species	Advantages	Disadvantages	Reference
RFLP	*Homo sapiens*	Inherited as simple Mendelian codominant markers	Labor-intensive and timc consuming = costly; low polymorphisms.	Botstein et al. (1980)
	Triticum aestivum			Sharp and Soltes-Rak (1988); Sharp et al. (1989); Chao et al. (1989)
SSR	human, *Drosophila*, sea urchin, *Stylonychia* (protozoan) andyeast	Abundance, high rate of polymorphism, ubiquitous distribution throughout the genome, codominant inheritance, high extent of allelic diversity, and ease of assay by PCR	Time-consuming and costly to develop and identify polymorphic markers	Tautz and Renz (1984)
	Triticum aestivum			Röder et al. (1995); Devos et al. (1995)
RAPD	*Homo sapien; Staphylococcus spp.*	No knowledge of the DNA sequence of the targeted genome is required; simple and fast = low costs.	Low polymorphisms and reproducibility; markers are dominant; not suitable for forming a cDNA databank.	Williams et al. (1990); Welsh and McClelland (1990)
	Triticum aestivum			Devos and Gale (1992)
DAF	*Homo sapien; Glycine max*	Same as RAPD	Same as RAPD; but no success in converting to SCAR	Caetano-Anolles et al. (1991)
	Triticum aestivum			Sen et al. (1997)
EST	*Homo sapien*	Easily developed from cDNA library by PCR, can map ESTs to specific chromosome locations, being expressed functional genes		Adams et al. (1991)
	Triticum aestivum			Qi et al. (2003); Lazo et al. (2004)

(*Continued*)

Table 7.7 (Continued)

Molecular marker	Species	Advantages	Disadvantages	Reference
STS	*Homo sapien* *Triticum aestivum*	quick and easy to use, high reproducibility, locus-specific, only low quantities of template DNA are required.	the need for sequence data to design the PCR primers	Green and Green (1991) Schachermayr et al. (1994)
SCAR	*Lactuca sativa L.* *Triticum aestivum*	Dominant or codominant, high reproducibility and are locus-specific. Due to the use of PCR, only low quantities of template DNA are required.	the need for sequence data to design the PCR primers	Paran and Michelmore (1993) Dedryver et al. (1996)
CAPS	*Arabidopsis* *Triticum aestivum*	Codominant, high reproducibility and are locus-specific. Due to the use of PCR, only low quantities of template DNA are required, more polymorphisms than STS detected by restriction digestions before or after amplification.	the need for sequence data to design the PCR primers	Konieczny and Ausube (1993) Mohler and Jahoor (1996)
AFLP	virus, bacteria, yeast, human, and plants *Triticum aestivum*	Higher reproducibility, resolution, and sensitivity, capability to amplify between 50 and 100 fragments at one time, and no prior sequence information is needed.	Time-consuming data analysis, fragments need to be isolated, cloned, and sequenced	Vos et al. (1995) Barrett and Kidwell (1998); Ma and Lapitan (1998)

(*Continued*)

Table 7.7 (Continued)

Molecular marker	Species	Advantages	Disadvantages	Reference
EST-SSR	*Triticum aestivum*	Greater transferability of these EST-derived SSR markers between species/	lower level of polymorphism than gSSR.	Kantety et al. (2002); Holton et al. (2002)
DArT	*Oryza sativa* *Triticum aestivum*	Rapid, low cost genotyping of any genome or complex genomic mixture.		Jaccoud et al. (2001) Akbari et al. (2006)
PLUG	*Triticum aestivum*	Similar to IT in producing intron sequences from PCR to detect SNPs and Indels.	Require the use of single-copy rice gene sequences for alignment to find exon-exon junctions	Ishikawa et al. (2007)
GBS	*Zea mays, Hordeum vulgare* *Triticum aestivum*	Low cost due to reduced sample handling, few PCR and purification steps, no DNA size fractionation, efficient barcoding system, simultaneous marker discovery and genotyping, avoiding repetitive regions, can discover millions of SNPs	a high rate of missing data, require reference genome sequences for analysis,	Elshire et al. (2011) Poland et al. (2012}
SLAF	carp (*Cyprinuscarpio* L.) *Triticum aestivum*	high throughput, high accuracy, low cost and short cycle,	Does not cover whole genome, require reference genome sequences for analysis	Sun et al. (2013) Hu et al. (2016)
STARP	*Oryza sativa L, Helianthus annuus L, Aegilops tauschii* *Triticum aestivum*	improved accuracy, flexible throughputs, low operational costs, and high platform compatibility	require SNP and indel information, require reference genome sequences for analysis	Long et al. (2017) Faris et al. (2020)

(*Continued*)

Table 7.7 (Continued)

Molecular marker	Species	Advantages	Disadvantages	Reference
IT	*Haynaldia villosa, Triticum aestivum*	are based on the sequence conservation of orthologous genes, therefore are specific to target genes and have higher transferability between species.	requires detailed sequence information on exon-exon junctions of targeted genes to design primers	Wang et al. (2017); Zhang et al. (2017a)

Abbreviations

RFLP = Restriction fragment length polymorphism
AFLP = Amplified fragment length polymorphism
RAPD = Random amplified polymorphic DNA
STS = Sequence tagged sites
DAF = DNA amplification fingerprinting
SCAR = sequence characterized amplified regions
CAPS = cleaved amplified polymorphic sequences
ALP = amplicon length polymorphisms
ISSA = inter-simple sequence repeat amplification
AP-PCR = arbitrarily primed PCR
AS-PCR = allele-specific PCR
MP-PCR = microsatellite-primed polymerase chain reaction
DAMD-PCR = direct amplification of minisatellite DNA by PCR
SSR or STMS = simple sequence repeats or microsatellites
EST = expressed sequence tags
SNP = single nucleotide polymorphisms
DArT = Diversity arrays technology
PLUG = PCR-based Landmark Unique Gene
GBS = Genotyping by sequencing
SLAF = specific-locus amplified fragment
STARP = semi-thermal asymmetric reverse PCR
IT = Intron targeting
GS = Genomic selection.
GWAS = Genome-wide association study

cereal genomes (Devos and Gale 2000; Gale and Devos 1998) is allowing molecular geneticists working with bioinformatics to develop and exploit these new molecular markers for marker-assisted selection (MAS) in wheat breeding.

These resources facilitated and enhanced the speed and the overall efficiency for mapping of quantititave trait loci (QTL) and gene discovery through linkage mapping using bi-parental recombinant inbred lines (RILs) and doubled haploid (DH) populations to identify markers that are associated with agronomic traits, disease resistance, quality attributes and abiotic stress tolerance using various platforms paving the way for the implementation of molecular markers in breeding programs to improve the rate of genetic gain. MAS is now common in wheat breeding programs and provide information to breeders on potential resistance to diseases and tolerance to major abiotic stress early in the breeding cycle when phenotypic screens are impractical.

The development and implementation of wheat molecular markers had been reviewed since 1999 (Eagles et al. 2001; Gupta et al. 1999; Liu et al. 2012; Rasheed et al. 2018; Rasheed and Xia 2019; Semagn et al. 2014; Toth et al. 2019; Vagndorf et al. 2018). More than 150 functional markers were initially developed and available for important genes in wheat (Liu et al. 2012), and were subsequently converted into high-throughput Kompetitive Allele Specific PCR (KASP) assays (Rasheed et al. 2016a). Some of the desirable genes for which MAS has been applied include disease resistance, agronomic and quality traits (Toth et al. 2019; Vagndorf et al. 2018). Other reported studies include powdery mildew (Huang and Röder, 2004), *Pm21* (Liu et al. 1999), a new unknown Pm gene (Wang et al. 2005), *Pm57* (Dong et al. 2020), leaf rust (Pinto Da Silva et al. 2018), Lr47 (Helguera et al. 2000), stem rust (Elangbam et al. 2018), Sr52 (Li et al. 2019), quality traits (Gale 2005) and pre-harvest sprouting (Imtiaz et al. 2008; Ogbonnaya et al. 2008). However, the use of RILs and DH populations is limited by the fact that such QTL mapping studies only locate associated genomic regions with low resolution limiting their utility across diverse genetic background.

In recent years, efforts have been directed at genome-wide association studies (GWAS) to identify marker-trait associations for desirable traits of agronomic importance. Ogbonnaya et al. (2017) and references cited therein, reported that GWAS based on random, high-density genotyping can help to "backfill" regions of the chromosome, where unknown genes with major effects are located. Furthermore, they indicated that the ability to survey large gene pools that are more representative of the breeding pool within any given country or geographical area lends itself to the detection and mapping of multiple traits in a single panel of genotypes (Jighly et al. 2016; Mulki et al. 2013; Neumann et al. 2011). GWAS has been bolstered by the availability of high-density molecular markers made possible by advances in the development of low-cost high-throughput genotyping resources (Wang et al. 2014a; Zegeye et al. 2014). The limitations of GWAS include its inability to detect rare alleles and the effects of unaccounted (Zhao et al. 2007) or overcorrected (Segura et al. 2012) population structure. However, GWAS is still useful to detect robust QTL that have an effect across different genetic backgrounds and environments (Jannink 2007; Wang et al. 2014b). Several studies have reported the identification of many QTL important in wheat traits: 7 QTLs for tan spot and 5 QTLs for Septoria Nodorum Blotch resistance (Singh et al. 2019), 18 QTLs for resistance to Karnal bunt (Gupta et al. 2019b), 3 QTLs for stripe rust resistance (Yang et al. 2019), multiple biotic stresses (Bhatta et al. 2019; Jighly et al. 2016), TaWRKY13 transcription factor for tolerance to abiotic stresses (Zhou et al. 2019), stay-green QTL for tolerance to drought and heat (Kamal et al. 2019) and salinity tolerance (Oyiga et al. 2019; 2018). Furthermore, the availability of IT markers where intron length variations measured in bp among genes on homoeologous chromosomes of wheat are advantageous for designing primers (Wang et al. 2017; Zhang et al. 2017b) for use in MAS.

Using 35,798 single nucleotide polymorphisms in 125 synthetic hexaploid wheats (SHWs; $2n = 6x = 42$, AABBDD), Bhatta et al. (2019) characterized their resistance to three rust diseases (leaf; *Puccinia triticina*, stem; *P. graminis f.sp. tritici* and stripe; *P. striiformis f.sp. tritici*); crown rot (*Fusarium* spp.); cereal cyst nematode (*Heterodera* spp.) and Hessian fly (*Mayetiola destructor*). Then they conducted a genome-wide association study (GWAS) to identify 124 significant marker-trait associations (MTAs) for multiple biotic stresses. Of these, 33 MTAs were found within genes that were playing certain roles in biotic resistances. Similarly Emebiri et al. (2020) reported the evaluation of 300 SHWs for aluminum (Al^{3+}) tolerance. Results from GWAS of the 300 SHW identified a set of 24 loci located to chromosomes 1B, 1D, 2A, 2B, 4A, 4D, 5A, 5B,6A, 6D and 7A that showed statistically significant association with Al^{3+} tolerance. In a similar vein, Joukhadar et al. (2020) used 2300 wheat accessions including landraces, cultivars and SHWs-derived germplasm with three Australian cultivars: Annuello, Yitpi and Correll, to investigate field-based resistance to Lr rust, Sr rust and Yr rust diseases across a range of Australian wheat agro-ecological zones. They reported that a number of synthetic backcross derived lines used in the study outperformed their recurrent parent Australian cultivar and displayed high levels of resistance to the three rust diseases across all environments. The presence of the synthetic derivatives in the germplasm allowed the detection of two major QTL inherited from the SHWs, Sr QTL at 29.4 cM on 2D and Lr/Sr QTL at 143 cM on 3D. Furthermore, they allowed the detection of a QTL that was almost fixed in the worldwide germplasm and less frequent in the SHWs derivative germplasm, Lr QTL at 27.2 cM on 7D. Rasheed and Xia (2019) reviewed recent technological advances in wheat genomics that provide new opportunities to uncover genetic variation in traits of breeding interest and enable genome-based breeding to deliver wheat cultivars for the projected food requirements for 2050. Their listings of open-source genomics/database resources for wheat and genome-based prediction studies in wheat are very useful. Saini et al. (2020) compiled information from QTL mapping studies on wheat and discussed the present status of MAS and different prospects of genomic selection (GS) for wheat biofortification of grain protein, zinc, iron, selenium, yellow pigment content, phytic acid and anthocyanin content.

Although SSRs, FMs and GBS have contributed greatly to gene discovery and MAS, whole-genome sequencing of bread wheat reference genome has resulted in the design of SNP arrays based on different densities or application targets. The solid chip-based arrays developed so far include the Illumina Wheat 9 K iSelect SNP array (Cavanagh et al. 2013), the Illumina Wheat 90 K iSelect SNP genotyping array (Wang et al. 2014a), the Wheat 15 K SNP array (Boeven et al. 2016), the Axiom® Wheat 660 K SNP array, the Wheat 55 K SNP array, the Axiom® HD Wheat genotyping (820 K) array (Winfield et al. 2016a), the Wheat Breeders' 35 K Axiom array (Allen et al. 2017) and the Wheat 50 K Triticum TraitBreed array (Rasheed and Xia 2019). Sun et al. (2020) evaluated seven types of wheat SNP arrays in terms of their SNP number, distribution, density, associated genes, heterozygosity and application. Their results suggested that the Wheat 660 K SNP array: i) contained the highest percentage (99.05%) of genome-specific SNPs with reliable physical positions; ii) the SNPs were almost evenly distributed across the whole genome; and iii) its 229 266 SNPs were located in 66,834 annotated gene or promoter intervals. The annotated genes in the Wheat 660 K SNP array almost covered all genes revealed in the Wheat Breeders' 35 K Axiom array (97.44%), 55 K SNP array with 99.73%, the Illumina Wheat 90 K iSelect SNP genotyping array (86.9%) and the Axiom® HD Wheat genotyping 820 K array (85.3%). Being both accurate and cost-effective, the Wheat 660 K SNP array may be the best choice for targeted genotyping and marker-assisted selection in wheat breeding.

Genes regulating tolerance to abiotic stresses are often involved in complex physiological and metabolic pathways, which are more likely located in proximal sections of chromosomes, in contrast to the higher distal distribution of genes conferring resistance to biotic stresses (Wang 2020). Resistance gene analogs (RGAs; including NBS-LRR) of the E genome and NBS-LRR genes of bread wheat had been shown to locate mainly in the distal end of chromosomes of *Th. elongatum* (Figure S12 in Wang et al. 2020) and *T. aestivum* (Xingfeng Li, unpublished data).

The E genome of *Thinopyrum* species and the St genome of *Pseudoroegneria* species are closely related to the D genome, slightly more than to A and B genomes of *T. aestivum*, based on the numbers of shared gene clusters (Figure S10 in Wang et al. 2020a) and GBS markers (Wang et al. 2020a), respectively. However, the E genome is more closely related to wheat genomes than the St genome (Liu et al. 2007). Thus, there were more desirable genes successfully transferred from the E genome of *Thinopyrum* species than the St genome of *Pseudoroegneria* into wheat (Wang 2011). Nevertheless, geneticists can attempt to transfer desirable traits from both E and St genomes to bread wheat by using the *ph1b* gene (Sears 1977) or the Ph inhibitor (Ph^I) line (Chen et al. 1994). Many genes for disease resistance had been translocated from the hexaploid intermediate wheatgrass (*Th. intermedium*) that has both the E- and St-genome chromosomes to D-genome chromosomes in bread wheat (Wang 2011). Now the *Fhb7* locus encoding a glutathione S-transferase (GST) enzyme to confer resistance to *Fusarium* head blight (FHB) has been shown to originate from the distal end of 7EL chromosome arm, which had been translocated to 7DL in wheat without yield penalty (Wang et al. 2020a). Molecular markers *XSdauK79* and *XSdauK80* flanking the *Fhb7* would be useful in monitoring the gene in breeding wheat with scab resistance.

Hayes et al. (2013) earlier suggested that QTLs for yield component traits and tolerance to abiotic stresses are better improved through genomic selection (GS) and the resistance to biotic stresses where genes are controlled by major gene can be incorporated by backcrossing (Vogel 2009). Genomic selection leverages genome-wide molecular marker information to select individuals based on their predicted genetic merit (Meuwissen et al. 2001) is a promising tool for accelerating crop genetic gains. Since then, the development of whole-genome sequencing resources in wheat and its relative species is facilitating the deployment of GS and genome-based breeding to develop high yielding, climate-resilient wheat cultivars with high nutritional quality (Li et al. 2018a; Rasheed and Xia 2019).

Jiang et al. (2020) and references therein, indicated that for most existing GS tools, linear mixed models, such as the ridge-regression best linear unbiased prediction (rrBLUP) model, are used to perform regression-based predictions. The GS model first needs to be trained using a reference population (training set), in which the genotypes and phenotypes of each individual sample are precisely measured. During training, the GS model infers the correlation between genotypes and phenotypes in the population and derives necessary parameters. Subsequently, the trained GS model is applied to a candidate population (testing set), and the genotype is used as the input to predict the phenotype outcome for each sample. Consequently, there is now increased use of GS for genetic improvement of various traits in wheat (Borrenpohl et al. 2020; Charmet et al. 2020; He et al. 2016; Heffner et al. 2010; 2009; Huang et al. 2016; Lozada et al. 2019; Poland et al. 2012a; 2012b; Rapp et al. 2018; Verges et al. 2020). Ramirez-Villegas et al. (2020) and references cited therein, reported that CGIAR has done extensive research to evaluate the genomic predictabilities of several traits including phenology, grain yield and its components, disease resistance, quality and micronutrients. The accuracy of forward predictions for grain yield (using a previous nursery year to predict the next nursery year) is, however, low and highly influenced by the environment, thereby highlighting the importance of incorporating environmental data in genomic prediction models for grain yield.

Concurrent with improvement in genotyping, over the last two decades, continuous improvements in genotyping technology have resulted in a quantum leap in the amount of genotypic data available. In contrast, methods for rapid characterization of plant phenotypes, in particular field phenotyping, have evolved little over the past 30 years, and accurate measurement of physiological and developmental traits on large sets of germplasm remains challenging (Furbank and Tester 2011). This is especially true for complex traits that vary over the life history of a crop species. Complex traits are of primary interest to crop improvement because most important agronomic traits are quantitative in nature (Pazhamala et al. 2021). As breeders and farmers often put it, yield is the king. As such, the integration of genomic selection and high throughput phenomics will accelerate a multiple-fold increase in the rate of genetic gain for yield, the primary determinant of economic value of most crops. The implementation of genomic selection and high throughput phenomics will shorten the breeding cycle through early stage selection as well as multi-parent crossing. This will deliver a more responsive dynamic path for breeding to adjust to changing environments, disease pressure and increase the quality and sustained yield improvement in an ongoing fashion.

Furthermore, gene banks in various countries hold extensive collections of plant genetic resources including wheat. The International Maize and Wheat Improvement Center (CIMMYT) hosts the most extensive collection of wheat germplasm, while the International Center for Agricultural Research in Dry Areas (ICARDA) holds a significant amount of accessions of wheat wild relatives. These genetic resources are an excellent source of new and novel trait alleles for crop improvement, since they capture genetic diversity that has been lost through bottlenecks imposed by domestication and breeding. The phenomenal advances in next-generation sequencing has created opportunities for pan-genome sequencing of various crops, resulting in the generation of multiple high-quality reference genome sequences than previously available. Milner et al. (2019) recently demonstrated how contemporary genomics and informatics technologies can be used to characterize almost all the gene banks accessions: 20,000 wild and domesticated barley genotypes at the German *ex-situ* genebank, IPK-Gatersleben. This offers not only the opportunity to interrogate genomic information of thousands of accessions but the creation of digital profile of gene banks accessions. The deep understanding of barley diversity revealed by this endeavor, provides the platform for both genome-enabled exploitation in plant breeding and accelerated domestication of an important and tractable inbred model crop (Langridge and Waugh 2019). They also argue that the pool of diversity for guided exploitation needs to be sensibly reduced and made available to the breeding and pre-breeding research communities.

Therefore, targeted research and developments in genotyping, phenotyping and population development underpinned by data driven approaches and decision including simulation would dramatically increase efficiencies for the management of the gene banks collection, as well as enhance germplasm accessibility and uptake in breeding. Voss-Fels et al. (2019) argued that combining beneficial, genome-wide haplotypes could help breeders to more efficiently exploit available genetic variation, optimizing future yield potential in more sustainable production systems. They indicated that a haplotype-based simulation approach, using haplotype blocks weighted according to their estimated trait variances, facilitated forecasts of genetic potential within our cultivar panel in the context of future sustainable production. Consideration of haplotype block effects can also facilitate design of complex crosses among complementary genotypes, to enrich breeding progenies with beneficial haplotypes. Combining accelerated breeding cycles with haplotype-based genomic selection methods could further accelerate the generation and recognition of superior recombinants to maximize exploitation of the available genetic potential.

We argue that if the estimated yield improvement for food and nutritional security is to be attained, deliberate efforts must be directed at effective mining of genetic resources uses genomics and phenomics coupled with envirotyping and improved analytics to:

i) Efficient mining of genetic resources: improve germplasm management and enrich passport information to improve accessibility of gene banks by breeders and researchers through identification and utilization of potentially new genes from exotic germplasm collections, including selection of germplasm set that maximizes genetic diversity.
ii) Employ the most effective phenotyping technologies, including high throughput phenomic (HTP) screening that are scalable, accurate, cost-effective and relevant in multiple target production environments. For example, any HTP that results to indirect selection for grain yield in early generations will be transformational, resulting in increasing the accuracy of breeding values for parent selection facilitates
iii) Data analysis pipeline that will involve genomic prediction and validation of prediction accuracies for traits. The generation of genomic estimated breeding values (GEBVs) for the population will inform the identification of lines enriched for both qualitative and quantitative traits, including the use of AI computing, breeding simulation to guide mating designs to select desirable cross combinations with the highest probability for maximizing identification of novel genes and alleles that can be deployed into elite varieties
iv) Finally, a rapid breeding cycle will be used to guide mating design, because many loci will need to be simultaneously introgressed into elite breeding germplasm. The logical approach is to use genomic prediction in a population that includes pre-breeding and breeding germplasm to guide this process. Rapid cycle breeding combined with genomic prediction (SpeedGS) is a powerful way to accomplish this quickly and deliver improved elite germplasm with a combination of desired traits in a fraction of the time it would take to achieve this with conventional approaches.

The combination of these approaches (Figure 7.12) underpinned by strong analyses will allow identification of novel genes and alleles that can be deployed to achieve improved rates of genetic gain in yield in elite varieties.

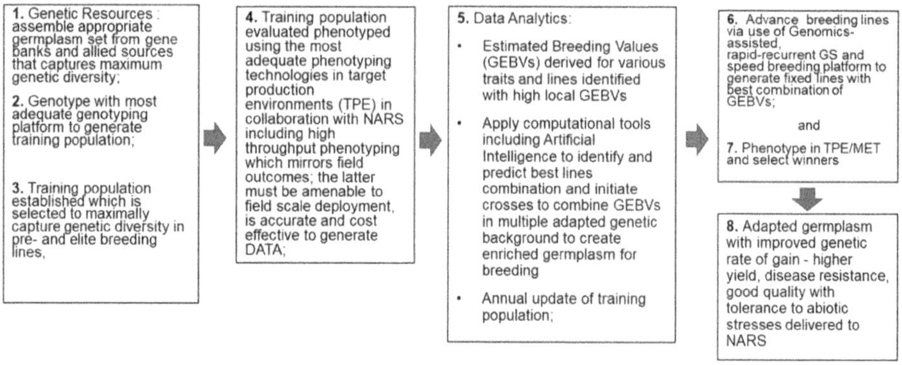

Figure 7.12 Schematic description of integrating R and D components from genebank to field to deliver improved rate of yield genetic gain to farmers.

7.15 The Way Forward and Wrap-Up

Toward attaining wheat productivity gains, it is paramount that genetic diversity is accessed and incorporated from the genetic resources of wheat as we are faced with a stagnancy in production levels. The practical way forward involves creative inroads where novelty is a definite gain and use of alien resources as providers of new allelic diversity are a priority. Within this umbrella, three areas are included that are intergeneric, interspecific and intraspecific hybridization. Recognizing the urgency of delivering productivity gains to meet the needs of a burgeoning population, use of the intergeneric route gets a back seat essentially because alien transfers that give significant returns are complex and very time-consuming. We have reviewed the huge arsenal of genes that are available and these can be enriched further with new non-conventional resources via adopting wise pre-breeding options. On their own, the balance is tilted strongly in favor of interspecific hybridization where the intraspecific area is automatically merged with it, thereby permitting all three wheat genomes to be enriched simultaneously via homologous chromosomal associations. This circumvents many of the limitations that prevail with the intergeneric crossing protocols. In our perception, we attach a high preference for giving maximum benefit to the D genome progenitor relative of wheat that had a limited role in its hexaploid origin around 6000 BC. These grass accessions are widely distributed and its allelic richness gives ample reason to attach to this genome a massive role in wheat improvement. The near 1500 synthetic stocks produced globally are a haven of new alleles and since, due to numerous promising reports in various fora and literature, stocks from this progenitor have become the starting point for addressing the difficulties of finding solutions to various aspects that govern wheat varietal improvement. The breeders have a select plateful of trait choices that through integration of conventional and non-conventional protocols can effectively chart the wheat breeding course.

The attributes of the D genome synthetics have been globally identified with testing data, reported in literature where details can be easily accessed. Several of these traits are summarized primarily to show the spectrum of allelic richness that is on hand for breeding across the following major productivity impacting categories:

1) general food security constraints
2) biotic stresses: spot blotch, Septoria, karnal bunt, head scab, all three rusts, powdery mildew, downy mildew, Alternaria spot, tan spot
3) barley yellow dwarf and wheat streak mosaic virus
4) abiotic stresses: tolerance to heat, drought, salinity, waterlogging, cold or frost, aluminum tolerance
5) quality components: various HMW and LMW units
6) nutritive value: iron and zinc
7) insect resistance: hessian fly, saw fly, aphids
8) mineral toxicities and desired level increases: boron and phytic acid
9) yield components: 1000 grain weight, efficient photosynthetic activity and its components, growth habit, leaf waxiness, pubescence, solid stem, robust root system, coleoptile emergence from deep seed planting, preharvest grain sprouting, stay green, growth habit, etc.
10) DNA polymorphisms for molecular application benefits, etc.
11) MSR (multiple stress resistances/tolerances, etc.) for broad coverage of multiple traits simultaneously instead of the slow individual trait exploitation

7.16 Concerns

By promoting the D genome synthetics as the preferred way forward, cognizance of some concerns is essential.

1) These D-genome synthetics are the storehouse of a wide variety of genes, but their having major genes may be considered a disadvantage. Resistances vary widely in this material and in order to combine the positive performance of two different synthetics, even with major genes for a trait, they are first crossed and their F_1 advanced to F_2. Field selections are then made for superior F_2's compared to either parent, tillers taken from such F_2 plants and doubled haploids produced with genes from both parents pyramided. An excellent example exists for Helminthosporium spot blotch, where many F_2 derivatives exhibited a strongly resistant 1-1 disease score versus the susceptible control 9-9 reading. The two parents had scores of 2-2 or 2-1. Pyramiding is a user option for harnessing major gene resistances for imparting durability and not to be viewed as an impediment in use of major genes.
2) *Aneuploidy*. The synthetics are hexaploids, i.e., $2n = 6x = 42$; AAABBDD, and each at C-0 stage had from 1 or 2 seeds to about 10. These were all somatically analyzed and there was a pre-dominance of progeny with the number 42. Subsequently, upon seed increase, aneuploidy was encountered (Figure 7.13) and only those with 42 were advanced of each combination. The first 42 somatic count plant with normal bivalent meiosis was selected for further study and purity maintenance. Normalcy of cytological status is a cumbersome exercise but a good precaution to maintain the stocks purity. It is doubtful if this earlier practice is continued by others handling primary synthetics any longer, but the protocol was followed in the CIMMYT program until 2004 prior to the retirement of the first author. Figure 7.13 demonstrates the somatically observed aneuploidy in the D genome synthetics.

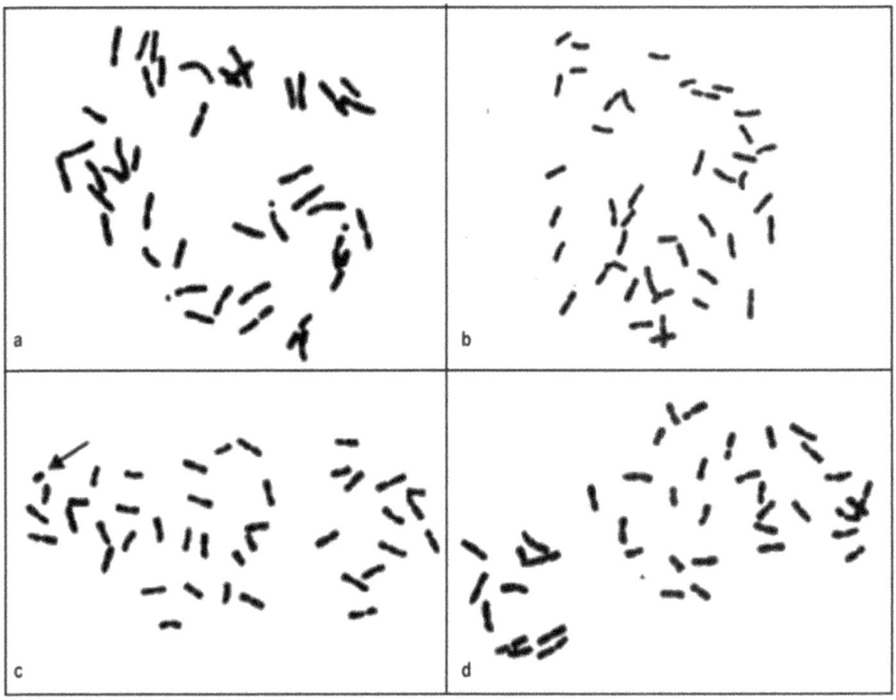

Figure 7.13 Somatic chromosome numbers in D genome synthetic hexaploids from *Triticum turgidum/Aegilops tauschii* combinations showing in: a) the expected 42 status, i.e., 2n = 6x = 42, AABBDD; b) hyperploid with 43 chromosomes; c) 42 with a telocentric chromosome; and d) hypoploid with 41 chromosomes.

3) *Seedling necrosis in pre-breeding stages.* Upon screening and identifying synthetics with desired values for wheat breeding were crossed with recipient wheat varieties and F_1's produced. When these seed were germinated, seedling necrosis was observed and so was a pitfall in their usage. The necrosis was due to the accumulation of the two necrotic genes and could be overcome if this knowledge was utilized and, if not on hand, then subjecting the seedling to high temperature exposure was the solution (Dhaliwal et al. 1986). Having knowledge of the necrotic genes was the safest option. Under other conditions, by trial and error, the non-necrotic wheats were identified and these formed the parents to be integrated within the initial crossing protocol followed by other cultivars that were crossed on further for getting backcross derivatives.

7.17 Conclusions

We have utilized the text below as expressed by the first author in a recent 2020 invited opinion document (Mujeeb-Kazi et al. 2020) with partial modifications to conclude this manuscript. It has approval to be freely used as we have done here.

Wheat has been unequivocally recognized as the major conduit to combat food security challenges envisioned for 2050. when the global populace is approximated to touch about 9.7 billion. Paramount for meeting annual wheat yield increases projected at 1.7%, there have to be ways and means that deliver swift progressive enhancement numbers that have been inconsistent so far. No matter what advancements occur in cutting edge science, if those outputs are not swiftly translated into tons per hectare via exploiting smooth multidisciplinary integrated crop improvement protocols, the desired applied targets will remain elusive. The enormous advances in molecular technology and the recent range of global quality publications offer us great impetus that the set targets covering food security, zero hunger and sustainability development goals shall be achieved. However, it is crucial that a "unified research effort exists and operates in tandem either at one location or structured around partners" that are target driven, as dictated by the 2050 timed output constraints of climate change. In essence, progress will be seen by integrating disciplines that execute the development program hand-in-hand via systematic crop breeding filial generation advance, where at each stage the technologies used permit efficient selective plant generation advance. The strategy envisioned as being the growing population non-destructive and molecular/conventional simultaneously would give a win-win picture is our contention. Such a well-integrated multidisciplinary wheat improvement program is a dire need, as only then shall we be able to see outputs that hopefully deliver gains in a timely manner. Have yields plateaued? How can this bottleneck be removed? No matter what the constraints, the ultimate measure of research progress will be yield enhancement on cultivation grounds that themselves will become limited versus time. When grounds become limiting, moving forward is a huge hurdle. A lot has been done using what had been available to give us the Green Revolution, whereby major biotic and abiotic stresses were well addressed. Quantum increases in productivity emerged with experimental creativity in the mid-1960s and mid-1970s. Stagnation since then necessitated that strategies to widen the genetic base were identified and exploited. Using novel genetic components paved the way for a surge in experimentation variations and by the early 1980s "wheat wide crosses" had gained prominence and fortuitously limited practical gains emerged from a few hybrid combinations, of which the most significant were materials that were superior in germplasm performance (Kishii 2019). These have been reported and reviewed in contributions of Gupta 2016; Gupta et al. 2019b, 2020; Gupta and Vasistha 2018; Mujeeb-Kazi et al. 2017, 2019; Sharma 1995). Missing was the need of a greater range of genetic diversity that could be exploited swiftly and cover a wide range of stress factors that are now linked with the hazards of climate change that penalize optimum food security projections (Wani et al. 2021).

Researchers have numerous options but we consider that in order to combat the constraints that are a threat to the food security issues of 2050, our emphasis here is to set priorities on *Ae. tauschii* or goat grass, of which a wide array of stocks were generated over the mid-1980s until recently. These are the arsenal of genes and a goldmine of diversity capable of combatting food production constraints adequately. These genetic stocks have resistances and tolerances to almost all biotic/abiotic stresses, as well as the resource for facilitating molecular tool development (Ogbonnaya et al. 2013; Rosyara et al. 2019). The stocks are housed in several gene banks, are freely distributed and show great promise to be able to address wheat production attributes perfectly geared to combat the adversities of climate change. Apart from stress traits, these stocks have enhanced yield per se by 22.7% based upon data gathered from field cultivation over numerous hectares through a cultivar CHAUNMAI 42 derived from a synthetic wheat cross combination (Li et al. 2018a).

7.18 Some Perceptions

Research is vital for generating quality platforms of outputs that is an agglomeration of three major facets encompassing basic, strategic and applied categories. These three facets require a common base where genetic diversity and its wise exploitation is paramount. Wise in the sense that researchers must recognize that experimentation cannot be open ended. The light at the end of the tunnel must be reached on time and impact seen if the demands for food security are well met and the complex environmental hazardous factors are circumvented. We feel that there is currently ample usable diversity generated and globally available that adequately covers the basic/strategic facets of a research program. From the outputs of these two segments, stringently focusing on the third "applied" component is the need. The available genetic stock data elucidates with clarity what materials should be selected for applied operations that can swiftly provide returns that mitigate all major environmental hazards jeopardizing food security projections. This is not the time to delve further into basic science but instead astutely move forward toward achieving positive practical outputs by using smart selective strategies that harness the needed genetic diversity stored in gene banks and well catalogued if the annual gains (Tadesse et al. 2019) up to 1.7% are achieved, leading us to the I billion tons production by the 2050 food security target.

Hence, the smart course would be to thoroughly exploit the over 1000 days genome synthetics that provide the answer of excellent diversity for almost all major stress factors. These stocks are freely available, are user-friendly for breeding, integral to applications of the current molecular technologies and genetically excellent as genetic recombination with bread wheat at a maximum. The literature is loaded with the resistance/tolerance information of these stocks and is profound. With this wealth we must ask the question: "What is the need to take a big step backward toward basic research instead of focusing on the over 1000 plus synthetics of the D genome that fit the applied scenario and is the answer to deliver swiftly the outputs needed for the 2050 food security mandate." We do not see the need to widen the D genome diversity as suggested by Mirzaghaderi and Mason (2019), as by using other higher ploidy D genome genera as the simple diploid level breeding effort will be confounded by the unrelated other genomes present and slow the breeding process. Our perception is that this abundant D genome diploid base created diversity in spring and winter habits, embracing wide pathogen and environmental stress coverage that governs global wheat productivity performance with supporting value from other traits. This germplasm has the variation that can also enrich the durum wheats across one genome by cytogenetic manipulation and many samples possess multiple stress attributes that can speed up agglomeration of needed attributes. This strength of accessibility to enormous diversity of practical value easily recombined with wheat renders this D genome progenitor genome resource as pivotal for wheat improvement and to combat food security is our stance. Integral to this strategy is to speed up varietal development by the appropriately timed under-utilized doubled haploid route to shorten the breeding cycle and exploiting its other practical additive values (Mujeeb-Kazi et al. 2006).

A concerted effort to harness this rich readily available progenitor genome resource warrants intensive exploitation for attaining time bound food security projected goals.

References

Abbasabad Zarei, E., Mohammadi, S.A., Moghaddam, M. et al. (2017). Analysis of genetic diversity, population structure and linkage disequilibrium in Iranian wheat landraces using SSR markers: Characterisation and utilisation. *Plant Genetic Resources* 15: 327–334.

Abu-Zaitoun, S., Chandrasekhar, K., Assili, S. et al. (2018). Unlocking the genetic diversity within a Middle-East panel of durum wheat landraces for adaptation to semi-arid climate. *Agronomy* 8: 233.

Adams, M.D., Kelley, J.M., Gocayne, J.D. et al. (1991). Complementary DNA sequencing: Expressed Sequence Tags and Human Genome Project. *Science* 252: 1651–1656.

Adugna, A., Nanda, G.S., Singh, K. et al. (2004). A comparison of cytoplasmic and chemically-induced male sterility systems for hybrid seed production in wheat (*Triticum aestivum*). *Euphytica* 135: 297–304.

Aghaee, M., Singh, H., and Dhaliwal, H. (2000). Ph^I gene derived from *Aegilops speltoides* induces homoeologous chromosome pairing in wide crosses of *Triticum aestivum*. *Journal of Heredity* 9: 417–421.

Ahmed, F. and Comeau, A. (1990). Wheat × pearl millet hybridization: Consequences and potential. *Euphytica* 50: 181–190.

Akbari, M., Wenzl, P., Caig, V. et al. (2006). Diversity arrays technology (DArT) for high-throughput profiling of the hexaploid wheat genome. *Theoretical and Applied Genetics* 113: 1409–1420.

Akhalkats, M., Ekhvaia, J., Mosulishvili, M. et al. (2010). Reasons and processes leading to the erosion of crop genetic diversity in mountainous regions of Georgia. *Mountain Research and Development* 3: 304.

Ali, N. (2012). Molecular markers, cytogenetics and epigenetics to characterize wheat-*Thinopyrum* hybrid lines conferring wheat streak mosaic virus resistance (Doctoral dissertation). University of Leicester.

Ali, N., Heslop-Harrison, J.P., Ahmad, H. et al. (2016). Introgression of chromosome segments from multiple alien species in wheat breeding lines with wheat streak mosaic virus resistance. *Heredity* 117 (2): 114e123.

Allen, A.M., Winfield, M.O., Burridge, A.J. et al. (2017). Characterization of a Wheat Breeders' Array suitable for high-throughput SNP genotyping of global accessions of hexaploid bread wheat (*Triticum aestivum*). *Plant Biotechnology Journal* 15: 390–401.

Anugrahwati, D.R., Shepherd, K.W., Verlin, D.C. et al. (2008). Isolation of wheat–rye 1RS recombinants that break the linkage between stem rust resistance gene *SrR* and secalin. *Genome* 51: 341–349.

Appels, R., Eversole, K., Stein, N. et al. (2018). Shifting the limits in wheat research and breeding using a fully annotated reference genome. *Science* 361 (6403).

Arabidopsis Genome Initiative (2000). Analysis of the genome sequence of the flowering plant *Arabidopsis thaliana*. *Nature* 408 (6814): 796–815.

Armstrong, J.M. (1936). Hybridization of *Triticum* and *Agropyron*. I: Crossing results and description of the first generation hybrids. *Canadian Journal of Research* 14 (C): 190–202.

Arora, S., Steuernagel, B., Gaurav, K. et al. (2019). Resistance gene cloning from a wild crop relative by sequence capture and association genetics. *Nature Biotechnology* 37: 139–143. doi:10.1038/s41587-018-0007-9.

Asiedu, R., Fisher, J.M., and Driscoll, C.J. (1990). Resistance to *Heterodera avenae* in the rye genome of triticale. *Theoretical and Applied Genetics* 79: 331–336.

Asseng, S., Martre, P., Maiorano, A. et al. (2018). Climate change impact and adaptation for wheat protein. *Global Change Biology* 25: 155–173.

Avni, R., Nave, M., Barad, O. et al. (2017). Wild emmer genome architecture and diversity elucidate wheat evolution and domestication. *Science* 357: 93–97.

Ayala, M., Guzmán, C., Peña, R.J. et al. (2016). Diversity of phenotypic (plant and grain morphological) and genotypic (glutenin alleles in Glu-1 and Glu-3 loci) traits of wheat landraces (*Triticum aestivum*) from Andalusia (Southern Spain). *Genetic Resources and Crop Evolution* 63: 465–475.

Ayed-Slama, O., Bouhaouel, I., Chamekh, Z. et al. (2018). Genetic variation of salt-stressed durum wheat (*Triticum turgidum* subsp. durum Desf.) genotypes under field conditions and gynogenetic capacity. *Journal of Genetic Engineering and Biotechnology* 16: 161–167.

Baboev, S.K., Buranov, A.K., Bozorov, T.A. et al. (2017). Biological and agronomical assessment of wheat landraces cultivated in mountain areas of Uzbekistan. *Sel'skokhozyaistvennaya Biologiya* 52: 553–560.

Baenziger, P.S. and DePauw, R.M. (2009). Wheat breeding: Procedures and strategies. In: *Wheat: Science and Trade* (ed. B.F. Carver), 275–308. Ames, IA, USA: Wiley-Blackwell Publishing.

Bai, D., Scoles, G.J., and Knott, D.R. (1995). Rust resistance in *Triticum cylindricum* Ces. (4x, CCDD) and its transfer into durum and hexaploid wheats. *Genome* 38: 8–16.

Bakshi, J.S. and Schlehuber, A.M. (1957). Identification of a substituted chromosome pair in a *Triticum-Agropyron* line. *Proceedings of the Oklahoma Academy of Science* 39: 16–21.

Banks, P.M., Xu, S.J., Wang, R.R.-C. et al. (1993). Varying chromosome composition of 56-chromosome wheat × *Thinopyrum intermedium* partial amphiploids. *Genome* 36: 207–215.

Barbu, S.P.D., Giura, A., and Lazăr, C. (2018). Potential sources of new genetic variability in mutant and mutant/ ecombinant wheat dh-lines. *Romanian Agricultural Research* 35: 81–87.

Barclay, I.R. (1975). High frequencies of haploid production in wheat (*Triticum aestivum*) by chromosome elimination. *Nature* 256: 410–411.

Barkworth, M. (2020). Genera of the *Triticeae* as recognized by various taxonomists after 1930. https://www.k-state.edu/wgrc/wheat_tax/post_1930.html (accessed 26 March 2020).

Barrett, B.A. and Kidwell, K.K. (1998). AFLP-based genetic diversity assessment among wheat cultivars from the Pacific Northwest. *Crop Science* 38: 1261–1271.

Barrett, J.A. (1978). A model for epidemic development in varietal mixtures. In: *Plant Disease Epidemiology* (ed. P.R. Scott and A. Bainsbridge), 129–137. Blackwells.

Bartos, P. (1993). Chromosome 1R of rye in wheat breeding. *Plant Breeding Abstracts* 63: 1203–1211.

Basil, E. and Blumwald, E. (2014). The ins and outs of intercellular ion homeostasis: NHX-type cation/H^+ transporters. *Current Opinion in Plant Biology* 22: 1–6.

Bates, L.S. and Deyoe, C.W. (1973). Wide hybridization and cereal improvement. *Economic Botany* 27: 401–412.

Bauer, E., Schmutzer, T., Barilar, I. et al. (2017). Towards a whole-genome sequence for rye (*Secale cereale* L.). *The Plant Journal* 89: 853–869. doi:10.1111/tpj.13436.

Bektas, H., Hohn, C.E., and Waines, J.G. (2016). Root and shoot traits of bread wheat (*Triticum aestivum* L.) landraces and cultivars. *Euphytica* 212: 297–311.

Bell, L.W., Byrne, F., Ewing, M.A. et al. (2008). A preliminary whole-farm economic analysis of perennial wheat in an Australian dryland farming system. *Agricultural Systems* 96: 166–174.

Bell, L.W., Wade, L.J., and Ewing, L.J. (2010). Perennial wheat: A review of environmental and agronomic prospects for development in Australia. *Crop & Pasture Science* 61: 679–690.

Berezhnoi, P. (1987). Development of the ideas of N.I. Vavilov on distant hybridization in wheat breeding. *Selektsiya I Semenovodstvo, Moscow* 6: 49–52.

Berzonsky, W.A. and Kimber, G. (1986). Tolerance of *Triticum* species to aluminium. *Plant Breeding* 97: 275–278.

BGRI. (2020). http://globalrust.org/about-bgri. (accessed 26 March 2020).

Bhatta, M., Morgounov, A., Belamkar, V. et al. (2019). Genome-wide association study for multiple biotic stress resistance in synthetic hexaploid wheat. *International Journal of Molecular Sciences* 20: 3667. doi:10.3390/ijms20153667.

Bhowal, J.G., Guha, G., and Vari, A.K. (1993). A cytogenetic study to transfer disease resistance genes from *Triticum timopheevi* to bread wheat. *Journal of Genetics and Breeding* 47: 71–76.

Biffen, R.H. (1905). Mendel's Law of Inheritance and Wheat Breeding. *The Journal of Agricultural Science* 1: 4–48.

Bishaw, Z., Struik, P.C., and Van Gastel, A.J.G. (2014). Wheat and barley seed system in Syria: How diverse are wheat and barley varieties and landraces from farmer's fields? *International Journal of Plant Production* 9: 117–150.

Blinkova, M.V. (1937). Winter-hardiness of the *Triticum-Agropyrum* hybrids. *Plant Breeding Abstracts* 9 (1939. Item 189): 49.

Blum, A., Golan, G., Mayer, J. et al. (1989). The drought response of landraces of wheat from the northern Negev Desert in Israel. *Euphytica* 43: 87–96.

Boeven, P.H., Longin, C.F.H., Leiser, W.L. et al. (2016). Genetic architecture of male floral traits required for hybrid wheat breeding. *Theoretical and Applied Genetics* 129: 2343–2357.

Bonjeau, A.P. and Angus, W.J. (2001). *The World Wheat Book: A History of Wheat Breeding*. Paris, France: Lavoisier Publishing.

Bonman, J.M., Babiker, E.M., Cuesta-Marcos, A. et al. (2015). Genetic diversity among wheat accessions from the USDA national small grains collection. *Crop Science* 55: 1243–1253.

Bonnin, I., Bonneuil, C., Goffaux, R. et al. (2014). Explaining the decrease in the genetic diversity of wheat in France over the 20th century. *Agriculture, Ecosystems & Environment* 195: 183–192.

Borlaug, N.E. (1958). The use of multilineal or composite varieties to control airborne epidemic diseases of self-pollinated crop plants. In *Proceedings of the First International Wheat Genetics Symposium* (ed. B. C. Jenkins), 12–27. Winnipeg, MB: University of Manitoba.

Borrenpohl, D., Huang, M., Olson, M.E. et al. (2020). The value of early-stage phenotyping for wheat breeding in the age of genomic selection. *Theoretical and Applied Genetics* 133: 2499–2520. https://doi.org/10.1007/s00122-020-03613-0.

Botstein, D., White, R.L., Skolnick, M. et al. (1980). Construction of a genetic linkage map in man using restriction fragment length polymorphisms. *American Journal of Human Genetics* 32: 314–331.

Brim, C.A. (1966). A modified pedigree method of selection in soybeans. *Crop Science* 6: 220.

Buerkert, A., Oryakhail, M., Filatenko, A.A. et al. (2006). Cultivation and taxonomic classification of wheat landraces in the upper Panjsher valley of Afghanistan after 23 years of war. *Genetic Resources and Crop Evolution* 53: 91–97.

Burr, B., Burr, F.A., Thompson, K.H. et al. (1988). Gene mapping with recombinant inbreds in maize. *Genetics* 118: 519–526.

Burt, C., Griffe, L.L., Ridolfini, A.P. et al. (2014). Mining the Watkins collection of wheat landraces for novel sources of eyespot resistance. *Plant Pathology* 63: 1241–1250.

Bux, H., Ashraf, M., and Chen, X. (2012). Expression of high-temperature adult-plant (HTAP) resistance against stripe rust (*Puccinia striiformis* f. sp. *tritici*) in Pakistan wheat landraces. *Canadian Journal of Plant Pathology* 34 (1): 68–74.

Caballero, L., Peña, R.J., Martín, L.M. et al. (2010). Characterization of Mexican Creole wheat landraces in relation to morphological characteristics and HMW glutenin subunit composition. *Genetic Resources and Crop Evolution* 57: 657–665.

Caetano-Anolles, G., Basam, J., and Gresshoff, P.M. (1991). DNA amplification finger-printing using very short arbitrary oligonucleotide primers. *Bio/Technology* 9: 553–557.

Cai, X., Xu, S.S., Oliver, R.E. et al. (2008). Alien introgression for FHB resistance in wheat-challenges and strategies. In: *Proceedings of the 11th International Wheat Genetics Symposium* (ed. R. Appels, R. Eastwood, E. Lagudah, et al.), 24–29 August 2008. Brisbane, QLD, Australia.

Cakmak, I., Derici, R., Torun, B. et al. (1997). Role of rye chromosomes in improvement of zinc efficiency in wheat and triticale. *Plant and Soil* 196: 249–253.

Cao, X.Y., Ming, C.H., Xu, Z.S. et al. (2012). Isolation and functional analysis of the bZIP transcription factor gene TaABP1 from a Chinese wheat landrace. *Journal of Integrative Agriculture* 11: 1580–1591.

Carvalho, M., Bebeli, P., Bettencourt, E. et al. (2012). Cereal landraces genetic resources in worldwide GeneBanks. A review. *Agronomy for Sustainable Development, Springer Verlag/ EDP Sciences/INRA* 33: 177–203. doi:10.1007/s13593-012-0090-0.hal-01201379.

Castiglioni, P., Ajmone-Marsan, P., Van Wijk, R. et al. (1999). AFLP markers in a molecular linkage map of maize: Codominant scoring and linkage group distribution. *Theoretical and Applied Genetics* 99: 425–431. https://doi.org/10.1007/s001220051253.

Cavanagh, C.R., Chao, S., Wang, S. et al. (2013). Genome-wide comparative diversity uncovers multiple targets of selection for improvement in hexaploid wheat landraces and cultivars. *Proceedings of the National Academy of Sciences of the United States of America* 110: 8057–8062.

Ceoloni, C. (1984). Transfer of a y mildew resistance gene from *Triticum longissimum* to common wheat by induced homoeologous recombination. *Genetica Agraria* 38: 326–327.

Ceoloni, C., Del Signore, G., and Bitti, O. (1989). Use of high pairing wheat mutants for the transfer of useful traits from alien species into cultivated wheats. In: *Current Options for Cereal Improvement. Doubled Haploids, Mutants and Heterosis. Advances in Agricultural Biotechnology* (ed. M. Maluszynski), 19–30. Dordrecht; Netherlands: No. 24. Kluwer Academic Publishers. CAB Accession 911697,26.

Ceoloni, C., Del Signore, G., Pasquini, M. et al. (1988). Transfer of mildew resistance from *Triticum longissimum* into wheat by *ph1* induced homoeologous recombination. In: *Proceedings of the Seventh International Wheat Genetics Symposium* (ed. T.E. Miller and R.M.D. Koebner), 221–226. 13–19 July 1988. Cambridge, UK: Institute of Plant Science Research.

Ceoloni, C., Kuzmanovic, L., Forte, P. et al. (2015). Wheat-perennial triticeae introgressions: Major achievements and prospects. In: *Alien Introgression in Wheat* (ed. M. Molnár-Láng, C. Ceoloni, and J. Doležel), 273–313. doi:10.1007/978-3-319-23494-6_11. Chapter 11.

Ceoloni, C., Signore, G.D., Ercoli, L. et al. (1992). Locating the alien chromatin segment in common wheat: *Aegilops longissima* mildew resistant transfers. *Hereditas-Landskrona* 116: 239–245.

Chaichi, M., Sanjarian, F., Razavi, K. et al. (2019). Analysis of transcriptional responses in root tissue of bread wheat landrace (*Triticum aestivum L.*) reveals drought avoidance mechanisms under water scarcity. *PloS One* 14 (3): e0212671.

Chan, K.Y., Bowman, A.M., Smith, W. et al. (2001). Restoring soil fertility of degraded hardsetting soils in semi-arid areas with different pastures. *Australian Journal of Experimental Agriculture* 41: 507–514.

Chao, S.P., Sharp, P.J., Worland, A.J. et al. (1989). RFLP-based genetic maps of wheat homoeologous group 6 chromosomes. *Theoretical and Applied Genetics* 78: 493–504.

Charmet, G., Tran, L.-G., Auzanneau, J. et al. (2020). BWGS: A R package for genomic selection and its application to a wheat breeding programme. *PLoS ONE* 15 (4): e0222733. https://doi.org/10.1371/journal.pone.0222733.

Chen, P., Wang, X., Zhang, R., et al. (2013). Development and utilization of wheat-alien small fragment translocations for wheat improvement. *Proceedings of the 12th International Wheat Genetics Symposium*, Pacifico Yokohama, Japan, 8–14 September 2013. pp 157.

Chen, P.D., Gao, Y., Zhou, B., et al. (2003). Development of wheat: *Haynaldia villosa* translocations using Ph^1 genes from *Aegilops speltoides*. *Proceedings of the Tenth International Wheat Genetics Symposium*, 1–6 September 2003, Paestum, Italy. Instituto Sperimentale per la Cerealicoltura. 2L893–895.

Chen, P.D., Tsujimoto, H., and Gill, B.S. (1994). Transfer of Ph^I genes promoting homoeologous pairing from *Triticum speltoides* to common wheat. *Theoretical and Applied Genetics* 88: 97–101.

Chen, S., Xia, G., Quan, T. et al. (2004). Introgression of salt-tolerance from somatic hybrids between common wheat and *Thinopyrum ponticum. Plant Science* 167: 773–779.

Chen, Y., Sidhu, H.S., Kaviani, M. et al. (2019). Application of image-based pheotyping tools to identify QTL for in-field witnter survival of winter wheat (*Triticum aestivum* L.). *Theoretical and Applied Genetics* 132: 2591–2604.

Chentoufi, L., Sahri, A., Arbaoui, M. et al. (2014). Anchoring durum wheat diversity in the reality of traditional agricultural systems: Varieties, seed management, and farmers' perception in two Moroccan regions. *Journal of Ethnobiology and Ethnomedicine* 10: art. no. 58.

Chhuneja, P., Dhaliwal, H.S., and Bains, N.S. (2006). *Aegilops kotschyi* and *Aegilops tauschii* as sources for higher levels of grain iron and zinc. *Plant Breeding* 125: 529–531.

Ciferri, R. (1955). The first interspecific wheat hybrids. *Journal of Heredity* 46 (2): 81–83.

CIMMYT (2020). https://www.cimmyt.org/news/farmers-in-india-embrace-high-zinc-wheat-for-its-nutritional-benefits. (accessed 26 March 2020).

Cisar, G. and Cooper, D. (2002). Hybrid wheat. In: *Wheat Improvement and Production* (ed. B.C. Curtis, S. Rajaram, and M.H. Gomez), 157–174. Rome: Food and Agriculture Organization.

Cohen, S., Chang, A., Boyer, H. et al. (1973). Construction of biologically functional bacterial plasmids *in vitro. Proceedings of the National Academy of Sciences of the United States of America* 70 (11): 3240–3244.

Colmer, T.D., Flowers, T.J., and Munns, R. (2006). Use of wild relatives to improve salt tolerance in wheat. *Journal of Experimental Botany* 57: 1059–1078.

Cox, C.M., Garrett, K.A., and Bockus, W.W. (2005). Meeting the challenge of disease management in perennial grain cropping systems. *Renewable Agriculture and Food Systems* 20: 15–24.

Cox, C.M., Murray, T.D., and Jones, S.S. (2002). Perennial wheat germplasm lines resistant to eyespot, Cephalosporium stripe, and wheat streak mosaic. *Plant Disease* 86: 1043–1048.

Cox, T.S., Glover, J.D., Van Tassel, D.L. et al. (2006). Prospects for developing perennial-grain crops. *Bioscience* 56: 649–659.

Cox, T.S., Van Ta, D.L., Cox, C.M. et al. (2010). Progress in breeding perennial grains. *Crop and Pasture Science* 61: 513–521.

Crespo-Herrera, L.A., Singh, R.P., Sabraoui, A. et al. (2019). Resistance to insect pests in wheat-rye and *Aegilops speltoides* Tausch translocation and substitution lines. *Euphytica* 215: 123. https://doi.org/10.1007/s10681-019-2449-7.

Crespo-Herrera, L.A., Smith, C.M., Singh, R.P. et al. (2013). Resistance to multiple cereal aphids in wheat-alien substitution and translocation lines. *Arthropod-Plant Interactions* 7 (2013): 535–545. doi:10.1007/s11829-013-9267-y.

Crossa, J., Jarquín, D., Franco, J. et al. (2016). Genomic prediction of gene bank wheat landraces. *G3: Genes, Genomes, Genetics* 6: 1819–1834. https://doi.org/10.1534/g3.116.029637.

Cui, L., Ren, Y.K., Murray, T.D. et al. (2018). Development of perennial wheat through hybridization between wheat and wheatgrasses: A review. *Engineering* 4: 507–513.

Çukadar, B., Peña, R.J., Dunphy, D. et al. (1997). The potential of hybrid wheat under irrigated conditions in Mexico. In: *The Genetics and Exploitation of Heterosis in Crops: An International Symposium* (D.F. Mexico 17–22 Aug. 1997. Book of abstracts), 190–191. Mexico, D.F.: CIMMYT.

Culman, S.W., Snapp, S.S., Ollenburger, M. et al. (2013). Soil and water quality rapidly responds to the perennial grain kernza wheatgrass. *Agronomy Journal* 105: 735–744.

Curwen-mcadams, C. and Jones, S.S. (2017). Breeding perennial grain crops based on wheat. *Crop Science* 57: 1172–1188.

Dedryver, F., Jubier, M.F., Thouvenin, J. et al. (1996). Molecular markers linked to the leaf rust resistance gene Lr13 in different wheat cultivars. *Genome* 39: 830–835.

DeHaan, L.R., Van Tassel, D.L., Anderson, J.A. et al. (2016). A pipeline strategy for grain crop domestication. *Crop Science* 56: 917–930.

DeHaan, L.R., Van Tassel, D.L., and Cox, T.S. (2005). Perennial grain crops: A synthesis of ecology and plant breeding. *Renewable Agriculture and Food Systems* 20: 5–14.

Delacy, I.H., Skovmand, B., and Huerta, J. (2000). Characterization of Mexican wheat landraces using agronomically useful attributes. *Genetic Resources and Crop Evolution* 47: 591–602.

Delgado, R., Cano, S., Cortes, A. et al. (2001). *Durum wheat/A and B Genome Amphiploids (2n = 6x = 42). Agronomy Abstracts.* American Society of Agronomy. CD-ROM.

Delibes, A., Romero, D., Aguaded, S. et al. (1993). Resistance to the cereal cyst nematode (*Heterodera avenae* Woll.) transferred from the wild grass *Aegilops ventricosa* to hexaploid wheat by a "stepping-stone" procedure. *Theoretical and Applied Genetics* 87: 402–408.

Devos, K.M., Bryan, G.L., Stephenson, P. et al. (1995). Application of two microsatellite sequences in wheat storage proteins as molecular markers. *Theoretical and Applied Genetics* 90: 247–252.

Devos, K.M. and Gale, M.D. (1992). The use of random amplified polymorphic DNA markers in wheat. *Theoretical and Applied Genetics* 84: 567–572. https://doi.org/10.1007/BF00224153.

Devos, K.M.D. and Gale, M.G. (2000). Genome relationships: The grass model in current research. *The Plant Cell* 12: 637–646.

Dewey, D.R. (1960). Salt tolerance of twenty-five strains of *Agropyron*. *Agronomy Journal* 52: 631–635.

Dewey, D.R. (1984). The genomic system of classification as a guide to intergeneric hybridization with the perennial *Triticeae*. In: *Gene Manipulation in Plant Improvement. 16th Stadler Genetics Symposium* (ed. J.P. Gustafsson), 209–279. New York: Plenum Press.

Dhaliwal, A.S., Mars, D.J., and Marshall, D.R. (1987). Effect of 1B/1R chromosome translocation on milling and quality characteristics of bread wheats. *Cereal Chemistry* 64: 72–76.

Dhaliwal, H.S., Sharma, S.K., and Ranbdhawa, A.S. (1986). How to overcome hybrid necrosis in wheat. *Wheat Information Service* 61/62: 27.

Dhaliwal, H.S., Singh, H., Singh, I., et al. (1998). Transfer of resistance to diseases and nematode from *Aegilops triuncialis* into wheat. *Proceedings of the Ninth International Wheat Genetics Symposium*. Saskatoon, Saskatchewan, Canada. 2–7 August 1998, 3, 245–247. University Extension Press, University of Saskatchewan.

Dodig, D., Zorić, M., Kandić, V. et al. (2012). Comparison of responses to drought stress of 100 wheat accessions and landraces to identify opportunities for improving wheat drought resistance. *Plant Breeding* 131: 369–379.

Donald, C.M. (1968). The breeding of crop ideotypes. *Euphytica* 17: 385–403.

Dong, Z.J., Tian, X.B., Ma, C. et al. (2020). Physical mapping of *Pm57*, a powdery mildew resistance gene derived from *Aegilops searsii*. *International Journal of Molecular Sciences* 21: 322.

Dowell, F.E., Maghirang, E.B., and Baenziger, P.S. (2009). Automated single-kernel sorting to select for quality traits in wheat breeding lines. *Cera-Chem* 86: 527–533.

Driscoll, C.J. (1972). XYZ system of producing hybrid wheat. *Crop Science* 12: 516517.

Driscoll, D.J. (1985). Modified XYZ system of producing hybrid wheat. *Crop Science* 25: 1115–1116.

Du Toit, F. (1989). Components of resistance in three bread wheat lines to Russian wheat aphid (*Homoptera: Aphididae*) in South Africa. *Journal of Economic Entomology* 6: 1779–1783.

Dubin, H.J. and Rajaram, S. (1996). Breeding disease-resistant wheats for tropical highlands and lowlands. *Annual Review of Phytopathology* 34: 503–526.

Dundas, I., Verlin, D., Zhang, P. et al. (2015a). Transfer to wheat of potentially new stem rust resistance genes derived from *Aegilops speltoides*. In: *Advances in Wheat Genetics: From Genome to Field* (ed. Y. Ogihara, S. Takumi, and H. Handa), 47–52. *Proceedings of the 12th International Wheat Genetics Symposium*, Yokohama, Japan. 8–14 September 2013. Springer. Chapter 5.

Dundas, I., Zhang, P., Verlin, D. et al. (2015b). Chromosome engineering and physical mapping of the *Thinopyrum ponticum* translocation in wheat carrying the rust resistance gene *Sr26*. *Crop Science* 55: 648–657.

Dundas, I.S., Frappell, D.E., Crack, D.M. et al. (2001). Deletion mapping of a nematode resistance gene on rye chromosome 6R in wheat. *Crop Science* 41: 1771–1778.

Dwivedi, S.L., Britt, A.B., Tripathi, L. et al. (2015). Haploids: Constraints and opportunities in plant breeding. *Biotechnology Advances* 33: 812–829.

Dyck, P.L., Kerber, E.R., and Martens, J.W. (1990). Transfer of a gene for stem rust resistance from *Aegilops caudata* to common wheat. *Canadian Journal of Plant Science* 70: 931–934.

Eagles, H.A., Bariana, H.S., Ogbonnaya, F.C. et al. (2001). Implementation of molecular markers in Australian wheat breeding. *Australian Journal of Agricultural Research* 52: 1349–1356.

Ehdaie, B., Whitkus, R.W., and Waines, J.G. (2003). Root biomass, water-use efficiency and performance of wheat-rye translocations of chromosomes 1 and 2 in spring bread wheat "Pavon." *Crop Science* 43: 710–717.

Elangbam, P.D., Deepshikha,, and Jaiswal, J.P. (2018). Revealing of brown rust resistance genes by molecular marker in w heat: A review. *Indian Journal of Pure & Applied Biosciences* 6: 407–417. http://dx.doi.org/10.18782/2320-7051.4095.

Elings, A. (1993). Durum wheat land races from Syria. *Euphytica* 70: 85. https://doi.org/10.1007/BF00029645.

Elshire, R.J., Glaubitz, J.C., Sun, O. et al. (2011). A robust, simple genotyping-by-sequencing (GBS) approach for high diversity species. *PLoS ONE* 6 (5): e19379. doi:10.1371/journal.pone.0019379.

Emebiri, L.C., Raman, H., and Ogbonnaya, F.C. (2020). Synthetic hexaploid wheat as a source of novel genetic loci for aluminium tolerance. *Euphytica* 216: 135. https://doi.org/10.1007/s10681-020-02669-9.

Eticha, F., Belay, G., and Bekele, E. (2006). Species diversity in wheat landrace populations from two regions of Ethiopia. *Genetic Resources and Crop Evolution* 53: 387–393.

Eucarpia (2020). Erich von Tschermak-Seysenegg. Austria, 1956. https://eucarpia.org/tschermak-seysenegg.html. (accessed 26 March 2020).

The European Wheat Database (EWDB) (2012). http://genbank.vurv.cz/ewdb/default.htm.

Falk, D.E. and Kasha, K.J. (1981). Comparison of the crossability of rye (*Secale cereale*) and *Hordeum bulbosum* onto wheat (*Triticum aestivum*). *Canadian Journal of Genetics and Cytology* 23: 81–88.

Fan, L., Han, J.Y., and Deng, J.Y. (1995). Transfer of gene for resistance to powdery mildew from *Ae. triuncialis* to common wheat by using Chinese Spring *Ta1* kr ph1b plants. In: *Proceedings of the Eighth International Wheat Genetics Symposium*, 1 (ed. Z.S. Li and Z.Y. Xin), 489–493. 20–25 July 1993. Beijing, China: China Agricultural Scientech Press.

Fanning, J., Linsell, K., McKay, A. et al. (2018). Resistance to the root lesion nematodes *Pratylenchus thornei* and *P. neglectus* in cereals: Improved assessments in the field. *Applied Soil Ecology* 132: 146–154.

FAO (2010). The second report on the state of the Worlds Plant Genetic Resources for Food and Agriculture Rome. ISBN 978-92-5-106534-1, 243–249.

Faris, J.D., Megan, E., Gayan, O., Kariyawasamet, K. et al. (2020). Identification of a major dominant gene for race-nonspecific tan spot resistance in wild emmer wheat. *Theoretical and Applied Genetics* 133: 829–841.

Farrer, W. (1904). Some notes on the wheat "Bobs"; its peculiarities, economic value and origin. *Agricultural Gazette of NSW* 15: 849–854.

Fedak, G. (1979). Source of meiotic abnormalities in barley-wheat hybrids. *Wheat Information Service* 50: 10–11.

Fedak, G. (1980). Production, morphology and meiosis of reciprocal barley wheat hybrids. *Canadian Journal of Genetics and Cytology* 22: 117–123.

Fedak, G. (1998). Procedures for transferring agronomic traits from alien species to crop plants. In: *Proceedings of the Ninth International Wheat Genetics Symposium*, 1 (ed. A.E. Slinkard), 1–7. 2–7 August 1998. Saskatoon, Saskatchewan, Canada: University Extension Press, University of Saskatchewan.

Fedak, G. and Jui, P.Y. (1982). Chromosomes of Chinese Spring wheat carrying genes for crossability with Betzes barley. *Canadian Journal of Genetics and Cytology* 24: 227–233.

Fedak, G. (1977). Increased homoeologous pairing in *Hordeum vulgare* L.x *Triticum aestivum* l. hybrids. *Nature* 266: 5229–5530.

Feldman, M. and Levy, A.A. (2012). Genome evolution due to allopolyploidization in wheat. *Genetics* 192: 763–774.

Ferrahi, M., Friebe, B., Hatchett, I.H. et al. (2017). Transfer and molecular mapping of *Aegilops tauschii* derived Hessian fly resistance genes (H22, H23, H24 and H26) from D genome of *Triticum aestivum* onto A genome chromosomes of *Triticum turgidum* by induced homoeologous recombination. *International Journal of Advanced Research* 5: 728–750.

Forster, B.P., Miller, T.E., and Law, C.N. (1988). The potential for transferring genes conferring salt tolerance from *Thinopyrum bessarabicum* into wheat. In: *Proceedings of the Seventh International Wheat Genetics Symposium*, 1 (ed. T.E. Miller and R.M.D. Koebner), 267–270. Cambridge, UK. 13–19 July 1988.

Forster, B.P. and Thomas, W.T.B. (2010). Doubled haploids in genetics and plant breeding. In: *Plant Breeding Reviews*, 25 (ed. J. Janick). Oxford, UK: John Wiley & Sons.

Friebe, B., Jiang, J., Raupp, W.J. et al. (1996). Characterization of wheat-alien translocations conferring resistance to diseases and pests: Current status. *Euphytica* 91: 59–87.

Furbank, R.T. and Tester, M. (2011). Phenomics: Technologies to relieve the phenotyping bottleneck. *Trends in Plant Science* 16: 635–644.

Gale, K.R. (2005). Diagnostic DNA markers for quality traits in wheat. *Journal of Cereal Science* 41: 181–192.

Gale, M.D. and Devos, K.M. (1998). Comparative genetics in the grasses. *Proceedings of the National Academy of Sciences of the United States of America* 95: 1971–1974.

Gessese, M., Bariana, H., Wong, D. et al. (2019). Molecular mapping of stripe rust resistance Gene Yr81 in a common wheat Landrace Aus27430. *Plant Disease* Apr 17: PDIS-06.

Ghimire, S.K., Akashi, Y., Masuda, A. et al. (2006). Genetic diversity and phylogenetic relationships among East Asian common wheat (*Triticum aestivum L.*) populations, revealed by the analysis of five isozymes. *Breeding Science* 56: 379–387.

Ghini, R. and Morandi, M.A.B. (2006). Biotic and abiotic factors associated with soil suppressiveness to *Rhizodonia solani*. *Scientia Agricola* 63: 153–160.

Gill, B.S., Browder, L.E., Hatchett, J.H. et al. (1983). Disease and insect resistance in wild wheats. In: *Proceedings of the Sixth International Wheat Genetics Symposium* (ed. S. Sakamoto), 785–792. 28 November–3 December 1983. Kyoto, Japan: Faculty of Agriculture, Kyoto University.

Gill, B.S. and Raupp, W.J. (1987). Direct gene transfers from *Aegilops squarrosa* L. to hexaploid wheat. *Crop Science* 27: 445–450.

Gill, B.S., Raupp, W.J., and Friebe, B. (2014). Dual threats of imperiled native agrosystems and climate change to world food security: Genomic perspectives. *Journal of Crop Improvemen* 28: 88–98.

Gill, K.S., Nanda, G.S., Singh, G. et al. (1981). *Indian Multilines*. Mexico: CIMMYT.

Gill, K.S., Nanda, G.S., Singh, G. et al. (1982). Sonalika multiline 1: The first multiline of "Sonalika." In: *Prog. Farm*, 9–10.

Glover, J.D., Culman, S.W., DuPont, S.T. et al. (2010a). Harvested perennial grasslands provide ecological benchmarks for agricultural sustainability. *Agriculture, Ecosystems & Environment* 137: 3–12.

Glover, J.D., Reganold, J.P., Bell, L.W. et al. (2010b). Increased food and ecosystem security via perennial grains. *Science* 328: 1638–1639.

Godron, D. (1858). On the natural and artificial fertilization of *Aegilops* by *Triticum*. *Journal of the Royal Agricultural Society of England* 19: 104–122.

Goel, S., Singh, B., Grewal, S. et al. (2018). Variability in Fe and Zn content among indian wheat landraces for improved nutritional quality. *Indian Journal of Genetics and Plant Breeding* 78: 426–432.

Goff, S.A., Ricke, D., Lan, T.H. et al. (2002). A draft sequence of the rice genome (*Oryza sativa* L. ssp. japonica). *Science* 296: 92–100.

Golden, C.H. (1939). Problems in plant selection. In: *Proceedings of the 7th International Genetics Congress (Edinburgh)* (ed. R.C. Burnett), 132–133. Cambridge, UK: Cambridge University Press.

Gonzelez, F.G., Rigalli, N., Miranda, P.V. et al. (2018). An interdisciplinary approach to study the performance of second-generation genetically modified crops in field trials: A case study with soybean and wheat carrying the sunflower *HaHB4* transcription factor. *Frontiers in Plant Science* 11: 178. doi:10.3389/fpls.2020.001178.

Gowda, M., Longin, C.F.H., Lein, V. et al. (2012). Relevance of specific versus general combining ability in winter wheat. *Crop Science* 52: 2494–2500.

Graham, R.D., Ascher, J.S., Ellis, P.A.E. et al. (1987). Transfer to wheat of the copper efficiency factor carried on rye chromosome arm 5RL. *Plant and Soil* 99: 107–114.

GrainGenes. 2020. Catalogue of gene symbols for wheat. https://wheat.pw.usda.gov/GG3/wgc (accessed 5 July 2020).

Graybosch, R.A., Peterson, C.J., Baenziger, P.S. et al. (2009). Registration of °Mace" hard red winter wheat. *Journal of Plant Registrations* 3: 51–56.

Graybosch, R.A., Peterson, C.J., Hansen, L.E. et al. (1990). Relationship between protein solubility characteristics, 1BL/1RS, high molecular weight glutenin composition, and end-use quality in winter wheat germ plasm. *Cera-Chem* 67: 342–349.

Green, E.D. and Green, P. (1991). Sequence-tagged Site (STS) content mapping of human chromosomes: Theoretical considerations and early experiences. *Genome Research* 1991 (1): 77–90.

Griffiths, S., Wingen, L., Pietragalla, J. et al. (2015). Genetic dissection of grain size and grain number trade-offs in CIMMYT wheat germplasm. *PLoS ONE* 10 (3): e0118847. doi:10.1371/journal.pone.0118847.

Guo, X., Li, H., Xiang, J. et al. (2011). Genetic variation of high molecular weight glutenin subunits in wheat accessions in China. *Crop Science* 51: 2423–2429.

Gupta, P.K. (2016). Use of alien genetic variation for wheat improvement. In: *Molecular Breeding for Sustainable Crop Improvement* (ed. V.R. Rajpal), 1–30. Switzerland: Springer International Publishing.

Gupta, P.K., Balyan, H.S., Gahlaut, V. et al. (2019a). Hybrid wheat: Past, present and future. *Theoretical and Applied Genetics* 132: 2463–2483. https://doi.org/10.1007/s00122-019-03397-y.

Gupta, P.K., Balyan, H.S., Sharma, S. et al. (2020). Genetics of yield abiotic stress tolerance and biofortification in wheat. *Theoretical and Applied Genetics* 133: 1569–1602.

Gupta, P.K., Varshney, R.K., Sharma, P.C. et al. (1999). Molecular markers and their applications in wheat breeding. *Plant Breeding* 118: 369–390.

Gupta, P.K. and Vasistha, N.K. (2018). Wheat cytogenetics and cytogenomics: The present status. *Nucleus* 61 (3): 195–212K.

Gupta, V., He, X.Y., Kumar, N. et al. (2019b). Genome wide association study of Karnal bunt resistance in a wheat germplasm collection from Afghanistan. *International Journal of Molecular Sciences* 20: 3124.

Guzmán, C., Medina-Larqué, A.S., Velu, G. et al. (2014). Use of wheat genetic resources to develop biofortified wheat with enhanced grain zinc and iron concentrations and desirable processing quality. *Journal of Cereal Science* 60: 617–622.

Habib, I., Rauf, M., Ahmad, M. et al. (2014). Optimization of somatic embryogenesis and Agrobacterium mediated transformation of elite wheat (*Triticum aestivum* L.). *International Journal of Agriculture and Biology* 16: 1098–1104.

Hamada, H., Linghu, Q., Nagira, Y. et al. (2017). An in planta biolistic method for stable wheat transformation. *Scientific Reports* 7: 11443.

Han, F.P., Liu, B., Fedak, G. et al. (2004). Genomic constitution and variation in five partial amphiploids of wheat – *Thinopyrum intermedium* as revealed by GISH, multicolor GISH and seed storage protein analysis. *Theoretical and Applied Genetics* 109: 1070–1076.

Haq, R.F.U., Saeed, N.A., Ahmed, M. et al. (2019). Barley vacuolar pyrophosphatase (*HVP1*) gene confers salinity tolerance in locally adapted wheat (*Triticum aestivum*). *International Journal of Agriculture and Biology* 22:1338–1346.

Harlan, H.V. and Pope, M.N. (1922). The use and value of back crosses in small-grain breeding. *The Journal of Heredity* 13: 319-322.

HarvestPlus 2020. https://www.harvestplus.org/search/node/zinc%20wheat. (accessed 26 March 2020).

Hatchett, J.H. and Gill, B.S. (1983). Expression and genetics of resistance to Hessian fly in *Triticum tauschii* (Coss) Schmal. In: *Proceedings of the Sixth International Wheat Genetics Symposium* (ed. S. Sakamoto), 807–811. 28 November–3 December 1983. Kyoto, Japan: Faculty of Agriculture, Kyoto University.

Hayes, B.J., Cogan, N.O.I., Pembleton, L.W. et al. (2013). Prospects for genomic selection in forage plant species. *Plant Breeding* 132: 133–143. doi:10.1111/pbr.12037.

Hayes, H.K., Immer, F.R., and Smith, D.C. (1955). *Methods of Plant Breeding*, 551. New York: McGraw-Hill.

Hayes, H.K., Parker, J.H., and Kurtzweil, C. (1920). Genetics of rust resistance in crosses of varieties of *Triticum vulgare* with varieties of *T. durum* and *T. dicoccum*. *Journal of Agricultural Research* 19: 523–542.

Hayes, R.C., Newell, M.T., DeHaan, L.R. et al. (2012). Perennial cereal crops: An initial evaluation of wheat derivatives. *Field Crops Research* 133: 68–89.

Hayes, R.C., Wang, S.W., Newell, M.T. et al. (2018). The performance of early-generation perennial winter cereals at 21 sites across four continents. *Sustainability* 10: 1124–1124.

Hayta, S., Smedly, M.A., Demir, S.U. et al. (2019). An efficient and reproducible Agrobacterium-mediated transformation method for hexaploid wheat (*Triticum aestivum* L.). *Plant Methods* 15: 121. doi:10.1186/s13007-019-0503-z.

He, S., Schulthess, A.W., Mirdita, V. et al. (2016). Genomic selection in a commercial winter wheat population. *Theoretical and Applied Genetics* 129: 641–651. https://doi.org/10.1007/s00122-015-2655-1.

He, X., Singh, P.K., Duveiller, E. et al. (2013). Identification and characterization of international Fusarium head blight screening nurseries of wheat at CIMMYT, Mexico. *European Journal of Plant Pathology* 136: 123–134.

Hede, A.R., Skovmand, B., Reynolds, M.P. et al. (1999). Evaluating genetic diversity for heat tolerance traits in Mexican wheat landraces. *Genetic Resources and Crop Evolution* 46: 37–45.

Heffner, E.L., Lorenz, A.J., Jannink, J.-L. et al. (2010). Plant breeding with genomic selection: Gain per unit time and cost. *Crop Science* 50: 1681–1690. doi:10.2135/cropsci2009.11.0662.

Heffner, E.L., Sorrells, M.E., and Jannink, J.-L. (2009). Genomic selection for crop improvement. *Crop Science* 49: 1–12. doi:10.2135/cropsci2008.08.0512.

Helguera, M., Khan, I., and Dubcovsky, J. (2000). Development of PCR markers for the wheat leaf rust resistance gene Lr47. *Theoretical and Applied Genetics* 100: 1137–1143. https://doi.org/10.1007/s001220051397.

Henry, R.J. (2012). Next-generation sequencing for understanding and accelerating crop domestication. *Briefings in Functional Genomics* 11: 51–56.

Henry, Y. and De Buyser, J. (1990). Wheat anther culture: Agronomic performance of doubled haploid lines and the release of a new variety Florin. In: *Biotechnology in Agriculture and Forestry: Wheat*, 13 (ed. Y.P.S. Bajaj), 285–352. Heidelberg, Germany: Springer-Verlag.

Hohn, C.E. and Lukaszewski, A.J. (2016). Engineering the 1BS chromosome arm in wheat to remove the *Rf* multi locus restoring male fertility in cytoplasms of *Aegilops kotschyi, Ae. uniaristata* and *Ae. mutica*. *Theoretical and Applied Genetics* 129: 1769–1774.

Holton, T.A., Christopher, J.T., McClure, L. et al. (2002). Identification and mapping of polymorphic SSR markers from expressed gene sequences of barley and wheat. *Molecular Breeding : New Strategies in Plant Improvement* 9: 63–71.

Hong, M.J., Kim, D.Y., Na, B.M. et al. (2019). Characterization of novel mutants of hexaploid wheat (*Triticum aestivum* L.) with various depths of purple grain color and antioxidant capacity. *Journal of the Science of Food and Agriculture* 99: 55–63. https://doi.org/10.1002/jsfa.9141.

Howard, A. and Howard, G.L.C. (1910). *Wheat in India: Its Production, Varieties and Improvement*, 128 pp. Calcutta: Thacker Spink & Co.

Hu, M.-J., Zhang, H.P., Liu, K. et al. (2016). C-Xvia SLAF-seq-BSA. *Frontiers in Plant Science* 7: 1902. doi:10.3389/fpls.2016.01902.

Huang, M., Cabrera, A., Hoffstetter, A. et al. (2016). Genomic selection for wheat traits and trait stability. *Theoretical and Applied Genetics* 129: 1697–1710. https://doi.org/10.1007/s00122-016-2733-z.

Huang, X.Q. and Roder, M.S. (2004). Molecular mapping of powdery mildew resistance genes in wheat: A review. *Euphytica* 137: 203–223.

Husenov, B., Otambekova, M., Morgounov, A. et al. (2015). *Wheat Landraces in Farmers' Fields in Tajikistan: National Survey, Collection, and Conservation, 2013–2015*, 29 p. Ankara, Turkey: FAO.

Hussain, M., Iqbal, M.A., Till, B.J. et al. (2018). Identification of induced mutations in hexaploid wheat genome using exome capture assay. *PLoS ONE* 13 (8): e0201918. https://doi.org/10.1371/journal.pone.0201918.

Imtiaz, M., Ogbonnaya, F.C., Oman, J. et al. (2008). Characterization of QTLs controlling genetic variation for pre-harvest sprouting in synthetic backcross derived wheat lines. *Genetics* 178: 1725–1736.

Inagaki, M.N., Varughese, G., Rajaram, S. et al. (1998). Comparison of bread wheat lines selected by doubled haploid, single-seed descent and pedigree selection methods. *Theoretical and Applied Genetics* 97: 550–556.

Ishida, Y., Hiei, Y., and Komari, T. (2015). High efficiency wheat transformation mediated by Agrobacterium tumefaciens. Chapter 18. In: *Proceedings of the 12th International Wheat Genetics Symposium* (ed. Y. Ogihara, S. Takumi, and H. Handa). Springer.

Ishikawa, G., Yonemaru, J., Saito, M. et al. (2007). PCR-based landmark unique gene (PLUG) markers effectively assign homoeologous wheat genes to A, B and D genomes. *BMC Genomics* 8: 135. doi:10.1186/1471-2164-8-135.

Islam, A.K.M.R., Shepherd, K.W., and Sparrow, D.H.B. (1978). Production and characterization of wheat-barley addition lines. In: *Proceedings of the 5th Internastional Wheat Geneics. Symposium*, 365–371. New Delhi, India.

Islam, A.K.M.R., Shepherd, K.W., and Sparrow, D.H.B. (1981). Isolation and characterization of euplasmic wheat-barley chromosomal addition lines. *Journal of Heredity* 46: 161–174.

Islam, S., Malik, A.I., Islam, A.K.M.R. et al. (2007). Salt tolerance in a *Hordeum marinum–Triticum aestivum* amphiploid, and its parents. *Journal of Experimental Botany* 58: 1219–1229.

Jaccoud, D., Peng, K., Feinstein, D. et al. (2001). Diversity Arrays: A solid state technology for sequence information independent genotyping. *Nucleic Acids Research* 29: e25.

Jaikumar, N.S., Snapp, S.S., Murphy, K. et al. (2012). Agronomic assessment of perennial wheat and perennial rye as cereal crops. *Agronomy Journal* 104: 1716–1726.

Jannink, J.-L. (2007). Identifying quantitative trait locus by genetic background interactions in association studies. *Genetics* 176: 553–561.

Jauhar, P.P. (1995). Morphological and cytological characteristics of some wheat × barley hybrids. *Theoretical and Applied Genetics* 90: 872–877.

Jensen, N.F. (1952). Intra-varietal diversification in oat breeding. *Agronomy Journal* 44: 30–34.

Jia, J.Z., Zhao, S.C., Kong, X.Y. et al. (2013). *Aegilops tauschii* draft genome sequence reveals a gene repertoire for wheat adaptation. *Nature* 496: 91–95.

Jiang, J., Friebe, B., and Gill, B.S. (1995). Recent advances in alien gene transfer in wheat. *Euphytica* 73: 199–212.

Jiang, S., Cheng, Q., Yan, J. et al. (2020). Genome optimization for improvement of maize breeding. *Theoretical and Applied Genetics* 133: 1491–1502. https://doi.org/10.1007/s00122-019-03493-z.

Jighly, A., Alagu, M., Makdis, F. et al. (2016). Genomic regions conferring resistance to multiple fungal pathogens in synthetic hexaploid wheat. *Molecular Breeding: New Strategies in Plant Improvement* 36: 127. doi:10.1007/s11032-016-0541-4.

Johannsen, W.L. (1903). Uber Erblickeit in Populationen und in reinen Leinen. Gustav Fischer, Jena. (Heredity in populations and pure lines. In: *Classic Papers in Genetics*, 1959 (ed. J.A. Peters), 20–26. Englewood Cliffs, NJ: Prentice-Hall.

Jones, E.T., Heyne, E.G., and Schmidt, J.W. (1952). A new source of Wheat Joint worm resistance in wheat. *Transactions of the Kansas Academy of Science* 55: 427–430.

Jones, H.D., Doherty, A., and Wu, H. (2005). Review of methodologies and a protocol for the Agrobacterium-mediated transformation of wheat. *Plant Methods* 1: 1–9. doi:10.1186/1746-4811-1-5.

Jones, T.A., Zhang, X.Y., and Wang, R.R.C. (1999). Genome characterization of MT-2 perennial and OK-906 annual wheat × intermediate wheatgrass hybrids. *Crop Science* 39: 1041–1043.

Joshi, B.C. and Singh, D. (1978). Introduction of alien variation into bread wheat. In: *Proceedings of the Fifth International Wheat Genetics Symposium*, 1 (ed. S. Ramanujam),

342–348. 23–28 February 1978. New Delhi, India: Indian Society of Genetics and Plant Breeding.

Joukhadar, R., Hollaway, G., Shi, F. et al. (2020). Genome-wide association reveals a complex architecture for rust resistance in 2300 worldwide bread wheat accessions screened under various Australian conditions. *Theoretical and Applied Genetics* 133: 2695–2712.

Kamal, N.M., Gorafi, Y.S.A., Abdelrahman, M. et al. (2019). Stay-green trait: A prospective approach for yield potential, and drought and heats stress adaptation in globally important cereals. *International Journal of Molecular Sciences* 20: 5837.

Kandel, J.S., Huang, M., Zhang, Z. et al. (2018). Genetic diversity of clinal freezing tolerance variation in winter wheat landraces. *Agronomy* 8: 95.

Kantety, R.V., La Rota, M., Matthews, D.E. et al. (2002). Data mining for simple sequences repeats in expressed sequence tags from barley, maize, rice, sorghum and wheat. *Plant Molecular Biology* 48: 501–510.

Karagoz, A. (2014). Wheat landraces of Turkey. *Emirates Journal of Food and Agriculture* 26: 149–156.

Kempe, K., Rubtsova, M., and Gils, M. (2014). Split-gene system for hybrid wheat. *Proceedings of the National Academy Sciences* 111: 9097–9102.

Kertho, A., Mamidi, S., Bonman, J.M. et al. (2015). Genome-wide association mapping for resistance to leaf and stripe rust in winter-habit hexaploidy wheat landraces. *PLoS One* 10 (6): e0129580. 15 June.

Khan, A.A., Iqbal, A., Awan, F.S. et al. (2010). Genetic diversity in wheat germplasm collections from Balochistan province of Pakistan. *Pakistan Jo0Ournal of Botany* 42: 89–96.

Khan, I.A. (2000). Molecular and agronomic characterization of wheat – *Agropyron intermedium* recombinant chromosomes. *Plant Breeding* 119: 25–29.

Kibirige-Sebunya, I. and Knott, D.R. (1983). Transfer of stem rust resistance to wheat from an *Agropyron* chromosome having a gametocidal effect. *Canadian Journal of Genetics and Cytology* 25: 215–221.

Kidane, Y.G., Hailemariam, B.N., Mengistu, D.K. et al. (2017). Genome-wide association study of *Septoria tritici* blotch resistance in Ethiopian durum Wheat Landraces. *Frontiers in Plant Science* 8: art. no. 1586.

Kim, D., Alptekin, A.B., and Budak, H. (2018). CRISPR/Cas9 genome editing in wheat. *Functional & Integrative Genomics* 18: 31–41.

King, I.P., Law, C.N., Cant, K.A. et al. (1997). Tritipyrum, a potential new salt-tolerant cereal. *Plant Breeding* 116: 127–132.

King, J., Grewal, S., Yang, C.Y. et al. (2017). A step change in the transfer of interspecific variation into wheat from *Amblyopyrum muticum*. *Plant Biotechnology Journal* 15: 217–226. doi:10.1111/pbi.12606.

Kishii, M. (2019). An update of recent use of *Aegilops* species in wheat breeding. *Frontiers in Plant Science* 10: 585. doi:10.3389/fpls.2019.00585.

Knott, D.R. (1961). The inheritance of rust resistance. VI: The transfer of stem rust resistance from *Agropyron elongatum* to common wheat. *Canadian Journal of Plant Research* 41: 109–123.

Knott, D.R. and Kumar, J. (1975). Comparison of early generation yield testing and a single seed descent in wheat breeding. *Crop Science* 15: 295–299.

Koebner, R.M.D. and Martin, P.K. (1996). High levels of salt tolerance revealed in triticale. In: *Triticale: Today and Tomorrow. Developments in Plant Breeding*, 5 edition (ed. H. Guedes-Pinto, N. Darvey, and V.P. Carnide), 429–436. Dordrecht: Springer.

Koebner, R.M.D. and Shepherd, K.W. (1986). Controlled introgression to wheat of genes from rye chromosome arm 1RS by induction of allosyndesis. 1: Isolation of recombinants. *Theoretical and Applied Genetics* 73: 197–208.

Koebner, R.M.D. and Shepherd, K.W. (1987). Induction of recombination between homoeologous chromosomes of wheat and rye. *Heredity* 59: 314–315.

Koebner, R.M.D. and Shepherd, K.W. (1988). Isolation and agronomic assessment of allosyndetic recombinants derived from wheat/rye translocation 1DL.1RS, carrying reduced amounts of rye chromatin. In: *Proceedings of the Seventh International Wheat Genetics Symposium* (ed. T.E. Miller and R.M.D. Koebner), 343–348. 13–19 July 1988. Cambridge; UK: Institute of Plant Science Research.

Kolmer, J.A., Garvin, D.F., Hayden, M. et al. (2018). Adult plant leaf rust resistance derived from the wheat landrace cultivar Americano 44d is conditioned by interaction of three QTL. *Euphytica* 214: art. no. 59.

Konieczny, A. and Ausubel, F.M. (1993). A procedure for mapping Arabidopsis mutations using co-dominant ecotype-specific PCR-based markers. *The Plant Journal* 4: 403–410. PMID: 8106085.

Kotvics, G., Krisztian, J., and Heszky, L. (2001). Perennial Rye: A new forage crop for the world, registered in Hungary. *Hungarian Agricultural Research* 10: 4–5.

Krugman, T., Chagué, V., and Peleg, Z. et al. (2008). Differential gene expression in wild emmer wheat genotypes contrasting in drought resistance. In: R. Appels, R. Eastwood, E.P. Lagudah, et al., *Proceedings of the 11th International Wheat Genetics Symposium*, 24–29 August 2008, 3, 910–912. Brisbane, QLD, Australia.

Kruse, A. (1967). Intergeneric hybrids between *Hordeum vulgare* L. ssp. *distichum* (v. Pallas, 2n=14) and *Secale cereale* L. (v. Petkus, 2n=14). In: *Royal Veterinary Agricultural College Yearbook 1967*, 82–92. Copenhagen.

Kruse, A. (1969). Intergeneric hybrids between *Triticum aestivum* L. (v. Koga II. 2¼42) and *Avena sativa* L. (v. Stal., 2n=42) with pseudogamous seed formation. In: *Royal Veterinary Agricultural College Yearbook 1969*, 188–200. Copenhagen.

Kruse, A. (1973). *Hordeum* × *Triticum* hybrids. *Hereditas* 73: 157–161.

Kruse, A. (1974). *Hordeum vulgare* ssp. *distichum* (var Bomi) × *Triticum aestivum* (var. Koga). An F1 hybrid with generative seed formation. *Hereditas* 78: 319.

Kulkarni, M., Soolanayaknahally, R., Ogawa, S. et al. (2017). Drought response in wheat: Key genes and regulatory mechanisms controlling root system architecture and transcription efficiency. *Frontiers in Chemistry* 5: 106. doi:10.3389/fchem.2017.00106.

Kulshrestha, V.P. (1989). A modified pedigree method of selection. *Theoretical and Applied Genetics* 78: 173–176.

Kyzeridis, N., Biesantz, A., and Limberg, P. (1995). Comparative trials with durum wheat land races and cultivars in different ecological environments in Meditterranean regions. *Journal of Agronomy and Crop Science* 174: 133–1.

Lahsaiezadeh, M., Ting, I.P., and Waines, J.G. (1983). Drought resistance in Chinese Spring wheat/Imperial rye addition and substitution lines. In: *Proceedings of the Sixth*

International Wheat Genetics Symposium (ed. S. Sakamoto), 945–950. 28 November–3 December 1983. Kyoto, Japan: Faculty of Agriculture, Kyoto University.

Langridge, P. and Waugh, R. (2019). Harnessing the potential of germplasm collections. *Nature Genetics* 51: 200–201. https://doi.org/10.1038/s41588-018-0340-4.

Lanza, L.L.B., Souza, C.L., Jr., Ottoboni, L.M.M. et al. (1997). Genetic distance of inbred lines and prediction of maize single-cross performance using RAPD markers. *Theoretical and Applied Genetics* 94: 1023–1030.

Larkin, P.J., Newell, M.T., Hayes, R.C. et al. (2014). Progress in developing perennial wheats for grain and grazing. *Crop & Pasture Science* 65: 1147–1164.

Laurie, D.A. and Bennett, M.D. (1986). Wheat × maize hybridization. *Canadian Journal of Genetics and Cytology* 28: 313–316.

Laurie, D.A. and Bennett, M.D. (1988). Cytological evidence for fertilization in hexaploidy wheat × sorghum crosses. *Plant Breeding* 100: 73–82.

Lazo, G.R., Chao, S., Hummel, D.D. et al. (2004). Development of an expressed sequence tag (EST) resource for wheat (*Triticum aestivum* L.): EST generation, unigene analysis, probe selection and bioinformatics for a 16,000-locus bin-delineated map. *Genetics* 168: 585–593. doi:10.1534/genetics.104.034777.

Leach, R.C., Dundas, I.S., and Houben, A. (2006). Construction of comparative genetic maps of two 4Bs.4Bl-5Rl translocations in bread wheat *Triticum aestivum* L. *Genome* 49: 729–734.

Lee, S., Choi, Y.-M., Lee, M.-C. et al. (2018). Geographical comparison of genetic diversity in Asian landrace wheat (*Triticum aestivum* L.) germplasm based on high-molecular-weight glutenin subunits. *Genetic Resources and Crop Evolution* 65: 1591–1602.

Leighty, C.E. and Sando, W.J. (1927). A trigeneric hybrid of *Aegilops, Triticum* and *Secale. The Journal of Heredity* 18: 433–442.

Li, A., Liu, D., Yang, W. et al. (2018a). Synthetic hexaploid wheats; Yesterday, today and tomorrow. *Engineering* 4: 552–558.

Li, H., Singh, R.P., Braun, H.J. et al. (2013). Doubled haploids versus conventional breeding in CIMMYT wheat breeding programs. *Crop Science* 53: 74–83.

Li, H.H., Dong, Z.J., Ma, C. et al. (2019). Physical mapping of stem rust resistance gene *Sr52* from *Dasypyrum villosum* based on ph1b-Induced homoeologous recombination. *International Journal of Molecular Sciences* 20: 4887.

Li, J., Dundas, I., Dong, C. et al. (2020). Identification and characterization of a new stripe rust resistance gene *Yr83* on rye chromosome 6R in wheat. *Theoretical and Applied Genetics* 133: 1095–1107. https://doi.org/10.1007/s00122-020-03534-y.

Li, T., Bai, G., Wu, S. et al. (2011). Quantitative trait loci for resistance to fusarium head blight in a Chinese wheat landrace Haiyanzhong. *Theoretical and Applied Genetics* 122: 1497–1502.

Li, X., Chen, W., Bao, Y., and Wang, H. (2013a). Response to phosphorus insufficiency of Trititrigia germplasm Shannong 303. *Proceedings of the 12th International Wheat Genetics Symposium*, Pacifico Yokohama, Japan, 8–14 September 2013, p. 100.

Li, Z., Li, B., Zheng, Q., and Li, H. (2013b). New progress in wheat wide hybridization for improving the resistance to biotic and abiotic stresses. *Proceedings of the 12th*

International Wheat Genetics Symposium, Pacifico Yokohama, Japan, 8–14 September 2013. p. 86.

Liang, Z., Chen, K., Li, K. et al. (2017). Efficient DNA-free genome editing of bread wheat using CRISPR/Cas9 ribonucleoprotein complexes. *Nature Communications* 8: 14261.

Ling, H.Q., Ma, B., Shi, X. et al. (2018). Genome sequence of the progenitor of wheat A subgenome *Triticum urartu*. *Nature* 557: 424–428. doi:10.1038/s41586-018-0108-0.

Ling, H.Q., Zhao, S.C., Liu, D.C. et al. (2013). Draft genome of the wheat A-genome progenitor *Triticum urartu*. *Nature* 496: 87–90.

Liu, W., Zheng, M.Y., Polle, E.A. et al. (2002). Highly efficient doubled-haploid production in wheat (*Triticum aestivum* L.) via induced microspore embryogenesis. *Crop Science* 42: 686–692.

Liu, X.L., Lu, B.Y., Wang, C.Y. et al. (2018). Identification of *Sitobion avenae* F. resistance and genetic diversity of wheat landraces from Qinling Mountains, China. *Cereal Research Communications* 46: 104–113.

Liu, Y., He, Z.H., Appels, R. et al. (2012). Functional markers in wheat: Current status and future prospects. *Theoretical and Applied Genetics* 125: 1–10. doi:10.1007/s00122-012-1829-3.

Liu, Z., Li, D., and Zhang, X.Y. (2007). Genetic relationships among five basic genomes St, E, A, B and D in Triticeae revealed by genomic Southern and *in situ* hybridization. *Journal of Integrative Plant Biology* 49: 1080–1086. doi:10.1111/j.1672-9072.2007.00462.x.

Liu, Z., Sun, Q., Ni, Z. et al. (1999). Development of SCAR markers linked to the *Pm21* gene conferring resistance to powdery mildew in common wheat. *Plant Breeding* 118: 215–219.

Ljubimova, V.F. (1936). A *Secale-Agropyrum* hybrid. In: *Plant Breeding Abstracts 9, 1937*, Item 189, 50.

Lodhi, S.S., John, P., Bux, H. et al. (2018). Resistance potential of Pakistani wheat landraces (*Triticum aestivum* l.) against stripe rust (*Puccinia striformis*) and Karnal bunt (*Tilletia indica*). *Pakistan Journal of Botany* 50: 801–806.

Lodhi, S.S., Maryam, S., Rafique, K. et al. (2020). Overview of the prospective strategies for conservation of genomic diversity in wheat land races. In: *Climate Change and Food Security with Emphasis on Wheat*, 293–309. Elsevier Press Inc.

Long, Y.M., Chao, W.S., Ma, G.J. et al. (2017). An innovative SNP genotyping method adapting to multiple platforms and throughputs. *Theoretical and Applied Genetics* 130: 597–607. doi:10.1007/s00122-016-2838-4.

Longin, C.F.H. (2016). Future of wheat breeding is driven by hybrid wheat and efficient strategies for pre-breeding on quantitative traits. In: *Plant Biotechnology and Its Applications* (ed. D. Zhang). e-ISSN:2320-0189 and p-ISSN:2347-2380.

Longin, C.F.H., Mühleisen, J., Maurer, H.P. et al. (2012). Hybrid breeding in autogamous cereals. *Theoretical and Applied Genetics* 125: 1087–1096.

Lopes, M.S., El-Basvoni, I., Baenziger, P.S. et al. (2015). Exploring genetic diversity from land races in wheat breeding for adaptation to climate change. *Journal of Experimental Botany* 66: 2377–3486.

Love, H.H. (1927). A program for selecting and testing small grains in successive generations following hybridization. *Agronomy Journal* 19: 705–712.

Lozada, D.N., Mason, R.E., Sarinelli, J.M. et al. (2019). Accuracy of genomic selection for grain yield and agronomic traits in soft red winter wheat. *BMC Genetics* 20: 82. https://doi.org/10.1186/s12863-019-0785-1.

Lukaszewski, A.J. (2015). Introgressions between wheat and rye. In: *Alien Introgression in Wheat*, Chapter 7 (ed. M. Molnár-Láng, C. Ceoloni, and J. Doležel), 163–189. doi:10.1007/978-3-319-23494-6_11.

Lukaszewski, A.J. (2017). Chromosomes 1BS and 1RS for control of male fertility in wheats and triticales with cytoplasms of *Aegilops kotschyi, Ae. mutica* and *Ae. uniaristata*. *Theoretical and Applied Genetics* 130: 2521–2526.

Luo, M.C., Gu, Y.O., Puiu, D. et al. (2017). Genome sequence of the progenitor of the wheat D genome *Aegilops tauschii*. *Nature* 551: 498–502. doi:10.1038/nature24486.

Ma, H., Singh, R.P., and Mujeeb-Kazi, A. (1995). Suppression/expression of resistance to stripe rust in synthetic hexaploid wheat (*Triticum turgidum* × *T. tauschii*). *Euphytica* 83: 87–93.

Ma, Z. and Lapitan, N.L.V. (1998). A comparison of amplified and restriction fragment length polymorphism in wheat. *Cereal Research Communications* 26: 7–13. https://doi.org/10.1007/BF03543462.

Mago, R., Verlin, D., Zhang, P. et al. (2013). Development of wheat – *Aegilops speltoides* recombinants and simple PCR-based markers for stem rust resistance genes on the 2S#1 chromosome. *Theoretical and Applied Genetics* 126: 2943–2955. doi:10.1007/s00122-013-2184-8.

Mago, R., Zhang, P., Bariana, H.S. et al. (2009). Development of wheat lines carrying stem rust resistance gene *Sr39* with reduced *Aegilops speltoides* chromatin and a simple PCR marker for marker-assisted selection of the gene. *Theoretical and Applied Genetics* 119: 1441–1450.

Mago, R., Zhang, P., Xia, X. et al. (2019). Transfer of stem rust resistance gene *SrB* from *Thinopyrum ponticum* into wheat and development of a closely linked PCR-based marker. *Theoretical and Applied Genetics* 132: 371–382. https://doi.org/10.1007/s00122-018-3224-1.

Manickavelu, A., Joukhadar, R., Jighly, A. et al. (2016). Genome wide association mapping of stripe rust resistance in Afghan wheat landraces. *Plant Science* 252: 222–229.

Manickavelu, A., Niwa, S., Ayumi, K. et al. (2014). Molecular evaluation of Afghan wheat landraces. *Plant Genetic Resources: Characterisation and Utilisation* 12 (SUPPL. 1): S31–S35.

Manyowa, N.M., Miller, T.E., and Forster, B.P. (1988). Alien species as sources for aluminium tolerance genes for wheat, *Triticum aestivum*. In: *Proceedings of the Seventh International Wheat Genetics Symposium*, 2 (ed. T.E. Miller and R.M.D. Koebner), 851–857. Cambridge, UK. 13–19 July 1988.

Mao, H., Li, S., Wang, Z. et al. (2020). Regulatory change in TaSNAC8-6A are associated with drought tolerance in wheat seedlings. *Plant Biotechnology Journal* 18: 1078–1092.

Marais, G.F. and Marais, A.S. (2003). Attempt to translocate *Thinopyrum distichum* chromosome arms carrying salt tolerance genes to triticale and wheat chromosomes. *Proceedings of the Tenth International Wheat Genetics Symposium*, 1–6 September 2003, 2, 893–895, Paestum, Italy: Instituto Sperimentale per la Cerealicoltura.

Martin, A.C., Atienza, S.G., Ramırez, M.C. et al. (2008). Male fertility restoration of wheat in *Hordeum chilense* cytoplasm is associated with 6HchS chromosome addition. *Australian Journal of Agricultural Research* 59: 206–213.
Martin, A.C., Atienza, S.G., Ramírez, M.C. et al. (2010). Molecular and cytological characterization of an extra acrocentric chromosome that restores male fertility of wheat in the msH1 CMS system. *Theoretical and Applied Genetics* 121: 1093–1101.
Martin, D.J. and Stewart, B.G. (1990). Dough stickiness in rye-derived wheat cultivars. *Euphytica* 51: 77–86.
Martin-Sanchez, J.A., Romero, M.D., Sin, E., et al. (1998). Relationship of the gene for resistance to *Heterodera avenae* derived from *Aegilops triuncialis* to other sources of resistance: Characterization of lines with the *Cre Aet* resistance gene. *Proceedings of the Ninth International Wheat Genetics Symposium*. Saskatoon, Saskatchewan, Canada. 2–7 August 1998, 3, 290–291. University Extension Press, University of Saskatchewan.
Marza, F., Bai, G.H., Carver, B.F. et al. (2006). Quantitative trait loci for yield and related traits in the wheat population Ning 7840 × Clark. *Theoretical and Applied Genetics* 112: 688–698.
Mascher, M., Gundlach, H., Himmelbach, A. et al. (2017). A chromosome conformation capture ordered sequence of the barley genome. *Nature* 544: 427–433. doi:10.1038/nature22043.
McCouch, S., Bante, G.J., Bradeen, J. et al. (2013). Agriculture: Feeding the future. *Nature* 499: 23–25.
McFadden, E.S. (1917). Wheat-rye hybrids. *The Journal of Heredity* 8: 335–336.
McFadden, E.S. (1930). A successful transfer of emmer characters to *vulgare* wheat. *Journal of the American Society of Agronomy* 22: 1020–1034.
McFadden, E.S. and Sears, E.R. (1947). The genome approach in radical wheat breeding. *Journal of the American Society of Agronomy* 39: 1011–1026.
McGuire, P.E. and Dvořák, J. (1981). High salt-tolerant potential in wheatgrasses. *Crop Science* 21: 702–705.
McIntosh, R.A. (1991). Alien sources of disease resistance in bread wheats. In: *Proceedings of Dr H. Kihara Memorial International Symposum on Cytoplasmic Engineering in Wheat* (ed. T. Sasakoma and T. Kinoshita), 320–332. 3–6 July 1991. Plant Breeding Institute, Hokkaido University.
McIntosh, R.A., Yamazaki, Y., Dubcovsky, J., et al. (2013). Catalogue of gene symbols for wheat. Proceedings of the 12th International Wheat Genetics Symposium, 8–13 September 2013. Yokohama, Japan.
McKinney, H.H. and Sando, W.J. (1951). Susceptibility and resistance to the wheat streak-mosaic virus in the genera *Triticum*, *Agropyron*, Secale, and certain hybrids. *Plant Disease Reporter* 35 (11): 476–479.
Mengistu, F., Pè Dell'Acqua, J.S. et al. (2014). Genetic diversity for stripe rust resistance in wheat landraces and identification of accessions with resistance to stem rust and stripe rust. *Crop Science* 54: 2131–2139.
Meuwissen, T.H.E., Hayes, B.J., and Goddard, M.E. (2001). Prediction of total genetic value using genome-wide dense marker maps. *Genetics* 157: 1819–1829.
Miller, T.E., Iqbal, N., Reader, S.M. et al. (1997). A cytogenetic approach to the improvement of aluminium tolerance in wheat. *The New Phytologist* 137: 93–98.

Miller, T.E., Reader, S.M., Mahmood, A. et al. (1995). Chromosome 3N of *Aegilops uniaristata*- a source of tolerance to high levels of aluminium for wheat. *Proceedings of the Eighth International Wheat Genetics Symposium*, Beijing, China. 20–25 July 1993. 2, 1037–1042.

Milner, S.G., Jost, M., Taketa, S. et al. (2019). Genebank genomics highlights the diversity of a global barley collection. *Nature Genetics* 51: 319–326. https://doi.org/10.1038/s41588-018-0266-x.

Mirzaghaderi, G. and Mason, A.S. (2019). Broadening the wheat D genome. *Theoretical and Applied Genetics* 132: 1295–1307.

Mohammadi, R., Haghparast, R., Sadeghzadeh, B. et al. (2014). Adaptation patterns and yield stability of durum wheat landraces to highland cold rainfed areas of Iran. *Crop Science* 54: 944–954.

Mohammed, Y.S.A., Eltayeb, A.E., and Tsujimoto, H. (2013). Enhancement of aluminium tolerance in wheat by addition of chromosomes from the wild relative *Leymus racemosus*. *Breeding Science* 63: 407–416.

Mohler, V. and Jahoor, A. (1996). Allele-specific amplification of polymorphic sites for the detection of powdery mildew resistance loci in cereals. *Theoretical and Applied Genetics* 93: 1078–1082.

Morgounov, A., Keser, M., Kan, M. et al. (2016). Wheat landraces currently grown in Turkey: Distribution, diversity, and use. *Crop Science* 56: 3112–3124.

Mujeeb-Kazi, A. (2003). New genetic stocks for durum and bread wheat improvement. In: *Tenth International Wheat Genetics Symposium*, 772–774. Paestum, Italy.

Mujeeb-Kazi, A. (2006). Utilization of genetic resources for bread wheat improvement. In: *CRC Series* (ed. R.J. Singh and P.P. Jauhar), 61–97. CRC Press.

Mujeeb-Kazi, A., Ali, N., Ibrahim, A. et al. (2017). Tissue culture mediated allelic diversification and genomic enrichment of wheat to combat production constraints and address food security. *Plant Tissue Culture and Biotechnology* 27: 89–140.

Mujeeb-Kazi, A., Ali, N., Tariq, M. et al. (2020). Perceptions and practicality of combating wheat based 2050 based food security challenges. *Modern Concepts & Developments in Agronomy* 6 (5): 683–685. doi:10.31031/mcda.2020.06.000649.

Mujeeb-Kazi, A. and Asiedu, R. (1990). Wide hybridization; Potential of alien genetic transfers for *Triticum aestivum* improvement. In: *Wheat. Biotech in Agriculture and Forestry*, 13 (ed. Y.P.S. Bajaj), 111–127. Berlin, Heidelberg: Springer.

Mujeeb-Kazi, A. and Bernard, M. (1985). Intergeneric hybridization to induce alien genetic transfers into *Triticum aestivum*. *Pakistan Journal of Botany* 17: 271–28.

Mujeeb-Kazi, A., Cano, S., Rosas, V. et al. (2001b). Registration of five synthetic hexaploid wheat and seven bread wheat germplasm lines resistant to wheat spot blotch. *Crop Science* 41: 1653–1654.

Mujeeb-Kazi, A., Fuentes-Davila, G., Villareal, R.L. et al. (2001a). Registration of 10 synthetic hexaploid wheat and six bread wheat germplasms resistant to karnal bunt. *Crop Science* 41: 1652–1653.

Mujeeb-Kazi, A., Gul, A., Ahmad, J. et al. (2006). A simplified and effective protocol for production of bread wheat haploids ($n = 3x = 21$, ABD) with some application areas in wheat improvement. *Pakistan Journal of Botany* 38: 393–406.

Mujeeb-Kazi, A. and Hettel, G.P. (1995). Utilizing wild grass biodiversity in wheat improvement: 15 years of Wide Cross Research at CIMMYT. CIMMYT REPORT 2, pp. 1–140.

Mujeeb-Kazi, A., Kazi, A.G., Dundas, I. et al. (2013). Genetic diversity for wheat improvement as a conduit for food security. *Advances in Agronomy* 122: 179-257.

Mujeeb-Kazi, A. and Kimber, G. (1985). The production, cytology and practicality of wide hybrids in the *Triticeae*. *Cereal Research Communications* 13: 111–124.

Mujeeb-Kazi, A., Munns, R., Rasheed, A. et al. (2019). Breeding strategies for structuring salinity tolerance in wheat. *Advances in Agronomy* 155: 121–187.

Mujeeb-Kazi, A. and Riera-Lizarazu, O. (1996). Polyhaploid production in the *Triticeae* by sexual hybridization. In: *In Vitro Haploid Production in Higher Plants* (ed. S.M. Jain et al.), 275–296.

Mujeeb-Kazi, A. and Rodríguez, R. (1980). Some intergeneric hybrids in the Triticeae. *Cereal Research Communications* 8: 469-475.

Mujeeb-Kazi, A. and Rodríguez, R. (1981). Cytogenetics of intergeneric hybrids involving genera within the *Triticeae*. *Cereal Research Communications* 9: 39-45.

Mujeeb-Kazi, A., Roldan, S., Suh, D.Y. et al. (1987). Production and cytogenetic analysis of hybrids between *Triticum aestivum* and some caespitose *Agropyron* species. *Genome* 29: 537–553.

Mujeeb-Kazi, A., Roldan, S., Suh, D.Y. et al. (1989). Production and cytogenetics of *Triticum aestivum* L. hybrids with some rhizomatous *Agropyron* species. *Theoretical and Applied Genetics* 77: 162–168.

Mulki, M.A., Jighly, A., Ye, G.Y. et al. (2013). Association mapping for soilborne pathogen resistance in synthetic hexaploid wheat. *Molecular Breeding: New Strategies in Plant Improvement* 31: 299–311.

Müller, T., Schierscher-Viret, B., Fossati, D. et al. (2018). Unlocking the diversity of genebanks: Whole-genome marker analysis of Swiss bread wheat and spelt. *Theoretical and Applied Genetics* 131: 407–416.

Mullis, K.B. (1990). The unusual origin of the polymerase chain reaction. *Scientific American* 262 (4): 56–61, 64–65. doi:10.1038/scientificamerican0490-56.

Munns, R. and Gillihan, M. (2015). Salinity tolerance of crops: What is the cost. *New Phytologist* 208: 668–673.

Munns, R., Rebetzke, G.J., Husain, S. et al. (2003). Genetic control of sodium exclusion in durum wheat. *Australian Journal of Agricultural Research* 54: 627–635.

Munns, R. and Tester, M. (2008). Mechanisms of salinity tolerance. *Annual Review of Plant Biology* 59: 651–681.

Nachit, M.M. and Tahir, M. (1983). The effect of rye chromosomes in triticale on moisture stress tolerance and its yield potential in North Africa and West Asia. In: *Proceedings of the Sixth International Wheat Genetics Symposium* (ed. S. Sakamoto), 925–931. 28 November–3 December 3 1983. Kyoto, Japan: Faculty of Agriculture, Kyoto University.

Nadolska-Orczyk, A., Rajchel, K., Orczyk, W. et al. (2017). Major genes determining yield related traits in wheat and barley. *Theoretical and Applied Genetics* 130: 1081–1098.

Neumann, K., Kobiljski, B., Dencic, S. et al. (2011). Genome-wide association mapping: A case study in bread wheat (*Triticum aestivum* L.). *Molecular Breeding: New Strategies in Plant Improvement* 27: 37–58.

Newcomb, M., Acevedo, M., Bockelman, H.E. et al. (2013). Resistance to the Ug99 race group of the stem rust pathogen in spring wheat landraces. *Plant Disease* 97: 882–890.

Newell, M.T., Hayes, R.C., Virgona, J.M. et al. (2015). Summer dormancy in *Elymus scaber* and its hybridity with wheat. *Euphytica* 204: 535–556.

Newman, L.H. (1912). *Plant Breeding in Scandinavia*. Ottawa, Canada: Canadian Seed Growers' Association.

Ni, Z., Li, H., Zhao, Y. et al. (2018). Genetic improvement of heat tolerance in wheat: Recent progress in understanding the underlying molecular mechanisms. *The Crop Journal* 6: 32–41. https://doi.org/10.1016/j.cj.2017.09.005.

O'Mara, J.G. (1940). Cytogenetic studies on *Triticale*. I. A method for determining the effects of individual *Secale* chromosomes in *Triticum*. *Genetics* 25: 401–408.

Ogbonnaya, F.C., Abdalla, O., Mujeeb-Kazi, A. et al. (2013). Synthetic hexaploid in wheat improvement. In: *Plant Breeding Reviews*, 37, 1st edition (ed. J. Janick), 35–122. John Wiley & Sons, Inc.

Ogbonnaya, F.C., Imtiaz, M., Ye, G. et al. (2008). Genetic and QTL analyses of seed dormancy and preharvest sprouting resistance in the wheat germplasm CN10955. *Theoretical and Applied Genetics* 116: 891–902.

Ogbonnaya, F.C., Rasheed, A., Okechukwu, E.C. et al. (2017). Genome-wide association study for agronomic and physiological traits in spring wheat evaluated in a range of heat prone environments. *Theoretical and Applied Genetics* 130: 1819–1835. doi:10.1007/s00122-017-2927-z.

Ogutu, E.A. and Charimbu, M.K. (2020). Assessment of adult plant resistance to stem rust (*Puccinia graminis* f.sp *tritici*) in wheat (*Triticum aestivum* L.) mutant lines. *American Journal of Agriculture and Forestry* 8 (1): 30–39. doi:10.11648/j.ajaf.20200801.15.

Ohkawa, Y., Suenaga, K., and Ogawa, T. (1992). Production of hexaploid wheat plants through pollination of Sorghum pollen. *Japanese Journal of Breeding* 42: 891–894.

Okamoto, M. (1957). Asynaptic effect of chromosome V. *Wheat Information Service* 5: 6.

Ortiz, R., Sayre, K.D., Govaerts, B. et al. (2008). Climate change: Can wheat beat the heat? *Agriculture, Ecosystems & Environment* 126: 46–58.

Ortiz, R., Trethowan, R., Ferrara, G.O. et al. (2007). High yield potential, shuttle breeding, genetic diversity, and a new international wheat improvement strategy. *Euphytica* 157: 365–384.

Osborne, L.D. and Rengel, Z. (2002). Screening cereals for genotypic variation in efficiency of phosphorus uptake utilization. *Australian Journal of Agricultural Research* 53: 295–303.

Oyiga, B.C., Ogbonnaya, F.C., Sharma, R.C. et al. (2019). Genetic and transcriptional variations in NRAMP-2 and OPAQUE1 genes correlate to salt-stress response in wheat. *Theoretical and Applied Genetics* 132: 323–246. https://doi.org/10.1007/s00122-018-3220-5.

Oyiga, B.C., Sharma, R.C., Baum, M. et al. (2018). Allelic variations and differential expressions detected at quantitative trait loci for salt stress tolerance in wheat. *Plant, Cell & Environment* 4: 919–935. doi:10.1111/pce.12898.

Paran, I. and Michelmore, R.W. (1993). Development of reliable PCR based markers linked to downy mildew resistance genes in lettuce. *Theoretical and Applied Genetics* 85: 985–993.

Pasam, R.K., Bansal, U., Daetwyler, H.D. et al. (2017). Detection and validation of genomic regions associated with resistance to rust diseases in a worldwide hexaploid wheat landrace collection using BayesR and mixed linear model approaches. *Theoretical and Applied Genetics* 130: 777–793.

Patial, M., Pal, D., Thakur, A. et al. (2017, also 2019). Doubled haploidy techniques in wheat (*Triticum aestivum* L.). *Proceedings of the National Academy of Sciences, India. B. Biological Sciences* 89: 27–41.

Pazhamala, L.T., Kudapa, H., Weckwerth, W. et al. (2021). Systems biology for crop improvement. *The Plant Genome (TSI)* 1–23.

Pena, R.J., Amaya, A., Rajaram, S. et al. (1990). Variation in quality characteristics associated with some spring 1B/1R translocation wheat. *Journal of Cereal Science* 12: 105–112.

Peng, Y., Yu, K., Zhang, Y. et al. (2015). Two novel Y-type high molecular weight glutenin genes in Chinese wheat landraces of the Yangtze-River region. *PLoS ONE* 10 (11): art. no. e0142348.

Percival, J. (1921). *The Wheat Plant. Hybridization and Wheat Hybrids*, 377–394. London: Duckworth and Co.

Phan, H.T.T., Rybak, K., Bertazzoni, S. et al. (2017). Novel sources of resistance to *Septoria nodorum* blotch in the Vavilov wheat collection identified by genome-wide association studies. *Theoretical and Applied Genetics* 131: 1223–1238.

Pickett, A.A. and Galwey, N.W. (1997). A further evaluation of hybrid wheat. *Plant Varieties and Seeds* 10: 15–32.

Pinto Da Silva, G.B., Zanella, C.M., Martinelli, J.A. et al. (2018). Quantitative trait loci conferring leaf rust resistance in hexaploid wheat. *Phytopath* 108: 1344–1354. https://doi.org/10.1094/PHYTO-06-18-0208-RVW.

Poland, J., Endelman, J., Dawson, J. et al. (2012b). Genomic selection in wheat breeding using genotyping-by-sequencing. *Plant Gene* 5: 103–113. doi:10.3835/plantgenome2012.06.0006.

Poland, J.A., Brown, P.J., Sorrells, M.E. et al. (2012a). Development of high-density genetic maps for barley and wheat using a novel two-enzyme genotyping-by-sequencing approach. *PLoS ONE* 7: e32253. doi:10.1371/journal.pone.0032253.

Porfiri, O., Torricelli, R., Silveri, D.D. et al. (2002). The *Triticeae* genetic resources of central Italy: Collection, evaluation and conservation. *Hereditas* 135: 187–192.

Pretorius, Z.A., Singh, R.P., Wagoire, W.W. et al. (2000). Detection of virulence to wheat stem rust resistance gene *Sr31* in *Puccinia graminis*. f. sp. tritici in Uganda. *Plant Diesease* 84: 203–203.

Qamar, Z.U., Bansal, U.K., Dong, C.M. et al. (2014). Detection of puroindoline (Pina-D1 and Pinb-D1) allelic variation in wheat landraces. *Journal of Cereal Science* 60: 610–616.

Qi, L.L., Echalier, B., Friebe, B. et al. (2003). Molecular characterization of a set of wheat deletion stocks for using in chromosome bin mapping of ESTs. *Functional & Integrative Genomics* 3: 39–55.

qi, s.y., yu, s.x., zhang, y.h. et al. (1979). studies on distant hybridization between spring wheat and *Agropyron glaucum*. *Scientia Agricultura Sinica* 2: 1–11.

Rabinovich, S.V. (1998). Importance of wheat-rye translocations for breeding modern cultivars of *Triticum aestivum* L. *Euphytica* 100: 323–340.

Rajaram, S. (1999). Approaches for breaching yield stagnation in wheat. *Genome* 42: 629–634.

Rajaram, S., Kohli, M.M., and Lopez-Cesati, J. (1991). Breeding for tolerance to aluminium toxicity in wheat. In: *Plant-Soil Interactions at Low pH. Developments in Plant and Soil Sciences*, 45 (ed. R.J. Wright, V.C. Baligar, and R.P. Murrmann), 1019–1028. Dordrecht: Springer.

Rajaram, S., Mann, C.E., Ortiz-Ferrara, G. et al. (1983). Adaption, stability and high yield potential of certain 1B/1R CIMMYT wheats. In: *Proceedings of the 6th International Wheat Genetics Symposium* (ed. S. Sakamoto), 613–621. 28 November–3 December 1983. Kyoto, Japan: Plant Germ-plasm Institute, Faculty of Agriculture, Kyoto University.

Rajaram, S. and Torres, E. (1986). An integrated approach to breeding for disease resistance: The CIMMYT wheat experience. In: *Genetic Improvement in Yield of Wheat* (ed. E.L. Smith). CSSA spec. publ. 13. Crop Sci. Soc. Am., Am. Soc. Agron. Madison, WI.

Rames, R.A., Von Caemmerer, S., Condon, A.T. et al. (2008). Genetic variation in tolerance to the osmotic stress component of salinity stress in durum wheat. *Functional Plant Biology* 35: 111–123.

Ramirez-Villegas, J., Milan, A.M., Alexandrov, N. et al. (2020). CGIAR modeling approaches for resource-constrained scenarios. I: Accelerating crop breeding for a changing climate. *Crop Science* 60: 547–567. https://doi.org/10.1002/csc2.20048.

Rapp, M., Lein, V., Lacoudre, F. et al. (2018). Simultaneous improvement of grain yield and protein content in durum wheat by different phenotypic indices and genomic selection. *Theoretical and Applied Genetics* 131: 1315–1329. https://doi.org/10.1007/s00122-018-3080-z.

Rasheed, A., Mujeeb-Kazi, A., Ogbonnaya, F.C. et al. (2018). Wheat genetic resources in the post-genomics era: Promise and challenges. *Annals of Botany* 121: 603–616. doi:10.1093/aob/mcx148.

Rasheed, A., Wen, W., Gao, F. et al. (2016a). Development and validation of KASP assays for functional genes underpinning key economic traits in wheat. *Theoretical and Applied Genetics* 129: 1843–1860. doi:10.1007/s00122-016-2743-x.

Rasheed, A. and Xia, X. (2019). From markers to genome-based breeding in wheat. *Theoretical and Applied Genetics* 132: 767–784. https://doi.org/10.1007/s00122-019-03286-4.

Rasheed, A., Xia, X., Mahmood, T. et al. (2016b). Comparison of economically important loci in landraces and improved wheat cultivars from Pakistan. *Crop Science* 56: 287–301.

Rauf, M., Saeed, N.A., Imran Habib, I. et al. (2016). Isolation and in silico analysis of a novel H^+-pyrophosphatase gene orthologue from the halophytic grass *Leptochloa fusca*. *Journal of Molecular Structure* 119: 179–187.

Rauf, M., Shahzad, K., Ali, K.,.R. et al. (2014). Cloning and characterization of Na^+/H^+ antiporter (*LfNHX1*) gene from a halophyte grass *Leptochloa fusca* for drought and salt tolerance. *Molecular Biology Reports* 41: 1669–1682.

Raupp, W.J., Gill, B.S., Browder, L.E. et al. (1988). Genetic diversity in wheat relatives for disease and insect resistance. In: *Proceedings of the Seventh International Wheat Genetics Symposium*, 2 (ed. T.E. Miller and R.M.D. Koebner), 879–884. Cambridge, UK. 13–19 July 1988.

Rawashdeh, N.K., Haddad, N.I., Al-Ajlouni, M.M. et al. (2007). Phenotypic diversity of durum wheat (*Triticum durum* Desf.) from Jordan. *Genetic Resources and Crop Evolution* 54: 129–138.

Rawat, N., Tiwari, V.K., Singh, N. et al. (2009). Evaluation and utilization of *Aegilops* and wild *Triticum* species for enhancing iron and zinc content in wheat. *Genetic Resources and Crop Evolution* 56: 53–64. doi:10.1007/s10722-008-9344-8.

Ray, B., Noren, S.K., Mandal, A.B. et al. (2011). Genetic engineering for abiotic stress tolerance in agricultural crops. *Biotech* 10: 1–22.

Ray, D.K., West, P.C., Clark, M. et al. (2019). Climate change has likely already affected global food production. *PLoS ONE* 145: e217148.

Reimann-Philipp, R. (1995). Breeding perennial rye. *Plant Breeding Reviews* 13: 265–292.

Reitz, L.P., Johnston, C.O., and Anderson, K.L. (1945). New combinations of genes in wheat x wheatgrass hybrids. *Transactions of the Kansas Academy of Science* 48: 151–159.

Ren, J., Sun, D., Chen, L. et al. (2013). Genetic diversity revealed by single nucleotide polymorphism markers in a worldwide germplasm collection of durum wheat. *International Journal of Molecular Sciences* 14 (2013): 7061–7088.

Reynolds, M., Dreccer, F., and Trethowan, R. (2007). Drought-adaptive traits derived from wheat wild relatives and landraces. *Journal of Experimental Botany* 58: 177–186.

Reynolds, M.P., Calderini, D.F., Condon, A.G. et al. (2001). Physiological basis of yield gains in wheat associated with the *Lr19* translocation from *Agropyron elongatum*. *Euphytica* 119: 137–141.

Riaz, A., Athiyannan, N., Periyannan, S. et al. (2017b). Mining Vavilov's treasure chest of wheat diversity for adult plant resistance to *Puccinia triticina*. *Plant Disease* 101: 317–323.

Riaz, A., Hathorn, A., Dinglasan, E. et al. (2017a). Into the vault of the Vavilov wheats: Old diversity for new alleles. *Genetic Resources and Crop Evolution* 64: 531–544.

Riera-Lizarazu, O. and Mujeeb-Kazi, A. (1990). Maize (*Zea mays* L) mediated polyhaploid production using various crossing methods. *Cereal Research Communications* 18: 339–346.

Riera-Lizarazu, O. and Mujeeb-Kazi, A. (1993). Maize mediated wheat (*Triticum aestivum* l.) in the Triticee: Wheat × *Tripsacum*. *Crop Science* 33: 973–976.

Riley, R. and Chapman, V. (1958). Genetic control of the cytologically diploid behaviour of hexaploid wheat. *Nature* 182: 713–715.

Riley, R. and Chapman, V. (1967). The inheritance in wheat of crossability with rye. *Genetical Research* 9: 259–267.

Riley, R., Chapman, V., and Johnson, R. (1968). The incorporation of alien resistance in wheat by genetic interference with the regulation of meiotic chromosome synapsis. *Genetical Research, Cambridge* 12: 199–219.

Robbana, C., Kehel, Z., Ben Naceur, M. et al. (2019). Genome-wide genetic diversity and population structure of tunisian durum wheat landraces based on dartseq technology. *International Journal of Molecular Sciences* 20: art. no. 1352.

Rockefeller Foundation. 1965. Ann. Rep. 1964/1965.

Röder, M.S., Plaschke, J., Konig, S.U. et al. (1995). Abundance, variability and chromosomal location of microsatellite wheat. *Molecular Genetics and Genomic : MGG* 246: 327–333.

Rosen, H.R. (1949). Oat percentage and procedures for combining resistance to crown rust, including race 45 and *Helminthosporium* blight. *Phytopathol* 39: 20.

Rosyara, U., Kishii, M., Payne, T. et al. (2019). Genetic contribution of synthetic hexaplod wheat to CIMMYT breeding germplasm. *Scientific Reports* 9: 1235.

Royo, C., Nazco, R., and Villegas, D. (2014). The climate of the zone of origin of Mediterranean durum wheat (*Triticum durum* Desf.) landraces affects their agronomic performance. *Genetic Resources and Crop Evolution* 61: 1345–1358. doi:10.1007/s10722-014-0116-3.

Royo, C., Villegas, D., Rharrabti, Y. et al. (2006). Grain growth and yield formation of durum wheat grown at contrasting latitudes and water regimes in a Mediterranean environment. *Cereal Research Communications* 34: 1021–1028. doi:10.1556/CRC.34.2006.2-3.233.

Ruiz, M., Giraldo, P., and González, J.M. (2018). Phenotypic variation in root architecture traits and their relationship with eco-geographical and agronomic features in a core collection of tetraploid wheat landraces (*Triticum turgidum* L.). *Euphytica* 214: art. no. 54.

RustTracker (2020a). Ug99 Race Group Effective Sr genes. https://rusttracker.cimmyt.org/?page_id=6745. (accessed 26 March 2020).

RustTracker (2020b). Pathotype Tracker: Where is Ug99? https://rusttracker.cimmyt.org/?page_id=22.

Saeed, N.A., Pandey, S., Xiong, L. et al. (2011). Identification and cloning of drought related genes in wheat (*T. aestivum*). Pak-US project on wheat biotechnology. https://sites.nationalacademies.org/PGA/Pakistan/phase2/PGA_052870.

Saini, D.K., Devi, P., and Kaushik, P. (2020). Advances in genomic interventions for wheat biofortificaion: A review. *Agronomy* 10(1): 62. https://doi.org/10.3390/agronomy10010062.

Sánchez-Martín, J., Steuernagel, B., Ghosh, S. et al. (2016). Rapid gene isolation in barley and wheat by mutant chromosome sequencing. *Genome Biology* 17: 221. doi:10.1186/s13059-016-1082-1.

Sando, W.J. (1935). Hybrids of wheat, rye, *Aegilops* and *Haynaldia*. *The Journal of Heredity* 26: 229–232. https://doi.org/10.1093/oxfordjournals.jhered.a104084.

Saranga, Y., Peleg, Z., Abbo, S. et al. (2003). Genetic diversity for drought resistance in wild emmer wheat, *Triticum dicoccoides*. In: *Proceedings of the Tenth International Wheat Genetics Symposium* (ed. N.E. Pogna, M. Romanò, E.A. Pogna, and G. Galterio). 1–6 September 2003. Paestum, Italy: Instituto Sperimentale per la Cerealicoltura.

Sasanuma, T., Takata, K., Kawaguchi, K. et al. (2008). Characterization of *Elymus humidus* as a candidate for genetic resource of wheat water tolerance. In: *Proceedings of the 11th International Wheat Genetics Symposium* (ed. R. Appels, R. Eastwood, E. Lagudah, et al.), 24–29 August 2008. Brisbane, QLD, Australia.

Sax, K. (1922). Sterility in wheat hybrids. II.: Chromosome behavior in partially sterile hybrids. *Genetics* 7: 513–552.

Sax, K. (1923). The relation between chromosome number, morphological characters and rust resistance in segregates of partially sterile wheat hybrids. *Genetics* 8: 301–321.

Schachermayr, G.M., Siedler, H., Gale, M.D. et al. (1994). Identification and localization of molecular markers linked to the *Lr9* leaf rust resistance gene of wheat. *Theoretical and Applied Genetics* 88: 110–115.

Schlegel, R. (2020). Current list of wheats with rye and alien introgression. http://www.rye-gene-map.de/rye-introgression. (accessed 6 July 2020).

Schlegel, R. and Korzun, V. (1997). About the origin of 1RS.1BL wheat-rye chromosome translocations from Germany. *Plant Breeding* 116: 537–540.

Schlegel, R., Kynast, R., Schwarzacher, T. et al. (1993). Mapping of gens for copper efficiency in rye and the relationshop between copper and iron efficiency. *Plant and Soil* 154: 61–65.

Schlehuber, A.M. and Sebesta, E.E. (1958). Progress in wheat-grass breeding. *Proceedings of the Oklahoma Academy of Science* 39: 6–16. https://ojs.library.okstate.edu/osu/index.php/OAS/issue/view/269.

Schmidt, J.W., Heyne, E.G., Johnston, C.O. et al. (1953). Progress of *Agrotricum* breeding in Kansas. *Transactions of the Kansas Academy of Science* 56 (1): 29–45. http://www.jstor.org/stable/3626185.

Schulz-Schaeffer, J. and Haller, S. (1987). Registration of Montana-2 perennial × *Agrotriticum intermediodurum* Khizhnyak. *Crop Science* 27: 822–823.

Sears, E.R. (1956). The transfer of leaf-rust resistance from *Aegilops umbellulata* to wheat. *Brookhaven Symposia in Biology* 9: 1–22.

Sears, E.R. (1972). Chromosome engineering in wheat. *Stadler Symposium* 4: 23–38. University of Missouri, Columbia.

Sears, E.R. (1973). Agropyron-wheat transfers induced by homoeologous pairing. In: *Proceedings of the Fourth International Wheat Genetics Symposium* (ed. E.R. Sears and L.M.S. Sears), 191–207. Columbia: University of Missouri. 6–11 August 1973.

Sears, E.R. (1977). An induced mutant with homoeologous pairing in common wheat. *Canadian Journal of Genetics and Cytology* 19: 585–593.

Sears, E.R. (1978a). Analysis of wheat-*Agropyron* recombinant chromosomes. In: *Interspecific Hybridization in Plant Breeding. Proceedings of the 8th Eucarpia Congress* (ed. E. Sanchez-Monge and F. Garcia-Olmedo), 63–72. Madrid, Spain.

Sears, E.R. (1978b). *Agropyron* – wheat transfers induced by homoeologous pairing. In: *Proceedings of the Fourth International Wheat Genetics Symposium* (ed. E.R. Sears and L.M.S. Sears), 191–199. Columbia; USA: University of Missouri. 8–11 August 1973.

Sears, E.R. (1983). The transfer to wheat of interstitial segments of alien chromosomes. In: *Proceedings of the 6th International Wheat Genetics Symposium* (ed. S. Sakamoto), 5–12. 28 November 28–3 December 1983. Kyoto, Japan: Plant Germ-plasm Institute, Faculty of Agriculture, Kyoto University.

Sears, E.R. and Okamoto, M. (1958). Intergenomic chromosome relationships in hexaploid wheat. *Proceedings of the 10th International Congress on Genetics*, 2, 258–259. McGill University, Montreal. 20–27 August 1958.

Segura, V., Vilhjalmsson, B.J., Platt, A. et al. (2012). An efficient multi-locus mixed-model approach for genome-wide association studies in structured populations. *Nature Genetics* 44: 825–830.

Sehgal, D., Dreisigacker, S., Belen, S. et al. (2016). Mining centuries old *in situ* conserved Turkish wheat landraces for grain yield and stripe rust resistance genes. *Frontiers in Genetics* 7 (Nov.): art. no. 201.

Semagn, K., Babu, R., Hearne, S. et al. (2014). Single nucleotide polymorphism genotyping using Kompetitive Allele Specific PCR (KASP): Overview of the technology and its application in crop improvement. *Molecular Breeding: New Strategies in Plant Improvement* 33: 1–14.

Sen, A., Balyan, H.S., Sharma, P.C. et al. (1997). DNA amplification fingerprinting (DAF) as a new source of molecular markers in bread wheat. *Wheat Inf. Ser.* 85: 35–42.

Shahzad, K., Rauf, M., Ahmed, M. et al. (2014). Functional characterization of an intron retaining K^+ transporter of barley reveals intron mediated alternate splicing. *Plant Biology* 17: 840–851.

Shamaya, N.J., Shavrukov, Y., Langridge, P. et al. (2017). Genetics of Na^+ exclusion and salinity tolerance in Afghani durum wheat landraces. *BMC Plant Biology* 17: 1–8.

Shan, Q., Wang, Y., Li, J. et al. (2014). Genome editing in rice and wheat using the CRISPR/Cas system. *Nature Protocols* 9: 2395–2410. doi:10.1038/nprot.2014.157.

Sharma, H.C. (1995). How wide can a wide cross be? *Euphytica* 82: 43–64.

Sharma, H.C. and Gill, B.S. (1983a). Current status of wide hybridization in wheat. *Euphytica* 32: 17–31.

Sharma, H.C. and Gill, B.S. (1983b). New hybrids between *Agropyron* and wheat. II: Production, morphology and cytogenetic analysis of F1 hybrids and backcross derivatives. *Theoretical and Applied Genetics* 66: 111–121.

Sharma, H.C. and Gill, B.S. (1983c). New hybrids between Agropyron and wheat. III: Backcross derivatives, effect of *Agropyron* cytoplasm, and production of addition lines. In: *Proceedings of the 6th International Wheat Genetics Symposium*, pp. 213–221.

Sharp, P.J., Chao, S., Desai, S. et al. (1989). The isolation, characterization and application in the *Triticeae* of a set of wheat RFLP probes identifying each homoeologous chromosome arm. *Theoretical and Applied Genetics* 78: 342–348.

Sharp, P.J. and Soltes-Rak, E. (1988). Homoeologous relationships between wheat group 2 chromosome arms as determined by RFLP analysis. In: *Proceedings of the 7th International Wheat Genetics Sympopsium* (ed. T.E. Miller and R.M.D. Koebner), 635–638. Cambridge, UK: IPSR.

Shebeski, L.H. and Wu, Y.S. (1952). Inheritance in wheat of stem rust resistance derived from *Agropyron elongatum*. *Agri Sciences* 32: 26–32.

Shewry, P.R. (2018). Do ancient types of wheat have health benefits compared with modern bread wheat? *Journal of Cereal Science* 79: 469–476.

Simeonovska, E., Gadžo, D., Jović, Z. et al. (2013). Collecting local landraces of maize and cereals in South Eastern Europe during 2009 and 2010. *Romanian Agricultural Research* 30: 37–43.

Simons, K.J., Fellers, J.P., Trick, H.N. et al. (2006). Molecular characterization of the major wheat domestication gene Q. *Genetics* 172: 547–555.

Singh, B., Thakur, K., Kukreja, S. et al. (2019). Functional genomic approaches to improve crop plant heat stress tolerance. *F1000 Research* 8: 1721. https://doi.org/10.12688/f1000research.19840.1.

Smit, H.A., Tolmay, V.L., Barnard, A. et al. (2010). An overview of the context and scope of wheat (*Triticum aestivum*) research in South Africa from 1983 to 2008. *South African Journal of Plant and Soil* 27: 81–96.

Smith, D.C. (1942). Intergeneric hybridization of cereals and other grasses. *Journal of Agricultural Research* 64: 33–47.

Sohail, Q., Manickavelu, A., and Ban, T. (2015). Genetic diversity analysis of Afghan wheat landraces (*Triticum aestivum*) using DArT markers. *Genetic Resources and Crop Evolution* 62: 1147–1157.

Soriano, J.M., Villegas, D., Sorrells, M.E. et al. (2018). Durum wheat landraces from east and west regions of the Mediterranean basin are genetically distinct for yield components and phenology. *Frontiers in Plant Science* 9: art. no. 80.

Sparks, C.A. and Jones, H.D. (2016). Wheat biotechnology: Recent developments and future trends. Chapter 6, 88–102. In: *Book: Biotechnology of Major Cereal* (ed. H.D. Jones). CAB International.

Steuernagel, B., Periyannan, S.K., Hernández-Pinzón, I. et al. (2016). Rapid cloning of disease-resistance genes in plants using mutagenesis and sequence capture. *Nature Biotechnology* 34: 652–655. doi:10.1038/nbt.3543.

Su, P., Jiang, C., Qin, H. et al. (2019a). Identification of potential genes responsible for thermotolerance in wheat under high temperature stress. *Genes* 10: 174. doi:10.3390/genes10020174.

Su, Z., Bernardo, A., Tian, B. et al. (2019b). A deletion mutation in TaHRC confers Fhb1 resistance to Fusarium head blight in wheat. *Nature Genetics* 51: 1099–1105. https://doi.org/10.1038/s41588-019-0425-8.

Suenaga, K. and Nakajima, K. (1989). Efficient production of haploid wheat (*Triticum aestivum*) through crosses between wheat maize (*Zea mays*). *Plant Cell Reports* 8: 263–266.

Sukumaran, S., Dreisigacker, S., Lopes, M. et al. (2015). Genome-wide association study for grain yield and related traits in an elite spring wheat population grown in temperate irrigated environments. *Theoretical and Applied Genetics* 128: 353–363.

Sukumaran, S., Lopes, M.S., Dreisigacker, S. et al. (2016). Identification of earliness Per Se flowering time locus in spring wheat through a genome-wide association study. *Crop Science* 56: 2962–2672. doi:10.2135/cropsci2016.01.0066.

Sukumaran, S., Reynolds, M.P., and Sansaloni, C. (2018). Genome-wide association analyses identify QTL hotspots for yield and component traits in durum wheat grown under yield potential, drought, and heat stress environments. *Frontiers in Plant Science* 9: 81.

Sun, C., Dong, Z., Zhao, L. et al. (2020). The Wheat 660K SNP array demonstrates great potential for marker assisted selection in polyploid wheat. *Plant Biotechnology Journal* 18: 1354–1360.

Sun, X., Liu, D., Zhang, X. et al. (2013). SLAF-seq: An efficient method of large-scale *de novo* SNP discovery and genotyping using high-throughput sequencing. *PLoS ONE* 8 (3): e58700. doi:10.1371/journal.pone.0058700.

Szakács, É., Schneider, A., and Rakszegi, M. et al. (2017). Transfer of rust resistance from perennial rye (*Secale cereanum*) to bread wheat via recombinant 1RS.1BL translocation. In: *Proceedings 13th International Wheat Genetics Symposium* (ed. H. Buerstmayr, C. Lang-Mladek, B. Steiner, et al.). 23–29 April 2017. Tulln, Austria: BOKU – University of Natural Resources and Life Sciences, Vienna. Department of Agrobiotechnology (IFA-Tulln) & Department of Crop Science.

Tadesse, W., Sanchez-Garcia, M., Assefa, S.G. et al. (2019). Genetic gains in wheat breeding and its role in feeding the world. *Crop Breeding, Genetics and Genomics* 1: e190005. https://doi.org/10.20900/cbgg20190005.

Tautz, D. and Renz, M. (1984). Simple sequences are ubiquitous repetitive components of eukaryotic genomes. *Nucleic Acids Research* 12: 4127–4138.

Tee, T.S. and Qualset, C.O. (1975). Bulk populations in wheat breeding: Comparison of single-seed descent and random bulk methods. *Euphytica* 14: 282–394.

Thépot, S., Gwendal, R., Frédéric, H. et al. (2015). Interest of a multiparental and outcrossing wheat population for fine mapping. In: *Advances in Wheat Genetics: From Genome to Field* (ed. Y. Ogihara, S. Takumi, and H. Handa), 331–338. Tokyo: Springer.

Thompson, A.L., Smiley, R.W., Paulitz, T.C. et al. (2016). Identification of resistance to *Pratylenchus neglectus* and *Pratylenchus thornei* in Iranian landrace accessions of wheat. *Crop Science* 56: 654–672.

Thompson, W.P. (1926). Chromosome behaviour in a cross between wheat and rye. *Genetics* 11: 317–332.

Tiara, T. and Larter, E.N. (1978). Factors influencing development of wheat-rye-hybrid embryos *in vitro*. *Crop Science* 18: 348–350.

Tiwari, V.K., Rawat, N., Neelam, K. et al. (2010). Substitutions of 2S and 7U chromosomes of *Aegilops kotschyi* in wheat enhance grain iron and zinc concentration. *Theoretical and Applied Genetics* 121: 259–269. doi:10.1007/s00122-010-1307-8.

Tiwari, V.K., Rawat, N., Singh, N. et al. (2008). Evaluation and utilization of *Aegilops* germplasm for biofortification of wheat for high grain iron and zinc content. In: *Proceedings of the 11th International Wheat Genetics Symposium*, (R. Appels, R. Eastwood, E. Lagudah, et al.), 24–29 August 2008. Brisbane, QLD, Australia.

Tohidinejad, E., Hassani, H.S., Madani, H. et al. (2008). Investigation of the relationship between grain yield and quantitative traits in T*ritipyrum*, a new salt tolerant amphiploid, in comparison with Triticale and Iranian wheat by factor analysis. In: *Proceedings of the 11th International Wheat Genetics Symposium*, (R. Appels, R. Eastwood, E. Lagudah, et al.), 24–29 August 2008. Brisbane, QLD, Australia.

Toth, J., Pandurangan, S., Burt, A. et al. (2019). Marker-assisted breeding of hexaploid spring wheat in the Canadian prairies. *Canadian Journal of Plant Science* 99: 111–127. http://doi.org/10.1139/cjps-2018-0183.

Trethowan, R.M. and Mujeeb-Kazi, A. (2008). Novel germplasm resources for improving environmental stress tolerance of hexaploid wheat. *Crop Science* 48: 1255–1265.

Tripathi, K., Chauhan, S.K., Gore, P.G. et al. (2017). Evaluation of wheat landraces of north-western Himalaya against rice weevil, *Sitophilus oryzae* L. vis-à-vis physical seed parameters. *Plant Genetic Resources: Characterisation and Utilisation* 15: 321–326.

Tschermak-Seysenegg, E.V. (1951). The rediscovery of Gregor Mendel's work. An historical retrospect. *Journal of Heredity* 42 (4): 163–171. https://doi.org/10.1093/oxfordjournals.jhered.a106195.

Tsegaye, B. and Berg, T. (2007). Genetic erosion of Ethiopian tetraploid wheat landraces in Eastern Shewa, Central Ethiopia. *Genetic Resources and Crop Evolution* 54: 715–726.

Tsunewaki, K. (2015). Fine mapping of the first multi-fertility-restoring gene, *Rfmulti*, of wheat for three *Aegilops plasmons*, using 1BS- 1RS recombinant lines. *Theoretical and Applied Genetics* 128: 723–732.

Tucker, E.J., Baumann, U., Kouidri, A. et al. (2017). Molecular identification of the wheat male fertility gene *Ms1* and its prospects for hybrid breeding. *Nature Communications* 8: 869. https://doi.org/10.1038/s41467-017-00945-2.

Tzitzin, N.V. (1936). The problem of perennial wheat. In: *Plant Breeding Abstracts*, 7 (1936. Item 594): 184.

Tzitzin, N.V. (1937). What does crossing wheat with *Agropyrum* give? *Plant Breeding Abstracts* 8 (1937/38): Item 1156, 350.

Uauy, C., Wulff, B.B.H., and Dubcovsky, J. (2017). Combining traditional mutagenesis with new high-throughput sequencing and genome editing to reveal hidden variation in polyploid wheat. *Annual Review of Genetics* 51: 435–454. https://doi.org/10.1146/annurev-genet-120116-024533.

Udol'skaja, N.L. (1937). Production of drought-resistant forms of *Triticum-Agropyrum* hybrids. *Plant Breeding Abstracts* 9 (1939): Item 189, 49.

Ullah, S., Bramley, H., Mahmood, T. et al. (2019). A strategy of ideotype development for heat-tolerant wheat. *Journal of Agronomy and Crop Science* 206. https://doi.org/10.1111/jac.12378.

United Nations (2019). Department of Economics and Social Affairs. Population Division. Available online: https://www.un.org/en/development/desa/population/index.asp. URL: https://www.jstor.org/stable/3625992.

Ushiyama, T., Shimizu, T., and Kuwabara, T. (1991). High frequency haploid production of wheat through intergeneric crosses with *Teosinte*. *Japanese Journal of Breeding* 41: 353–357.

Vagndorf, N., Kristensen, P.S., Andersen, J.R. et al. (2018). Marker-assisted breeding in wheat. In: *Next Generation Plant Breeding*, 1–22. IntechOpen. http://dx.doi.org/10.5772/intechopen.74724.

Validya, A.S., Helander, J.M.D., Peterson, F.C. et al. (2019). Dynamic control of plant water use using designed ABA receptor agonists. *Science* 366: 446.

Van Dijk, K.V., Parlevliet, J.E., Kema, G.H.J. et al. (1988). Characterization of the durable resistance to yellow rust in old winter wheat cultivars in the Netherlands. *Euphytica* 38: 149–158.

Van Ginkel, M. and Ortiz, R.R. (2018). Cross the best with the best, and select the best: HELP in breeding selfing crop. *Crop Science* 58: 1–14. https://doi.org/10.2135/cropsci2017.05.0270.

Van Ginkel, M., Trethowan, R., Ammar, K., et al. (2002). Guide to bread wheat breeding at CIMMYT (rev). Wheat Special Report No. 5. CIMMYT, Mexico, D.F.

Van Niekerk, H.A. (2001). Southern Africa Wheat Pool. In: *The World Wheat Book: The History of Wheat Breeding* (ed. A.P. Bonjean and W.J. Angus), 923–936. Paris: Lavoisier Publishing.

Van Tassel, D.L., DeHaan, L.R., and Cox, T.S. (2010). Missing domesticated plant forms: Can artificial selection fill the gap? *Evol Applications* 3: 434–452.

Varella, A.C., Weaver, D.K., Cook, J.P. et al. (2017). Characterization of resistance to the wheat stem saw fly in spring wheat landrace accessions from targeted geographic regions of the world. *Euphytica* 213: art. no. 153.

Vavilov, N.I. (1914). Immunity to fungous diseases as a physiological test in genetics and systematics, exemplified in cereals. *Journal of Genetics* 4: 49–65.

Velu, G., Crespo Herrera, L., Guzman, C. et al. (2019). Assessing genetic diversity to breed competitive biofortified wheat with enhanced grain Zn and Fe concentrations. *Frontiers in Plant Science* 9: 1971. doi:10.3389/fpls.2018.01971.

Velu, G., Singh, R.P., Crespo-Herrera, L. et al. (2018). Genetic dissection of grain zinc concentration in spring wheat for mainstreaming biofortification in CIMMYT wheat breeding. *Scientific Reports* 8: Article number: 13526.

Verges, V.L. and Van Sanford, D.A. (2020). Genomic selection at preliminary yield trial stage: Training population design to predict untested lines. *Agronomy* 2020 (10): 60.

Verma, S.K., Kumar, S., Sheikh, I. et al. (2016). Transfer of useful variability of high grain iron and zinc from *Aegilops kotschyi* into wheat through irradiation approach. *International Journal of Radiation Biology* 92 (3): 132–139. doi:10.3109/09553002.2016.1135263.

Verushkine, S. and Shechurdine, A. (1933). Hybrids between wheat and couch grass. Fertile *Triticum-Agropyrum* hybrids of great scientific and practical interest. *The Journal of Heredity* 24: 329–335.

Vico, G., Manzoni, S., Nkurunziza, L. et al. (2016). Trade-offs between seed output and life span. A quantitative comparison of traits between annual and perennial congeneric species. *New Phytologist* 209: 104–114.

Vikram, P., Franco, J., Burgueño-Ferreira, J. et al. (2016). Unlocking the genetic diversity of Creole wheats. *Scientific Reports* 6: art. no. 23092.

Villa, T.T.C., Maxted, N., Scholten, M. et al. (2006). Defining and identifying crop landraces. *Plant Genetic Resources* 3: 373–384.

Villareal, R.L., Banuelos, O., and Mujeeb-Kazi, A. (1997). Agronomic performance of related durum (*Triticum turgidum* L) stocks possessing the chromosome substitution T1BL.1RS. *Crop Science* 37: 1735–1740.

Villareal, R.L., Banuelos, O., Mujeeb-Kazi, A. et al. (1998). Agronomic performance of chromosome 1B and T1BL.1RS near-isolines in the spring bread wheat Seri M82. *Euphytica* 103: 195–202.

Vilmorin, M.H. (1880). Essais de croisement entre blés différents. *Bulletin De La Société Botanique De France* 27 (9): 356–361. doi:10.1080/00378941.1880.10827932.

Vogel, K.E. (2009). Backcross breeding. *Methods in Molecular Biology (Clifton, N.)* 526: 161–169. doi:10.1007/978-1-59745-494-0_14.

Vos, P.R., Hogers, M., Bleeker, M. et al. (1995). AFLP: A new technique for DNA fingerprinting. *Nucleic Acids Research* 23: 4407–4414.

Voss-Fels, K.P., Stahl, A., Wittkop, B. et al. (2019). Breeding improves wheat productivity under contrasting agrochemical input levels. *Nature Plants* 5: 706–714. https://doi.org/10.1038/s41477-019-0445-5.

Wagoner, P. (1990). Perennial grain development – past efforts and potential for the future. *Critical Reviews in Plant Sciences* 9: 381–408.

Waines, J.G. and Ehdaie, B. (2005). Optimizing root characters and grain yield in wheat. *Czech Journal of Genetics and Plant Breeding* 41: 326–330.

Waines, J.G. and Ehdaie, B. (2007). Domestication and crop physiology: Roots of Green Revolution wheat. *Annals of Botany* 100: 991–998.

Wang, H., Dai, K., Xiao, J. et al. (2017). Development of intron targeting (IT) markers specific for chromosome arm 4VS of *Haynaldia villosa* by chromosome sorting and next-generation sequencing. *BMC Genomics* 18: 167.

Wang, H.W., Sun, S.L., Ge, W.Y. et al. (2020a). Horizontal gene transfer of *Fhb7* from fungus underlies *Fusarium* head blight resistance in wheat. *Science* 368: 6493. doi:10.1126/science.aba5435.

Wang, J., Singh, R.P., Braun, H. et al. (2009). Investigating the efficiency of the single backcrossing breeding strategy through computer simulation. *Theoretical and Applied Genetics* 118: 683–694.

Wang, J., Van Ginkel, M., Podlich, D., Ye, G. et al. (2003). Comparison of two breeding strategies by computer simulation. *Crop Science* 43: 1764–1773.

Wang, R.R.-C. (1989). Intergeneric hybrids involving perennial Triticeae. *Genetics (Life Sci. Adv.)* 8: 57–64.

Wang, R.R.-C. (2011). *Agropyron* and *Psathyrostachys*. In: *Wild Crop Relatives: Genomic and Breeding Resources, Cereals* (ed. C. Kole), 77–108. Berlin and Heidelberg, Germany: Springer-Verlag. (ISBN 10: 3642142273; ISBN 13: 9783642142277). doi:10.1007/978-3-642-14228-4_2.

Wang, R.R.-C. (2020). Chromosomal distribution of genes conferring tolerance to abiotic stresses versus that of genes controlling resistance to biotic stresses in plants. *International Journal of Molecular Sciences* 21: 1820.

Wang, R.R.-C., Li, X.-F., Robbins, M.D. et al. (2020b). DNA sequence-based mapping of the **St** genome of *Pseudoroegneria spicata* (Pursh) Á. Löve. versus wheat (*Triticum aestivum* L.) and barley (*Hordeum vulgare* L.). *Genome* 63: 445–457.

Wang, S., Wong, D., Forrest, K. et al. (2014a). Characterization of polyploid wheat genomic diversity using a high-density 90,000 single nucleotide polymorphism array. *Plant Biotechnology Journal* 12: 787–796.

Wang, S.C., Wong, D.B., Forrest, K. et al. (2014b). Sequencing IWG (2014) Characterization of polyploid wheat genomic diversity using a high-density 90,000 single nucleotide polymorphism array. *Plant Biotechnology Journal* 12: 787–796.

Wang, W., Pan, Q., He, F. et al. (2018). Transgenerational CRISPR-Cas9 activity facilitates multiplex gene editing in allopolyploid wheat. *The CRISPR Journal* 1 (1). doi:10.1089/crispr.2017.0010.

Wang, Z., Ren, J., Du, Z. et al. (2019). Identification of a major QTL on chromosome arm 2AL for reducing yellow rust severity from a Chinese wheat landrace with evidence for durable resistance. *Theoretical and Applied Genetics* 132: 457–471.

Wang, Z.Y., Zhao, P., Chen, H. et al. (2005). Identification of RAPD markers and development of SCAR markers linked to a powdery mildew resistance gene, and their location on chromosome in wheat cultivar brock. *Plant Production Science* 8 (5): 578–585. doi:10.1626/pps.8.578.

Wani, S.H., Khan, H., Riaz, A. et al. (2021). Genetic diversity for developing climate resilient wheats to achieve food security goals. *Advances in Agronomy*. 10.1016/bs.agron.2021.08.006.

Warburton, M.L., Crossa, J., Franco, J. et al. (2006). Bringing wild relatives back into the family: Recovering genetic diversity in CIMMYT improved wheat germplasm. *Euphytica* 149: 289–301.

Watson, J.D. and Crick, F.H.C. (1953). A structure for deoxyribose nucleic acid. *Nature* 171: 737–738.

Welch, R.M. and Graham, R.D. (1999). A new paradigm for world agriculture: Meeting human needs. Productive, sustainable, nutritious. *Field Crops Research* 60: 1–10.

Weller, D.M., Raaijmakers, J.M., Gardener, B.B.M. et al. (2002). Microbial populations responsible for specific soil suppressiveness to plant pathogens. *Annual Review of Phytopathology* 40: 309–348.

Welsh, J. and McClelland, M. (1990). Fingerprinting genomes using PCR with arbitrary primers. *Nucleic Acids Research* 18: 7213–7218.

Weruschkin, S.M. (1937). The relationship between the genera *Agropyrum* and *Triticum*. *Plant Breeding Abstracts* 9 (Item 191): 50.

Whelan, E.D.P. and Thomas, J.B. (1988). Resistance to colonization by the wheat curl mite in *Triticum-Agropyron* hybrids and Robertsonian translocations. In: *Proceedings of the Seventh International Wheat Genetics Symposium*, 2 (ed. T.E. Miller and R.M.D. Koebner), 913–916. Cambridge, UK. 13–19 July 1988.

Whitford, R., Fleury, D., Reif, J.C. et al. (2013). Hybrid breeding in wheat: Technologies to improve hybrid wheat seed production. *Journal of Experimental Botany* 64: 5411–5428.

Williams, J.G.K., Kubelik, A.R., Livak, K.J. et al. (1990). DNA polymorphisms amplified by arbitrary primers are useful as genetic markers. *Nucleic Acids Research* 18: 6531–6535.

Wilson, J.A. and Ross, W.M. (1962). Male sterility interaction of the *Triticum aestivum* nucleus and *Triticum timopheevi* cytoplasm. *Wheat Information Service* 14: 29–30.

Winfield, M.O., Allen, A.M., Burridge, A.J. et al. (2016a). High-density SNP genotyping array for hexaploid wheat and its secondary and tertiary gene pool. *Plant Biotechnology Journal* 14: 1195–1206.

Winfield, M.O., Allen, A.M., Wilkinson, P.A. et al. (2016b). Development of pre-breeding technology for root system study and selection of Kihara Afghan wheat landraces (KAWLR) to enhance wheat breeding in the rain-fed region. *Breeding Science* 66: 808–822.

Winfield, M.O., Allen, A.M., Wilkinson, P.A. et al. (2018). High-density genotyping of the A.E. Watkins Collection of hexaploid landraces identifies a large molecular diversity compared to elite bread wheat. *Plant Biotechnology Journal* 16: 165–175.

Wingen, L.U., Orford, S., Goram, R. et al. (2014). Establishing the *A. E. Watkins Landrace Cultivar Collection as a Resource for Systematic Gene Discovery in Bread Wheat. Theor. Appl. Genet.* 127: 1831–1842.

Wingen, L.U., West, C., Waite, M.L. et al. (2017). Wheat landrace genome diversity. *Genetics* 205: 1657–1676.

Wulff, B.H. and Moscou, M.J. (2014). Strategies for transferring resistance into wheat: From wide crosses to GM cassettes. *Frontiers in Plant Science* Article 692. doi:10.3389/fpls.2014.00692.

Xu, Y., An, D., Li, H. et al. (2011). Breeding wheat for enhanced micronutrients. *Canadian Journal of Plant Research* 91: 231–237.

Yang, M.Y., Li, G.R., Wan, H.S. et al. (2019). Identification of QTLs for stripe rust resistance in a recombinant inbred line population. *International Journal of Molecular Sciences* 20: 3410.

Youliang, Z., Mingcheng, L., Chi, Y. et al. (1992). Chromosome location of a new crossability gene in common wheat. *Wheat Information Service* 75: 36–40.

Yu, J., Hu, S., Wang, J. et al. (2002). A draft sequence of the rice genome (*Oryza sativa* L. ssp. indica). *Science* 296: 79–92.

Yuan, H.X., Zhang, F.X., Zhang, J.J. et al. (2011). Resistance of CIMMYT wheat germplasm to *Heterodera filipjevi* Xuchang population from Henan Province, China. *Acta Agronomica Sinica* 37 (11): 1956–1966. doi:10.1016/S1875-2780(11)60052-7.

Zadoks, J.C. (1985). Cereal rusts, dogs and stars in antiquity. *Cereal Rusts Bull* 13 (1): 1–9.

Zaharieva, M., Bonjean, A., and Monneveux, A.P. (2014). Saharan wheats: Before they disappear. *Genetic Resources and Crop Evolution* 61: 1065–1084.

Zegeye, H., Rasheed, A., Makdis, F. et al. (2014). Genome-wide association mapping for seedling and adult plant resistance to stripe rust in synthetic hexaploid wheat. *PLoS ONE* 9 (8): e105593. doi:10.1371/journal.pone.0105593.

Zeller, F.J. (1973). 1B/1R wheat-rye chromosome substitutions and translocations. In: *Proceedings of the Fourth International Wheat Genetics Symposium* (ed. E.R. Sears and L.M.S. Sears), 209–221. 6–11 August 1973. Columbia: University of Missouri.

Zeller, F.J. and Hsam, S.L.K. (1983). Broadening the genetic variability of cultivated wheat by utilizing rye chromatin. In: *Proceedings of the 6th International Wheat Genet Symposium* (ed. S. Sakamoto), 161–173. Kyoto: Japan.

Zeven, A.C. (1998). Landraces: A review of definitions and classifications. *Euphytica* 104: 127–139.

Zhang, H., Mittal, N., Leamy, L.J. et al. (2017a). Back into the wild: Apply untapped genetic diversity of wild relatives for crop improvement. *Evolutionary Applications* 10: 5–24. doi:10.1111/eva.12434.

Zhang, H.B. and Dvorák, J. (1990). Isolation of repeated DNA sequences from *Lophopyrum elongatum* for detection of *Lophopyrum* chromatin in wheat genomes. *Genome* 33: 283–293.

Zhang, W., Li, T., Qin, L. et al. (2018). TaDRO, A gene associated with wheat root architectures, its global distribution and evolution in breeding. *Scientia Agricultura Sinica* 51: 1815–1829.

Zhang, X.D., Wei, X., Xiao, J. et al. (2017b). Whole genome development of intron targeting (IT) markers specific for *Dasypyrum villosum* chromosomes based on next-generation sequencing technology. *Molecular Breeding : New Strategies in Plant Improvement* 37: 115.

Zhang, Y., Liang, Z., Zong, Y. et al. (2016). Efficient and transgene-free genome editing in wheat through transient expression of CRISPR/Cas9 DNA or RNA. *Nature Communications* 7: 12617. doi:10.1038/ncomms12617.

Zhao, K., Aranzana, M.J., Kim, S. et al. (2007). An Arabidopsis example of association mapping in structured samples. *PLoS Genetics* 3: e4.

Zheng, M.Y., Weng, Y., Liu, W. et al. (2002). The effect of ovary-conditioned medium on microspore embryogenesis in common wheat (*Triticum aestivum* L.). *Plant Cell Reports* 20: 802–807.

Zhou, S., Zheng, W.J., Liu, B.H. et al. (2019). Characterizing the role of *TaWRKY13* in salt tolerance. *International Journal of Molecular Sciences* 20: 5712.

Part II

Bioresource and Future Energy Security

8

Waste-to-Energy

Potential of Biofuels Production from Sawdust as a Pathway to Sustainable Energy Development

Oyebanji Joseph Adewumi[1], *Oyedepo Sunday Olayinka*[2], *Kilanko Oluwaseun*[2] *and Dunmade Israel Sunday*[3]

[1] *Department of Mechanical Engineering, Bells University of Technology, Ota, Nigeria*
[2] *Department of Mechanical Engineering, Covenant University, Ota, Nigeria*
[3] *Earth and Environmental Sciences Department at the Mount Royal University, Calgary, Canada*

CHAPTER MENU

8.1 Introduction

Global warming drifts and energy crunches call for immediate eradication of fossil fuel usage. Moreover, as the global energy demand has increased at a very large rate and in parallel, wastes (e.g., agricultural, food, industrial) are generated in extensive quantities

Bioresource Technology: Concept, Tools and Experiences, First Edition. Tanveer Bilal Pirzadah, Bisma Malik, Rouf Ahmad Bhat, and Khalid Rehman Hakeem.
© 2022 John Wiley & Sons Ltd. Published 2022 by John Wiley & Sons Ltd.

day-by-day with significant problems in their management and disposal. All these factors are posing enormous technological challenges to the world's sustainable growth and development (Arancon et al 2013; Hoornweg and Bhada-Tata 2012). Rising in energy demand is possibly linked to increasing population growth and technological development. The notion of energy from biomass waste is accepted by the populace worldwide, based on its capacity to generate energy from more than one fuel source and value-added products from different fractions of biomass waste (Hoornweg and Bhada-Tata 2012; Nemerow, 2009). Waste-to-energy technologies (WTE-T) are optimistic techniques, especially for the developing countries, to turn waste into a functional form of energy (Ahmadi et al. 2020; Kothari et al. 2010). Waste-to-energy (WTE) and sustainable development concepts are the new and current global trends for sustaining renewable energy and for the production of environmentally-friendly fuels. Over the years, various investigations have been carried out on biofuel production from biomass (especially from agricultural crops, residues or wastes), as well as energy recovery from municipal solid waste (MSW). The shortage of limited available fossil fuel in conjunction with environmental pollution due to human activities has raised global concerns in this last decade. Agro waste (AGW), especially biomass, contributes to these environmental impact issues, yet it acts as a potential energy source for energy recovery (Tabasova et al. 2012). Conversion of biomass waste to energy can abate environmental degradation and can replace some part of the primary energy currently supplied by fossil fuels with proper management practice and waste handling (Fodor and Klemes 2012). Effective utilization of AGW for WTE acts as one of the potential solutions for modern AGW management in urban areas. An AGW management system aims to handle health, environment, aesthetic, land-use resources, economic and employment concerns related to improper disposal of waste (Nemerow 2009; Ouda and Cekirge 2014; Allen et al. 2000). Generation of AGW in developing nations was due to the increase in the population, urbanization growth and the rising standards of living (Chiaramonti et al. 2015; Colesca and Ciocoiu 2013). Disposal methods have been a major problem in developing countries due to increase in AGW generation growth, open burning of waste and open landfills causing environmental pollution (Ouda 2013; Iwasaki et al. 2014). Environmental health hazards result due to inadequate management of these wastes. Energy poverty is relieved from the waste and biomass energy to give employment opportunities, increase national income and improve the national economy instead of constituting environmental pollutants (Oyedepo et al. 2019; Akorede et al. 2017). One-seventh of the global energy consumption (43%) derived from renewable energy through biomass has great potential to replace the conventional fossil fuels in both developed and developing countries (Searchinger et al. 2008; Anupama et al. 2015). Fossil fuel greenhouse gas (GHG) emissions offsets and GHG emission reductions occur by biomass energy's significant role (Kumar et al. 2015). Compared to fossil fuel sources (e.g., natural gas, coal, oil, etc.), it has the advantage of being neutral in regard to the emissions of GHG.

Biomass residues and waste was encouraged by the International Energy Agency (IEA 2007) for energy generation, GHG emissions mitigation, national energy sources diversification and to achieve energy independency without threatening the world's food supply. Biomass resources are classified into three types, as follows:

- *Agriculture and forestry residues, and organic waste, including MSW*,giving energy production per year between 50 and 150 EJ yr^{-1} with 100 EJ yr^{-1} as the mean estimate per year.
- *Surplus forestry*, as forestry residues energy of 60–100 EJ yr^{-1} is available from surplus forest growth of trees, wood sawdust, bark, logging residues, etc.
- *Biomass produced via cropping systems*.Reliable studies revealed that a more conservative estimate of energy crop production on possible surplus good-quality agricultural and pasture lands reserves amounts to an estimated value of 120 EJ yr^{-1}. Examples of energy crops include woody crops, herbaceous woody crops, grasses, starch crops, sugar crops, oil seed crops, etc. (Kumar et al. 2015).

Thermal-chemical conversion of biomass consists of gasification, pyrolysis and direct combustion to produce mainly liquid, char and gaseous fuels to be processed further for energy production. The prime objectives of this study are to: i) carry out a comprehensive review on WTE technologies; ii) assess WTE technology development in developing countries; and iv) Gas Chromatography-Mass Spectrometry (GC-MS) characterization and determination of bio-oils functional groups produced from Africana birch (Al) and West African cordia (Cm) sawdust pyrolysis.

8.2 Overview of Potential WTE Technologies for Biomass Wastes

Biomass waste potentials are realized from available technologies ranging from very simple systems for disposing of dry waste to more complex technologies capable of dealing with large amounts of biomass waste for steam, fuel or electricity generation (Lombardi and Corti 2009). Biological and thermal are the major two paths for energy recovery from waste. The biological route produces biogas as an energy source from organic wastes. Waste refuse-derived-fuel for combustion, gasification or pyrolysis to produce electricity, heating or both can also be obtained via the thermal route (Nordi et al. 2017; David and Ragauskas 2010). Figure 8.1 illustrates the energy recovery routes from biomass wastes. Presently, different technologies have been developed to convert biomass waste to bio-oil, biogas or bio-char, which compete favorably with by-products of fossil fuels (Lee et al. 2019).

8.2.1 Thermo-Chemical Conversion Technologies

Thermo-chemical processes aid the conversion of biomass wastes over a massive range. The biomass wastes can either be from agricultural waste, crop residues or wastes that are traditionally discarded and have no apparent value (Jameel et al. 2010). These processes are generally characterized by elevated temperatures and chemical reactions occurring at different temperatures, and may require partial oxidation as in gasification or proceed in the absence of oxygen as in pyrolysis (Gumisiriza et al. 2017). The thermo-chemical conversion process is determined by the end-user requirements, source and nature of the biomass and environmental circumstances for production purposes (Goyal et al. 2008).

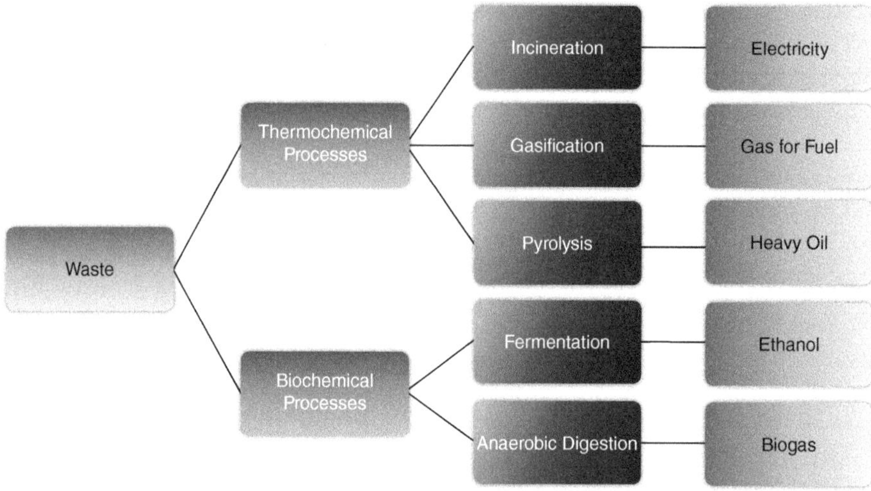

Figure 8.1 Wastes to Energy Conversion Routes (*source*: based on David and Ragauskas 2010).

8.2.1.1 Gasification

The gasification technique is a thermo-chemical conversion at high temperatures (500–1400°C) in the absence of oxygen (Ahmad et al. 2016). It converts carbonaceous materials into syngas consisting of CO, H_2, CH_4 and small amounts of impurities (e.g., tar, sulfur, nitrogen oxides, particulate matter and alkali metals), and production of gaseous fuel (producer gas) for energy generation and value-added chemicals for chemical industries (Messerle et al. 2018; Watson et al. 2018). The gas products are used directly or indirectly in running of internal combustion engines (both compression and spark ignition), furnace oil substitutes, application in heat engines, production of methanol and also serve as chemical feedstock for industries (Bhatia 2014; Thakkar et al. 2018). The biomass gasification process is a more efficient and attractive route for fuel production from any biomass materials than from ethanol production.

8.2.1.2 Pyrolysis

Pyrolysis can be defined as the thermal decomposition of carbon-based materials in an oxygen-deficient atmosphere at temperatures of around 200–400°C using heat to produce syngas (Mazzoni and Janajreh 2017). No air or oxygen is present and no direct burning takes place. The process is endothermic. Pyrolysis process product can be summarized as char, liquids (tar) and gases (H_2, CO, CO_2, CH_4 and N_2), as shown in the equation below (Young 2010):

$$\text{Biomass} \rightarrow \text{Bio-char} + \text{Bio-oil} + \text{bio-Gases}\left(H_2 + CO + CO_2 + CH_4\right)$$

Pyrolysis processes are classified as slow, fast and flash pyrolysis (Figure 8.2) that produce fuel oil as alternatives to fossil fuel for electricity generation and heating applications, which can be influenced by the operating conditions to give liquid production more attention due to high pyro-oil yield (75 wt. %), environmentally-friendly energy production, and

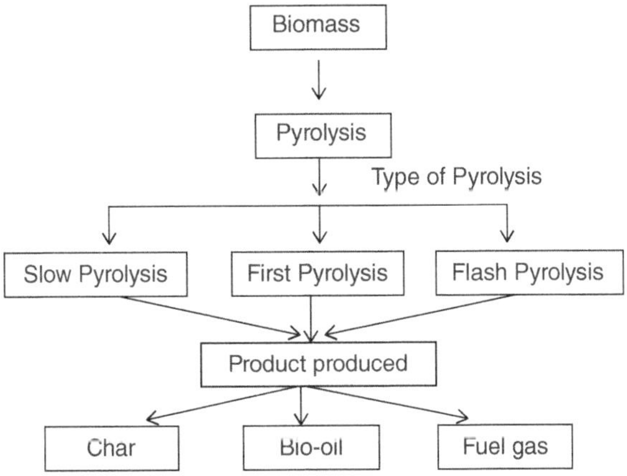

Figure 8.2 Pyrolysis process (Lee et al. 2019).

cost-effective and highly efficient technology in production (Dhyani and Bhaskar 2018; Bridgwater 2012; Jahirul et al. 2012). Recently, improvements and enhancement of pyro-oil properties have become a serious issue which requires further processing, so that it can be utilized as a substitute for fossil fuel. In the pyrolysis process, gases are produced, particularly syngas, which was capable of producing energy due to its high heating value. It mainly contains carbon monoxide, carbon dioxide, hydrogen, nitrogen, ethane, propane, butane and methane and also can be used as fuel for generating electricity and heat (Brownsort 2009; Cannon and Sperlin 2009; Li et al. 2019; IEA 2007; Zhang et al. 2013).

8.2.1.3 Liquefaction

Biomass is a desirable fuel ecologically because it contains very little sulfur and generally much less ash than coal. The liquefaction of biomass is an important technology and through thermo-chemical routes it converts biomass waste into valuable biofuel (Zhang et al. 2019; Soudham 2015). Indirect liquefaction and direct liquefaction are the two liquefaction classifications. Indirect liquefaction involves the presence of catalysts for pyro-oil cracking to produce clean fuel oil, while direct liquefaction refers to pyrolysis, solvolysis liquefaction and high-pressure hydrothermal/solvolytic liquefaction to produce liquid tars and oils and/or condensable organic vapors. In the indirect liquefaction process, first, biomass is gasified to syngas in the presence of partial O_2/air at very high temperatures of more than 1000°C and then this gas mixture undergoes a catalytic reaction. In this reaction, Fe-, Co-, Ru- and Ni-based catalysts are mostly used at high temperatures of around 340°C or low temperatures of around 230°C to produce olefins and gasoline or diesel and linear waxes, respectively (Lee et al. 2019).

8.3 Biochemical Conversion Technologies

Biochemical transformation of biomass waste encompasses techniques for conversion of crop residues or chemical biomass sources into chemicals and clean fuel through hydrolysis and fermentation of, for example, cellulose and hemicellulose, into simple sugars to serve as feedstock to the industries (Soudham 2015). Biochemical conversion performance regarding GHG emissions and energy balance are better than thermo-chemical conversion of biomass wastes according to comparative life-cycle assessment. Biomass biochemical transformation is divided into three kinds: technology of anaerobic fermentation biogas, bio-ethanol and biodiesel (Li et al. 2017; Sialve et al. 2009; Brennan and Owende 2010; Cantrell et al. 2008).

8.4 Potential Feedstocks for Waste-to-Energy

8.4.1 Agricultural Residues

Quantities of agricultural residues and waste produced annually are estimated to be 4.6 billion tons dry matter and estimated energy to be 78 EJ yr^{-1}, and are widely under-utilized (Bentsen et al. 2014). The residue referred to as waste is the utilized portion of crop residue for energy production at low material cost volumes. The most common biomass residues from agricultural products are rice husk, wood sawdust, sugar cane fiber (known as bagasse), coconut husks and shells, groundnut shells, cereal straw, etc. These residues serve as potential sources of fuels for production of heat and electricity when undergoing biomass conversion processes (Punia et al. 2017).

8.4.2 Animal Waste

Animal wastes are the source wastes for production of biomass energy which are derived from poultry and animal manures. The manures are processed as fertilizers and were spread on farm land in the past. Recently, introduction of waste management and tighter environmental controls is now vital to provide motivations for WTE conversion. Waste materials conversion gives rise to biogas that serves as a potential source of fuel used for space and water heating, small gas turbines and internal combustion engines (Ben-Iwo et al. 2016).

8.4.3 Forestry Residues

Forestry residues are usually generated wood and grasses and natural attrition in which wood processing contributes a large percentage of waste in the form of sawdust, off-cuts and bark (IRENA 2019). This serves as raw materials for production of steam and hot gasses with the help of a biomass gasifier (Panel et al. 2016). Nigerian forests generate 100 tons of logs and sawdust and 42 tons from split timber log and ply wood annually, which can serve as potential sources of energy generation and biofuel feedback instead of constituting environmental problems (Ohunakin 2010).

8.4.4 Industrial Wastes

A large number of residues and by-products are from industries such as solid wastes, including food that does not meet quality control standards, peelings and scraps from fruit and vegetables, filter sludge and coffee grounds, sugar pulp and starch fiber. Liquid wastes generated from fruit, fruit and vegetable blanching, fish and poultry, washing of meat, processing and cleaning operations, as well as wine-making, constitute industrial waste to generate energy. These industrial wastes serve as potential sources for the production of biogas anaerobically to produce electricity and process heat from steam (IRENA 2018).

8.4.5 Municipal Solid Waste (MSW)

According to the International Renewable Energy Agency, millions of tons of household waste is collected each year (IRENA 2018). Presently, there are over 800 WTE plants across 40 countries. These plants use 11% of MSW around the world, and produce 429 TWh. Approximately 10% of these plants are used for electricity generation. In Germany, only 1% of the waste is landfilled, while the share of WTE is 35%, which is 24% higher than the average for European countries and 11,540 MW power capacity was generated from municipal wastes as from 2018 (Llanes and Kalogirou 2019). If wastes are properly processed. either through direct combustion or natural anaerobic digestion, they will serve as potential sources of energy generation for both developed and developing countries (Ahmadi et al. 2020; Akorede et al. 2017).

8.4.6 Black Liquor

Black liquor is a kraft pulp waste liquor that contains various chemicals, water and very fine organics. It contains most of the original cooking inorganic elements and the degraded, dissolved wood substances used to fire boilers. It is obtained from pyrolysis of paper waste from paper manufacturing, producing liquor of 13.5–14.5 MJ kg^{-1} heat content per pulp ton. A study of the prospect of bioenergy production from biomass for GHG emission reduction and possibilities of sustaining energy development was embarked on in this study, as shown in Figure 8.3.

8.5 Waste-to-Energy and Sustainable Energy Development

Worldwide, the energy production and utilization pattern is tending toward sustainable development. Moreover, the WTE postulation is being encouraged as a component in an attempt to achieve a sustainable energy system that is environmentally-friendly, cost-efficient and reliable, which is based on effective utilization of local resources and networks (Laura et al. 2020; Oyedepo 2012). Research reveals that the energy output from WTE power plants is about 50% renewable (biomass, etc.) by proportion.

The above estimate is based on the fact that the significant constituents of biomass waste is organic matter derived from plants and trees, which are renewable resources and form major parts of the natural ecosystem (Qin et al. 2014). The WTE technique is a powerful approach and an effective substitute energy option to reduce CO_2 emissions; hence, it is capable of preserving the limited available fossil fuel resources used in conventional power

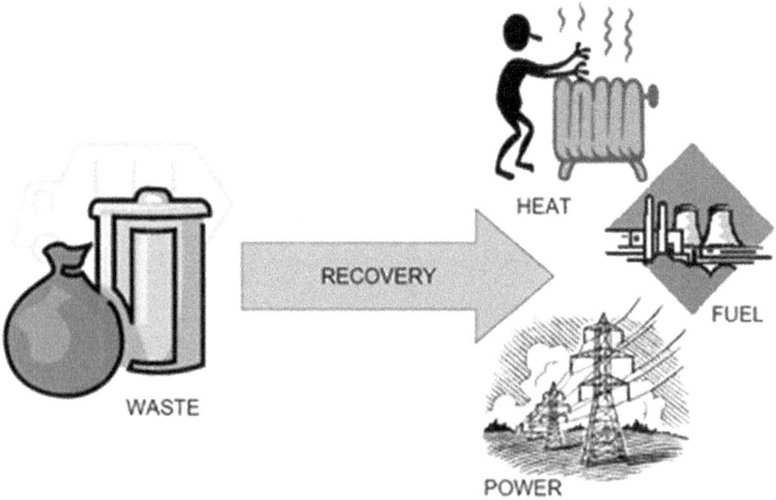

Figure 8.3 Waste to Energy process.

plants (Gumbo 2014). Among the prime advantages of the WTE conversion system is that the emitted GHGs during the energy generation process from biomass waste are recycled in that process in order to reduce GHG emissions into the atmosphere (Massaruto 2015). This is much better than the burning of fossil fuels such as coal, which releases carbon dioxide that remains in the atmosphere for several centuries. Degrees of industrialization, socio-economic development and climate all influence waste generation rates, the amount of solid waste produced depending on the urban population percentage and economic prosperity. However, sustainable energy generation and development is a major concern to meet the present needs without compromising future energy generation (Ryu 2010). Environmental degradation and energy security issues have raised global concerns that call for alternative sources for energy generation and recovery. MSW is one of the potential sources that needs appropriate management practice and waste handling, and its efficient mitigation has environmental impacts which can serve as replacement for fossil fuels. This utilization of MSW for WTE acts as one of the potential solutions for modern MSW management and sustainable energy development (Oyedepo 2012).

8.6 Challenges and Future Prospects of Waste-to-Energy Technologies

Promising feedstocks for bioenergy are available across the globe but challenges are posed by energy production such as conversion of biomass residues, and waste to immediate energy resources and technologically economical viabilities for generation of heat and electricity. Waste-to-bioenergy production cost-competitiveness is very low compared to fossil-based fuels in this generation, due to a high increase in bioenergy production costs, operational costs and high cost of conversion technologies such aspyrolysis, gasification and incineration accordingly (Oyedepo 2012; Ryu 2010). High energy requirements, reactor maintenance and operations hinder waste-to-bioenergy technology commercialization, thus the adaptation of process optimization to maximized production yield and process efficiency is a current research focus. Presently, in developing countries like Nigeria, wood residues constitute environmental pollution in which crude methods are used in disposing of such waste by open burning of sawdust and fuel wood in domestic heating and cooking, which needs waste-to-bioenergy modern technology (Tan et al. 2015). Adequate utilization of wastes to energy brings about clean energy generation and better waste management. Numerous African researchers (Oyebanji et al. 2017, 2018; Ogunjobi and Lajide 2013; Tan et al. 2015; Adegoke et al. 2014) study bio-oil production of different feedstocks (palm fruit bunches, wood residue and corn cobs) using pyrolysis conversion processes.

8.7 Case Study: Application of Fast Pyrolysis for Conversion of Sawdust to Bio-Oil

8.7.1 Samples Collection and Experimental Analysis

In this case study, characterization of West African cordia (Cm) and Africana birch (Al) sawdust were study and experimentally investigated. The wood samples were obtained from a sawmill in Ota, southwestern Nigeria. Tree lumber collected was turned into wood chips (15 × 10 × 10 mm), air dried for 3 days and ground to a particle size of 500 μm using the milling machine at Bells University of Technology, Ota, Ogun State, Nigeria. Samples composition were determined according to American Standard of Testing and Materials (ASTM) standards (Ogunsina et al. 2014; American Standard of Testing and Materials 2007).

8.7.2 Instrumentation and Experimental Set-up

The instrumentation for this study (pyrolysis unit) was carried out in a fixed-bed reactor with an automatic temperature control 4-kW furnace. At a pre-set temperature, the volatile stream was give off and driven through the product pipes channel to the condenser, which was immersed in an ice-bath (tar trapper) for the collection of the condensate (bio-oil or tar). The pyrolysis unit was fabricated to separate bio-char and biogas in the volatile stream. The bio-oil were transferred to the condenser while the bio-char are collected at room temperature in the retort and the biogas was trapped in a gas collector. Figure 8.4 shows the assembly view of the pyrolysis unit used for this study (Tan et al. 2015; Adegoke et al. 2014; American Standard of Testing and Materials 2004).

8.7.3 GCMS Analysis

The bio-oil was analyzed using Shimadzu GCMS-QP2010 Plus (Japan) with 30 m × 0.25 mm internal diameter (di) and 0.25 mm in film thickness. Analysis conditions were as follows: column oven temperature (60–280°C), injection mode Split ratio of 41.6, injection temperature 250°C, linear velocity (36.2 cm sec^{-1}), purge flow (3.0 ml min^{-1}), pressure at 56.2 kPa and helium carrier gas flow rate of 45.0 ml min^{-1}, and identification of mass

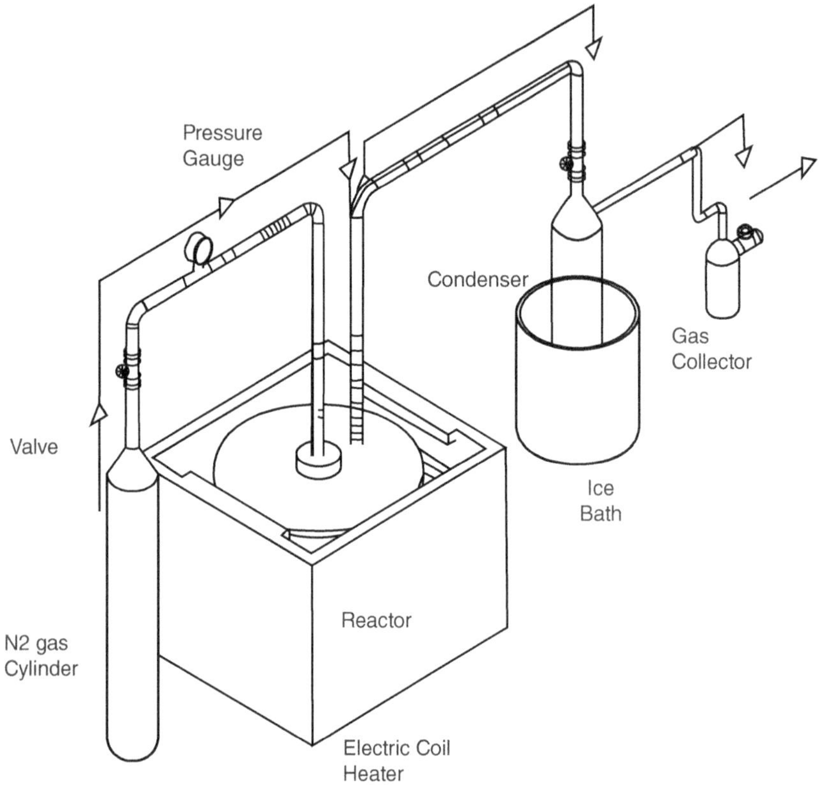

Figure 8.4 Assembly view for pyrolysis unit (Oyebanji et al. 2017).

spectra and fragmentation pattern is according to the NIST computer data bank with those reported in the literature (Oyebanji and Oyedepo 2018; Ololade et al. 2014).

8.7.4 Chemical and Physical Composition of Biofuel Yield

Generally, wood bio-oil comes with a dark-brownish color and a slight pungent odor. The, density, ash content, elemental composition, viscosity, pH, lower heating value (LHV), moisture content and higher heating values (HHV) of the feed materials and bio-oil are the physical and chemical properties that show how high energy values can be obtained from feed material used in this study. Compared to results achieved from analyses of bio-oils from the other sources, bio-oil yield results indicated more carbon (43.56%) and oxygen (48.89%) in the study analysis. The low nitrogen (0.62%) and sulfur (0.68%) showed that the bio-oil has low pollutant effect, which indicates that they are an eco-friendly bio-oil (Adegoke et al. 2014; Ololade et al. 2014). Moreover, HHV and LHV were greater than other bio-oils reported (Ng et al. 2014; Okoroigwe et al. 2012; Rahim et al. 2013). The moisture content and viscosity of the bio-oil were similar to those of the other bio-oils (Ng et al. 2014), indicating that the bio-oil could be easily handled. The moisture content could be traced to moisture in feed sample and water of dehydration reactions, bio-oils pH and ash content was at normal range for biomass and pH cracking or neutralization of the produced bio-oil to prevent corrosion and other reactions must be carried out before adequate utilization in internal combustion engines, furnaces and boilers (Ng et al. 2014; Okoroigwe et al. 2012; Rahim et al. 2013; Hossain et al. 2014; Kato et al. 2016).

8.7.5 Characterization of Bio-Oil Yield from Sawdust Samples

Tables 8.1 and 8.2 present the chemical composition of *Anogeissue Leiocarpa* and *Cordia Milleni*, found in bio-oil from wood samples, while bio-oil yield chromatograms are shown in Figures 8.5 and 8.6 at 250°C injection temperature. The presence of carboxylic acid, ketones, esters, ethers, alcohol and aromatic hydrocarbons in pyrolysis liquids were indicated by MS and this result was in agreement with the study of Czernik and Bridgwater (Sukhbaatar et al. 2009).

Table 8.1 Chemical composition of *Anogeissue Leiocarpa (AL)*.

Compounds	Retention Index	Percentage Composition	MF	MW
Luprosil	676	3.0	$C_3H_6O_2$	74
2,5-dimethylfuran	732	2.4	C_6H_8O	96
1,4-dimethyl-1 hour-pyrazole	804	0.8	$C_5H_8N_2$	96
2-methylbutyric acid	811	1.0	$C_5H_{10}O_2$	102
trans-3-Octene	824	0.4	C_8H_{16}	112
cis-3-octene	824	0.4	C_8H_{16}	112
5-methyl-3-methylene-5-hexen-2-one	887	0.5	$C_8H_{12}O$	124
4,5-dimethyl-1 hour-imidazol	927	0.8	$C_5H_8N_2$	96
Cyclooctane	959	0.4	C_8H_{16}	112
Corylone	972	0.4	$C_6H_8O_2$	112
3-methylcyclopentane-1,2-dione	1003	0.4	$C_6H_8O_2$	112
methyl-1-cyclohexenyl ketone	1027	1.6	$C_8H_{12}O$	124
o-guaiacol	1090	0.5	$C_7H_8O_2$	124
2-nitroethylpropionate	1067	1.0	$C_5H_9\ NO_4$	147
cis-1,9-hexadecadiene	1610	12.0	$C_{16}H_{30}$	222
cis-10-pentadecen-1-ol	1763	12.0	$C_{15}H_{30}O$	226
myristic acid	1769	5.0	$C_{14}H_{28}O_2$	228
palmitic acid	1968	15.0	$C_{16}H_{32}O_2$	256
9-octadecenal	2007	12.0	$C_{18}H_{34}O$	266
trans-2-octadecadecen-1-ol	2061	12.0	$C_{18}H_{36}O$	268
stearic acid	2167	5.0	$C_{18}H_{36}O_2$	284
oleic acid	2175	12.3	$C_{18}H_{36}O_2$	284
Percentage Total		98.9		

Table 8.2 Chemical composition of *Cordia Milleni* (Oyebanji et al. 2017).

Compounds	Retention Index	Percentage Composition	MF	MW
Tetramethylmethane	434	0.4	C_5H_{12}	72
Dimethylvinylmethanol	600	0.5	$C_5H_{10}O$	86
2-methylfuran	642	1.0	C_5H_6O	82
3-methylfuran	643	1.0	C_5H_6O	82
Trimethylnitromethane	716	0.5	$C_4H_9NO_2$	103
2-pentynal	724	1.6	C_5H_6O	82
4-ethyl-1,3-dioxolane	738	0.5	$C_5H_{10}O_2$	102
vinyl butyrate	775	0.5	$C_6H_{10}O_2$	114
1-hydroxy-2-butanone	798	1.0	$C_4H_8O_2$	88
cis-3-octene	823	1.0	C_8H_{16}	112
trans-3-octene	824	2.0	C_8H_{16}	112
2-methoxymethyltetrahydrofuran	825	0.5	$C_6H_{12}O$	116
2,4-pentadienoic acid	873	1.0	$C_5H_6O_2$	98
trans-2-furanmethanol	885	3.5	$C_5H_6O_2$	98
5-methyl-3-methylene-5-hexen-2-one	887	2.0	$C_8H_{12}O$	124
tetrahydro-2-furanmethanol	892	0.5	$C_5H_{10}O_2$	
Corylone	972	1.0	$C_6H_8O_2$	112
Butylglyoxylate	973	0.5	$C_6H_{10}O_3$	130
3-hydroxy-6-methylpyridazine	975	1.0	$C_5H_6N_2O$	110
3-methyl-1,2-cyclopentanedione	1003	1.0	C_6H_8O	112
methyl-1-cyclohexenyl ketone	1027	6.0	$C_8H_{12}O$	124
methyl 2-butyl-2-cyclopropene-1-carboxylate	1057	1.0	$C_9H_{14}O_2$	154
o-guaiacol	1090	2.0	$C_7H_8O_2$	124
pelargic acid	1272	2.0	$C_9H_{18}O_2$	158
Syringol	1279	1.0	$C_5H_8N_2$	96
2,4-dimethoxyphenol	1279	1.0	$C_8H_{10}O_3$	154
3-nonynoic acid	1290	1.0	$C_9H_{14}O_2$	154
1-methyl-3-nitro-2(1 hour)-pyridinone	1376	1.0	$C_6H_6N_2O_3$	154
cis-10-pentadecen-1-ol	1763	11.0	$C_{15}H_{30}O$	226
n-pentadecanoic acid	1869	2.0	$C_{15}H_{30}O_2$	242
palmitic acid	1968	6.2	$C_{16}H_{32}O_2$	256
9-octadecenal	2007	11.0	$C_{18}H_{34}O$	266
trans-2-octadecadecen-1-ol	2061	11.0	$C_{18}H_{36}O$	268
oleic acid	2175	22.0	$C_{18}H_{34}O_2$	282
Percentage Total		99.2		

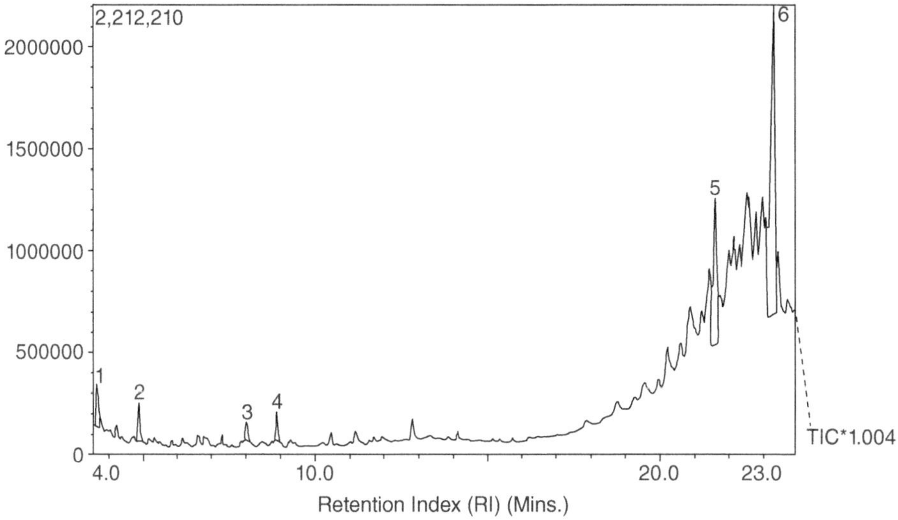

Figure 8.5 GC-MS chromatogram for bio-oil from Africana birch. Labeled peaks are: (1) Cis-10-pentadecen-1-ol, 9-octadecenal, trans-2-octadecen-1-ol,cis-1,9-hexadecadiene; (2) stearic acid, myristic acid; (3) 2,5-dimethylfuran; (4) luprosic; (5) oleic acid; and (6) palmitic acid (Oyebanji et al. 2017).

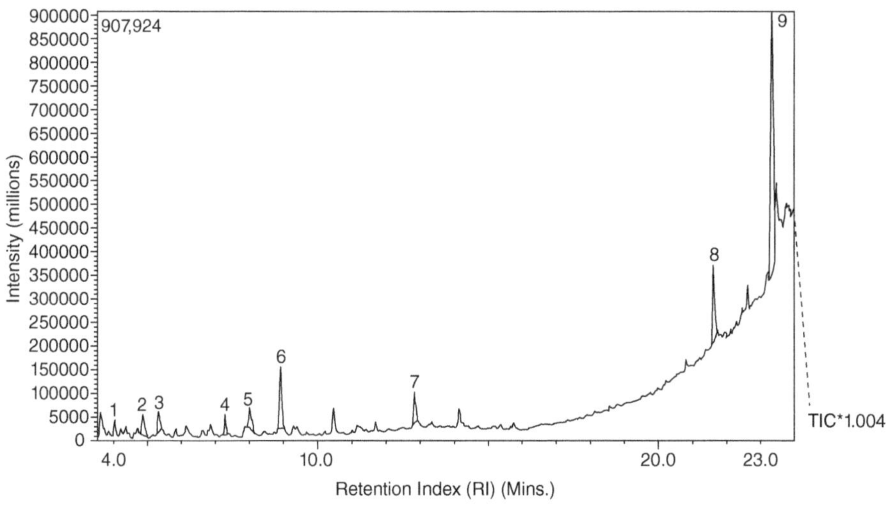

Figure 8.6 GC-MS chromatogram for bio-oil from West African cordia. Labeled peaks are: (1) tetramethylmethane; (2) tetrahydro-2- furanmetheanol; (3) 5-methl-3-methylene-5-hexen-2-2one; (4) 2, 4dimethoxyphenol; (5) trans-2-furanmethanol; (6)palmitic acid; (7) Methyl-1-cyclohexenyl ketone; (8) trans-2-octadecadecen-1-ol,9-octadecenal,cis-10-pentadencen-1-ol; and (9) oleic acid (Oyebanji et al. 2017).

8.8 Economic and Environmental Benefits of Biofuel

8.8.1 Economics Benefits

Feedstock value addition, increased income taxes opportunities, low GHG emissions and a reduced country's reliance on crude oil imports are the major economic benefit of biofuel industries in these present times. Biofuels substituting for fossil fuels has the potential to reduce dependence on unstable foreign suppliers of fossil fuel, exhaustible resource depletion, conventional and GHG emissions, and also increase farm income (Czernik and Bridgwater 2004).

Waste biomass, algae, and non-food crops are the sources of cellulose to produce 2nd- and 3rd-generation biofuels, or cellulosic biofuels. Biofuels from both 2nd- and 3rd-generation biofuels can be produced domestically to lower fossil fuel imports and reduce adverse impacts of supply disruptions (US Department of Energy 2013; Huang et al. 2013).

8.8.2 Environmental Benefits of Biofuel

The biofuel alternative is viable to provide a net energy gain to enhance environmental satisfaction and economic competitiveness without food supplies quantities reduction. Sufficient environmental advantages to merit subsidies can be harnessed through biofuel production and energy generation, if produced from low-input biomass and residue grown on agricultural marginal land or from waste biomass (US Environmental Protection Agency 2010). Globally, land degradation, climate change and loss of biodiversity have both positive and negative effects on biomass-derived energy today. The extensive and increasing areas of degraded lands provide an opportunity for woody biomass species to be used economically and rehabilitated and sustainable establishment of multi-purpose bioenergy plantations returns land to productive use. Rehabilitation incentives require adequate and functional waste management strategies and policies to ensure future modes of land use sustainability. The sustainable production of biomass fuels and the use of present agricultural and forestry-derived residues play an important role in reducing the need for GHG and acid rain emitting fossil fuels (He and Lin 2019).

8.9 Conclusion and Recommendations

Considerable attention in recent years has been paid to plant biomass waste conversion, which in contrast to the fossil sources is continuously renewed in nature. Prospective realization and assessment of biofuels production from sawdust has been carried out in this research. The outcome of this study revealed availability of feedstock for biofuels production, most especially in developing countries where biomass wastes constitutes environmental pollutants. Moreover, among the recent biomass waste-to-energy (WTE) conversion technologies, the fast pyrolysis technique was experimentally investigated to convert sawdust samples to bio-oils. Both GC-MS analysis and chemical and physical composition of biofuel yields were investigated. Results of the case study experimentally evaluated and showed that the maximum yield of bio-oil was obtained at pyrolysis temperature in the range of 500 to 550°C and bio-oil energy content produced from fast pyrolysis was in the range of 64.35 to 66.63 MJ kg^{-1}, which can be combusted in an existing heating application system or as a mixture with other fuels. The dehydration of bio-oil can be used to improve the quality, increasing the heating value and reducing the water content of the bio-oil. Bio-oil GC-MS analysis discovered 22 compounds containing majorly aromatic hydrocarbons (benzene, toluene, indene, methylindene, naphthalene and methylnaphthalene). Considering a future perspective with most probably rising crude oil prices, continued research on the production of bio-oil from biomass wastes and its utilization both as a combustible/fuel and source of valuable chemical products seems highly advisable. It is recommended that government policy on development of waste-to-energy technology should be favorable. This is because, presently, the government policy in developing countries does not favor the development of WTE conversion technologies.

References

Adegoke, O.A., Fuwape, J.A., and Fabiyi, J.S. (2014). Combustion properties of some tropical wood species and their pyrolytic products characterization. *Energy Power* 4 (3): 54–57.

Ahmad, A.A., Zawawi, N.A., Kasim, F.H. et al. (2016). Assessing the gasification performance of biomass: A review on biomass gasification process conditions, optimization and economic evaluation. *Renewable and Sustainable Energy Reviews* 53: 1333–1347.

Ahmadi, A., Esmaeilion, F., Esmaeilion, A. et al. (2020). Benefits and limitations of waste-to-energy conversion in Iran. *Renewable Energy Research and Application* 1 (1): 27–45.

Akorede, M.F., Ibrahim, O., Amuda, S.A. et al. (2017). Current status and outlook of renewable energy development in Nigeria. *Nigerian Journal of Technology* 36 (1): 196–212.

Allen, M.J., Dai, X., Shimizu, B. et al. (2000). Attainment of the theoretical yield of carbon from biomass. *Industrial & Engineering Chemistry Research* 39 (11): 4024–4031.

American Standard of Testing and Materials. ASTM D5865-04. (2004). *Standard Test Method for Gross Calorific Value of Coal and Coke*. West Conshohocken (PA): ASTM International.

American Standard of Testing and Materials. ASTM D1102-84. (2007). *Standard Test Method for Ash in Wood*. West Conshohocken, PA: ASTM International.

Anupama, R., Niju, S., Meera Sheriffa Begum, K.M. et al. (2015). Agro-residues as fuel and as a feedstock for other products. In: *Advances in Bioprocess Technology* (ed. P. Ravindra), 247–256. Switzerland: Springer International Publishing.

Arancon, R.A.D., Lin, C.S.K., Chan, K.M. et al. (2013). Advances on waste valorisation: New horizons for a more sustainable society. *Energy Science and Engineering* 1 (2): 53–71.

Ben-Iwo, J., Vasilije, M., and Philip, L. (2016). Biomass resources and biofuels potential for the production of transportation fuels in Nigeria. *Renewable and Sustainable Energy Reviews* 63: 172–192.

Bentsen, N.S., Felby, C., and Thorsen, B.J. (2014). Agricultural residue production and potentials for energy and materials services. *Progress in Energy and Combustion Science* 40 (1): 59–73.

Bhatia, S.C. (2014). *Advanced Renewable Energy Systems (Part 1)*, 473–488. New Delhi: Woodhead Publishing India Pvt. Ltd.
Brennan, L. and Owende, P. (2010). Biofuels from microalga: A review of technologies for production, processing, and extractions of biofuels and co-products. *Renewable and Sustainable Energy Review* 14: 557–577.
Bridgwater, A.V. (2012). Review of fast pyrolysis of biomass and product upgrading. *Biomass Bioenergy* 38: 68–94.
Brownsort, P.A. (2009) Biomass pyrolysis processes: Review of scope, control and variability. UK Bioresearch Centre, UKBRC Working paper 5.
Cannon, J.S. and Sperling, D. (2009). *Reducing Climate Impacts in the Transportation Sector*. Dordrecht: Springer Netherlands. doi:10.1007/978-1-4020-6979-6
Cantrell, K.B., Ducey, T., Ro, K.S. et al. (2008). Livestock waste-to-bioenergy generation opportunities. *Bioresource Technology* 99: 7941–7953.
Chiaramonti, D., Prussi, M., Buffi, M. et al. (2015). Thermochemical conversion of microalgae: Challenges and opportunities. *Energy Procedia* 75: 819–826.
Colesca, S.E. and Ciocoiu, C.N. (2013). An overview of the Romanian renewable energy sector. *Renewable and Sustainable Energy Reviews* 24: 149–158.
Czernik, S. and Bridgwater, A.V. (2004). Overview of applications of biomass fast pyrolysis oil. *Energy Fuel* 18 (2): 590–59.
David, K. and Ragauskas, A.J. (2010). Switchgrass as an energy crop for biofuel production: A review of its ligno-cellulosic chemical properties. *Energy Environment and Science* 3: 1182–1190.
Dhyani, V. and Bhaskar, T. (2018). A comprehensive review on the pyrolysis of lignocellulosic biomass. *Renewable Energy* 129: 695–716.
Fodor, Z. and Klemes, J.J. (2012). Waste as alternative fuel: Minimising emissions and effluents by advanced design. *Process Safety and Environmental Protection* 90: 263–284.
Goyal, H.B., Seal, D., and Saxena, R.C. (2008). Bio-fuels from thermochemical conversion of renewable resources: A review. *Renewable and Sustainable Energy Reviews* 12: 504–517.
Gumbo, T. (2014). Scaling up sustainable renewable energy generation from municipal solid waste in the African Continent: Lessons from eThekwini, South Africa. *The Journal of Sustainable Development* 12 (1): 46–62.
Gumisiriza, R., Hawumba, J.F., Okure, M. et al. (2017). Biomass waste-to-energy valorisation technologies: A review case for banana processing in Uganda. *Biotechnol Biofuels* 10 (11): 1–29.
He, J. and Lin, B. (2019). Assessment of waste incineration power with considerations of subsidies and emissions in China. *Energy Policy* 126: 190–199.
Hoornweg, D. and Bhada-Tata, P. (2012). What a waste: A global review of solid waste management. In: *Urban Development Series Knowledge Papers* (ed. World Bank), 8–12. Washington DC: World Bank.
Hossain, A., Hasan, R., and Islam, R. (2014). Design, fabrication and performance study of a biomass solid waste pyrolysis system for alternative liquid fuel production. *Global Journal of Researches in Engineering: A Mechanical and Mechanics Engineering* 14 (5): 25–34.
Huang, H., Khanna, M., Onal, H. et al. (2013). Stacking low carbon policies on the renewable fuels standard: Economic and greenhouse gas implications. *Energy Policy* 56: 5–15.

IEA. (2007). *Bioenergy: Potential Contribution of Bioenergy to the World's Future Energy Demand*. Paris: OECD/IEA.

IRENA. (2018). *Renewable Capacity Statistics 2018, Edited by International Renewable Energy Agency*. Abu Dhabi: IRENA.

IRENA. (2019). *Renewable Capacity Statistics 2018, Edited by International Renewable Energy Agency*. Abu Dhabi: IRENA.

Iwasaki, T., Suzuki, S., and Kojima, T. (2014). Influence of biomass pyrolysis temperature, heating rate and type of biomass on produced char in a fluidized bed reactor. *Energy and Environment Research* 4 (2): 64–72.

Jahirul, M.I., Rasul, M.G., Chowdhhury, A.A. et al. (2012). Biofuel production through biomass pyrolysis. A technological review. *Energies* 5 (12): 4952–5001.

Jameel, H., Keshwani, D.R., Carter, S.F. et al. (2010). Thermochemical conversion of biomass to power and fuels. In: *Biomass to Renewable Energy Processes* (ed. J. Cheng), 438–488. New York: Taylor & Francis Group.

Kato, Y., Enomoto, R., Akazawa, M., and Kojima, Y. (2016). Characterization of Japanese cedar bio-oil produced using a bench-scale auger pyrolyzer. *Springer Plus* 5 (177): 1–11.

Kothari, R., Tyagi, V.V., and Pathak, A. (2010). Waste-to-energy: A way from renewable energy sources to sustainable development. *Renewable and Sustainable Energy Reviews* 14: 3164–3170.

Kumar, A., Singh, J., Baskar, C. et al. (2015). Bioenergy: Biofuels process technology. In: *Advances in Bioprocess Technology* (ed. P. Ravindra), 165–207. Switzerland: Springer International Publishing.

Laura, L., Rosella, L., Carmen, M. et al. (2020). Waste-to-energy in the EU: The effects of plant ownership, waste mobility, and decentralization on environmental outcomes and welfare. *Sustainability* 12: 5743.

Llanes, J.L. and Kalogirou, E. (2019). Waste-to-energy conversion in Havana: Technical and economic analysis. *Social Science* 8 (119): 1–17.

Lee, S.Y., Sankaran, R., Chew, K.W. et al. (2019). Waste to bioenergy: A review on the recent conversion technologies. *BMC Energy* 1: 4–7.

Li, M., Luo, N., and Lu, Y. (2017). Biomass energy technological paradigm (BETP): Trends in this sector. *Sustainability* 9 (567): 1–28.

Li, R., Liu, D., Zhang, Y. et al. (2019). Improved methane production and energy recovery of post-hydrothermal liquefaction waste water via integration of zeolite adsorption and anaerobic digestion. *Science of the Total Environment* 651: 61–69.

Lombardi, L. and Corti, A. (2009). Energy recovery from waste: Comparison of different technology combinations. In: *Energy Recovery* (ed. E. Dubois and A. Mercier), 213–228. New York: Nova Science Publishers, Inc.

Massarutto, A. (2015). Economic aspects of thermal treatment of solid waste in a sustainable WM system. *Waste Management* 37: 45–57.

Mazzoni, L. and Janajreh, I. (2017) Plasma gasification of municipal solid waste with variable content of plastic solid waste for enhanced energy recovery. Proceedings of the 2016 International Renewable and Sustainable Energy Conference IRSEC, 42, 907–912.

Messerle, V.E., Mosse, A.L., and Ustimenko, A.B. (2018). Processing of biomedical waste in plasma gasifier. *Waste Management* 79: 791–799.

Nemerow, N. (2009). *Environmental Health and Safety for Municipal Infrastructure, Land Use and Planning, and Industry*, 6th ed., Hoboken, NJ: Wiley.

Ng, W.P.Q., Lam, H.L., Varbanov, P.S. et al. (2014). Waste-to-energy (WTE) network synthesis for municipal solid waste (MSW). *Energy Conversion Management* 85: 866–874.

Nordi, G.H., Reynaldo, P.B., Antonio, G.G. et al. (2017). Electricity production from municipal solid waste in Brazil. *Waste Management & Research* 35: 709–720.

Ogunjobi, J.K. and Lajide, L. (2013). Characterisation of bio-oil and bio-char from slow pyrolysed Nigerian yellow and white corn cobs. *Journal of Sustainable Energy and Environment* 4: 77–84.

Ogunsina, B.S., Ojolo, S.J., and Ohunakin, O.S. (2014). Potentials for generating alternative fuels from empty palm bunches by pyrolysis. *JOMAR* 8: 50–60.

Ohunakin, S.O. (2010). Energy utilisation and renewable energy sources in Nigeria. *Journal of Engineering and Applied Sciences* 5 (2): 171–177.

Okoroigwe, E., Li, Z., Struecken, T., Saffron, C. et al. (2012). Pyrolysis of *Gmelina arborea* wood for Bio-oil/Bio-char production: Physical and chemical characterization of products. *Journal of Applied Sciences* 12 (4): 369–374.

Ololade, Z.S., Fakankun, O.A., Alao, F.O. et al. (2014). Phytochemical and therapeutic studies of the fruit essential oil of *Thuja orientalis* from Nigeria. *Global Journal of Science Frontier Research* 14 (7): 15–20.

Ouda, O. (2013). Assessment of the environmental values of waste-to-energy in the Gaza Strip. *Current World and Environment* 8 (3): 335–364.

Ouda, O.K.M. and Cekirge, H.M. (2014). Potential environmental values of waste-to-energy facilities in Saudi Arabia. *Arabic Journal of Science and Engineering* 39 (2): 7525–7533.

Oyebanji, J.A. and Oyedepo, S.O. (2018). Compositional and experimental investigation of the effect of reactor temperature on softwood and hardwood pyrolysis. *World Journal of Engineering* 15 (1): 21–26.

Oyebanji, J.A., Okekunle, P.O., Lasode, O.A. et al. (2017). Chemical composition of bio-oils produced by fast pyrolysis of two energy biomass. *Biofuels* 9 (4): 1–9. doi:10.1080/17597269.2017.1284473

Oyebanji, J.A., Oyedepo, S.O., Okekunle, P.O. et al. (2018). Product yield distribution and essential oil composition of Eucalyptus Terminalis Sawdust Pyrolysis. *IOP Conference Series: Material Science and Engineering* 391: 012008.

Oyedepo, S.O. (2012). On energy for sustainable development in Nigeria. *Renewable Sustainable Energy Review* 16: 2583–2598.

Oyedepo, S.O., Dunmade, I.S., Adekeye, T. et al. (2019). Bioenergy technology development in Nigeria: Pathway to sustainable energy development. *International Journal of Environment and Sustainable Development* 18 (2): 175–205.

Panel, S.M., Saurabh, J.S., Satinder, K.B. et al. (2016). Agro-industrial wastes as feedstock for sustainable bio-production of butanol by *Clostridium beijerinckii*. *Food and Bioproducts Processing* 98: 217–226.

Punia, R., Marlar, R.R., Kuma, R.S. et al. (2017). Potential of bioelectricity production in India through thermochemical conversion of lignocellulosic biomass. In: *Sustainable Biofuels Development in India*, 189–206. Cham: Springer International Publishing.

Qin Ng, W.P., Lam, H.L., Varbanov, P.S. et al. (2014). Waste-to-Energy (WTE) network synthesis for municipal solid waste (MSW). *Energy Conversion and Management* 85: 866–874.

Rahim, M.A., Babu, T., and Rana, S. (2013). Possibility study of pyrolysis oil produced from pine seeds as an alternative of fossil oil and its comparison with pyrolysis oil produced from other sources. *Global Journal of Researches in Engineering. A: Mechanical and Mechanics Engineering* 13 (10): 43–48.

Ryu, C. (2010). Potential of municipal solid waste for renewable energy production and reduction of greenhouse gas emissions in South Korea. *Journal of Air and Waste Management and Association* 60: 176–183.

Searchinger, T., Heimlich, R., Houghton, R.A. et al. (2008). Use of US croplands for biofuels increases greenhouse gases through emissions from land-use change. *Science* 319: 1238–1240.

Sialve, B., Bernet, N., and Bernard, O. (2009). Anaerobic digestion of microalgae as a necessary step to make microalgal biodiesel sustainable. *Biotechnology Advances* 27: 409–416.

Soudham, V.P. (2015) Biochemical conversion of biomass to biofuels: Pretreatment–Detoxification–Hydrolysis–Fermentation, PhD Thesis, Department of Chemistry Umeå University, Sweden, pp: 17–38.

Sukhbaatar, B., Steele, P.H., Ingram, L.L. et al. (2009). An exploratory study on the removal of acetic and formic acids from bio-oil. *BioResources* 4 (4): 1319–1329.

Tabasova, A., Kropac, J., Kermes, V. et al. (2012). Waste-to-energy technologies: Impact on environment. *Energy* 44: 146–155.

Tan, S.T., Ho, W.S., Hashim, H. et al. (2015). Energy, economic and environmental (3E) analysis of waste-to-energy (WTE) strategies for municipal solid waste (MSW) management in Malaysia. *Energy Conversions and Management* 102: 111–120.

Thakkar, M., Mohanty, P., Shah, M. et al. (2018). An overview of biomass gasification. In: *Recent Advancements in Biofuels and Bioenergy Utilization* (ed. P.K. Sarangi, S. Nanda, and P. Mohanty), 147–176. Springer Nature Singapore Pte Ltd.

US Department of Energy (DOE) Alternative Fuels Data Center. Ethanol Fueling StationLocations. http://www.afdc.energy.gov/fuels/ethanol_locations.html (accessed 10 September 2013).

US Environmental Protection Agency (2010) Renewable Fuel Standard Program (RFS2) Regulatory Impact Analysis. (accessed 10 September 2013).

Watson, J., Zhang, Y., Si, B. et al. (2018). Gasification of biowaste: A critical review and outlooks. *Renewable and Sustainable Energy Reviews* 83: 1–17.

Young, G.C. (2010). *Municipal Solid Waste to Energy Conversion Processes: Economic, Technical, and Renewable Comparisons*, 1–10. Hoboken, NJ and Canada: John Wiley & Sons, Inc.

Zhang, W., Yu, E., Rozelle, S. et al. (2013). The impact of biofuel growth on agriculture: Why is the range of estimates so wide? *Food Policy* 38: 227–239.

Zhang, S., Yang, X., Zhang, H. et al. (2019). Liquefaction of biomass and upgrading of bio-oil: A review. *Molecules* 24 (2250): 1–30.

9

Biogas Production and Processing from Various Organic Wastes in Anaerobic Digesters and Landfills

Setareh Heidari[1], *David A. Wood*[2], *Birendra K. Rajan*[3] *and Ahmad Fauzi Ismail*[4]

[1] *Faculty of Chemical Engineering, Sahand University of Technology, Tabriz, Iran*
[2] *DWA Energy Limited, Bassingham, Lincoln, LN5 9JP, United Kingdom*
[3] *Environmental Science & Engineering Department, IIT Bombay*
[4] *Advanced Membrane Technology Research Center (AMTEC), Universiti Teknologi Malaysia 81310, Skudai, Johor Bahru, Johor, Malaysia*

CHAPTER MENU

9.1 Introduction

Generically, the term "biomass" is used to cover all biologically-generated compounds. Biomass is broadly exploited, providing about fifty exajoules (50 EJ) of energy worldwide, constituting some 10% of energy currently consumed globally (Clarke and Trinnaman 2010). A broad range of novel technologies and their applications, introduced in recent years, play a substantial role in improving standards of living and access to energy and resources such as: enhanced oil recovery (Sadatshojaei et al. 2016, 2019; Choubineh et al. 2017; Dehshibi et al. 2019; Nezhad et al. 2021); fossil energy technologies (Sadatshojaei et al. 2021; Bahrami et al. 2020; Khajehesamedini et al. 2018); corrosion (Seddiqi et al. 2019); biofuels (Sadatshojaei

Bioresource Technology: Concept, Tools and Experiences, First Edition. Tanveer Bilal Pirzadah, Bisma Malik, Rouf Ahmad Bhat, and Khalid Rehman Hakeem.
© 2022 John Wiley & Sons Ltd. Published 2022 by John Wiley & Sons Ltd.

et al. 2020b, 2020d, 2020e, 2020f; Jafari et al. 2021); biocatalysis (Sadatshojaei et al. 2020a, 2020c; Heidari et al. 2020a, 2020d); solvents (Heidari et al. 2020b; Sadatshojaei and Wood, 2020a; 2020b); carbon dioxide (Sadatshojaei and Rahimpour 2020; Sadatshojaei et al. 2018, 2020g); and handling and utilization and wastewater treatments (Heidari et al. 2017, 2019, 2021, 2020c; Heidari and Yegani 2018; Shokri et al. 2015). The efficient transformation of biomass into useful forms of energy also contributes to eco-friendly solutions that improve living conditions. According to the World Energy Council (WEC), there remains considerable opportunity to further develop biomass resources to generate energy and manufacture useful chemicals. However, this needs to be done in eco-friendly ways that avoid deforestation or using agricultural food crops that places price pressures on cereal, sugar crops and water resources. Generating biogas from the problematic quantities of organic wastes and waste byproducts produced from many manufacturing processes offers eco-friendly processing of environmentally-damaging feedstocks. Substantial biomass is generated as waste products from agriculture and forestry, industrial and urban effluents. The waste biomass typically contains large quantities of fats, proteins, starches and/or sugars making them readily digestible by anaerobic-digestion (AD). If handled and processed appropriately, all these sources of biomass can be employed as raw materials for producing energy and other useful products. Such resources may be slurries or solids of consistent composition (e.g., farm wastes) or complex mixtures of damp solids (e.g., municipal solid waste – MSW). Many can be effectively processed using AD techniques.

9.2 Urban Waste as a Raw Material for Biogas Production

9.2.1 Independent-Source Organic Waste

Independent-source organic waste is composed of organic content amended in the domestic-waste materials that included unused or general waste produced during food preparation and consumption. It also includes various wastes generated from gardens. Such domestic wastes can vary substantially in their compositions and consistencies. This makes them difficult to handle, particularly as consumption patterns and the wastes generated from them change in space and time. The need to manage and process such waste is growing as population, consumption and urbanization continue to grow at an alarming rate. Isolating waste types, pre- or post-collection, so that they can be more easily gathered and prepared as suitable feedstocks for anaerobic digesters avoids them being processed inefficiently from an energy perspective by incineration or ending up in landfill sites (Briassoulis et al. 2019; Favoino 2002; Rutz et al. 2012). Such approaches facilitate to an extent more efficient nutrient recovery and recycling. In large urban areas there are many opportunities to enhance AD domestic waste processing embedded within communities. The utilization of domestic-waste materials as raw feedstock for biogas-production plants offers sustainable solutions because it does not compete for land usage or generate toxic byproducts (Rutz et al. 2012). However, to be appropriate for recycling as AD raw materials, organic domestic wastes do need to be segregated to establish a high-purity feedstock. Mixed waste, with compositions varying over time and various contaminant concentrations pose a substantial obstacle to the efficient operation of AD plants. The quantity of contaminants in organic wastes cannot be more than about

0.1% by weight (Amon and Boxberger 1999). Efficaciously, this is only feasible if organic wastes are gathered individually at their location of generation or usage and with care being taken not to mix wastes from different sources.

Individual gathering or isolation of suitable wastes plays a key role in effective waste-management systems (Figure 9.1). It offers high-purity feedstock of appropriate quality for optimized AD production. Gathering of mixed wastes transported to a centralized-separation plant is more costly and poses a higher risk of the separated wastes being contaminated with pollutants (Favoino 2002). Consequently, the utilization of digestate derived from centrally gathered multiply-sourced wastes is generally not suitable for use as a fertilizer or organic soil supplement for many purposes. Organic-domestic waste can be readily biodegraded to generate methane and other gases and the microorganisms are well suited to carry out the process in an efficient manner (Hwang and Kim 2019; Zhang et al. 2007). However, domestic waste does contain some pathogenic materials, including bacteria, viruses, fungi and other microbes that act as hindrances in the AD process and thus need to be pre-treated to enhance the efficiency and reduce the cost. Favoino (2002) demonstrated that, indeed the costs and expenses associated with discrete gathering of waste are comparable with those of gathering mixed waste. This is typically the case, even when the gathering logistics are optimally implemented and/or if waste components are gathered individually as part of an integrated-waste-management plan to a dedicated AD system. Additionally, a considerable source of waste supply for many biogas-generation plants comes from deliveries paid for a consignment weight basis, making it difficult for them to check and guarantee the quality of each delivery. Purity remains a key confining variable when trying to rely upon source-independent wastes for biogas feedstocks. The amount of foreign compounds present within such supplies tends to be governed by human actions rather than technological factors (Favoino 2002), rendering multi-supplier gathering systems often unreliable with respect to feedstock quality. Favoino (2002) also demonstrated that even when supply purity exceeded about 89%, organic-domestic waste could still contain damaging quantities of contaminants such as glass, metal, plastic and sand. These annoying materials can be eliminated by digestion in dry digestion systems or in plug flow digesters.

9.2.2 Sewage Sludge

Anaerobic-digestion (AD) is globally employed to process various grades of sewage and effluent-related sludge derived from aerobically-treated urban-waste water. Sewage-derived sludge offers the opportunity to produce methane on a commercially viable scale. On the other hand, waste-activated sludge offers less potential than untreated sludge. An activated sludge is one that has been biologically treated aerobically, often as part of water purification processes. There are a huge range of pre-treatments that can be effective in improving methane yield. These mainly involve decomposition of biomass by bacteria and microbes to produce organic materials more conducive to AD processes. These pre-treatments involve enzymatic decomposition, different types of hydrolysis and mechanical decomposition (Rulkens 2008; Fox et al. 2019). Again, the main risks associated with sewage-derived sludge used for AD feedstocks is unacceptable levels of contaminants erratically impacting the process and rendering the digestates produced

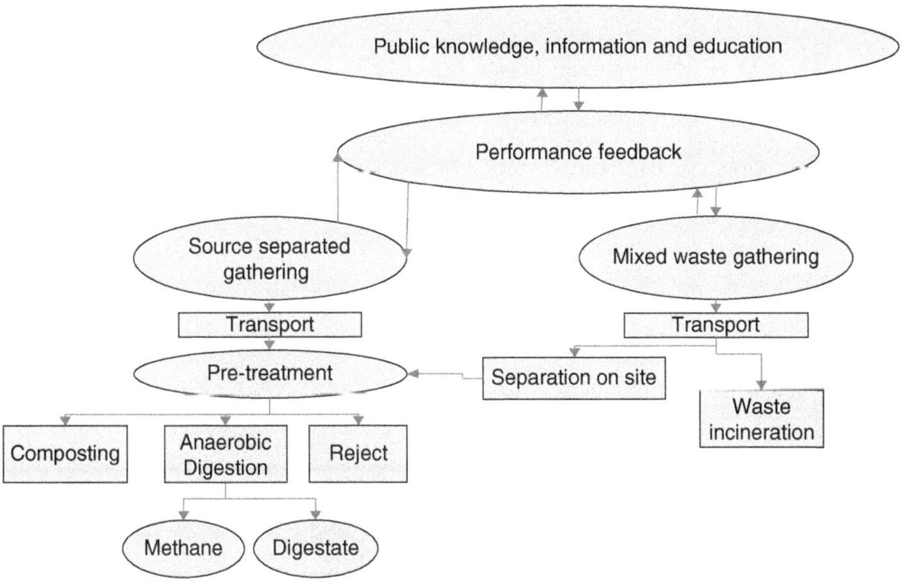

Figure 9.1 Gathering systems for anaerobically-digestible organic wastes as part of integrated-waste-management systems (*source*: modified from Biosantech et al. 2013).

unreliable fertilizers. Unfortunately, sewage sludge involves considerable amounts of chemical and biological contaminants, depending upon its source. The application of sewage sludge as raw materials for AD is consequently constrained and controlled, taking into account local regulations and quality-related standards regarding the distribution and utilization of digestates as fertilizers (Lamastra et al. 2018; Aso 2020).

On the other hand, some countries do allow the application of digested sewage sludge on agricultural land in the form of fertilizer or for other agricultural uses. It is deemed to meet the provisions that set upper limits on heavy-metal concentrations and non-biodegradable organic contaminants, while satisfying the health requirements to deactivate pathogens and other harmful biological components (Sezai et al. 2020; Smith 2009). However, concerns exist about how rigorously the regulation limits are checked for and adhered to. In some countries its utilization for such purposes is forbidden. Sewage-based sludge is routinely digested together with animal and plant agricultural wastes plus organic industrial byproducts and domestic waste. Such feedstock mixtures tend to degrade biogas-production efficiency and destabilize digester processes (Pyae et al. 2019; Kuglarz and Mrowiec 2009).

9.3 Biogas Feedstock Properties

9.3.1 Suitability and Availability

When it comes to biogas generation, suitability and availability of substrates are important. Availability refers to the accessibility and cost-off supply for biogas plant operators.

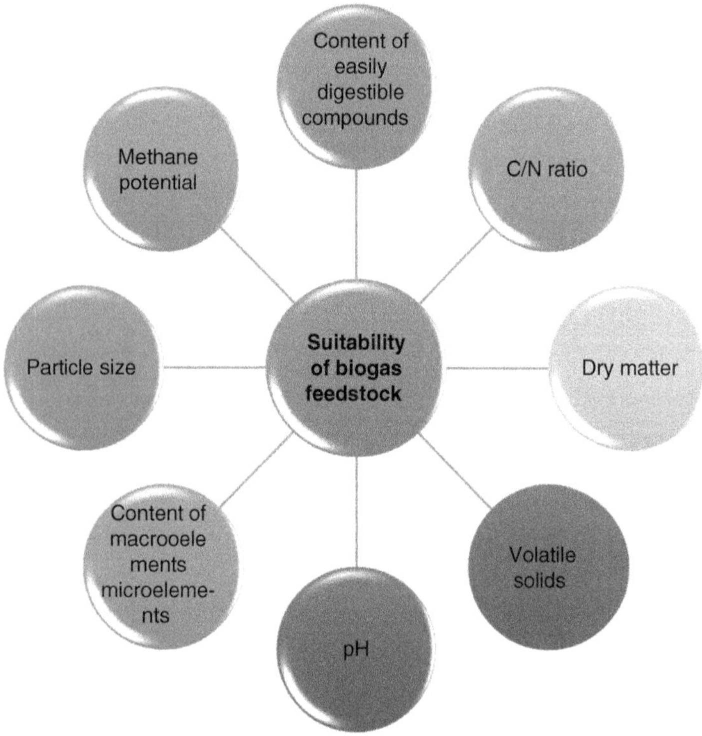

Figure 9.2 Parameters to consider when evaluating the suitability of biogas feedstocks.

Suitability reflects various compositional factors such as the proportions of digestible and non-digestible organic matters, its potential to produce methane and the expected methane yield, plus the concentrations of major and trace elements in the gaseous and digestate phases (Figure 9.2) (Wellinger et al. 2013). Key compositional factors associated with various biogas feedstocks delivered to European biogas plants are listed in Table 9.1.

Table 9.1 identifies that a wide variety of materials can be used to produce biogas. The compositional variations identified in Table 9.1 are significant as they are determining factors in the quality and efficiency of methane generation. Moreover, they do influence the overall performance and stability of AD processes. For instance, the maximum yield from AD processes tends to occur at neutral pH conditions and when the carbon/nitrogen ratio ranges from 20:1 to 30:1 (Biosantech et al. 2013). Furthermore, to improve AD performance, feedstocks typically need to be mixed with additives, sometimes derived from other feedstocks or their byproducts, such that they are able to provide the basic nutrients required by the anaerobic micro-organisms within the digesters (Angelidaki et al. 2002). The selection of the feedstocks and nutrient additives, if any, directly influences the methane yield. Figure 9.3 compares the methane yields expressed as volume of CH_4 produced (m^3 per unit volume of feedstock). Figure 9.3 highlights that methane yield from fish oil and soya oil is substantially higher than the other AD feedstocks evaluated. Both fish oil and soya oil have the capability of yielding more than 600 m^3/m^3 of

Table 9.1 Compositional characteristics of typical feedstocks used in the biogas industries (*source*: modified from Biosantech et al. 2013).

Type	Source	Composition					C/N ratio in substrates	Dry matter (%)	Volatile solids of Dry matter (%)
		Carbohydrates	Proteins	Lipids	Lactose	Alcohol			
Animal wastes	Pig slurry	√	√	√			7	5	80
	Cattle slurry	√	√	√			13	8	80
	Cattle manure	√	√	√				20	80
	Poultry droppings	√	√	√			7	5	80
	Poultry manure	√	√	√				20	80
Plant wastes	Straw	√		√			90	90-70	90-80
	Garden wastes	√		√			125	70-60	90
	Grass	√		√			18	25-20	90
	Fruit wastes	√		√			35	20-15	75
Organic wastes obtained from industries	Flotation sludge		√	√					
	Concentrated whey		√		√		–	10	90
	Fish oil			√			–	90	90
	Soya oil						–	95	90
	Alcohol					√	–	40	95
	Bleach clay						–	98	40
	Olive pulp						–	24	96
	Brewers spent						–	20	90
	Thin grain-based silage							8.5	86

(*Continued*)

Table 9.1 (Continued)

		Composition							
Type	Source	Carbohydrates	Proteins	Lipids	Lactose	Alcohol	C/N ratio in substrates	Dry matter (%)	Volatile solids of Dry matter (%)
Sewage sludge	Waste-water sludge						–	5	75
	Concentrated wastewater sludge						–	10	75
	Food remains						–	10	80
Energy-generating crops	Grass-based silage						17	40-15	90

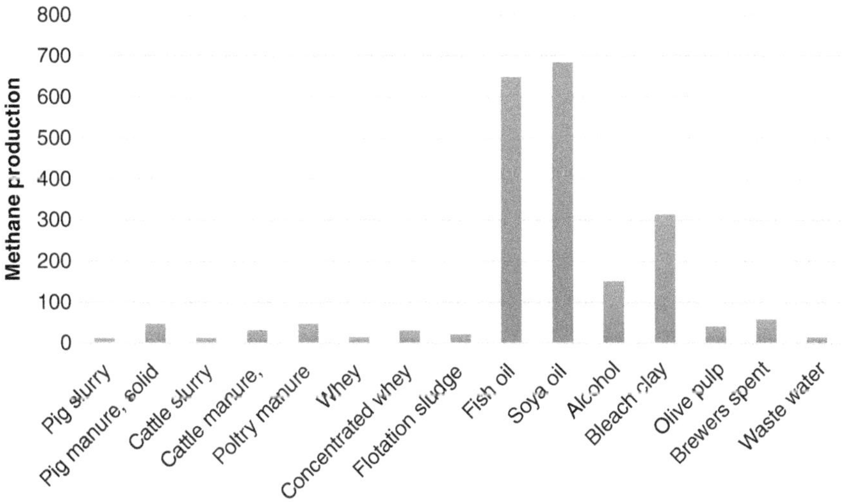

Figure 9.3 Methane yield CH_4 m^3/m^3 of some biogas feedstocks. This figure is drawn based on data reported in (Biosantech et al. 2013).

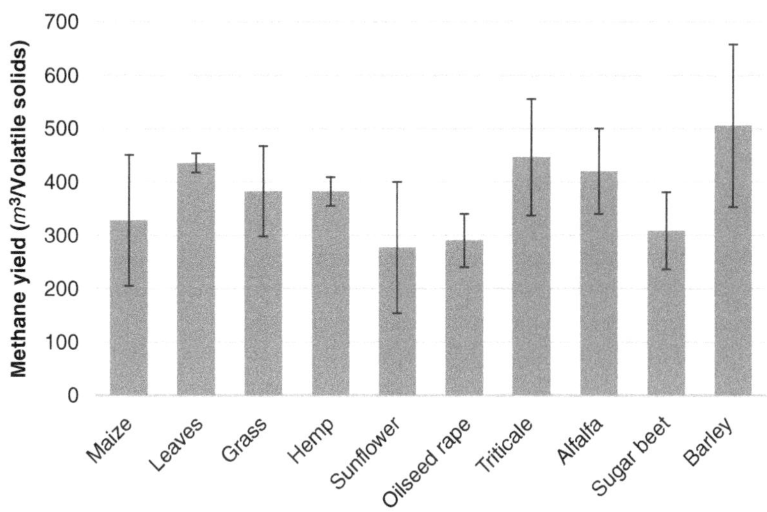

Figure 9.4 Methane yields, with ranges shown, from agricultural crops processed as AD feedstocks for methane production. This figure is drawn based on data reported in (Steffan et al. 1998).

methane compared to yields of less than 50 m^3/m^3 from animal wastes and byproducts as feedstock (Figure 9.4). For this reason, fish oil and soya feedstock are often used as methane production boosters (Biosantech et al. 2013).

9.3.2 Digestibility

Digestibility is the major factor in AD of feedstocks, having a direct influence on the methane yield. The higher the digestibility, the higher the rate of decomposition of the substrate during anaerobic digestion and rate of methane production. On the other hand, the higher the amount of recalcitrant matter, e.g., lignocellulose contained in a feedstock, the lower its digestibility and slower the rate of methane production (Steffen et al. 1998). The anaerobic degradation rate varies substantially depending on feedstock type and its composition (Figure 9.4). The reaction time, i.e., the time required for the decomposition of feedstock, is also influenced by the composition, quality and condition of the feedstock introduced into a digester. The estimated decomposition time for feedstock including proteins, lipids and hemicelluloses is considerably higher than those of low molecular weight carbohydrates, alcohol and volatile fatty acids (Murphy et al. 2011; Rutz and Janssen 2013). In terms of feedstock containing substantial oils and fats, the reaction rates can be much slower; perhaps several weeks would be required for their full conversion (Rutz and Janssen 2013). For economic consideration, those feedstocks with lower digester-retention times, that is faster decomposition rates, are favored.

9.3.3 Impurities with Digester-Disrupting Effects

Many digestible feedstocks, unfortunately, contain some undesirable impurities that have the potential to disturb the normal AD conversion procedures. These can result in digester instability and process interruptions. They tend to gradually reduce active-digestion anaerobic processes within the digesters due to physical or chemical effects. For instance, the build-up of foams or floating undigestible debris, such as straw and woody materials, can act to inhibit the anaerobic processes (Biosantech et al. 2013; Steffan et al. 1998). The presence of metals and metallic compounds as well as sand commonly found in supplied feedstocks may also harm and gradually abrade operating equipment such as pumps and filters. Table 9.2 summarizes the typical disrupting processes encountered during AD operations associated with specific feedstocks. When such contamination-related disruptions occur in digesters, it requires substantial efforts and incremental operating and maintenance cost to control and stabilize the anaerobic digestion processes. Pre-treatments of feedstocks in order to achieve efficient, trouble-free anaerobic digestion is, therefore, a common strategy adopted (Wellinger et al. 2013).

9.3.4 Feedstocks Acting as AD Biogas Boosters

For economic reasons, the addition of high-quality methane boosters, either in small intermittently added doses or continuously as specified fractions of the main feedstock form an integral part of biogas plant operations. Their addition is designed to improve the potential methane yield and/or rate of production. The most used methane-yield boosters in the biogas industry include: fatty compounds like fish oils, obtained from different fish industries; alcohol residues, generally derived from the brewing industry; glycerin, which is a byproduct of biodiesel trans-esterification processes; residues from the sugar processing and drinks manufacturing industries; and food oils like margarine and soya bean oil

Table 9.2 Feedstocks and their disturbing effects during anaerobic digestion procedure (Wellinger et al. 2013; Steffen et al. 1998).

Composition	Feedstock source	Disturbing impacts
Fats	Slaughterhouses Rendering plants	Foaming
Volatile fatty acids (VFA)	Rendering plants Oil mills	Poor water solubility
Organic Pollutants (Antibiotics, Pesticides Detergents)	Animal manures Organic wastes Crop and crops residues	Foaming
Carbohydrates (Starch, Sugars, Cellulose)	Agro-industries Animal manures Crop residues	Foaming Lignin Encrustation
Proteins	Dairy and Pharmaceutical industries	Foaming

(Wellinger et al. 2013). The addition of excessive fractions of biogas boosters can limit achieving high methane yield or fast reaction rates (Biosantech et al. 2013; Wellinger et al. 2013). It is therefore necessary to carefully establish upper limits for methane-booster additives to AD processes and this requires rigorous laboratory and pilot-scale plant testing.

9.4 Biogas Production Technology Applied to Landfills

Biogas production from landfills is an alternate approach to burning, composting or anaerobically digesting in purpose-built equipment. Landfill materials are diverse and tend to be dominated by municipal solid wastes (MSWs) and are converted slowly and progressively to methane under prevailing anaerobic conditions in the near-sub-surface. MSW consists of various and to some extent unpredictable organic materials interspersed unevenly with diverse inorganic components. A typical poorly-regulated landfill site consists of a mélange of food wastes, rotting vegetation and garden wastes, paper and plastic packing materials, wood and other waste building materials, non-biodegradable plastics and polymers, glass, metals and clothing and other natural and man-made fibers. In some countries, landfills are better regulated and materials are segregated into zones before being dumped into landfills. However, it is almost impossible to ensure single-source sectors within most landfills. Biogas production depends on multiple variables impacting landfills sites. These include the dominant compositions within the buried waste mélange, its water content and movement of ground and surface water through it, governed by its porosity, permeability and connectivity to the water table. The density of the MSW and how that varies over time and in different landfill strata, due to the changing consumption trends and the nature of materials purchased and discarded, the consumption of materials varies over the scale of decades also having an impact on the way it degrades. Additionally, variations of composition and temperature of landfill strata influence gasification rates. Anaerobic reactions driven by microorganism metabolism that generate biogas are to a large extent driven by temperature-related factors. The biogas components are dominated by CH_4, CO_2 and to a lesser extent by N_2. However, there are multiple trace and minor constituents that include ammonia, carbon monoxide, halogenated-volatile-organic compounds, hydrocarbons (C_2 to C_5 gaseous molecules), hydrogen sulfide, methyl siloxanes, oxygen and water vapor. The actual compositions of biogases generated in landfills with the mentioned constituents can vary substantially. Typical ranges of component concentrations are: methane (35–65%); carbon dioxide (15–40%); nitrogen (5–40%); water vapor (1–5%); oxygen (0–5%); hydrogen (0–3%); carbon monoxide (0–3%); halogenated hydrocarbons (20–200 ppmv Cl^-/F^-); hydrogen sulfide (0–100 ppmv); ammonia (0–5 ppmv); volatile organic compound (0–4500 mg/m^3); and siloxanes (0–50 mg Si m^{-3}) (Awe et al. 2017). This gaseous cocktail is typically processed to provide a biomethane product of a consistent composition: CH_4 varying from about 95 to 99%, CO_2 varying from about 1 to 5% and H_2S diminished to trace levels. Biomethane of such compositions could be blended into many natural gas pipeline streams but tends to be consumed locally in gas engines to generate electricity in plants adjacent to the landfills (Barragán-Escandón et al. 2020). Some landfill sites, particularly in urban areas, are dominated by food wastes. Food waste is all biodegradable but some types of food can be problematic for landfills and digesters, for example, meats, noodles and rice tend to form volatile acids as part of their degradation processes (Clarke and Trinnaman 2010). Extremely low-pH fluids of about pH = 3.5 are a consequence of anaerobic decay of such foodstuffs and not only destabilize digesters but they lead to acidic ground waters emanating from landfill sites rich in these acid-producing foods (Awe et al. 2017). Moreover, low-pH conditions (acidification) degrade methanogenic reactions leading to reduced methane yields and slower methane production rates (Ghosh et al. 2020). One metric ton of organic waste left to decay would emit between 50 and 110 m^3 of carbon dioxide, and between 90 and 140 m^3 of methane (Ghosh et al. 2020; Singh et al. 2020). If not recovered, that would dissipate into the atmosphere as greenhouse gases. It makes sense to either control this decay in a dedicated AD process or to gather and process the gaseous emissions from landfill sites. Table 9.3 shows the biogas content and bio-methane production potential of carbohydrate, concentrated protein,

Table 9.3 Biogas content and bio-methane potential of some raw organic components making up MSW (*source*: modified from Labatut et al. 2011).

	Biogas (Nm^3/TS (metric ton of substrate))	CH_4 (%)	CO_2 (%)
Carbohydrate	790–800	50	50
Concentrated Protein	700–790	70–71	29–30
Lignin	Nil	Nil	Nil
Concentrated Animal Fat	1200–1250	67–68	32–33

lignin and concentrated animal fat. AD consists of a complex set of biogenic reactions initiated and sustained by microbial actions that involves various forms of hydrolysis and is fundamentally associated with significant methane generation. This is true for controlled digestion or for the uncontrolled digestion that occurs in landfill sites. It is beneficial to find ways to accelerate AD reactions and to do so requires enhancing microbial metabolism and growth. By doing so, it leads to more rapid degradation in landfills of the organic components of MSW and substantial efforts are being made to identify technologies and handling strategies that can achieve this. Technology is focused not just on increasing reaction rates, but also on improving the quality of the biogas produced, particularly the quantity of methane content in the biogas.

9.4.1 Anaerobic Digester Pre-Treatments

Digestibility of wastes by microorganisms is a major issue for AD feedstocks. Therefore, pretreatments are applied to various organic wastes to remove the contaminants that tend to retard microbial reactions. This can substantially increase biogas-production quantities and reduce unwanted volatiles. Figure 9.5 illustrates that pre-treatment is one of the distinctive strategies applied to enhance biogas production. Hydrolysis reactions can act to limit biogas production in AD processors. These reactions tend to form volatile-fatty acids and in some cases toxins that have detrimental effects on AD reaction rates. Consequently, pre-treatments often aim to inhibit hydrolysis while at the same time enhancing methane production. Pre-treatment methods can be classified into several groups (Pilli et al. 2020):

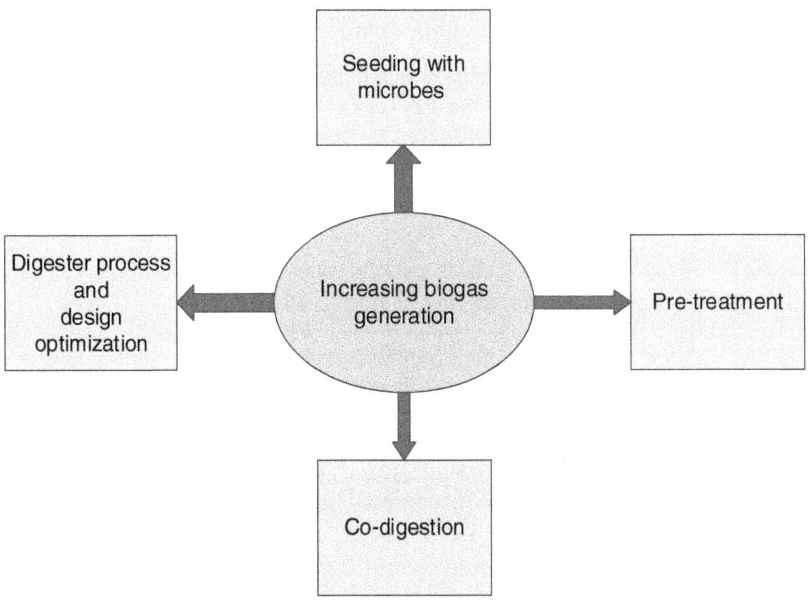

Figure 9.5 Various approaches to enhance biogas production in AD (*source*: modified from Andriani et al. 2015).

- Particle-size reduction of substrates to about two millimeters tends to optimize methane production.
- Hot-water soaking tends to increase the solubility of cellulose- and lignin-based materials, thereby restricting the development of aldehydes (e.g., furfural) that tend to inhibit some AD reactions. The energy cost to heat the water can be prohibitive.
- Acid soaking can enhance the outcomes of hot-water soaking, but incrementally increases operating costs and has the associated risks of corrosion and producing detrimental byproducts.
- Alkali soaking can accelerate the breakdown of paper-related waste, but the concentrations of sodium hydroxide used need to be selected carefully to avoid saponification-type reactions occurring and negatively impacting digester stability.
- Pre-heating substrates with additives before introducing them to the digester can improve methane-producing reaction rates and for some additives tailored to suit the substrate compositions, both acids and alkalis can improve methane yields from certain substrates.
- Microwave irradiation can assist in substrate fragmentation and increase biogas yield by up to 7%.
- Ultrasonic irradiation is worthwhile to assist in the mechanical fragmentation of some substrates before they are introduced into the AD.
- Hydrolytic-type enzyme additions, particularly to food-based wastes, act to catalyze hydrolysis reactions and ease hydrolysis constraints, thereby enhancing overall methane production.

9.4.2 Digester Design and Process Optimization

There are mainly two types of digester designs: a) Single-stage systems b) Multi-stage systems. Studies show that the ability to enhance biogas production in single-stage systems is limited. The limitations are due mainly to the nutrient requirements of active microbes being constrained, making it difficult to adjust their rates of growth and digester reaction rates. However, adopting more costly and complex, two-stage AD systems can accelerate reactions and increase AD methane yield (Markphan et al. 2020). The first stage of such systems focuses on one or more of the pre-treatments mentioned; the second stage focuses on applying digester reactions to optimized substrates. Table 9.4 summarizes the pros and cons of these systems (Figure 9.5).

9.4.3 Hydrolysis Enhancements

Sludge and slurries have different biogas yields in AD, depending upon their degree of activations. Primary sludges are accumulated during sedimentation due to the forces of gravity and are mechanically formed slurries comprising mainly wet solid and organic-rich particles and fragments of various sizes. On the other hand, secondary sludges are distinct in that active microbes have been introduced to consume some of the organic matter. Secondary sludges are commonly derived from settlement tanks in water and sewage treatment plants. A wide range of enzymes act as catalysts to accelerate

Table 9.4 Advantages and disadvantages of single-stage AD systems and multi-stage AD systems (Moedinger et al. 2012; Massanet-Nicolau et al. 2013).

	Single-Stage AD Systems	Multi-Stage AD Systems
Advantages	Low cost Easy maintenance	Higher biogas production Increased process stability High organic loading rate Offers process optimization Low retention time
Disadvantages	Low biogas production Low organic loading rate High retention time Low process stability Process optimization problem	Expensive Maintenance complex Skilled persons required to operate

hydrolysis reactions in secondary sludges; some of these are naturally occurring, while others are selected and added as pretreatments to suit the slurry compositions, for example, various types of lipase work well with fat-rich substrates (Ferreira-Dias et al. 2013). Tongco et al. (2020) applied various fractions of lipase plus protease to hydrolyze of primary sludge to assess its impact on biogas production. They found that such treatments increased methane production by nearly 90% by increasing reaction rates and enhance CH_4 yields by almost 10%. Lignin contents in MSW tend to slow down and otherwise inhibit AD and various methods had been adopted in attempts to improve the digestion of lignin-containing wastes. Rahimi et al. (2019) found termite-gut symbionts to be highly effective in the rapid breakdown of lignin. The bacterium *Opitutaceae*, strain TAV5 that can metabolize at low oxygen levels (<2%), was shown to be particularly effective in breaking down lignin in conditions not dissimilar from those encountered in landfills (Rahimi 2019).

9.4.4 Bacterial Clean-up of AD Digester Effluent

Most AD effluent cannot be used as fertilizer or otherwise released to the environment because of its inappropriate composition. It tends to be too rich in carbon, nitrogen and/or phosphorus with problematic carbon to nitrogen (C:N) ratios and high chemical-oxygen demand (COD) (Foresti et al. 2006). Various biological processes have been applied to adjust concentration of these nutrients in AD effluent. Some of these involve additional AD stages in a series of several digesters arranged in series. For instance, anaerobic sludge from one vessel is mixed with further waste material (e.g., a mixture of food-based, paper-based, and crop-based waste) in the next digester in the sequence. For instance, the Bardenpho process developed in the 1970s involves four stages of digestion (Foresti et al. 2006). Bio-electrochemical systems exploiting direct electron transfer techniques have also been tested in the laboratory to treat AD effluent by converting some of the carbon dioxide produced into methane (Liu et al. 2019). However, so far such systems have proved difficult to scale up. Bacterial community analysis, based on their 16S rRNA and 18S rRNA gene levels, was performed by Xie et al. (2020), with the objective of effectively producing biodiesel from AD effluent using micro-algae. *Acinetobacter* sp. and/or *Citrobacter* sp. were effective at breaking down the organic material and reducing nitrogen and phosphorus levels, whereas *Brevundimonas* sp. improved algal metabolism and biomass growth, essential for maximizing lipid and biodiesel yields in photo-biotic reactors. *Chlorella* sp. made up more than 99% of the bacterial community present in the photo-biotic reactor tests evaluated, and that acted to stabilize the bacterial community. The results suggest that AD effluent could be used to effectively generate biodiesel using micro-algal processes.

9.4.5 Additives to Enhance Methane Yield

Rasapoor et al. (2020) identified three groups of additives which can be exploited in large-scale AD plants to improve the methane content in the biogas generated. These include major-element adjustment, trace-element balancing and direct-electron transfer. Major element adjustment is primarily focused on additives that adjust the C:N ratio and the nitrogen and phosphorus nutrient levels. Co-digestion of appropriate mixture of carbohydrates, lipids and proteins in the substrates, combined with enzyme-based additives are effective in major element adjustments. Traces of iron, nickel and/or molybdenum have proved effective in creating an appropriate nutrient balance in AD systems (Choong et al. 2016). Some methanogens are able to directly accept electrons from an electrode or from specific microbes. Liu et al. (2019) showed a microbial electrolysis cell could convert about 8% of its carbon dioxide into methane by exploiting direct electron transfer technique.

9.4.6 Biogas Upgrading Technologies

Biogas produced from AD and landfill sites once generated needs to be conditioned and upgraded to make it suitable for customers. Water/moisture reduction is an essential process to ensure that water does not drop out during transportation or storage. For most uses, scrubbers are required to remove hydrogen sulfide and carbon dioxide. Heavier hydrocarbons may be removed by chilling, whereas separation of nitrogen from methane requires cryogenic-separation techniques (Stewart and Arnold 2011; Carreon 2018). Adsorption and membranes may be required to remove certain contaminants, such as metal ions and aqueous salts. Upgrading of the methane content of the raw biogas produced can be conducted, for example carbon dioxide conversion to methane as described by Nelabhotia and Dinamarca (2019).

9.4.6.1 Carbon Dioxide Removal Technologies

It is typically necessary to remove substantial amounts of carbon dioxide from the biogas. This is necessary to meet gas quality specifications with respect to energy content, calorific values and/or Wobbe-Index quality measures. Various methods are applied including chemical adsorption in a scrubber facility causing the carbon dioxide to form a liquid absorbent; water and organic solvents (e.g., methanol or glycols) are commonly used for this purpose (Table 9.5).

Table 9.5 Comparison of carbon dioxide removal technologies applied biogas streams (Wei 2016; Rasapoor et al. 2020).

Technology	Disadvantages	Advantages
Absorption technology employing water scrubbing	Costly 0.4–0.5 kWh/m^3 of gas Bacterial growth Feasibility of foaming	Higher efficiency (>97% methane) Hydrogen sulfide is eliminated simultaneously (when H_2S < 300 cm^3/m^3) Regulatable temperature or pressure Low methane losses Reproduction feasible
Adsorption technology employing organic solvent-polyethylene glycol	Costly 0.21 kWh/m^3 of gas Not easy to operate	Higher efficiency (>97% methane) Removal of H_2S, NH_3, HCN, and water Energetic more favorable than water Low methane losses regeneration possible with low-temperature waste heat
Chemical absorption employing Amines	Costly Breaking down and poisoning of the amines by oxygen or other chemicals	Higher efficiency (>99% methane) Cheap operation, regenerative, Carbon dioxide dissolved per unit of volume compared with water Very low methane losses (<0.1%) Power cost: 0.05–0.25 kWhm^{-3} of gas
PSA/VSA (Carbon Molecular sieves, Zeolite Molecular sieves Alumina silicates)	Costly Extensive procedure control is needed Methane losses during malfunctioning of valves	H_2S and water are eliminated Construction and operation is easy Highly reliable
Membrane technology employing gas or liquid	Purity of methane and amount of biogas upgradation is compromised Multiple stages needed	Gas/gas elimination efficiency >96% methane (1 step) or >96% methane, water is eliminated Gas/liquid removal efficiency >96% methane, cheap investment and operation, Pure carbon dioxide can be generated
Cryogenic separation	Costly Carbon dioxide stay in methane High energy needed to cool	90–98% CH_4 can be attained carbon dioxide and methane in high purity Low extra energy cost to reach liquid biomethane (LBM)
Biological elimination	Addition of hydrogen gases, still on experimental basis (not on large-scale)	Hydrogen sulfide and carbon dioxide is eliminated
Air/Oxygen doing digester slurry		Enriched methane
Biofilter/ Biotrickling filter		No unwanted end products

9.4.7 Hydrogen Sulfide and Ammonia Removal

Several solvent-based scrubbing techniques are employed to remove hydrogen sulfide from biogas streams. The actual technique applied is often determined by the concentration of H_2S in the biogas; the higher the concentration, the more costly it is to remove. For low concentration, simple chemical scavengers can be dosed into the biogas streams to remove H_2S. It is difficult to remove all the sulfur-based compounds present, which include mercaptans and ammonia in a single scrubber unit. This is due to sulfur compounds being more easily extracted in alkaline conditions, whereas ammonia, and other nitrogen-based compounds, requires acidic conditions (Okoro and Sun 2019; Ghyselbrecht et al. 2018). Chemical costs to operate such scrubbers to remove large quantities of hydrogen sulfide and ammonia from biogas can be high and negatively impact commercial returns.

9.4.8 Siloxane Removal

Siloxanes are a problem in biogas because when they are combusted they are converted into silicon dioxide (i.e., sand or glass) which is highly abrasive and can damage engines, turbines, fuel cells and their component parts and thus should be removed (Nyamukamba et al. 2020). Siloxanes are a type of silicone composed of silicon-oxygen bonds. They are used in consumer products such as detergents and cosmetics but do not decompose in activated-sludge processes applied in water purification processes. However, in AD. as temperatures reach 50°C to 60°C, siloxanes incorporated into substrate sludge tend to vaporize and enter the raw biogas streams (Beauchemin 2020). A number of absorbents are used to remove siloxanes from biogas. These include activated carbon, activated alumina, synthetic resins and silica gels. Chilling and membrane technologies can also be used to remove siloxanes. The problem with absorbents is that their absorption capacity degrades as they are reused. This means that high operating costs tend to be associated with replacing absorbents in siloxane-removal plants.

9.5 Conclusions

Biogas production in AD and landfills is associated with a set of proven technologies. Biogas compositions can vary substantially due to the wide range of wastes that provide feedstocks for AD or end up in landfills. Various techniques and materials are available that can potentially boost biogas production, including fish oil and brewers' wastes. These are often added to AD feedstocks or introduced as pre-treatment stages to improve methane productivity. Once produced, the biogas typically needs to be processed to remove contaminants and often upgraded to improve its methane content as per the international ASTM standard. Biogas upgrading technologies involve a range of additives to adjust major- and trace element compositions. They also include physical processing to adjust the AD substrate composition by blending feedstocks to achieve specific carbohydrate, protein and lipid fractions. Also, substrate particle size and texture can be optimized using microwave and ultrasonic treatments making it mode susceptibility to microbial actions. Additions of enzymes and certain bacteria that facilitate direct electron transfer are also able to improve the efficiency of AD processes. The raw biogas produced contains a range of contaminants that need to be removed, particularly carbon dioxide, ammonia, hydrogen sulfide and siloxanes.

References

Amon, T. and Boxberger, J. (1999) Organic wastes for co-digestion in agricultural biogas plants: Guidelines and legal conditions in Austria. *Proceedings of the IEA Workshop: Hygienic and Environmental Aspects of Anaerobic Digestion: Legislation and Experiences in Europe.*

Andriani, D., Wresta, A., Saepudin, A. et al. (2015). A review of recycling of human excreta to energy through biogas generation: Indonesia case. *Energy Procedia* 68 (April): 219–225.

Angelidaki, I., Ellegaard, L., Sørensen, A.H. et al. (2002). *Environmental Biotechnology.* Denmark: Department of Environment and Resources, DTU.

Aso, S.N. (2020). Digestate: The co-product of biofuel production in a circular economy, and new results for cassava peeling residue digestate. In: *Renewable Energy Technologies and Applications.* (ed. T. Taner, A. Tiwari and T.S. Ustun, IntechOpen. doi:10.5772/intechopen.91340.

Awe, O.W., Zhao, Y., Nzihou, A. et al. (2017). A review of biogas utilisation, purification and upgrading technologies. *Waste and Biomass Valorization* 8 (2): 267–283.

Bahrami, B., Sadatshojaie, A., and Wood, D.A. (2020). Assessing wellbore stability with a modified lade failure criterion. *Journal of Energy Resources Technology* 142 (8): 083004 (12 pages).

Barragán-Escandón, A., Ruiz, J.M.O., Tigre, J.D.C. et al. (2020). Assessment of power generation using biogas from landfills in an equatorial tropical context. *Sustainability* 12 (7): 2669.

Beauchemin, D. (2020). *Sample Introduction Systems in ICPMS and ICPOES.* Newnes.

Biosantech, T.A.S., Rutz, D., Janssen, R., and Drosg, B. (2013). Biomass resources for biogas production. In: *The Biogas Handbook*, 19–51. Elsevier.

Briassoulis, D., Pikasi, A., and Hiskakis, M. (2019). End-of-waste life: Inventory of alternative end-of-use recirculation routes of bio-based plastics in the European Union context. *Critical Reviews in Environmental Science and Technology* 49 (20): 1835–1892.

Carreon, M.A. (2018). Molecular sieve membranes for N_2/CH_4 separation. *Journal of Materials Research* 33 (1): 32–43.

Choong, Y.Y., Norli, I., Abdullah, A.Z. et al. (2016). Impacts of trace element supplementation on the performance of anaerobic digestion process: A critical review. *Bioresource Technology* 209: 369–379.

Choubineh, A., Ghorbani, H., Wood, D.A. et al. (2017). Improved predictions of wellhead choke liquid critical-flow rates: Modelling based on hybrid neural network training learning based optimization. *Fuel* 207: 547–560.

Clarke, A. and Trinnaman, J. (2010). *Survey of Energy Resources.* United Kingdom: World Energy Council.

Dehshibi, R., Sadatshojaie, A., Mohebbi, A. et al. (2019). A new insight into pore body filling mechanism during waterflooding in a glass micro-model. *Chemical Engineering Research & Design: Transactions of the Institution of Chemical Engineers Part A* 151.

Favoino, E. (2002). Drivers for separate collection in the EU, optimisation and cost assessment of high capture schemes. *Proceedings of EC Conference on Biological Treatment of Biodegradable Waste*, Brussels.

Ferreira-Dias, S., Sandoval, G., Plou, F.J. et al. (2013). The potential use of lipases in the production of fatty acid derivatives for the food and nutraceutical industries. *Electronic Journal of Biotechnology* 16 (3): 12–42.

Foresti, E., Zaiat, M., and Vallero, M. (2006). Anaerobic processes as the core technology for sustainable domestic wastewater treatment: Consolidated applications, new trends, perspectives, and challenges. *Reviews in Environmental Science and Bio-Technology* 5 (1): 3–19.

Fox, J.T., Zook, A.N., Freiss, J. et al. (2019). Thermal conversion of blended food production waste and municipal sewage sludge to recoverable products. *Journal of Cleaner Production* 220: 57–64.

Ghosh, P., Shah, G., Sahota, S. et al. (2020). Biogas production from waste: Technical overview, progress, and challenges. In: *Bioreactors*, 89–104. Elsevier.

Ghyselbrecht, K., Monballiu, A., Somers, M.H. et al. (2018). Stripping and scrubbing of ammonium using common fractionating columns to prove ammonium inhibition during anaerobic digestion. *International Journal of Energy and Environmental Engineering* 9 (4): 447–455.

Heidari, S. and Yegani, R. (2018) Experimental study on fouling characteristics of HDPE and HDPE/PE-g-MA/EVA blend membranes in microfiltration of Collagen: Effects of surface morphology and physicochemical membrane features. In: *Proceedings of the 10th International Chemical Engineering Congress & Exhibition (IChEC)*. Isfahan University, Iran.

Heidari, S., Jafarzadeh, Y., Samarin, M.S. et al. (2017). Study on the fouling behavior of HDPE/PE-g-MA/EVA blend membrane fabricated via thermally induced phase separation method. *Polyolefins Journal* 4 (2): 235–251.

Heidari, S., Amirinejad, M., and Jahangirian, H. (2019). Investigation of fouling mechanisms using surface morphology and physicochemical membrane features. *Chemical Engineering & Technology* 42 (6): 1310–1320.

Heidari, S., Sadatshojaei, E., and Wood, D.A. (2020a). *Ionic Liquids and Their Beneficial Contributions to Enzyme-catalyzed Reactions, Catalytic Biomass Conversion and Energy Conversion and Storage Systems*. Elsevier.

Heidari, S., Sadatshojaei, E., and Wood, D.A. (2020b). Large-scale molecular solvents for environmentally sustainable applications. In: *Green Sustainable Process for Chemical and Environmental Engineering and Science: Environmental Remediation*. Elsevier.

Heidari, S., Sadatshojaei, E., Wood, D.A. et al. (2020c). *Algae-based Membrane Bioreactor for Wastewater Treatment*. Switzerland AG: Springer-Nature.

Heidari, S., Sadatshojaei, E., Foroozesh, J. et al. (2020d). *Application of Supercritical Fluid for Biocatalysis with an Emphasis on Supercritical Water*. Elsevier.

Heidari, S., Etemadi, H., and Yegani, R. (2021). A comprehensive analysis of membrane fouling in microfiltration of complex linear macromolecules based on theoretical modeling and FESEM images. *Chemical Technology and Biotechnology*. 96 (2): 360–373. https://doi.org/10.1002/jctb.6548.

Hwang, K. and Kim, D. (2019). Comparison of anaerobic digestion efficiency with different temperature of food wastes. *Journal of Korean Society on Water Environment* 35 (4): 332–339.

Jafari, A., Esmaeilzadeh, F., Mowla, D. et al. (2021). New insights to direct conversion of wet microalgae impregnated with ethanol to biodiesel exploiting extraction with supercritical carbon dioxide. *Fuel* 285 (1): 119199.

Khajehesamedini, A., Sadatshojaie, A., Parvasi, P. et al. (2018). Experimental and theoretical study of crude oil pretreatment using low-frequency ultrasonic waves. *Ultrasonics Sonochemistry* 48: 383–395.

Kuglarz, M. and Mrowiec, B. (2009). *Co-digestion of Municipal Biowaste and Sewage Sludge for Biogas Production*. University of Bielsko-Biala.

Labatut, R.A., Angenent, L.T., and Scott, N.R. (2011). Biochemical methane potential and biodegradability of complex organic substrates. *Bioresource Technology* 102 (3): 2255–2264.

Lamastra, L., Suciu, N.A., and Trevisan, M. (2018). Sewage sludge for sustainable agriculture: Contaminants' contents and potential use as fertilizer. *Chemical and Biological Technologies in Agriculture* 5 (1): 10.

Liu, C., Sun, D., Zhao, Z., Dang, Y. et al. (2019). Methanothrix enhances biogas upgrading in microbial electrolysis cell via direct electron transfer. *Bioresource Technology* 291: 121877.

Markphan, W., Mamimin, C., Suksong, W. et al. (2020). Comparative assessment of single-stage and two-stage anaerobic digestion for biogas production from high moisture municipal solid waste. *Peer Journal* 8. doi:10.7717/peerj.9693.

Massanet-Nicolau, J., Dinsdale, R., Guwy, A. et al. (2013). Use of real time gas production data for more accurate comparison of continuous single-stage and two-stage fermentation. *Bioresource Technology* 129: 561–567.

Moedinger, F., Ast, F., Ragazzi, M. et al. (2012). Innovative biogas multi-stage biogas plant and novel analytical system: First project experiences. *Energy Procedia* 18: 672–680.

Murphy, J., Braun, R., Weiland, P. et al. (2011). Biogas from crop digestion. In: *IEA Bioenergy Task*, 1–23.

Nelabhotla, A.B.T. and Dinamarca, C. (2019). Bioelectrochemical CO_2 reduction to methane: MES integration in biogas production processes. *Applied Sciences* 9 (6): 1056.

Nezhad, M.S., Wood, D.A., Sadatshojaei, E. et al. (2021). New insight to experimental study of ionic solutions with a nonionic surfactant on wettability, interfacial tension and micro-model flooding. *Fuel* 285 (1): 119126.

Nyamukamba, P., Mukumba, P., Chikukwa, E.S. et al. (2020). Biogas upgrading approaches with special focus on siloxane removal: A review. *Energies* 13 (22): 6088.

Okoro, O.V. and Sun, Z. (2019). Desulphurisation of biogas: A systematic qualitative and economic-based quantitative review of alternative strategies. *Chemical Engineering* 3 (3): 76.

Pilli, S., Pandey, A.K., Katiyar, A. et al. (2020). Pre-treatment technologies to enhance anaerobic digestion. In: *Sustainable Sewage Sludge Management and Resource Efficiency*. doi:10.5772/intechopen.93236.

Pyae, H.A., Aye, W.W., Yossapol, C. et al. (2019). Comparative study on the performance of iron-amended cassava pulp feed bio-methanation in CSTRs. *Engineering and Applied Science Research* 46 (3): 219–226.

Rahimi, H. (2019) Effect of termite gut microorganism Tav5 ratios on lignin degradation and methane generation from paper, yard, and wood waste. PhD Thesis, University of Texas Arlington, USA. http://hdl.handle.net/10106/28141

Rasapoor, M., Young, B., Brar, R. et al. (2020). Recognizing the challenges of anaerobic digestion: Critical steps toward improving biogas generation. *Fuel* 261: 116497.

Rulkens, W. (2008). Sewage sludge as a biomass resource for the production of energy: Overview and assessment of the various options. *Energy & Fuels* 22 (1): 9–15.

Rutz, D. and Janssen, R. (2013). Biomass resources for biogas production. In: *The Biogas Handbook: Science, Production and Applications*, 19. doi: 10.1533/9780857097415.1.19

Rutz, D., Janssen, R., Ramanauskaite, R. et al. (2012) The use of bio-waste for biomethane production in European cities. *Proceedings of the 20th European Biomass Conference and Exhibition*, pp. 1481–1490; ISBN 978-88-89407-54-7. doi:10.5071/20thEUBCE2012-3CO.2.2

Sadatshojaei, E., Esmaeilzadeh, F., Fathikaljahi, J. et al. (2018). Regeneration of the midrex reformer catalysts using supercritical carbon dioxide. *Chemical Engineering Journal* 343: 748–758.

Sadatshojaei, E. and Wood, D.A. (2020a). Water, the most accessible green solvent, extraction and separation agent. In: *Green Sustainable Process for Chemical and Environmental Engineering and Science: Environmental Remediation*. Elsevier.

Sadatshojaei, E. and Wood, D.A. (2020b). Applications of supercritical fluids in environmental remediation. In: *Green Sustainable Process for Chemical and Environmental Engineering and Science: Environmental Remediation*. Elsevier.

Sadatshojaie, A. and Rahimpour, M.R. (2020). CO_2 emission and air pollution (volatile organic compounds, etc.)–related problems causing climate change. In: *Current Trends and Future Developments on (Bio-) Membranes*, 1–30. Elsevier.

Sadatshojaei, E., Jamialahmadi, M., Esmaeilzadeh, F. et al. (2016). Effects of low-salinity water coupled with silica nanoparticles on wettability alteration of dolomite at reservoir temperature. *Petroleum Science and Technology* 34 (15): 1345–1351.

Sadatshojaei, E., Jamialahmadi, M., Esmaeilzadeh, F. et al. (2020). The impacts of silica nanoparticles coupled with low-salinity water on wettability and interfacial tension: Experiments on a carbonate core. *Journal of Dispersion Science and Technology* 41 (8): 1–15.

Sadatshojaei, E., Heidari, S., and Haghniaz, R. (2020a). *The Potential Use of Supercritical CO_2 as a Sustainable Solvent in Biocatalytic Reactions*. Switzerland AG: Springer-Nature.

Sadatshojaei, E., Heidari, S., and Wood, D.A. (2020b). Biodiesel production methods and feedstocks. In: *Biodiesel Technology and Applications*. USA: Wiley-Scrivener.

Sadatshojaei, E., Heidari, S., and Wood, D.A. (2020c). *Sustainable Approach in Biocatalytic Preparation of Antibiotic Peptide*. Elsevier.

Sadatshojaei, E., Mowla, D., and Wood, D.A. (2020d). Review of progress in microalgal biotechnology applied to wastewater treatment. In: *Sustainable Green Chemical Process and Their Allied Applications*. Switzerland AG: Springer Nature.

Sadatshojaei, E., Wood, D.A., and Mowla, D. (2020e). Third generation of biofuels exploiting microalgae. In: *Sustainable Green Chemical Process and Their Allied Applications*. Switzerland AG: Springer Nature.

Sadatshojaei, E., Wood, D.A., and Rahimpour, M.R. (2020f). Progress towards the sustainable generation of biofuels exploiting microalgae. In: *Biodiesel Technology and Applications*. USA: Wiley-Scrivener.

Sadatshojae, E., Wood, D.A., and Rahimpour, M.R. (2020g). Potential and challenges of carbon sequestration in soils. In: *Applied Soil Chemistry*. USA: Wiley-Scrivener.

Sadatshojaie, A., Wood, D.A., Jokar, S.M. et al. (2021). Applying ultrasonic fields to separate water contained in medium-gravity emulsions and determining emulsion adhesion coefficients, 2020. *Ultrasonics Sonochemistry* 70: 105303.

Seddiqi, H., Sadatshojaie, A., Vaferi, B. et al. (2019). Mathematical model for iron corrosion that eliminates chemical potential parameters. *Chinese Journal of Chemical Engineering* 28: 603–612.

Sezai, D., Voronina, L., Morachevskaya, E. et al. (2020). Use of sewage sludge in agricultural soils: Useful or harmful. *Eurasian Journal of Soil Science* 9 (2): 126–139.

Shokri, E., Yegani, R., Heidari, S. et al. (2015). Effect of PE-g-MA compatibilizer on the structure and performance of HDPE/EVA blend membranes fabricated via TIPS method. *Chemical Engineering Research & Design* 100: 237–247.

Singh, L., Yousuf, A., and Mahapatra, D.M. (2020). *Bioreactors: Sustainable Design and Industrial Applications in Mitigation of GHG Emissions*. Elsevier. ISBN-13 978-0128212646.

Smith, S. (2009). Organic contaminants in sewage sludge (biosolids) and their significance for agricultural recycling. *Philosophical Transactions of the Royal Society A: Mathematical, Physical and Engineering Sciences* 367 (1904): 4005–4041.

Steffen, R., Szolar, O., and Braun, R. (1998). *Feedstocks for Anaerobic Digestion*. Vienna: Institute of Agrobiotechnology Tulin, University of Agricultural Sciences.

Stewart, M. and Arnold, K. (2011). *Gas Sweetening and Processing Field Manual*. Gulf Professional Publishing.

Tongco, J.V., Kim, S., Oh, B-R. et al. (2020). Enhancement of hydrolysis and biogas production of primary sludge by use of mixtures of protease and lipase. *Biotechnology and Bioprocess Engineering* 25: 132–140. doi: 10.1007/s12257-019-0302-4

Wei, S. (2016). The application of biotechnology on the enhancing of biogas production from lignocellulosic waste. *Applied Microbiology and Biotechnology* 100 (23): 9821–9836.

Wellinger, A., Murphy, J.D., and Baxter, D. (2013). *The Biogas Handbook: Science, Production and Applications*. Elsevier.

Xie, C., Goi, C. L., Huson, D. H. et al. (2016). RiboTagger: Fast and unbiased 16S/18S profiling using whole community shotgun metagenomic or metatranscriptome surveys. *BMC Bioinformatics* 17(Suppl 19): 508. doi:10.1186/s12859-016-1378-x

Zhang, R., El-Mashad, H.M., Hartman, K. et al. (2007). Characterization of food waste as feedstock for anaerobic digestion. *Bioresource Technology* 98 (4): 929–935.

10

Extremophiles as Gold Mines for Bioprospecting

Sheikh Tanveer Salam[1], *Mukhtar Ahmad Malik*[2] *and Tanveer Bilal Pirzadah*[3,4]

[1] *Department of Zoology, Government Degree College Sopore, Baramulla, J&K, India 193201*
[2] *Department of Environmental Science, University of Kashmir, Srinagar, J&K, India 190006*
[3] *University Center for Research and Development (UCRD), Chandigarh University, Punjab, India 140413*
[4] *Department of Bioresources, Cluster University, Srinagar*

CHAPTER MENU

10.1 Introduction

Extremophiles are the organisms that have unusual potential of surviving, adapting and thriving in such hostile conditions that were previously thought to be unsuitable or lethal for life, such as waters with extreme pH, geothermal geysers, tar pits, volcanic magma and even the extreme cold of Antarctica. Moreover, the polyextremophiles and extremotolerant organisms reside in distinct ecological niches, as reported by various authors (Dumorne et al. 2017; Gundala and Chinthala et al. 2017; Singh et al. 2019b). Scientists are very curious to understand the metabolic cycles of extremophiles so as to utilize them for various potential applications in the industry because they possess high stability at extreme temperatures, they favor good solubility of the substrates, they have a high mass transfer rate and the most important being their lowering of the risk of pollution during any industrial process (Chen and Jiang 2018). Considering the complexity of reaching the extreme ecological niches and isolating these microbes, the study of extremophiles is a highly challenging field. However, recent technological advances have allowed research on extreme environments to provide valuable insights on how extremophiles can survive such challenges. Extremophiles can be named on the basis of the extreme conditions in which they grow, like a thermophilic organism having an optimum temperature of 50°C for growth, a hyperthermophilic organism having an optimum growth temperature of about 80–110°C, a halophilic organism residing in about 3–30% salinity conditions, acidophilic (optimum growth at pH 2 or below), alkaliphile

Bioresource Technology: Concept, Tools and Experiences, First Edition. Tanveer Bilal Pirzadah, Bisma Malik, Rouf Ahmad Bhat, and Khalid Rehman Hakeem.
© 2022 John Wiley & Sons Ltd. Published 2022 by John Wiley & Sons Ltd.

(optimal growth at pH > 10), psychrophilic (grown at < 15^0C or lower temperature) and a piezophile or barophile organism that lives optimally at high hydrostatic pressure ranging from 10–130 MPa. Piezophilic bacteria possess an unusually high proportion of fatty acids in their cytoplasmic membrane, which prevent its gelling at higher pressure and metallophiles that can tolerate environments with high metal concentrations (Klenk et al. 1997; Suzuki et al. 2002). Extremophiles are an unexplored and promising bioresource that possesses potential industrial applications like biopolymers, biofuels, value added products and other environmental services like bioremediation and biomining, as reported by several authors (Bhalla et al. 2013; Niehaus et al. 1999; Coker 2016; Pirzadah et al. 2018). Moreover, with the evolution of novel technologies, microbes have been largely explored for various industrial applications including food processing, biofertilizers, biopolymers, beverages, pharmaceuticals, bioremediation, biomining and biofuels (Adrio and Demain 2014, Yamada and Shimizu 1988). According to Economic Impact Analysis of the US Bio-based Products Industry sponsored by the US Department of Agriculture (USDA), in 2013 the bio-based products industry supported about 4 million jobs with a total contribution of $369 billion to the US economy (Golden et al. 2015). The data also revealed that such bio-based products are currently displacing about 300 million gallons of petroleum per year. The report "US Bio-based Products: Market Potential and Projections Through 2025" predicted that in the global chemical industry, market share of bio-based chemicals will show an exponential increase from 2% in 2008 to 22% in 2025. However, the challenges remain to grow these microorganisms in the laboratory without having knowledge of the optimum growth media and conditions. However, new technological advances have allowed research on extreme environments so as to provide valuable insights into how extremophiles can thrive in such hostile environments.

Various classes of extremophiles, their growth conditions, dwelling places and their representative examples are given in the Table 10.1

10.2 Bioprospecting of Extremophiles

Bioprospecting can be defined as the systematic exploration of biological material for the genetic and biochemical information to develop commercially valuable products for agricultural, pharmaceutical, cosmetic and other applications. In other words, it is the scientific research that looks for any useful product, application or process in nature (Jorquera et al. 2019). Bioprospecting involves systematic identification, evaluation and exploitation of the diversity of life in a certain place to seek genetic resources for commercial purposes. Extremophiles are proving to be gold mines for bioprospecting involving the rush for collecting the biological materials and the bio-prospectors are chemists, biologists and the corporations. Although it is extremely challenging for microbiologists worldwide to hunt for microbes in extreme environments, its returns can be innumerable by providing microorganisms, enzymes and biomolecules for diverse applications in biotechnology, agriculture, biomedicine and other allied industries (Table 10.2). Cid et al. (2016) reported the cold-active enzymes from Arctic and Antarctic extremophiles that are promising alternatives for biotechnologists, as they find great application in the food industry like food processing, improving milk fermentation and ice-cream production

Table 10.1 Extremophile classes, their growth conditions, dwelling places and their representative examples.

Extremophile class	Growth condition	Environment/ Source	Representative organisms
Thermophile	60–80°C	Hot spring	*Synechococcus lividus, Sulfolobus.*
Hyperthermophile	>80°C	Submarine hydrothermal vents	*Pyrolobus fumarii, Methanopyrus kandleri, Pyrococcus furiosus, Geothermobacterium ferrireducens, Aquifex aeolicus*
Halophile	2–5 M NaCl	Salt lakes, salt mines	Halobacteriaceae, *Dunaliella salina, Halanaerobacter* sp.
Acidophile	pH < 5	Acid mine drainage, volcanic springs	*Picrophilus oshimae/torridus, Stygiolobus azoricus*
Alkaliphile	pH > 9	Soda lakes	*Bacillus firmus OF4, Haloanaerobium alcaliphilum*
Psychrophile	<15°C	Ice, snow	*Psychrobacter,Methanogenium, Arthrobacter* sp., *Halomonas, Pseudomonas, Hyphomonas, Sphingomonas*, Lichens like *Umbilicaria antarctica* and *Xanthoria elegans*
Piezophile	High Pressure (10–130 MPa)	Deep ocean like Mariana Trench	*Thermococcus piezophilus, Methanopyrus kandleri, Pyrococcus* sp., *Colwellia* sp
Xerophile	Anhydrobiotic	Deserts, rocks	*Artemia salina, Deinococcus* sp., Lichens, *Methanosarcina barkeri, Desulfovibrio cavernae*
Metallophiles	High conc. of metals	Metal mines	*Ferroplasma acidarmanus* (Cu, As, Cd, Zn), *Ralstonia sp* (Zn, Co, Cd,Hg, Pb)

and storing frozen yogurt. Ureolytic bacteria from desert salt lakes could be used as new reservoirs for bio-precipitation of ion crystals like calcium and magnesium. Similarly, the core microbiome of Arctic fish is a possible tool in aquatic biotechnology to assist in improving fish health and waste management practices in aquaculture. The bioprospecting of airborne microbial communities can give clues to their application in the degradation of toxic compounds and increased tolerance of heavy metal stress in plants. Similarly, epoxide hydrolases found in the metagenome from hot springs have been proposed for use in the detoxification of xenobiotics and as a defense against pathogen attack and stress response in plants. The examples of various commercially available cold-active enzymes are depicted in Table 10.3.Bioprospecting of extremophiles for various technological applications are discussed throughout this chapter (Figure 10.1).

Table 10.2 Applications of extremophiles and extremozymes in industry.

Species	Source	Enzyme	Optimum reaction temperature (°C)	Use (potential)	References
Bacillus mojavensis SO-10	No information	α-amylase	70	Biorefinery, food, detergent	(Ozdemir et al. 2018)
Thermophilic Anoxybacillus sp. GXS-BL	Hot spring	α-amylase	60	Food, pharmaceutics, textile, detergent and bioenergy industries	(Liao et al. 2019)
Anoxybacillus thermarum FRM-RBK02	Hot spring	α-amylase	80	Biorefinery, food, detergent	(Mantiri et al. 2019)
Erwinia sp. E602	Frozen soil	Cold-adapted β-galactosidase	40	Dairy industry	(Xia et al. 2018)
Anoxybacillus flavithermus Bacillus licheniformis	Hot spring	β-galactosidase	60	Food, bioremediation, biosensor	(Rani et al. 2019)
Alteromonas sp. ML117	Marine	β-galactosidase	10	Food	(Yao et al. 2019)
Bacillus tequilensis ARMATI	Feces soil	Thermo-alkali stable xylanase	60	Biorefinery, food	(Khusro et al. 2016))
Bacillus subtilis Lucky9	No information	Alkali tolerant xylanase	60	Biofuel and food	(Chang et al. 2017)
Dictyoglomus turgidum	No information	Thermostable β-glucosidase	80	Biorefinery, food	(Fusco et al. 2018)
Exiguobacterium antarcticum B7	Antarctica soil	Cold-adapted β-glucosidase	30	Biorefinery, ethano	(Crespim et al. 2016))
Micrococcus antarcticus	Antarctica	Psychrophilic β-glucosidase	35	Detergents, textiles, bioremediation	(Miao et al. 2016)
Metagenome	Marine	Thermoactive endoglucanase	115	Biorefinery	(Suleiman et al. 2019)
Cellulomonas fimi ATCC484	Soil	Thermostable endoglucanase	65	Biorefinery	(Saxena et al. 2018)
Bacillus cereus FT 1	Soil	Alkaline protease	35	Detergents, pharmaceutical, leather, food, bioremediation	(Asha and Palaniswamy 2018)
Janibacter sp. strain R02	Antarctic soil	Thermophilic and halophilic esterase	80	Detergents, pharmaceutical, leather, food, bioremediation	(Castilla et al. 2017)

(*Continued*)

Table 10.2 (Continued)

Species	Source	Enzyme	Optimum reaction temperature (°C)	Use (potential)	References
Thermophilic Anoxybacillus sp. HBB16	Hot spring	Alkaline lipase	50	Organic synthesis, detergent, wastewater treatment, biodiesel	(Burcu Bakir and Metin, 2017)
Paenicibacillus barengoltzii	Marine	Chitinase	55	Conversion of cellulose to ethanol	(Yang et al. 2016; Yang et al. 2020)

Table 10.3 Commercially available cold-active enzymes.

Market	Enzyme	Commercially available	Uses
Molecular Biology	Alkaline phosphatases	Antarctic phosphatase (New England Biolabs Inc.)	Dephosphorylation of 5′ end of a linearized fragment of DNA
	Uracil-DNA N-glycosylases (UNGs)	Uracil-DNA N-glycosylase (UNG) (ArcticZymes), Antarctic Thermolabile UDG (New England Biolabs Inc.)	Release of free uracil from uracil-containing DNA
	Nucleases	Cryonase (Takara-Clontech)	Digestion of all types of DNA and RNA
Detergent	Lipases	Lipoclean®, Lipex®, Lipolase® Ultra, Kannase, Liquanase®, Polarzyme®, (Novozymes)	Breaking down of lipid stains
	Proteases	Purafect® Prima, Properase®, Excellase (Genencor)	Breaking down of protein stains
	Amylases	Stainzyme® Plus (Novozymes), Preferenz™ S100 (DuPont), Purafect® OxAm (Genencor)	Breakdown starch-based stains
	Cellulases	Rocksoft™ Antarctic, Antarctic LTC (Dyadic), UTA-88 and UTA-90 (Hunan Youtell Biochemical), Retrocell Recop and Retrocell ZircoN (EpyGen Biotech), Celluzyme®, Celluclean® (Novozymes)	Wash of cotton fabrics
	Mannanases	Mannaway® (Novozymes), Effectenz™ (DuPont)	Degradation of mannan or gum
	Pectate lyases	XPect® (Novozymes	Pectin-stain removal activity

(Continued)

Table 10.3 (Continued)

Market	Enzyme	Commercially available	Uses
Textile	Amylases	Optisize® COOL and Optisize NEXT (Genencor/DuPont)	Desizing of woven fabrics
	Cellulases	Primafast® GOLD HSL IndiAge® NeutraFlex, PrimaGreen® EcoLight 1 and PrimaGreen® EcoFade LT100 (Genencor/DuPont)	Bio-finishing combined with dyeing of cellulosic fabrics
Food and beverages	Pectinases	Novoshape® (Novozymes), Pectinase 62l (Biocatalysts), Lallzyme® (Lallemand)	Fermentation of beer and wine, breadmaking, and fruit juice processing
Other	Catalase	Catalase (CAT), (Swissaustral)	Textile, research, and cosmetic applications

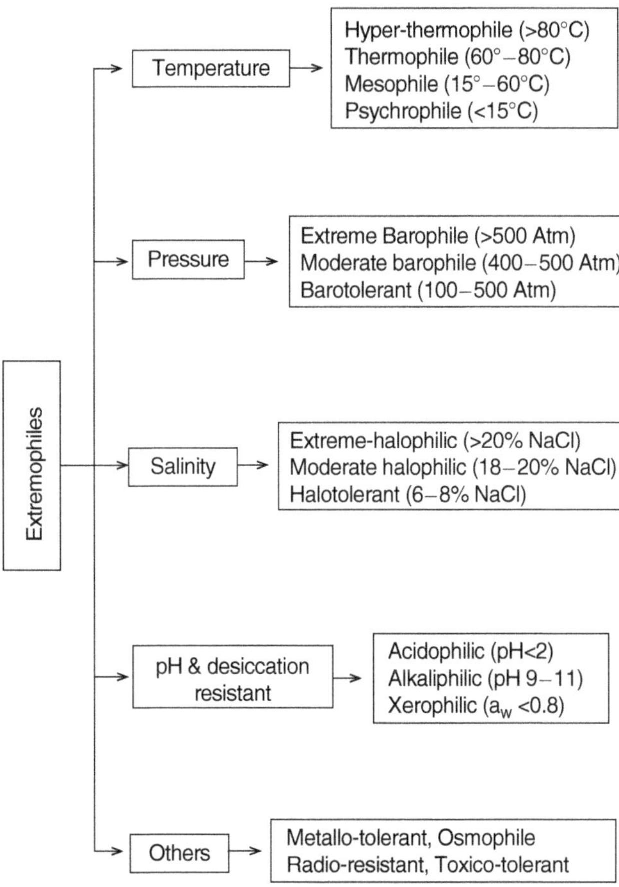

Figure 10.1 Flowchart depicting the classification of extremophiles.

10.3 Bioprospecting of Thermophiles

The thermophiles produce industrially important enzymes which are reported to be stable at higher temperatures. These include cellulases, glucosidases, gelatin hydrolyses, α-amylases, alkaline proteases, endoglucanases and cellobiohydrolases. Additionally, some thermophilic isolates are capable of using sugarcane bagasse as a carbon source to produce ethanol (Chaudhary and Qazi 2008). Some thermophilic isolates having antimicrobial properties could be further investigated as a potent source of various therapeutic compounds (Muhammad et al. 2009). Some thermophiles are even capable of depyritizing coal, oxidizing iron and sulfur (Munawar et al. 2007) and some have the potential of leaching out various metals (Ilyas et al. 2007). Thermophilic processes have been correlated with better production of biogas which increases the likelihood of utilizing pure cultures of thermophiles to enhance the biogas plant efficiency (Angelidaki and Ellegaard 2003). Thermophilic keratinases and chitinases are used in conversion of wool, feathers and chitin into feed. Lipids from hyperthermophiles are used in the formation of liposomes for drug administration. Strains of *Fishcherella* have been isolated from nuclear reactor water reservoirs at the Savannah River Site in Aiken, South Carolina, which have been found to be viable even after exposure to a high temperature of about 60°C for 2 hours or a low temperature of 15°C for 2 days. Thermostable DNA polymerase like *Taq polymerase* is a thermostable restriction enzyme obtained from thermophiles used in molecular biology, genomics and DNA amplification for polymerase chain reaction (PCR) and is considered to be the most successful application of a product which is derived from an extremophile (hyperthermophile), with annual sales of about half a billion dollars (Podar and Reysenbach 2006).

10.4 Bioprospecting of Acidophiles

The acidophilic extremophiles have tremendous economic potential in the well-established dairy industry for production of fermented dairy products. Extremophiles with probiotic characteristics and low or neutral pH stable enzymes in dairy fermentation processes may improve the quality and availability of cost-effective fermented dairy products like cheese, yogurt, etc. Some acidophilic isolates possess bioleaching capabilities, e.g., strains of *Thiobacillus thiooxidans*, *T. ferrooxidans* and *Azotobacter* have the potential of leaching out a variety of metals (Rawlings 2001). These strains play an important role in hyperaccumulation of metals, nitrogen fixation and metal sulfide ore solubilization in their natural environments like waste water and soil (landfills) (Mehmood et al. 2009). In yet another study, it was observed that acidophilic heterotrophs like *Sulfobacillus thermosulfidooxidans* have the capability of recovering metals from electronic scrap with leaching abilities of about 80% for Ni, Al, Cu and Zn (Ilyas et al. 2007). Coal bio-depyritization studies carried out with pure and mixed consortia of some of the local acidophilic isolates showed bio-depyritization rates were much higher when such microbial consortia were employed (Munawar et al. 2007). Such local isolates can therefore be potentially utilized in different bioleaching processes of various low or high-grade ores. The potentials of these acidophilic extremophiles may prove to be the potential candidates for recycling of local electronic wastes. Lipases and cellulases are used as additives in laundry and dishwashing detergents, and proteases are used for dehairing of hides and skins and to improve smoothness and dye affinity of wool (Eichler 2001; Wiegel and Kevbrin 2004; Horikoshi 1999) and also as additives in detergent industries (Kumar et al. 2011; Shukla et al. 2009), cellulase free xylanases help in bio-bleaching of pulp and paper, pectinases are used in degumming of ramie fibers and catalases are used to remove hydrogen peroxide residue from the textile processing industry effluents.

10.5 Bioprospecting of Psychrophiles

The enzymes obtained from psychrophilic bacteria have unique catalytic activity and stability. These enzymes have low activation energy which results in their high catalytic efficiency, even at lower temperatures (Podar and Reysenbach 2006). Therefore, these bacteria can serve as illustrative model systems to help us in understanding the molecular basis of adaptations to low temperature (Podar and Reysenbach 2006). Antifreeze proteins from psychrophiles can be used in freezing protection and alkaline phosphatases obtained from psychrophiles can be routinely used in molecular biology for recombinant DNA technology (Hoshino et al. 2003). Alga, namely *Mesotaenium berggrenii*, isolated from open glacier surfaces in alpine and polar regions, produce brownish vacuolar pigments and purpurogalin carboxylic acid-6-O-β-D-glucopyranoside which appears to have lot of eco-physiological relevance like analogous protective pigments (Remias et al. 2012).

10.6 Bioprospecting of Halophiles

Bacteriorhodopsin from halophiles can be used in making "Optical Switch" 500.000 Ghz. (500 × 10–15 s) and re-writeable films of high resolution (5 μm). Halophilic membranes can be used in surfactants. γ-linoleic acid, β-carotene and other pigments from halophiles are used in dietary supplements and pigments (Sarmiento et al. 2015). Haloarchea produce peptide or protein antibiotics called bacteriocins and halocins to kill or inhibit the sensitive haloarchaeal strains present in the same niche to avoid competition. *Halobacterium salinarum* facilitates the degradation of wastes during water treatment in oxidation ponds (Oren 2010). Many explorations are carried out for possible applications of halophiles in increasing crude oil extraction and insertion of genetically engineered halophilic enzyme coding genes into crops to make them salt tolerant.

10.7 Bioprospecting of Metallophiles

Bioprospecting of metallophiles involves their utilization locally in a variety of commercial bioleaching operations and a wide range of bioremediation projects. Engineered bioleaching for mining of various metals like gold, cobalt and copper is already being used in many countries like Australia, South Africa, Chile, Myanmar, Ghana, Brazil and

Peru (Brierley and Brierley 2001). Endogenous metallophiles can be employed for bioleaching of low-grade ore heaps of iron and uranium, which can boost the development of bio-hydrometallurgy in the country. On the other hand, bioremediation of various metal contaminants is of great interest among scientific communities (Xiong et al. 2008; Gadd 2010; Guo et al. 2010; Monachese et al. 2012; Rajkumar et al. 2012; Basha and Rajaganesh 2014). The Industrial Revolution has led to the contamination of both ground as well as surface waters with pesticides, metals and other toxic metals, resulting in dispensing toxic levels of Ca^{2+}, Cl^{-}, Ag^{+}, Na^{+}, K^{+}, Mg^{2+}, HCO_3^{-}, arsenic, mercury, iron, copper, lead, chromium, zinc, cadmium, cobalt and nickel (Azizullah et al. 2011). This problem has further been exacerbated by improper disposal of industrial effluents and indiscriminate use of agrochemicals. Only a small percent of these pollutants undergo proper treatment prior to their disposal (Azizullah et al. 2011). There are a number of reports that clearly reveal that these pollutants have directly or indirectly affected the health of people (Arain et al. 2009; Farooqi 2015). So there is an immediate need to develop new, ecofriendly approaches for effluent treatments to minimize the environmental contamination. Bioprocessing of the effluents can be a potential approach to address the metal contamination. Implementation of mini-scale biological waste water treatment plants aided with metallophiles in the local industries may substantially reduce the metal load of waste water effluents.

10.8 Conclusion and Future Perspective

Many products and processes that stem from bioprospecting are already abundant in areas of agriculture, pharmaceutical, environmental and industrial biotechnology, but extremophiles thriving under the adverse conditions are the highly sought after due to their wide range of applications. Bioprospecting of extremophiles is proving to be the gold mine for extraction of more novel biological products with extreme applications. A multi-interdisciplinary approach of bioprospecting, omics technology and synthetic biology has a promising role to explore extremophilic sites for the development of products of commercial importance and is advancing the field along a path of development in which designing proteins for stability and activity under seemingly impossible conditions may some day become a reality.The field of bioprospecting of extremophiles is vast and open for research to explore more biological products of great industrial, agricultural and environmental importance.

References

Adrio, J.L. and Demain, A.L. (2014). Microbial enzymes: Tools for biotechnological processes. *Biomolecules* 4 (1): 117–139. https://doi.org/10.3390/biom4010117

Angelidaki, I. and Ellegaard, L. (2003). Codigestion of manure and organic wastes in centralized biogas plants. *Applied Biochemistry and Biotechnology* 109: 95–105.

Arain, M.B., Kazi, T.G., Baig, J.A. et al. (2009). Determination of arsenic levels in lake water, sediment, and food stuff from selected area of Sindh, Pakistan: Estimation of daily dietary intake. *Food and Chemical Toxicology* 47: 242–248.

Asha, B. and Palaniswamy, M. (2018). Optimization of alkaline protease production by *Bacillus cereus* FT 1 isolated from soil. *Journal of Applied Pharmaceutical Science* 8: 119–127. doi:10.7324/JAPS.2018.8219.

Azizullah, A., Khattak, M.N., Richter, P. et al. (2011). Water pollution in Pakistan and its impact on public health: A review. *Environment International* 37: 479–497.

Basha, S.A. and Rajaganesh, K. (2014). Microbial bioremediation of heavy metals from textile industry dye effluents using isolated bacterial strains. *International Journal of Current Microbiology and Applied Sciences* 3: 785–794.

Bhalla, A., Bansal, N., Kumar, S. et al. (2013). Improved lignocellulose conversion to biofuels with thermophilic bacteria and thermostable enzymes. *Bioresource Technology* 128: 751–759.

Brierley, J.A. and Brierley, C.L. (2001). Present and future commercial applications of biohydrometallurgy. *Hydrometallurgy* 59: 233–239.

Burcu Bakir, Z. and Metin, K. (2017). Production and characterization of an alkaline lipase from thermophilic *Anoxybacillus* sp. HBB16. *Chemical & Biochemical Engineering Quarterly* 31: 303–312. doi:10.15255/CABEQ.2016.99.

Castilla, A., Panizza, P., Rodríguez, D. et al. (2017). A novel thermophilic and halophilic esterase from *Janibacter* sp. R02, the first member of a new lipase family (Family XVII). *Enzyme and Microbial Technology* 98: 86–95. doi:10.1016/j.enzmictec.2016.12.010.

Chang, S., Guo, Y., Wu, B. et al. (2017). Extracellular expression of alkali tolerant xylanase from *Bacillus subtilis* Lucky9 in *E. coli* and application for xylooligosaccharides production from agro-industrial waste. *International Journal of Biological Macromolecules* 96: 249–256. doi:10.1016/j.ijbiomac.2016.11.032.

Chaudhary, N. and Qazi, J.I. (2008). Ethanologenic potential of thermophilic bacterium and yeast from acid saccharified sugarcane bagasse. *Pakistan Journal of Zoology* 40: 99–103.

Chen, G.Q. and Jiang, X.R. (2018). Next generation industrial biotechnology based on extremophilic bacteria. *Current Opinion in Biotechnology* 50: 94–100. doi:10.1016/j.copbio.2017.11.016

Cid, F.P., Rilling, J.I., Graether, S.P. et al. (2016). Properties and biotechnological applications of ice-binding proteins in bacteria. *FEMS Microbiology Letters* 363: fnw099. doi:10.1093/femsle/fnw099.

Coker, J.A. (2016) Extremophiles and biotechnology: Current uses and prospects. F1000Research 5, F1000 Faculty Rev-396. https://doi.org/10.12688/f1000research.7432.1.

Crespim, E., Zanphorlin, L.M., De Souza, F.H. et al. (2016). A novel cold-adapted and glucose-tolerant GH1 β-glucosidase from *Exiguobacterium antarcticum* B7. *International Journal of Biological Macromolecules* 82: 375–380. doi:10.1016/j.ijbiomac.2015.09.018.

Dumorne, K., Camacho Córdova, D., Astorga-Eló, M. et al. (2017). Extremozymes: A potential source for industrial applications. *Journal of Microbiology and Biotechnology* 27: 649–659. 10.4014/jmb.1611.11006

Eichler, J. (2001). Biotechnological uses of archaeal extremozymes. *Biotechnology Advances* 19: 261–278.

Farooqi, A. (2015). Status of As and F-groundwater and soil pollution in Pakistan. In: *Arsenic and Fluoride Contamination*, pp. 21–33. India: Springer.

Fusco, F.A., Fiorentino, G., Pedone, E. et al. (2018). Biochemical characterization of a novel thermostable βglucosidase from *Dictyoglomus turgidum. International Journal of Biological Macromolecules* 113: 783–791. doi:10.1016/j.ijbiomac.2018.03.018.

Gadd, G.M. (2010). Metals, minerals and microbes: Geomicrobiology and bioremediation. *Microbiology* 156: 609–643.

Golden, J.S., Handfield, R.B., Daystar, J. et al. (2015) An economic impact analysis of the US biobased products industry. A report to the congress of the United States of America, a joint publication of the Duke Center for Sustainability & Commerce and the Supply Chain Resource Cooperative at North Carolina State University.

Guo, H., Luo, S., Chen, L., et al. (2010). Bioremediation of heavy metals by growing hyperaccumulator endophytic bacterium *Bacillus* sp. L14. *Bioresource Technology* 101: 8599–805.

Gundala, P.B. and Chinthala, P. (2017). Extremophilic pectinases. In: *Extremophilic Enzymatic Processing of Lignocellulosic Feedstocks to Bioenergy* (ed. R.K. Sani and R.N. Krishnaraj), 155–180. Cham: Springer International Publishing. doi:10.1007/978-3-319-54684-1_9

Horikoshi, K. (1999). Alkaliphiles: Some applications of their products for biotechnology. *Microbiology and Molecular Biology Reviews* 63: 735–750.

Hoshino, T., Kiriaki, M., Ohgiya, S. et al. (2003). Antifreeze proteins from snow mold fungi. *Canadian Journal of Botany* 81: 1175–1181.

Ilyas, S., Anwar, M.A., Niazi, S.B. et al. (2007). Bioleaching of metals from electronic scrap by moderately thermophilic acidophilic bacteria. *Hydrometallurgy* 88: 180–188.

Jorquera, M.A., Graether, S.P., and Maruyama, F. (2019). Editorial: Bioprospecting and biotechnology of extremophiles. *Frontiers in Bioengineering and Biotechnology* 7: 204. 10.3389/fbioe.2019.00204

Khusro, A., Kaliyan, B.K., Al-Dhabi, N.A. et al. (2016). Statistical optimization of thermo-alkali stable xylanase production from *Bacillus tequilensis* strain ARMATI. *Electronic Journal of Biotechnology* 22: 16–25. doi:10.1016/j.ejbt.2016.04.002

Klenk, H., Clayton, R., Tomb, J. et al (1997). The complete genome sequence of the hyperthermophilic, sulphate reducing archaeon *Archaeoglobus fulgidus. Nature* 390: 364–370.

Kumar, L., Aggarwal, T., Singh, B. et al. (2011). Bioprospecting for L-asparaginase producing soil bacterial isolates of Sahastradhara, Doon Valley. *Journal of Engineering and Technology Management* 3: 18–23.

Liao, S.-M., Liang, G., Zhu, J. et al. (2019). Influence of calcium ions on the thermal characteristics of α-amylase from thermophilic *Anoxybacillus* sp. GXS-BL. *Protein and Peptide Letters* 26: 148–157. 10.2174/0929866526666190116162958

Mantiri, F., Rumende, R., and Sudewi, S. (2019). Identification of α-amylase gene by PCR and activity of thermostable α-amylase from thermophilic *Anoxybacillus thermarum* isolated from Remboken hot spring in Minahasa, Indonesia. In: *IOP Conference Series: Earth and Environmental Science*. Bristol: IOP Publishing. doi:10.1088/1755-1315/217/1/012045.

Mehmood, T., Malik, S.A., and Hussain, S.T. (2009). Role of microbes in nitrogen and metalhyperaccumulation by taxilaion *Eichhorniacrassipes*. *African Journal of Microbiology Research* 3: 914–924.

Miao, -L.-L., Hou, Y.-J., Fan, H.-X. et al. (2016). Molecular structural basis for the cold adaptedness of the psychrophilic β-glucosidase BglU in *Micrococcus antarcticus*. *Applied and Environmental Microbiology* 82: 2021–2030. doi:10.1128/AEM.03158-15.

Monachese, M., Burton, J.P., and Reid, G. (2012). Bioremediation and tolerance of humans to heavy metals through microbial processes: A potential role for probiotics? *Applied and Environmental Microbiology* 18: 6397–6404.

Muhammad, S.A., Ahmad, S., and Hameed, A. (2009). Antibiotic production by thermophilic *Bacillus* species SAT-4. *Pakistan Journal of Pharmaceutical Sciences* 22: 339–345.

Munawar, N., Anwar, M.A., Saleem, M. et al. (2007). Biodepyritization of indigenous coal using acidophilic mesophilic and moderately thermophilic bacteria. *Pakistan Journal of Zoology* 39: 147–151.

Niehaus, F., Bertoldo, C., Kähler, M. et al. (1999). Extremophiles as a source of novel enzymes for industrial application. *Applied Microbiology and Biotechnology* 51 (6): 711–729.

Oren, A. (2010). Industrial and environmental applications of halophilic microorganisms. *Environmental Technology* 31 (8–9): 825–834.

Ozdemir, S., Fincan, S.A., Karakaya, A. et al. (2018). A novel raw starch hydrolyzing thermostable α-amylase produced by newly isolated *Bacillus mojavensis* SO-10: Purification, characterization and usage in starch industries. *Brazilian Archives of Biology and Technology* 61: 399. 10.1590/1678-4324-2018160399

Pirzadah, T.B., Malik, B., and Hakeem, K.R. (2018). Metagenomic analysis of uncultured microorganisms and their enzymatic attributes. *Journal of Microbiological Methods* 155: 65–69.

Podar, M. and Reysenbach, A.L. (2006). New opportunities revealed by biotechnological explorations of extremophiles. *Current Opinion in Biotechnology* 17 (3): 250–255.

Rajkumar, M., Sandhya, S., Prasad, M.N.V. et al. (2012). Perspectives of plant-associatedmicrobes in heavy metal phytoremediation. *Biotechnology Advances* 30: 1562–1574.

Rani, V., Sharma, P., and Dev, K. (2019). Characterization of thermally stable β galactosidase from *Anoxybacillus flavithermus* and *Bacillus licheniformis* isolated from tattapani hotspring of North Western Himalayas, India. *International Journal of Current Microbiology and Applied Sciences* 8: 2517–2542. doi:10.20546/ijcmas.2019.801.266

Rawlings, D.E. (2001). The molecular genetics of *Thiobacillus ferrooxidans* and other mesophilic, acidophilic, chemolithotrophic, iron or sulfur oxidizing bacteria. *Hydrometallurgy* 59: 187–201.

Remias, D., Schwaiger, S., Aigner, S. et al. (2012). Characterization of an UV AND VIS-absorbing, purpurogallin-derived secondary pigment new to algae and highly abundant in *Mesotaeniumberggrenii* (*Zygnematophyceae, Chlorophyta*), an extremophyte living in glaciers. *FEMS microbiology Ecology* 79 (3): 638–648.

Sarmiento, F., Peralta, R., and Blamey, J.M. (2015). Cold and hot extremozymes: Industrial relevance and current trends. *Front. Bioeng. Biotechnol.* 3: 148. 10.3389/fbioe.2015.00148

Saxena, H., Hsu, B., De Asis, M. et al. (2018). Characterization of a thermostable endoglucanase from *Cellulomonas fimi* ATCC484. *Biochemistry and Cell Biology* 96: 68–76. doi:10.1139/bcb-2017-0150.

Shukla, A., Rana, A., Kumar, L. et al. (2009). Assessment of detergent activity of *Streptococcus* sp. AS02 protease isolated from soil of Sahastradhara, Doon Valley, Uttarakhand. *Asian Journal of Microbiology, Biotechnology and Environmental Sciences* 11: 587–591.

Singh, P., Jain, K., Desai, C., Tiwari, O. et al. (2019b). Chapter 18: Microbial community dynamics of extremophiles/extreme environment. In: *Microbial Diversity in the Genomic Era* (ed. S. Das and H.R. Dash), 323–332. Cambridge, MA: Academic Press.

Suleiman, M., Schröder, C., Klippel, B. et al. (2019). Extremely thermoactive archaeal endoglucanase from a shallow marine hydrothermal vent from Vulcano Island. *Applied Microbiology and Biotechnology* 103: 1267–1274. doi:10.1007/s00253-018-9542-z.

Suzuki, T., Iwasaki, T., Uzawa, T. et al. (2002). *Sulfolobus tokodaii* sp. nov. (f. *Sulfolobus* sp. strain 7), a new member of the genus *Sulfolobus* isolated from Beppu Hot Springs, Japan. *Extremophiles* 6 (1): 39–44.

Wiegel, J. and Kevbrin, V.V. (2004). Alkalithermophiles. *Biochemical Society Transactions* 32: 193–198.

Xia, Y., He, L., Mao, J. et al. (2018). Purification, characterization, and gene cloning of a new cold-adapted β-galactosidase from *Erwinia* sp. E602 isolated in northeast China. *Journal of Dairy Science* 101: 6946–6954. doi:10.3168/jds.2018-14605.

Xiong, J., He, Z., Liu, D. et al. (2008). The role of bacteria in the heavy metals removal and growth of *Sedum alfredii* Hance in an aqueous medium. *Chemosphere* 70: 489–494.

Yamada, H. and Shimizu, S. (1988). Microbial and enzymatic processes for the production of biologically and chemically useful compounds. *Angewandte Chemie* 27 (5): 622–642.

Yang, S., Fu, X., Yan, Q. et al. (2016). Cloning, expression, purification and application of a novel chitinase from a thermophilic marine bacterium *Paenibacillus barengoltzii*. *Food Chemistry* 192: 1041–1048. doi:10.1016/j.foodchem.2015.07.092.

Yang, X., Wu, Y., Zhang, Y. et al. (2020). A thermo-active laccase isoenzyme from *Trametes trogii* and its potential for dye decolorization at high temperature. *Frontiers in microbiology* 11: 241. doi:10.3389/fmicb.2020.00241

Yao, C., Sun, J., Wang, W. et al. (2019). A novel cold-adapted β-galactosidase from *Alteromonas* sp. ML117 cleaves milk lactose effectively at low temperature. *Process Biochemistry* 82: 94–101. doi:10.1016/j.procbio.2019.04.016

Part III

Bioresource Technology: Solution to Sustainable Environment and Management Policies

11

Algal-based Membrane Bioreactor for Wastewater Treatment

Setareh Heidari[1], *David A. Wood*[2] *and Ahmad Fauzi Ismail*[3]

[1] Faculty of Chemical Engineering, Sahand University of Technology, Tabriz, Iran
[2] DWA Energy Limited, Bassingham, Lincoln, LN5 9JP, United Kingdom
[3] Advanced Membrane Technology Research Center (AMTEC), Universiti Teknologi Malaysia, 81310, Skudai, Johor Bahru, Johor, Malaysia

CHAPTER MENU

11.1 Introduction

As water scarcity intensifies due to the stresses of increasing water demand, groundwater depletion and global warming effects, the development of sustainable and environmentally-friendly technologies for reusing secondary municipal effluent becomes the need of the hour. Such technologies require effluents to be effectively treated without stages of oxidation that often lead to compounds being formed that are harmful to human and animal health and can result in severe pollution. In developing appropriate technologies, it is necessary to seek: i) compact plant designs with low surface footprints; ii) plants that are reliable and can efficiently treat a wide range of effluent compositions; and iii) sustainable technologies in terms of energy consumption and recycling or disposal of associated chemical effluents and other waste products (Wisniewski 2007).

Bioresource Technology: Concept, Tools and Experiences, First Edition. Tanveer Bilal Pirzadah, Bisma Malik, Rouf Ahmad Bhat, and Khalid Rehman Hakeem.
© 2022 John Wiley & Sons Ltd. Published 2022 by John Wiley & Sons Ltd.

Typically, untreated and partially treated effluents represent health risks for humans, terrestrial animals and aquatic habitats (Das et al. 2017). Effective handling and recycling of effluents often becomes an obstacle for the sustainable urban and industrial developments. However, the availability of sustainable wastewater treatments offers the potential to convert such effulents into a valuable resource (Prieto 2011). For example, some of the chemicals that are abundant in many effluents and wastewaters, including nitrogen (N) and phosphorus (P), are beneficial, justifying their reclamation. It is also imperative to eliminate these substances from effluent discharge to reduce their detrimental environmental impacts. Currently, most common water/wastewater treatment systems rely on the chemical and physical reactions that are not easy to combine with the reliability and sustainability objectives for effluent treatment economically (Silveira et al. 2017). The global shortages of water and energy and increasing demand for sustainable water resources are prompting many nations and regional communities to consider alternative technologies for water/wastewater treatment. Correspondingly, recent findings have confirmed that algae can be successfully used to treat various effluents and various valuable resources can be recovered from them. The uptake of N and P by algae is particularly significant in this regard and, indeed, microalgae are photosynthetic microorganisms (3–30 μm), living in aquatic environments, which consume carbon dioxide and nutrients (particularly N and P) to generate organic matter (Kumar and Buddolla 2019; Chen et al. 2012; Ardekani and Stocker 2010). These precious, small-sized organisms have recently drawn attention of the scientific community because of their promising applications in the generation of bioenergy (Chen et al. 2019; Lam et al. 2019; Zebed et al. 2019; Mathimani et al. 2019; Darda et al. 2019; Nagappan et al. 2019; Shokravi et al. 2019; Sudhakar and Viswanaathan 2019; Singhai et al 2020; Patel et al. 2002; Chandra et al. 2020; Aravind et al. 2020; Bharadwaj et al. 2020). Besides, due to their great efficiency in the elimination of nutrients and toxic metals from effluents, they have opened up new horizons to develop state-of-the-art water treatment processes with minimum energy consumption (Valizadeh and Davarpanal 2020; Pachés et al. 2020; Suryawan and Sofiyah 2020; Leng et al. 2020; Shahid et al. 2020; Cheng et al. 2020). An additional benefit is that they uptake carbon dioxide (CO_2), consuming it as energy for their metabolism, thereby limiting its release to the atmosphere as a greenhouse gas and assisting in the mitigation of climate change (Shahid et al. 2020). Nevertheless, algal wastewater treatments on a commercial scale have been limited by slow nutrient removal rates and the difficulty of harvesting the algae before effluent discharge. These issues can be overcome by the entrapment of the algae in Ca-alginate beads. These items facilitate their retention and increased concentrations within the algal reactor, enabling rapid nutrient removal and simple algal cell removal by settling (Kube et al 2019.). Recovery of the algal biomass produced during wastewater treatment enables resource recovery of energy and nutrient-rich fertilizers (Cai et al 2013). Furthermore, an established and mature technology available for waste management, providing sustainability and pollution control in the drinking and wastewater sectors, involves the utilization of membrane bioreactors (MBRs). Currently, algae-based membrane bioreactor (AMBR) systems are attracting significant research attention. This is because they offer the potential advantage of combining activated sludge processing with membrane filtration. Compared to the conventional microalgae processes applied in wastewater treatments, membrane-based systems are able to cost-effectively meet demands for algae cultivation, harvesting, dewatering and processing (Liao et al. 2018). This system has the potential to provide a single step, compact process, producing a high-quality effluent, providing better biomass treatment, with less biofouling and waste issues, making it overall a more environmentally-friendly option.

11.2 Algal Treatment System: Requirements and Complications

A broad range of novel technologies and their applications introduced in recent years play a substantial role in improving standards of living and access to energy and resources, such as novel technologies of the hydrocarbon industry (Sadatshojaei et al. 2016a; 2019; Choubineh et al. 2017; Dehshibi et al. 2019; Nezhad et al. 2020; Bahrami et al. 2020; Sadatshojaie and Wood, 2020a), corrosion (Seddiqui et al. 2020), ultrasonic technology (Khajehesamedini et al. 2018; Sadatshojaie et al. 2020a), biofuels (Jafari et al. 2020), solvents (Heidari et al. 2020; Sadatshojaei and Wood 2020b), nanomaterials (Sadatshojaei et al. 2016b), carbon dioxide handling and utilization and wastewater treatments. In the field of wastewater treatment technologies, algae are recognized as one of the most important microbial organisms that have the capability to impact this sector on a commercial basis. Macro- and microalgae commercial application represents an industrial sector with an annual turnover of several billion US dollars (Pulz and Gross 2004). There are a wide range of algal species, with variable concentrations of polyunsaturated fatty acids, pigments, minerals, carbohydrates, protein and vitamins at the molecular level (Duarte et al. 2020; Gujar et al. 2019; Morineau-Thomas et al. 2002). Several algal species are known to be effective in the processes associated with biofuel production (Sadatshojaei et al. 2020b, 2020c; 2020d) and for a wide range of products used in food processing, nutrient harvesting for animal and fish food, fertilizers, cosmetics and pharmaceuticals. The fundamental requirements for the growth and development of most algae are carbon dioxide, water, sunlight and nutrients, particularly phosphorus and nitrogen. In general, algal species are flexible to cultivate because they exploit a broad range of feedstocks as their energy source as they can be either mixotrophic, autotrophic or heterotrophic (Liang et al. 2009; Bansal 2019; Mangal et al. 2020). For instance, there are certain algal species that are ideally suited to exploit the varied compositions of nutrients, water and carbon dioxide present in industrial flue gases and wastewaters (Mata et al. 2010; Upadhyay et al. 2020). Certain algae are consequently able to treat a broad range of wastewater types including those that are generated from industrial, agricultural or urban sources. Additionally, some algae are able to digest the cultivated biomass anaerobically, as well as convert it into liquid fuels (Chisti 2007; Kimev et al. 2020). Microalgae generation fed by carbon dioxide also have a potentially wide range of applications (Sadatshojaei et al. 2018, 2021; Sadatshojaei and Wood 2020c; Sadatshojaei and Rahimpour 2020). However, balanced delivery of carbon dioxide to microalgal systems can represent an obstacle when the microalgae are being cultivated. Furthermore, during the harvesting stages of microalgae, dewatering algal cells, bioproducts and separation process represents additional challenges to overcome. It is difficult to accomplish these processes on a cost-effective basis, because microalgae are of very small size, with single-celled varieties ranging from 3 to 30 micrometers. They also typically exist in very low concentrations with developed cultures, i.e,. around 0.5 g dry biomass per liter of culture (Grima et al. 2003). There are a wide range of common technologies including centrifugation, precipitation, flocculation and screening that need to be applied during the various steps of algal cultivation and harvesting. Each of these processes is associated with advantages and disadvantages. Being able to recycle at least some of the algal-treated materials, including treated water and extracted nutrients, substantially improves the economic viability and sustainability of these applications.

11.3 Elements of Microalgae Cultivation

Attached microalgae cultivations (Figure 11.1) are defined as microalgae growing on solid substrates to enhance biofilm expansion (Zhuang et al. 2018). Figure 11.1c shows attached microalgae cultivation and the biofilm expansion procedure. Acid–base

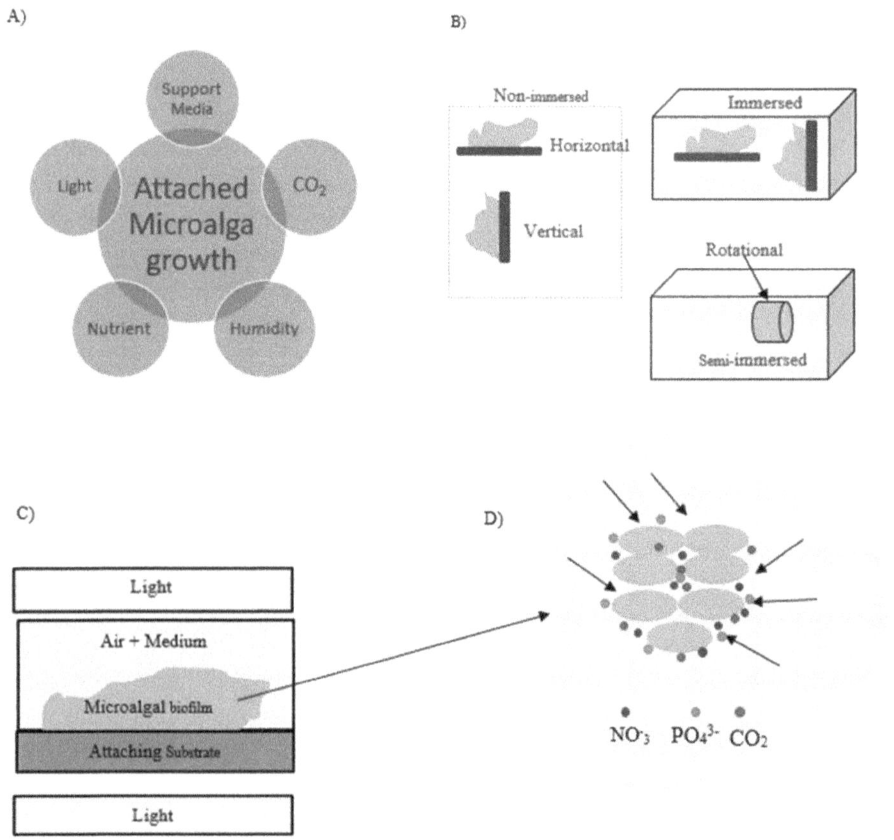

Figure 11.1 The diagram of cultivation of microalgae and its elements parts (Ishmail et al. 2018).

interactions between the support media and the microalgae and hydrophobic interplays are responsible for the connections to the support media (Ozkan and Bergeroglu 2013). There are some essential requirements that are associated with the cultivation of microalgae such as carbon dioxide, light and water which must be available. These primary factors apply for both attached and suspended microalgae and the attached microalgal networks also depend upon the availability of strong supporting elements (Figure 11.1a). The outer biofilm areas are exposed to light and a culture medium. The biofilm surfaces adjacent to the supporting elements are referred to as the "internal sections," and the exposed biofilm surfaces far from the supporting elements in contrast, referred to as the "external sections." Furthermore, light is commonly accessible only through the external sections of a microalgae biofilm and the culture broth component surrounding the biofilm is utilized in their growth. Regarding the location of the support media compared to the culture broth, there are three types of attached microalgae cultures that include immersed, semi-immersed or non-immersed. Additionally, according to the positioning of the support media, they can be configured horizontally, vertically or rotationally (Figure 11.1b). The most commonly-used configuration is a semi-immersed microalgae/bacteria culture system arranged as a rotational system. All microalgae consortia are contained within and influenced by water-based systems. The outside-cellular organic materials (OOMs) and nutrients soluble in the OOM are represented diagrammatically in Figure 11.1d. The detailed compositions of the outside-cellular substances present in typical cultivation reactors fluctuate considerably during their operation, especially the OOMs (Shen et al. 2016). Microalgae densities tend to be much greater in attached systems than in suspended systems.

11.4 Membranes and Their Application in Water and Wastewater Treatments

A membrane is a selectively permeable barrier, used to remove unwanted constituents from both gas and liquid streams (Ismail et al. 2015). According to the driving forces, membranes are typically classified as pressure-driven, concentration-driven, temperature-driven and ion-exchanger types (Ismail et al. 2018). Pressure-driven membranes can be of microfiltration type (MF), nanofiltration type (NF), reverse-osmotic type (RO) and ultrafiltration type (UF). Concentration-driven membranes are divided intodialysis type (DL), forward-osmosis type (FO) and gas-separation type (GS). Temperature-driven membranes work via distillation or and pervaporation (i.e., separating liquids by partial vaporization). Membrane separation is therefore achieved by a range of process mechanisms depending upon the type of membrane involved. This includes sieving, charge effects, pervaporation and solution-diffusion (Van Der Bruggen et al. 2003). Based on the size of separated molecules, synthetic membranes are typically classified as asymmetric, dense or porous (Alen et al 2019). Membranes are applied at a commercial scale in various industry sectors including water/wastewater treatment, pharmaceutical processes, food industry, biomedical processes, desalination and environmental and energy applications (Ismail et al. 2018, 2019). Indeed, one of the ground-breaking technologies applied in the drinking and wastewater sector involves membrane technology, most notably pressure-driven membranes (Ismail et al. 2018; Heidari et al. 2017, 2019; Shokri et al, 2015). Membranes are fabricated from many different materials including polymers, metals and ceramics. Polymeric membranes tend to be made of polyethersulfone (PES), polyethylene (PE), polyvinyl chloride (PVC) and polyvinylidene fluoride (PVDF), due to the low cost of production, packing densities and ease of installation, replacement and operational procedures (Heidari et al. 2017, 2019; Shokri et al. 2015; Singh and Purkait 2019). Numerous studies have also confirmed that inorganic nanomaterials, such as TiO_2, Fe_2O_3, GO and Fum-A nanoparticles, can be effectively embedded in the synthesis of the polymeric membranes. These mixed-matrix membranes display a marked improvement in properties and performances (Heidari et al. 2019; Moslehyani et al 2019; Homayoonfal et al. 2013, 2015; Emadzadeh et al. 2014a, 2014b; Nasir et al. 2019, 2020; Chong et al. 2019; Zulhairun et al. 2020; Chee et a. 2020; Bakhsheshi-Rad et al. 2020; Said et al. 2020; Jumnaidi et al. 2020). Regarding porous membranes, the membrane pore size and its pore distribution are the key factors that determine the effectiveness and efficiency of their purification processes (Pourbozorg 2017). Figure 11.2 illustrates the removal capacity of MF, UF and RO membranes used in water/wastewater treatment application, testifying to the importance of pore size to their effectiveness. Recent research and testing have confirmed that reverse osmosis (RO) is a viable and effective technology to address the issue of micro- and nano-pollutants in water. RO is capable of removing a wide range of contaminants often present in water supplies that cannot be otherwise eliminated sufficiently by conventional water treatment technologies (Alen et al. 2019; Goh et al. 2019). Despite the promising results obtained in applying

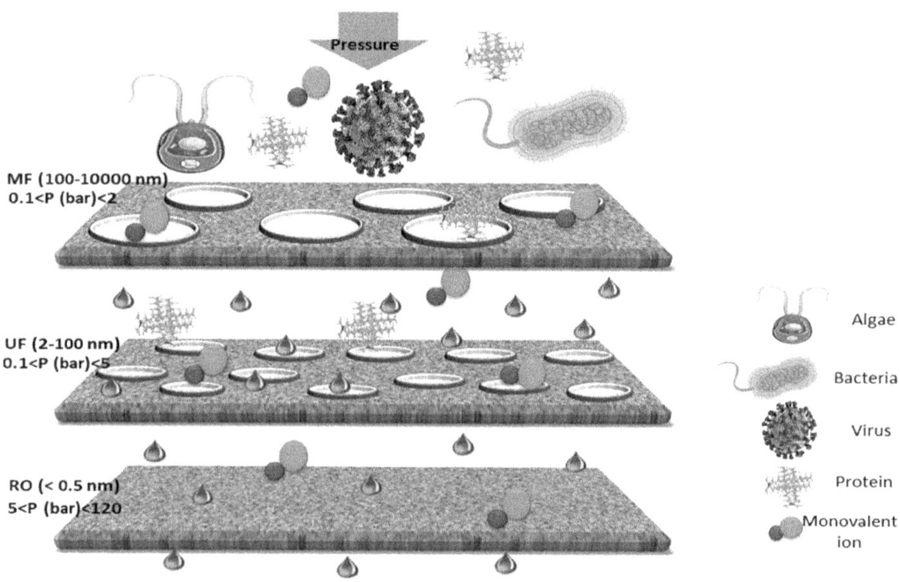

Figure 11.2 Cut-offs and operating pressures for various pressure-driven membranes (source: Centers for Disease Control and Prevention (CDC)/Wikipedia/Public domain; based on Pourbozorg 2017.

RO membranes for water treatment, their application is much more energy-consuming as they demand a much higher transmembrane pressure (TMP) compared with other pressure-driven membrane-based treatment systems such as MF and UF. Transmembrane pressure should be greater for small-pore-size (tighter) membranes, if they have to achieve high-flux throughputs (Ramaswami et al. 2018). Moreover, biofouling can block some of the small pores in such membranes, leading to an increase in transmembrane-pressure requirements. Most successful applications of RO membranes are to separate mineral salts and protect them from exposure and potential fouling from larger contaminant particles potentially blocked their small pores and this avoids the need to frequently clean and unblock their pores. If such protection is not possible, then alternative membrane technologies need to be exploited and applied. Unfortunately, this is often the case in water treatment plants where the water and effluent sourccs, including sea water, are often contaminated with a wide range of macro particles. In this regard, integrated bioreactors membrane systems (MF/UF) with a biological treatment component together offer a sustainable technology to potentially overcome membrane limitations. The ultimate objective of such integrated algal/membrane systems is to effectively treat highly contaminated sources of wastewater and effluents, including those derived from urban sewage systems.

11.5 Algal Membrane Photobioreactors

MF or UF membranes are utilized in algal membrane photobioreactors to provide a closed algal cultivation process. There is a priority to use closed systems such as raceways or ponds rather than open systems due to their greater productivity (Mata et al. 2010; Chisti 2007; Kimev et al. 2020; Fithriani and Melanie 2020). In detail, closed systems display productivities of about 4 gm l^{-1} compared to about 0.14 gm l^{-1} for open systems (Chisti 2007). Regarding the cultivation stage, membrane systems possess the ability to utilize high algal cell densities. This leads to a reduction in the land area needed (i.e., reduced footprint) and an enhanced nutrient elimination efficiency for such treatment systems (Singh and Thomas, 2012; Mandotra et al. 2020). Algal MBRs are now utilized in some water treatment plants, particularly to treat the effluents from digesters. More commonly, they are exploited to control algal cultivation on a commercial scale (Silveira et al. 2017; Singh and Thomas 2012; Ruiz-Martinez et al. 2012; Singh and Mishra 2019). Algal-membrane photobioreactors (MPBR) can eliminate phosphates with 60 to 100% effectiveness, nitrates with about 35% effectiveness and ammonia with 50 to 100% effectiveness. The average algal cell density is about 0.5 gm l^{-1} with transmembrane pressures varying from about 0.06 bar to 0.34 bar (Silveira et al. 2017; Kitani et al. 2020; Danquah et al. 2009) and the more elevated end of that pressure scale tends to enhance the algal culture density. An MPBR system to treat a combination of household wastewater and industrial wastewater (from a plant manufacturing electrical components) enhances cell operating time (Zhen-Feng et al. 2011). Figure 11.3 shows a schematic diagram of the MPBR set-up for such a treatment system. Xu et al. (2014) evaluated a steady-flow photobioreactor utilizing synthetic wastewater and attained a value of 4.8 gm l^{-1} for algal cell density and 66% effectiveness for

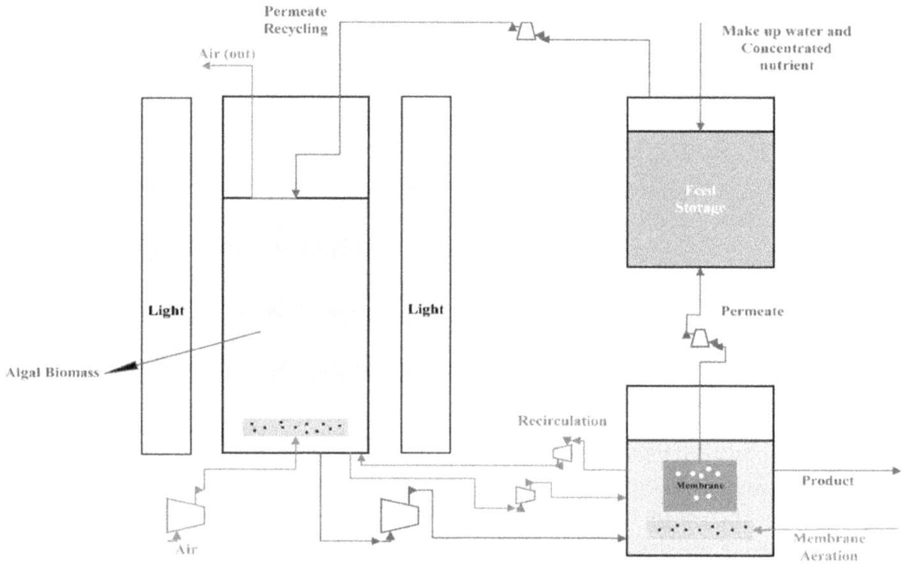

Figure 11.3 Schematic diagram of MPBR set up (*source*: based on Zhen-Feng et al. 2011).

phosphorous elimination (Xu et al. 2014). The combination of wastewater treatment and a photobioreactor provides the capability to establish an inexpensive closed system feeding on the intrinsic nutrient resource to cultivate high algal cell densities. Table 11.1 summarizes the latest applications of algal-based membrane systems in water/wastewater industry.

Table 11.1 Characteristics of various algal-based membrane systems.

Applications	Characteristics of Membranes	Aim	Algae	References
ASB-MBR	PVDF/hollow fiber/0.2 μm	$NH4^+$-N, Nitrogen and $PO4^{3-}$-P removal	Algae collected from a purifier wall located in a municipal wastewater treatment plant	(Zhang et al. 2018)
MADMB	PE/flat sheet/0.4 μm	Mercury removal from simulated dental wastewater	*Chlorella Vulgaris*	(Fard and Mehrnia 2017)
Integrated OPPBR and MBR	PES/flat sheet/0.2 μm	Nitrogen, NH_4-N, NO_3^-N, phosphorus and PO_4^-P removal	*C. Vulgaris* (FC-16)	(Choi and Lee 2015)

(*Continued*)

Table 11.1 (Continued)

Applications	Characteristics of Membranes	Aim	Algae	References
Pressure-driven	PES, PES-SPEEK-TiO_2, PES-PEG-TiO_2, PES-$ZnCl_2$-TiO_2, PES-PEI-TiO_2/Cross flow	Pore-former enhancement on TiO_2 incorporated PES MMMs for algae harvesting	*Coelastrella* sp. FI69	(Yogarathinam et al. 2018)
MPBRs	PVDF/hollow fiber/0.1 μm	To investigate the effects of HRT on algal biomass production and nutrient removal	*Chlorella Vulgaris*	(Peng et al. 2020)
Pressure-driven MF and UF	PVDF, PES and regenerated cellulose/crossflow/0.2 μm and 150, 50 and 30 kDa	To harvest microalgae biomass	*Chlorella Vulgaris*	(Elcik and Cakmakci 2017b)
HMPBR and MPBR	PVDF/hollow fiber/0.1 μm	To compare HMPBR and MPBR efficacy in the removal of atrazine and CNP	not mentioned	Derakhshan et al. 2019)
AMPBR	PVDF/hollow fiber/0.45 μm	To evaluate SRT impacts on biomass production	mixed microalgae collected from the effluent	(Solmaz and Işik 2019)
OM-MPBR	PVDF/0.2 μm	NH_3-N removal	*Chlorella pyrenoidosa* (FACHB-28)	(Parakh et al. 2019)
AMPBR	PVDF/hollow fiber/0.2 μm	to evaluate fouling control strategies	*Chlorella Vulgaris* (UTEX 259)	(Fortunato et al. 2020)
IPB and an algal harvesting	Polyester and nylon MM/o.1–5.31 μm	algal growth/harvesting	algae inocula	(Luo et al. 2017)
MPBR and NPBR	PVDF/hollow fiber /0.1, 0.5, and 1 μm	nitrogen and phosphorus removal	*S. dimorphus*, *S. quadricauda*, *C. sorokiniana*, *C. vulgaris ESP-6*	(Chen et al. 2018)

11.6 Factors Affecting the Performance of Membrane Photobioreactors

11.6.1 Operating Factors

11.6.1.1 Temperature

Alteration in temperature may considerably influence enzyme activity as well as microalgal metabolic processes (Ru et al. 2020; Chawla et al. 2020; Chinnasamy et al. 2009). Most experimental MPBR tests have been performed at approximately 25°C. However, it is important to evaluate MPBR efficiency over a range of operating temperatures. The optimal temperature range to achieve sustained growth of a wide range of microalgae species is reported to be between 25 and 30°C (Chinnasamy et al. 2009). Temperatures higher than 35°C have a negative influence on the growth of some species, such as *C. vulgaris* (Chinnasamy et al. 2009; Conveeti et al. 2009; Choi et al. 2019). However, certain algal species are known to thrive in extreme conditions including high temperatures and hypersaline conditions (Wood 2019).

11.6.2.2 Acidity-Alkalinity (Ph)

Regarding MPBR operations, pH influences the cultivation processes of microalgal biomass and controls their growth rates (Yoo et al. 2015). Microalgae's photosynthetic activities play a substantial role as they act to increase the pH in the cultivation media (Munoz and Guieysse 2006; Pascual et al. 2020). There is a preference to utilize MPBR rather than PBR because the MPBR typically provide a reduced operating pH than PBR. Consequently, it is possible to attain more benefits regarding microalgae cultivation in MPBR (Gao et al. 2014; Mao et al. 2020). The value for pH that optimizes microalgae cultivation varies significantly among algal species. For example, *Chlorella sp.* was observed to be most productive in PBRs, treating wastewater derived from dairies, occurring in the pH range of 6.3 to 7.5 (Aguer et al. 2020; {Posadas et al. 2015).

11.6.3.3 Flux and Permeate Flux Through the Reactors

In order to optimize algal cultivation in large-scale bioreactors, high flux (rate per unit area) and permeate flux (taking into account the number of pores per unit area) of about 40 liters/square meter/hour are often employed (Praveen and Loh 2019; Boonchai and Seo 2015). However, the higher the flux, the greater will be the impacts of bio-fouling. Consequently, many smaller-scale laboratory tests involving MPBR are conducted at much lower flux in the range of 2.6 to 15 liters/square meter/hour to constrain membrane-fouling rates.

11.6.4.4 Hydraulic and Solids Retention Time

Hydraulic retention time (HRT) plays a key role as an operational condition affecting algal reactor productivity (Matamoros et al 2015; Larsen et al 2019). MPBR analyses have revealed that an HRT reduction from 24 to 12 hours resulted in the chemical-oxygen demand (COD) of *C. vulgaris* increasing from 895 to 1473 milligrams per liter (Novon et al. 2020; Xu et al. 2015). Solid retention time (SRT) also influences the cultivation rates of biomass and certain rates tend to favor specific microalgal species (Bilad et al. 2014; Hou et al. 2019). The SRT is equivalent to HRT for an ordinary PBR with no biomass recycling. On the other hand, in MPBR the membrane enhances biomass-retention time leading to a high SRT and low HRT system. Such conditions often act to encourage the development of slow-growth algal species in preference to fast-growth species (Iqbal et al. 2020; Ahmed et al. 2007). The SRT/HRT balance of reactor systems tends to play a substantial role in the cultivation rates of algal organic matter that go on to influence the overall efficiency of the MPBR. Some studies have provided unexpected outcomes with respect to the effects of HRT on algal organic matter development. For example, some studies, by increasing HRT from 2.7 to 5 days, observed that algal organic matter increased from 2 to 8 milligrams per liter due to decreased nutrient-load rate (Discart et al. 2014; Ashok et al. 2019).

11.6.5.5 Lighting

The intensity and access to light is a substantial influencing factor, because microalgae depend on a light-energy source to exist and metabolize. It is apparent that optimizing the efficiency of light usage will enhance biomass productivity and subsequently nutrient uptake efficiency (Voltolina et al. 2005). A common strategy in microalgae cultivation systems is to maximize the ratio of surface/volume in order to enhance light availability to the algae. MPBR research was conducted utilizing artificial light sources generally made up of banks of multi-LED bulbs and/or fluorescent lamps (Yew et al. 2019; Chen et al. 2011). Several aspects of light availability influence microalgae growth in bioreactors, including light component wavelengths, the light to darkness (day length versus night length) cycle, as well as intensity.

11.6.6.6 Aeration

Regarding MPBR operation, it is essential to have sufficient aeration in order to penetrate the algal biomass efficaciously, disperse carbon dioxide and reduce fouling within the system. A degree of air circulation within the system is also required to assist in eliminating ammonia (NH_3) from the system and to avoid the build-up of oxygen concentrations (Zhou et al. 2020; Park et al. 2010). Aeration needs to be carefully controlled in order to optimize algal biomass cultivation. Furthermore, unnecessary excess aeration requires more energy usage and leads to increased operating costs. To operate MPBRs successfully, paying careful attention to aeration is essential.

11.7 Biomass Properties Impacting MPBR Performance

11.7.1 Microorganisms

There are two main classes for microalgal cultivation systems. These are homogenous (monoculture) and heterogeneous (mixed). Most historic MPBR research has concentrated on pure cultures. However, it needs to probe more widely algae/bacteria consortia in MPBR. When treating wastewater with an algal monoculture, the conditions are optimized to increase the growth of that one particular species. This makes it possible to attain high productivity while, at the same time, removing contaminants efficaciously. Several *Chorella* species are known to thrive in many water-based effluents and wastewater generally, such as *Chorella vulgaris*. Ongoing research is evaluating their effectiveness in MPBR systems. *C. saccharophila*, *C. kessler* and *C. zofingiensis* have been used to sustainably cultivated biomass in wastewaters derived from a pig farm and a carpet-manufacturing plant (Cheng et al. 2019; Zhu et al. 2013).

11.7.2 Wastewater Properties

MPBR set-ups need to be customized to suit the compositional ranges of the wastewater under treatment. Types of nutrients in the wastewater and their concentrations will determine which algal species/consortia are most suitable for the task and the optimal operating conditions for the system as a whole. Wastewater feeds into MPBR are likely to have total nitrogen (TN) contents in the range of 7.5 to 40.0 milligrams/liter (Solmaz and Işik 2019; Honda et al. 2012) and vary from source to source and from a particular source from time to time. Total phosphorus (TP) can be made up of a wide range of phosphorus-containing molecules including orthophosphate, polyphosphate, pyrophosphate, metaphosphate and organophosphorus compounds with a range of oxidation states (Markou et al. 2014; Zuccaro et al. 2020). The proportion of TP absorption by algae is a function of the algal species in the consortia, other nutrient levels, the specific phosphorus molecules present in the wastewater, and the generic environmental parameters influencing algal cultivation including acidity-alkalinity, light characteristics and system temperature (Choi and Lee 2015; Nguyen et al. 2020). The nitrogen/phosphorus ratio (N/P) in wastewaters was also a substantial influencing factor on the development and sustainability of microalgae. Generally, a similarity was observed between the N:P ratios of microalgal biomass in the MPBR, the N:P ratios of the wastewaters being treated and the secondary effluent from the MPBR (i.e., 7.5–9.6 milligrams/liter), although occasionally phosphorus fluctuations may occur (Chisti 2007; Bilad et al. 2014; Hou et al. 2019).

11.8 Challenges and Limitations

Although algae-based membrane reactors have advantages over conventional water treatment processes, the main problem impeding wider use of this effective technology is membrane fouling. This leads to permeate flux losses, coupled with increased energy consumption (under constant-flux mode), and a significant rise in hydraulic resistance and membrane biodegradation with an increase in operating costs due to frequent requirements to clean (de-foul) the membranes (Liao et al. 2018). Based on the relative size of pores and solutes, three different fouling mechanisms are possible:

1) Cake filtration increases the concentration of microalgae cells and high molecular weight organic/inorganic particles on the membrane surface forming a cake or layer that grows during filtration operations. In this case, particle diameter (d_p) is presumed to be larger than membrane pores diameter (d), so they are incapable of moving into or through the membrane pores. The build-up of the cake layer leads to an additional resistance that combines with the membrane resistance, and consequently decreases water flux.

2) Pore blockage assumes that each particle approaching the surface of the membrane settles on it and blocks perfectly or fractionally some of the membrane pores, in this case ($d_p \approx d$).
3) Pore narrowing can also lead to membrane fouling (Liao et al. 2018; Heidari et al. 2017; Zhang and Fu 2018). This was attributed to biopolymer adsorption primarily involving soluble microbial products (SMPs) and/or extracellular polymeric substances (EPSs) originating from some microalgae that exist within the membrane pores. This process assumes that the diameter of the particles is much smaller than the membrane pores ($d_p << d$), so that particles are easily able to penetrate the membrane pores and when pore narrowing occurs, the effective pore radii decrease and the total hydraulic resistance to permeate flow increases considerably.

Figure 11.4 provides a schematic diagram illustrating the three proposed membrane fouling mechanisms that have been identified. Fouling can be also categorized, based on the types of foulants involved, as bio-fouling or organic fouling. Biofouling is attributed to the microorganism metabolism, resulting in the growth of biofilms either within the membrane's pores or on its surface. This degrades the water flux by creating additional resistance against the water stream, leading to a substantial reduction in efficiency of algal-based membrane bioreactors. Organic fouling is, however, of greater concern in algal-based membrane bioreactors. It involves fouling caused by organic substances derived from the feed wastewaters or secreted by the microalgae active within a reactor. The type of substances involved include humic materials and other naturally occurring organic materials, together with lipid, protein and polysaccharide molecular accumulations (Liao et al.2018). Fouling reduces the flux, selectivity, effectiveness and operating life of membranes. Consequently, it increases the costs associated with membrane cleaning and membrane replacement. To improve performance and efficiency of MPBR processes, it is essential to prevent or minimize membrane-fouling processes. To achieve this, it is first necessary to fully understand the parameters affecting membrane fouling. There are many variables that influence the fouling behavior within MPBR systems. The most influential ones are identified as reactor configurations (Praveen et al. 2019),

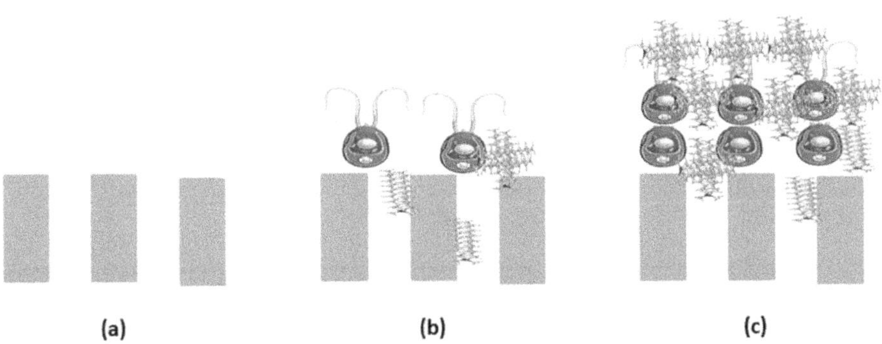

Figure 11.4 Fouling mechanisms that potentially occur during MF/UF process; (a): clean membrane before filtration process, (b): pore blockage and pore narrowing, occurred at the initial phase of filtration (c) fouling during long filtration runs including pore blockage, pore narrowing and cake filtration.

hydrodynamic and operation conditions (e.g., temperature) (Gonzalez-Camejo et al. 2018), cross-flow rate (Ahmad et al. 2012; Elcik and Cakmakci 2017a), transmembrane pressure (Wang et al. 2019), vibrations (Praveen et al. 2019), hydraulic-retention times (HRT) and solid/sludge-retention times (SRT) (Praveen et al. 2019; Gonzalez-Camejo et al. 2018; 2019; Yang et al. 2018; Low et al. 2016; Luo et a. 2017), microalgal characteristics (e.g., size and its particle size distributions (PSD) (Low et al. 2016), nature and composition (Chen et al. 2018), shape (Marbelia et al. 2014), concentration (Kang 2005), zeta potential (Low et al. 2016)), membrane characteristics (e.g., configuration) (Yogarathinam et al. 2018), material (Yogarathinam et al. 2018), hydrophilicity/hydrophobicity (Liao et al. 2018; Yogarathinam et al. 2018; Elcik et al. 2017), charge (Drexler and Yeh 2014), roughness (Guo et al. 2019) and pore size (Chen et al 2018). Hence, in order to construct efficient and sustainable algal-based membrane photobioreactors, detailed analysis is required to identify how best to mitigate membrane fouling (Castaing et al. 2011). Such analysis is likely to include considering ways of improving membrane design by modifying its materials and properties, adjusting operating methods and conditions, and/or considering the introduction of advanced pre-treatment techniques to diminish or alter those components in the feed wastewaters most likely to cause fouling.

11.9 Future Directions for Algal-based Membrane Bioreactors

Today, there is a wide range of research on medical subjects (Ghashghaei et al. 2018; Sadatshojaei et al. 2020e) that can improve the lives of human beings. As an initial objective, microalgae must efficiently be harvested such that it reaches and is sustained at about 0.5% w/w concentration in a reactor utilizing algal-based membrane technology. Greater concentrations of microalgae typically require additional energy due to the higher viscosities of their algal broths. However, utilization of algae to treat bulk household wastewater has historically not performed competitively with other systems, because such feed wastewaters contain extremely low levels of nutrients that result in low productivities of ordinary PBR systems. Modern MPBRs are designed to overcome these limitations. Moreover, anaerobic digestion plants can potentially be coupled with MPBR to extract the carbon dioxide from the biogas effluents from these reactors. Future research is required to assess the performance of such systems and to optimize their commercial-scale operating conditions and, in particular, identify how membrane fouling of such systems can be minimized. Additionally, it is essential for more analysis to be focused in an integrated way on the environmental impacts, sustainability and cost-effectiveness of industrial-scale MPBR.

11.10 Conclusions

The cultivation of microalgae within membrane photo-bioreactors (MPBR) for use in treating industrial effluents and wastewaters is now well studied and understood based upon the results of published results from laboratory-scale experiments. It has been demonstrated that microalgal MPBRs are capable of working efficaciously at lower hydraulic retention times in comparison with conventional photobioreactors. This means that MPBR have the potential to be built at commercial scales with a lower overall system footprint. MPBR are also well suited to treat the secondary effluents from other photobioreactors because such effluents generally consist of low levels of organic material, thereby avoiding bacterial contamination of the microalgal cultures with the MPBR and preventing potential biomass washout from the combined system. Published test results suggest that MPBR technology has the potential to attain algal productivities in the range of 50 to 100 milligrams/liter/day. Moreover, such systems can reach elimination efficiencies for nitrogen ranging from 30 to 97% and for phosphorus ranging from 46 to 93%. The exact efficiency achieved by specific systems depends on the composition of the wastewaters being treated, the algal species or consortia present in the MPBR, the reactor design and operating conditions.

References

Aguera, A., Plaza-Bolaños, P., and Fernández, F.A. (2020). *Removal of Contaminants of Emerging Concern by Microalgae-based Wastewater Treatments and Related Analytical Techniques, Current Developments in Biotechnology and Bioengineering*, 503–525. *Elsevier.*

Ahmad, A., Yasin, N.M., Derek, C. et al. (2012). Crossflow microfiltration of microalgae biomass for biofuel production. *Desalination* 302: 65–70.

Ahmed, Z., Cho, J., Lim, B.R. et al. (2007). Effects of sludge retention time on membrane fouling and microbial community structure in a membrane bioreactor. *Journal of Membrane Science* 287 (2): 211–218.

Alen, S.K., Nam, S., and Dastgheib, S.A. (2019). Recent advances in graphene oxide for gas separation applications. *International Journal of Molecular Sciences* 20 (22): 5609.

Aravind, S., Kumar, P.S., Kumar, N.S. et al. (2020). Conversion of green algal biomass into bioenergy by pyrolysis. A review. *Environmental Chemistry Letters* 18 (3): 829–849. doi:10.1007/s10311-020-00990-2

Ardekani, A.M. and Stocker, R. (2010). Stratlets: Low Reynolds number point-force solutions in a stratified fluid. *Physical Review Letters* 105 (8): 084502.

Ashok, V., Gupta, S.K., and Shriwastav, A. (2019). *Photobioreactors for Wastewater Treatment, Application of Microalgae in Wastewater Treatment*, 383–409. *Springer.*

Bahrami, B., Sadatshojaie, A., and Wood, D.A. (2020). Assessing wellbore stability with a modified lade failure criterion. *Journal of Energy Resources Technology* 142 (8): 083004. Paper No: JERT-20-1075 https://doi.org/10.1115/1.4046387

Bakhsheshi-Rad, H.R., Ismail, A.F., Aziz, M. et al. (2020). Co-incorporation of graphene oxide/silver nanoparticle into poly-L-lactic acid fibrous: A route toward the development of cytocompatible and antibacterial coating layer on magnesium implants. *Materials Science and Engineering*: 111: 110812. doi: 10.1016/j.msec.2020.110812.

Bansal, S. (2019). Mixotrophic growth of *Chlorella* Sp. using glycerol for the production of biodiesel: A review. *Mapan – Journal of Sciences* 18 (2): 1–12.

Bharadwaj, S.V., Ram, S., Pancha, I. et al. (2020). Recent trends in strain improvement for production of biofuels from microalgae, microalgae cultivation for biofuels production. In: *Microalgae Cultivation for Biofuels Production* (ed. A. Yousuf), 211–225. Elsevier.

Bilad, M., Discart, V., Vandamme, D. et al. (2014). Coupled cultivation and pre-harvesting of microalgae in a membrane photobioreactor (MPBR). *Bioresource Technology* 155: 410–417.

Boonchai, R. and Seo, G. (2015). Microalgae membrane photobioreactor for further removal of nitrogen and phosphorus from secondary sewage effluent. *Korean Journal of Chemical Engineering* 32 (10): 2047–2052.

Cai, T., Park, S.Y., and Li, Y. (2013). Nutrient recovery from wastewater streams by microalgae: Status and prospects. *Renewable and Sustainable Energy Reviews* 19: 360–369.

Castaing, J.B., Masse, A., Sechet, V. et al. (2011). Immersed hollow fibres microfiltration (MF) for removing undesirable micro-algae and protecting semi-closed aquaculture basins. *Desalination* 276 (1–3): 386–396.

Chandra, R., Vishal, G., Sánchez, C.E.G. et al. (2020). Bioreactor for algae cultivation and biodiesel production. In: *Bioreactors: Sustainable Design and Industrial Applications in Mitigation of GHG Emissions* (ed. L. Singh, A. Yousuf, and D.M. Mahapatra), 289–307. Elsevier.

Chawla, P., Malik, A., Sreekrishnan, T.R. et al. (2020). Selection of optimum algae via wastewater combination. *Environmental Technology & Innovation* 18: 100758. https://doi.org/10.1016/j.eti.2020.100758

Chee, D.N.A., Amin, M.A.M., Aziz, F. et al. (2020). *Development of Advanced Nanocomposite Membranes by Molecular Sieving Nanomaterials (Zeolite and MOF), Nanocomposite Membranes for Water and Gas Separation*, 163–198. *Elsevier*.

Chen, X., Huang, C., and Liu, T. (2012). Harvesting of microalgae *Scenedesmus* sp. using polyvinylidene fluoride microfiltration membrane. *Desalination and Water Treatment* 45 (1–3): 177–181.

Chen, C.Y., Yeh, K.L., Aisyah, R. et al. (2011). Cultivation, photobioreactor design and harvesting of microalgae for biodiesel production: A critical review. *Bioresource Technology* 102 (1): 71–81.

Chen, X., Li, Z., He, N. et al. (2018). Nitrogen and phosphorus removal from anaerobically digested wastewater by microalgae cultured in a novel membrane photobioreactor. *Biotechnology for Biofuels* 11 (1): 190.

Chen, H., Li, T., and Wang, Q. (2019). Ten years of algal biofuel and bioproducts: Gains and pains. *Planta* 249 (1): 195–219.

Cheng, D., Ngo, H., Guo, W. et al. (2019). Microalgae biomass from swine wastewater and its conversion to bioenergy. *Bioresource Technology* 275: 109–122.

Cheng, X., Delanka-Pedige, H.M., Munasinghe-Arachchige, S.P. et al. (2020). Removal of antibiotic resistance genes in an algal-based wastewater treatment system employing *Galdieria sulphuraria*: A comparative study. *Science of the Total Environment* 711: 134435.

Chinnasamy, S., Ramakrishnan, B., Bhatnagar, A. et al. (2009). Biomass production potential of a wastewater alga *Chlorella vulgaris* ARC 1 under elevated levels of CO_2 and temperature. *International Journal of Molecular Sciences* 10 (2): 518–532.

Chisti, Y. (2007). Biodiesel from microalgae. *Biotechnology Advances* 25 (3): 294–306.

Choi, H.J. and Lee, S.M. (2015). Effect of the N/P ratio on biomass productivity and nutrient removal from municipal wastewater. *Bioprocess and Biosystems Engineering* 38 (4): 761–766.

Chong, C.Y., Lau, W.J., Yusof, N. et al. (2019). Roles of nanomaterial structure and surface coating on thin film nanocomposite membranes for enhanced desalination. *Composites Part B: Engineering* 160: 471–479.

Choubineh, A., Ghorbani, H., Wood, D.A. et al. (2017). Improved predictions of wellhead choke liquid critical-flow rates: Modelling based on hybrid neural network training learning based optimization. *Fuel* 207: 547–560.

Converti, A., Casazza, A.A., Ortiz, E.Y. et al. (2009). Effect of temperature and nitrogen on the growth and lipid content of *Nannochloropsis oculata* and *Chlorella vulgaris* for biodiesel poduction. *Chemical Engineering and Processing: Process Intensification* 48 (6). 1146–1151.

Danquah, M.K., Gladman, B., Moheimani, N. et al. (2009). Microalgal growth characteristics and subsequent influence on dewatering efficiency. *Chemical Engineering Journal* 151 (1–3): 73–78.

Darda, S., Papalas, T., and Zabaniotou, A. (2019). Biofuels journey in Europe: Currently the way to low carbon economy sustainability is still a challenge. *Journal of Cleaner Production* 208: 575–588.

Das, R., Vecitis, C.D., Schulze, A. et al. (2017). Recent advances in nanomaterials for water protection and monitoring. *Chemical Society Reviews* 46 (22): 6946–7020.

Dehshibi, R., Sadatshojaie, A., Mohebbi, A. et al. (2019). A new insight into pore body filling mechanism during waterflooding in a glass micro-model. *Chemical Engineering Research & Design: Transactions of the Institution of Chemical Engineers Part A* 151: 100–107.

Derakhshan, Z., Ehrampoush, M.H., Mahvid, A.H. et al. (2019). A comparative study of hybrid membrane photobioreactor and membrane photobioreactor for simultaneous biological removal of atrazine and CNP from wastewater: A performance analysis and modeling. *Chemical Engineering Journal* 355 (2019): 428–438.

Discart, V., Bilad, M., Marbelia, L. et al. (2014). Impact of changes in broth composition on Chlorella vulgaris cultivation in a membrane photobioreactor (MPBR) with permeate recycle. *Bioresource Technology* 152: 321–328.

Drexler, I.L. and Yeh, D.H. (2014). Membrane applications for microalgae cultivation and harvesting: A review. *Reviews in Environmental Science and BioTechnology* 13 (4): 487–504.

Duarte, J.H., Cardoso, L.G., De Souza, C.O. et al. (2020). Brackish groundwater from Brazilian backlands in spirulina cultures: Potential of carbohydrate and polyunsaturated fatty acid production. *Applied Biochemistry and Biotechnology* 190 (3): 907–917.

Elcik, H. and Cakmakci, M. (2017a). Harvesting microalgal biomass using crossflow membrane filtration: Critical flux, filtration performance, and fouling characterization. *Environmental Technology* 38 (12): 1585–1596.

Elcik, H. and Cakmakci, M. (2017b). Harvesting of microalgae via submerged membranes: Flux, fouling and its reversibility. *Membrane Water Treatment* 8 (5): 499–515.

Elcik, H., Cakmakci, M., and Ozkaya, B. (2017). Preparation and characterisation of novel polysulfone membranes modified with Pluronic F-127 for reducing microalgal fouling. *Chemical Papers* 71 (7): 1271–1290.

Emadzadeh, D., Lau, W.J., Matsuura, T. et al. (2014a). A novel thin film composite forward osmosis membrane prepared from PSf–TiO_2 nanocomposite substrate for water desalination. *Chemical Engineering Journal* 237: 70–80.

Emadzadeh, D., Lau, W.J., Matsuura, T. et al. (2014b). Synthesis and characterization of thin film nanocomposite forward osmosis membrane with hydrophilic nanocomposite support to reduce internal concentration polarization. *Journal of Membrane Science* 449: 74–85.

Fard, G.H. and Mehrnia, M.R. (2017). Investigation of mercury removal by Micro-Algae dynamic membrane bioreactor from simulated dental wastewater. *Journal of Environmental Chemical Engineering* 5 (1): 366–372.

Fithriani, D. and Melanie, S. (2020) Potential utilization of *Botriococcus braunii* and *Nannochloropsis microalgae* as a biodiesel source in Indonesia. *IOP Conference Series: Earth and Environmental Science*, IOP Publishing, pp. 012005.

Fortunato, L., Lampresa, A.F., and Leiknes, T.O. (2020). Evaluation of membrane fouling mitigation strategies in an algal membrane photobioreactor (AMPBR) treating secondary wastewater effluent. *Science of the Total Environment* 15 (708): 134548.

Gao, F., Yang, Z.H., Li, C. et al. (2014). Concentrated microalgae cultivation in treated sewage by membrane photobioreactor operated in batch flow mode. *Bioresource Technology* 167: 441–446.

Ghashghaei, S., Ghashghaei, S., Dastsooz, H. et al. (2018). Haplotype analysis for five unrelated patients with a novel ATP7B gene mutation. *European Journal of Human Genetics* 26: 1020–1020.

Goh, P.S., Ismail, A.F., Ng, B.C. et al. (2019). Recent progresses of forward osmosis membranes formulation and design for wastewater treatment. *Water* 11 (10): 2043.

Gonzalez-Camejo, J., Barat, R., Ruano, M. et al. (2018). Outdoor flat-panel membrane photobioreactor to treat the effluent of an anaerobic membrane bioreactor. Influence of operating, design, and environmental conditions. *Water Science and Technology* 78 (1): 195–206.

González-Camejo, J., Jiménez-Benítez, A., Ruano, M. et al. (2019). Optimising an outdoor membrane photobioreactor for tertiary sewage treatment. *Journal of Environmental Management* 245: 76–85.

Grima, E.M., Belarbi, E.-H., Fernandez, F.A. et al. (2003). Recovery of microalgal biomass and metabolites: Process options and economics. *Biotechnology Advances* 20 (7–8): 491–515.

Gujar, A., Cui, H., Ji, C. et al. (2019). Development, production and market value of microalgae products. *Appli Microbiol Open Access* 5: 162.

Guo, C., Duan, D., Sun, Y. et al. (2019). Enhancing *Scenedesmus obliquus* biofilm growth and CO_2 fixation in a gas-permeable membrane photobioreactor integrated with additional rough surface. *Algal Research* 43: 101620.

Heidari, S. and Wood, D.A. (2020). Biodiesel production methods and feedstocks. In: *Biodiesel Technology and Applications*. (ed. Inamuddin, Ahamed, M.I., Boddula, R., and Rezakazemi, M.). Chapter 17, 447–464. Wiley Online Library. https://doi.org/10.1002/9781119724957.ch17

Heidari, S., Jafarzadeh, Y., Seyfollahi Samarin, M. et al. (2017). Study on the fouling behavior of HDPE/PE-g-MA/EVA blend membrane fabricated via thermally induced phase separation method. *Polyolefins Journal* 4 (2): 235–251.

Heidari, S., Amirinejad, M., and Jahangirian, H. (2019). Investigation of fouling mechanisms using surface morphology and physicochemical membrane features. *Chemical Engineering & Technology* 42 (6): 1310–1320.

Heidari, S., Sadatshojaei, E., and Wood, D.A. (2020). Large-scale molecular solvents for environmentally sustainable applications. In: *Green Sustainable Process for Chemical and Environmental Engineering and Science: Environmental Remediation* (ed. I. Inamuddin, R. Boddula, and A. Asiri). Elsevier. doi:10.1016/B978-0-12-821884-6.00001-2

Homayoonfal, M., Mehrnia, M.R., Mojtahedi, Y.M. et al. (2013). Effect of metal and metal oxide nanoparticle impregnation route on structure and liquid filtration performance of polymeric nanocomposite membranes: A comprehensive review. *Desalination and Water Treatment* 51 (16–18): 3295–3316.

Homayoonfal, M., Mehrnia, M.R., Shariaty-Niassar, M. et al. (2015). Fabrication of magnetic nanocomposite membrane for separation of organic contaminant from water. *Desalination and Water Treatment* 54 (13): 3603–3609.

Honda, R., Boonnorat, J., Chiemchaisri, C. et al. (2012). Carbon dioxide capture and nutrients removal utilizing treated sewage by concentrated microalgae cultivation in a membrane photobioreactor. *Bioresource Technology* 125: 59–64.

Hou, C., Qiao, S., Yang, Y. et al. (2019). A novel sequence batch membrane carbonation photobioreactor developed for microalgae cultivation. *Frontiers of Environmental Science & Engineering* 13 (6): 92.

Ismail, A.F., Khulbe, K.C., and Matsuura, T. (2015). *Gas Separation Membrane Materials and Structures, Gas Separation Membranes*, 37–192. *Springer.*

Ismail, A.F., Rahman, M.A., Othman, M.H.D. et al. (2018). *Membrane separation principles and applications from material selection to mechanisms and industrial uses* (ed. A. Ismail, M.R. Rahman, M.H.D. Othman, et al.). Elsevier. ISBN: 9780128128152.

Ismail, A.F., Hilal, N., Jaafar, J. et al. (2019). *Nanofiber Membranes for Medical, Environmental, and Energy Applications.* CRC Press. doi:10.1201/9781351174046

Iqbal, T., Weerasekara, N.A., Lee, K. et al. (2020). Impact of encapsulated quorum-quenching bacterial dose and feed type on biofouling control in membrane bioreactors. *Journal of Environmental Engineering* 146 (2): 04019109.

Jafari, A., Esmaeilzadeh, F., Mowla, D. et al. (2020). New insights to direct conversion of wet microalgae impregnated with ethanol to biodiesel exploiting extraction with supercritical carbon dioxide. *Fuel* 285: 119199.

Junaidi, N.F.D., Othman, N.H., Shahruddin, M.Z. et al. (2020). Fabrication and characterization of graphene oxide–polyethersulfone (GO–PES) composite flat sheet and hollow fiber membranes for oil–water separation. *Journal of Chemical Technology & Biotechnology* 95 (5): 1308–1320.

Kang, P.S. (2005). Hollow Fibre Fouling in Submerged Membrane Bioreactor (MBR) Process. University of New South Wales. http://unsworks.unsw.edu.au/fapi/datastream/unsworks:36018/SOURCE01?view=true

Khajehesamedini, A., Sadatshojaie, A., Parvasi, P. et al. (2018). Experimental and theoretical study of crude oil pretreatment using low-frequency ultrasonic waves. *Ultrasonics Sonochemistry* 48: 383–395.

Kirnev, P., Carvalho, J., Vandenberghe, L. et al. (2020). Technological mapping and trends in photobioreactors for the production of microalgae. *World Journal of Microbiology and Biotechnology* 36 (3): 1–9.

Kitani, H., Demura, M., Morisada, S. et al. (2020). Dewatering of algal suspension using microfiltration with cross flow in the presence of magnetite as a filter aid. *Biotechnology Progress* 36 (4): e2979.

Kube, M., Mohseni, A., Fan, L. et al. (2019). Impact of alginate selection for wastewater treatment by immobilised *Chlorella vulgaris*. *Chemical Engineering Journal* 358: 1601–1609.

Kumar, M.S. and Buddolla, V. (2019). Future prospects of biodiesel production by microalgae: A short review. In: *Recent Developments in Applied Microbiology and Biochemistry* (ed. V. Buddolla), 161–166. Elsevier.

Lam, M.K., Khoo, C.G., and Lee, K.T. (2019). Scale-up and commercialization of algal cultivation and biofuels production. In: *Biofuels from Algae. A Volume in Biomass, Biofuels, Biochemicals* (ed. A. Pandey, J. Chang, and Y. Chisti), 475–506. Elsevier.

Larsen, C., Yu, Z.H., Flick, R. et al. (2019). Mechanisms of pharmaceutical and personal care product removal in algae-based wastewater treatment systems. *Science of the Total Environment* 695: 133772.

Leng, L., Wei, L., Xiong, Q. et al. (2020). Use of microalgae based technology for the removal of antibiotics from wastewater: A review. *Chemosphere* 238: 124680.

Liang, Y., Sarkany, N., and Cui, Y. (2009). *Biomass and lipid productivities of Chlorella vulgaris* under autotrophic, heterotrophic and mixotrophic growth conditions. *Biotechnology Letters* 31 (7): 1043–1049.

Liao, Y., Bokhary, A., Maleki, E. et al. (2018). A review of membrane fouling and its control in algal-related membrane processes. *Bioresource Technology* 264: 343–358.

Low, S.L., Ong, S.L. et al. (2016). Characterization of membrane fouling in submerged ceramic membrane photobioreactors fed with effluent from membrane bioreactors. *Chemical Engineering Journal* 290: 91–102.

Luo, Y., Le-Clech, P., and Henderson, R.K. (2017). Simultaneous microalgae cultivation and wastewater treatment in submerged membrane photobioreactors: A review. *Algal Research* 24: 425–437.

Mandotra, S., Lolu, A.J., Kumar, S. et al. (2020). *Integrated Approach for Bioremediation and Biofuel Production Using Algae, Restoration of Wetland Ecosystem: A Trajectory Towards A Sustainable Environment*, 145–160. Springer.

Mangal, V., Phung, T., and Guéguen, C. (2020). An estimation of sulfur concentrations released by three algae (*Chlorella vulgaris*, *Chlamydomonas reinhardtii*, *Scenedesmus obliquus*) in response to variable growth photoperiods. *Environmental Science and Pollution Research* 27 (11): 12491–12498. doi: 10.1007/s11356-020-07812-6

Mao, X., Myavagh, P.H., Lotfikatouli, S. et al. (2020). Membrane bioreactors for nitrogen removal from wastewater: A review. *Journal of Environmental Engineering* 146 (5): 03120002.

Marbelia, L., Bilad, M.R., Passaris, I. et al. (2014). Membrane photobioreactors for integrated microalgae cultivation and nutrient remediation of membrane bioreactors effluent. *Bioresource Technology* 163: 228–235.

Markou, G., Vandamme, D., and Muylaert, K. (2014). Microalgal and cyanobacterial cultivation: The supply of nutrients. *Water Research* 65: 186–202.

Mata, T.M., Martins, A.A., and Caetano, N.S. (2010). Microalgae for biodiesel production and other applications: A review. *Renewable and Sustainable Energy Reviews* 14 (1): 217–232.

Matamoros, V., Gutiérrez, R., Ferrer, I. et al. (2015). Capability of microalgae-based wastewater treatment systems to remove emerging organic contaminants: A pilot-scale study. *Journal of Hazardous Materials* 288: 34–42.

Mathimani, T., Baldinelli, A., Rajendran, K. et al. (2019). Review on cultivation and thermochemical conversion of microalgae to fuels and chemicals: Process evaluation and knowledge gaps. *Journal of Cleaner Production* 208: 1053–1064.

Morineau-Thomas, O., Jaouen, P. et al. (2002). The role of exopolysaccharides in fouling phenomenon during ultrafiltration of microalgae (*Chlorella* sp. and *Porphyridium purpureum*): Advantage of a swirling decaying flow. *Bioprocess and Biosystems Engineering* 25 (1): 35–42.

Moslehyani, A., Ismail, A.F., Matsuura, T. et al. (2019). Recent progresses of ultrafiltration (UF) membranes and processes in water treatment. In: *Membrane Separation Principles and Applications from Material Selection to Mechanisms and Industrial Uses* (ed. A. Ahmad Ismail, M.A. Rahman, M.H.D. Othman, et al.), 85–110. Elsevier.

Munoz, R. and Guieysse, B. (2006). Algal–bacterial processes for the treatment of hazardous contaminants: A review. *Water Research* 40 (15): 2799–2815.

Nagappan, S., Devendran, S., Tsai, P.-C. et al. (2019). Passive cell disruption lipid extraction methods of microalgae for biofuel production: A review. *Fuel* 252: 699–709.

Nasir, A.M., Goh, P.S., Abdullah, M.S. et al. (2019). Adsorptive nano-composite membranes for heavy metal remediation: Recent progresses and challenges. *Chemosphere* 232: 96–112.

Nasir, A.M., Goh, P.S., and Ismail, A.F. (2020). Synthesis route for the fabrication of nanocomposite membranes. In: *Nanocomposite Membranes for Water and Gas Separation*, 69–89. *Elsevier*.

Nezhad, S. M., Wood, D.A., Sadatshojaei, E. et al. (2020). New insight to experimental study of ionic solutions with a non-ionic surfactant on wettability, interfacial tension and micro-model flooding. *Fuel* 285: 119126.

Nguyen, L.N., Truong, M.V., Nguyen, A.Q. et al. (2020). A sequential membrane bioreactor followed by a membrane microalgal reactor for nutrient removal and algal biomass production. *Environmental Science: Water Research & Technology* 6 (1): 189–196.

Novoa, A.F., Fortunato, L., Rehman, Z.U. et al. (2020). Evaluating the effect of hydraulic retention time on fouling development and biomass characteristics in an algal membrane photobioreactor treating a secondary wastewater effluent. *Bioresource Technology* 309: 123348. https://doi.org/10.1016/j.biortech.2020.123348

Ozkan, A. and Berberoglu, H. (2013). Cell to substratum and cell to cell interactions of microalgae. *Colloids and Surfaces B: Biointerfaces* 112: 302–309.

Pachés, M., Martínez-Guijarro, R., Gonzalez-Camejo, J. et al. (2020). Selecting the most suitable microalgae species to treat the effluent from an anaerobic membrane bioreactor. *Environmental Technology* 41 (3): 267–276.

Parakh, S.K., Praveen, P., Loh, K.C. et al. (2019). Wastewater treatment and microbial community dynamics in a sequencing batch reactor operating under photosynthetic aeration. *Chemosphere* 215: 893–903.

Park, J., Jin, H.F., Lim, B.R. et al. (2010). Ammonia removal from anaerobic digestion effluent of livestock waste using green alga *Scenedesmus* sp. *Bioresource Technology* 101 (22): 8649–8657.

Pascual, C., Akmirza, I., Pérez, R. et al. (2020). Trimethylamine abatement in algal-bacterial photobioreactors. *Environmental Science and Pollution Research* 27: 9028–9037. https://doi.org/10.1007/s11356-019-07369-z

Patel, A.K., Choi, Y.Y., and Sim, S.J. (2020). Emerging prospects of mixotrophic microalgae: Way forward to sustainable bioprocess for environmental remediation and cost-effective biofuels. *Bioresource Technology* 300: 122741. doi: 10.1016/j.biortech.2020.122741

Peng, -Y.-Y., Gao, F., Yang, H.-L. et al. (2020). Simultaneous removal of nutrient and sulfonamides from marine aquaculture wastewater by concentrated and attached cultivation of *Chlorella vulgaris* in an algal biofilm membrane photobioreactor (BF-MPBR). *Science of the Total Environment* 725: 138524.

Posadas, E., Del Mar Morales, M., Gomez, C. et al. (2015). Influence of pH and CO_2 source on the performance of microalgae-based secondary domestic wastewater treatment in outdoors pilot raceways. *Chemical Engineering Journal* 265: 239–248.

Pourbozorg, M. (2017) Effect of turbulence on hollow fiber membrane filtration in multiphase reactors. Doctoral dissertation. https://hdl.handle.net/10356/70645.

Praveen, P. and Loh, K.C. (2019). Nutrient removal in an algal membrane photobioreactor: Effects of wastewater composition and light/dark cycle. *Applied Microbiology and Biotechnology* 103 (8): 3571–3580.

Praveen, P., Xiao, W., Lamba, B. et al. (2019). Low-retention operation to enhance biomass productivity in an algal membrane photobioreactor. *Algal Research* 40: 101487.

Prieto, A.L. (2011) Sequential anaerobic and algal membrane bioreactor (A2MBR) system for sustainable sanitation and resource recovery from domestic wastewater. Graduate School Theses and Dissertations. https://www.google.com/url?sa=t&rct=j&q=&esrc=s&source=web&cd=&ved=2ahUKEwiDm_-Y5vvzAhXCi1wKHdWWCHAQFnoECA4QAQ&url=https%3A%2F%2Fdigitalcommons.usf.edu%2Fcgi%2Fviewcontent.cgi%3Farticle%3D4491%26context%3Detd&usg=AOvVaw1OLNfWJ43CpzEbHTwWySED (accessed 3 November 2021).

Pulz, O. and Gross, W. (2004). Valuable products from biotechnology of microalgae. *Applied Microbiology and Biotechnology* 65 (6): 635–648.

Ramaswami, S., Behrendt, J., and Otterpohl, R. (2018). Comparison of NF-RO and RO-NF for the treatment of mature landfill leachates: A guide for landfill operators. *Membranes* 8 (2): 17.

Ru, I.T.K., Sung, Y.Y., Jusoh, M. et al. (2020). *Chlorella vulgaris*: A perspective on its potential for combining high biomass with high value bioproducts. *Applied Phycology* 1 (1): 2–11. https://doi.org/10.1080/26388081.2020.1715256

Ruiz-Martinez, A., Garcia, N.M., Romero, I. et al. (2012). Microalgae cultivation in wastewater: Nutrient removal from anaerobic membrane bioreactor effluent. *Bioresource Technology* 126: 247–253.

Sadatshojaie, A. and Rahimpour, M.R. (2020). *CO_2 Emission and Air Pollution (Volatile Organic Compounds, etc. – Related Problems Causing Climate Change, Current Trends and Future Developments on (Bio-) Membranes*, 1–30. *Elsevier*.

Sadatshojaie, A. and Wood, D.A. (2020a) *Effectiveness of Colloidal Gas*. doi:10.26434/chemrxiv.12764708.v1

Sadatshojaei, E. and Wood, D.A. (2020b). Water, the most accessible green solvent, extraction and separation agent. In: *Green Sustainable Process for Chemical and*

Environmental Engineering and Science: Environmental Remediation (ed. I. Inamuddin, R. Boddula, and A. Asiri). Elsevier. doi:10.1016/B978-0-12-821884-6.00012-7

Sadatshojaei, E. and Wood, D.A. (2020c). *Applications of Supercritical Fluids in Environmental Remediation, Green Sustainable Process for Chemical and Environmental Engineering and Science: Environmental Remediation*. Elsevier. doi:10.1016/B978-0-12-821884-6.00003-6

Sadatshojaei, E., Jamialahmadi, M., Esmaeilzadeh, F. et al. (2016a). Effects of low-salinity water coupled with silica nanoparticles on wettability alteration of dolomite at reservoir temperature. *Petroleum Science and Technology* 34 (15): 1345–1351.

Sadatshojaei, E., Sadatshojaie, A., and Fakhar, M.H. (2016b). Differential quadrature method for nonlocal nonlinear vibration analysis of a boron nitride nanotube using sinusoidal shear deformation theory. *Mechanics of Advanced Materials and Structures* 23 (11): 1278–1283.

Sadatshojaei, E., Esmaeilzadeh, F., Fathikaljahi, J. et al. (2018). Regeneration of the midrex reformer catalysts using supercritical carbon dioxide. *Chemical Engineering Journal* 343: 748–758.

Sadatshojaei, E., Jamialahmadi, M., Esmaeilzadeh, F. et al. (2019). The impacts of silica nanoparticles coupled with low-salinity water on wettability and interfacial tension: Experiments on a carbonate core. *Journal of Dispersion Science and Technology* 41 (8): 1159–1173. https://doi.org/10.1080/01932691.2019.1614943

Sadatshojaie, A., Wood, D.A., Jokar, S.M. et al. (2020a). Applying ultrasonic fields to separate water contained in medium-gravity crude oil emulsions and determining crude oil adhesion coefficients. *Ultrasonics Sonochemistry* 70: 105303.

Sadatshojaei, E., Wood, D.A., and Mowla, D. (2020b). Third generation of biofuels exploiting microalgae. In: *Sustainable Green Chemical Process and Their Allied Applications*. Switzerland AG: Springer Nature. doi:10.1007/978-3-030-42284-4_21

Sadatshojaei, E., Mowla, D., and Wood, D.A. (2020c). Review of progress in microalgal biotechnology applied to wastewater treatment. *Sustainable Green Chemical Process and Their Allied Applications, Springer Nature*. doi:10.1007/978-3-030-42284-4_19

Sadatshojaei, E., Wood, D.A., and Rahimpour, M.R. (2020d). *Progress Towards the Sustainable Generation of Biofuels Exploiting Microalgae, Biodiesel Technology and Applications*. Wiley-Scrivener (in press).

Sadatshojae, E., Wood, D.A., and Rahimpour, M.R. (2021). *Potential and challenges of carbon sequestration in soils, In: Applied Soil Chemistry* (ed. Inamuddin, M.I.A., Boddula, R., and Altalhi, T.). Wiley-Scrivener. https://doi.org/10.1002/9781119711520.ch1

Sadatshojaei, E., Heidari, S., and Wood, D.A. (2020e). Sustainable approach in biocatalytic preparation of antibiotic peptide. In: *Green Sustainable Process for Chemical and Environmental Engineering and Science* (ed. I. Inamuddin, A.M. Ahmed Asiri, and S. Kanchi). Elsevier. doi:10.1016/B978-0-12-819721-9.00003-0

Said, N., Abidin, M.N.Z., Hasbullah, H. et al. (2020). Polysulfone hemodialysis membrane incorporated with Fe_2O_3 for enhanced removal of middle molecular weight uremic toxin. *Structure* 38: 43.

Seddiqi, H., Sadatshojaie, A., Vaferi, B. et al. (2020). Mathematical model for iron corrosion that eliminates chemical potential parameters. *Chinese Journal of Chemical Engineering* 28 (2): 603–612.

Shahid, A., Malik, S., Zhu, H. et al. (2020). Cultivating microalgae in wastewater for biomass production, pollutant removal, and atmospheric carbon mitigation; A review. *Science of the Total Environment* 704: 135303.

Shen, Y., Zhu, W., Chen, C. et al. (2016). Biofilm formation in attached microalgal reactors. *Bioprocess and Biosystems Engineering* 39 (8): 1281–1288.

Shokri, E., Yegani, R., Heidari, S. et al. (2015). Effect of PE-g-MA compatibilizer on the structure and performance of HDPE/EVA blend membranes fabricated via TIPS method. *Chemical Engineering Research and Design* 100: 237–247.

Shokravi, H., Shokravi, Z., Aziz, M.M.A. et al. (2019). IIIII: Algal Biofuel: A promising alternative for fossil fuel. In: *Fossil Free Fuels Trends in Renewable Energy* (ed. M.B.A. Aziz, K.A. Kassim, W.A.W.A. Bakar, et al.), 187–211. CRC Press.

Silveira, E.O., Moura, D., Rieger, A. et al. (2017). Performance of an integrated system combining microalgae and vertical flow constructed wetlands for urban wastewater treatment. *Environmental Science and Pollution Research* 24 (25): 20469–20478.

Singh, V. and Mishra, V. (2019). *Bioremediation of Nutrients and Heavy Metals from Wastewater by Microalgal Cells: Mechanism and Kinetics, Microbial Genomics in Sustainable Agroecosystems*, 319–357. *Springer*.

Singh, R. and Purkait, M.K. (2019). *Microfiltration Membranes, Membrane Separation Principles and Applications*, 111–146. *Elsevier*.

Singh, G. and Thomas, P.B. (2012). Nutrient removal from membrane bioreactor permeate using microalgae and in a microalgae membrane photoreactor. *Bioresource Technology* 117: 80–85.

Singhal, G., Verma, V., Bhagyawant, S.S. et al. (2020). Production of biofuel through metabolic engineering: Processing, types, and applications. In: *Genetic and Metabolic Engineering for Improved Biofuel Production from Lignocellulosic Biomass* (ed. A. Kuila and V. Sharma), 155–169. Elsevier.

Solmaz, A. and Işik, M. (2019a). Effect of sludge retention time on biomass production and nutrient removal at an algal membrane photobioreactor. *BioEnergy Research* 12 (1): 197–204.

Sudhakar, M. and Viswanaathan, S. (2019). Algae as a sustainable and renewable bioresource for bio-fuel production. In: *New and Future Developments in Microbial Biotechnology and Bioengineering* (ed. J.S. Singh and D.P. Singh), 77–84. Elsevier.

Suryawan, I.W.K. and Sofiyah, E.S. (2020). Cultivation of *Chlorella* Sp. and Algae Mix for NH_3-N and PO_4-P domestic wastewater removal. *Civil and Environmental Science Journal* 3 (1): 31–36. doi: https://doi.org/10.21776/ub.civense.2020.00301.4 https://chemrxiv.org/engage/chemrxiv/article-details/60c74e7e337d6c10b9e27fad (accessed 3 November 2021).

Tabzar, A., Sadatshojaie, A., and Wood, D.A. (2020) *Effectiveness of colloidal gas aphron fluids formulated with a biosurfactant enhanced by silica nanoparticles*. ChemRxiv. Cambridge: Cambridge Open Engage (preprint). https://chemrxiv.org/engage/chemrxiv/article-details/60c74e7e337d6c10b9e27fad. [Accessed 3 Nov 2021]

Upadhyay, A.K., Singh, R., and Singh, D. (2020). *Phytoremediation and Sustainable Developmental Policies and Practices, Restoration of Wetland Ecosystem: A Trajectory Towards A Sustainable Environment*, 17–29. *Springer*.

Valizadeh, K. and Davarpanah, A. (2020). Design and construction of a micro-photo bioreactor in order to dairy wastewater treatment by micro-algae: Parametric study. *Energy Sources. Part A: Recovery, Utilization, and Environmental Effects* 42 (5): 611–624.

Van Der Bruggen, B., Vandecasteele, C., Van Gestel, T. et al. (2003). A review of pressure-driven membrane processes in wastewater treatment and drinking water production. *Environmental Progress* 22 (1): 46–56.

Voltolina, D., Gómez-Villa, H. et al. (2005). Nitrogen removal and recycling by *Scenedesmus obliquus* in semicontinuous cultures using artificial wastewater and a simulated light and temperature cycle. *Bioresource Technology* 96 (3): 359–362.

Wang, L., Pan, B., Gao, Y. et al. (2019). Efficient membrane microalgal harvesting: Pilot-scale performance and techno-economic analysis. *Journal of Cleaner Production* 218: 83–95.

Wisniewski, C. (2007). Membrane bioreactor for water reuse. *Desalination* 203 (1–3): 15–19.

Wood, D.A. (2019). Microbial improved and enhanced oil recovery (mieor): Review of a set of technologies diversifying their applications. *Advances in Geo-Energy Research* 3 (2): 122–140.

Xu, M., Bernards, M., and Hu, Z. (2014). Algae-facilitated chemical phosphorus removal during high-density *Chlorella emersonii* cultivation in a membrane bioreactor. *Bioresource Technology* 153: 383–387.

Xu, M., Li, P., Tang, T. et al. (2015). Roles of SRT and HRT of an algal membrane bioreactor system with a tanks-in-series configuration for secondary wastewater effluent polishing. *Ecological Engineering* 85: 257–264.

Yang, J., Gou, Y., Fang, F. et al. (2018). Impacts of sludge retention time on the performance of an algal-bacterial bioreactor. *Chemical Engineering Journal* 343: 37–43.

Yew, G.Y., Lee, S.Y., Show, P.L., et al. (2019). Recent advances in algae biodiesel production: From upstream cultivation to downstream processing. *Bioresource Technology Reports* 7: 100227.

Yogarathinam, L.T., Gangasalam, A., Ismail, A.F. et al. (2018). Harvesting of microalgae *Coelastrella* sp. FI69 using pore former induced TiO_2 incorporated PES mixed matrix membranes. *Journal of Chemical Technology & Biotechnology* 93 (3): 645–655.

Yoo, C., La, H.J., Kim, S.C. et al. (2015). Simple processes for optimized growth and harvest of *Ettlia sp.* by pH control using CO_2 and light irradiation. *Biotechnology and Bioengineering* 112 (2): 288–296.

Zabed, H.M., Akter, S., Yun, J. et al. (2019). Recent advances in biological pretreatment of microalgae and lignocellulosic biomass for biofuel production. *Renewable and Sustainable Energy Reviews* 105: 105–128.

Zhang, Y. and Fu, Q. (2018). Algal fouling of microfiltration and ultrafiltration membranes and control strategies: A review. *Separation and Purification Technology* 203: 193–208.

Zhang, H., Li, B., Sun, D. et al. (2018). SiO_2-PDMS-PVDF hollow fiber membrane with high flux for vacuum membrane distillation. *Desalination* 429 (1): 33–43.

Zhen-Feng, S., Xin, L., Hong-Ying, H. et al. (2011). Culture of *Scenedesmus* sp. LX1 in the modified effluent of a wastewater treatment plant of an electric factory by photo-membrane bioreactor. *Bioresource Technology* 102 (17): 7627–7632.

Zhou, W., Lu, Q., Han, P. et al. (2020). *Microalgae Cultivation and Photobioreactor Design, Microalgae Cultivation for Biofuels Production*, 31–50. *Elsevier*.

Zhu, L., Wang, Z., Shu, Q. et al. (2013). Nutrient removal and biodiesel production by integration of freshwater algae cultivation with piggery wastewater treatment. *Water Research* 47 (13): 4294–4302.

Zhuang, L.L., Yu, D., Zhang, J. et al. (2018). The characteristics and influencing factors of the attached microalgae cultivation: A review. *Renewable and Sustainable Energy Reviews* 94: 1110–1119.

Zuccaro, G., Yousuf, A., Pollio, A. et al. (2020). *Microalgae Cultivation Systems, Microalgae Cultivation for Biofuels Production*, 11–29. *Elsevier*.

Zulhairun, A., Wijiyanti, R., Widiastuti, N. et al. (2020). Prospects of nanocomposite membranes for natural gas treatment. In: *Nanocomposite Membranes for Water and Gas Separation*, 355–378. *Elsevier*.

12

Engineering Plants for Metal Tolerance and Accumulation

Fernanda Maria Policarpo Tonelli[1], *Flávia Cristina Policarpo Tonelli*[2], *Moline Severino Lemos*[3], *Helon Guimarães Cordeiro*[4] *and Danilo Roberto Carvalho Ferreira*[5]

[1] *Department of Cellular Biology, Institute of Biological Science, Federal University of Minas Gerais, Belo Horizonte, Brazil*
[2] *Department of Biochemistry, Federal University of São João del Rei, Divinópolis, Brazil*
[3] *Department of Cellular Biology, Institute of Biological Science, Federal University of Minas Gerais, Belo Horizonte, Brazil*
[4] *Department of Biochemistry and Tissue Biology, University of Campinas (UNICAMP-SP), Campinas, Brazil*
[5] *Department of Materials, CDTN/CNEN, Federal University of Minas Gerais, Belo Horizonte, Brazil*

CHAPTER MENU

12.1 Introduction

Human action, ignoring sustainability principles, can cause enormous damage to ecosystems, resulting in polluting of soil, surface water, groundwater and air. It can present serious consequences, resulting in the death of living beings (Chaoua et al. 2019; Zheng et al. 2020). Inorganic and/or organic pollutants are the result of activities such as industrial ones (that generate toxic wastes), mine exploitation, pesticide use, etc. (Jafari et al. 2013; Barrios-Estrada et al. 2018; Carpenter 2018; Bilal et al. 2019; Mendes et al. 2019; Pesantes et al. 2019; Pu et al. 2019; Rosculete et al. 2019; Vázquez-Luna and Cuevas-Díaz 2019; Zhang et al. 2019).

The major problems faced by humanity regarding environmental pollution arise from persistent contaminants that are able to remain for a very long period of time in the polluted areas without undergoing any form of transformation, threatening further generations (Fulekar and Pathak 2020; Gavrilescu et al. 2015).

Bioresource Technology: Concept, Tools and Experiences, First Edition. Tanveer Bilal Pirzadah, Bisma Malik, Rouf Ahmad Bhat, and Khalid Rehman Hakeem.
© 2022 John Wiley & Sons Ltd. Published 2022 by John Wiley & Sons Ltd.

Polluted water, for example, is an enormous problem worldwide. Providing water of good quality to the population is an important concern in order to allow people healthy lives. However, conventional methods applied to treat water (such as reverse osmosis) need urgent improvements, since they tend to have elevated costs and generate more pollution by their processing (Albering et al. 2016; Bilal et al. 2018; Dasgupta et al. 2015).

When it comes to heavy metals, they are highly toxic and persistent contaminants, difficult to eliminate from contaminated areas (Tonelli and Tonelli 2020). The term "heavy metals" is used to refer to metals and metalloids possessing atomic density above 4 g/cm^3, such as manganese, arsenic, nickel, chromium, vanadium, cadmium, lead, cobalt, iron, zinc, mercury and copper (Hawkes 1997). These heavy metals may reach living organisms from domestic, chemical, industrial, agricultural and technological sources (Aprile and De Bellis 2020). Some of them, known as essential heavy metals (i.e., Fe, Cu, Fe, Mn, Ni and Zn) are required in small quantities for plant survival (Cempel and Nikel 2006). However, when present in abundance they can do as much damage as non-essential heavy metals (i.e., As, Cd, Pb and Hg) that are highly threatening to the survival of human beings (Fasani et al. 2018).

They can severely impact human health as a result of acute or chronic exposure, even in small amounts, by poisoning people, inducing oxidative stress that causes diseases (even cancer) and/or causing death. They may accumulate in living beings performing trophic level transfer in the food chain; by doing damage to the first poisoned organism and subsequently to the ones that will eat it or eat its predator (Ali et al. 2019; Alkorta et al. 2004; Ashraf et al. 2019; Jacob et al. 2018; Jaishankar et al. 2014).

It is necessary to develop and execute efficient and eco-friendly protocols to promote the extraction of these toxins from contaminated environments. When addressing these strategies, bioremediation-based protocols deserve attention.

12.2 Metals' Bioremediation

The remediation of environmental pollution can be performed by physicochemical methods, biological methods or protocols that combine both. The suitable protocol is chosen by: i) considering the polluted area's characteristics; ii) evaluating how polluted the environment is; iii) taking into consideration the available financial resources; and iv) considering the type of pollution present (Azubuike et al. 2016). However, the physicochemical methods, such as incineration, photocatalysis, chemical extraction, soil washing and electrokinetics, when applied on the large scale, may present higher costs and even generate more toxic waste (making the previously existing problem worse) (Baldissarelli et al. 2019; Bilal et al. 2018; Dasgupta et al. 2015; Ye et al. 2017).

Bioremediation strategies, in their turn, involve the use of living oganisms like microbes (e.g., fungi and bacteria) and plants; are environmentally-friendly and generally incur lower costs (Azubuike et al. 2016; Strong and Burgess 2008). The environmental biotechnology field offers safe, economic, feasible and efficient protocols as an alternative to classical physicochemical treatments that are generally used to fulfill this goal (Fulekar and Pathak 2020; Guerra et al. 2018; Koźmińska et al. 2018).

Microbes can remediate metals through processes such as volatilization, extracellular chemical precipitation, metal binding and valence transformation (Huang et al. 2008; Wu et al. 2010). *Pseudomonas aeruginosa* strain KUCd1 can accumulate high levels of cadmium, thus remediating industrial wastewater containing this metal (Sinha and Mukherjee 2009). *Streptomyces* sp. MC1 can act in polluted soil by converting Cr^{6+} into a less toxic form (Polti et al. 2009) and the methane-oxidizing bacteria *Methylococcus capsulatus* has also proved to be able to remediate this pollutant (Hasin et al. 2010)

Microbes acting as plant endophytes and also plant growth promoting rhizobacteria can influence the plant's capacity to remediate, improving it, in a process known as microbe-assisted phytoremediation, especially with a phytoremediation mechanism known as phytodegradation (a mechanism in which a microorganism associated with the plant's roots takes part in the process) (Cheng et al. 2019; Gong et al. 2018, 2019; He et al. 2020). The fungi *Funneliformis mosseae* can assist the plant *Robinia pseudoacacia* to immobilize Pb ions and attenuate their toxicity to the vegetable's roots during phytoremediation (Huang et al. 2019). The fungi *Rhizophagus intraradices* avoids the overload of rice plants during cadmium remediation by promoting the proliferation of Actinobateria (Chen et al. 2019). Due to this, there are researchers that consider these microorganism, and not only the plant species, as agents to promote phytoremediation (He et al. 2020).

Not only microorganisms can influence a plant's capacity to remediate polluted environments. Abiotic factors such as pH, temperature and chemical composition of the pollutants themselves and also substances that may sequester or react with contaminants can also have an influence on phytoremediation potential (He et al. 2020; Xu et al. 2019). For example, the leachates from Italy's landfills can be remediated by *Chrysopogon zizanioides*. However, the efficiency of the process depends on the sample's composition (Fasani et al. 2019). Plant–chemical combined remediation protocols have been developed aiming at process quality improvement. For example, the capacity of the radish to remediate cadmium and lead polluted environments can be improved by citric acid's presence (that can reduce these metals' toxicity and consequently facilitates Cd transportation inside the plant) (Chen et al. 2003). Biochar, for example, is an interesting tool, which is obtained through different protocols (varying mainly according to biomass type) of biomass pyrolysis and exists in multifunctional carbonaceous material that can efficiently sequestrate carbon (Woolf et al. 2010; Zheng et al. 2020). However, it can also help to regulate soil pH and reduce the stress, especially from metal pollutants, which may affect plant species during phytoremediation. Biochar can help *Ricinus communis* to deal with oxidative stress during Pb phytoremediation, increasing the plant's tolerance to the contaminant, nutrient intake, plant growth and consequently the efficiency of the whole process (Kiran and Prasad 2019).

Vegetable species can promote *in situ* and *ex situ* (through different mechanisms such as rhizofiltration) restoration of polluted environments in an auto-sustained manner (preventing erosion and improving local fertility), which is an interesting feature of phytoremediation (Ai et al. 2016; Brunner et al. 2008; Muthusaravanan et al. 2018; Nwoko 2010; Panesar et al. 2019; Strong and Burgess 2008).

12.2.1 Metal Phytoremediation

There are plant species that are naturally capable of phytoremediating polluted environments. For example, *Eichhornia crassipes*, the water hyacinth, can remediate diverse kinds of pollutants such as organic ones like dyes, or inorganic ones like phosphate, radioactive isotopes and also heavy metals (Jayaweera and Kasturiarachchi 2004; Nie et al. 2015; Priya and Selvan 2017; Thapa et al. 2016).

When compared to microorganisms, plants are more suitable for use as remediate metals. Instead of just converting them into a less toxic form, there are mechanisms of phytoremediation that allow plants to remove these contaminants from contaminated areas (Garbisu et al. 2002; Ojuederie and Babalola 2017). It is also interesting to highlight that there are phytoremediation protocols that require low initial investment and the remediation process can be maintained in an auto-sustained and safe way, being a well accepted strategy (Muthusaravanan et al. 2018).

Phytoaccumulation (also known as phytoextraction), phytodegradation (also known as phytotransformation), phytohydraulic control, phytostabilization,phytovolatilization, rhizofiltration and rhizodegradation are important mechanisms of phytoremediation (Abdullahi 2015; Dzantor 2007; Muthusaravanan et al. 2018). Besides phytodegradation and rhizodegradation, which are performed generally for organic contaminants, metals can be efficiently removed by these methods (Dominguez et al. 2019; Muthusaravanan et al. 2018) and especially by phytostabilization, phytovolatilization and phytoextraction, which involve pollutant sequestration (Nwoko 2010). The methods vary depending on plant species, whether it is soil or water to be remediated, and which pollutant should be removed (Panesar et al. 2019).

Hyperaccumulators are plant species that can efficiently perform phytoextraction of metals, removing them from polluted areas while accumulating the pollutants in certain parts of the plant (Lajayer et al. 2019; Muthusaravanan et al. 2018). Plants from the Brassicaceae family, for example, are able to hyperaccumulate metals by transporting them from root to shoot. *Brassica juncea* is such an example and can have the potential (increased by the presence of thiosulfate) to accumulate mercury. The thiosulfate can damage the plant cell's plasma membrane and complex with Hg from the contaminated environment. The complex can easily penetrate the structurally damaged part, increasing the heavy metal accumulation capacity (Wang et al. 2018). However, an important disadvantage related to this species' use in phytoremediation is how time-consuming the process can be due to the slow growth some of these plants present when growing in areas containing high levels of pollutants (Jafari et al. 2013).

Rhizofiltration involves plant roots immobilization of contaminants, for example, as with *Arundo donax* dealing with copper pollution (Oustriere et al. 2017). Phytostabilization involves the immobilization of pollutants in the rhizosphere region and is a mechanism that allows *Erica australis* to deal with heavy metals in the soil (Monaci et al. 2019). Through phytovolatilization, metals such as mercury can be converted to a less toxic form and released to the environment by *Brassica juncea* (Moreno et al. 2008).

Genetic modification in plants can also be used to make it possible to enhance metal remediation. Genes from vacuolar transporters, for example, are interesting to be used in these strategies, playing an important role in the development of plants with a high capacity for accumulating metals. Vacuolar metal transporters have proved to be efficient in the generation of transgenic plants with improved phytoremediation capacity, especially in soils polluted with heavy metals such as Cd and As (Zhang et al. 2018). The *Sacharomyces cerevisae* gene that codes for the YCF1 protein, that deals with cadmium pollution by transporting the pollutant to the interior of the vacuoles, allows *Arabidopsis* plants to dramatically increase their tolerance and accumulation rate of Pb^{2+} and Cd^{2+} (Song et al. 2003). Besides vacuoles, heavy metal ions can also be sequestrated and compartmentalized into other locations inside the plant, such as leaf petioles, leaf sheathes and trichomes (Eapen and D'souza 2005; Robinson et al. 2003), where heavy metals cause less damage to the plant.

12.2.2 Non-Target Specific Engineered Plants to Metal Phytoremediation

Phytoremediation is an important strategy to mitigate the impact of heavy metal pollution on the environment. It consists of green low-cost technology, powered by solar energy (Mendez and Maier 2008). Plant species most suitable to be used in this kind of protocol should present fast growth rate, high biomass yield, robustness, high tolerance to significant levels of pollutants, large root systems and the ability to immobilize and/or absorb large amounts of contaminants (Padmavathiamma and Li 2007). Several species of plants have been identified as possessing a high potential to be explored for phytoremediation, also being targets to become even more improved by scientists through transgenesis (Cherian and Oliveira 2005).

The advances in the field of molecular biology and genetics regarding genome elucidation and genome editing technologies have allowed the development of genetic engineering to generate organisms optimized to perform desirable tasks such as dealing with environmental pollution (Liu et al. 2019; Zhao et al. 2017).

Living organisms can be genetically modified to improve an already existing desirable characteristic or to confer to it a new desirable characteristic to promote bioremediation (Liu et al. 2019; Prasad 2019; Zhao et al. 2017).

Hence, to deal with metal pollution, recombinant DNA technology possesses elegant solutions to alter the DNA content of organisms. It offers tools to introduce gene(s) generating transgenic organisms, to ablate the expression of genes generating knockout organisms or to alter target DNA sequences (Doyle et al. 2012).

The basis for the creation of popular known transgenic organisms (technically referred to as allotransgenic) involves the transfer of genetic material, not from the organism itself, but from a different species of organism. The foreign gene is called the transgene and the main idea is to confer a new characteristic to the transgenic organism (Ozyigit et al. 2020). However, it is also possible to perform autotransgenesis to improve, for example, already existing features by using DNA sequences from the same species (Reis et al. 2006).

There are also important differences between genetic engineering and genome editing technologies. Genome editing techniques are more recent and allow DNA editing at targeted locations. The genetic engineering discussed in this chapter uses molecular biology tools to integrate recombinant DNA (DNA sequence of interest from the same species or from a different species) into an organism's genome (Mackelprang and Lemaux 2020).

Plants can be genetically engineered to convert a non-remediating vegetable species into organisms capable of promoting phytoremediate, or also to increase a previously present natural remediation ability. For example, a species can already be capable of dealing with a pollutant but when the concentration of the pollutant reaches high levels, the plant suffers stress, and the toxicity may even limit the success of phytoremediation by inducing the plant's death (Tiwari and Lata 2018). By genetically modifying this plant, the problem can be solved.

The main problems researchers aim to solve by genetically engineering plants to perform efficient phytoremediation are to reduce the risk of food chain contamination occurrence (Nesler et al. 2017), reduce the remediation time (Feng et al. 2017), increase the plant's biomass and growth rate (Pérez-Palacios et al. 2017) and enhance tolerance to stressful conditions such as the one induced by the presence of pollutants (Charfeddine et al. 2017). Therefore, strategies commonly focus on overexpressing genes encoding metal transporters, metal chelators and/or components of antioxidant machinery to achieve a more efficient metal tolerance and remediating capacity (Koźmińska et al. 2018).

A large variety of genes can be used to optimize plants to phytoremediate metals. They can be, for example, DNA sequences from the same species that will be modified. In this kind of strategy, an autotransgenic organism is generated overexpressing the transgene, enhancing a phytoremediation capacity that already exists. StDREB transcription factors, when overexpressed in potato plants, allowed the transgenic to deal with Cd pollution better than the wild type (Charfeddine et al. 2017). The gene *SpMTL* from *Sedum plumbizincicola* (Cd/Zn hyperaccumulator) contains the information to synthesize a metallothionein-like protein with the capacity to offer the plants increased capacity to tolerate Cd and accumulate it in its leaves (Peng et al. 2017).

Plant species can be modified to contain, for example, genes from microbes that code for proteins involved in metal remediation. The *merApe9* gene is a modified version of *Escherichia coli* gene *merA* that allows the synthesis of a reductase from mercury ions. When inserted into plants, this DNA fragment allows mercury detoxification through phytoremediation of mercury-contaminated regions (He et al. 2001; Rugh et al. 1996). The *Pseudomonas fluorescens copC* gene allows transgenic tobacco plants with hairy roots to hyperaccumulate copper (Pérez-Palacios et al. 2017). This gene that encodes for a periplasmic Cu-binding protein could also modify *Arabidopsis thaliana* to induce a 5-fold increase in root and shoot Cu accumulation (Rodríguez-Llorente et al. 2012). *Pseudomonas putida* genes *CzcB* or *CzcA* allow the same vegetel species to enhance Cd accumulation in roots when developing in a cadmium contaminated area when compared to wild type (Nesler et al. 2017). *Brassica juncea* engineered to overexpress *gshI* gene from *E. coli* (coding a glutamylcysteine synthetase) presents enhanced tolerance to cadmium, growing better in its presence, and can also increase Cd's accumulation capacity in its shoots (Zhu et al. 1999a). A similar result is observed when the same species (*Brassica juncea*) is modified to overexpress the *gshII* gene (Zhu et al. 1999b). A sterile line of poplar (*Populus alba* X *P. tremula* var. glandulosa) could be successfully engineered to contain yeast cadmium factor 1 (*ScYCF1*) gene (a heavy metal resistance gene) and remediate soil from a closed mine. The genetically modified plant (GMP) grew better than wild type organisms in this area, and increased Cd, Zn and Pb accumulation in the root, thus better tolerating the stress induced by these pollutants (Shim et al. 2013). An interesting study used genes from *Pseudomonas putida* CzcCBA efflux system to produce desirable effects on *Nicotiana tabacum*, choosing the site of Cd accumulation during phytoremediation, aiming to control toxicity of feed crops cultivated in Cd presence. GMPs modified with *CzcB* or *CzcA* accumulated more Cd in roots and less in shoots when compared to wild type. Plants modified with *CzcB* and *CzcA* accumulated less Cd in both plant parts, whereas organisms modified with *CzcB*, *CzcC* and *CzcA* at the same time accumulated less Cd in shoots (Nesler et al. 2017). To induce hairy roots on *Scirpus americanus*, the agropine-type plasmid pRiA4 of *Agrobacterium rhizogenes* was used to modify the plant species. As a result, besides a faster growth rate, the GMP presented increased capacity to accumulate Pb and Cr, remediating polluted water (Alfaro-Saldaña et al. 2016).

Genes from plant species can also be used to engineer plants to phytoremediate. Cadmium accumulation in tobacco plants was made possible by inserting the *OsMTP1* gene from *Oryza sativa* L. (indica rice) to be overexpressed. It resulted in metal tolerance/transport protein family's production that allowed reduction of Cd-induced stress over tobacco plants during phytoremediation. Transgenic plants grew better and could also better tolerate arsenic pollution in soil samples (Das et al. 2016). The *O. sativa* gene *OsMT2c* encoding a metallothionein, when overexpressed in *A. thaliana*, offers enhanced capacity of Cu remediation and increased tolerance to Cu stress (Liu et al. 2015). Cassava could also be engineered to accumulate metal ions such as zinc. *ZIP1* and *MTP1* genes from *Arabidopsis thaliana*, when overexpressed, could induce Zn accumulation in roots (Gaitán-Solís et al. 2015). The transporter coded by the AtATM3 gene is a member of the ATP-binding cassette transporter family. After being obtained from *A. thaliana* and inserted into *Brassica juncea*, it showed enhanced Cd and Pb tolerance during phytoremediation (Bhuiyan et al. 2011). The CYP450-like gene (Os08g01480) from rice could enhance *A. thaliana*'s tolerance to abiotic stresses allowing, for example, remediation of As, Cr and Cd (Rai et al. 2015).

Genes from hyperaccumulators, plants that can naturally survive in contaminated environments, are very useful in generating transgenic plants to phytoremediate metals. The coding sequence for selenocysteine methyltransferase from *Astragalus bisulcatus* can be used to convert non-hyperaccumulator species into organisms capable of remediating metals. An interesting feature is that this kind of strategy allows the solution of the problem previously mentioned regarding the hyperaccumulator: they grow slowly. By using the gene from *A. bisulcatus*, the non-hyperaccumulator *Arabidopsis* (that grows fast) could be efficiently engineered to remediate selenium in a rapid way (Ali et al. 2013; LeDuc et al. 2004).

Arabidopsis thaliana, an important plant model, became more capable of accumulating Zn and Cd from polluted samples due to the modification to overexpress a gene from *Noccaea cearulescens* (a hyperaccumulator from the Brassicaceae family), coding the protein ZNT1 (a zinc transporter) (Lin et al. 2016). *Psychotria gabriellae* is a species capable of hyperaccumulating nickel and the gene of the metal transported PgIREG1 obtained from this species, after being expressed in *A. thaliana*, could confer Ni tolerance (Merlot et al. 2014). The *SnYSL3* gene (that contains the information to produce a metal transporter) from *Solanum nigrum* (cadmium hyperaccumulator) could allow *A. thaliana* to increase Fe and Cd transportation (root to shoot) in the presence of pollutant Cd, making remediation faster and increasing the plant's tolerance (Feng et al. 2017). *Brassica campestris* L. genes of metallothioneins MT1 and MT2 offered *A. thaliana* a better phytoremediaiton capacity and better tolerance to Cd and Cu pollution, reducing reactive oxygen species generation in response to their presence (Lu et al. 2015).

Salix matsudana could offer increased Cu tolerance and consequently remediation capacity after being engineered to express the *Tamarix hispida ThMT*3 gene that allows the synthesis of a metallothionein (Yang et al. 2015).

Tobacco plants (*Nicotiana tabacum*) can be converted into copper phytoremediators by overexpression of metallothionein 1 (MT1) from *Elsholtzia haichowensis* (a copper accumulator species) (Xia et al. 2012) or by being engineered to express *Sedum alfredii*'s (a plant species known as a Cd/Zn co-hyperaccumulator) *MT2* gene that allows the generation of a cysteine-rich metal-binding protein (Zhang et al. 2014). Both modifications allow plants to accumulate large amounts of copper in their roots (increasing accumulation capacity) by dealing well with the oxidative stress it caused (increasing plant's tolerance). When expressing the *MT1* gene, for example, the transgenic plant presented reduced levels of lipid peroxidation (Xia et al. 2012).

Mammalian DNA sequences can also be used to engineer plants. *Linum usitatissimum* L. is naturally able to phytoremediate Cd. However, this species' manipulation to overproduce a Cd-binding peptide, that consists in a fusion of the β-glucuronidase *gus* gene and the α-domain of mammalian metallothionein 1a (αMT1a), improved remediation efficiency (Vrbová et al. 2015). Alfalfa plants co-expressing glutathione *S*-transferase and human cytochrome P450 2E1 genes could remediate mixed pollution, tolerating better not only organic pollutants, but also Cd and Hg (Zang et al. 2013).

12.2.3 Target Specific Genomic Engineering Technique to Enhance Plants Metal Tolerance and Accumulation

Bioremediation strategies can also benefit from site-directed genome editing tools such as Transcription activator-like effector nuclease (TALEN), zinc finger nuclease (ZFN) and the clustered regularly interspaced short palindromic repeats (CRISPR)/Cas system to generate genetically-modified organisms optimized to perform remediation of environmental pollutants, reducing the problems associated with off-target integration of transgenes (e.g., gene silencing) (Jaiswal et al. 2019) (Figure 12.1).

Zinc finger nuclease works as a dimer containing each FokI domain (to generate a site-directed double-strand break after recognizing 3 to 4 base-pair sequence on DNA) and DNA sequencing recognition domains, the same as TALENs (both methods use artificial fusion proteins) (Bortesia and Fischer 2015; Kim et al. 1996; Petolino 2015; Urnov et al. 2010).

The nucleases used in TALEN methodology are generated through a combination of the endonuclease domain of the FokI restriction enzyme with artificial DNA sequences (TAL effector repeats). Dimers of these structures are able to insert the transgene after generating a target-specific double-strand break on the DNA. The discovery of TAL

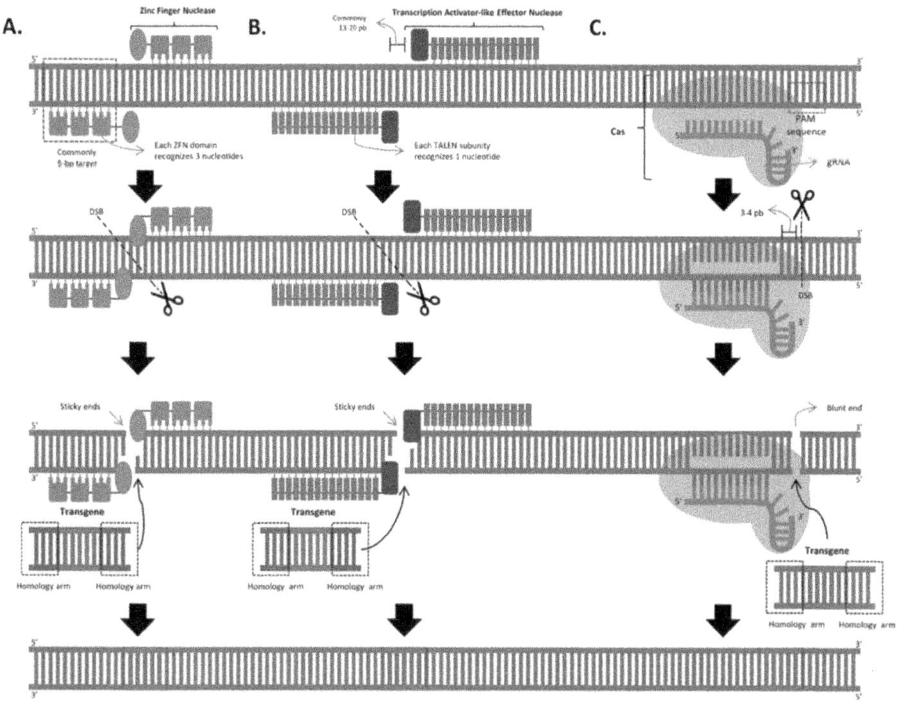

Figure 12.1 Logical steps of technologies of site-directed genome editing that can be applied to promote the integration of a particular DNA sequence of interest (the transgene represented in red) in a plant's genome. (A) Zinc finger nuclease (ZFN) system (FokI domain in orange and recognition domains in green). (B) TALEN system (FokI domain in purple and recognition domains in blue). (C) CRISPR/Cas system (gRNA in green and Cas in pink) (*source*: based on Jaiswal et al. 2019).

effectors was performed on plant pathogens that possessed these 10 to 30 tandem repetitive sequences of approximately 35 amino acids each, containing two adjacent residues responsible for conferring recognition specificity to the system (Bogdanove and Voytas 2011; Christian et al. 2010). In order to generate TALEN to be used in genetic engineering protocols reducing off-target integration, the recommendation is to design them to bind to a specific 18-base-pair or longer sequences (Gupta and Musunuru 2014). TALEN, the same as ZFN (that is also a designed nuclease), can perform from single nucleotide mutations to large sequence deletions, rearrangements and/or integrations in the genome (if the sequence containing the transgene is also provided with homology arms) once they are capable of cleaving specific DNA sequences, thus allowing the recruitment of different DNA repair mechanisms that will allow various types of changes in the genome (Curtin et al. 2012).

In the CRISPR/Cas system, Cas9 is commonly the chosen nuclease from *Streptococcus pyogenes*. However, other nucleases can also be used, such as Cpf1 from *Prevotella* and *Francisella* (Zaidi et al. 2017) to be guided by a guide RNA (gRNA – designed to be complementary and pair to the desired DNA sequence) to the target region of DNA in which the transgene's insertion is desired (through homologous recombination due to homology arms on both sides of transgene sequence also provided by the protocol). After reaching the location that gRNA is guided to, the nuclease can break the DNA and a double-strand break is generated. gRNA binding to DNA consists of approximately 20 base-pair regions located 3 base-pair distant (in an upstream direction) of the proto-spacer-adjacent motif (PAM). If no transgene is placed in the system to be inserted, a double-strand break is generated which will be repaired by error-prone mechanism non-homologous end-joining, that brings together the broken ends of DNA, resulting in small insertions or deletions (indels) near to the cleavage site, which may cause frameshift mutations or premature termination codons (Cong et al. 2013; Jinek et al. 2012; Mali et al. 2013). The CRISPR-Cas9 technique allows performing knockouts, targeting transcriptional regulation or making substitutions in the genome or insertion of new sequences. CRISPR allows generation of several simultaneous modifications by simply using more than one gRNA (Basharat et al. 2018; Sander and Joung 2014; Yin et al. 2017).

The CRISPR/Cas system is promising (Basharat et al. 2018) and advantageous when compared to the TALEN and ZFN strategies, since it does not depend on the generation of fusion protein capable of recognizing DNA sequences, making it an easier tool to be designed (Holkers et al. 2013; Jinek et al. 2012; Mali et al. 2013; Sander et al. 2011). However, the use of this method in phytoremediation is still underrepresented (Basharat et al. 2018).

Researchers are, through technological platforms such as omics ones, looking for candidate genes to promote transgenesis in plants to enhance abiotic stress tolerance using site-direct genome editing. In the field of phytoremediation, CRISPR/Cas can be used to increase or allow the synthesis of metal ligands (such as metallothioneins and phytoquelatins), metal transport proteins (from the CDF, HMA, MATE, YSL and ZIP families) and plant growth hormones and root exudates (particularly LMWOA and siderophores) (Basharat et al. 2018). It is important to highlight that many factors contribute to the use of CRISPR-Cas9 genome editing in various plant species, such as the availability of plant genome sequences, bioinformatics-based approaches and the availability of optimized versions of Cas9 codons for monocots and dicots (Lowder et al. 2015). The reprogramming of the plant's genome, mediated by CRISPR, can contribute considerably to phytoremediation, including its use in the regulation of gene expression by repression or activation of transcription (CRISPRi and CRISPRa). The simplicity, low cost and capabilities of this gene editing technique may soon potentiate the ability of plants to phytoremediate (Basharat et al. 2018).

In the near future, good results are expected in this field (Anwar and Kim 2020). However, the CRISPR/Cas system has already been successfully applied to plants to generate mutations and deletions, for example to study gene function or to improve desirable characteristics and to promote interference (Ali et al. 2015; DalCorso et al. 2019; Didovyk et al. 2016; Khatodia et al. 2016; Ma et al. 2015; Wang et al. 2015; Xie et al. 2015; Xing et al. 2014; Zhang et al. 2015). *Catharanthus roseus MDS1* to *MDS4* genes were studied through CRISPR/Cas multiplex mutagenesis and revealed to be involved in response to different metals: Ni^{2+} during hypocotyl elongation and Cd^{2+} and Zn^{2+} during root growth (Richter et al. 2018).

Thus, not only gene insertion can be used to increase the tolerance of plants to pollutants. Besides mutations, gene knock-out also presents an interesting potential. Rice that suffered knock-out through CRISPR/Cas of gene *NRAMP5*, that codes for a metal transporter, was able to better tolerate Cd pollution, reducing this pollutant's accumulation in aerial parts, roots and grains in hydroponic cultivation. It is noteworthy that the biomass remained similar to the wild type, proving that the GMP generation protocol can be employed in order to achieve an organism suitable for environmental sustainability, thus reducing the risk of metal pollutants insertion into the food chain (Tang et al. 2017).

It is also important to mention that target-genome editing technologies can also contribute to phytoremediation by allowing the generation of more competent bacterial strains, since some phytoremediation mechanism can also benefit from the association of plants with plant growth-promoting rhizobacteria (PGPR) (Basharat et al. 2018). Therefore, their application to plants and PGPR could increase the production and/or tolerance of biomass, accumulating heavy metals and detoxifying polluted areas.

12.2.4 Important Methodologies to Engineer Plants to Metals Phytoremediation

There are different methodologies to generate transgenic plants. However, most research works use bacteria from the *Agrobacterium* gender: *Agrobacterium rhizogenes* and more commonly, *Agrobacterium tumefaciens.*

A. tumefaciens is a bacterium from the soil that naturally acts as a plant pathogen capable of infecting dicots. It is capable of transforming plants with a DNA content called T-DNA present on a Ti, a large virulence plasmid. As a consequence of phytohormones action, it induces a tumor (gall) in the plant. The pathogen can also make the plant produce opines for its nutrition. In Ti plasmid, the virulence genes code for proteins that process and translocate T-DNA to the nucleus to be integrated into the plant's chromosome. If the integration is not performed, the expression will be only transient (Thompson et al. 2020).

A biotechnological procedure was developed based on a natural process to engineer the Ti plasmid to contain desirable transgene in T-DNA replacing some virulence genes, so the transgene can be inserted into the host plant's DNA to generate the transgenic plant (Cunningham et al. 2018) (Figure 12.2).

Although strategies using bacteria from the Agrobacterium gender are the most used by researchers to genetically modify plants, there are also other strategies capable of achieving this goal. It is possible, for example, to generate a GMP by using viral vectors to deliver the transgene (Keshavareddy et al. 2018). The virus can contain genetic material DNA (e.g., Wheat dwarf virus) or RNA (e.g., Tobacco rattle virus) that are being tested as vectors to engineer diverse plant species (Zaidi and Mansoor 2017).

Biolistics is also a viable technique. It consists of microprojectile bombardment of plants (commonly callus tissues) to deliver transgenes (Ismagul et al. 2018), not only for metal remediation. Tobacco plants overexpressing the gene of maize glutathione S-transferase I could be programmed, after biolistic method performing, to remediate chloroacetanilide herbicides (Karavangeli et al. 2005). Metallothionein genes *AtMT2a*

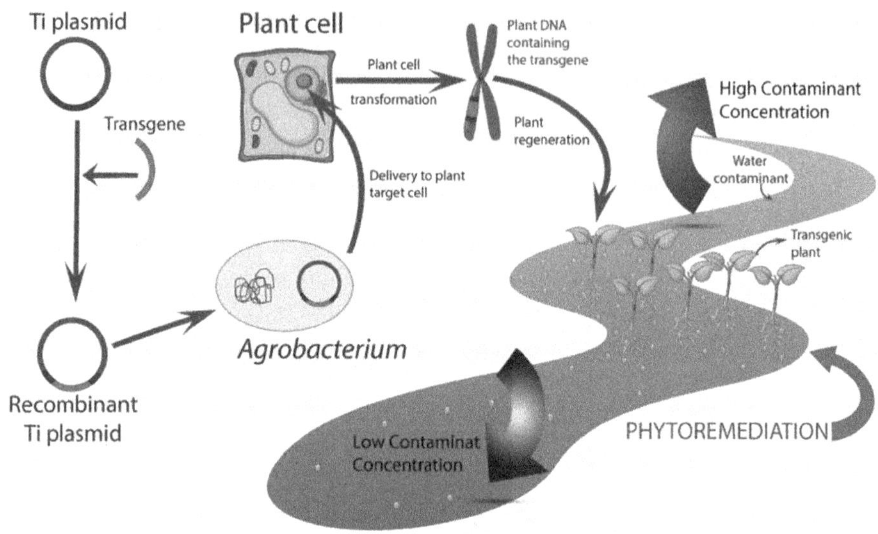

Figure 12.2 The genetic engineering ofplants to phytoremediation by using bacterium from *Agrobacterium* gender. The transgene (e.g., a gene from a protein that binds to heavy metal ions) is inserted into the Ti-plasmid at the T-DNA site. The transgene is integrated into the host plant's DNA from target cells and after vegetative regeneration, the transgenic organism suitable for phytoremediation is obtained (in this example, as the protein is bound to heavy metals ions, the GMP can be used to remediate the pollution caused by these inorganic contaminants).

and *AtMT3* from *A. thaliana* were delivered to *Vicia faba* by biolistic bombardment to confer Cd tolerance to the transgenic plant (Lee et al. 2004).

There are also strategies that involve plastid genome transformation (Chen et al. 2014), aiming to surpass some difficulties associated with genomic manipulation such as wrong insertion site causing effects like gene silencing (Daniell et al. 2016) and also to contain the risk of gene transfer into the environment (Maliga 2002). To differentiate the genetically-modified organism from the one obtained through genomic modification, the plants generated through plastid genome engineering are generally called transplastomic plants, and are interesting, especially when protocols require engineering with multiple genes. Tobacco plants, for example, can be successfully engineered to remediate mercury in organic and inorganic forms by expressing both mercuric ion reductase and organomercuriallyase genes present in transgenic chloroplasts (Hussein et al. 2007).

12.3 Omics as Tools to Elucidate Important Genes to Plants Engineering

Plants exposed to pollutants such as heavy metals suffer physiological and biochemical alterations, and in order to survive need to be able to deal with abiotic stress. Signaling pathways involving a large array of molecules need to be triggered to allow plants to respond to this situation, and the response can alter the profile of, for example, RNA, metabolites and proteins (Han et al. 2019; Liu et al. 2013; Nakabayashi and Saito 2015) that can be accessed through technological platforms know as omics.

So, in order to develop strategies to make plants more capable of dealing with this stressful exposure to pollutants, it is interesting to analyze, for example, transcripts of tolerant species under the contaminant's stress compared to the transcripts generated in the absence of pollutants. Some RNAs will present increased levels being possibly involved in tolerance. If they code for proteins, it is interesting to study them to understand the tolerance mechanisms involved. qRT-PCR studies in *Oryza sativa* indicated that OsMSR3 transcript's level is enhanced under Cd stress, coding for a low-molecular-weight heat shock protein. Researchers are also dedicated to investigate if the protein also helps a plant to deal with cadmium stress and decided to express it in *Arabidopsis thaliana*. The correspondent gene increased the plant's tolerance to Cd through stress-related genes expression induction (Cui et al. 2013). The mechanism by which tolerance improvement happens was later investigated, which involved antioxidative defense mechanisms (enhancing expression of superoxide dismutase and peroxidase), abscisic acid synthesis and production of this hormone's responsive genes (Cui et al. 2019). So, technological platforms such as transcriptomic, proteomic and metabolomic are interesting tools, not only helping to understand better the response process to the stress induced by pollutants but also to reveal candidate genes to plant engineering protocols to an efficient phytoremediation.

Metabolomics allow low-molecular weight metabolite identification and quantification and consequently the data generated reflects the plant's physiological activities, which can be highly influenced by environmental conditions such as the presence of the pollutant (Arbona et al. 2013; Bundy et al. 2005). By studying metabolites that have their production increased as a consequence of metallic pollutant-induced stress (Table 12.1), it is possible to find candidate genes for transgenic plant generation aiming to deal better with this kind of pollution. In this case, the genes of enzymes responsible for this metabolite generation are good candidates: as genes of enzymes involved in amino acid proline synthesis. Proline, which can chelate metal ions and contribute to antioxidant defense, commonly presents increased levels in plants exposed to these contaminants. In sunflower, for example, heavy metals induce proline synthesis (Belhaj et al. 2016; Szabados and Savouré 2010). In *Alyssum lesbiacum*, Ni stress induces the synthesis of the nitrogen compound putrescine (Singh et al. 2016).

Transcriptomic analysis consists of studying the transcribed RNAs: coding ones (messenger RNAs) and non-coding ones (Srivastava et al. 2019). Cotton tolerance to Cd can be successfully studied through this approach that shows a difference in the expression levels of 4627 genes involved in metal binding and redox processes. From this analysis, the coding sequence of GhHMAD5 protein stood out as a candidate for plant engineering. Gene-silencing revealed increased sensitivity to Cd stress and *Arabidopsis thaliana*, overexpressing this protein, developed longer roots dealing better with Cd contamination (Han et al. 2019). *Lolium multiflorum* exhibited 2367 differentially expressed genes under Cd stress. The gene *LmAUX1* was among these and when overexpressed in *A. thaliana* allowed enhanced accumulation of cadmium (Hu et al. 2020). Comparing the transcriptome of *Viola baoshanensis* (a cadmium hyperaccumulator of metalliferous

Table 12.1 Examples of increased metabolites in plant species as a result of metallic pollutants.

Plant species	Pollutant	Increased metabolite or group of metabolites	Reference
Arabidopsis thaliana	Cd and Cu	alpha-tocopherols	Collin et al. (2008)
A.s thaliana	Cd	Glutathione	Cui et al. (2013)
Helianthus annuus	Cr, Cu, Ni, and Zn	Glutathione	Belhaj et al. (2016)
H. annuus	Cr, Cu, Ni, and Zn	Proline	Belhaj et al. (2016)
Alyssum lesbiacum	Ni	putrescine	Singh et al. (2016)
Solanum nigrum	Ni	Ascorbate	Soares et al. (2016)
S. nigrum	Ni	Carotenoids	Soares et al. (2016)
Miscanthus sacchariflorus	Cd	Malate	Guo et al. (2017)
A. thaliana	Cu	Abscisic acid	Cui et al. (2019)
A.thaliana	Cu	Hydrogen peroxide	Cui et al. (2019)
A. thaliana	Cu	Carotenoids	Cui et al. (2019)
Zea mays	Cd and Pb	Auxin	Małkowski et al. (2020)

areas) and *V. inconspicua* (non-tolerance species), it was possible to notice that Cd enhances transcription of RNA from proteins involved in the ubiquitin proteosome system pathway, sucrose metabolism and tonoplast transport (Shu et al. 2019). It helped researchers to understand better how the tolerance can occur at the molecular level in *V. baoshanensis*, involving respectively degradation of misfolded proteins, cell wall modification and participation in antioxidant system, and Cd compartmentalization in the vacuoles, and highlighting key genes involved in this process for plant engineering strategies' development aiming at Cd remediation (Shu et al. 2019). The antioxidant system is also triggered in *Brassica juncea* exposed to metallic pollutants. The mRNA and protein level (the same as activity) of superoxide dismutase, catalase and ascorbate peroxidase, involved in dealing with oxidative stress generated through reactive oxygen species, were higher after exposure to Cu, Zn, Cd and Pb (Małecka et al. 2019).

Proteomics investigates proteins' large-scale expression, allowing the study of changes in their generation in an organism as a whole when compared to other or different expression profiles in different parts (cells, tissues, organs) of an organism (Ahsan et al. 2009; Anderson and Anderson 1998; Liu et al. 2013).

Proteome of *Oenothera glazioviana* roots under Cu stress presents 36 upregulated proteins involved in protein metabolism (i.e., Elongation factor Tu), carbohydrate metabolism (i.e., Fructokinase), signal transduction (i.e., GDP dissociation inhibitor), detoxification and stress defense (i.e., Isopentenyl-diphosphate Delta-isomerase II), development (i.e., Methylmalonate-semialdehyde dehydrogenase) and oxidoreduction (i.e., Phosphomannomutase) (Wang et al. 2017). *Brassica napus* proteomics revealed 73 proteins unique to Cd-exposed plants being involved in cell wall modification, redox processes and metabolism of lipids and proteins. Researchers developed transgenic *Escherichia coli* and *Arabidopsis thaliana* overexpressing BnPDFL (a defensin-like protein) and confirmed an enhancement in Cd tolerance acting as a Cd-chelating molecule (Luo and Zhang 2019). *Sorghum bicolor* leaf proteome revealed increased levels of (i.e., Double-stranded RNA-binding protein 1), transport (i.e., Proton-pump interactor 1), transcriptional regulation (i.e., zinc finger CCCH domain-containing protein 14), stress response (i.e., cytochrome P450), carbohydrate metabolism (i.e., glyceraldehyde-3-phosphate dehydrogenase) and redox homeostase and defense response (i.e., aspartate aminotransferase 3 and cytochrome c oxidase subunit 5b-2). In this study, ion levels were also analyzed showing decreased concentrations of $Zn^{2+}$and Ca^{2+} and increased concentration of Fe^{2+} after cadmium exposure (Roy et al. 2016).

When applied alone, each of these strategies already possess a good information potential. However, when applied simultaneously (multi-omics approaches), they are able to provide a more holistic view (Jia et al. 2016). In order to study cadmium influence on plant cells, the important physical barrier that is the cell wall, metabolomics and proteomics analyses, for example, can be performed providing data regarding gene expression profile analyses results. Cells from the stem of *Medicago sativa* L. exposed to cadmium were analyzed and compared to non-exposed plants, allowing researchers to comprehend that the pollutant's presence enhanced the presence of proteins involved in redox processes, defense responses, remodeling of cell walls and carbohydrate metabolism. Isoflavone biosynthesis is enhanced at long-term exposure through the phenylpropanoid pathway (Gutsch et al. 2019).

12.4 Conclusion

The possibility to engineer plants for phytoremediation of heavy metals indubitably has been presenting elegant solutions to the limitations present in conventional methods of environmental remediation. Significant advancements have already been made in the field, offering elegant solutions to deal with heavy metals environmental pollution, thus reducing their negative effects on living beings.

12.5 Future Perspectives

Recombinant DNA technology is an essential tool provided by the biotechnology field that allows strategic design of genetically-modified organisms to perform specific tasks. This is a promising strategy when it comes to environmental pollution bioremediation, since there is an urgency to deal with contaminants, especially persistent ones (such as heavy metals) that threaten human health and life on Earth in general.

Plants are interesting organisms to be engineered to perform eco-friendly phytoremediation, restoring polluted areas.

An enormous advancement has been performed since the beginning of phytoremediation protocols proposed; however, there are still important aspects to be explored in detail.

It is expected that the number of studies involving field experiments with GMPs will increase, the same as strategies to assure safety (avoiding, for example, uncontrolled transgene transfer); this evolution has to occur together with updates in countries' regulations regarding this kind of activity.

Future developments regarding determining the role of genes elucidated in genome projects from various plant species, especially the ones related to dealing with heavy metals, are expected to point to new candidate genes for DNA engineering aiming for efficient phytormediation. The omics studies are strategical tools to contribute to this.

Finally, site-specific genome editing tools applied to generating GMPs to phytoremediate heavy metals studies are also expected to be developed by multidisciplinary research groups worldwide in the near future in the field, to access the possible aspects that require more attention and the success rate gain in comparison to non-site-specific strategies of genetic modification.

References

Abdullahi, M.S. (2015). Soil contamination, remediation and plants. In: *Soil Remediation and Plants* (ed. K. Rehman, H. Muhammad, S. Münir, et al.), 525–546. Cambridge: Academic Press.

Ahsan, N., Renaut, J., and Komatsu, S. (2009). Recent developments in the application of proteomics to the analysis of plant responses to heavy metal. *Proteomics* 9: 2602–2621.

Ai, S., Guo, R., Liu, B. et al. (2016). A field study on the dynamic uptake and transfer of heavy metals in Chinese cabbage and radish in weak alkaline soils. *Environmental Science and Pollution Research* 23: 20719–20727.

Albering, H.J., Rila, J.P., Moonen, E.J. et al. (2016). Human health risk assessment in relation to environmental pollution of two artificial freshwater lakes in the Netherlands. *Research Journal of Medical Sciences* 107: 27–35.

Alfaro-Saldaña, E.F., Pérez-Molphe-Balch, E., and Santos-Díaz, M.S. (2016). Generation of transformed roots of *Scirpus americanus Pers.* and study of their potential to remove Pb^{2+} and Cr^{3+}. *Plant Cell, Tissue and Organ Culture* 127 (1): 15–24.

Ali, H., Khan, E., and Ilahi, I. (2019). Environmental chemistry and ecotoxicology of hazardous heavy metals: Environmental persistence, toxicity, and bioaccumulation. *Journal of Chemistry* (2019): 1–14.

Ali, H., Khan, E., and Sajad, M.A. (2013). Phytoremediation of heavy metals: Concepts and applications. *Chemosphere* 91: 869–881.

Ali, Z., Abulfaraj, A., Idris, A., Ali, S. et al. (2015). CRISPR/Cas9-mediated viral interference in plants. *Genome Biology* 16: 238.

Alkorta, I., Hernandez-Allica, J., Becerril, J.M. et al. (2004). Recent findings on the phytoremediation of soils contaminated with environmentally toxic heavy metals and metalloids such as zinc, cadmium, lead and arsenic. *Reviews in Environmental Science and Biotechnology* 3: 71–90.

Anderson, N.L. and Anderson, N.G. (1998). Proteome and proteomics: New technologies, new concepts, and new words. *Electrophoresis* 19: 1853–1861.

Anwar, A. and Kim, J.-K. (2020). Transgenic breeding approaches for improving abiotic stress tolerance: Recent progress and future perspectives. *International Journal of Molecular Sciences* 21 (8): 2695.

Aprile, A. and De Bellis, L. (2020). Editorial for special issue "Heavy metals accumulation, toxicity, and detoxification in plants". *International Journal of Molecular Sciences* 21: 4103.

Arbona, V., Manzi, M., Ollas, C.D. et al. (2013). Metabolites as a tool to investigate abiotic stress tolerance in plants. *International Journal of Molecular Sciences* 14: 4885–4911.

Ashraf, S., Ali, Q., Zahir, Z.A. et al. (2019). Phytoremediation: Environmentally sustainable way for reclamation of heavy metal polluted soils. *Ecotoxicology and Environmental Safety* 174: 714–727.

Azubuike, C.C., Chikere, C.B., and Okpokwasili, G.C. (2016). Bioremediation techniques–classification based on site of application: Principles, advantages, limitations and prospects. *World Journal of Microbiology and Biotechnology* 32 (180): 1–18. https://doi.org/10.1007/s11274-016-2137-x.

Baldissarelli, D.P., Vargas, G.D.L.P., Korf, E.P. et al. (2019). Remediation of soils contaminated by pesticides using physicochemical processes: A brief review. *Planta Daninha* 37: e019184975.

Barrios-Estrada, C., De Jesús Rostro-alanis, M., Muñoz-Gutiérrez, B.D. et al. (2018). Emergent contaminants: Endocrine disruptors and their laccase-assisted degradation: A review. *Science of the Total Environment* 612: 1516–1531.

Basharat, Z., Novo, L.A.B., and Yasmin, A. (2018). Genome editing weds CRISPR: What is in it for phytoremediation? *Plants (Basel)* 7 (3): 51.

Belhaj, D., Elloumi, N., Jerbi, B. et al. (2016). Effects of sewage sludge fertilizer on heavy metal accumulation and consequent responses of sunflower (*Helianthus annuus*). *Environmental Science and Pollution Research* 23 (20): 20168–20177.

Bhuiyan, M.S.U., Min, S.R., Jeong, W.J. et al. (2011). Overexpression of AtATM3 in *Brassica juncea* confers enhanced heavy metal tolerance and accumulation. *Plant Cell, Tissue and Organ Culture* 107 (1): 69–77.

Bilal, M., Rasheed, T., Sosa-Hernández, J. et al. (2018). Biosorption: An interplay between marine algae and potentially toxic elements: A review. *Marine Drugs* 16: 65–70.

Bogdanove, A.J. and Voytas, D.F. (2011). TAL effectors: Customizable proteins for DNA targeting. *Science* 333: 1843–1846.

Bortesia, L. and Fischer, R. (2015). The CRISPR/Cas9 system for plant genome editing and beyond. *Biotechnology Advances* 33: 41–52.

Brunner, I., Luster, J., Gunthardt-Goerg, M.S. et al. (2008). Heavy metal accumulation and phytostabilisation potential of tree fine roots in a contaminated soil. *Environmental Pollution* 152: 559–568.

Bundy, J.G., Willey, T.L., Castell, R.S. et al. (2005). Discrimination of pathogenic clinical isolates and laboratory strains of *Bacillus cereus* by NMR-based metabolomic profiling. *FEMS Microbiology Letters* 242: 127–136.

Carpenter, A. (2018). Oil pollution in the North Sea: The impact of governance measures on oil pollution over several decades. *Hydrobiologia* 2018: 1–19.

Cempel, M. and Nikel, G. (2006). Nickel: A review of its sources and environmental toxicology. *Polish Journal of Environmental Studies* 15: 375–382.

Chaoua, S., Boussaa, S., El Gharmali, A. et al. (2019). Impact of irrigation with wastewater on accumulation of heavy metals in soil and crops in the region of Marrakech in Morocco. *Journal of the Saudi Society of Agricultural Sciences* 18 (4): 429–436.

Charfeddine, M., Charfeddine, S., Bouaziz, D. et al. (2017). The effect of cadmium on transgenic potato (*Solanumtuberosum*) plants overexpressing the StDREB transcription factors. *Plant Cell, Tissue and Organ Culture* 128 (3): 521–541.

Chen, P.J., Senthilkumar, R., Jane, W.N. et al. (2014). Transplastomic *Nicotiana benthamiana* plants expressing multiple defence genes encoding protease inhibitors and chitinase display broad-spectrum resistance against insects, pathogens and abiotic stresses. *Plant Biotechnology Journal* 12 (4): 503–515.

Chen, Y.X., Lin, Q., Luo, Y.M. et al. (2003). The role of citric acid on the phytoremediation of heavy metal contaminated soil. *Chemosphere* 50: 807–811. https://doi.org/10.1016/S0045-6535(02)00223-0

Cheng, L., Zhou, Q., and Yu, B. (2019). Responses and roles of roots, microbes, and degrading genes in rhizosphere during phytoremediation of petroleum hydrocarbons contaminated soil. *International Journal of Phytoremediation* 21 (12): 1161–1169.

Cherian, S. and Oliveira, M.M. (2005).Transgenic plants in phytoremediation: Recent advances and new possibilities. *Environmental Science & Technology* 39 (24): 9377–9390.

Christian, M., Cermak, T., Doyle, E.L. et al. (2010). Targeting DNA double-strand breaks with TAL effector nucleases. *Genetics* 186: 757–761.

Collin, V.C., Eymery, F., Genty, B. et al. (2008). Vitamin E is essential for the tolerance of *Arabidopsis thaliana* to metal induced oxidative stress. *Plant, Cell & Environment* 31: 244–257.

Cong, L., Ran, F.A., Cox, D. et al. (2013). Multiplex genome engineering using CRISPR/Cas systems. *Science* 339: 819–823.

Cui, Y., Wang, M., Yin, X. et al. (2019). OsMSR3, a small heat shock protein, confers enhanced tolerance to copper stress in *Arabidopsis thaliana*. *International Journal of Molecular Sciences* 20: 6096.

Cui, Y., Xu, G., Wang, M. et al. (2013). Expression of OsMSR3 in *Arabidopsis* enhances tolerance to cadmium stress. *Plant Cell, Tissue and Organ Culture* 113: 331–340.

Cunningham, F.J., Goh, N.S., Demirer, G.S. et al. (2018). Nanoparticle-mediated delivery towards advancing plant genetic engineering. *Trends in Biotechnology* 36: 882–897.

Curtin, S.J., Voytas, D.F., and Stupar, R.M. (2012). Genome engineering of crops with designer nucleases. *The Plant Genome* 5 (2): 42–50.

DalCorso, G., Fasani, E., Manara, A. et al. (2019). Heavy metal pollutions: State of the art and innovation in phytoremediation. *International Journal Of Molecular Sciences* 20 (14): 3412.

Daniell, H., Lin, C., Yu, M., and Chang, W. (2016). Chloroplast genomes: Diversity, evolution and applications in genetic engineering. *Genome Biology* 17: 134.

Das, N., Bhattacharya, S., and Maiti, M.K. (2016). Enhanced cadmium accumulation and tolerance in transgenic tobacco overexpressing rice metal tolerance protein gene OsMTP1 is promising for phytoremediation. *Plant Physiology and Biochemistry* 105: 297–309.

Dasgupta, J., Sikder, J., Chakraborty, S. et al. (2015). Remediation of textile effluents by membrane based treatment techniques: A state of the art review. *Journal of Environmental Management* 147: 55–72.

Didovyk, A., Borek, B., Tsimring, L., and Hasty, J. (2016). Transcriptional regulation with CRISPR-Cas9: Principles, advances, and applications. *Current Opinion in Biotechnology* 40: 177–184.

Dominguez, J.J.A., Inoue, C., and Chien, M.F. (2019). Hydroponic approach to assess rhizodegradation by sudangrass (*Sorghum x drummondii*) reveals pH- and plant age-dependent variability in bacterial degradation of polycyclic aromatic hydrocarbons (PAHs). *Journal of Hazardous Materials* 387: 121695.

Doyle, A., McGarry, M.P., Lee, N.A. et al. (2012). The construction of transgenic and gene knockout/knockin mouse models of human disease. *Transgenic Research* 21 (2): 327–349.

Dzantor, E.K. (2007). Phytoremediation: The state of rhizosphere "engineering" for accelerated rhizodegradation of xenobiotic contaminants. *Journal of Chemical Technology & Biotechnology* 82 (3): 228–232.

Eapen, S. and D'souza, S. (2005). Prospects of genetic engineering of plants for phytoremediation of toxic metals. *Biotechnology Advances* 23: 97–114.

Fasani, E., DalCorso, G., Zerminiani, A. et al. (2019). Phytoremediatory efficiency of *Chrysopogon zizanioides* in the treatment of landfill leachate: A case study. *Environmental Science and Pollution Research* 26: 10057–10069.

Fasani, E., Manara, A., Martini, F. et al. (2018). The potential of genetic engineering of plants for the remediation of soils contaminated with heavy metals. *Plant, Cell & Environment* 41: 1201–1232.

Feng, S., Tan, J., Zhang, Y. et al. (2017). Isolation and characterization of a novel cadmium-regulated Yellow Stripe-Like transporter (SnYSL3) in *Solanum nigrum*. *Plant Cell Reports* 36: 281–296.

Fulekar, M.H. and Pathak, B. (2020). *Bioremediation Technology: Hazardous Waste Management*, 366. CRC Press.

Gaitán-Solís, E., Taylor, N.J., Siritunga, D. et al. (2015). Overexpression of the transporters AtZIP1 and AtMTP1 in cassava changes zinc accumulation and partitioning. *Frontiers in Plant Science* 6: 492.

Garbisu, C., Hernandez-Allica, J., Barrutia, O. et al. (2002). Phytoremediation: A technology using green plants to remove contaminants from polluted areas. *Reviews on Environmental Health* 17: 173–188.

Gavrilescu, M., Demnerová, K., Aamand, J. et al. (2015). Emerging pollutants in the environment: Present and future challenges in biomonitoring, ecological risks and bioremediation. *Biotechnology* 32 (1): 147–156.

Gong, Y., Chen, J., and Pu, R. (2019). The enhanced removal and phytodegradation of sodium dodecyl sulfate (SDS) in wastewater using controllable water hyacinth. *International Journal of Phytoremediation* 2019: 1–10.

Gong, Y., Zhou, X., Ma, X., and Chen J (2018) Sustainable removal of formaldehyde using controllable water hyacinth. Author links open overlay panel. *Journal of Cleaner Production* 181: 1–7.

Guerra, F.D., Attia, M.F., Whitehead, D.C. et al. (2018). Nanotechnology for environmental remediation: Materials and applications. *Molecules* 23 (7): 1760.

Guo, H., Feng, X., Hong, C. et al. (2017). Malate secretion from the root system is an important reason for higher resistance of *Miscanthus sacchariflorus* to cadmium. *Physiologia Plantarum* 159 (3): 340–353. https://doi.org/10.1111/ppl.12526

Gupta, R.M. and Musunuru, K. (2014). Expanding the genetic editing tool kit: ZFNs, TALENs, and CRISPR-Cas9. *The Journal of Clinical Investigation* 124: 4154–4161.

Gutsch, A., Vandionant, S., Sergeant, K. et al. (2019). Systems biology of metal tolerance in plants: A case study on the effects of Cd exposure on two model plants. In: *Plant Metallomics and Functional Omics* (ed. G. Sablok), 330.

Han, M., Lu, X., Yu, J. et al. (2019). Transcriptome analysis reveals cotton (*Gossypium hirsutum*) genes that are differentially expressed in cadmium stress tolerance. *International Journal of Molecular Sciences* 20: 1479.

Hasin, A.A., Gurman, S.J., Murphy, L.M. et al. (2010). Remediation of chromium (VI) by a methane-oxidizing bacterium. *Environmental Science & Technology* 44: 400–405.

Hawkes, J.S. (1997). Heavy metals. *Journal of Chemical Education* 74: 1369–1374.

He, W., Megharaj, M., Wu, C.Y. et al. (2020). Endophyte-assisted phytoremediation: Mechanisms and current application strategies for soil mixed pollutants. *Critical Reviews in Biotechnology* 40 (1): 31–45.

He, Y.K., Sun, J.G., Feng, X.Z. et al. (2001). Differential mercury volatilization by tobacco organs expressing a modified bacterial merA gene. *Cell Research* 11 (3): 231–236.

Holkers, M., Maggio, I., Liu, J. et al. (2013). Differential integrity of TALE nuclease genes following adenoviral and lentiviral vector gene transfer into human cells. *Nucleic Acids Research* 41: e63.

Hu, Z., Zhang, Y., He, Y. et al. (2020). Full-length transcriptome assembly of Italian ryegrass root integrated with RNA-seq to identify genes in response to plant cadmium stress. *International Journal of Molecular Sciences* 21: 1067.

Huang, D.L., Zeng, G.M., Feng, C.L. et al. (2008). Degradation of lead-contaminated lignocellulosic waste by *Phanerochaete chrysosporium* and the reduction of lead toxicity. *Environmental Science & Technology* 42: 4946–4951.

Huang, L., Chen, D., Zhang, H. et al. (2019). *Funneliformis mosseae* enhances root development and Pb phytostabilization in *Robinia pseudoacacia* in Pb-contaminated soil. *Frontiers in Microbiology* 10: 2591.

Hussein, H.S., Ruiz, O.N., Terry, N. et al. (2007). Phytoremediation of mercury and organomercurials in chloroplast transgenic plants: Enhanced root uptake, translocation to shoot and volatization. *Environmental Science & Technology* 41: 8439–8446.

Ismagul, A., Yang, N., Maltseva, E. et al. (2018). A biolistic method for high-throughput production of transgenic wheat plants with single gene insertions. *BMC Plant Biology* 18: 135.

Jacob, J.M., Karthik, C., Saratale, R.G. et al. (2018). Biological approaches to tackle heavy metal pollution: A survey of literature. *Journal of Environmental Management* 217: 56–70.

Jafari, M., Danesh, Y.R., Goltapeh, E.M. et al. (2013). Bioremediation and genetically modified organisms. In: *Fungi as Bioremediators* (ed. E. Goltapeh, Y. Danesh, and A.M. Varma), 433–450. Berlin: Springer.

Jaishankar, M., Tseten, T., Anbalagan, N. et al. (2014). Toxicity, mechanism and health effects of some heavy metals. *Interdisciplinary Toxicology* 7: 60–72.

Jaiswal, S., Singh, D.K., and Shukla, P. (2019). Gene editing and systems biology tools for pesticide bioremediation: A review. *Frontiers in Microbiology* 10: 87.

Jayaweera, M. and Kasturiarachchi, J. (2004). Removal of nitrogen and phosphorus from industrial wastewaters by phytoremediation using water hyacinth (*Eichhornia crassipes* (Mart.) Solms). *Water Science and Technology* 50 (6): 217–225.

Jia, X., Sun, C., Zuo, Y. et al (2016). Integrating transcriptomics and metabolomics to characterise the response of *Astragalus membranaceus* Bge. var. mongolicus (Bge.) to progressive drought stress. *BMC Genomics* 17: 188.

Jinek, M., Chylinski, K., Fonfara, I. et al. (2012). A programmable dual-RNA-guided DNA endonuclease in adaptive bacterial immunity. *Science* 337 (6096): 816–821.

Karavangeli, M., Labrou, N.E., Clonis, Y.D. et al. (2005). Development of transgenic tobacco plants overexpressing maize glutathione S-transferase I for chloroacetanilide herbicides phytoremediation. *Biomolecular Engineering* 22 (4): 121–128.

Keshavareddy, G., Kumar, A.R.V., and Ramu, V.S. (2018). Methods of plant transformation: A review. *Journal of Current Microbiology and Applied Sciences* 7: 2656–2668.

Khatodia, S., Bhatotia, K., Passricha, N. et al. (2016). The CRISPR/Cas genome-editing tool: Application in improvement of crops. *Frontiers in Plant Science* 7: 506.

Kim, Y.G., Cha, J., and Chandrasegaran, S. (1996). Hybrid restriction enzymes: Zinc finger fusions to Fok I cleavage domain. *Proceedings of the National Academy of Sciences USA* 93: 1156–1160.

Kiran, B.R. and Prasad, M.N.V. (2019). Biochar and rice husk ash assisted phytoremediation potentials of *Ricinus communis* L. for lead-spiked soils author links open overlay panel. *Ecotoxicology and Environmental Safety* 183: 109574.

Koźmińska, A., Wiszniewska, A., Hanus-Fajerska, E. et al. (2018). Recent strategies of increasing metal tolerance and phytoremediation potential using genetic transformation of plants. *Plant Biotechnology Reports* 12 (1): 1–14.

Lajayer, B.A., Moghadam, N.K., Maghsoodi, M.R. et al. (2019). Phytoextraction of heavy metals from contaminated soil, water and atmosphere using ornamental plants: Mechanisms and efficiency improvement strategies. *Environmental Science and Pollution Research* 26: 8468–8484.

LeDuc, D.L., Tarun, A.S., Montes-Bayon, M. et al. (2004). Overexpression of selenocysteine methyltransferase in *Arabidopsis* and Indian Mustard increases selenium tolerance and accumulation. *Plant Physiology* 135: 377–383.

Lee, J., Donghwan, S., Won-Yong, S. et al. (2004). *Arabidopsis* metallothioneins 2a and 3 enhance resistance to cadmium when expressed in *Vicia faba* guard cells. *Plant Molecular Biology* 54: 805–815.

Lin, Y.F., Hassan, Z., Talukdar, S. et al. (2016). Expression of the ZNT1 zinc transporter from the metal hyperaccumulator *Noccaea caerulescens* confers enhanced zinc and cadmium tolerance and accumulation to *Arabidopsis thaliana*. *PLoS One* 11: e0149750.

Liu, J., Shi, X., Qiana, M. et al. (2015). Copper-induced hydrogen peroxide upregulation of a metallothionein gene, OsMT2c, from *Oryza sativa* L. confers copper tolerance in *Arabidopsis thaliana*. *Journal of Hazardous Materials* 294: 99–108.

Liu, L., Bilal, M., Duan, X. et al. (2019). Mitigation of environmental pollution by genetically engineered bacteria: Current challenges and future perspectives. *Science of the Total Environment* 667: 444–454.

Liu, X., Wu, H., Ji, C. et al. (2013). An integrated proteomic and metabolomic study on the chronic effects of mercury in *Suaeda salsa* under an environmentally relevant salinity. *PLoS ONE* 8: e64041.

Lowder, L.G., Zhang, D., Baltes, N.J. et al. (2015). A CRISPR/Cas9 toolbox for multiplexed plant genome editing and transcriptional regulation. *Plant Physiology* 169 (2): 971–985.

Lu, Y., Deng, X., Quan, L. et al. (2015). Metallothioneins BcMT1 and BcMT2 from *Brassica campestris* enhance tolerance to cadmium and copper and decrease production of reactive oxygen species in *Arabidopsis thaliana*. *Plant and Soil* 367: 507–519.

Luo, J.S. and Zhang, Z. (2019). Proteomic changes in the xylem sap of Brassica napus under cadmium stress and functional validation. *BMC Plant Biology* 19: 280.

Ma, X., Zhang, Q., Zhu, Q. et al. (2015). A robust CRISPR/Cas9 system for convenient, high-efficiency multiplex genome editing in monocot and dicot plants. *Molecular Plant* 8: 1274e1284.

Mackelprang, R. and Lemaux, P.G. (2020). Genetic engineering and editing of plants: An analysis of new and persisting questions. *Annual Review of Plant Biology* 71: 659–687.

Małecka, A., Konkolewska, A., Han´c, A. et al. (2019). Insight into the phytoremediation capability of *Brassica juncea* (v. Malopolska): Metal accumulation and antioxidant enzyme activity. *International Journal of Molecular Sciences* 20: 4355.

Mali, P., Yang, L., Esvelt, K.M. et al. (2013). RNA-guided human genome engineering via Cas9. *Science* 339: 823–826.

Maliga, P. (2002). Engineering the plastid genome of higher plants. *Current Opinion in Plant Biology* 5 (2): 164–172.

Małkowski, E., Sitko, K., Szopiński, M. et al. (2020). Hormesis in plants: The role of oxidative stress, auxins and photosynthesis in corn treated with Cd or Pb. *International Journal of Molecular Sciences* 21 (6): 2099.

Mendes, K.F., Apj, R., Takeshita, V. et al. (2019) Water Resource Pollution by Herbicide Residues. IntechOpen https://doi.org/10.5772/intechopen.85159 (accessed 2 November 2021).

Mendez, M.O. and Maier, R.M. (2008). Phytostabilization of mine tailings in arid and semiarid environments: An emerging remediation technology. *Environmental Health Perspectives* 116 (3): 278–283.

Merlot, S., Hannibal, L., Martins, S. et al. (2014). The metal transporter PgIREG1 from the hyperaccumulator *Psychotriagabriellae* is a candidate gene for nickel tolerance and accumulation. *Journal of Experimental Botany* 65 (6): 1551–1564.

Monaci, F., Trigueros, D., Mingorance, M.D. et al. (2019). Phytostabilization potential of *Erica australis* L. and *Nerium oleander* L.: A comparative study in the Riotinto mining area (S.W. Spain). *Environmental Geochemistry and Health* 2019: 1–16.

Moreno, F.N., Anderson, C.W.N., Stewart, R.B. et al. (2008). Phytofiltration of mercury-contaminated water: Volatilisation and plant-accumulation aspects. *Environmental and Experimental Botany* 62: 78–85.

Muthusaravanan, S., Sivarajasekar, N., Vivek, J.S. et al. (2018). Phytoremediation of heavy metals: Mechanisms, methods and enhancements. *Environmental Chemistry Letters* 16: 1339–1359.

Nakabayashi, R. and Saito, K. (2015). Integrated metabolomics for abiotic stress responses in plants. *Current Opinion in Plant Biology* 24: 10–16.

Nesler, A., DalCorso, G., Fasani, E. et al. (2017). Functional components of the bacterial CzcCBA efflux system reduce cadmium uptake and accumulation in transgenic tobacco plants. *New Biotechnology* 35: 54–61.

Nie, X., Dong, F., Ding, D.X. et al. (2015). Ability of *Pistia stratiotes* L. and *Eichhornia crassipes* for remediation of uranium-contaminated waste water. *Yuanzineng Kexue Jishu/ Atomic Energy Science and Technology* 49 (11): 1946–1953.

Nwoko, C.O. (2010). Trends in phytoremediation of toxic elemental and organic pollutants. *African Journal of Biotechnology* 9 (37): 6010–6016.

Ojuederie, O.B. and Babalola, O.O. (2017). Microbial and plant-assisted bioremediation of heavy metal polluted environments: A review. *International Journal of Environmental Research and Public Health* 14: 1504–1511.

Oustriere, N., Marchand, L., Roulet, E. et al. (2017). Rhizofiltration of a Bordeaux mixture effluent in pilot-scale constructed wetland using *Arundo donax* L. coupled with potential Cu-ecocatalyst production. *Ecological Engineering* 105: 296–305.

Ozyigit, I.I., Can, H., and Dogan, I. (2020). Phytoremediation using genetically engineered plants to remove metals: A review. *Environmental Chemistry Letters*. https://doi.org/10.1007/s10311-020-01095-6

Padmavathiamma, P.K. and Li, L.Y. (2007). Phytoremediation technology: Hyper-accumulation metals in plants. *Water, Air, and Soil Pollution* 184: 105–126.

Panesar, A.S., Kumar, A., and Kalpana. (2019). Phytoremediation: An ecofriendly tool for *in-situ* remediation of contaminated soil. *Journal of Pharmacognosy and Phytochemical* SP1: 311–316.

Peng, J.S., Ding, G., Meng, S. et al. (2017). Enhanced metal tolerance correlates with heterotypic variation in SpMTL, a metallothionein-like protein from the hyperaccumulator *Sedum plumbizincicola*. *Plant Cell & Environment*. https://doi.org/10.1111/pce.12929

Pérez-Palacios, P., Agostini, E. et al. (2017). Removal of copper from aqueous solutions by rhizofiltration using genetically modified hairy roots expressing a bacterial Cu-binding protein. *Environmental Technology* 28 (22): 2877–2888.

Pesantes, A.A., Carpio, E.P., Vitvar, T. et al. (2019). A Multi-Index Analysis approach to heavy metal pollution assessment in river sediments in the Ponce Enríquez area, Ecuador. *Water* 11: 590–596.

Petolino, J.F. (2015). Genome editing in plants via designed zinc finger nucleases. *Vitro Cellular & Developmental Biology* 51: 1–8.

Polti, M.A., García, R.O., Amoroso, M.J. et al. (2009). Bioremediation of chromium(VI) contaminated soil by *Streptomyces* sp. MC1. *Journal of Basic Microbiology* 49: 285–292.

Prasad, M.N.V. (2019). *Transgenic Plant Technology for Remediation of Toxic Metals and Metalloids*. 564. New York: Academic Press.

Priya, E.S. and Selvan, P.S. (2017). Water hyacinth (*Eichhornia crassipes*) :An efficient and economic adsorbent for textile effluent treatment: A review. *Arabian Journal of Chemistry* 10 (2): S3548–S3558.

Pu, Q., Sun, J.Q., Zhang, F.F. et al. (2019). Effects of copper mining on heavy metal contamination in a rice agrosystem. *Acta Geochimica* 2019: 1–21.

Rai, A., Singh, R., Shirke, P.A. et al. (2015). Expression of rice CYP450-like gene (Os08g01480) in *Arabidopsis* modulates regulatory network leading to heavy metal and other abiotic stress tolerance. *PloS One* 10 (9): e0138574.

Reis, L.F.L., Van Sluys, M.A., Garratt, R.C. et al. (2006). GMOs: Building the future on the basis of past experience. *Anais Da Academia Brasileira De Ciências* 78 (4): 667–686.

Richter, J., Watson, J.M., Stasnik, P. et al. (2018). Multiplex mutagenesis of four clustered CrRLK1L with CRISPR/Cas9 exposes their growth regulatory roles in response to metal ions. *Scientific Reports* 8: 12182.

Robinson, B.H., Lombi, E., Zhao, F.J. et al. (2003). Uptake and distribution of nickel and other metals in the hyperaccumulator *Berkheya coddii*. *New Phytologist* 158: 279–285.

Rodríguez-Llorente, I.D., Lafuente, A., Doukkali, B. et al. (2012). Engineering copper hyperaccumulation in plants by expressing a prokaryotic cop C gene. *Environmental Science & Technology* 46 (21): 12088–12097.

Rosculete, C.A., Bonciu, E., Rosculete, E. et al. (2019). Determination of the environmental pollution potential of some herbicides by the assessment of cytotoxic and genotoxic effects on *Allium cepa*. *International Journal of Environmental Research and Public Health* 16: 75–85.

Roy, S.K., Cho, S.-W., Kwon, S.J. et al. (2016). Morpho-physiological and proteome level responses to cadmium stress in sorghum. *PLoS ONE* 11 (2): e0150431.

Rugh, C.L., Wilde, H.D., Stacks, N.M. et al. (1996). Mercuric ion reduction and resistance in transgenic *Arabidopsis thaliana* plants expressing a modified bacterial merA gene. *Proceedings of the National Academy of Sciences* 93: 3182–3187.

Sander, J.D., Dahlborg, E.J., Goodwin, M.J. et al. (2011). Selection-free zinc-finger-nuclease engineering by context-dependent assembly (CoDA). *Nature Methods* 8: 67–69.

Sander, J.D. and Joung, J.K. (2014). CRISPR-Cas systems for editing, regulating and targeting genomes. *Nature Biotechnology* 32 (4): 347–355.

Shim, D., Kim, S., Choi, Y.I. et al. (2013). Transgenic poplar trees expressing yeast cadmium factor 1 exhibit the characteristics necessary for the phytoremediation of mine tailing soil. *Chemosphere* 90 (4): 1478–1486.

Shu, H., Zhang, J., Liu, F. et al. (2019). Comparative transcriptomic studies on a cadmium hyperaccumulator *Viola baoshanensis* and its non-tolerant counterpart *V. inconspicua*. *International Journal of Molecular Sciences* 20: 1906.

Singh, S., Parihar, P., Singh, R. et al. (2016). Heavy metal tolerance in plants: Role of transcriptomics, proteomics, metabolomics and ionomics. *Frontiers in Plant Science* 6: 1143.

Sinha, S. and Mukherjee, S.K. (2009). *Pseudomonas aeruginosa* KUCD1, a possible candidate for cadmium bioremediation. *Brazilian Journal of Microbiology* 40: 655–662.

Soares, C., De Sousa, A., Pinto, A. et al. (2016). Effect of 24-epibrassinolide on ROS content, antioxidant system, lipid peroxidation and Ni uptake in *Solanum nigrum* L. under Ni stress. *Environmental and Experimental Botany* 122: 115–125.

Song, W.Y., Ju Sohn, E., Martinoia, E. et al. (2003). Engineering tolerance and accumulation of lead and cadmium in transgenic plants. *Nature Biotechnology* 21 (8): 914–919.

Srivastava, A., George, J., and Karuturi, R.K.M. (2019). Transcriptome analysis. *Encyclopedia of Bioinformatics and Computational Biology* 3: 792–805.

Strong, P.J. and Burgess, J.E. (2008). Treatment methods for wine-related ad distillery wastewaters: A review. *Bioremediation Journal* 12: 7087–7095.

Szabados, L. and Savouré, A. (2010). Proline: A multifunctional amino acid. *Trends in Plant Science* 15: 89–97.

Tang, L., Mao, B., Li, Y. et al. (2017). Knockout of OsNramp5 using the CRISPR/Cas9 system produces low Cd-accumulating indica rice without compromising yield. *Scientific Reports* 7: 14438.

Thapa, G., Das, D., Gunupuru, L.R. et al. (2016). Endurance assessment of *Eichhornia crassipes* (Mart.) Solms, in heavy metal contaminated site: A case study. *Cogent Environmental Science* 2 (1): 1215280.

Thompson, M.G., Moore, W.M., Hummel, N.F.C. et al. (2020). *Agrobacterium tumefaciens*: A bacterium primed for synthetic biology. *BioDesign Research* 2020: 1–16.

Tiwari, S. and Lata, C. (2018). Heavy metal stress, signaling, and tolerance due to plant-associated microbes: An overview. *Frontiers in Plant Science* 9: 452.

Tonelli, F.C.P. and Tonelli, F.M.P. (2020). Causes and effects of pesticide and metal pollution on different ecosystems. In: *Bioremediation and Biotechnology*, 2 (ed. R.A. Bhat, K.R. Hakeem and M.A. Dervash), 1–11. Berlin: Springer.

Urnov, F.D., Rebar, E.J., Holmes, M.C. et al. (2010). Genome editing with engineered zinc finger nucleases. *Nature Reviews Genetics* 11 (9): 636–646.

Vázquez-Luna, D. and Cuevas-Díaz, M.C. (2019). Soil contamination and alternatives for sustainable development. In: *Soil Contamination and Alternatives for Sustainable Development* (ed. D. Vázquez-Luna and M.C. Cuevas-Díaz). IntechOpen. https://doi.org/10.5772/intechopen.83720

Vrbová, M., Kotrba, P., Horáček, J. et al. (2015). Enhanced accumulation of cadmium in *Linumusitatissimum* L. plants due to overproduction of metallothionein a-domain as a fusion to b-glucuronidase protein. *Plant Cell Tissue and Organ Culture* 112: 321–330.

Wang, C., Shen, L., Fu, Y. et al. (2015). A simple CRISPR/Cas9 system for multiplex genome editing in rice. *Journal of Genetics and Genomics* 42: 703.

Wang, C., Wang, J., Wang, X. et al. (2017). Proteomic analysis on roots of *Oenothera glazioviana* under copper-stress conditions. *Scientific Reports* 7: 10589.

Wang, J., Anderson, C.W.N., Xing, Y. et al. (2018). Thiosulphate-induced phytoextraction of mercury in *Brassica juncea*: Spectroscopic investigations to define a mechanism for Hg uptake. *Environmental Pollution* 242: 986–993. https://doi.org/10.1016/j.envpol.2018.07.065

Woolf, D., Amonette, J.E., Street-Perrott, F.A. et al. (2010). Sustainable biochar to mitigate global climate change. *Nature Communications* 1: 56.

Wu, G., Kang, H., Zhang, X. et al. (2010). A critical review on the bio-removal of hazardous heavy metals from contaminated soils: Issues, progress, eco-environmental concerns and opportunities. *Journal of Hazardous Materials* 174: 1–8.

Xia, Y., Qi, Y., Yuan, Y. et al. (2012). Overexpression of *Elsholtzia haichowensis* metallothionein 1 (EhMT1) in tobacco plants enhances copper tolerance and accumulation in root cytoplasm and decreases hydrogen peroxide production. *Journal of Hazardous Materials* 234: 65–71.

Xie, K., Minkenberg, B., and Yang, Y. (2015). Boosting CRISPR/Cas9 multiplex editing capability with the endogenous tRNA-processing system. *Proceedings of the National Academy of Sciences* 112: e3570–e3575.

Xing, H.L., Dong, L., Wang, Z.P. et al. (2014). A CRISPR/Cas9 toolkit for multiplex genome editing in plants. *BMC Plant Biology* 14: 327.

Xu, L., Xing, X., Liang, J. et al. (2019). *In situ* phytoremediation of copper and cadmium in a co-contaminated soil and its biological and physical effects. *RSC Advances* 9: 993–1003.

Yang, J.L., Chen, Z., Wu, S. et al. (2015). Overexpression of the *Tamarix hispida* ThMT3 gene increases copper tolerance and adventitious root induction in *Salix matsudana* Koidz. *Plant Cell, Tissue and Organ Culture* 121 (2): 469–479.

Ye, S., Zeng, G., Wu, H. et al. (2017). Biological technologies for the remediation of co-contaminated soil. *Critical Reviews in Biotechnology* 37 (8): 1062–1076.

Yin, K., Gao, C., and Qiu, J.L. (2017). Progress and prospects in plant genome editing. *Nature Plants* 3: 17107.

Zaidi, S.S., Mahfouz, M.M., and Mansoor, S. (2017). CRISPR-Cpf1: A new tool for plant genome editing. *Trends in Plant Science* 22 (7): 550–553.

Zaidi, S.S. and Mansoor, S. (2017). Viral vectors for plant genome engineering. *Frontiers in Plant Science* 8: 539–545.

Zhang, J., Martinoia, E., and Lee, Y. (2018). Vacuolar transporters for cadmium and arsenic in plants and their applications in phytoremediation and crop development. *Plant and Cell Physiology* 59 (7): 1317–1325. https://doi.org/10.1093/pcp/pcy006

Zhang, J., Zhang, M., Tian, S. et al. (2014). Metallothionein 2 (SaMT2) from *Sedum alfredii* Hance confers increased Cd tolerance and accumulation in yeast and tobacco. *PLoS One* 9: e102750.

Zhang, Q., Yu, R., Fu, S. et al. (2019). Spatial heterogeneity of heavy metal contamination in soils and plants in Hefei, China. *Scientific Reports* 1049: 2045–2051.

Zhang, Y., Liu, J., Zhou, Y. et al. (2013). Enhanced phytoremediation of mixed heavy metal (mercury)–organic pollutants (trichloroethylene) with transgenic alfalfa co-expressing glutathione S-transferase and human P450 2E1. *Journal of Hazardous Materials* 260: 1100–1107.

Zhang, Z., Mao, Y., Ha, S. et al. (2015). A multiplex CRISPR/Cas9 platform for fast and efficient editing of multiple genes in *Arabidopsis*. *Plant Cell Reports* 7: 1519e1533.

Zhao, Q., Yue, S., Bilal, M. et al. (2017). Comparative genomic analysis of 26 *Sphingomonas* and *Sphingobium* strains: Dissemination of bioremediation capabilities, biodegradation potential and horizontal gene transfer. *Science of the Total Environment* 609: 1238–1247.

Zheng, H., Zhang, C., Liu, B. et al. (2020). Biochar for water and soil remediation: Production, characterization, and application. In: *A New Paradigm for Environmental Chemistry and Toxicology* (ed. G. Jiang and X. Li), 153–196. Singapore: Springer.

Zhu, Y., Eah, P.-S., Jouanin, L. et al. (1999b). Overexpression of glutathione synthetase in *Brassica juncea* enhances cadmium tolerance and accumulation. *Plant Physiology* 119: 73–79.

Zhu, Y., Pilon-Smits, E.A., Tarun, A.S. et al. (1999a). Cadmium tolerance and accumulation in Indian Mustard is enhanced by overexpressing γ-glutamylcysteine synthetase. *Plant Physiology* 121: 1169–1177.

13

Recent Advances in Enzymatic Membranes and Their Sustainable Applications Across Industry

Setareh Heidari[1], *David A. Wood*[2] *and Ahmad Fauzi Ismail*[3]

[1] *Faculty of Chemical Engineering, Sahand University of Technology, Tabriz, Iran*
[2] *DWA Energy Limited, Bassingham, Lincoln, LN5 9JP, UK*
[3] *Advanced Membrane Technology Research Center (AMTEC), UniversitiTeknologi Malaysia, 81310, Skudai, Johor Bahru, Johor, Malaysia*

CHAPTER MENU

13.1 Introduction

Long-term success in the current industrial world requires overtly employing efficient, eco-friendly and sustainable strategies (Prather 2020). Such strategies received the attention and support of governments, and the broad scientific and industrial communities seeking innovation without damaging the environment, eco-systems or human health (Ismail et al. 2022; Liu et al 2021). This emphasis is shifting the focus to interdisciplinary approaches involving the development of environmentally-friendly and renewable processes and avoiding the use and generation of hazardous materials (Heidari and Wood 2021; Sadatshojaei et al. 2021a, 2021b, 2021c). The scarcity of potable water and ever-increasing endemic quality problems necessitate the development of

Bioresource Technology: Concept, Tools and Experiences, First Edition. Tanveer Bilal Pirzadah, Bisma Malik, Rouf Ahmad Bhat, and Khalid Rehman Hakeem.
© 2022 John Wiley & Sons Ltd. Published 2022 by John Wiley & Sons Ltd.

a new generation of large-scale but reliable and sustainable treatment units (Imbulana et al. 2020; Managheb et al. 2021; Kim et al. 2021). Correspondingly, micro-pollutants consisting of microorganisms, steroid hormones, pesticides, natural organic matter, micro-plastics derived from discarded man-made materials, pharmaceuticals, personal care products, heavy metals and toxic anions are increasingly detected in water bodies (Heidari et al. 2020; Zhang and Jiang 2021). These present a growing threat to aquatic habitats, all established food chains and ultimately human health. Importantly, these waste substances in many cases behave synergistically with other contaminants, exacerbating the detrimental effects (Silva et al. 2017). These issues represent some of the most significant global environmental concerns requiring urgent attention (Ojajuni et al. 2015). A broad range of novel technologies and their applications introduced in recent years play a substantial role in improving standards of living like medical research (Sadatshojaei et al. 2021a, 2021b, 2021c) and access to energy and resources, such as new improving methods (Sadatshojaei et al. 2016a, 2016b, 2021a, 2021b, 2021c; Choubineh et al. 2017; Dehshibi et al. 2019; Nezhad et al. 2021; Khajehesamedini et al. 2018; Bahrami et al. 2020), biofuels (Heidari and Wood 2020; Sadatshojaei et al. 2020a, 2020b; Jafari et al. 2021), solvents (Sadatshojaei and Wood 2020; Heidari et al. 2021, 2021b), nanoparticles (Sadatshojaei et al. 2016a, 2016b), corrosion (Seddiqi et al. 2019), carbon dioxide (Sadatshojaei and Rahimpour 2020; Sadatshojaei et al. 2018, 2021a, 2021b, 2021c) handling and utilization, and wastewater treatments. Many traditional treatment technologies are exploited to extract contaminants from a wide range of materials and systems; typically, these treatments require solvents (Daoud et al. 2019; Faragmandfar et al. 2019; Alara et al. 2020; Jain and Verma 2011). However, many widely used extraction technologies continue to consume significant quantities of organic solvents. These are effective but many are associated with carcinogenic and hazardous impacts and are responsible for long-term negative health consequences for humans and ecosystems, and lead to substantial pollution of the environment (Heidari and Haghniaz 2020; Heidari and Wood 2020; Heidari et al. 2021a, 2021b). These organic reagents find their way into landfills, water supplies and food chains (Epinoza-Acosta et al. 2014). Tackling this issue necessitates the development of the new generations of sustainable alternative technologies for concentration, extraction, purification and separation processes. Key beneficial technologies that have emerged over recent decades for use in purification, waste management and sustainability and pollution control in the drinking and food sectors, involve membranes. These offer substantial advantages over other conventional separation systems (Huang and Feng 2019; Omidvar et al. 2019; Scott 2019; Wei et al. 2018; Cao et al. 2020). The development and application of biotechnology solutions in these sectors has grown rapidly (Jochems et al. 2011). Combining membrane technologies with biotechnologies offers complementary benefits. Enzymatic membranes have done this by coupling membrane-separation processes with enzymatic reactions. Their potential has been successfully tested in a wide spectrum of applications in dairy, drinks and food industries alongside treatment, chemical production and pharmaceutical industries (Singh et al. 2018; Asif et al. 2018a, 2018b). In enzymatic-membrane systems, the active enzymes are able to circulate freely on the retentate side. On the other hand, the enzymes can be immobilized either on the membrane surfaces or inside their porous structures, preventing the removal of the enzyme from the medium (Kim et al. 2021). To achieve maximum efficiency in an enzymatic membrane, focus is required on both components to avoid degrading their productivity (Jochems et al. 2011). In this regard, careful selection of membrane surfaces is essential to optimize enzyme immobilization (Espinoza-Acosta et al. 2014; Zeuner et al. 2020; Zdarta et al. 2018a, 2018b). Membranes require enzyme-binding sites to retain enzymes and if these are not present on the membrane materials used, they need to be added. Various surface-modification methods, including wet chemical, ultra-violet (UV)-induced and γ-irradiation plasma techniques are applied to modify membrane surfaces (Jochems et al. 2011). Such membrane-modification techniques are employed to induce functional groups on a membrane's surface without disrupting its primary characteristics (Deng et al. 2009). These functional groups, in fact, immobilize the enzymes on the surface of the membranes. Another key issue for enzymatic membranes is maintaining effective activity levels of the enzyme. For doing so, it requires accurate control of the quantities of introduced functional molecules onto the membrane surfaces. The effectiveness of the functional groups can be substantially influenced by the enzyme characteristics, and vice versa. Indeed, maintenance of enzyme activity is strongly dependent on membrane-enzyme interactions (Liu and Jiang 2016). Furthermore, membrane modifications can have additional advantages. Typically, membrane modifications are conducted with the objectives of enhancing materials and their performances. A combination of key properties, such as permeability, selectivity, fouling resistance (by reducing undesired, secondary interactions), biocompatibility, chemical and thermal stability, cost-effectiveness and easy manufacturing are generally considerations for optimizing membrane performance and determining their uptake. Multiple methods are employed to modify membrane properties. These include photo-irradiation, surface plasma irradiation, blending with suitable polymers, atomic layer deposition, chemical cross-linking, surfactants and nanoparticles (Kim and Van Der Bruggen 2010; Sun et al. 2013). Each of these techniques has its specific benefits, drawbacks and application areas. In multilayer-membrane composites, membranes are integrated with a range of biocatalytic activities and/or perm-selective characteristics associated with individual enzymatic films (Jochems et al. 2011). Enzymatic-membrane reactors (EMR) intensify component interactions, combining desired biological reactions and separation in one device. Some of the exceptional features that make EMRs a leading technology in drink and food industries are their compactness, large surface area, excellent separation performance, antifouling potential and high-water flux. These features act in conjunction with selective mass transport through the membrane, sustaining enzyme activity and avoiding the use of toxic catalysts (Merz et al. 2014). For these reasons, EMRs offer environmentally sustainable solutions for food production and water supply and quality issues in the near future.

13.2 Enzymes

Enzymes are proteins that serve as biological catalysts in chemical reactions (Sahni and Goel 2015; Bautista-Cruz and Ortiz-Hernández 2015). They are, in turn, biomolecules that are able to increase the rate of chemical reactions. They do so by decreasing their activation energies, without undergoing any permanent alterations to their own structures. In enzymatic reactions, the chemical species with which enzymes may react are referred to as substrates. Such reactions occur in the region of an enzyme known as an active site. These sites are comprised of amino-acid domains that can establish temporary bonds with the substrates and ultimately convert them into desired products (Moraes et al. 2017). Enzymes are now routinely used in many industries, including food, pharmacy, water treatment, energy, agriculture, chemicals and textiles (Cen et al. 2019). Enzyme-mediated processes offer numerous, multifaceted advantages including time savings, cost-effectiveness, low energy consumption, high product selectivity, non-toxicity, biodegradability and eco-friendly sustainability. Furthermore, the introduction of protein engineering in the late twentieth century revolutionized the development of industrial enzymes. This facilitated the production of tailor-made enzymes exhibiting new catalytic capabilities and a high level of adaptation to a wide range of operating conditions. This enabled and promoted the multiplicity and expansion of their industrial use (Lobedanz et al. 2016). The growing demand for food and energy with less potential detrimental effects on human health, ecosystems and the wider environment, growing demand for affordable consumer goods, while conserving natural resources for a sustainable future, are all adriving forces behind the expanding uptake of enzyme catalysis in commercial-scale applications. Figure 13.1 summarizes the most exploited enzymes for industrial applications and many of these applications are in the food sector. These include, among many others, applications in drink and food processing, dairy products, bakery products, fruit-juice manufacture, nutraceuticals, vegetable oil processing, etc. On the other hand, they are also widely used in pharmaceutical, water/wastewater treatment, biofuels, textile, paper and chemical detergent industries.

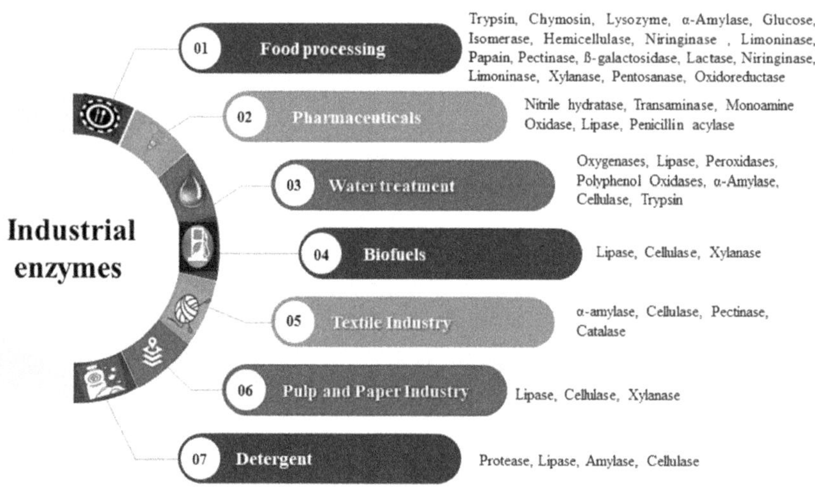

Figure 13.1 The most industrially exploited enzymes and their application sectors in multifaceted industries. This figure is drawn based on data reported in Kumar et al. (2014).

13.3 Global Demand for Commercial Enzymes

The synthesis of commercial enzymes has soared in recent years to meet the demand for expanding markets in conjunction with producing new biocatalysts to reduce the environmental burden. According to global enzymes market research, demand for enzymes globally (both the industrial and specialty enzymes) were projected to experience a significant increase of approximately 7% from $5.8 bn to $11.3 bn between 2010 and 2020. Specialty enzymes, however, have seen higher growth than that of their counterparts in industrial enzymes – 7.5% vs. 6.5% growth (Kumar et al. 2014). Figure 13.2 illustrates the proportion of projected global demand for the enzyme market in different areas in 2020. North America is the leading enzyme market accounting for 41% of global demand. Europe is the second-largest market with a 24% share, rivalled closely by Asia-Pacific with a 23% share. Central and South America, Africa and the Middle East have not yet become extensive markets for enzyme products, and collectively they account for less than 15% of the global enzyme market. Figure 13.3 illustrates the demand for enzymes in various industries from 2010 to 2020. In 2020, Enzyme use in the food industry was valued at about US$2.5 bn which is a considerably greater value than enzyme use by other industries (Figure 13.4). Moreover, the food sector has experienced substantial growth in enzyme demand over the past decade, valued at about US$1.3 trillion. This rising trend coincided with growing demand for better quality food products containing more natural flavors (Kumar et al. 2014). To some extent, this can be attributed to increased health and nutrition awareness among a more educated public, triggering the demand for food products that can demonstrate that they are produced by more sustainable techniques. The past decade has also witnessed substantial growth in enzyme use by the pharmaceutical industry. This demand approximately doubled from $1.2 trillion to $2.2 trillion from 2010 to 2020 (Figure 13.3). Food and pharmaceuticals currently

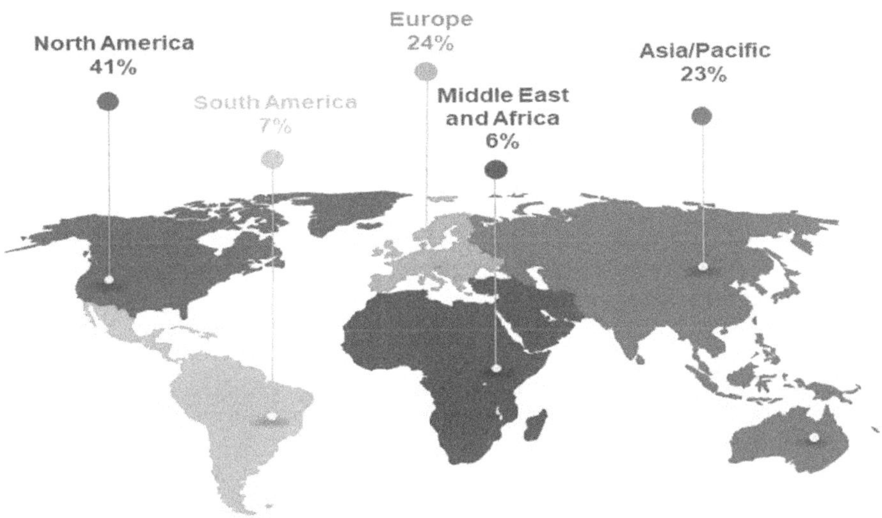

Figure 13.2 The proportion of estimated global demand in the enzyme markets geographically for 2020. This figure is drawn based on data reported in Kumar et al. (2014).

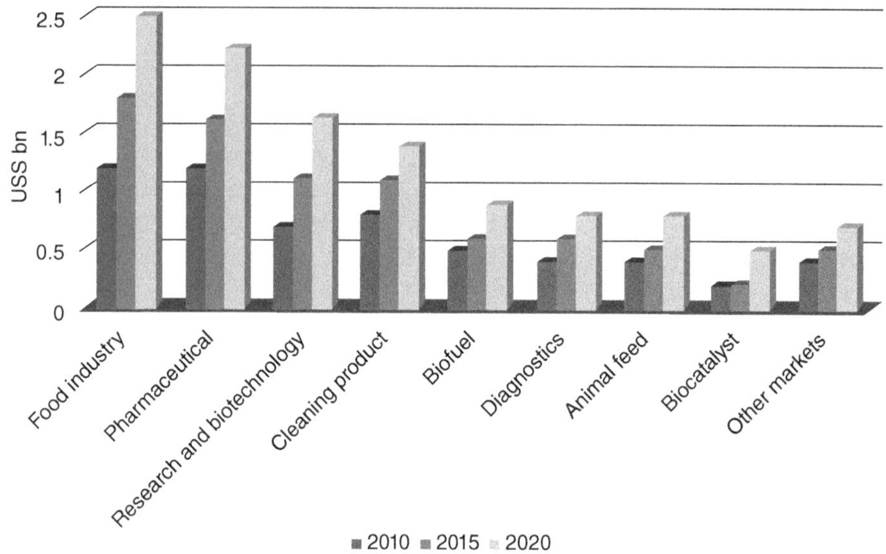

Figure 13.3 Recent relative growth in demand for enzymes in a multi-trillion-dollar market incorporating diverse industries for 2010, 2015 and 2020. This figure is drawn based on data reported in Kumar et al. (2014).

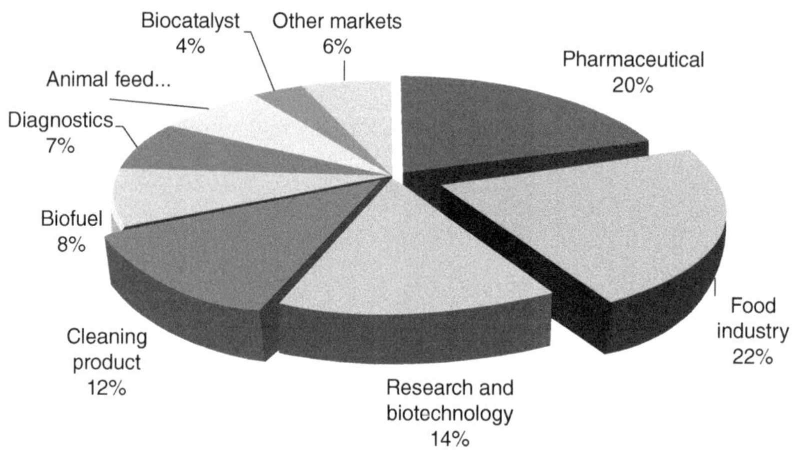

Figure 13.4 Approximate percentage shares of enzyme market by industrial sectors in 2020. This figure is drawn based on data reported in Kumar et al. (2014).

account for about half of the global enzyme market. However, research and biotechnology and cleaning products also constitute substantial current markets for enzymes. Some 22% of the global enzyme market is associated with the food industry (Figure 13.5). The food industry itself can be divided into main sub-sectors, namely, baked goods, dairy products and drinks and other foods. All three sub-sectors have rapidly increased enzyme consumption over the past two decades. The drinks industry constitutes a substantial proportion of the total food enzyme market, which was about US $1010 million in 2020.

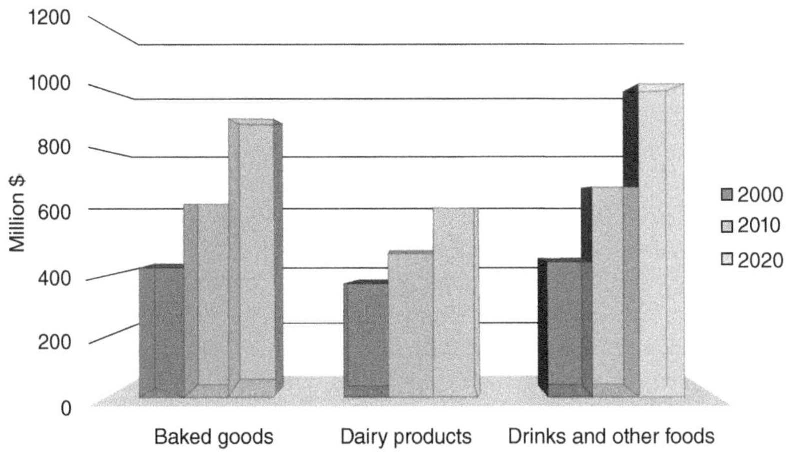

Figure 13.5 Worldwide baked food, dairy and drinks sectors enzyme demand (US$ millions) between 2000 and 2020 (*source*: based on Kumar et al. 2014).

13.4 Membrane Technology

A membrane is defined as a selectively permeable boundary, facilitating the elimination of unwanted constituents from both gas and liquid streams (Ismail et al. 2015). According to the driving forces, membranes can be categorized as pressure-driven, temperature-driven, concentration-driven and ion-exchanger types (Ismail et al. 2018). Pressure-driven membranes are distinguished as microfiltration (MF), ultrafiltration (UF), nanofiltration (NF) and reverse-osmotic (RO) types. Concentration-driven membranes can be of dialysis-type, forward-osmosis type and gas-separation-type. Temperature-driven membranes work via distillation or and pervaporation (i.e., separating liquids by partial vaporization) (Van Der Bruggen et al. 2003). Membrane separation is therefore achieved by the above-mentioned process mechanisms depending upon the type of membrane involved. Generically, these function by sieving, charge differentials, pervaporation and solution-diffusion. Based on the size of separated molecules, synthetic membranes can also be classified as asymmetric, dense or porous (Alen et al. 2019; Heidari et al. 2021a, 2021b). Membranes are widely exploited in industrial applications, especially the food industry, water/wastewater treatment, pharmaceutical processes, biomedical processes, desalination, environmental remediation and energy applications (Alen et al. 2019; Heidari et al. 2017a, 2017b; Zhang et al 2016; Hong et al. 2020). Typically, membranes can be synthesized from a broad range of materials, both organic and inorganic. Among organic membranes, polymeric membranes made of polyethylene (PE), polyethersulfone (PES), polysulfone (PSU), polyvinyl chloride (PVC) and polyvinylidene fluoride (PVDF) are widely employed. This is due to their low manufacturing costs, durability, packing densities, ease of installation, replacement and operation. Moreover, they display high thermal/chemical stability (Heidari et al. 2017a, 2017b; Shokri et al. 2015; Samsami et al. 2020; Heidari and Yegani 2018). Inorganic nanomaterials, such as titanium dioxide (TiO_2), ferric oxide (Fe_2O_3), raphene oxide (GO) and fumaratealumoxane (Fum-A) nanoparticles, can also be successfully incorporated into polymeric matrices, markedly improving their properties and performances (Ma et al. 2019; Huang et al. 2020; Halakoo and Feng 2020; Heidari et al. 2019; Sheikh and Sheikh 2020; Chong et al. 2019). For porous membranes, pore size and pore distributions represent key variables determining the effectiveness and efficiency of their purification/separation processes (Go et al. 2019). Figure 13.6 depicts the removal capacity of MF, UF, NF and RO membranes utilized in separation processes, testifying to the significance of pore size to their effectiveness. Recent research and testing have confirmed that membranes can be effectively used as a support for ensyme immobilization (Cen et al. 2019). Such biocatalytic membranes can be effectively integrated to form enzymatic membrane reactors (EMR) combining bio-catalysis and separation. Figure 13.7 provides a schematic diagram describing the functional steps of an EMR. The main objective of such integrated enzyme/membrane systems is to efficiently filter and treat highly contaminated aqueous/non-aqueous streams.

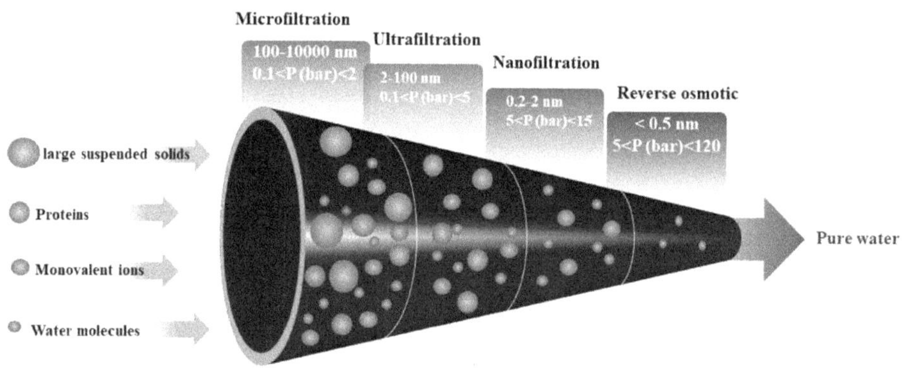

Figure 13.6 Filtration particle-size effectiveness, cut-off pressures and operating pressures for various pressure-driven membranes (*source*: based on Guo et al. 2012.

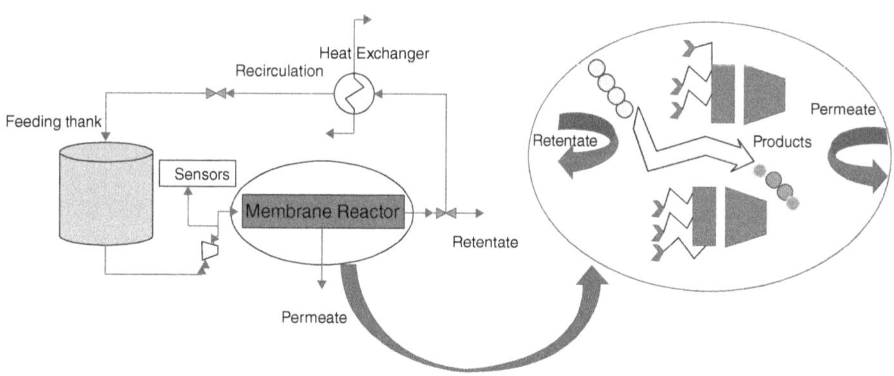

Figure 13.7 Schematic representation of enzymatic membrane reactor components and systems (*source*: De Cazes et al. 2014).

13.5 Fouling-Type Immobilization Membranes

Membranes, and their porous components, are repeatedly and unavoidably faced with a wide range of drawbacks during filtration operations. Membrane fouling reduces the flux and gradually imposes irreversible damage to membranes (Guo et al. 2012). Consequently, it is essential to identify and understand the mechanisms responsible for membrane fouling in order to mitigate them (Luo et al. 2014). There are a broad range of potential pollutants of membranes and these include humic compounds, proteins, microbes and polysaccharides (He et al. 2019). There are many mitigation techniques that are designed to prevent the fouling of membranes. These include membrane area alteration, enhanced cake-layer resistors, pore blocking and adsorption fouling (Luo et al. 2013). On the other hand, membrane fouling provides a mechanism to immobilize enzymes. Consequently, some membrane-fouling enzyme-immobilization procedures have been beneficially exploited (Luo et al. 2013). Adsorption processes by membranes tend to be analogous to adsorption processes operating in soils. One type of membrane area alteration is covalent linkage, and another consists of cross-linkages of combined inorganic and organic fouling, referred to as "cake" layering (Luo et al. 2013). Four distinct procedures are available to immobilize enzymes in and/or on membranes such as cross-linking, covalent linkage, trapping and adsorption.

13.6 Physical Procedures that Immobilize Enzyme in/on Membranes

Membranes focused on immobilization involve substantial quantities of functional groups and require sizeable surface areas (Jesionowski et al. 2014). Physical adsorption is a facilitated and easily available membrane-immobilization procedure for enzymes. Electrostatic forces and van der Waals forces, as well as hydrogen bonds between the carrier and the enzyme, play a substantial role in physical adsorption (Sassolas et al. 2012). The physical interplays between enzymes and membranes can vary. The chemical and physical factors of the membrane area, with different kinds of active functional groups, pore structures and surface-area dimensions, play a crucial role in achieving successful physical adsorption.

13.7 Covalent Bonds that Immobilize Enzymes in/on Membranes

Covalent procedures join the enzymes to carriers and their functional molecules. Chemical bonding occurs either within the membrane pores or on their surfaces. Certain functional groups present in membranes often directly bond covalently (Ariaeenejad et al. 2021). The membrane pre-treatments and modifications associated with functionalizing reagents are usually required to enhance covalent immobilization. Such procedures include surface plasma irradiation, the UV irradiation, γ-irradiation and wet chemical applications (Jochems et al. 2011). By applying these pre-treatments, the pores or surface areas of the membrane are covered with active groups including amides, sulfhydryl, tyrosine, carboxyls, hydroxyls, azidos and aminos. Subsequently, components including enzymes become covalently bonded to these chemical groups (Zdarta et al. 2018a, 2018b). It is crucial to apply such membrane modification in order to immobilize enzymes covalently. The modifications can be customized to accommodate covalent bonding of particular enzymes and membrane-surface chemistry. There are two distinct functions for membrane modifiers: i) to prevent the binding to the enzyme-active center of the membrane in order to sustain high recovery activity; and ii) to constantly bind enzymes to the membrane surface.

13.7.1 Amino Groups that Modify Membranes

1,12-diaminododecane (Sulaiman et al. 2017), 1,6-hexamethylenediamine (1,6-HMDA) (Vasileva et al. 2016) and 3-aminopropyltriethoxysilane (APTES) (Hou et al. 2014) are the mostly employed modification reagents capable of producing membrane amino groups. There are interactions between amino and carboxyl groups and enzyme molecules that facilitate their bonding with a range of modifiers. Indeed, the three reagents mentioned are routinely employed to couple with glutaraldehyde (GLU) in order to create bio-catalytic membranes. TiO_2 nanoparticles can also be introduced into membranes through phase-inversion techniques (Razmjou et al. 2011) during fabrication to create titania-functionalized membranes. APTES functionalize the titania nanoparticles, which go on to disperse within the GLU. Such APTES-induced modifications enable membranes to hold amino groups. These are able to bond with the aldehyde component of the GLU molecule group forming a Schiff-base compound. The enzyme amino groups are subsequently able to bond covalently, exploiting this mechanism (Hou et al. 2014). The consecutive reaction steps involved result in about 79% recovery and 0.135 U mg^{-1} support for laccase enzymatic activity (Hou et al. 2014). This mechanism also works for 1,6-HMDA and is exploited to immobilize β-galactosidase (β-Gal).

13.7.2 Carboxylic Groups that Modify Membranes

Carboxyl groups are also capable of functionalizing membrane support and once present, they can effectively bond covalently with amino groups present in enzymes (Meller et al. 2017). (Poly)acrylic acids modify films, enabling them to attain a carboxyl group. Acrylic acid can be used as a raw compound to generate a polymer film. Acrylic acid combined with a carboxyl andvinyl group enhances the rate of polymerization. Ultrafiltration membranes relying on the acrylic-acid, graft-copolymerization method (AAc-g) were invented by Ying et al. (2002). This technique is used to manufacture polyvinylidenefluoride (PVDF) copolymer membranes. The amino groups contained within the enzyme molecules bond covalently bound carboxyl side-chains of PVDF-membrane surfaces.

13.7.3 Epoxy Groups that Modify Membranes

An epoxy group can react directly with sulfhydryl, carboxyl or amino groups within enzyme molecules to bond covalently in the absence of additional chemical activators (Pithawala et al. 2010). This modification procedure is more widely employed in covalent immobilization because the epoxy group is able to form bonds with miscellaneous components of enzymes (Pujari et al. 2006). Glycidyl methacrylate (GMA) is a commonly employed modifier in the covalent immobilization procedure, due to its ionic reactive epoxy group and active vinyl group. Hollow-fiber membranes link GMA and polypropylene utilizing polymerization, typically induced by radiation to immobilize a lipase. The epoxy is transformed into a diethylamino group during this process. The diethylamino group can bind the molecules of a lipase to form an anion-exchange medium (Abrol et al. 2007). Ultimately, an enzyme membrane that is active in terms of bioactivity is attained and typically offers high-storage stability and reaction stability.

13.7.4 Azido Groups that Modify Membranes

1-Fluoro-2-nitro-4-azidobenzene (FNAB) is a photo coupler frequently employed to provide the azido functional group for activating polymeric membranes. The carbon–hydrogen bond can accommodate the extremely reactive nitrene. An azido at the photo-sensitive surfaces of cellulose films creates the nitrene in the presence of UV irradiation. The nitrene is then available to covalently bond with enzyme components via insertion reactions (Bora et al. 2006). As irradiation by UV light is required for these reactions to occur, immobilization of enzymes and other bio-molecules is restricted to the surface of modified membranes (Kumar et al. 2016).

13.8 Cross-linkage Procedures

Cross-linkage can produce a reactor with Enzymes Immobilized on Membranes (REIM) in the presence of bi-functional molecules involving bis-diazotized benzidine, glyoxal, hexamethylenediamine and GLU. The catalytic enzymes are able to directly crosslink among themselves (Razib et al. 2020). Alternatively, the enzymes can become immobilized by cross-linking with selected non-functional proteins, such as bovine serum albumin. The load of enzymes at a membrane can be substantially enhanced by cross-linkage. On the downside, cross-linkages often induce or contribute to membrane fouling. This leads to restricted diffusion through the membrane, particularly as the membranes become progressively overloaded with enzymes (Palai and Bhattacharya 2013). The cross-linkages created often include complex biochemical reactions, which sometimes provide exploitable opportunities to beneficially amend an enzyme's structure. Bioactive membranes can be expanded through cross-linkages to enhance their performance. A polysulfone-membrane surface, cross-linked with GLU, can successfully immobilize a lipase of *Candida rugosa* (Zhu et al. 2016).

13.9 Applications of Enzymatic Membrane Reactors

13.9.1 Treatment of Milk or Cheese Whey

Membrane reactors provide sustainable systems for hydrolyzing the lactose available in cheese whey or milk. Furthermore, by employing beta-galactosidase immobilized in membrane reactors, the recovery of large proteins and lactose transformation can be performed in a single step (Illanes et al. 1990; Asif et al. 2018a, 2018b). Hydrolyzed milk is suitable for consumers who are sensitive to lactose and other milk components. Additionally, generating hydrolyzed cheese whey effectively provides environmental and economic benefits (Ahmed et al. 2020; Marwaha and Kennedy 1988; Ngoc et al. 2018). Cheese whey is made up of about 93% water and salts, 0.7% proteins and 5% lactose, constituting a major contaminant to be removed. EMRs make it feasible and cost-effective to remove or reduce lactose content during cheese production and at the same time reduce waste through recycling of separated materials. For instance, some whey proteins, such as alpha-lactalbumin, a protein that regulates mammalian milk production, offer attractive functional characteristics. These can be easily recovered through hydrolysis and ultrafiltration. Among other uses, such extracts are exploited as pharmaceutical intermediates.

13.9.2 Treatments of Animal, Plant, and Waste Oils and Fats

Progress is being made in substituting traditional chemical methods for oils and fats (DePietro et al. 2020) with biotechnological methods (Giorno and Drioli 2000). The exploitation of lipase-immobilized membrane reactors for this purpose are well documented (Gonawan 2019). The operating mechanisms of various reactors used for this purpose varies according to the physicochemical characteristics of the reactions involved. Lipase is able to efficiently hydrolyze triglyceride-containing vegetable oils to a range of fatty acids and glycerol. Different types of reactors can be employed to achieve such conversions (Giorno et al. 1998), which include: i) an EMR exploiting an emulsified-organic–aqueous enzyme membrane, referred to as an E-EMR, performing ultrafiltration of its emulsified throughput; ii) an ordinary emulsion stirring tank reactor (STR); and iii) an E-EMR with two-phase throughputs (B-EMR, the "B" referring to its biphasic nature) that are split by the membrane via an enzyme immobilization process. The reaction rate and the volumetric extent of the reactions are greater for the free-enzyme phase than for the immobilized-enzyme phase. Moreover, the immobilized enzyme's catalytic activity remains significantly more stable in such biphasic configurations.

13.9.3 Pharmaceutical Production Employing Biocatalytic Membrane Reactors

There are miscellaneous systems that are adapted to produce pure enantiomers using optical techniques. Additionally, membrane-technology strategies exploited to optically generate pure isomers are successfully utilized (Vladisavljević 2016). There is a broad range of research focusing on the generation of carboxylic acids, aryl propionic acids, amino acids and amines for use in such systems (Vishnoi et al. 2020). The main drawbacks in generating such materials for pharmaceutical production on a large scale involve: i) their need for costly ancillary processes; ii) the low level of water-solubility of many of the associated substrates; and iii) the extraction, refinement and separation of the products from complex solutions. Although many such systems have been demonstrated to work on laboratory scales, it is challenging to scale these up for commercial production. Membrane reactor designs and developing efficient membranes tailored to specific products are the key hurdles that are yet to be fully resolved for commercially-viable applications of these systems.

13.9.4 Biocatalytic-Membrane Reactors for Biomedical Applications

Some artificial organs are now driven by membrane reactors (Losappio et al. 2020), and this has broadened the possible applications of extracorporeal treatments and organ substitutions. Membranes also play a substantial role in many other medical applications and tools. For instance, artificial-kidney haemodialysis equipment involves rapid counter-current circulation of buffer solutions on one side of a membrane, while the recipient's blood passes along the other side of the dialyser membrane (Losappio et al. 2020). Membranes are also employed to enhance blood oxygenation in artificial lungs and respirators. Intracorporeal and extracorporeal large-scale and micro-membrane oxygenators are being developed (Mortensen 1993).

13.9.5 Biocatalytic-Membrane Reactors for Agricultural Applications

Effective membrane applications methods and biocatalytic reactors exploit microfiltration, ultrafiltration, reverse osmosis and membrane extraction. These are widely used in fermentation processes to produce organic acids, antibiotics and the processing of various drink and food items. Biocatalytic membrane reactors are particularly exploited in agriculture and food sectors to: a) decrease juice viscosities in pectin hydrolysis; b) decrease lactose in milk and/or whey by transforming it into more digestible sugars; c) transforming polyphenol and anthocyanin components of wine to improve its quality and durability; and d) peroxide removal from various diary items (Zhu et al. 2020; Ottone et al. 2020).

13.9.6 Biocatalytic-Membrane Reactors for Waste-Water Treatment

Membrane technologies have been applied enthusiastically to wastewater and water treatments (Zdarta et al. 2019). Membrane bioreactors and membranes are successfully applied to treat activated sludge (Mohammad et al. 2019) and to remove nitrogen from wastewater (Yuan et al. 2018). Some highly activated sludge methods employ a rotary-disk-ultrafiltration membrane in order to treat wastewater with concentrated zymosis products (Natesan 2019). The operation of such systems over long test periods of about four months has demonstrated their effectiveness for zymosis-wastewater treatments. Bhanot et al. (2020) tested the combination of membrane bioreactors equipped with a ceramic-cross-flow-ultrafiltration component to effectively treat explosives-industry wastewater. This system hydrolyzes the highly explosive hexahydro-1,3,5-trinitro-1,3,5-triazine byproducts (RDX), which are potentially carcinogenic. Unfortunately, the hydrolysis of RDX produces additional byproducts, including acetate, formaldehyde, formates and nitrates. The nitrate is eliminated through utilizing an anoxic biochemical reaction that transforms hydrolysate into non-toxic nitrogen and carbon dioxide. The permeate of this process is purified water, while other effluents are passed through a ceramic membrane for recycling in the bioreactor (Bhanot et al. 2020).

13.10 Limitations, Challenges and Solution for EMR Applications

Despite major advances of enzyme-membrane technology over the past two decades, this highly promising technology still faces developmental issues for commercial applications. If these challenges can be overcome, it should lead to further market expansion for this technology. The main challenges are considered in detail with some mitigation actions identified:

1) While membrane separation processes (MSP) offer advantages over conventional water treatments, the main problems and barriers impeding the design of an efficient immobilized enzyme system is to obtain retention of high enzyme activity, enhanced stability and reusability (Sigudardóttir et al. 2018). However, to do so, it is difficult and time-consuming. Many variables are involved and their interactions are complex. Additionally, there are many methods and materials available for synthesizing enzymatic membranes and it is difficult to identify the most suitable for specific objectives. Despite the numerous immobilization methods used in scientific research, no general rules exist for choosing between them to ensure maximum efficiency and effectiveness. Hence, careful evaluations of enzyme-immobilization systems must be conducted for each membrane considered. These identify effects of the immobilization system on the enzyme properties upon immobilization, thereby quantifying their activity, stability and reusability.
2) Enzyme leakage presents a substantial challenge and such losses can be minimized by narrowing the membrane cut-offs in some cases. However, a decrease in membrane pore size inevitably causes drop in pressure, and reduces water flux through a membrane (Jochems et al. 2011). Careful optimization of membrane preparation variables and verification experiments can help to limit and control such impacts.
3) Concentration polarization and membrane fouling lead to reduction in permeate flux. They also increase energy consumption under constant-flux operating modes. This is due to significant increases in hydraulic resistance. Such conditions, in turn, accelerate membrane biodegradation and increase in operating costs due to subsequent requirements for frequent membrane cleaning and component replacements (Heidari et al. 2021a, 2021b; Nguyen et al. 2012). Membrane fouling is a complex process covering a wide range of scientific disciplines. In reality, this involves a complex interplay of additional variables. However, a rigorous theoretical and experimental analysis of membrane behavior and prediction of fouling mechanisms can help to identify and introduce an effective method to alleviate fouling in membrane-associated processes. Hence, finding sustainable fouling mitigation strategies to avoid reduced treated permeate productivity remains one of the main challenges (Ioprhemen et al. 2016).
4) Sustaining aseptic and/or sterile conditions is essential in enzymatically produced products, particularly those destined for use in food and pharmaceuticals (Jochems et al. 2011). A key challenge for the biotechnological industry is to find suitable and effective membranes that can be easily sterilized, preferably *in situ*.

5) Membrane materials and their characteristics can incur negative impacts on adsorption and retention of contaminants, such as nanofiltration of estrogenic MPs identified by Semião and Schäfer (2013). Systematic and comprehensive analysis was required to determine the complex interactions between these particles and different polymer materials to identify and quantify both relative contributions and strengths of each interaction. This is a complicated task given the intricacy of bioreactor systems and because efficient removal of contaminants is synchronously influenced by additional secondary factors (Silva et al. 2017).
6) Determining feed specifications also poses its own challenges. Ultrafiltration membrane enzymatic reactors are designed to remove contaminants in the range 0.01 to 0.1 μm. They are not configured to handle large particles outside that size range, and also they are not specifically designed to be defouled and cleaned frequently. This becomes particularly problematic when such systems are used to treat high-fouling feed water sources, such as surface water and wastewater. Integrated membrane systems (IMS), where microfiltration (MF) is introduced as a technically and cost-effective pre-treatment option for EMRs, may offer more sustainable strategies for efficient separation in such circumstances. In water treatment, such configurations become essential as low-fouling feed-water sources become less available and it becomes necessary to treat more difficult feed-water sources such as surface waters, industrial wastewaters and municipal sewer waters.
7) Regeneration of immobilized enzymes can be costly and time-consuming. Immobilized enzymes typically deactivate during their use and need to be replaced. Therefore, to minimize the associated regeneration costs, it is crucial to develop easy and cheap processes to restore enzyme activity. On the other hand, stability and reusability of the enzymes tend to be improved by immobilization (Jochems et al. 2011). The activity and enantio selectivity of the enzyme may also be enhanced by carefully tuning the process conditions. Hence, there is a trade-off of benefits and regeneration costs and operational obstacles to achieve with respect to enzyme immobilization.
8) Prediction of the enzyme behavior after the immobilization process is difficult to establish prior to plant scale-up. Thus, large quantities of trial-and-error experiments are required at different operational scales and conditions to achieve the desired recycled enzyme performance.
9) The presence of carry-over enzymes in the permeates can be problematic. If this occurs, it necessitates additional processing of the permeate phase to remove the carry-over substances to ensure that the final products meet quality standards. Such additional product purification steps tend to be costly.
10) Unexpected or difficult-to-deal-with impurities found in the feed may have a detrimental effect on biocatalytic membranes. Besides, changes in the operational conditions, such as shifts in pH or temperature, that they can lead to, may denature the active enzymes resulting in a dramatic decrease in system efficiency. There is, therefore, a need to systematically control the experimental conditions and importantly develop biocatalytic membranes that are able to withstand fluctuations within limits of feed compositions. This is essential in wastewater and sewage treatment systems.

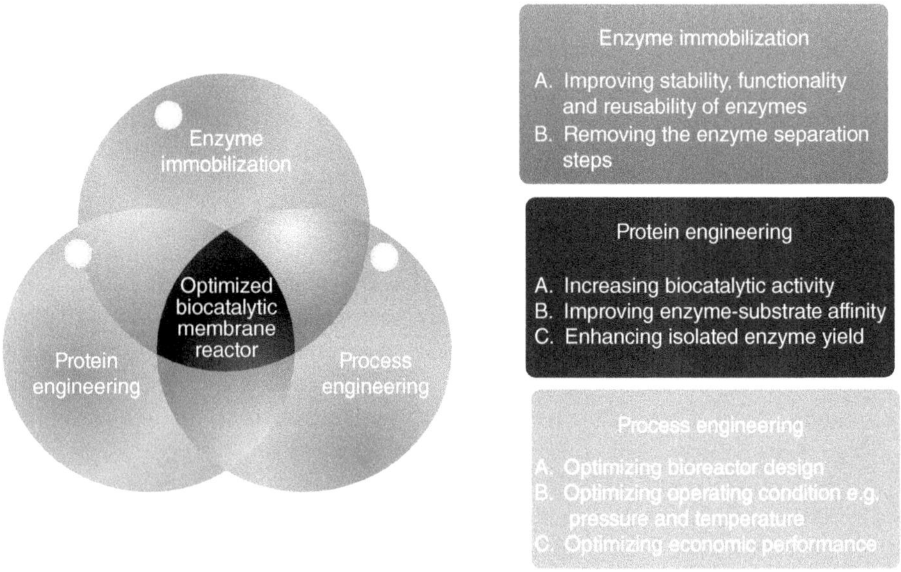

Figure 13.8 Techniques with potential to improve enzymatic membrane reactor efficiencies (*source*: modified from Chapman et al. 2018).

11) The service lives of many enzymatic membranes require improvement. Research and development of enzymatic membranes to extend mechanical and chemical life spans is essential for extending membrane efficiency, reducing costs and making such systems commercially sustainable (Wang et al. 2008).
12) EMRs have been extensively studied at laboratory scales; however, operational tests at pilot- and full-commercial scales are lacking for many potential designs. This represents uncertainty and risk regarding their true operational capabilities. With the current level of knowledge and understanding of EMR performance for the removal of specific contaminants, it is inappropriate to rely upon predictions regarding contaminant-removal efficiency in commercial-scale reactor implementations, based solely on laboratory-scale experiments with synthetic pollutants.

Figure 13.8 summarizes the available techniques and strategies currently employed in attempts to optimize and achieve maximum efficiency enzymatic membrane bioreactors. Although many obstacles remain, the successfully implementation of enzymatic membrane processes can be achieved by diligently addressing the issues identified. By doing so, EMR operating costs can be substantially reduced. Continuing improvements, plus ongoing research and development, should enable greater quantities of liquids to be more effectively treated by EMRs in the future.

13.11 Conclusions

Enzymatic-membrane-reactor (EMR) technologies now offer more efficient and low-cost alternatives to many, but not all, fluid treatment and purification systems. They are particularly successful where they are able to decrease equipment costs by removing some processing steps. The drink and food sector dominate EMR use and these are closely followed by the pharmaceutical and cleaning product sectors. Much potential remains to expand their usage in water, wastewater and sewage treatment systems. Emerging markets for EMRs include non-aqueous enzymology and pharmaceutical applications for the catalytic properties of antibodies. New applications for enzymes coupled with non-biological catalysts and the expansion of bio-sensors use for medical diagnosis also offer future potential. However, in order to achieve substantial expansion in many industrial sectors, the enzyme- and biocatalytic-membrane-reactor sector needs to focus more on process scale-up, improved handling of immobilized enzymes, membrane defouling and contaminant removal methods in order to commercially outperform established alternatives.

References

Abrol, K., Qazi, G., and Ghosh, A.K. (2007). Characterization of an anion-exchange porous polypropylene hollow fiber membrane for immobilization of ABL lipase. *Journal of Biotechnology* 128: 838–848.

Ahmed, S., Sharif, M.K., Butt, M.S. et al. (2020). Development of nutritious probiotic sportsman drink. *Pakistan Journal of Pharmaceutical Sciences* 33 (1): 191–197.

Alara, O., Abdurahman, N., and Olalere, O. (2020). Ethanolic extraction of flavonoids, phenolics and antioxidants from *Vernonia amygdalina* leaf using two-level factorial design. *Journal of King Saud University – Science* 32: 7–16.

Alen, S.K., Nam, S., and Dastgheib, S.A. (2019). Recent advances in graphene oxide membranes for gas separation applications. *International Journal of Molecular Sciences* 20: 5609.

Ariaeenejad, S., Motamedi, E., and Salekdeh, G.H. (2021). Application of the immobilized enzyme on magnetic graphene oxide nano-carrier as a versatile bi-functional tool for efficient removal of dye from water. *Bioresource Technology* 319: 124228.

Asif, M.B., Hai, F.I., Kang, J. et al. (2018a). Biocatalytic degradation of pharmaceuticals, personal care products, industrial chemicals, steroid hormones and pesticides in a membrane distillation-enzymatic bioreactor. *Bioresource Technology* 247: 528–536. https://doi.org/10.1016/j.biortech.2017.09.129

Asif, M.B., Hai, F.I., Jegatheesan, V. et al. (2018b). Applications of membrane bioreactors in biotechnology processes. In: *Current Trends and Future Developments on (Bio-) Membranes* (ed. A. Basuke and C. Charciosset), pp. 223–257. USA: Elsevier. http://dx.doi.org/10.1016/B978-0-12-813606-5.00008-7

Bahrami, B., Sadatshojaie, A., and Wood, D.A. (2020). Assessing wellbore stability with a modified lade failure criterion. *Journal of Energy Resources Technology* 142 (8): 083004 (12 pages).

Bautista-Cruz, A. and Ortiz-Hernández, Y.D. (2015). Hydrolytic soil enzymes and their response to fertilization: A short review. *Comunicata Scientiae* 6: 255–262.

Bhanot, P., Celin, S.M., Sreekrishnan, T. et al. (2020). Application of integrated treatment strategies for explosive industry wastewater: A critical review. *Journal of Water Process Engineering* 35: 101232.

Bora, U., Sharma, P., Kannan, K. et al. (2006). Photoreactive cellulose membrane: A novel matrix for covalent immobilization of biomolecules. *Journal of Biotechnology* 126: 220–229.

Cao, X., Lee, H.-S., and Feng, X. (2020). Extraction of dissolved methane from aqueous solutions by membranes: Modelling and parametric studies. *Journal of Membrane Science* 596: 117594.

Cen, Y.K., Liu, Y.X., Xue, Y.P. et al. (2019). Immobilization of enzymes in/on membranes and their applications. *Advanced Synthesis & Catalysis* 361: 5500–5515.

Chapman, J., Ismail, A.E., and Dinu, C.Z. (2018). Industrial applications of enzymes: Recent advances, techniques, and outlooks. *Catalysts* 8: 238. doi:10.3390/catal8060238

Chong, C.-Y., Lau, W.-J., Yusof, N. et al. (2019). Roles of nanomaterial structure and surface coating on thin film nanocomposite membranes for enhanced desalination. *Composites Part B: Engineering* 160: 471–479.

Choubineh, A., Ghorbani, H., Wood, D.A. et al. (2017). Improved predictions of wellhead choke liquid critical-flow rates: Modelling based on hybrid neural network training learning based optimization. *Fuel* 207: 547–560.

Daoud, A., Malika, D., Bakari, S. et al. (2019). Assessment of polyphenol composition, antioxidant and antimicrobial properties of various extracts of date palm pollen (DPP) from two Tunisian cultivars. *Arabian Journal of Chemistry* 12: 3075–3086.

de Cazes, M., Abejón, R., Belleville, M. et al. (2014). Membrane bioprocesses for pharmaceutical micropollutant removal from waters. *Membranes (Basel)* 4 (4): 692–729. doi:10.3390/membranes4040692

Dehshibi, R., Sadatshojaie, A., Mohebbi, A. et al. (2019). A new insight into pore body filling mechanism during waterflooding in a glass micro-model. *Chemical Engineering Research & Design: Transactions of the Institution of Chemical Engineers Part A* 151: 100–107. https://doi.org/10.1016/j.cherd.2019.08.019

Deng, J., Wang, L., Liu, L. et al. (2009). Developments and new applications of UV-induced surface graft polymerizations. *Progress in Polymer Science* 34: 156–193.

Di Pietro, M.E., Mannu, A., and Mele, A. (2020). NMR determination of free fatty acids in vegetable oils. *Processes* 8: 410.

Espinoza-Acosta, J.L., Torres-Chávez, P.I., Carvajal-Millán, E. et al. (2014). Ionic liquids and organic solvents for recovering lignin from lignocellulosic biomass. *BioResources* 9: 3660–3687.

Farahmandfar, R., Esmaeilzadeh Kenari, R., Asnaashari, M. et al. (2019). Bioactive compounds, antioxidant and antimicrobial activities of *Arum maculatum* leaves extracts as affected by various solvents and extraction methods. *Food Science & Nutrition* 7: 465–475.

Giorno, L. and Drioli, E. (2000). Biocatalytic membrane reactors: Applications and perspectives. *Trends in Biotechnology* 18: 339–349.

Giorno, L., Molinari, R., and Drioli, E. (1998). Experimental studies on enzyme membrane reactors in oil treatment. *Advances in Oils and Fats, Antioxidants, and Oilseed By-products* 2: 91–94.

Goh, P.S., Ismail, A.F., Ng, B.C. et al. (2019). Recent progresses of forward osmosis membranes formulation and design for wastewater treatment. *Water* 11: 2043.

Gonawan, F.N. (2019). *Immobilized B-Galactosidase-Mediated Conversion of Lactose: Process, Kinetics and Modeling Studies*. Springer.

Guo, W., Ngo, -H.-H., and Li, J. (2012). A mini-review on membrane fouling. *Bioresource Technology* 122: 27–34.

Halakoo, E. and Feng, X. (2020). Layer-by-layer assembly of polyethyleneimine/graphene oxide membranes for desalination of high-salinity water via pervaporation. *Separation and Purification Technology* 234: 116077.

He, Y., Zhuang, X., Lei, C. et al. (2019). Porous carbon nanosheets: Synthetic strategies and electrochemical energy related applications. *Nano Today* 24: 103–119. https://doi.org/10.1016/j.nantod.2018.12.004

Heidari, S. and Haghniaz, R. (2020). The potential use of supercritical CO_2 as a sustainable solvent in biocatalytic reactions. In: *Green Sustainable Process for Chemical and Environmental Engineering and Science Green Solvents for Biocatalysis* (ed. B.R. Inamuddin, M.I. Ahamed, and A.M. Asiri), Chapter 16. Elsevier. https://doi.org/10.1016/B978-0-12-819721-9.00013-3

Heidari, S. and Wood, D.A. (2020). Ionic liquids and their beneficial contributions to enzyme-catalyzed reactions, catalytic biomass conversion and energy conversion and storage systems. In: *Green Sustainable Process for Chemical and Environmental Engineering and Science: Green Solvents for Biocatalysis* (ed. B.R. Inamuddin, M.I. Ahamed, and A.M. Asiri), Chapter 18, 369–407. Elsevier. https://doi.org/10.1016/B978-0-12-819721-9.00013-3

Heidari, S. and Wood, D.A. (2021). Biodiesel production methods and feedstocks. In: *Biodiesel Technology and Applications* (ed. B.R. Inamuddin, R. Boddula, and M. Rezakazemi), Chapter 17, pp. 447–464. Wiley Online Library. https://doi.org/10.1002/9781119724957.ch17

Heidari, S. and Yegani, R. (2018) Experimental study on fouling characteristics of HDPE and HDPE/PE-g-MA/EVA blend membranes in microfiltration of collagen: Effects of surface morphology and physicochemical membrane features. *Proceedings of the 10th International Chemical Engineering Congress & Exhibition (IChEC)*. Iran.

Heidari, S., Jafarzadeh, Y., Seyfollahi Samarin, M.A. et al. (2017a). Study on the fouling behavior of HDPE/PE-g-MA/EVA blend membrane fabricated via thermally induced phase separation method. *Polyolefins Journal* 4: 235–251.

Heidari, S., Amirinejad, M., and Jahangirian, H. (2017b) Fouling identification for nanofiltration of whey: Resistance-in-series model. *Proceedings of the 5th International Conference on The New Findings of the Chemistry and Chemical Engineering*. Iran.

Heidari, S., Amirinejad, M., and Jahangirian, H. (2019). Investigation of fouling mechanisms using surface morphology and physicochemical membrane features. *Chemical Engineering & Technology* 42: 1310–1320.

Heidari, S., Etemadi, H., and Yegani, R. (2020). A comprehensive analysis of membrane fouling in microfiltration of complex linear macromolecules based on theoretical modeling and FESEM images. *Chemical Technology and Biotechnology* 96 (2): 360–373.

Heidari, S., Sadatshojaei, E., and Wood, D.A. (2021a). Large-scale molecular solvents for environmentally sustainable applications. In: *Green Sustainable Process for Chemical and Environmental Engineering and Science: Environmental Remediation* (ed. B.R. Inamuddin and A.M. Asiri), 267–282. Elsevier. https://doi.org/10.1016/B978-0-12-821884-6.00001-2.

Heidari, S., Foroozesh, J., and Latifid, M. et al. (2021a). Application of supercritical fluid for biocatalysis with an emphasis on supercritical water. In: *Green Sustainable Process for Chemical and Environmental Engineering and Science: Green Solvents for Biocatalysis* (ed.

B.R. Inamuddin, M.I. Ahamed, and A.M. Asiri), Chapter 15, 301–323. Elsevier. https://doi.org/10.1016/B978-0-12-819721-9.00015-7.

Heidari, S., Sadatshojaei, E., Wood, D.A. et al. (2021b). Algae-based membrane bioreactor for wastewater treatment. In: *Bioresource Technology: Concept, Tools and Experiences* (ed. T.B. Pirzadah, B. Malik, R.A, Bhat, and K.R. Hakeem), Chapter 11. Wiley. ISBN: 978-1-119-78938-3.

Hong, X., Xiong, J., Feng, Z. et al. (2020). Extracorporeal membrane oxygenation (ECMO): Does it have a role in the treatment of severe COVID-19? *International Journal of Infectious Diseases* 94: 78–80.

Hou, J., Dong, G., Ye, Y. et al. (2014). Laccase immobilization on titania nanoparticles and titania-functionalized membranes. *Journal of Membrane Science* 452: 229–240.

Huang, Y. and Feng, X. (2019). Polymer-enhanced ultrafiltration: Fundamentals, applications and recent developments. *Journal of Membrane Science* 586: 53–83.

Huang, X., Chen, Y., Feng, X. et al. (2020). Incorporation of oleic acid-modified Ag ZnO core-shell nanoparticles into thin film composite membranes for enhanced antifouling and antibacterial properties. *Journal of Membrane Science* 602: 117956.

Illanes, A., Ruiz, A., Zuniga, M. et al. (1990). Immobilization of lactase for the continuous hydrolysis of whey permeate. *Bioprocess Engineering* 5: 257–262.

Imbulana, S., Oguma, K., and Takizawa, S. (2020). Evaluation of groundwater quality and reverse osmosis water treatment plants in the endemic areas of Chronic Kidney Disease of Unknown Etiology (CKDu) in Sri Lanka. *Science of the Total Environment* 745, 140716.

Iorhemen, O., Hamza, R., and Tay, J. (2016). Membrane bioreactor (MBR) technology for wastewater treatment and reclamation: Membrane fouling. *Membranes* 6: 33.

Ismail, A.F., Khulbe, K.C., and Matsuura, T. (2015). Gas separation membrane materials and structures. In: *Gas Separation Membranes*, 37–192. Springer.

Ismail, A.F., Rahman, M.A., Othman, M.H.D. et al. (2018). *Membrane Separation Principles and Applications: From Material Selection to Mechanisms and Industrial Uses*, 1–494. Elsevier. eBook ISBN: 9780128128169

Ismail A.F., Heidari, S., Wood, D.A. et al. (in press, 2022) Biogas production and processing from various organic wastes in anaerobic digesters and landfills. In: *Bioresource Technology: Concept, Tools and Experiences* (ed. T.B. Pirzadah, B. Malik, R.A. Bhat, et al.). Wiley. ISBN 1119789389, 9781119789383

Jafari, A., Esmaeilzadeh, F., Mowla, D. et al. (2021). New insights to direct conversion of wet microalgae impregnated with ethanol to biodiesel exploiting extraction with supercritical carbon dioxide. *Fuel* 285 (1): 119199. https://doi.org/10.1016/j.fuel.2020.119199

Jain, A. and Verma, K.K. (2011). Recent advances in applications of single-drop microextraction: A review. *Anal. Chim. Acta.* 706 (1): 37–65. doi: 10.1016/j.aca.2011.08.022

Jesionowski, T., Zdarta, J., and Krajewska, B. (2014). Enzyme immobilization by adsorption: A review. *Adsorption* 20: 801–821.

Jochems, P., Satyawali, Y., Diels, L. et al. (2011). Enzyme immobilization on/in polymeric membranes: Status, challenges and perspectives in biocatalytic membrane reactors (BMRs). *Green Chemistry* 13: 1609–1623.

Khajehesamedini, A., Sadatshojaie, A., Parvasi, P. et al. (2018). Experimental and theoretical study of crude oil pretreatment using low-frequency ultrasonic waves. *Ultrasonics Sonochemistry* 48: 383–395.

Kim, J. and Van Der Bruggen, B. (2010). The use of nanoparticles in polymeric and ceramic membrane structures: Review of manufacturing procedures and performance improvement for water treatment. *Environmental Pollution* 158: 2335–2349.

Kim, S., Cho, H., and Hwang, W. (2021). Robust superhydrophilic depth filter and oil/water separation device with pressure control system for continuous oily water treatment on a large scale. *Separation and Purification Technology* 256: 117779.

Kumar, V., Sangwan, P., Singh, D. et al. (2014). Global scenario of industrial enzyme market. In: *Industrial Enzymes: Trends, Scope and Relevance*, 176–196. New York: Nova Science Publishers.

Kumar, S., Kumar, D., Ahirwar, R. et al. (2016). Exploring the flexible chemistry of 4-fluoro-3-nitrophenyl azide for biomolecule immobilization and bioconjugation. *Analytical and Bioanalytical Chemistry* 408: 6945–6956.

Liu, S. and Jiang, S. (2016). Chemical conjugation of zwitterionic polymers protects immunogenic enzyme and preserves bioactivity without polymer-specific antibody response. *Nano Today* 11: 285–291.

Liu, T., Chen, D., Cao, Y. et al. (2021). Construction of a composite microporous polyethylene membrane with enhanced fouling resistance for water treatment. *Journal of Membrane Science* 618: 118679.

Lobedanz, S., Damhus, T., Borchert, T.V. et al. (2016). Enzymes in industrial biotechnology. *Kirk-Othmer Encyclopedia of Chemical Technology* 1–73. doi:10.1002/0471238961.0914042114090512.a01.pub3

Losappio, V., Franzin, R., Infante, B. et al. (2020). Molecular mechanisms of premature aging in hemodialysis: The complex interplay between innate and adaptive immune dysfunction. *International Journal of Molecular Sciences* 21: 3422. https://doi.org/10.3390/ijms21103422

Luo, J., Meyer, A.S., Jonsson, G. et al. (2013). Fouling-induced enzyme immobilization for membrane reactors. *Bioresource Technology* 147: 260–268.

Luo, J., Marpani, F., Brites, R. et al. (2014). Directing filtration to optimize enzyme immobilization in reactive membranes. *Journal of Membrane Science*, 459: 1–11. https://doi.org/10.1016/j.memsci.2014.01.065

Ma, Y., Sun, Y., Yin, J. et al. (2019). An MOF membrane with ultrathin ZIF-8 layer bonded on ZIF-8 *in-situ* embedded PSf substrate. *Journal of the Taiwan Institute of Chemical Engineers* 104: 273–283. doi:10.1016/j.jtice.2019.08.012

Managheb, M., Zarghami, S., Mohammadi, T. et al. (2021). Enhanced dynamic Cu(II) ion removal using hot-pressed chitosan/poly (vinyl alcohol) electrospun nanofibrous affinity membrane (ENAM). *Process Safety and Environmental Protection* 146: 329–337.

Marwaha, S. and Kennedy, J. (1988). Whey: Pollution problem and potential utilization. *International Journal of Food Science &technology* 23: 323–336.

Meller, K., Szumski, M., and Buszewski, B. (2017). Microfluidic reactors with immobilized enzymes: Characterization, dividing, perspectives. *Sensors and Actuators. B, Chemical* 244: 84–106.

Merz, M., Ewert, J., Stressler, T. et al. (2014) Enzyme membrane reactor technology as a promising tool to increase enzyme efficiency. Poster University of Hohenheim. https://merzmichael.de/assets/Merz_Poster_2.pdf (accessed 9 January 2021)

Mohammad, A.W., Teow, Y.H., Chong, W.C. et al. (2019). Hybrid processes: Membrane bioreactor. In: *Membrane Separation Principles and Applications*, 401–470. Elsevier.

Moraes, J.P., Pappa, G.L., Pires, D.E. et al. (2017). GASS-WEB: A web server for identifying enzyme active sites based on genetic algorithms. *Nucleic Acids Research* 45: W315–W319.

Mortensen, J. (1993). Extrapulmonary blood gas exchange: Why? Who? When? How? *Perfusion* 8: 443–447.

Natesan, S. (2019). Extracellular fungal peroxidases and laccases for waste treatment: Recent improvement. Recent advancement in white biotechnology through fungi. *Perspective for Sustainable Environments* 3: 153.

Nezhad, M.S., Wood, D.A., Sadatshojaei, E. et al. (2021). New insight to experimental study of ionic solutions with a nonionic surfactant on wettability, interfacial tension and micro-model flooding. *Fuel* 285 (1): 119126.

Ngoc, T.P., Fatahi, B., and Khabbaz, H. (2018) Impact of liquid whey waste on strength and stiffness of cement treated clay. *Proceedings of the Civil Infrastructures Confronting Severe Weathers and Climate Changes Conference*: Springer, 1–10.

Nguyen, T., Roddick, F., and Fan, L. (2012). Biofouling of water treatment membranes: A review of the underlying causes, monitoring techniques and control measures. *Membranes* 2: 804–840.

Ojajuni, O., Saroj, D., and Cavalli, G. (2015). Removal of organic micropollutants using membrane-assisted processes: A review of recent progress. *Environmental Technology Reviews* 4: 17–37.

Omidvar, M., Nguyen, H., Huang, L. et al. (2019). Unexpectedly strong size-sieving ability in carbonized polybenzimidazole for membrane H_2/CO_2 separation. *ACS Applied Materials & Interfaces* 11 (50): 47365–47372. https://doi.org/10.1021/acsami.9b16966

Ottone, C., Romero, O., Aburto, C. et al. (2020). Biocatalysis in the winemaking industry: Challenges and opportunities for immobilized enzymes. *Comprehensive Reviews in Food Science and Food Safety* 19: 595–621.

Palai, T. and Bhattacharya, P.K. (2013). Kinetics of lactose conversion to galacto-oligosaccharides by β-galactosidase immobilized on PVDF membrane. *Journal of Bioscience and Bioengineering* 115: 668–673.

Pithawala, K., Mishra, N., and Bahadur, A. (2010). Immobilization of urease in alginate, paraffin and lac. *Journal of the Serbian Chemical Society* 75: 175–183.

Prather, K.L. (2020) Accelerating and expanding nature to address its greatest challenges. *Nature Catalysis*, 3: 181–183.

Pujari, N., Vaidya, B., Bagalkote, S. et al. (2006). Poly (urethane methacrylate-co-glycidyl methacrylate)-supported-polypropylene biphasic membrane for lipase immobilization. *Journal of Membrane Science* 285: 395–403.

Razib, M.S.M., Rahman, R.N.Z.R.A., Shariff, F.M. et al. (2020). Biochemical and structural characterization of cross-linked enzyme aggregates (CLEAs) of organic solvent tolerant protease. *Catalysts* 10: 55.

Razmjou, A., Mansouri, J., and Chen, V. (2011). The effects of mechanical and chemical modification of TiO_2 nanoparticles on the surface chemistry, structure and fouling performance of PES ultrafiltration membranes. *Journal of Membrane Science* 378: 73–84.

Sadatshojaie, A. and Rahimpour, M.R. (2020). CO_2 emission and air pollution (volatile organic compounds, etc.)-related problems causing climate change. In: *Current Trends and Future Developments on (Bio-) Membranes*, 1–30. Elsevier.

Sadatshojaei, E. and Wood, D.A. (2021). Water, the most accessible eco-friendly green solvent, extraction and separation agent. In: *Green Sustainable Process for Chemical and Environmental Engineering and Science: Environmental Remediation* (ed. B.R. Inamuddin and A.M. Asiri), 283–292. Elsevier. https://doi.org/10.1016/B978-0-12-821884-6.00012-7.

Sadatshojaei, E., Sadatshojaie, A., and Fakhar, M.H. (2016a). Differential quadrature method for nonlocal nonlinear vibration analysis of a boron nitride nanotube using sinusoidal shear deformation theory. *Mechanics of Advanced Materials and Structures* 23: 1278–1283.

Sadatshojaei, E., Jamialahmadi, M., Esmaeilzadeh, F. et al. (2016b). Effects of low-salinity water coupled with silica nanoparticles on wettability alteration of dolomite at reservoir temperature. *Petroleum Science and Technology* 34 (15): 1345–1351.

Sadatshojaei, E., Esmaeilzadeh, F., Fathikaljahi, J. et al. (2018). Regeneration of the midrex reformer catalysts using supercritical carbon dioxide. *Chemical Engineering Journal* 343: 748–758.

Sadatshojaei, E., Wood, D.A., and Mowla, D. (2020a). Third generation of biofuels exploiting microalga. In: *Sustainable Green Chemical Process and Their Allied Applications* (ed. B.R. Inamuddin), Chapter 21, 575–588. Switzerland AG: Springer Nature. https://doi.org/10.1007/978-3-030-42284-4_21

Sadatshojaei, E., Mowla, D., and Wood, D.A. (2020b). Review of progress in microalgal biotechnology applied to wastewater treatment. In: *Sustainable Green Chemical Process and Their Allied Applications* (ed. B.R. Inamuddin), Chapter 19, 539–557. Switzerland AG: Springer Nature. https://doi.org/10.1007/978-3-030-42284-4_19.

Sadatshojaei, E., Heidari, S., and Wood, D.A. (2021a). Sustainable approach in biocatalytic preparation of antibiotic peptide. In: *Green Sustainable Process for Chemical and Environmental Engineering and Science* (ed. B.R. Inamuddin and A.A.M. Ahamed), Chapter 17, 345–367. Elsevier. https://doi.org/10.1016/B978-0-12-819721-9.00017-0.

Sadatshojae, E., Wood, D.A., and Rahimpour, M.R. (2021b). Potential and challenges of carbon sequestration in soils. In: *Applied Soil Chemistry*, 1st ed. (ed. B.R. Inamuddin, R. Boddula, and T.A. Altalhi), Chapter 1, 1–22. Wiley-Scrivener. https://doi.org/10.1002/9781119711520.ch1.

Sadatshojaie, A., Wood, D.A., Jokar, S.M. et al. (2021c). Applying ultrasonic fields to separate water contained in medium-gravity emulsions and determining emulsionadhesion coefficients, 2020. *Ultrasonics Sonochemistry* 70: 105303.

Sahni, T.K. and Goel, A. (2015). Microbial enzymes with special reference to α-amylase. *Evolutionary Biology* 2: 19–25.

Samsami, S., Mohamadi, M., Sarrafzadeh, M.-H. et al. (2020). Recent advances in the treatment of dye-containing wastewater from textile industries: Overview and perspectives. *Process Safety and Environmental Protection* 143: 138–163.

Alara, A., Blum, L.J., and Leca-Bouvier, B.D. (2012). Immobilization strategies to develop enzymatic biosensors. *Biotechnology Advances* 30: 489–511.

Scott, A.J. (2019) Design of Polymeric Materials: Novel Functionalized Polymers for Enhanced Oil Recovery & Gas Sorption Applications. PhD Thesis, University of Waterloo.

Semião, A.J. and Schäfer, A.I. (2013). Removal of adsorbing estrogenic micropollutants by nanofiltration membranes. Part A; Experimental evidence. *Journal of Membrane Science* 431: 244–256.

Shokri, E., Yegani, R., Heidari, S. et al. (2015). Effect of PE-g-MA compatibilizer on the structure and performance of HDPE/EVA blend membranes fabricated via TIPS method. *Chemical Engineering Research & Design* 100: 237–247.

Seddiqi, H., Sadatshojaie, A., Vaferi, B. et al. (2019). Mathematical model for iron corrosion that eliminates chemical potential parameters. *Chinese Journal of Chemical Engineering* 28: 603–612.

Sheikh, E. and Sheikh, A. (2020). Effects of gelatin and Nano-ZnO on the performance and properties of cellulose acetate water membranes. *Chemical Engineering & Technology* 43 (9): 1749–1757. https://doi.org/10.1002/ceat.202000022

Sigurdardóttir, S.B., Lehmann, J., Ovtar, S. et al. (2018). Enzyme immobilization on inorganic surfaces for membrane reactor applications: Mass transfer challenges, enzyme leakage and reuse of materials. *Advanced Synthesis & Catalysis* 360: 2578–2607. https://doi.org/10.1002/adsc.201800307

Silva, L.L., Moreira, C.G., Curzio, B.A. et al. (2017). Micropollutant removal from water by membrane and advanced oxidation processes: A review. *Journal of Water Resource and Protection* 9: 411.

Alara, J., Saharan, V., Kumar, S. et al. (2018). Laccase grafted membranes for advanced water filtration systems: A green approach to water purification technology. *Critical Reviews in Biotechnology* 38: 883–901.

Sulaiman, S., Cieh, N.L., Mokhtar, M.N. et al. (2017). Covalent immobilization of cyclodextrin glucanotransferase on kenaf cellulose nanofiber and its application in ultrafiltration membrane system. *Process Biochemistry* 55: 85–95.

Sun, W., Liu, J., Chu, H. et al. (2013). Pretreatment and membrane hydrophilic modification to reduce membrane fouling. *Membranes* 3: 226–241.

Van Der Bruggen, B., Vandecasteele, C., Van Gestel, T. et al. (2003). A review of pressure-driven membrane processes in wastewater treatment and drinking water production. *Environmental Progress* 22: 46–56.

Vasileva, N., Ivanov, Y., Damyanova, S. et al. (2016). Hydrolysis of whey lactose by immobilized β-galactosidase in a bioreactor with a spirally wound membrane. *International Journal of Biological Macromolecules* 82: 339–346.

Vishnoi, N., Dixit, S., and Mishra, J. (2020). Microbial lipases and their versatile applications. In: *Microbial Enzymes: Roles and Applications in Industries*, 207–230. Springer.

Vladisavljević, G.T. (2016). Biocatalytic membrane reactors (BMR). In: *Saha, Catalytic Reactors* (ed. De Gruyter). ISBN 978-3-11-033296-4. https://dspace.lboro.ac.uk/2134/19508 (accessed 3 November 2021).

Wang, Z., Wu, Z., Mai, S. et al. (2008). Research and applications of membrane bioreactors in China: Progress and prospect. *Separation and Purification Technology* 62: 249–263. https://doi.org/10.1016/j.seppur.2007.12.014

Wei, Q., Oribayo, O., Feng, X. et al. (2018). Synthesis of polyurethane foams loaded with TiO_2 nanoparticles and their modification for enhanced performance in oil spill cleanup. *Industrial & Engineering Chemistry Research* 57: 8918–8926.

Ying, L., Kang, E., and Neoh, K. (2002). Covalent immobilization of glucose oxidase on microporous membranes prepared from poly (vinylidene fluoride) with grafted poly (acrylic acid) side chains. *Journal of Membrane Science* 208: 361–374.

Yuan, J., Dong, W., Sun, F. et al. (2018). Low temperature effects on nitrification and nitrifier community structure in V-ASP for decentralized wastewater treatment and its improvement by bio-augmentation. *Environmental Science and Pollution Research* 25: 6584–6595.

Zdarta, J., Meyer, A.S., Jesionowski, T. et al. (2018a). General overview of support materials for enzyme immobilization: Characteristics, properties, practical utility. *Catalysts* 8: 92.

Zdarta, J., Meyer, A.S., Jesionowski, T. et al. (2018b). Developments in support materials for immobilization of oxidoreductases: A comprehensive review. *Advances in Colloid and Interface Science* 258: 1–20.

Zdarta, J., Meyer, A.S., Jesionowski, T. et al. (2019). Multi-faceted strategy based on enzyme immobilization with reactant adsorption and membrane technology for biocatalytic removal of pollutants: A critical review. *Biotechnology Advances* 37 (7): 107401.

Zeuner, B., Ovtar, S., Persson, Å.H. et al. (2020). Surface treatments and functionalization of metal-ceramic membranes for improved enzyme immobilization performance. *Journal of Chemical Technology and Biotechnology* 95: 993–1007. https://doi.org/10.1002/jctb.6278

Zhang, S. and Jiang, J.Q. (2021). Comparative removal of imidacloprid, bisphenol-S, and azithromycin with ferrate and $FeCl_3$ and assessment of the resulting toxicity. *Journal of Chemical Technology and Biotechnology* 96: 99–112.

Zhang, Y., Feng, X., Yuan, S. et al. (2016). Challenges and recent advances in MOF–polymer composite membranes for gas separation. *Inorganic Chemistry Frontiers* 3: 896–909.

Zhu, X.-Y., Chen, C., Chen, P.-C. et al. (2016). High-performance enzymatic membrane bioreactor based on a radial gradient of pores in a PSF membrane via facile enzyme immobilization. *RSC Advances* 6: 30804–30812. https://doi.org/10.1039/C5RA25602J

Zhu, F., Xu, W., Li, X. et al. (2020). Lipase immobilization on UiO-66/poly (vinylidene fluoride) hybrid membranes and active catalysis in the vegetable oil hydrolysis. *New Journal of Chemistry* 44: 14379–14388. https://doi.org/10.1039/D0NJ03282D

14

Use and Manufacture of Biopesticides and Biofertilizers in Latin America

Luis Jesús Castillo-Pérez[1]*, Juan José Maldonado Miranda*[2] *and Candy Carranza-Álvarez*[2]

[1] *Multidisciplinary Graduate Program in Environmental Sciences, Autonomous University of San Luis Potosí, San Luis Potosí, SLP, Mexico*
[2] *Faculty of Professional Studies Huasteca Zone, Autonomous University of San Luis Potosí, S.L.P., Mexico*

CHAPTER MENU

14.1 Introduction

In the last 40 years, the increase in the world population has generated the need to increase food production around the world (Bartley et al. 2019; Desiere et al. 2018; FAO 2019). Due to this, the use of commercial pesticides and artificial fertilizers has played an essential role in increasing agricultural production (Scott et al. 2018). During agricultural processes, pesticides are used to control diseases and pests caused mainly by other plants, rodents, insects, fungi and other phytopathogenic organisms, while fertilizers contribute to improve the nutritional characteristics of soils to achieve higher crop yields. However, intensification of these products has generated strong environmental world-wide pollution (Da Costa-Chaulet et al. 2019).

The use of synthetic pesticides and fertilizers in agriculture has had a negative effect on several environmental matrices, particularly in soils and water ecosystems, in addition to accelerating the loss of biodiversity in different parts of the world (Meena et al. 2020; Rundlöf et al. 2016). The accumulation of these substances in ecosystems, together with their high degree of toxicity and difficult degradation, affect organisms and consequently originate genetic resistance to these synthetic products (Fodjo et al. 2018; Wachowska et al. 2018).

Bioresource Technology: Concept, Tools and Experiences, First Edition. Tanveer Bilal Pirzadah, Bisma Malik, Rouf Ahmad Bhat, and Khalid Rehman Hakeem.
© 2022 John Wiley & Sons Ltd. Published 2022 by John Wiley & Sons Ltd.

Therefore, products have been developed that fulfill the function of pesticides and fertilizers, but that have been created from molecules mainly obtained from plants and bacteria. These products are called biopesticides and biofertilizers and are a source of sustainable biological controllers and enhancers used in agriculture. Furthermore, they have the advantage of being biodegradable and thus reduce environmental pollution (Copping and Menn 2000; Kumar 2012).

Currently, the Latin American countries together with the United States register the highest consumption of pesticides and fertilizers world-wide due to the several transgenic crops developed in these regions (Hapeman et al. 2003). Additionally, the Pan American Health Organization reports that in 12 Latin American and Caribbean countries, lead poisoning and agrochemicals such as pesticides and fertilizers have caused 15% of the illnesses registered in these countries (García 1998; Quinteros et al. 2017). For this reason, Latin American countries have dedicated greater emphasis to manufacture and use of biopesticides and biofertilizers that contribute to sustainable agriculture.

This chapter contains evidence about the problems of pesticides and fertilizers in Latin America, as well as current technological developments related to the manufacture and use of biopesticides and biofertilizers and finally we present a case study carried out in the state of San Luis Potosí, Mexico. The information will be useful to provide some strategies that promote the manufacture and use of these products toward sustainable agriculture in Latin America.

14.2 Current Problems of Pesticides and Fertilizers in Latin America

In the current times, the exposure to agrochemicals is a critical threat to humanity and all ecosystems on the planet. The toxicity caused by these substances produces effects that can be acute or chronic and can be fatal for the organisms that suffer from such effects. Acute effects appear immediately in the body or shortly after exposure, while chronic effects can manifest themselves many years later and, in this case, the source of exposure is difficult to trace (Prior et al. 2019; Sánchez-Bayo and Tennekes 2017). The World Health Organization (WHO) estimates that every year 3 million people suffer from acute intoxication caused by some type of agrochemicals, the original people from developing countries being the most affected population in the world (Jørs et al. 2006; Kumar et al. 2013).

The use of pesticides and fertilizers has not decreased due to these products being necessary to obtain high agricultural productivity. Pesticides are used to control, repel or eliminate organisms that cause havoc during the production, storage, transportation, distribution and processing of agricultural products. These agricultural products include herbicides, insecticides and fungicides (Copping and Menn 2000). Concerning their use, Asia and America appear as the main continents in the use of these substances with 52.8% and 30.2%, respectively (Figure 14.1), putting the ecosystems and the health of the people at risk who are connected with its production and use.

With respect to fertilizers, the most used are those that provide nutrients like nitrogen and phosphorus, of which their production has increased in recent years (Vance 2001). The use of artificial fertilizers is necessary to increase agricultural production and meet

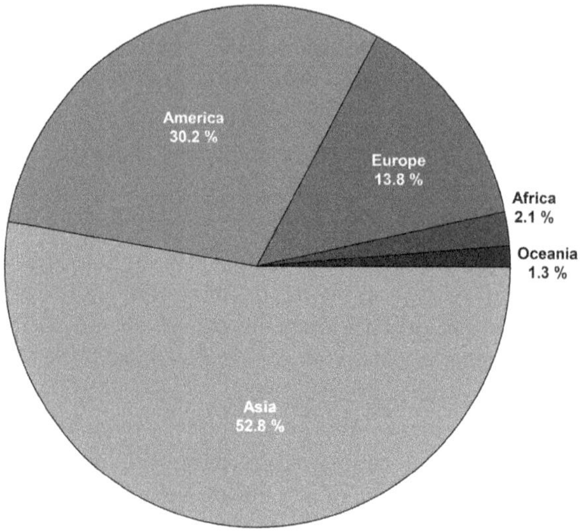

Figure 14.1 Percentage of pesticide use by different continents.

the food demands of the current population (Bartley et al. 2019). Nevertheless, synthetic fertilizers used on crops around the world have already been reported to cause the following problems: i) spoilage of soils and surface and groundwater; ii) increased air pollution; iii) reduced biodiversity and affecting the function of ecosystems; and iv) negatively impacting the capacity of ecosystems to generate environmental services (Atikur-Rahman and Zhang 2018; Chen et al. 2018; Tong et al. 2019; Yuan et al. 2018).

Until 1988, in Latin America, 1000 to 2000 intoxications occurred annually from agrochemicals (García 1998). Currently, in some Latin American countries such as Argentina, Brazil, Chile, El Salvador, Mexico, Nicaragua and Peru, about 4 million people related directly and indirectly to field work have been affected by this problem. Hendges et al. (2019) mentioned that between 1999 and 2014, men between 15 and 59 years of age were the most affected by intoxication caused by agrochemicals in the southern region of Brazil. In Mexico, the Single Information System for Epidemiological Surveillance (SUIVE, for its acronym in Spanish) reported an increase in intoxications of 3849 cases for the year 2003, which increased to 3902 for the year 2005 (Hernández-González et al. 2007). It is estimated that with the use of biopesticides and biofertilizers, the number of intoxicated people in Latin America will decrease significantly. Furthermore, these products will contribute to a decrease in contamination, promoting sustainable agriculture (Mishra and Dash 2014).

14.3 Manufacture and Use of Biopesticides and Biofertilizers in Latin America

The excessive use of commercial pesticides and fertilizers in the agricultural sector has generated high costs for farmers, and they cause adverse effects on the planet's ecosystems, as mentioned earlier. With the aim of minimizing environmental impacts and migration toward sustainable agriculture, several Latin American countries are working on the manufacture of biopesticides that act as weed or pest control agents, and biofertilizers that naturally stimulate growth and development of plants (Dhir 2017; Moshi and Matoju 2017).

Biopesticides are products created with complete microorganisms or biomolecules extracted from living organisms such as bacteria, fungi, nematodes and secondary plant metabolites (Figure 14.2). These molecules operate as an active ingredient and it is possible for them to contain the entire extracted substance or only a fraction of it. Some biopesticides have also been manufactured using viruses as an active ingredient.

On the other hand, biofertilizers are created from biomolecules produced during the secondary metabolism of plants or microorganisms that promote plant growth and are usually applied to soils or composting areas with the aim of increasing the degree of availability of nutrients that can be better assimilated by the plant (Mohammadi and Sohrabi 2012).

Currently, several countries in the Latin American region are migrating toward sustainable agriculture and have developed technology focused on the manufacture of biopesticides and biofertilizers. Table 14.1 presents 40 researches developed during the last 20 years in Latin American countries, of which 32.5% correspond to the development of biopesticides and 67.5% of biofertilizers.

According to the research reported in Table 14.1, Mexico is the Latin American country where the most biopesticides and biofertilizers are manufactured, followed by Colombia, Argentina, Peru, Chile, Brazil, Ecuador, Nicaragua and Bolivia. The crops that have benefited from the use of these biopesticides are chilli, beans, rice, sorghum, wheat and others. The bacteria *Azospirillum brasilense* and the fungus *Glomus*

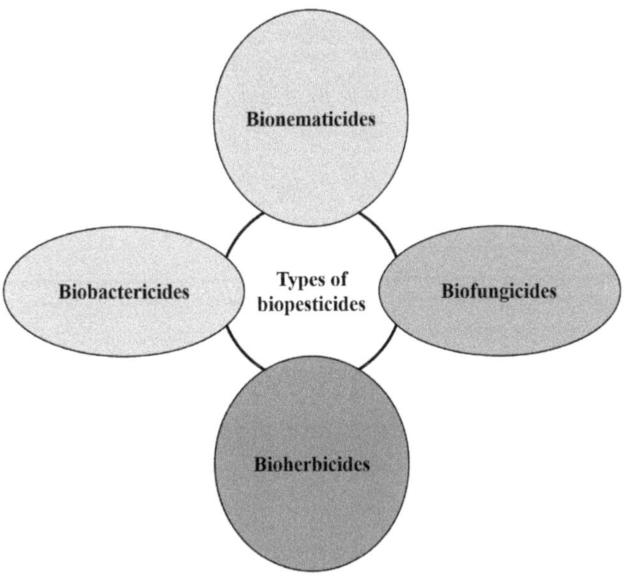

Figure 14.2 Diversity of biopesticide products.

Table 14.1 Some research on biopesticides and biofertilizers manufactured in Latin America during the last two decades (*source*: based on Haase et al. 2015).

Biomaterial used	Biopesticide/ Biofertilizer	Country of manufactured	Crop benefited	Reference
Organic waste (beef and chicken manure, rice husk, parchment and coffee pulp)	BF	Nicaragua	Radish	Laguna-Miranda and Cisne-Contreras 2001
Liquid pig excretia	BF	Mexico	NM	Soria-Fregoso et al. 2001
Neem extracts (*Azadirachta indica*)	BP	Mexico	Cotton	Rodríguez 2002
Lippia alba extracts	BP	Colombia	Bean	Guzmán-Rivera et al. 2004
Glomus intraradices and *Glomus fasciculatum*	BF	Mexico	Sorghum	Garza-Cano et al. 2005
Azospirillum brasilense and *Glomus intraradices*	BF	Mexico	Safflower	Díaz-Franco et al. 2006
Azospirillum brasilense and *Glomus intraradices*	BF	Mexico	Corn	Uribe-Valle y Dzib-Echeverría 2006
Azospirillum brasilense and *Glomus intraradices*	BF	Mexico	Cocoa	Aguirre-Medina et al. 2007
Glomus intraradices	BF	Mexico	Sorghum	Díaz-Moreno et al. 2007
Polygonum hydropiperoides, *Calliandra pittieri* and *Solanum nigrum*	BP	Colombia	Corn	Lizarazo et al. 2008
Trichothecium roseum	BP	Brazil	Rice, sorghum and wheat	Negri et al. 2010
Vermicompost	BP	Mexico	NM	Domínguez et al. 2010
Glomus intraradices, *Azotobacter* sp. and *Azospirillum* sp.	BF	Mexico	Coffee	Adriano-Anaya et al. 2011
Oil and neem seeds (*Azadirachta indica*)	BP	Mexico	Vegetables and cucurbits	González-Maldonado and García-Gutiérrez 2012
Arbuscular mycorrhizal fungi	BF	Mexico	Corn	Pérez-Luna et al. 2012
Bacillus thuringiensis	BP	Brazil	Against *Aedes aegypti*	Ernandes et al. 2013

(*Continued*)

Table 14.1 (Continued)

Biomaterial used	Biopesticide/ Biofertilizer	Country of manufactured	Crop benefited	Reference
Basil, rue, tobacco, chilli pepper and oregano	BP	Ecuador	Cucumber, zucchini, tomato, pepper, aubergine and lettuce	Franco-Rodríguez et al. 2014
Azospirillum spp.	BF	Mexico	Corn, sorghum and wheat	Rangel-Lucio et al. 2014
Cachanilla extract (*Pluchea sericea*)	BP	Mexico	NM	Osuna-Carreón et al. 2014
Ectomycorrhizal fungi *Suillus luteus* and *Rhizopogon luteolus*, and saprobic fungi *Coriolopsis rigida* and *Trichoderma harzianum*	BF	Chile	*Pinus radiata*	Chávez et al. 2014
Rhizobacteria from habanero pepper, corn, and pumpkin	BF	Mexico	Tomato	Noh-Medina et al. 2014
Serratia sp., *Pseudomonas koreensis* and *Plectosphaerella cucumerina*	BF	Colombia	Vanilla	Álvarez-López et al. 2014
Vinaza	BF	Chile	Vineyards	Callejas et al. 2014
Baculoviruses	BP	Argentina	NM	Haase et al. 2015
Azospirillum sp.	BF	Chile	Potato	Castillo et al. 2016
Cyanobacteria	BF	Ecuador	Grass	Freire et al. 2016
Vermicompost with pine sawdust	BF	Mexico	Apple	Orozco-Corral et al. 2016
Cry proteins of *Bacillus thuringiensis*	BP	Colombia	Potato, rice and cotton	Cerón et al. 2017
Pseudomonas fluorescens, *Pseudomonas* spp., *Bacillus subtilis* and *Bacillus* sp.	BF	Argentina	Chia	Krieger et al. 2017
Potato endophytic bacteria	BF	Bolivia	Potato	Ortuño et al. 2018
Bacillus subtilis subsp. *spizizenii*	BF	Argentina	Tomato	Sarti et al. 2019
Capsaicinoids from ají panca	BP	Peru	Potato	Gamarra-Mendoza et al. 2019
Cyanobacteria	BF	Mexico	Blue corn	Hernández-Reyes et al. 2019

(*Continued*)

Table 14.1 (Continued)

Biomaterial used	**Biopesticide/ Biofertilizer**	**Country of manufactured**	**Crop benefited**	**Reference**
Clibadium Peruvianum and *Petiveria alliacea*	BP	Peru	NM	Bracho-Pérez et al. 2019
Serratia sp. and *Pseudomonas* sp.	BF	Mexico	Chilli	Quiroz-Sarmiento et al. 2019
Bacillus subtilis and *Trichoderma harzanium*	BF	Mexico	Chilli	Gamboa-Angulo et al. 2020
Glomus mosseae	BF	Mexico	Wheat	Báez-Pérez et al. 2020
Pseudomonas tolaasii	BF	Mexico	Chilli	Cabanzo-Atilano et al. 2020
Rhizobium sp. and *Azotobacter* sp.	BF	Perú	Bean	Curi et al. 2020
Rhizophagus aggregatus and a consortium of *Glomus claroides*, *Rhizophagus diaphanus* and *Paraglomus albidum*	BF	Mexico	Coffee	Hernández-Acosta et al. 2020

intraradices are the two microorganisms that stand out the most in the manufacture of biopesticides and biofertilizers. Interestingly, research carried out in Argentina was the only one that used a type of virus called baculovirus, and with this microorganism created a biopesticide that neutralized the phytopathogenic effects of several insects (Table 14.1) (Haase et al. 2015).

As can be seen in Table 14.1, microorganisms are an ecological and sustainable alternative that contribute to improve the productivity and protection of agricultural crops. These organisms adapt to the rhizosphere of plants and promote plant growth by improving the bioavailability of nutritional elements such as nitrogen and phosphorus and encouraging the production of growth regulators such as auxins and gibberellins. Currently, biopesticides and biofertilizers that replace their artificial analogs and molecular processes to increase their efficacy are being investigated (Basu et al. 2017; Jiméncz-Gómez et al. 2020; Kadam et al. 2018).

14.4 Manufacture of A Natural Repellent: A Case Study

Tagetes erecta, commonly called tagete or Indian carnation, and known in Mexico as cempasúchil, cempoal or death flower, is a species of the Asteraceae family. The common name "cempasúchil" comes from the Nahuatl word "cempōhualxōchitl" which means "twenty flower." It is an annual plant that can reach a height of up to 110 cm. Its leaves are opposite, pinnate and subdivided into lanceolate, dentate or ciliated segments. Its flower is compound and very aromatic, whose color varies from orange to yellow (Figure 14.3). It has a long flowering period that extends throughout the summer and fall. The reproduction of *T. erecta* is carried out mainly by the seeds. This plant is native to Mexico and can be found in warm, semi-warm, dry and temperate weather regions, from 8 to 3900 meters altitude. It adapts easily to different habitats, and can be cultivated in orchards, cornfields or urban areas. It is associated with different types of vegetation such as tropical deciduous and sub-deciduous forests, xerophilous scrub and thorny forests and cloud mountain forest.

Cempasuchil leaves contain an essential oil in which monoterpenes geraniol, limonene, linalool, ocimeno, β-phelandrene, α-pinene, β-pinene and tagetone have been identified. Flavonoids comferitrin and camferol are also present. The flowers and petals are rich in carotenoids and lutein, xanthophyll and five esters of both components with fatty acids of 10, 16 and 18 carbons have been identified. Bitienyl and α-tertienyl sulfur components have been detected in flowers and roots.

In agriculture it is used to repel nematodes and it is effective especially against the species *Pratylenchus penetrans*, which attacks some crops such as tomatoes. In addition, it is used to make new natural and artificial fertilizers due to its high stability. In some rural communities, the flower of this plant is used to repel mosquitoes due to its aromatic properties. Therefore, in order to check its effects, its essential oils were extracted through steam distillation, and a bioinsecticide was developed.

Figure 14.3 Complete flower of T. erecta.

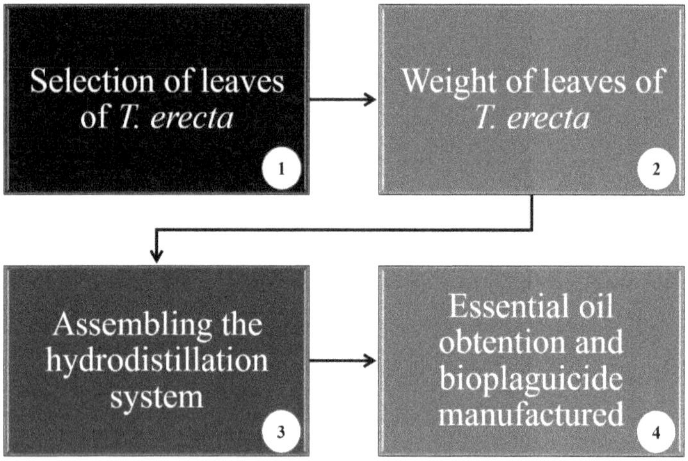

Figure 14.4 Stages of production of bioinsecticide from T. erecta.

For manufacture of this product, the following methodology was carried out: i) collection of the plant; ii) leaves and flowers separated from the plants; iii) weight determination of the sample (for this case 100 g leaves); and iv) assembly of the distillation system and start of the process using temperatures from 115 to 200°C. After 3 hours of hydrodistillation, 200 μl of essential oil was obtained. Finally, for the manufacturing of this bioinsecticide, a diluent oil was used in a proportion of 5 μl for every 20 μl of essential oil and it was completed to a volume of 50 ml with 70% ethanol. In Figure 14.4, the stages of production are presented.

After manufacturing the bioinsectide, its effectiveness on the *Aedes aegypti* (L.) mosquito was evaluated. Laboratory tests were performed using the standard protocol established by WHO/CTD/WHOPES/IC (Barnard 2005; WHO 2009). For this, the availability of the mosquitoes in sugar solution was evaluated 12 hours before the tests, then they were placed in different boxes and the product was added. Results shows 100% direct inhibition. However, it is necessary to carry out new tests to determine if the bioinsecticide in people is functional. Finally, bioinsecticide odor durability tests were carried out on 100 persons, finding that in the 68% of the participants the product presents a durability of 6 hours (Table 14.2).

Table 14.2 Durability of the bioinsecticide manufactured from *T. erecta* leaves.

Time elapsed	Frequency	Percentage (%)	Valid percentage (%)	Accumulated percentage (%)
1 hour	7	7	7	7
3 hours	25	25	25	32
6 hours	68	68	68	100
Total	100	100	100	

14.5 Biotechnological Interventions in Biopesticide Synthesis

The emergency to increased global agricultural productivity and the requirements of the current and future population on the planet, has caused the excessive use of fertilizers and pesticides. This situation has put the health of ecosystems and the human population at risk. In this context, current advances in science related to the manufacture of biopesticides have also increased throughout the world. The synthesis of these sophisticated bioproducts needs to be environmentally-friendly, and requires biotechnological interventions where a wide range of microbial and plant biomolecules are used to attack several threats. These new technologies use more sophisticated processes that involve the genetic incorporation of DNA in agricultural products (Keswani et al. 2016; Pavela and Benelli 2016; Chen et al. 2020).

In this context, Table 14.3 shows 10 scientific researches of the development of biotechnological processes established throughout the last 5 years for the synthesis of biopesticides. As can be seen, the use of microorganisms such as *Bacillus thuringiensis*, *Paenibacillus polymyxa* and *Pseudomonas chlororaphis* is one of the most used strategies for the synthesis of biopesticides.

On the other hand, the manufacture of nanoparticles has been another of the strategies that have started to be developed. However, applications of nanotechnology in agriculture are still in the early stages. The manufacture and use of nanoparticles have been developed with more emphasis because the release of the encapsulated biopesticide has higher efficiency, there is better uniformity of coverage for highly active compounds and they present lower risk of exposure for the people who use them (Sahayaraj 2014).

The isolation and purification of plant secondary metabolites is also a recurring method for the synthesis of biopesticides. An example of this methodology is the azaridactin obtained from neem (*Azadirachta indica*), as this biomolecule presents pesticidal properties and has been widely studied (Chaudhary et al. 2017). The effectiveness of these plant-based products lies in the presence of molecules such as terpenes, iridoids, flavonoids, naphthoquinones and phenylpropanoids. These biomolecules have been shown to have interesting biological activities (Céspedes et al. 2014). Finally, it is important to note that most of the synthesized biopesticides have mainly focused on destroying threats like flies, mosquitoes, ticks and beetles.

14.6 Biofertilizers Relevance and Plant Tolerance to Abiotic/Biotic Stress

Advances in current metabolomics and proteomics research have shown that abiotic factors related to osmotic processes such as salinity and drought are the main stress situations that affect agricultural crops and cause crop loss around the world (Coutinho et al. 2018; Guo et al. 2018). Some situations that activate osmotic stress and other types of stress can be observed in Figure 14.5. However, when the plants are in this situation, they are capable to adapt to and combat environmental stress by activating cascades of molecular networks. These molecular networks activate specific genes and metabolites

Table 14.3 Some biotechnological processes used for the synthesis of biopesticides.

Manufactured product	Threat	Biotechnological process development	Reference
Biopesticide	–	Production of *Bacillus thuringiensis* biopolymers called melanins	Sansinenea and Ortiz 2015
	Against *B. cereus*-associated diseases	Production of a new bacteriocin similar to a cold shock protein synthesized by *Bacillus thuringiensis*	Huang et al. 2016
	Some common plant pathogens like *Phytophthora sojae* (oomycete), *Rhizoctonia solani* (basidiomycete fungus), *Cylindrocarpon destructans* (ascomycete fungus), *Pseudomonas syringae* (bacterium) and *Xanthomonas campestris* (bacterium)	Isolation and production of *Paenibacillus polymyxa*, a plant growth-promoting rhizobacterium that can benefit agriculture through antimicrobial activity, nitrogen fixation, phosphate solubilization, plant hormone production or lignocellulose degradation	Weselowski et al. 2016
	Against phytophagous insects	Isolation and purification of azadirachtin for the manufacture of nanocarriers that provide a controlled release of this bioactive compound	Chaudhary et al. 2017
	Against the pulse beetle *Callosobruchus maculatus*	Development of a coated zinc oxide nanoparticle of the bacterium *Bacillus thuringiensis*	Malaikozhundan et al. 2017
	Mosquito, tick, and phytophagous mites	Manufacture of silver nanoparticles with *Saponaria officinalis* root extract	Pavela et al. 2017
	Against an array of microbial pathogens, insects and nematodes	Synthesis of *Pseudomonas chlororaphis* isolates metabolites	Anderson and Kim 2018
	Against several insect infestations	Nano-biopesticides synthesis	Lade and Gogle 2019
	Against serious invasive pest, South American leaf miner *Tuta absoluta* (Lepidoptera)	Synthesis of iron oxide nanoparticles (FeNPs) based on leaves of *Trigonella foenum-graecu*	Ramkumar et al. 2020
	Whiteflies (*Bemisia tabaci*)	Development of transgenic plants through nuclear or chloroplast transformation with toxins against whitefly	Suhag et al. 2020

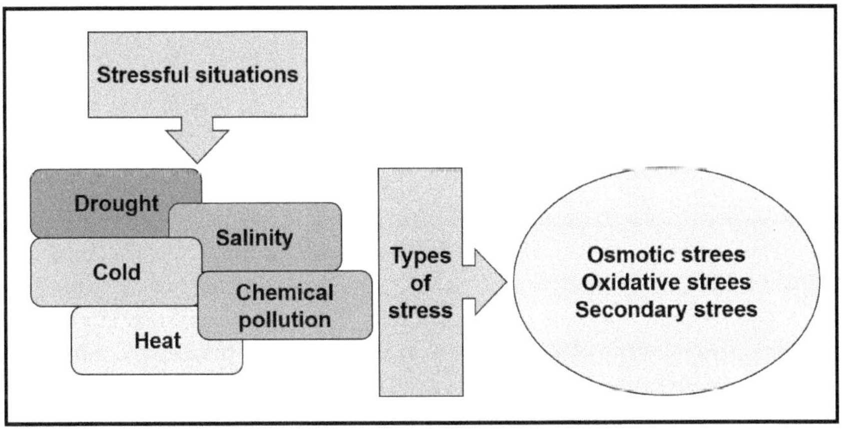

Figure 14.5 Some stress situations that occur in crops and the types of stress they generate.

related to stress and produce the response to reestablish homeostasis and protect the affected proteins and membranes in the plant organism (Evseeva et al. 2019; Osakabe et al., 2013; Xiong and Zhu, 2002).

Some molecules that have been studied that are related to abiotic stress are phytochelatins, cysteine and methionine amino acids, and glutarredoxins/thioredoxins related to the tripeptide glutathione. The molecules related to glutathione play an important role in the response to stress through oxidative and enzymatic signaling. Furthermore, evidence exists that several protein and non-protein thiols, in conjunction with sulfur-containing molecules, contribute to plant stress tolerance (Hasanuzzaman et al., 2017; Nianiou-Obeidat et al., 2017).

Other interesting molecules that have been shown to genetically improve the tolerance of plants to certain types of stress are microRNAs. MicroRNAs regulate gene expression in multiple development and signaling pathways. However, when a situation of abiotic stress is generated, they are expressed in an aberrant way, and produce problems in the affected plant. The plant responds by decreasing the number of negative regulators and at the same time allows the accumulation and operation of positive regulators (Bej and Basak 2014; Shukla et al. 2008; Sunkar et al. 2012).

For the biotic stress situations that decrease the agricultural crops' productivity, the main organisms that cause damage are fungi, bacteria, viruses, parasitic nematodes, insects, weeds and other plant species. In response to these attacks, plants have developed complex physiological, molecular and cellular adaptations (Moustafa-Farag et al. 2020). However, current studies suggest that a combination of abiotic and biotic stress can benefit the plant because it can then activate signaling pathways that encourage the production of phytohormones, transcription factors and kinase cascades. Unfortunately, antagonistic processes have been detected, mainly due to the production of reactive oxygen species (Suzuki et al. 2014; Saijo and Loo 2020).

The correct and accurate use of biofertilizers in agriculture will contribute to improving the well-being of the main crops on the planet, since they contribute to the reduction and tolerance of different types of stress. These new technological developments will also help improve food security globally.

14.7 Conclusions

In the last two decades, several countries in Latin America have developed research to begin the manufacture of biopesticides and biofertilizers. The use of these bioproducts has attempted to develop an integral pest management system which has increased soil productivity and reduced the environmental impact. Some advantages that are offered by the use of biopesticides and biofertilizers are: i) easy and major biodegradability of the product; ii) decrease the amount of toxic waste; iii decrease in production costs; and iv) decrease in the number of contaminated sites. Research in the development of molecular techniques can help improve the effectiveness of biopesticides and biofertilizers and increase their activity against a wide range of pathogens and diseases.

References

Adriano-Anaya, M.D., Jarquín-Gálvez, R., Hernández-Ramos, C. et al. (2011). Biofertilización de café orgánico en etapa de vivero en Chiapas, México. *Revista mexicana de ciencias agrícolas* 2: 417–431.

Aguirre-Medina, J.F., Mendoza-López, A., Cadena-Iñiguez, J. et al. (2007). Efecto de la biofertilización en vivero del cacao (*Theobroma cacao* L.) con *Azospirillum brasilense* Tarrand, Krieg et Döbereiner y *Glomus intraradices* Schenk et Smith. *Interciencia* 32: 541–546.

Álvarez-López, C., Osorio-Vega, W., Díez-Gómez, M.C. et al. (2014). Biochemical characterization of rhizosphere microorganisms from vanilla plants with potential as biofertilizers. *Agronomía Mesoamericana* 25: 226–241.

Anderson, A.J. and Kim, Y.C. (2018). Biopesticides produced by plant-probiotic *Pseudomonas chlororaphis* isolates. *Crop Protection* 105: 62–69.

Atikur-Rahman, K.M. and Zhang, D. (2018). Effects of fertilizer broadcasting on the excessive use of inorganic fertilizers and environmental sustainability. *Sustainability* 10: 759.

Báez-Pérez, A., Limón-Ortega, A., Ramírez-Barrientos, C.E. et al. (2020). Efecto de biofertilizantes y agricultura de conservación en la producción de trigo en un Vertisol. *Terra Latinoamericana* 38: 569–581.

Barnard, D.R. (2005). Biological assay methods for mosquito repellents. *Journal of the American Mosquito Control Association* 21 (sp. 1): 12–16.

Bartley, T.J., McCann, K.S., Bieg, C. et al. (2019). Food web rewiring in a changing world. *Nature Ecology and Evolution* 3: 345–354.

Basu, S., Rabara, R., and Negi, S. (2017). Towards a better greener future: An alternative strategy using biofertilizers. I: Plant growth promoting bacteria. *Plant Gene* 12: 43–49.

Bej, S. and Basak, J. (2014). MicroRNAs: The potential biomarkers in plant stress response. *American Journal of Plant Sciences* 5: 748–759.

Bracho-Pérez, J., Tacza-Valverde, I., and Vásquez-Castro, J. (2019). Peruvian botanical biopesticides for sustainable development and protection of the environment. *Peruvian Journal of Agronomy* 3: 126–133.

Cabanzo-Atilano, I., Rodríguez, M., García-Cué, J.L. et al. (2020). La biofertilización y nutrición en el desarrollo de plántulas de chile serrano. *Revista Mexicana de Ciencias Agrícolas* 11: 699–712.

Callejas, R., Silva, Á., Peppi, C. et al. (2014). Factibilidad agronómica del uso de vinaza, subproducto de la fabricación del pisco, como biofertilizante en viñedos. *Revista Colombiana de Ciencias Hortícolas* 8: 230–241.

Castillo, C., Huenchuleo, M.J., Michaud, A. et al. (2016). Micorrización en un cultivo de papa adicionado del biofertilizante Twin-N establecido en un Andisol de la Región de La Araucanía. *Idesia (Arica)* 34: 39–45.

Cerón, J.A., Velasquez, L.F., and Torres, D.R. (2017). Proteínas CRY con actividad tóxica hacia algunos insectos lepidópteros y coleópteros. *Revista Colombiana de Biotecnología* 25: 56–57.

Céspedes, C.L., Salazar, J.R., Ariza-Castolo, A. et al. (2014). Biopesticides from plants: Calceolaria integrifolia sl. *Environmental Research* 132: 391–406.

Chaudhary, S., Kanwar, R.K., Sehgal, A. et al. (2017). Progress on *Azadirachta indica* based biopesticides in replacing synthetic toxic pesticides. *Frontiers in Plant Science* 8: 610.

Chávez, D., Pereira, G., and Machuca, Á. (2014). Estimulación del crecimiento en plántulas de *Pinus radiata* utilizando hongos ectomicorrícicos y saprobios como biofertilizantes. *Bosque* 35: 57–63.

Chen, H., Chen, L., Shen, Z. et al. (2020). Synthesis of mesoporous silica post loaded by methyl eugenol as an environment-friendly slow-release bio pesticide. *Scientific Reports* 10: 1–12.

Chen, J., Lü, S., Zhang, Z., Zhao, X. et al. (2018). Environmentally friendly fertilizers: A review of materials used and their effects on the environment. *Science of the Total Environment* 613: 829–839.

Copping, L.G. and Menn, J.J. (2000). Biopesticides: A review of their action, applications and efficacy. *Pest Management Science: Formerly Pesticide Science* 56: 651–676.

Coutinho, I.D., Mertz-Henning, L.M., Döpp, S.A. et al. (2018). Flooded soybean metabolomic analysis reveals important primary and secondary metabolites involved in the hypoxia stress response and tolerance. *Environmental and Experimental Botany* 153: 176–187.

Curi, M.A., Medina, E.G., and Jiménez, V.H. (2020). Azotobacter y Rhizobium como biofertilizantes naturales en semillas y plantas de frijol caupí. *Avances* 22: 239–251.

Da Costa-Chaulet, F., De Alcantara-barcellos, H.H., Fior, D. et al. (2019). Glyphosate-and fipronil-based agrochemicals and their mixtures change zebrafish behavior. *Archives of Environmental Contamination and Toxicology* 77: 443–451.

Desiere, S., Hung, Y., Verbeke, W. et al. (2018). Assessing current and future meat and fish consumption in Sub-Sahara Africa: Learnings from FAO Food Balance Sheets and LSMS household survey data. *Global Food Security* 16: 116–126.

Dhir, B. (2017). Biofertilizers and biopesticides: Eco-friendly biological agents. In: *Advances in Environmental Biotechnology* (ed. R. Kumar, A. Sharma and S. Ahluwalia). Singapore: Springer.

Díaz-Franco, A., Ortegón-Morales, A.S., and Garza-Cano, I. (2006). Biofertilización del Cártamo (*Carthamus tinctorius* L.) en condiciones restringidas de humedad en el suelo. *Revista Fitotecnia Mexicana* 29: 175–180.

Díaz-Moreno, R., Díaz-Franco, A., Garza-Cano, I. et al. (2007). Brassinoesteroides e inoculación con *Glomus intraradices* en el crecimiento y la producción de sorgo en campo. *Terra Latinoamericana* 25: 77–83.

Domínguez, J., Aira, M., and Gomez-Brandon, M. (2010). Vermicomposting: earthworms enhance the work of microbes. In: *Microbes at Work* (ed. H. Insam and I. Franke-Whittle). Berlin: Springer.

Ernandes, S., Bianchi, V.L., and Oliveira-Moraes, I. (2013). Avaliação de dois diferentes meios de cultura para o desenvolvimento de biopesticidas à base de *Bacillus thuringiensis* e sua aplicação em larvas de *Aedes aegypti*. *Acta Scientiarum: Technology* 35: 11–18.

Evseeva, N.V., Tkachenko, O.V., Denisova, A.Y. et al. (2019). Functioning of plant-bacterial associations under osmotic stress *in vitro*. *World Journal of Microbiology and Biotechnology* 35: 195.

FAO, Food and Agriculture Organization of the United Nations (2019). FAOSTAT Statistical Database. Rome, Italy: FAO.

Fodjo, B.K., Koudou, B.G., Tia, E. et al. (2018). Insecticides resistance status of *An. Gambiae* in areas of varying agrochemical use in Côte D'Ivoire. *BioMed Research International* 2874160: 1–9.

Franco-Rodríguez, J.E., Monar, J.B., and Andrade, X.F. (2014). El uso de biocidas botánicos para el control de las plagas en agricultura urbana (II parte y final). *Alternativas* 15: 43–52.

Freire, E., Koch, A., and Salvador, L. (2016). Evaluación del potencial biofertilizante de consorcios de Cianobacterias en pasto Raygrass (*Lolium multiflorum*). *Revista Científica Ecuatoriana* 4: 1–7.

Gamarra-Mendoza, N.N., Gamarra-Poma, R.N., and Tito-León, R. (2019). Efecto de capsaicinoides en larvas de *Premnotrypes vorax* de papas (*Solanum tuberosum*) cosechadas y almacenadas en el Valle del Mantaro. Región Junín. *Journal of Agri-Food Science* 1: 83–89.

Gamboa-Angulo, J.J., Ruíz-Sánchez, E., Alvarado-López, C.J. et al. (2020). Efecto de biofertilizantes microbianos en las características agronómicas de la planta y calidad del fruto del chile xcatík (*Capsicum annuum* L.). *Terra Latinoamericana* 38: 817–826.

García, J.E. (1998). Intoxicaciones agudas con plaguicidas: Costos humanos y económicos. *Revista Panamericana de Salud Pública* 4: 383–387.

Garza-Cano, I., Pecina-Quintero, V., Díaz-Franco, A. et al. (2005). Sorghum cultivated with biofertilizers, fitohormones, and inorganic phosphorus. *Terra Latinoamericana* 23: 581–586.

González-Maldonado, M.B. and García-Gutiérrez, C. (2012). Uso de biorracionales para el control de plagas de hortalizas en el norte de Sinaloa. *Ra Ximhai* 8: 31–45.

Guo, S.H., Hu, N., Li, Q.S. et al. (2018). Response of edible amaranth cultivar to salt stress led to Cd mobilization in rhizosphere soil: A metabolomic analysis. *Environmental Pollution* 241: 422–431.

Guzmán-Rivera, S.P., Tróchez-Parra, A., Correa-Zapata, L.F. et al. (2004). Efecto insecticida y residual de tres extractos de *Lippia alba* para el control de *Acanthoscelides obtectus* en fríjol Diacol Calima. *Revista Guillermo de Ockham* 2: 187–199.

Haase, S., Sciocco-Cap, A., and Romanowski, V. (2015). Baculovirus insecticides in Latin America: Historical overview, current status and future perspectives. *Viruses* 7: 2230–2267.

Hapeman, C.J., McConnell, L.L., Rice, C.P. et al. (2003). Current United States Department of Agriculture – Agricultural Research Service research on understanding agrochemical fate and transport to prevent and mitigate adverse environmental impacts. *Pest Management Science* 59: 681–690.

Hasanuzzaman, M., Nahar, K., Anee, T.I. et al. (2017). Glutathione in plants: Biosynthesis and physiological role in environmental stress tolerance. *Physiology and Molecular Biology of Plants* 23: 249–268.

Hendges, C., Schiller, A.P., Manfrin, J. et al. (2019). Human intoxication by agrochemicals in the region of South Brazil between 1999 and 2014. *Journal of Environmental Science and Health, Part B* 54: 219–225.

Hernández-Acosta, E., Trejo-Aguilar, D., Rivera-Fernández, A. et al. (2020). La micorriza arbuscular como biofertilizante en cultivo de café. *Terra Latinoamericana* 38: 613–628.

Hernández-González, M.M., Jiménez-Garcés, C., Jiménez-Albarrán, F.R. et al. (2007). Caracterización de las intoxicaciones agudas por plaguicidas: Perfil ocupacional y conductas de uso de agroquímicos en una zona agrícola del Estado de México, México. *Revista Internacional de Contaminación Ambiental* 23: 159–167.

Hernández-Reyes, B.M., Rodríguez-Palacio, M.C., Castilla-Hernández, P. et al. (2019). Uso potencial de cianobacterias como biofertilizante para el cultivo de maíz azul en la Ciudad de México. *Revista Latinoamericana de Biotecnología Ambiental y Algal* 10: 13–27.

Huang, T., Zhang, X., Pan, J. et al. (2016). Purification and characterization of a novel cold shock protein-like bacteriocin synthesized by *Bacillus thuringiensis*. *Scientific Reports* 6: 1–15.

Jiménez-Gómez, A., García-Estévez, I. et al. (2020). Increase in phenolic compounds of *Coriandrum sativum* L. after the application of a *Bacillus halotolerans* biofertilizer. *Journal of the Science of Food and Agriculture* 100: 2742–2749.

Jørs, E., Aguilar, G.C., Huici, O. et al. (2006). Occupational pesticide intoxications among farmers in Bolivia: A cross-sectional study. *Environmental Health: A Global Access Science Source* 5: 1–9.

Kadam, R.V., Jadhav, S.W., and Abhang, P.B. (2018). Molecular characterization of MPKV biofertilizers. *Journal of Pharmacognosy and Phytochemistry* 7: 418–423.

Keswani, C., Sarma, B.K., and Singh, H.B. (2016). Synthesis of policy support, quality control, and regulatory management of biopesticides in sustainable agriculture. In: *Agriculturally Important Microorganisms* (ed. H. Singh, B. Sarma and C. Keswani). Singapore: Springer.

Krieger, S., Molina, S.G., Piquín, E. et al. (2017). Efecto promotor de un biofertilizante en plantas de chía (*Salvia hispanica* L.). *Agrotecnia* 25: 28.

Kumar, A., Verma, A., and Kumar, A. (2013). Accidental human poisoning with a neonicotinoid insecticide, imidacloprid: A rare case report from rural India with a brief review of literature. *Egyptian Journal of Forensic Sciences* 3: 123–126.

Kumar, S. (2012). Biopesticides: A need for food and environmental safety. *Journal of Biofertilizers & Biopesticides* 3: 1–3.

Lade, B.D. and Gogle, D.P. (2019). Nano-biopesticides: Synthesis and applications in plant safety. In: *Nanobiotechnology Applications in Plant Protection. Nanotechnology in the Life Sciences* (ed. K. Abd-Elsalam and R. Prasad). Cham: Springer.

Laguna-Miranda, R.J. and Cisne-Contreras, J. (2001). Efecto de biofertilizante (Em-Boskashi) sobre el crecimiento y rendimiento de Rábano (*Raphanus sativus*). *La Calera* 1: 26–29.

Lizarazo, K., Mendoza, C., and Carrero, R. (2008). Efecto de extractos vegetales de *Polygonum hydropiperoides, Solanum nigrum* y *Calliandra pittieri* sobre el gusano cogollero (*Spodoptera frugiperda*). *Agronomía colombiana* 26: 427–434.

Malaikozhundan, B., Vaseeharan, B., Vijayakumar, S. et al. (2017). *Bacillus thuringiensis* coated zinc oxide nanoparticle and its biopesticidal effects on the pulse beetle. *Callosobruchus maculatus. Journal of Photochemistry and Photobiology B: Biology* 174: 306–314.

Meena, R.S., Kumar, S., Datta, R. et al. (2020). Impact of agrochemicals on soil microbiota and management: A review. *Land* 9: 34.

Mishra, P. and Dash, D. (2014). Rejuvenation of biofertilizer for sustainable agriculture and economic development. *Consilience* 11: 41–61.

Mohammadi, K. and Sohrabi, Y. (2012). Bacterial biofertilizers for sustainable crop production: A review. *Journal of Agriculture and Biological Sciences* 7: 307–316.

Moshi, A.P. and Matoju, I. (2017). The status of research on and application of biopesticides in Tanzania: Review. *Crop Protection* 92: 16–28.

Moustafa-Farag, M., Almoneafy, A., Mahmoud, A. et al. (2020). Melatonin and its protective role against biotic stress impacts on plants. *Biomolecules* 10: 54.

Negri, G., May-de Mio, L.L., and Wordell-Filho, J.A. (2010). Produção e armazenamento de *Trichothecium roseum* para uso como biopesticida. *Scientia Agraria* 11: 247–254.

Nianiou-Obeidat, I., Madesis, P., Kissoudis, C. et al. (2017). Plant glutathione transferase-mediated stress tolerance: Functions and biotechnological applications. *Plant Cell Reports* 36: 791–805.

Noh-Medina, J., Yam-Chimal, C., Borges-Gómez, L. et al. (2014). Aislados bacterianos con potencial biofertilizante para plántulas de tomate. *Terra Latinoamericana* 32: 273–281.

Orozco-Corral, A.L., Valverde-Flores, M.I., Martínez-Téllez, R. et al. (2016). Propiedades físicas, químicas y biológicas de un suelo con biofertilización cultivado con manzano. *Terra Latinoamericana* 34: 441–456.

Ortuño, N., Córdoba, M., Claros, M. et al. (2018). Evaluación de bacteriasendófitas de papa nativa (*Solanum tuberosum* L.) y el desarrollo de un biofertilizante. *Revista Latinoamericana de la papa* 22: 12–37.

Osakabe, Y., Arinaga, N., Umezawa, T. et al. (2013). Osmotic stress responses and plant growth controlled by potassium transporters in *Arabidopsis*. *The Plant Cell* 25: 609–624.

Osuna-Carreón, A., Ail-Catzim, C., García-López, A.M. et al. (2014). Efecto insecticida de *Pluchea sericea* sobre adultos de *Bemisia tabaci*. *Entomología Mexicana* 1: 1018–1021.

Pavela, R. and Benelli, G. (2016). Essential oils as ecofriendly biopesticides? Challenges and constraints. *Trends in Plant Science* 21: 1000–1007.

Pavela, R., Murugan, K., Canale, A. et al. (2017). *Saponaria officinalis*-synthesized silver nanocrystals as effective biopesticides and oviposition inhibitors against *Tetranychus urticae* Koch. *Industrial Crops and Products* 97: 338–344.

Pérez-Luna, Y.C., Álvarez-Solís, J.D., Mendoza-Vega, J. et al. (2012). Diversidad de hongos micorrícicos arbusculares en maíz con cultivo de cobertura y biofertilizantes en Chiapas, México. *Gayana Botánica* 69: 46–56.

Prior, H., Casey, W., Kimber, I. et al. (2019). Reflections on the progress towards non-animal methods for acute toxicity testing of chemicals. *Regulatory Toxicology and Pharmacology* 102: 30–33.

Quinteros, E., Ribó, A., Mejía, R., López, A. et al. (2017). Heavy metals and pesticide exposure from agricultural activities and former agrochemical factory in a Salvadoran rural community. *Environmental Science and Pollution Research* 24: 1662–1676.

Quiroz-Sarmiento, V.F., Almaraz-Suarez, J.J., Sánchez-Viveros, G. et al. (2019). Biofertilizantes de rizobacterias en el crecimiento de plántulas de chile Poblano. *Revista Mexicana de Ciencias Agrícolas* 10: 1733–1745.

Ramkumar, G., Asokan, R., Ramya, S. et al. (2020). Characterization of *Trigonella foenum-graecum* derived iron nanoparticles and its potential pesticidal activity against *Tuta absoluta* (Lepidoptera). *Journal of Cluster Science* 32: 1–6.

Rangel-Lucio, J.A., Ramírez-Gama, R.M., Cervantes-Ortíz, F. et al. (2014). Biofertilización de *Azospirillum* spp. y rendimiento de grano de maíz, sorgo y trigo. *Revista de la Facultad de Ciencias Agrarias* 46: 231–238.

Rodríguez, H.C. (2002). Plantas contra plagas. In: *Potencial práctico de ajo, anona, nim, chile y tabaco* (ed. RAAPAN), 133 p. México.

Rundlöf, M., Smith, H.G., and Birkhofer, K. (2016). Effects of organic farming on biodiversity. *eLS* 1–7. https://doi.org/10.1002/9780470015902.a0026342

Sahayaraj, K. (2014). Nanotechnology and plant biopesticides: An overview. In: *Advances in Plant Biopesticides* (ed. D. Singh). New Delhi: Springer.

Saijo, Y. and Loo, E.P. (2020). Plant immunity in signal integration between biotic and abiotic stress responses. *New Phytologist* 225: 87–104.

Sánchez-Bayo, F. and Tennekes, H.A. (2017). Assessment of ecological risks of agrochemicals requires a new framework. *Environmental Risk Assessment and Remediation* 1: 20–28.

Sansinenea, E. and Ortiz, A. (2015). Melanin: A photoprotection for *Bacillus thuringiensis* based biopesticides. *Biotechnology Letters* 37: 483–490.

Sarti, G.C., Cristóbal-Miguez, A.E., and Curá, A.J. (2019). Optimización de las condiciones de cultivo para el desarrollo de una biopelícula bacteriana y su aplicación como biofertilizante en *Solanum lycopersicum* L. var. Río grande. *Revista de Protección Vegetal* 34: 1–13.

Scott, N.R., Chen, H., and Cui, H. (2018). Nanotechnology applications and implications of agrochemicals toward sustainable agriculture and food Systems. *Journal of Agricultural and Food Chemistry* 66: 6451–6456.

Shukla, L.I., Chinnusamy, V., and Sunkar, R. (2008). The role of microRNAs and other endogenous small RNAs in plant stress responses. *Biochimica et Biophysica Acta (BBA)-Gene Regulatory Mechanisms* 1779: 743–748.

Soria Fregoso, M.J., Ferrera-Cerrato, R., Etchevers Barra, J. et al. (2001). Biodigestion of hog slurry to produce biomanure. *Terra Latinoamericana* 19: 353–362.

Suhag, A., Yadav, H., Chaudhary, D. et al. (2020). Biotechnological interventions for the sustainable management of a global pest, whitefly (*Bemisia tabaci*). *Insect Science* 28: 1228–1252.

Sunkar, R., Li, Y.F., and Jagadeeswaran, G. (2012). Functions of microRNAs in plant stress responses. *Trends in Plant Science* 17: 196–203.

Suzuki, N., Rivero, R.M., Shulaev, V. et al. (2014). Abiotic and biotic stress combinations. *New Phytologist* 203: 32–43.

Tong, Z., Quan, G., Wan, L. et al. (2019). The effect of fertilizers on biomass and biodiversity on a semi-arid Grassland of Northern China. *Sustainability* 11: 2854.

Uribe-Valle, G. and Dzib-Echeverría, R. (2006). Vesicular arbuscular mycorrhizal (*Glomus intraradices), Azospirillum brasilense* and Brasinoesteroid on yield of maize in luvisol soil. *Agricultura técnica en México* 32: 67–76.

Vance, C.P. (2001). Symbiotic nitrogen fixation and phosphorus acquisition. Plant nutrition in a world of declining renewable resources. *Plant Physiology* 127: 390–397.

Wachowska, U., Irzykowski, W., and Jędryczka, M. (2018). Agrochemicals: Effect on genetic resistance in yeasts colonizing winter wheat kernels. *Ecotoxicology and Environmental Safety* 162: 77–84.

Weselowski, B., Nathoo, N., Eastman, A.W. et al. (2016). Isolation, identification and characterization of *Paenibacillus polymyxa* CR1 with potentials for biopesticide, biofertilization, biomass degradation and biofuel production. *BMC Microbiology* 16: 1–10.

WHO, World Health Organization (2009). Guidelines for efficacy testing of mosquito repellents for human skin (No. WHO/HTM/NTD/WHOPES/2009.4). World Health Organization.

Xiong, L. and Zhu, J.K. (2002). Molecular and genetic aspects of plant responses to osmotic stress. *Plant, Cell & Environment* 25: 131–139.

Yuan, J., Yuan, Y., Zhu, Y. et al. (2018). Effects of different fertilizers on methane emissions and methanogenic community structures in paddy rhizosphere soil. *Science of the Total Environment* 627: 770–781.

15

Carbon Sequestration Alternatives for Mitigating the Accumulation of Greenhouse Gases in the Atmosphere

Erfan Sadatshojaei[1], Setareh Heidari[2], Zahra Edraki[3], David A. Wood[4] and Ahmad Fauzi Ismail[5]

[1] *Department of Chemical Engineering, Shiraz University, Shiraz, 71345, Iran*
[2] *Faculty of Chemical Engineering, Sahand University of Technology, Tabriz, Iran*
[3] *Sharif University of Technology, Tehran, Iran*
[4] *DWA Energy Limited, Bassingham, Lincoln, LN5 9JP, UK*
[5] *Advanced Membrane Technology Research Center (AMTEC), Universiti Teknologi Malaysia, 81310, Skudai, Johor Bahru, Johor, Malaysia*

CHAPTER MENU

15.1 Introduction

The Earth's climate is a function of the energy flowing in and out from the surface of the Earth through the atmosphere and oceans. Progressively accumulating greenhouse gases (GHG) into the atmosphere creates an imbalance that gradually traps more radiation and heat energy within the atmosphere and oceans. A broad range of novel technologies and their applications, introduced in recent years, play a substantial role in improving standards of living and more efficient access to and use of available energy resources, such as enhanced oil recovery (Sadatshojaei et al. 2016, 2019; Choubineh et al. 2017; Dehshibi et al. 2019; Nezhad et al. 2020), fossil energy technologies (Sadatshojaie et al. 2020a), corrosion (Seddiqi et al. 2019), biofuels (Jafari et al. 2020; Sadatshojaei et al. 2020a, 2020b), solvents (Kozinski et al. 2018; Sadatshojaei and Wood 2020), carbon dioxide (Sadatshojaei et al. 2018), handling and utilization, and wastewater treatments

Bioresource Technology: Concept, Tools and Experiences, First Edition. Tanveer Bilal Pirzadah, Bisma Malik, Rouf Ahmad Bhat, and Khalid Rehman Hakeem.
© 2022 John Wiley & Sons Ltd. Published 2022 by John Wiley & Sons Ltd.

(Heidari et al. 2019, 2017; Shokri et al. 2015). Despite these breakthroughs in technology, carbon dioxide (CO_2) continues to accumulate in the atmosphere and oceans.

Trapped energy in the atmosphere and oceans is responsible for many recordable changes in global climate including average annual temperature rise, melting of polar ice and mountain glaciers, increased frequency of draughts, severe storms and cyclones and other more extreme weather events. CO_2-concentration changes in the oceans and biosphere is also impacting ecosystems on an unprecedented scale, as well as the universal climate.

From 1750 to 2015, the increase in atmospheric CO_2 concentration has been of anthropogenic origin. This resulted in a long-term greenhouse gases (GHG) increase of about 37%, with CO_2 recording approximately 79% growth from 1990 to 2015 (Ussiri and Lal 2017). The two fundamental resources of CO_2 emission are fossil-fuel consumption and changes in the type of land-use, particularly urbanization and expanded conversion of land into a range of agricultural uses.

Carbon capture and storage (CCS) as a potential approach for mitigating GHG emissions is defined as "the capture and secure storage of carbon that would otherwise be emitted to the atmosphere" (Herzog and Golomb 2004). This chapter discusses different approaches that could be exploited to implement CCS.

15.2 Impact of Greenhouse Gases

15.2.1 The Natural Greenhouse Impacts

The gases oxygen, argon and nitrogen that form the bulk of the Earth's atmosphere do not release or absorb thermal irradiance. There would not be any clouds in our atmosphere and no greenhouse impacts if those were the only gases present. In such conditions. radiative equilibrium would lead to an average surface temperature for the Earth of about −6°C (Houghton 2005), which is much lower than the 20°C that actually prevails. This highlights that traces of other gases, making up much less than 1% of the atmosphere by dry volume, have a major impact on controlling energy outflow from the atmosphere by absorbing infra-red (IR) radiation.

CO_2, methane (CH_4), nitrous oxide (N_2O) and water vapor are the key trace-component atmospheric gases contributing to greenhouse effects. These GHGs absorb some IR radiation and scatter long-wave radiation in their roles as atmospheric warming agents. Some of that scattered radiation is reflected downward, heating the lower atmospheric layers and the surface of Earth itself. The remaining IR radiation, not absorbed by the GHGs, is lost to space through the troposphere. Water vapor, CH_4 and CO_2 exert the strongest greenhouse erffects within the atmosphere.

The quantity of irradiance that GHGs emit depends on the temperatures that prevail in the upper atmosphere. Convection creates atmospheric temperature gradients that cause the Earth's surface to be as much as 50°C warmer than the upper atmosphere. GHG molecules therefore emit much less irradiance from the outer atmospheric layers to space due to their coldness acting like a thermal blanket. Clouds within the atmosphere tend to play a counter role by reflecting some of the incident radiation arriving from the Sun out of the upper atmosphere, thereby cooling the atmosphere. During geological periods of higher than usual volcanic activity, excess clouds have been responsible for lowering surface temperatures.

This GHG phenomenon is not a consequence in itself of anthropogenic actions but a natural consequence of the physical and chemical properties of the gases involved. Moreover, atmospheric water-vapor concentrations depend to a large extent on the temperature of the oceans. Higher ocean temperatures result in more water evaporating, adding to the GHG burden in the atmosphere and exacerbating the greenhouse impacts. Energy arriving from the Sun is concentrated in equatorial regions and subsequently transmitted to higher latitudes by oceanic currents as atmospheric weather and circulation systems. On the other hand, long-wave radiation leaves the Earth's surface more evenly from all latitudes cooling the polar regions (Fasullo and Trenberth 2008a, 2008b; Schwarz et al. 2019). It is this complex atmosphere-ocean GHG cycle that the increasing anthropogenic inputs, particularly of water vapor, CH_4 and CO_2, have progressively distorted.

The heating impacts of solar irradiance were first recognized and understood at a fundamental level by Fourier and Claude Pouillet based on widely distributed barometric measurements in the 1830s (Mudge 1997). Their observations lead to an explanation of the atmosphere's role in maintaining higher temperatures at the Earth's surface than if the atmosphere did not exist. Tyndall (1872) was the first to identify the role of the minor gases in atmospheric heating by configuring the first infra-red spectrometer able to directly measure heat intensity.

15.2.2 Anthropogenic Greenhouse Impacts

Recent information published by the Intergovernmental Panel on Climate Change (IPCC), reported that human-induced global warming reached approximately 1°C in 2017. Moreover, they estimated that planetary-scale warming is increasing by about 0.2°C per decade (Allen et al. 2019). GHG emissions have for more than two decades been identified as being closely linked to this progressive warming process (Cooper 1998; Herzog 2009; Mohajan 2011). GHGs, although mostly water vapor, CO_2, CH_4 and N_2O, also include man-made halogenated compounds such as chlorofluorocarbons (CFC) that accumulate in the upper atmosphere. Collectively, these GHGs, including the anthropogenic contributions, are recognized to be contributing to climate change (Bloomfield and Pearson 2000; Bolin and Doos 1989) and ecosystem damage (Botkin et al. 2007; Moore 1998; Shao et al. 2014). Thus, the United Nations Framework Convention on Climate Change (UNFCCC)'s ultimate objective was set to stabilize GHG concentrations in the atmosphere at a level that would prevent dangerous anthropogenic interference with the climate system. In this regard, in 2010, policy-makers amended the Kyoto Protocols of 1997 to a stated goal of limiting the global average temperature rise to 2°C. To achieve this. calculations suggest that cumulative anthropogenic CO_2 emissions from 2011 to 2050 should be limited to 1100 gigatonnes (Meinshausen et al. 2009; Van Der Ploeg and Rezai 2017).

One of the mechanisms limiting GHG accumulations in the atmosphere is the natural carbon cycle (NCC). This biogeochemical cycle is the natural flow of carbon through the

Earth's ecosystems (Olson et al. 1985). This carbon moves through the atmosphere, ocean, terrestrial biosphere and lithosphere (Schimel 1995). Prior to the contributions of anthropogenic emissions, the NCC was nearly balanced between animals and plants emitting CO_2 into the atmosphere through respiration, and plants absorbing it through photosynthesis. The extended usage of fossil fuels and global warming have altered this natural carbon cycle (Trabalka et al. 1985).

15.3 Soil's Role in the Sequestration of Carbon

Soil-carbon sequestration (SCS) may involve either the soil's organic-carbon content (SOC) or the soil's inorganic-carbon content (SIC). Organic carbon tends to play the dominant role in regions covered by an expansive biosphere. On the other hand, SIC, through the deposition of carbonate minerals, dominates in more barren regions with sparse biospheres. The SOC pools include calcite minerals accumulated from the remains of animals, plants and microbes, whereas the SIC pools consist mainly of primary and secondary carbonate minerals derived from non-biotic sources (Lal 2016; Greipsson and El-Mayas 2000).

15.3.1 Organic Carbon Sequestration

Typically, SCS is achieved where the carbon content in the soil is incrementally increased over time. This is typically achieved continuously by CO_2 being removing from the atmosphere by plants and some microbes via their photosynthesis. In dry and semi-arid regions, SCS processes may also involve transformation of CO_2 absorbed from the atmosphere into various inorganic carbon-containing minerals, mainly carbonates. However, the rate of inorganic carbon mineralization is much lower than organic carbon absorption by the soil (Ussiri and Lal 2017). A good deal of organic carbon accumulated in soils suffers erosion, causing its redistribution with much of it ending up back in the atmosphere or oceans.

Organic carbon accumulation in the soil generally takes place when a positive disequilibrium persists over considerable time periods on a scale of thousands of years. This enables carbon to be consistently abstracted from the CO_2 present in the air (Ussiri and Lal 2017). Removing CO_2 from the atmosphere not only counters global warming but also has many ecosystem benefits. It enriches the soil's organic content, increasing soil fertility, and helps to rehabilitate ecosystems associated with low-carbon soils. However, conversion of carbon from the air into the biosphere is in dynamic equilibrium at most locations and can be easily disrupted by small changes in conditions. Even such small changes can cause organic carbon stocks built up in soils over millennia to be eroded and returned to the atmosphere in short disruptive episodes.

The storage or residence time of carbon in the soil is paramount in determining the potential of soil under particular climatic and ecosystem conditions to act as a long-term viable carbon sink (Luo et al. 2003). Clearly, transferring some carbon periodically from the soil or biosphere back into the atmosphere has negative impacts on the atmosphere's greenhouse mechanisms (Powlson et al. 2011). Disruption of farmland soils by the agricultural sector, together with tree removal and soil turnover by the forestry industry and other soil disrupting activities, such as peat harvesting, are responsible for approximately a quarter of global anthropogenic GHG emissions (Vermeulen et al. 2012).

Organic carbon retention by soils is influenced by numerous factors such as the soil's characteristics, type of ecosystem, species diversity, climate and exposure to extreme weather events. Furthermore, various types and sources of organic carbon present as layers, or as distinctive carbon pools in a soil, have their own particular function and contribution to enabling that soil to accumulate and retain carbon (Ussiri and Lal 2017). The functionality of different types of organic carbon present in soils is summarized in Table 15.1.

Soil represents the main source of the global carbon stocks in the Earth's surface and near-surface (ecosystem) environments. It accounts for some 80% of the total carbon stocks (Figure 15.1). Figures 15.2 and 15.3 illustrate the world's biomass and carbon stocks in vegetation and soil, respectively.

Figure 15.2 highlights the key role of tropical forests as carbon sinks. They are the main source of carbon stocks in the terrestrial biosphere, containing some 212 Pg of carbon. Boreal forests are the second-largest source of carbon held by vegetation storing 88

Table 15.1 Soil organic components and their functions. Data from Ussiri and Lal (2017).

Type	Function
Crop residues	These species tend to be rapidly mineralized and provide a substratum for microorganisms living in the soil.
Particulate organic carbon	• The rate of its mineralization is quite fast, but lower than the rest of crop residues. • These species supply nutrients for vegetation growth.
Recalcitrant organic carbon	Because of its stability and insolubility form, it is not further decomposed. Yet it plays a vital role in soil protection via improving a soil's physical properties.

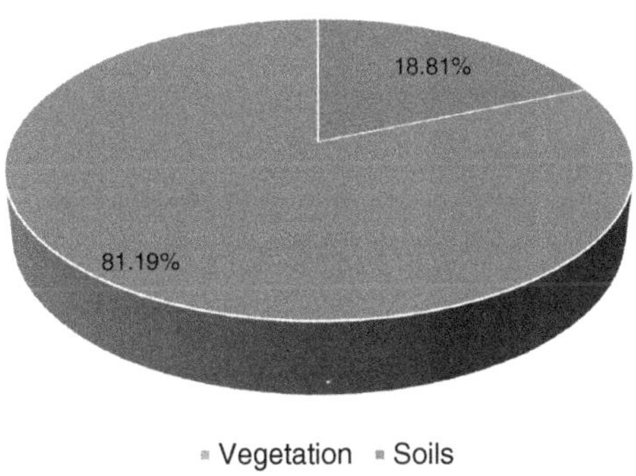

Figure 15.1 The global carbon stocks (expressed as percent) in the ecosystem. These data were reported by Ravindranath and Ostwald (2007) and Watson et al. (2000).

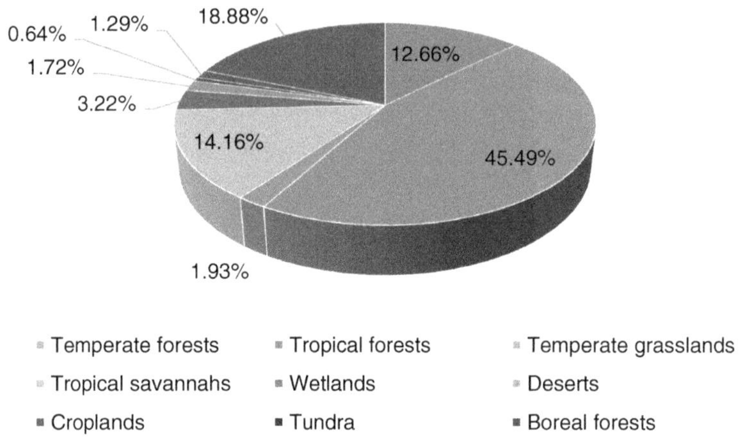

Figure 15.2 Global climatic environments (biomes) and their carbon stocks (%) held in vegetation. These data were reported by (Ravindranath and Ostwald (2007) and Watson et al. (2000).

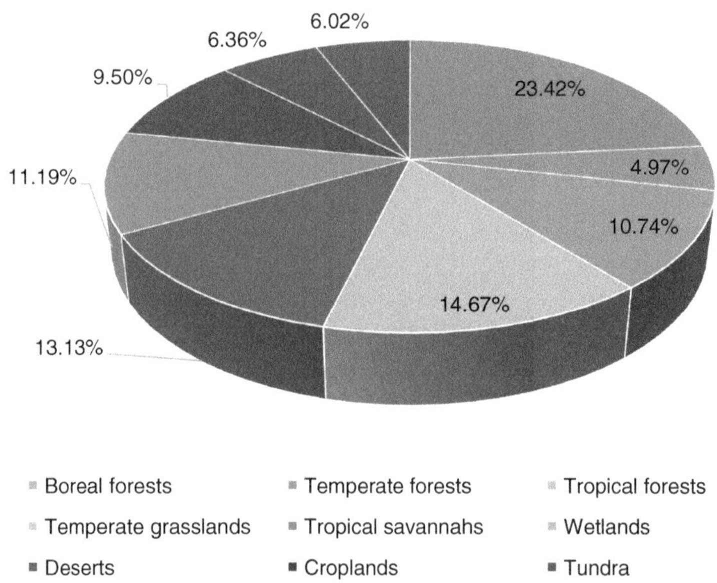

Figure 15.3 Global climatic environments (biomes) and their carbon stocks held in the top 1 meter of soil. These data were reported by Ravindranath and Ostwald (2007) and Watson et al. (2000).

Pg of carbon. Tropical savannahs and temperate forests also make valuable contributions of the carbon stocks of global biomes in vegetation by holding 66 Pg and 59 Pg of carbon, respectively. On the other hand, wetlands, deserts, croplands and tundra collectively contribute less than 10% of the total carbon stocks held in vegetation.

However, the carbon stocks in the uppermost meter of soils (Figure 15.3) is distributed quite differently under the various climatic environments compared to in their overlying vegetation. Boreal forests account for nearly a quarter of the global carbon stocks in the uppermost soil layers (~471 Pg of carbon). Another major source of carbon stock in the soil is associated with temperate grasslands (~295 Pg of carbon). Tropical-savannah soils are also an important carbon sink (~264 Pg of carbon). A comparison between Figures 15.2 and 15.3 reveals that the soils of the wetlands, deserts, croplands and tundra contain a much more significant portion of carbon with respect to the other climatic environments than is contained in the vegetation. This suggests that the carbon in the wetlands, deserts, croplands and tundra soils accumulates more slowly but is more likely to be retained.

15.3.2 Inorganic Carbon Sequestration in Soils

The conversion of atmospheric carbon into inorganic carbon in soils, mainly as carbonate minerals, takes a long time. However, once converted into inorganic carbon, the mean residence time of carbon in soil carbonates, dominated by calcium carbonate ($CaCO_3$), is approximately 80,000 years. That is a much longer-term carbon store than for organic carbon in soils. The global soil inorganic carbon pool is considerably lower at 940 Pg of carbon than in carbonate minerals present in oceanic sediments (~40,000 Pg of carbon) (Sarmiento and Gruber 2002). However, the inorganic carbon estimated to be present in soils is estimated to be more than half the amount of soil organic carbon (~1505 Pg of carbon) (Batjes 2014; Sarmiento and Gruber 2002). These values demonstrate the important role played by soil inorganic carbon in the long term as a carbon store. Based on the processes involved in soil inorganic carbon generation, the inorganic carbonate minerals can be divided into three distinctive groups: geogenic, biogenic and pedogenic (Table 15.2).

Table 15.2 Different types of soil inorganic carbonate minerals: their characteristics and formation processes.

Category	Characteristics	Reference
Geogenic carbonates (from speleothem fragments and bedrock)	• They account for 84% up to around 98% of the carbonate • They are formed by the chemical processes (e.g., stalagmites and stalactites) • Carbonates minerals formed indigenously or inherited • Indigenous: carbonate fragments from a soil's parent rock, e.g., limestone and other carbonate bedrocks • Inherited: carbonates redeposited in the soil but derived from other regions through dust/landslides/glaciation. These tend to be dominated by calcium carbonate	Nicosia and Stoops (2017); Ussiri and Lal (2017)
Biogenic carbonates (from biologic precursors)	Carbonates found in plants and/or in the skeletons of terrestrial animals. These include micro-flora and micro-fauna	Ussiri and Lal (2017)
Pedogenic carbonates (secondary diagenesis)	• A vadose zone carbonates deposited by diagenetic processes in the soil above the water table, particularly in arid and semi-arid climatic zones • This is a secondary carbonate deposited in soils by the dissolution and solution of soil organic carbon. It is a vital process in slowly mineralizing soil organic carbon pool	Verrecchia (2011); Ussiri and Lal (2017)

15.4 Terrestrial Carbon Sequestration

Terrestrial ecosystems are considered piotentially as major carbon sinks that could be exploited to accumulate and store CO_2. The photosynthesis and storage of CO_2 in live and/or dead organic matter sequesters approximately 28% of CO_2 emissions originating from anthropogenic activities into the biosphere and its underlying soils (Canadell et al. 2007; Lal 2008a; Le Quéré et al. 2014). The overall terrestrial carbon pool includes biotic and pedologic pools. The biotic pool photosynthesizes ~120 Pg of carbon per year, a good deal of which is returned to the atmosphere from the respiration of plants and microorganisms (Lal 2008b). The pedologic pool consists of soil organic (SOC) and inorganic (SIC) pools, as described in Section 15.3.

The anthropogenic processes, especially fossil-fuel based industries and land-use changes, continue to substantially disturb the Earth's natural carbon cycle and release CO_2 and other GHGs to the atmosphere. These actions progressively reduce the net uptake and increase the release rate of carbon from many terrestrial ecosystems. Assessments of land-use changes collectively in North America, Europe and Asia showed that an additional 55 ± 50 Pg of carbon is expected to be released from upper soil layers by 2050 (Crowther et al. 2016). Terrestrial carbon sequestration undertakes actions that attempt to maintain, sustain and promote carbon storage in the soil and biosphere by enhancing ecological factors and protecting valuable ecosystems. The actions undertaken on a large scale include promoting eco-friendly forest and land management practices, modifying agricultural land use and expanding the use of biofuels not generated from food crops or on existing agricultural land (Ussiri and Lal 2017).

15.4.1 Global Forest Management

Besides all the economic, environmental and eco-system benefits global forests provide, they act as global-scale "lungs" for the planet through the combination of their collective and continuous photosynthesis and respiration. Near 30% of global land is covered by forested areas. These forests absorb about 2 billion tonnes of CO_2 per year (Freer-Smith et al. 2007). Clearly, forests play a crucial role in the carbon cycle. They carry about 90% of the worldwide carbon flux between the atmosphere and terrestrial ecosystems. Forests are estimated to store more than 80% of land-based above-ground carbon, plus forest litter and soils account for some 70% of global SOC (Alemu 2014; Dixon 1995). It is the significant contributions of forests and the agroforest industries to the carbon cycle, that has made them the focus of targeted management initiatives to protect and/or enhance their carbon absorption and storage capacities.

Unfortunately, anthropogenic actions including forest degradation and clearance, land-use changes, expanding agricultural requirements, overgrazing, illegal logging, mineral extraction and recreational pressure continue to cause serious damage to the forested areas. The agriculture-forestry-land-use sector (AFOLU) contributes about 25% of global anthropogenic GHG emissions (Bustamante et al. 2014). Large-scale GHG emissions associated with deforestation and agricultural activities account for the largest negative environmental impacts on the atmosphere and global warming; they also wreak irreparable damage to eco-systems that have taken millennia to establish. On the other hand, natural phenomena, including wildfires, hurricanes, floods, bacterial/insect infestations and diseases, also pose considerable threats to forest sustainability. The deforestation caused by anthropogenic actions and natural disasters represent the most serious issue in terms of its direct influences on GHG emissions (Freer-Smith et al. 2007; Mills et al. 2011; Franklin et al. 2018).

Mitigating the impacts of various destructive human influences, planning and implementing actions to enhance the carbon storage capacity and increasing forest areas, are the main focus of forest management (d'Annunzio et al. 2015). Global forest management is usefully sub-divided into three distinct activities: conservation, substitution and carbon storage (Brown et al. 1996).

Conservation helps to curb GHG emissions by conserving and increasing the carbon pools, limiting deforestation, sustaining reserves, monitoring and controlling harvesting and preventing fire and pollution impacts. *Substitution* strives to replace harvested forest biomass with replacement plantations and to encourage the use of alternative materials to wood from endangered species for unsustainable use in building construction and wood-based manufactured goods. *Storage* is focused on expanding forested areas and their carbon pools. These distinctive forest management strategies include several complimentary actions. A key aim is to reduce carbon loss from biota and soil. This can be achieved partly by treating soils to encourage biomass growth that ultimately increases the concentration per unit area of carbon stored in vegetation, dead organic matter and soil. Another approach is to promote the increase of carbon uptake by biota and long-lived, fast-growing species through selective tree planting, reforestation and afforestation. Moreover, encouraging the use durable wood products for long-lived and recyclable wood-based manufactured goods and building materials helps to prevent stored carbon being released to the atmosphere in short-term cycles.

Global forest management can be successful only if public policies and private markets join together to formulate and execute effective forest-management strategies and plans (Siry et al. 2005). Environmental legislations needs to be implemented, accomplished and strictly supervised to prevent illegal deforestation (Jørgensen 2010). For example, road construction and harvesting patterns need to be carefully reviewed with forest conservation in mind. Increasing forest area through establishment of a forest in an area with no previous tree cover can make valuable long-term contributions to terrestrial carbon storage (Kanowski 1997). Applying reforestation techniques to compensate for the negative effects of deforestation or land-use change has been applied as a successful policy for more than two decades (Mather 2001). To prevent land-use change, especially clearing forest areas for agricultural purposes, can be achieved by offering farmers economic incentives and support for better agriculture practices supported by advanced technologies, reduced rental on land already available for agricultural use and incentives to protect ecosystems (Angelsen 2009). Agroforestry strategies that integrate the growth of trees and hedgerows alongside crop farming and/or animal grazing can help to maintain the organic content of soils, creating valuable ecosystems and, to some extent, mitigate the damage caused by land-use change (Dollinger and Jose 2018; Raj and Lal 2013; Young 1989).

15.4.1.1 Improving Agricultural Practices

About 11% of the global land-surface area, equivalent to about 1.5 billion hectares, is dedicated to the production of agricultural crops (Bruinsma 2003). Moreover, the land area dedicated to crop growth increases year-on-year as global population and food consumption continues to grow. This rapid growth in agriculture is a major anthropogenic source of carbon emissions that accounts for about 13.5% of global GHG emissions (Follett et al. 2011; Salinger 2007). Deep ploughing of land during the cultivation cycle acts to displace substantial organic carbon held within the upper layers of soil back into the atmosphere.

To mitigate negative agricultural impacts on GHG emissions from the soil, substantial changes are required to agricultural practices. These changes need to focus on increasing carbon uptake through photosynthesis of plants, stimulating that to be captured and retained in the soil and reducing carbon emissions from deep soil disturbance (Hutchinson et al. 2007). Besides the climatic benefits, carbon sequestration in terrestrial ecosystems has the potential to reduce other negative consequences of widespread poor agricultural practices including soil erosion, excess use of nitrogen fertilizers and atrazine pollution from agricultural runoff which can contaminate drinking water sources (Almberg et al. 2018). It also helps to improve biodiversity conservation, local ecology, water quality, landscaping aesthetics and flood or drought protection, etc. (Cantonati et al. 2020; Feng et al. 2004; Jindal et al. 2008; Plantinga and Wu 2003). SOC in agricultural land could be enhanced through land-use management and/or changes to agricultural practices. Changes in agricultural land has the potentially to sequester between 75 and 208 million metric tons of carbon per year (Byrne et al. 2007; Lal et al. 1998).

There are a number of technologies offering the capability to improve terrestrial carbon sequestration. Photosynthesis boosting can be achieved by manipulating growing conditions and by genetic engineering to increase a lighter uptake by certain crops (Evans 2013). Also, planting fast-growing trees to fringe agricultural land with forestry (Pereira and Landsberg 2012), and better recycling of agricultural wastes and residues are actions that can beneficially influence SOC stocks directly or indirectly (Sanaullah et al. 2019; Ussiri and Lal 2017a).

Biochar production, a process used to convert woody agricultural and forestry wastes by pyrolysis, yields organic material that can be retained by soils and serve as a long-term carbon sink (Verheijen et al. 2010). Conversion of woody biomass carbon into biochar carbon sequesters about 50% of the initial carbon present in the waste (Lehmann et al. 2006). Recycling biochar into the soil can enhance natural rates of carbon sequestration in soils, reduce farm waste, and, in some cases, be used as a solid-fuel substitute for fossil-derived fuels (McHenry 2009).

Sequestering carbon into the deeper subsoil presents another potential carbon sink. Deep-soil carbon sequestration is a promising option worthy of more attention (Rumpel and Kögel-Knabner 2011). Experimental studies on soil-depth impacts on the net carbon sequestration studies have revealed that more than half of the total carbon in soils is stored more than 30 cm below the surface (Paul et al. 2001; Tarnocai et al. 2009). Besides the higher capacity of carbon in deep soil, studies have shown that it is able to more reliably stabilize and retain carbon (Liao et al., 2020). However, some investigations on the stability of carbon in deep-soil layers have shown that adding fresh plant-derived carbon to the subsoil could disrupt the stability of the organic carbon pool and lead to some loss of older buried carbon (Fontaine et al. 2007). More studies are required to confirm the carbon sequestration benefits of adding agricultural wastes to the deep-soil layers.

15.4.1.2 Improving Biofuel Production Processes

Although fossil fuels dominate global energy supply, and in particular transportation fuels, health and environmental consequences associated with their extensive use have been documented for more than 30 years (Barbir et al. 1990; Kampa and Castanas 2008), and the finite nature of such fuels is also cause for sustainability concern (Kruger 2006). Thus, developing biomass-derived yet eco-friendly alternatives to fossil fuels for transportation continues to be a desirable objective (Lago et al. 2019). Eco-friendly sources of biomass to convert into biofuels include: i) agricultural residues and other organic waste materials; ii) forestry litter; and iii) specialist crops, fast-growing grasses and algae grown specifically as biofuel feedstocks (Ben-Iwo et al. 2016; IEA Bioenergy 2010). The energy produced can be converted into electricity, heat or transportation biofuels. Model-based studies have predicted that bioenergy has the potential to supply up to 35% of global primary energy by 2050 and 50% by 2100 (Rose et al. 2014). Clearly, biomass from a range of sources could be contributing more to energy supply, particularly transportation fuels, with more research and development investment. By doing so, the carbon it extracts from the atmosphere by photosynthesis would help to better balance GHG emissions during its combustion.

The life cycle of all biomass commences with naturally sequestering CO_2 from the atmosphere through photosynthesis to sustain cell growth. That intake of carbon supplies energy essential for the growth and metabolism of all organisms. Biomass is a renewable energy source, since it can be grown, harvested and regrown in a continuous short-term cycle, unlike fossil fuels that take millennia to form and are consumed faster than they can be replaced (Azar et al. 2006). CO_2 emitted during the combustion of biomass is to a large extent offset by the intake during biomass growth. As long as the cycle of regrowing biomass is perpetuated, efficient biomass to fuel supply chains are able to approach zero emissions on a life-cycle basis.

By applying CCS technologies to biomass fuel supply chains, the capture of CO_2 emissions can, in certain circumstances. achieve negative net CO_2 emissions over their life cycles. Bioenergy with carbon capture and storage (BECCS) generates energy via biomass burning and further integrates the emissions from that combustion with CCS exploiting geological or alternative long-term carbon sinks (Bellamy et al. 2019). Coupling biomass growth with CCS on a large scale offers a highly effective means of mitigating CO_2 accumulating in the atmosphere (Santos et al. 2019). The BECCS strategy has been credited with the potential to reduce atmospheric accumulation of carbon by 0.68–2.7 Gt C yr^{-1} (Haszeldine and Scott 2014).

Despite the potentially important role BECCS could play in atmospheric GHG mitigation, the technology remains in its early stages of research and development. One biomass to energy technology that has been tested on commercial scales is torrefaction. This process involves thermally upgrading biomass feedstocks to a coal-like but renewable solid biofuel. The solid material produced has improved fuel properties than raw biomass as it is brittle and can be ground with little energy consumed in its preparation for combustion. Torrefaction involves processing the biomass for some hours in an inert atmosphere (typically N_2) at atmospheric pressure at temperatures between 200°C and 300°C (Chen et al. 2015). Products of torrefaction display improved energy densities compared to raw biomass and are resistant to damage by moisture absorption (National Academies of Sciences and Medicine 2018). Crucially, these products can be readily consumed in existing coal power plants, either on their own or as a blend with conventional coal, thereby reducing emissions from such power plants.

Biofuel can be effective in mitigating GHG and other polluting emissions by improving environmental and health consequences compared to the consumption of fossil fuel-based transportation fuels. Biofuel production on small and large scales can provide opportunities for economic and social development in rural communities, at the same time as reducing biomass waste disposal problems from agriculture and forestry (Clarke et al. 2015). Although bioenergy production and utilization display many positive attributes, there are negative issues if these processes are not carefully planned and implemented. For instance, the impacts on land restoration and land-use change, water-use efficiency, biodiversity and ecosystem all require careful consideration and rigorous scientific evaluation. Although it may be profitable for farmers to produce biofuels from agricultural crops on pristine land areas or clearing tropical forests for palm oil plantations, it is imperative for regulators to take into account the negative environmental and eco-system impacts of such actions (Pulighe et al. 2019). Generally, bioenergy systems can cause both positive and negative effects. Hence, their environmental, eco-system and socio-economic impacts need to be rigorously analyzed in advance of sanctioning such projects (Edenhofer 2015) if genuine mitigation of GHG accumulations in the atmosphere is to be achieved.

15.5 Carbon Sequestration into Sub-Surface Geological Reservoirs

Geological carbon sequestration into sub-surface porous rock formations and fluid traps offers substantial potential to store over the long-term large-scale captured CO_2 from large industrial sites such as power plants, steel and glass works, cement production units and crude oil refineries. It offers opportunities for carbon capture sequestration and utilization (CCSU) for the coal, oil and gas industry and algal biomass applications. Large fossil-fuel producers and consumer (e.g., power plants) see underground CCSU as a means of at least partially mitigating their GHG emissions and at the same time using the sequestered CO_2 to boost oil and gas production from porous and tight reservoirs and shale, and coal-seam gas from *in-situ* coal seams. In oil reservoirs, such a strategy is referred to as CCS with enhanced oil recovery (CCS-EOR).

Geological carbon sequestration starts with adsorbing or absorbing the CO_2 from the exhaust and flue gases of industrial plants, piping concentrated CO_2 in its gaseous phase to the storage location, injecting it under pressure into the subsurface porous media and constant monitoring and analyzing the fluid's movements over time to ensure it is not escaping (Leung, Caramanna, and Maroto-Valer 2014). During sub-surface injection, supercritical CO_2, with high density but low viscosity characteristics, is stored at substantial depths under high temperatures and pressures that enable it to penetrate deep within the pore-scale distributions of multi-scale porous zones (Kaszuba et al. 2005).

Different subsurface and injection targets have been proposed for sub-surface CO_2 injection into deep saline aquifers, depleted oil and/or gas reservoirs, coal seams that are too deep to mine, and other deep-porous formations (Ajayi et al. 2019). Among these options, oil and gas reservoirs are most likely to offer safe and secure long-term entrapment of CO_2, since they already have trapped large amounts of oil and/or gas. Moreover, gas injection methodologies for EOR are well established and applied (Bickle 2009). Large-scale CCS-EOR plants are now operational, such as the UAE's Masdar project enhancing oil recovery from a mature oil field (Adnoc Website 2017).

Geological carbon sequestration can be effective in climate change mitigation, only if it is optimally implemented from the economic and environmental perspectives. Preventing leakage and sustaining permanency for the carbon sink represent the most challenging issues geological carbon sequestration, and these have been recognized for many years (Haugan and Joos 2004). If depletion rate of the injected CO_2 exceeds 0.1% annually, the process may not successfully achieve its goal to become a long-term carbon sink that can effectively mitigate CO_2 emissions to the atmosphere (Verkerke et al. 2014). Thus, leak detection and monitoring are essential aspects of subsurface CCSU projects. Carbon leakage can possibly migrate over time into overlying aquifers, other oil/gas zones, or back to the Earth's surface. Such outcomes pose substantial risks for ecosystem safety, health by altering pH and inorganic carbon concentrations of drinking-water supplies and/or increasing CO_2 contents of produced oil and gas (Carroll et al. 2014; Mahmoud and Elkatatny 2019). Such outcomes would be costly to rectify and pose significant economic risks for such projects. Also the chemical effects of dissolved CO_2 in sub-surface fluids can have secondary negative consequences (Wilson et al. 2003). For instance, the reactions of wellbore cements with the CO_2-rich and acid fluids can damage their matrix structures and cause fluid leakages into the annulus zones of cased wellbores (Mahmoud and Elkatatny 2019). The World Resources Institute (WRI) issued guidelines for conducting rigorous risk assessments associated with leakage of injected or displaced fluids of any type, including CO_2 (Forbes et al. 2008).

To prevent CO_2 leakage from sub-surface storage zones, multiple actions are required, such as establishing effective and durable cement seals around the CO_2-injection wells (Bagheri et al. 2018). There are multiple well-bore conditions that pose risks to the successful achievement of such an outcome. These include the grade of cement used, the thickness and integrity of the cement sheath installed, cementing procedures and conditions, the age and condition of the wellbore and its steel casing, and internal wellbore plug locations and materials (Li et al. 2017). The mechanical properties of the cement and steel tubes used, the physical properties of the rock formation into which the CO_2 is injected, and prevailing temperature and pressure at the point of injection into the reservoir influence CO_2-leakage likelihood (Roy et al. 2018).

A range of technologies can be implemented to detect and monitor the rate of CO_2 leakage from and through subsurface pressurized zones. At the surface, these include monitoring surface deformation and vegetative stress. Near-surface detection technologies include soil gas analysis, soil CO_2 flux changes, regular soil analyses, eddy covariance changes, CO_2 laser recording, noble gas tracer recordings, airborne hyperspectral and microbial soil genomics (Benson and Myer 2003; Feitz et al. 2014). Sub-surface simulations such as predictive modeling of the injected plume within the reservoir combined with regular well-pressure recordings, and seismic surveys to detect subsurface fluid movements, provide early warnings of adverse fluid movements deep within the subsurface (Benson and Myer 2003). Regular well-pressure monitoring at the reservoir level provides a simple and reliable source of data to history match CO_2-leakage simulations with real data observations (Arora et al. 2019). Reservoir pressure monitoring systems and procedures can be designed to minimize the cost of drilling, instrumenting and operating deep, CO_2-leakage-monitoring wells (Jeong et al. 2018).

15.6 Oceanic Carbon Sequestration

Two-thirds of the Earth is covered by oceans and they represent the largest long-term pool for carbon storage (Hossain 2019). Oceans ultimately uptake approximately 2 of 6 Pg of carbon per year associated with anthropogenically emitted CO_2 (Caldeira et al. 2001). They do this through the chemical reactions that continuously impact the CO_2 equilibrium between seawater and the atmosphere (Eric et al. 2008). This equilibrium plays an important role in controlling the dynamics of the Earth's carbon cycle. This has generated interest in strategies that exploit the oceans for carbon capture and storage. Proposed CCS strategies involving oceans include: i) direct injection of CO_2 into deep oceanic areas; and/or ii) accelerating the organic carbon uptake rate through "biological pumping" which effects phytoplankton's carbon uptake rate and thereby increasing the atmospheric CO_2-diffusion rate into the oceans by effectively shifting in the CO_2-equilibrium state between atmosphere and ocean (Divya et al. 2019).

Different implementation scenarios for potentially injecting CO_2 into the oceans have been proposed: i) release of dry ice at the ocean surface from a transport ship, which has substantial operating costs relating to the transportation and preservation of CO_2 in a solid state; ii) transport of CO_2 in liquid form from onshore capture locations via pipeline pumping and dispersing it through a manifold positioned on the seafloor or deep within the water column, creating a ring plume of CO_2 droplets; iii) generating dense CO_2-seawater mixture within the water column, ideally in the depth range 500 m to 1000 m, where CO_2 is in a liquid phase, thereby limiting its tendency to sink; and iv) tanker transport of liquid CO_2 with release through a discharge nozzle as the tanker sails around a selected discharge zone (Liro et al. 1991; Herzog 1998; Haugan and Drange 1992; Nakashiki 1991; Ohsumi 1995).

Accelerating phytoplankton bloom rate is achieved, in conjunction with CO_2 release into the ocean, by injecting micronutrient fertilizers such as iron and phosphate compounds (Buesseler et al. 2004). The addition of nutrients speeds up the phytoplankton regeneration rate. Once the necessary nutrient input is halted, the artificial phytoplankton bloom would terminate and the newly formed population would grow and sink into the deep layers of the ocean. The sinking biomass is expected to remain in the cold, high-density oceanic zones providing food for the oceanic ecosystem (Markels and Barber 2001).

Although the chemistry and dynamics of oceanic CO_2 storage are feasible, huge environmental and ecosystem risks are associated with all the proposed techniques. Foremost, the increased concentration of CO_2 in the deep layers of the oceans would over time lower the pH, which would very likely affect the ecosystem's behavior causing unpredictable imbalances in the oceanic food chain. Also, the anthropogenically induced increase in ocean acidity is likely to facilitate the dissolution of carbonate skeletons of some marine life such as corals, ultimately leading to their demise. Moreover, some carbonate ocean sediments would also dissolve, leading to increased CO_2 in the ocean and ultimately having an adverse influence on the atmosphere–ocean CO_2 balance. Such actions would likely have negative consequences for oceanic ecosystem diversity (Raj and Lal 2012). This CCS option is not being implemented because of the mentioned risks, adverse public and scientific opinions and uncertainties associated with all its possible impacts (Creutzig et al. 2015).

15.7 Conclusions

A range of technically feasible methods have been extensively evaluated to conduct carbon capture, sequestration (CCS), and in some cases accompanied by carbon utilization (CCSU), in large-scale attempts to reverse the trend of carbon dioxide (CO_2) accumulating in the atmosphere and mitigate its negative environmental and eco-system impacts. The Earth's natural carbon cycle involving the atmosphere, oceans, forests and other terrestrial biomass and soils, is complex and creates equilibrium concentrations of CO_2 in the atmosphere and oceans that vary over geological time scales as conditions vary within elements of that carbon cycle. Massive and growing anthropogenic CO_2 and other greenhouse gas (GHG) inputs to the atmosphere have disrupted the natural carbon cycle and threaten irreversible climate change and eco-system damage within a few decades if not reversed. CCS and CCSU offer techniques that could resolve these problems if implemented rapidly on a large scale, alongside efforts to reduce energy consumption and move from fossil fuel-dominated energy supply to more sustainable renewable energies with minimum environmental footprints.

Effective CCS options include: expanding forests of all kinds; reducing agricultural land and changing farming practices to become more eco-friendly, efficient and avoid deep ploughing; enhancing the carbon pools in soils and the biosphere by boosting biomass production, soil carbonate accumulation and preventing soil erosion; expanding biofuel production from organic waste materials and algal biomass avoiding crop production on agricultural land; geological CO_2 sequestration in subsurface reservoirs; and deep ocean CO_2 sequestration. All of these methods can make a valuable contribution as man-made carbon sinks, except deep ocean sequestration which poses huge risks to ecosystems and the global carbon cycle equilibriums. Some additional techniques can be beneficially implemented as CCSU systems that create both sustainable and economically beneficial outcomes. These include CCS combined with enhanced oil and gas recovery projects in the short term and medium term and bioenergy burning from processes, such as torrefaction, that reduce agricultural and forestry waste. International support policies and large-scale multi-lateral investments, supported by governments and industry, are required to fully exploit the multiple routes for CO_2-emission mitigation offered by CCS and CCSU.

References

Adnoc Website (2017). www.adnoc.ae/en/news-and-media/press-releases/2017/adnoc-and-masdars-carbon-capture-facility-holds-key-to-limiting-industrial-co2-emissions.

Ajayi, T., Gomes, J.S., and Bera, A. (2019). A review of CO_2 storage in geological formations emphasizing modeling, monitoring and capacity estimation approaches. *Petroleum Science* 16: 1028–1063. https://doi.org/10.1007/s12182-019-0340-8

Alemu, B. (2014). The role of forest and soil carbon sequestrations on climate change mitigation. *Research Journal of Agricultural and Environmental Management* 3: 492–505.

Allen, M., Antwi-Agyei, P., Aragon-Durand, F. et al. (2019). Technical Summary: Global warming of 1.5°C. An IPCC Special Report on the impacts of global warming of 1.5°C above pre-industrial levels and related global greenhouse gas emission pathways, in the context of strengthening the global response to the threat of climate change, sustainable development, and efforts to eradicate poverty.

Almberg, K.S., Turyk, M.E., Jones, R.M. et al. (2018). Atrazine contamination of drinking water and adverse birth outcomes in community water systems with elevated atrazine in Ohio, 2006–2008. *International Journal of Environmental Research and Public Health* 15: 1889.

Angelsen, A. (2009). *Realising REDD+: National Strategy and Policy Options*. Cifor.

Arora, V., Saran, R.K., Kumar, R. et al. (2019). Separation and sequestration of CO_2 in geological formations. *Materials Science for Energy Technologies* 2: 647–656.

Azar, C., Lindgren, K., Larson, E. et al. (2006). Carbon capture and storage from fossil fuels and biomass–costs and potential role in stabilizing the atmosphere. *Climatic Change* 74: 47–79.

Bagheri, M., Shariatipour, S.M., and Ganjian, E. (2018). A review of oil well cement alteration in CO_2-rich environments. *Construction and Building Materials* 186: 946–968.

Barbir, F., Veziroğlu, T.N., and Plass, H.J., Jr (1990). Environmental damage due to fossil fuels use. *International Journal of Hydrogen Energy* 15: 739–749.

Batjes, N.H. (2014). Total carbon and nitrogen in the soils of the world. *European Journal of Soil Science* 65: 10–21.

Bellamy, R., Lezaun, J., and Palmer, J. (2019). Perceptions of bioenergy with carbon capture and storage in different policy scenarios. *Nature Communications* 10: 1–9.

Ben-Iwo, J., Manovic, V., and Longhurst, P. (2016). Biomass resources and biofuels potential for the production of transportation fuels in Nigeria. *Renewable and Sustainable Energy Reviews* 63: 172–192.

Benson, S. and Myer, L. (2003). Monitoring to ensure safe and effective geologic sequestration of carbon dioxide. Workshop on Carbon Dioxide Capture and Storage Proceedings, p. 178. Netherlands: ECN.

Bickle, M.J. (2009). Geological carbon storage. *Nature Geoscience* 2: 815–818.

Bloomfield, J. and Pearson, H.L. (2000). Land use, land-use change, forestry, and agricultural activities in the clean development mechanism: Estimates of greenhouse gas offset potential. *Mitigation and Adaptation Strategies for Global Change* 5: 9–24.

Bolin, B., Jager Doos, B.R., Jager, J. et al. (1989). The greenhouse effect, climatic change and ecosystems. 574 pages. John Wiley & Sons. ISBN: 9780471926351.

Botkin, D.B., Saxe, H., Araujo, M.B. et al. (2007). Forecasting the effects of global warming on biodiversity. *Bioscience* 57: 227–236.

Brown, S., Sathaye, J., Cannell, M. et al. (1996). Mitigation of carbon emissions to the atmosphere by forest management. Special Issue: Forests And Climate. The Commonwealth Forestry Review 75 (1): 80–91. https://www.jstor.org/stable/42607279 (accessed 9 December 2021).

Bruinsma, J. (2003). *World Agriculture: Towards 2015/2030: An FAO Perspective*. Earthscan.

Buesseler, K.O., Andrews, J.E., Pike, S.M. et al. (2004). The effects of iron fertilization on carbon sequestration in the Southern Ocean. *Science* 304: 414–417.

Bustamante, M., Robledo-Abad, C., Harper, R. et al. (2014). Co-benefits, trade-offs, barriers and policies for greenhouse gas mitigation in the agriculture, forestry and other land use (AFOLU) sector. *Global Change Biology* 20: 3270–3290.

Byrne, J., Hughes, K., Rickerson, W. et al. (2007). American policy conflict in the greenhouse: Divergent trends in federal, regional, state, and local green energy and climate change policy. *Energy Policy* 35: 4555–4573.

Caldeira, K., Herzog, H.J., and Wickett, M.E. (2001). Predicting and evaluating the effectiveness of ocean carbon sequestration by direct injection. *Proceedings of First National Conference on Carbon Sequestration*, Washington DC, 14–17.

Canadell, J.G., Pataki, D.E., Gifford, R. et al. (2007). Saturation of the terrestrial carbon sink. In: *Terrestrial Ecosystems in a Changing World*. Springer.

Cantonati, M., Poikane, S., Pringle, C.M. et al. (2020). Characteristics, main impacts, and stewardship of natural and artificial freshwater environments: Consequences for biodiversity conservation. *Water* 12: 260.

Carroll, S.A., Keating, E., Mansoor, K. et al. (2014). Key factors for determining groundwater impacts due to leakage from geologic carbon sequestration reservoirs. *International Journal of Greenhouse Gas Control* 29: 153–168.

Chen, W.-H., Lin, B.-J., Huang, M.-Y. et al. (2015). Thermochemical conversion of microalgal biomass into biofuels: A review. *Bioresource Technology* 184: 314–327.

Choubineh, A., Ghorbani, H., Wood, D.A. et al. (2017). Improved predictions of wellhead choke liquid critical-flow rates: Modelling based on hybrid neural network training learning based optimization. *Fuel* 207: 547–560.

Clarke, L.E., Jiang, K., Akimoto, K. et al. (2015). Assessing transformation pathways. In: *Climate Change 2014: Mitigation of Climate Change. Contribution of Working Group III to the Fifth Assessment Report of the Intergovernmental Panel on Climate Change*. Richland, WA): Pacific Northwest National Lab (PNNL).

Cooper, R.N. (1998). Toward a real global warming treaty. *Foreign Affairs* 77: 66.

Creutzig, F., Ravindranath, N.H., Berndes, G. et al. (2015). Bioenergy and climate change mitigation: An assessment. *GCb Bioenergy* 7: 916–944.

Crowther, T.W., Todd-Brown, K.E.O., Rowe, C.W. et al. (2016). Quantifying global soil carbon losses in response to warming. *Nature* 540: 104–108.

d'Annunzio, R., Sandker, M., Finegold, Y. et al. (2015). Projecting global forest area towards 2030. *Forest Ecology and Management* 352: 124–133.

Dehshibi, R., Reza, A.S., Mohebbi, A. et al. (2019). A new insight into pore body filling mechanism during waterflooding in a glass micro-model. *Chemical Engineering Research*

& Design: Transactions of the Institution of Chemical Engineers Part A 151: 100–107. https://doi.org/10.1016/j.cherd.2019.08.019

Divya, M., Dinesh Kumar, S., Krishnaveni, N. et al. (2019). A study of carbon sequestration by phytoplankton. In: *Basic and Applied Phytoplankton Biology*. Springer.

Dixon, R.K. (1995). Agroforestry systems: Sources of sinks of greenhouse gases? *Agroforestry Systems* 31: 99–116.

Dollinger, J. and Jose, S. (2018). Agroforestry for soil health. *Agroforestry Systems* 92: 213–219.

Edenhofer, O. (2015). *Climate Change 2014: Mitigation of Climate Change*. Cambridge University Press.

Eric, S., Robert, B., Stephen, F., Robert, G. et al. (2008). Carbon sequestration to mitigate climate change. In: *US Geological Survey, Fact Sheet*, 3097.

Evans, J.R. (2013). Improving photosynthesis. *Plant Physiology* 162: 1780–1793.

Fasullo, J.T. and Trenberth, K.E. (2008a). The annual cycle of the energy budget. Part I: Global mean and land–ocean exchanges. *Journal of Climate* 21: 2297–2312.

Fasullo, J.T. and Trenberth, K.E. (2008b). The annual cycle of the energy budget. Part II: Meridional structures and poleward transports. *Journal of Climate* 21: 2313–2325.

Feitz, A., Jenkins, C., Schacht, U. et al. (2014). An assessment of near surface CO_2 leakage detection techniques under Australian conditions. *Energy Procedia* 63: 3891–3906.

Feng, H., Kling, C.L., and Gassman, P.W. (2004). Carbon sequestration, co-benefits, and conservation programs. *Choices* 19: 19–24.

Follett, R., Mooney, S., Morgan, J. et al. (2011). Carbon sequestration and greenhouse gas fluxes in agriculture: Challenges and opportunities. In: *Council for Agricultural Science and Technology (CAST)*. Ames.

Fontaine, S., Barot, S., Barré, P. et al. (2007). Stability of organic carbon in deep soil layers controlled by fresh carbon supply. *Nature* 450: 277–280.

Forbes, S.M., Verma, P., Curry, T.E. et al. (2008). Guidelines for carbon dioxide capture, transport and storage. World Resources Institute (27 October 2008). https://www.wri.org/research/guidelines-carbon-dioxide-capture-transport-and-storage (accessed 9 December 2021).

Franklin, J.F., Norman Johnson, K., and Johnson, D.L. (2018). *Ecological Forest Management*. Waveland Press.

Freer-Smith, P.H., Broadmeadow, M.S.J., and Lynch, J.M. (2007). *Forestry and Climate Change*. CAB International.

Greipsson, S. and El-Mayas, H. (2000). Arbuscular Mycorrhizae of *Leymus arenarius* on coastal sands and reclamation sites in Iceland and response to inoculation. *Restoration Ecology* 8: 144–150. doi:10.1046/J.1526-100X.2000.80021.X

Haszeldine, R.S. and Scott, V. (2014). Storing carbon for geologically long timescales to engineer climate. *Geoengineering of the Climate System* 38: 22.

Haugan, P.M. and Drange, H. (1992). Sequestration of $C0_2$ in the deep ocean by shallow injection. *Nature* 357: 318–320.

Haugan, P.M. and Joos, F. (2004). Metrics to assess the mitigation of global warming by carbon capture and storage in the ocean and in geological reservoirs. *Geophysical Research Letters* 31 (18): L18202 (4 pages). https://doi.org/10.1029/2004GL020295

Heidari, S., Amirinejad, M., and Jahangirian, H. (2019). Investigation of fouling mechanisms using surface morphology and physicochemical membrane features. *Chemical Engineering & Technology* 42: 1310–1320.

Heidari, S., Jafarzadeh, Y., Samarin, M.S. et al. (2017). Study on the fouling behavior of HDPE/PE-g-MA/EVA blend membrane fabricated via thermally induced phase separation method. *Polyolefins Journal* 4: 235–251.

Herzog, H.J. (1998). Ocean sequestration of CO_2: An overview. *Proceedings of the AWMA's Second International Specialty Conference*, 13–15 October 1998, Washington DC.

Herzog, T. (2009). *World Greenhouse Gas Emissions in 2005*. World Resources Institute.

Herzog, H. and Golomb, D. (2004). Carbon capture and storage from fossil fuel use. *Encyclopedia of Energy* 1: 277–287.0

Hossain, M.M. (2019). Future importance of healthy oceans: Ecosystem functions and biodiversity, marine pollution, carbon sequestration, ecosystem goods and services. *Journal of Ocean and Coastal Economics* 6: 4.

Houghton, J. (2005). Global warming. *Reports on Progress in Physics* 68: 1343.

Hutchinson, J.J., Campbell, C.A., and Desjardins, R.L. (2007). Some perspectives on carbon sequestration in agriculture. *Agricultural and Forest Meteorology* 142: 288–302.

IEA Bioenergy (2010). Bioenergy, land use change and climate change mitigation. *Report for Policy Advisors and Policy Makers. IEA Bioenergy: ExCo* 3.

Jafari, A., Esmaeilzadeh, F., Mowla, D. et al. (2020). New insights to direct conversion of wet microalgae impregnated with ethanol to biodiesel exploiting extraction with supercritical carbon dioxide. *Fuel*.

Jeong, H., Sun, A.Y., and Zhang, X. (2018). Cost-optimal design of pressure-based monitoring networks for carbon sequestration projects, with consideration of geological uncertainty. *International Journal of Greenhouse Gas Control* 71: 278–292.

Jindal, R., Swallow, B., and Kerr, J. (2008). Forestry-based carbon sequestration projects in Africa: Potential benefits and challenges. In: *Natural Resources Forum*, 116–130. Wiley Online Library.

Jørgensen, S.E. (2010). *Global Ecology*. Academic Press.

Kampa, M. and Castanas, E. (2008). Human health effects of air pollution. *Environmental Pollution* 151: 362–367.

Kanowski, P. (1997). Afforestration and plantation forestry, plantation forestry for 21st century. *World Forestry Congress XI*: 25–37.

Kaszuba, J.P., Janecky, D.R., and Snow, M.G. (2005). Experimental evaluation of mixed fluid reactions between supercritical carbon dioxide and NaCl brine: Relevance to the integrity of a geologic carbon repository. *Chemical Geology* 217: 277–293.

Kozinski, J., Łukasik, R.M., Matsuda, T. et al. (2018). *Supercritical and Other High-Pressure Solvent Systems: For Extraction, Reaction and Material Processing*. Royal Society of Chemistry.

Kruger, P. (2006). *Alternative Energy Resources: The Quest for Sustainable Energy*. Hoboken: Wiley.

Lago, C., Herrera, I., Caldés, N., and Yolanda, L. (2019). Nexus bioenergy–bioeconomy. In: *The Role of Bioenergy in the Bioeconomy*. Elsevier.

Lal, R. (2008a). Carbon sequestration. *Philosophical Transactions of the Royal Society B: Biological Sciences* 363: 815–830.

Lal, R. (2008b). Sequestration of atmospheric CO_2 in global carbon pools. *Energy & Environmental Science* 1: 86–100.

Lal, R. (2016). Soil health and carbon management. *Food and Energy Security*, 5 (4): 212–222. https://doi.org/10.1002/fes3.96

Lal, R., Kimble, J.M., Follett, R. et al. (1998). *Management of Carbon Sequestration in Soils.* Boca Raton FL: CRC.

Le Quéré, C., Peters, G.P., Andres, R.J. et al. (2014). Global carbon budget 2013. *Earth System Science Data* 6: 235–263.

Lehmann, J., Gaunt, J., and Rondon, M. (2006). Bio-char sequestration in terrestrial ecosystems: A review. *Mitigation and Adaptation Strategies for Global Change* 11: 403–427.

Leung, D.Y.C., Caramanna, G., and Mercedes Maroto-Valer, M. (2014). An overview of current status of carbon dioxide capture and storage technologies. *Renewable and Sustainable Energy Reviews* 39: 426–443.

Li, B., Li, H., Zhou, F., Guo, B. et al. (2017). Effect of cement sheath induced stress on well integrity assessment in carbon sequestration fields. *Journal of Natural Gas Science and Engineering* 46: 132–142.

Liao, C., Li, D., Huang, L., Yue, P. et al. (2020). Higher carbon sequestration potential and stability for deep soil compared to surface soil regardless of nitrogen addition in a subtropical forest. *Peer Journal* 8: e9128.

Liro, C.R., Adams, E., and Herzog, H.J. (1991). Modeling the release of CO_2 in the deep ocean. Energy Laboratory Report MIT-EL-91-002, Cambridge USA: Massuchussetts Institute of Technology, 105 pages. http://hdl.handle.net/1721.1/60614 (accessed 9 December 2021).

Luo, Y., White, L.W., Canadell, J.G. et al. (2003). Sustainability of terrestrial carbon sequestration: A case study in Duke Forest with inversion approach. *Global Biogeochemical Cycles* 17 (10): 10211–10213. doi: 10.1029/2002GB001923

Mahmoud, A.A. and Elkatatny, S. (2019). Mitigating CO_2 reaction with hydrated oil well cement under geologic carbon sequestration using nanoclay particles. *Journal of Natural Gas Science and Engineering* 68: 102902.

Markels, M. and Barber, R. (2001). Sequestration of CO_2 by ocean fertilization. *Proceedings of the Ist National Conference on Carbon Sequestration*, Washington DC.

Mather, A. (2001). The transition from deforestation to reforestation in Europe. In: *Agricultural Technologies and Tropical Deforestation* (ed. A. Angelsen and D. Kaimowitz), 35–52. Wallingford UK: CAB International. doi: 10.1079/9780851994512.0035

McHenry, M.P. (2009). Agricultural bio-char production, renewable energy generation and farm carbon sequestration in Western Australia: Certainty, uncertainty and risk. *Agriculture, Ecosystems & Environment* 129: 1–7.

Meinshausen, M., Meinshausen, N., Hare, W. et al. (2009). Greenhouse-gas emission targets for limiting global warming to 2°C. *Nature* 458: 1158–1162.

Mills, R.T., Hoffman, F.M., Kumar, J. et al. (2011). Cluster analysis-based approaches for geospatiotemporal data mining of massive data sets for identification of forest threats. *Procedia Computer Science* 4: 1612–1621.

Mohajan, H. (2011). Greenhouse gas emissions increase global warming. MPRA Paper 50839, University Library of Munich, Germany. https://ideas.repec.org/p/pra/mprapa/50839.html (accessed 9 December 2021).

Moore, T.G. (1998). Health and amenity effects of global warming. *Economic Inquiry* 36: 471–488.

Mudge, F.B. (1997). The development of the "greenhouse" theory of global climate change from Victorian times. *Weather* 52: 13–17.

Nakashiki, N. (1991). Sequestering of Co_2 in a deep-ocean: Fall velocity and dissolution rate of solid CO_2 in the ocean. *CRIEPI (Central Research Institute of Electoric Power Industry) Report* 91003: 1–19.

National Academies of Sciences, Engineering, and Medicine (2018). *Bioenergy with Carbon Capture and Storage Approaches for Carbon Dioxide Removal and Reliable Sequestration: Proceedings of a Workshop: in Brief.* Washington DC: The National Academies Press.

Nezhad, S.M., Wood, D.A., Sadatshojaei, E. et al. (2020). New insight to experimental study of ionic solutions with a nonionic surfactant on wettability, interfacial tension and micro-model flooding. *Fuel* 285: 119126. https://doi.org/10.1016/j.fuel.2020.119126

Nicosia, C. and Stoops, G. (2017). *Archaeological Soil and Sediment Micromorphology*. John Wiley & Sons.

Ohsumi, T. (1995). CO_2 disposal options in the deep sea. *Marine Technology Society Journal* 29: 58–66.

Olson, J.S., Garrels, R.M., Berner, R.A. et al. (1985). The natural carbon cycle. *Atmospheric Carbon Dioxide and the Global Carbon Cycle* 239: 175–214.

Paul, E.A., Collins, H.P., and Leavitt, S.W. (2001). Dynamics of resistant soil carbon of Midwestern agricultural soils measured by naturally occurring ^{14}C abundance. *Geoderma* 104: 239–256.

Pereira, J.S. and Landsberg, J.J. (2012). *Biomass Production by Fast-Growing Trees.* Springer Science & Business Media.

Plantinga, A.J. and Wu, J. (2003). Co-benefits from carbon sequestration in forests: Evaluating reductions in agricultural externalities from an afforestation policy in Wisconsin. *Land Economics* 79: 74–85.

Powlson, D.S., Whitmore, A.P., and Goulding, K.W.T. (2011). Soil carbon sequestration to mitigate climate change: A critical re-examination to identify the true and the false. *European Journal of Soil Science* 62: 42–55.

Pulighe, G., Bonati, G., Colangeli, M. et al. (2019). Ongoing and emerging issues for sustainable bioenergy production on marginal lands in the Mediterranean regions. *Renewable and Sustainable Energy Reviews* 103: 58–70.

Raj, A.J. and Lal, S.B. (2012). *Forestry Principles and Applications.* India: Scientific Publisher.

Raj, A.J. and Lal, S.B. (2013). *Forestry Principles and Applications.* Scientific Publishers.

Ravindranath, N.H. and Ostwald, M. (2007). *Carbon Inventory Methods: Handbook for Greenhouse Gas Inventory, Carbon Mitigation and Roundwood Production Projects.* Springer Science & Business Media.

Rose, S.K., Kriegler, E., Bibas, R. et al. (2014). Bioenergy in energy transformation and climate management. *Climatic Change* 123: 477–493.

Roy, P., Morris, J.P., Walsh, S.D.C. et al. (2018). Effect of thermal stress on wellbore integrity during CO_2 injection. *International Journal of Greenhouse Gas Control* 77: 14–26.

Rumpel, C. and Ingrid, K.-K. (2011). Deep soil organic matter: A key but poorly understood component of terrestrial C cycle. *Plant and Soil* 338: 143–158.

Sadatshojaei, E. and Wood, D.A. (2020). Water, the most accessible green solvent, extraction and separation agent. In: *Green Sustainable Process for Chemical and Environmental Engineering and Science: Environmental Remediation*. Elsevier.

Sadatshojaei, E., Mowla, D., and Wood, D.A. (2020a). Review of progress in microalgal biotechnology applied to wastewater treatment. In: *Sustainable Green Chemical Process and Their Allied Applications*. Springer Nature.

Sadatshojaei, E., Wood, D.A., and Mowla, D. (2020b). Third generation of biofuels exploiting microalgae. In: *Sustainable Green Chemical Process and Their Allied Applications*. Switzerland AG: Springer Nature.

Sadatshojaei, E., Esmaeilzadeh, F., Fathikaljahi, J. et al. (2018). Regeneration of the midrex reformer catalysts using supercritical carbon dioxide. *Chemical Engineering Journal* 343: 748–758.

Sadatshojaei, E., Jamialahmadi, M., Esmaeilzadeh, F. et al. (2016). Effects of low-salinity water coupled with silica nanoparticles on wettability alteration of dolomite at reservoir temperature. *Petroleum Science and Technology* 34: 1345–1351.

Sadatshojaei, E., Jamialahmadi, M., Esmaeilzadeh, F. et al. (2019). The impacts of silica nanoparticles coupled with low-salinity water on wettability and interfacial tension: Experiments on a carbonate core. *Journal of Dispersion Science and Technology* 41 (8): 1159–1173. https://doi.org/10.1080/01932691.2019.1614943

Sadatshojaie, A., Wood, D., and Rahimpour, M.R. (2020). Applying ultrasonic fields to separate water contained in medium-gravity crude oil emulsions and determining crude oil adhesion coefficients. *Ultrasonics Sonochemistry* 70: 105303. https://doi.org/10.1016/j.ultsonch.2020.105303

Salinger, M.J. (2007). Agriculture's influence on climate during the Holocene. *Agricultural and Forest Meteorology* 142: 96–102.

Sanaullah, M., Afzal, T., Shahzad, T. et al. (2019). Carbon sequestration for sustainable agriculture. In: *Innovations in Sustainable Agriculture*. Springer.

Santos, F.M., Gonçalves, A.L., and Pires, J.C.M. (2019). Negative emission technologies. In: *Bioenergy with Carbon Capture and Storage*. Elsevier.

Sarmiento, J.L. and Gruber, N. (2002). Sinks for anthropogenic carbon. *Physics Today* 55: 30–36.

Schimel, D.S. (1995). Terrestrial ecosystems and the carbon cycle. *Global Change Biology* 1: 77–91.

Schwarz, M., Folini, D., Yang, S. et al. (2019). The annual cycle of fractional atmospheric shortwave absorption in observations and models: Spatial structure, magnitude, and timing. *Journal of Climate* 32: 6729–6748.

Seddiqi, H., Sadatshojaie, A., Vaferi, B. et al. (2019). Mathematical model for Iron corrosion that eliminates chemical potential parameters. *Chinese Journal of Chemical Engineering* 28 (2): 603–612. https://doi.org/10.1016/j.cjche.2019.09.007

Shao, W., Keim, B.D., Garand, J.C. et al. (2014). Weather, climate, and the economy: Explaining risk perceptions of global warming, 2001–10. *Weather, Climate, and Society* 6: 119–134.

Shokri, E., Yegani, R., Heidari, S. et al. (2015). Effect of PE-g-MA compatibilizer on the structure and performance of HDPE/EVA blend membranes fabricated via TIPS method. *Chemical Engineering Research and Design* 100: 237–247.

Siry, J.P., Cubbage, F.W., and Ahmed, M.R. (2005). Sustainable forest management: Global trends and opportunities. *Forest Policy and Economics* 7: 551–561.

Tarnocai, C., Canadell, J.G., Schuur, E.A.G. et al. (2009). Soil organic carbon pools in the northern circumpolar permafrost region. *Global Biogeochemical Cycles* 23 (2): GB2023 pages 1 to 11. https://doi.org/10.1029/2008GB003327

Trabalka, J.R., Edmonds, J.A., Reilly, J.M. et al. (1985). Human alterations of the global carbon cycle and the projected future. In: *Atmospheric Carbon Dioxide and the Global Carbon Cycle* (ed. J.R. Trabalka), 247–287. DOE/ER-0239. Washington DC: US Department of Energy.

Tyndall, J. (1872). *Contributions to Molecular Physics in the Domain of Radiant Heat.* Longmans, Green & Company.

Ussiri, D.A.N. and Lal, R. (2017). Greenhouse gas mitigation under agriculture and livestock landuse. In: *Carbon Sequestration for Climate Change Mitigation and Adaptation.* Springer.

Van Der Ploeg, F. and Rezai, A. (2017). Cumulative emissions, unburnable fossil fuel, and the optimal carbon tax. *Technological Forecasting and Social Change* 116: 216–222.

Verheijen, F., Jeffery, S., Bastos, A.C. et al. (2010). Biochar application to soils. A critical scientific review of effects on soil properties, processes, and functions. *EUR* 24099: 162.

Verkerke, J.L., Williams, D.J., and Thoma, E. (2014). Remote sensing of CO_2 leakage from geologic sequestration projects. *International Journal of Applied Earth Observation and Geoinformation* 31: 67–77.

Vermeulen, S.J., Campbell, B.M., and Ingram, J.S.I. (2012). Climate change and food systems. *Annual Review of Environment and Resources* 37: 195–222. https://doi.org/10.1146/annurev-environ-020411-130608

Verrecchia, E.P. (2011). Pedogenic carbonates. In: *Encyclopedia of Geobiology* (ed. J. Reitner and V. Theil), 721–725. Encyclopedia of Earth Sciences Series. Springer, Dordrecht. https://doi.org/10.1007/978-1-4020-9212-1_161

Watson, R.T., Noble, I.R., Bolin, B. et al. (2000). *Land Use, Land-use Change and Forestry: A Special Report of the Intergovernmental Panel on Climate Change.* Cambridge University Press.

Wilson, E.J., Johnson, T.L., and Keith, D.W. (2003). *Regulating the Ultimate Sink: Managing the Risks of Geologic CO_2 Storage.* ACS Publications.

Young, A. (1989). Agroforestry for soil conservation. 276 pages. Wallingford UK: CAB International. ISBN 0 85198 648 X

16

Nanotechnology for Future Sustainable Plant Production Under Changing Environmental Conditions

Ayesha Tahir[1], *Jun Kang*[2], *Jaffer Ali*[1], *Ameer Bibi*[3] *and Shabbar Abbas*[1]

[1] *Department of Biosciences, Park Road, COMSATS University Islamabad (CUI), Islamabad, Pakistan*
[2] *School of Life Sciences, Tianjin University, Nankai District, Tianjin, China*

CHAPTER MENU

16.1 Introduction

The agriculture sector is important throughout the world because it provides for key and basic needs for food and related food industries. A number of challenges are posed to the

Bioresource Technology: Concept, Tools and Experiences, First Edition. Tanveer Bilal Pirzadah, Bisma Malik, Rouf Ahmad Bhat, and Khalid Rehman Hakeem.
© 2022 John Wiley & Sons Ltd. Published 2022 by John Wiley & Sons Ltd.

agricultural sector due to injudicious use of nonrenewable energy resources, runoff of nutrients, the greenhouse effect and global warming. Climate change is causing unpredictable losses to agricultrural production, a problem which requires immediate attention owing to the increased demand for food by an ever-increasing population (Roychoudhury et al. 2008). However, due to the post-green revolution era and overutilization of high yielding hybrid crops, the current attainable yield has already reached a plateau. Unfortunately, conditions have been worsened exponentially by excessive use of chemical fertilizers, which have caused deterioration of soil quality by loss of minerals and elements in the soil. Therefore, environmentally-friendly technology is required to improve agricultural productivity and quality, and to meet the growing demand for food is of utmost importance (Fischer et al. 2014; Ray et al. 2013). Nowadays, many modern approaches are in practice that help to enhance the sustainability of the agricultural system.

Nanotechnology is a revolutionary field, with unique characteristics of NPs, have opened up new doors for extensive and precise research on animals, microorganisms and plants. NPs can be defined as any materials with a size of less than 100 nm in at least one dimension (Thomas et al. 2012). Scientists are optimizing different synthetic techniques for NPs of different nature, i.e., organic, physical, metallic, inorganic and hybrid, with a set of desired characteristics, i.e., optical, physical, chemical or biological (Nanjwade et al. 2011; Thomas et al. 2012) to increase their implication in food security. Key objectives of nanotechnology application in the agricultural sector under the present climate change scenario include limited spreading of chemicals, minimizing nutrient loss in fertilization and increased yield through pest and nutrient management. Other aspects may include monitoring the soil quality and plant health through the use of nano sensors. This chapter's main concern is about summarizing the synthesis of nanomaterials and their role in productivity enhancement under rapidly-changing environmental conditions.

16.2 Nanotechnology and Synthesis of Nanomaterials

In the literature, synthesis of NPs is based on two principles, i.e., top down and bottom up approaches. The top down approach refers to milling and sputtering methods adopted to break down the large blocks of material into the smallest particles (Arulmani et al. 2018). In the bottom up approach, chemical vapor deposition, sol–gel processes, spray pyrolysis, laser pyrolysis and atomic/molecular condensation are used to grow NPs from simpler molecules, as described by Arulmani et al. (2018).

NPs can be categorized on the basis of shape, size, chemical nature and mode of action, etc. (Khan et al. 2019). Different types of NPs may include phyto-NPs, metallic NPs, polymeric (organic) NPs and carbon-based NPs, etc. Carbon (nanotubes, fullerenes, etc.) and metal (metal oxides, salts)-based NPs have a broad range of uses, whereas those forged by green synthesis are becoming popular owing to their environmentally-friendly properties. NPs can be synthesized chemically, physically and biologically.

16.2.1 Chemical Methods

It is possible to prepare diverse materials through variation in chemical composition and the reaction environment in the chemical synthesis of nanomaterials. Different methods like sol–gel, polysynthesis, hydro-solvo thermal and microemulsion techniques are major among chemical-based nanomaterials synthesis approaches. The chemicals of the nanosynthesis include metallic precursors, and organic and inorganic stabilizing agents. Reducing agents such as sodium citrate, ascorbate, sodium borohydride ($NaBH_4$), elemental hydrogen, polyol process, tollens reagent, N,N-dimethylformamide (DMF) and poly(ethylene glycol)-block copolymers are commonly used (Arulmani et al. 2018; Zhang et al. 2016).

16.2.2 Physical Methods

Physical methods involve the top down approach that includes different strategies to break down the bulk materials to the small particles. This has the advantage of being solvent-free but consumes time and energy for synthesis if NPs. Electricalchemical methods such as ultrasonication and microwave irradiation, etc. sre opted for size reduction. Laser removal and evaporation condensation (performed in a tube heater) are the most essential physical methods to develop NPs. For example, the source material is first vaporized into the bearer gas and utilized to synthesize NPs of Ag, PbS, Au and Cd through the dissipation build-up process (Mathur et al. 2018). Methods other than laser pyrolysis include mechanical methods, inert gas techniques and pulse vapor deposition.

16.2.3 Biological Methods (Green Synthesis)

Green synthesis, a term relating the NPs to plants and their metabolites, is identified as the safe route to orchestrate the NPs for their application as compared to the more toxic traditional methods. This approach uses ideal solvents of organic nature and metallic NPs encompass biological materials, e.g., bacteria, fungi and plant extracts, etc. Accordingly, the use of vitamins, amino acids and plants extracts is being becoming popular nowadays. The process is efficient for being easy and simple to produce NPs on a large scale. The methodology is based on biological precursors and reaction conditions such as pH, solvent used, temperature and pressure. For the synthesis of metal-based NPs, a diversity of plant species with effective phytochemicals are considered, such as ketones, aldehydes, flavones, amides, terpenoids, carboxylic acids, phenols and ascorbic acids.

16.2.3.1 Plant Extract-based Synthesis of Nanomaterials

Plants have biomolecules like carbohyderates, proteins and coenzymes which facilitate the conversion of metal ions into NPs. Gold and silver nanoparticles were first investigated in plant extract assisted synthesis using oat (*Avena sativa*), alfalfa (*Medicago sativa*) and neem (*Azadirachta indica*), *In vivo* synthesis of NPs like zinc, nickel, cobalt and copper was also observed in mustard (*Brassica juncea*), alfalfa (*Medicago sativa*) and sunfower (*Helianthus annuus*) (Marchiol 2012). ZnO NPs have been prepared using a great variety of plant leaf extracts such as coriander (*Coriandrum sativum*) (Anastas and Warner 1998) and crown flower (*Calotropis gigantean*) (Din et al. 2018). The phytochemical composition of leaf extracts such as flavones, terpenoids, sugars, ketones, aldehydes, carboxylic acids and amides are NPs reducers. In *Ocimum basilicum*, this method was used to construct silver NPs (Ahmad et al. 2010). Different sizes and shapes of metallic NPs were modulated by glucose sugar, while fructose has been used to synthesize gold and silver NPs (Panigrahi et al. 2004). Similarly, nearly all 20 types of amino acids are observed to reduce Au metal ions, while cystine, lysine, arginine and methionine have the ability to bind silver ions (Tan et al. 2010). Additionally, functional groups such as -NH_2, -C-O-, -C = C-, available in the phytochemicals, can also generate metallic NPs (Li et al. 2007). The role of different NPs in productivity enhancement is presented in the Table 16.1.

16.2.3.2 Microorganism-based Synthesis of Nanomaterials

Cell walls of the microorganisms can interact with the metals during their synthesis, which is a complex interaction (Bhattacharya and Mukherjee 2008). Microorganism-assisted biosynthesis includes the withholding of the ions from their surroundings and transforming them into their elemental metal through enzymatic activity. This formation can be grouped into inter- and extra-cellular synthesis in the presence of the enzymes (Zhang et al. 2011).

The properties of bacteria to reduce metal ions, and feasibility in bioleaching and genetic engineering, have made it a momentous candidate in NPs preparation. Bacterial

Table 16.1 Green synthesized NPs in Plant productivity enhancement.

NPs	Source of NPs	Treated plant	Trait Improved	Reference
Fe_2O_3	Pistachio leaf extract	Tomatoes	Vigor, Shoot length, Biomass	(Abusalem et al. 2019)
AgNPs	Kafr lime leaf extract	Rice	Germination performance, Seedling vigor	(Mahakham et al. 2017)
AgNPs	Onion peel extract	Watermelon	Germination rate	(Acharya et al. 2020)
AuNP	Onion extract-based synthesis	Onion	Seedling growth, Yield, Quality	(Acharya et al. 2019)
TiO_2	Leaf extract of *Melia azedarach*	*Pennisetum glaucum* seed	Germination	(Rafique et al. 2019)
ZnO	Root extract of *Sphagneticola trilobata*	Fenugreek seeds	Germination and growth promotion	(Shaik et al. 2020)
Ag	*Pseudomonas rhodesiae*	Sweet potato	Plant disease management	(Hossain et al. 2019)
Ag	*Penicillium duclauxii*	Sorghum	*Bipolaris sorghicola*	(Almaary et al. 2020)
Ag	*Trichoderma harzianum*	Cabbage	*Sclerotinia sclerotiorum*	(Guilger et al. 2017)
Cu	*Streptomyces griseus*	Tea	Antifungal agent	(Ponmurugan et al. 2016)

strains such as *Escherichia coli, Lactobacillus casei, Bacillus cereus, Aeromonas sp. SH10 Phaeocystis antarctica, Pseudomonas proteolytica, B. amyloliquefaciens, B. indicus, B. cecembensis, Enterobacter cloacae, Geobacter spp., Arthrobacter gangotriensis, Corynebacterium sp. SH09* and *Shewanella oneidensis* have been exploited for the synthesis of silver NPs. Likewise, other bacterial species such as *B. megaterium D01, E. coli, B. subtilis 16*, and *Plectonema boryanum UTEX 485* have been used in preparation of gold NPs (Gericke and Pinches 2006; Iravani 2014; Thakkar et al. 2010).

Yeast, a single-celled microorganisms of almost 1500 species, has also been identified for its potential in the synthesis of NPs. Silver and gold NPs are synthesized from a *Saccharomyces cerevisiae* broth (Niknejad et al. 2015). Fungi are easy to use and produce metabolites excessively, enabling their use in production of myco-NPs (Parkash et al. 2020). Myco-NPs have a number of applications in the fields of medicine, energy, industry, environment and agriculture, accrediting nanotechnology as their important and integral part. Competent fungi due to the presence of a variety of intracellular enzymes can synthesize larger amounts of NPs compared to bacteria. The important thing in myco-synthesis is that the production of NPs can be controlled to generate NPs of required size. Its mycelial body can better withstand the conditions of the bioreactor than bacteria and plants (Guilger-Casagrande and Lima 2019).

16.3 Potential Applications of Nanotechnology in Agriculture for Climate Resilient Crops

Controlled release of NPs improves plant growth, productivity and yield. Precision farming, being a smart diagnostic tool due to successful application of nanotechnology, can also promote stress tolerance (Shang et al. 2019). Nowadays, NP application in agriculture and other environmental fields are widely accepted in the diversity of scientific fields, including agriculture and environment, as presented in Figure 16.1.

16.3.1 Nanotechnology and Efficient Use of Input Resources

Yield increase is the ultimate target of any breeding program. In the absence of biotic and abiotic stresses, yield is maximum for a given crop. Under changing climatic conditions, limited radiation, water, nutrients and enhanced pressure due to biotic agents is negatively impacting yields (Lobell et al. 2009). The beneficial role of nanotechnology to efficiently utilize the input resources (water, light, nutrients), enhances vigor, productivity and tolerance to stress conditions.

16.3.1.1 Water Use Efficiency Enhancement

Water plays a key role in crop yield, and water use efficiency (WUE) in crop plants can be enhanced in the cropping systems, such as with a controlled environment agricultural system. The following options have near-term potential to improve WUE in crops, i.e., improved plant resilience, soil water-holding capacity and reuse of water through alternative water sources. NPs have the ability to enhance the water retention ability of the crops. As discussed by Giraldo et al. (2019), plant-based sensors can be used to monitor

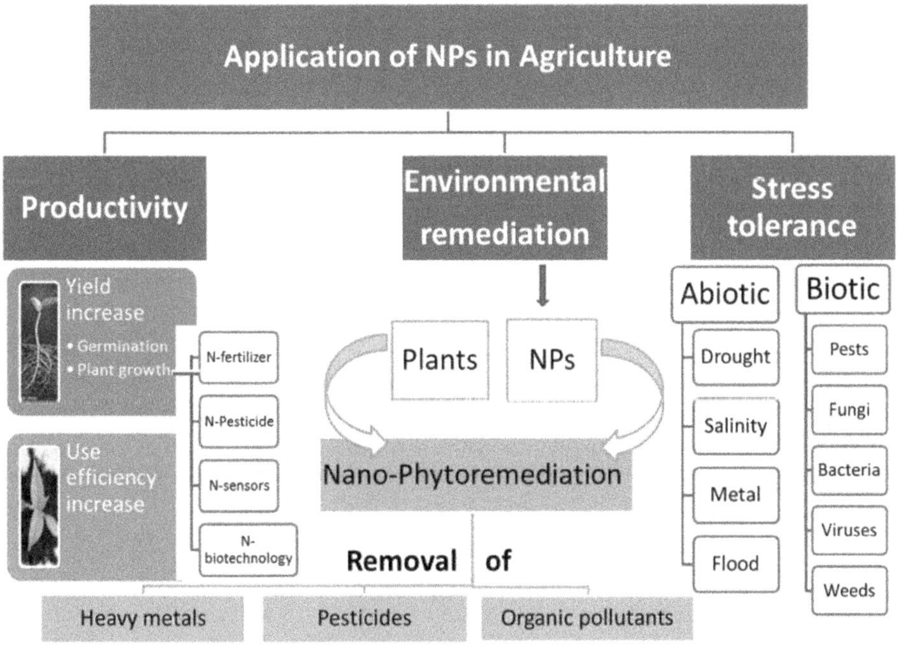

Figure 16.1 Application of nanotechnology in agriculture and environmental science.

the individual plant needs that can help to save water. NPs help to make the coupling production possible, which can minimize crop loss due to water-related stress. The use of nanocarbon made from coconut shell has been found to increase the water-holding capacity of soil on the Chinese Loess Plateau (Zhou and Chen 2017).

16.3.1.2 Light Use Efficiency Enhancement

The yield of a crop is the product of several components, including the proportion of photosynthetically active radiations (PARS), with conversion efficiency as the major one (Monteith et al. 1977). Foliar applied engineered nanomaterials (ENMs) can increase plant photosynthesis efficency by various mechanisms: i) increasing chlorophyll content (Willett et al. 2019); ii) enhancing light absorption; and iii) photon use (Xiong et al. 2018) assisting in electron transfer (Raliya et al. 2015). The single-walled nanotubes (SWNTs) have discrete optical and electronic properties and a broad range of absorption spectra (ultraviolet, visible and near-infrared). The enhancement of light reaction after the insertion of SWNTs in chloroplasts isolated from commercially available baby spinach leaves (*Spinacia oleracea L.*) has been observed (Giraldo et al. 2014).

16.3.1.3 Nutrient Use Efficiency Enhancement

The nanobiotechnology (NBT) platform provides new delivery strategies that ensure controlled release of nutrients. NPs can serve as a nutrient or as a platform to deliver the nutrient. Nanofertlizer is a product made with NPs to improve crop nutrition more efficiently (Priyanka et al. 2020; Subramanian et al. 2015), catalyze soil fertility (Varma and

Uma 2017), promote enzymatic activity in plants (Yang et al. 2017) and increase chlorophyll content and seed germination. Nanofertilizers are of three types: nanoscale fertilizer (NPs which contain nutrients), nanoscale additives (traditional fertilizers with nanoscale additives) and nanoscale coating (traditional fertilizers coated or loaded with NPs).

Nanofertilizers may contain nano zinc, silica, iron and titanium dioxide, ZnCdSe/ZnS core shell QDs, InP/ZnS core shell QDs, Mn/ZnSe QDs, gold nanorods, core shell QDs, etc., and should facilitate in controlled release of nutrients with improved quality (Prasad et al. 2014). Zeolite-based NPs are also capable of releasing the nutrients slowly, thereby preventing the loss of nutrients through denitrification and volatilization, etc. Encapsulation of fertilizer elements and compounds inside the inorganic nanomaterials have enhanced ferillizer availability to the crops. CNTs, SiO_2 and synthetic zeolite are NPs used for this purpose (Ditta and Arshad 2016). NPs used as nano-fertillizers are absorbed by plant roots and uptaken by xylem and throughout the vascular tissues, subsequently transferring them to other plant parts, e.g., mesoporous silica NPs, and SiO_2 is transported inside the plants through apoplast and symplast pathways.

Plant growth-promoting rhizobacteria (PGPR) are bacterial species which have a symbiotic relation with roots of leguminous plants to develop nodules for nitrogen fixation. PGPR is a supreme technique to deliver suitable microbes to soil but their maintenance in the soil causes problems. Carrier nanomaterials have the ability to encapsulate inoculum, protect it from environmental stress and allow control dispersion upon the degradation of these polymers by soil microbes. The effect of nanoencapsulation (silica NPs and CNTs) of the PGPR enhancing the root and shoot propagation was observed by Pour et al. (2019). Methodology of seed coating by micro polymers containing immobilized microorganisms can increase the production of soybean in an environmentally-friendly manner by the application of microbial inoculation (De Gregorio et al. 2017).

16.3.2 Nanomaterials and Plant Growth Enhancement

16.3.2.1 Germination and Vigor Enhancement

NPs are found to have positive effects on germination as well as plant growth and development. These properties are further affected by factors such as moisture availability and soil fertility. The crop-wise optimization of priming agent (pure water, polyethylene glycol, nutrients and inorganic salts) and its treatment time are crucial to improve seed vigor. The enhancement in germination and reduction in aging of rice seeds was observed through application of Ag-NPs (Mahakham et al. 2017). Nanomaterials (e.g., nano-TiO_2, CNT) help plants to absorb water, regulate permeability of water, activate water channel protein genes and improve seed germination (Khodakovskaya et al. 2011; Zaytseva and Neumann 2016).

The rice seeds primed in FeS-NFs showed a germination percentage of more than 90% with a significant increase in shoot length (Ahuja et al. 2020). Nanoiron pyrite treatment (FeS_2) of spinach hastened germination was observed by Das et al. (2016). Nano ZnO has also been reported to enhance germination in various plant species (Faizan et al. 2020; Reddy Pullagurala et al. 2018). Early germination was observed after seed treatment with MWCNTs, while seed and seedling tolerance was increased in *Alnus miller* and wheat

(Joshi et al. 2018; Rahimi et al. 2016). Single-walled carbon nano-horns (SWCNHs) have enhanced seed germination in tomato, soybean and switchgrass (Milewska-Hendel et al. 2016). The enhancement in dry weights/plant biomass and sulfur content was observed in the plantlets developed from the seeds pretreated with FeS_2 (Srivastava et al. 2014). Seed treatment with low concentrations of SiO_2 NPs has enhanced the germination of tomato seeds along with an increase in the vigor, fresh weight and dry weight of tomato seedlings (Siddiqui and Al-Whaibi 2014).

NPs have been found to increase plant growth and photosynthetic quantum under various environmental conditions. Nano-TiO_2 enhanced growth by enhancing nitrogen metabolism, photosynthesis and an increase in cell growth through cell cycle regulation (Khodakovskaya et al. 2011). The mixture SiO_2 and TiO_2 NPs applied to *Glycine max* enhanced the absorption potential of fertilizer, water and nitrate reductase, encouraged the mechanism of antioxidant defense, and accelerated both seed germination and seedling growth (Lin and Xing 2007).

16.3.3 Nanoparticles to Mitigate Biotic Stresses

16.3.3.1 Nano-Pesticides

Measures against damage by insects, pests and plant protection have always been a challenge to plant protection departments. The environmental problems that arise due to the unmonitored and exccessive use of pesticides is of serious concern for the scientist and the public. Pesticides encapsulated in the nanomaterial are considered environmentally-friendly as they help to avoid premature degradation, reduce dosage and increased efficacy for longer periods of time (Prasad et al. 2017). Red mite, one of the major pests of agricultural crops, is significantly controlled by the zein NPs encompassing citronella essential oils (EOs) (JLd et al. 2018). Similarly, the *Sitophilus oryzae* adult showed significant mortality after exposure to neem oil base emulsion encapsulated azadirachtin due to increased efficacy and stability of coated insecticide (Choupanian Choupanian et al. (2017).

16.3.3.2 Nano-Fungicides

Globally, plant diseases developed by pathogenic fungi cause $45 billions worth of crop loss every year. Among devastating soil-borne fungi, *Rhizoctonia solani, Macrophomina phaseolina, Sclerotium rolfsii* and *Fusarium oxysporum* are prominent to reduce the crop yield. Nanoformulations of clotrimazole were found effective against *R. solani*, while miconazole nitrate nanoformulations were effective against *M. phaseolina* and *F. oxysporum* (Tippannanavar et al. 2020). ZnO-NPs synthesized by chemical and green synthesis methods extracted from garlic (*Allium sativum*) and their antifungal potential were tested on the two pathogenic fungi (*Mycena citricolor* and *Colletotrichum sp.*). The nanomaterials showed 97% and 93% inhibition in growth of *M. citricolor*, and *Colletotrichum* sp., respectively (Arciniegas-Grijalba et al. 2019).

Copper NPs also have the potential to protect plants from diseases in a sustainable manner and act as anti-fungal and anti-bacterial agents (Kasana et al. 2017). Green-synthesized copper NPs are reported as potential fungicides against *Neofusicoccum* sp., *F. solani*, and *F. oxysporum* (Pariona et al. 2019).

16.3.4 Nanomaterials to Mitigate Abiotic Stresses

16.3.4.1 Nanoparticles to Mitigate Drought Stress

Water stress conditions like flood or drought, affect various aspects of the metabolism of plant physiology (Aqaei et al. 2020a). Nanotechnology promises a significant role to mitigate drought stress. Several mechanisms of abiotic stress tolerance may include H^+ATPase activity triggered by nanozerovalent iron (nZVI). The process in turn increases leaf area, stomatal aperture and decreased pH, and enhances drought tolerance (Stefaniuk et al. 2016). In plants, NPs of ZnO may enhance the salinity and drought tolerance. Potassium performs a key role in the reduction of detrimental effects caused by drought, with its central function in enzyme reactions, respiration and its advantageous impact on plant resistance under a stressed environment (Wang et al. 2015). Drought stress results in reduced concentrations of Si, P, Zn, Ca, Mn and Fe within the plant seed and N, Cu, P, Zn, Ca, Si, Fe and Mn concentrations within the plant shoot (Zörb et al. 2014). The usage of potassium nano-silica (PNS) under drought stress enhanced plant growth and mineral concentrations in seed and shoot of maize plants that were reduced in control (Aqaei et al. 2020). The Fullerol NPs reduce sugarbeet's oxidative stress by functioning like an additional intercellular supply of water (Shafiq et al. 2019). The specific roles of Si and TiO_2 NPs were also identified by Pei et al. (2009) and Jaberzadeh et al. (2013), respectively.

16.3.4.2 Nanoparticles to Mitigate Metal Stress

Numerous NPs replenish or mitigate the heavy metals and inorganic and organic soil contaminants (Yadav et al. 2017). Environmental application-based nanomaterials were created for their wide surface range and usefulness in remediation and reactivity as compared to their bulky shape, enabling their effective passage through the sites of contamination (Sherry Davis et al. 2017).

NPs are unique due to their greater adsorption abilities, leading to increased usage in heavy metal elimination from the soil (Cai et al. 2019). Alumina NPs-induced antioxidant enzymes resulted in enhanced seedling growth and helped to mitigate the phytotoxic effects of the aluminium ions through metabolic processes (Amist et al. 2017). A significant decrease in the growth of barley occurred by exposure to nano-NiO. The co-exposure to nano-NiO and nano-SiO_2 experimentally confirmed lower reduction in fresh and dry biomass. Nano-SiO_2 is suspected to have reduced the anions produced as a result of nano-NiO through reversion (Soares et al. 2018).

For the removal of chromium in soils contaminated by tannery waste, the nano-scale zerovalent iron (nZVI) combined with carboxy-methyl cellulose (CMC) is used because of its properties to enhance reduction efficiency of Cr. Nanosized zerovalent iron plays a substantial role in heavy metal degradation and lowers Zn and Pb concentrations compared to untreated soils (Gil-Díaz et al. 2014). It is also stated that 10% of nano-scale zerovalent iron substantially lowers the arsenic uptake in a heavily polluted brownfield soil by decreasing the arsenic concentration in the available portion while enhancing the residual portion (Gil-Díaz et al. 2017).

Nano-phytoremediation combines nanotechnology and phytoremediation to remediate the contaminants in the environment such as for heavy metals mitigation from the polluted soil. Some studies show that the NP applications, under both *in-situ* and *ex-situ* conditions, can enhance plant stress tolerance, subsequently promoting potential of phytoremediation

(Pillai and Kottekottil 2016). Combined nano-phytoremediation treatments have potential to remediate the contaminated soils (Figure 16.2). Living plants utilized for the extraction and elimination of pollutants from the soil, air or water is known as phytoremediation (Figure 16.2). The utilization of the plant may be either direct or indirect for the elimination or degradation of water and soil pollutants (Cunningham et al. 2018).

Silver NPs greatly increase the production of phytohormones (ABA and GA), assisting plants to cope with stresses and increasing water and nutrient absorption for growth improvement (Khan and Bano 2016). Fipronil (phenyl pyrazole insecticide)-contaminated water and soil was phytoremediated by a combination of *Plantago major* and green synthesized AgNPs (Romeh 2018). The application of ZnONPs carrying a novel plant growth promoter to increase plant productivity resulted in an increase in growth, biomass and *chlorophyll a* as compared to control in cotton (Venkatachalam et al. 2017).

The application of Mn-NPs in the soil enhanced the uptake and assimilation of nitrogen, and consequently there were positive impacts on shoot and root elongation of plants (Pradhan et al. 2014).

In addition to the above-described metal stress remediation strategies, reduced mobility of heavy metals in contaminated soil can decrease potential plant bioavailability of heavy metals (Chang et al. 2014) and reduce their risk of entering the food chain (Lu et al. 2014). Biochar (Wang et al. 2020), fly ash (Dang et al. 2019) and phosphate minerals (Saxena and D'Souza 2006) have been investigated for immobilization of heavy metals in soils. The acute toxicity of Hg has made it a priority control pollutant (Clarkson and Magos 2006). Conversion of Hg to MeHg (form available to the plants) is one of the solutions to reduce its damage to the ecosystem as well as to human health (Beckers and Rinklebe 2017; Liu et al. 2019; Ullrich et al. 2001). Low concentrations of SeO_3^{2-} and

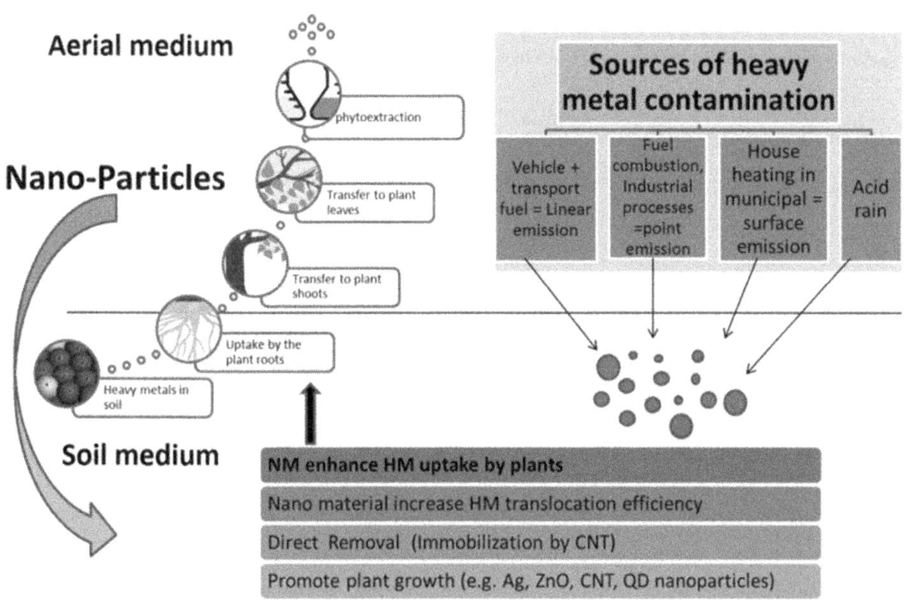

Figure 16.2 Heavy metal contamination and process of phytoremediation.

SeO_4^{2-} (10 ppb) can promote plant growth and mitigate Hg toxicity to plant systems via formation of HgSe compounds. Secondly, the translocation of Hg to plant systems is greatly reduced due to HgSeNPs and HgSe formation (Gong et al. 2014; Hofacker et al. 2013). The silica shell containing iron oxide core was used to prepare a functional magnetic polymer Fe_3O_4/SiO_2/glycidyl methacrylate (GMA)-S-SH nanocomposite. The material has high adsorption capacity for Cd^{+2} ions with an additional property to degrade anti-inflammatory drugs. The immobilization of the laccase enzyme is one of such strategies for drugs and dye removal from wastewater. Therefore, Fe_2O_4 NPs, a novel composite, are tested for their possible role in the removal of heavy metal and pharmaceutical pollutants (Mohammadi et al. 2018).

16.3.4.3 Nanoparticles to Mitigate Salinity Stress

Salinity is badly affecting the quality, growth and productivity of crops worldwide (Sharifi et al. 2007). Under saline stress, a large number of chloride ions and sodium ions are accumulated in plants (Farooq et al. 2017).

The adverse effects of salinity were alleviated by nanosized SiO_2 treatment through increased root, shoot and seedling dry weight as compared to bulk application of SiO_2 and control conditions (Ivani et al. 2018) (Table 16.2). All the agronomic traits of Moldavian balm (*Dracocephalum moldavica L.*) were negatively affected under saline

Table 16.2 Nanoparticles utilized for different objectives in tissue culture.

NPs	Objective	Reference
ZnO/ CuO	Role in tissue culture regenerants	Ahmad et al. (2020)
Se	Role in growth, growth mechanism and callogenesis	Rajaee Behbahani et al. (2020)
Ag	Putative role in characterization and antibacterial activities via callus culture	Aref and Salem (2020)
Se	Role in epigenetic DNA methylation	Sotoodehnia-Korani et al. (2020)
Ag	Role in micropropagation	CdO et al. (2019)
Si	Role in growth, physiology and anatomy	Moghanloo et al. (2019)
AgNO3	Enhance plant growth and development	Mahendran et al. (2019)
Fe NPs	The absorption, distribution and accumulation of the Fe NPs in *C. annuum*	Yuan et al. (2018)
nZVI	Root elongation	Kim et al. (2014)
Fe3O4	Compare callus morphology, and to estimate the genotoxicity	Kokina et al. (2017)
Fe3O4	Increased root length and chlorophyll inflourescence	Kokina et al. (2020)
AuNPs	Regulation of size and easier recovery of NPs	Bhaskaran et al. (2019)

conditions. The application of titanium dioxide NPs (TiO^2 NPs) significantly ameliorated the salinity effects on oil profile and biochemical attributes (Gohari et al. 2020) (Table 16.2). Nano-ZnO in sunflower plants increased growth under saline stress (Mahil et al. 2019). Another study on the sunflower depicted that ZnO NPs had positive effects on biomass production, enabling Olsion and Hysun33 cultivars to be suitable for saline conditions (Torabian et al. 2016). Betaine serves as an organic osmolyte to withstand different environmental hazards such as drought, salinity, heavy metals, UV radiation, heat and cold (Khalifa et al. 2016) through its antioxidant mechanism enhancing growth and development under stress (Adibah et al. 2020).

16.3.4.4 Nanoparticles to Mitigate Flooding Stress

Flooding decreases the supply of oxygen to roots, resulting in root dysfunction, with inadequate respiration and uptake of nutrients (Aqaei et al. 2020). The NPs of Si have tremendous potential in the field of agriculture and can perform better than bulk materials to alleviate different abiotic stresses including flooding (Cui et al. 2017). In flooding stress, the effects of Nano-Ag particles were studied on *Crocus sativus* plant species by promoting root growth and through blocking ethylene signaling (Rezvani et al. 2012). Mustafa et al. (2015) noticed an increase in seedling growth through application of nano-Ag particles to soybean plants.

16.3.4.5 Nanoparticles to Mitigate Heat Stress

Heat stress showed reduction in various plant growth parameters, i.e., yield, dry biomass, etc. At molecular levels, high temperature activity heat shock proteins (HSPs) were mediated by nitric oxide (NO) (Piterková et al. 2013), which reduces relative water content and the activities of antioxidant enzymes (Shah et al. 2013). Ag-NPs have shown increase in plant growth in wheat under heat stress (Iqbal et al. 2019). Low Se-NPs concentration also showed increase in chlorophyll content, level of hydration and growth in the case of tomato plants and sorghum under high temperature conditions (Haghighi and Abolghasemi 2014; Djanaguiraman et al. 2018). In another study to evaluate the effect of selenium (Se) as a stress mediator and nano selenium (N-Se) on growth in response to high and low temperatures, promoted growth in tomato plants. Shoot fresh weights, dry weights, chlorophyll content and root fresh weights, dry weights and volumes were increased at different durations of stress treatments. Moreover, Se and N-Se increased relative water content results in improving plant growth parameters (Haghighi and Abolghasemi 2014). Zhao et al. (2012) reported the stress response by CeO-NPs, which lead to the overproduction of H_2O_2 along with HSP70 upregulation in corn plants. Qi et al. (2013) identified that TiO_2-NPs mitigated heat stress by regulating the opening of stomata.

16.3.4.6 Nanoparticles to Mitigate Cold Stress

Crop yields and quality are also reduced due to reduced carbohyderates metabolism and photosynthetic rates at low temperatures. Cold stress or chilling injury causes electrolyte leakage of the membranes leading to poor germination affecting overall productivity (Suzuki et al. 2008; Welti et al. 2002) and chilling may cause crystal formation leading to killing of plants (Hasanuzzaman et al. 2013). NMs application has shown improvement in the growth as well as physio-biochemical features of crop plants under cold stress (Azimi et al. 2014; Haghighi and Abolghasemi 2014; Kohan-Baghkheirati and Geisler-Lee 2015). The TiO NPs showed reduced or no oxidative damage to chickpea variants owing to positively affecting the physiology and improving the redox status of genotypes under cold treatment (Mohammadi et al. 2013). Use of TiO_2 is found to have increased Rubisco enzyme and chlorophyll-binding protein gene expression (Hasanpour et al. 2015), along with ascorbate peroxidase (APX), CAT and SOD activities (Mohammadi et al. 2014). Upregulation of the MeCu/ZnSOD and MeAPX2 genes and enhancement of monodehydroascorbate reductase, dehydroascorbate reductase and glutathione reductase activities can improve ROS scavenging, leading to suppression of lipid peroxidation and hydrogen peroxide synthesis that leads to the stress resistance (Xu et al. 2014).

16.4 Advances in Nanotechnology

Nanotechnology is immensely emerging and is being applied or is on the verge of application in many advance agriculture applications.

16.4.1 Nanotechnology in Tissue Culture

Plant tissue culture is a combination of *in vitro* techniques to produce thousands of agricultural, horticultural and other commercial plantations by providing nutrient medium under lab conditions. The techniques may involve callus, hybrid, protoplast and cell suspension cultures (Park et al. 2008). Climate change and global warming are threats to a number of plant species that become endangered directly due to the pressing need of tissue culture technology to conserve the germplasm (*in vitro* storage) owing to its ability to duplicate rapidly and protect hereditary assets as compared to field quality banks.

NP application in tissue culture is at its infancy stage, yet several achievements have been reported. Microbial infection in tissue culture is a common issue, which has been resolved by metals and metals oxide NPs including ZnO, Ag and TiO_2 (Rajput et al. 2018). Several other NPs, i.e., ZnO, Fe_2O_3, MgO, Al_2O_3, CuO, Ni and SiO_2 have also been successfully utilized as sterilizing agents of explants. However, concentration and treatment durations should be optimized well to avoid toxic damage and enhance the efficiency of pathogenic control in explants. Some studies related to NPs utilization in tissue culture are listed in Table 16.2.

Tecomella undulate has been identified as an ideal example of NPs having increased root and shoot formation of stem explants enhancement when silver NPs were added in the medium (Aghdaei et al. 2012). Number, percentage and length of shoots also increased with addition of silver NPs in an MS medium containing nodal explants (Sarmast and Salehi 2016). Zinc oxide NPs (ZnO NPs) also have positive effects on callus proliferation. NPs have affected mitotic activity, DNA stabilization and protein modification in plants. Carbon-based NPs (CNPs) in an MS medium containing callus culture of *L. usitatissimum* improved the ploidy levels. Possible role of C, Ag and Au-based NPs in DNA methylation and somaclonal variation has also been reported by Kokina et al. (2017).

Genetic transformation for targeted objectives has been adopted by many biotechnologists in versatile areas including plant sciences. Plant transformations through NPs and development of resistant plants will help to reduce losses on the one hand and minimize chemical use on the other, leading to a reduction in pollution (Alghuthaymi et al. 2015). Each and every step of genetic transformation is complex and crucial, thus needs to be performed in a controlled environment. These traditional methods do have certain limitations, namely excessive energy consumption and damage to DNA, etc. Nanotechnology importance in transformation is evident from the publications and patents of the past decade. This technology is clearly the smartest one for delivery of desired chemicals along with its precision (Emeka et al. 2019). Various systmes for transformation have been adopted, such as:

- *Calcium phosphate based NPs* (CaPNP) are 80.7% efficient in transformation as compared to 54.4% of *Agrobacterium tumefaciens* and 8% of naked DNA. The plasmid DNA can protect the reporter gene (DNA) against nucleases (Roy Chowdhury et al. 2018). A complex of (CaPNP) and pDNA (Plasmid DNA) can act as the best tool to NP-mediated genetic engineering in plants (Naqvi et al. 2012).
- *Magnetic NPs* have been used in *Brassica juncea* (canola) cells, where 95% transformation efficiency of gold magnetic NPs was observed in the presence of a magnetic field (Hao et al. 2013). However, a sustainable method of loading foreign DNA to magnetic NPs and finally delivering to the pollen and their delivary into pollen was done by Zhao et al. (2017). The genetic information delivered to the pollen was transferable to the next generations.
- *Dendrimers* are also being used. Branching in polyamidoamine dendrimers has enahanced its ability to transfer the gene. In alfalfa, ssDNA conjugated dendrimers transferred with a transfection efficiency of 36% through ultrasonication (Amani et al. 2018). Supramolecular complexes of G4 poly(amidoamine) dendrimers were coupled with plasmid DNA and introduced electrostatically into *Agrostis stolonifera L.*, as observed by Pasupathy et al. (2008).
- *Liposomes* are most unique NPs, because they have a close resemblance to the cell membrane. The negatively-charged liposomes carrying the plasmid (pLGV23neo) with the kanamycin-resistant gene were integrated into tobacco protoplasts. Kanamycin-resistant offspring with 3–5 copies of genes highlighted the efficacy of the method in successful transformation (Ditta and Arshad 2016). The experiment demonstrated both transient gene expression and silencing of gene expression by the use of CNT complexes. Demirer et al. (2019) exploited the potential of SWCNTs and MWCNTs by covalently binding them to DNA and delivering to arugula protoplasts with 85% transfection efficiency.

16.4.2 NPs in Genome Editing

RNA interference or RNAi is a process in eukaryotic cells that involves plant growth as well as a defense mechanism against viruses and transposons, which have the ability of inhibiting the target insects, viruses, fungi and weeds (Worrall et al. 2018). Although the method is attractive, its application faces issues like its short life and specific targeting.

DNA nanostructures, being powerful for nanostructure fabrication, have been extensively used for different objectives including cell internalization, intracellular delivery and downstream diagnostic and therapeutic applications due to biocompatibility (Wang et al. 2018). Though the work in animal-based research has been well conducted, analogous work in plant systems is lacking (Kang et al. 2021). A few studies have reported the biolistic or nonmechanical uptake, translocation or localization of engineered NPs (carbon nanotubes, SiO_2, quantum dots, TiO NPs) to plants (Cunningham et al. 2018; Demirer et al. 2019; Mitter et al. 2017). Recently, siRNA have been successfully delivered and silenced gene expression in *Nicotiana benthamiana* leaves by adopting DNA nanostructures as mediators (Zhang et al. 2019). CRISPR/cas9 is a time-efficient measure of gene editing using short RNA fragments to locate, target and break the DNA double strands, thus avoiding the need of protein engineering (Adhikari et al. 2020). A number of challenges to CRISPR/cas9 delivery includes its formulation stability, large loading capacity, long retention times and protection from degradation (Wan et al. 2019). Nanotechnology can play an indispensable role in nonviral gene delivery. Polymeric and lipid NPs are observed to be successful in delivary of CRISPR/cas9 to the target site (Deng et al. 2019).

16.4.3 Nanosensors/Smart Plant Sensors

Precision farming involves the combination of information technology, nano biosensors, remote sensing and global positioning system (GPS) at one platform. Nanobiosensors act as diagnostic tools by providing precise information to the growers, which help them make better decisions to increase crop yield (Shang et al. 2019). The smart sensor devices help in better phenotyping of desired traits. These electronic sensors provide real-time data about the plant's condition via wireless signals by converting the chemical information to digital signals. By this technique, data about stress levels and available resources of every individual plant can be collected and analyzed, which helps in selecting the best phenotypic species of plants for optimum production (Giraldo et al. 2014). Nanobiotechnology (NBT)-based plant sensors enable the use of electronics in agriculture for monitoring and communication purposes.

Secondary metabolites of plants are expensive and their synthesis is also time-consuming. On the other hand, the worth of natural products cannot be denied. The traditional methods (high performance liquid chromatography and gas chromatography) of detecting plant metabolites are not economical and time-efficient, suggesting the need for nanobiosensors (Bharali et al. 2019). Recently, a fluorescence-based sucrose sensor was used by Bagal-Kestwal et al. (2015), declared to be potentially utilized as a region-specific nanosensor for sucrose detection. Nanobiosensors can also detect the presence of herbicides (Usman et al. 2020), and can be applied to the plants in combination with herbicides (Lin and Xing 2007). Such herbicides are available to the plants, depending upon the need of leading to better crop production.

Growth and development of plants under normal as well as stressed conditions need to be phenotyped for a plethora of objectives. Single-chirality Single-Walled Nanotube (SWNTs) are the sensors able to detect NO, H_2O_2 and dopamine under such conditions

(Giraldo et al. 2014, 2015). Recently, leaf tissues of spinach plants were implanted with CoPhMoRe-based Bombolitin II nanosensors (BSWNTs). The spinach plant acted like a self-powered auto sampler, which was transferred to the user's smartphone (Wong et al. 2017).

Encapsulation of inoculum, fungicides and pesticides using NPs were discussed in Section 16.2. All these applications are based on nano-sensors and smart delivery systems that are based on the need and site-specificity of making this system cost-efective (Kumar and Arora 2020; Kwak et al. 2017). The unique properties of the nanosensors to detect the chemicals, pathogens, aromas and even the changes in environmental conditions, has enabled their extensive use in food packaging. Usually consumers have to rely on the expiry dates mentioned on the packaging materials. So, transport and storage conditions, if violated for a specific period of time, may cause food quality deterioration unknown to the consumer until the food packaging is opened. CM5 biosensor chips are used to detect the specific binding proteins and vitamins complex. IPs are packaging technologies that, through internal and external indicators, monitor the interaction between the food, the packaging and the environment (Biji et al. 2015; Brockgreitens and Abbas 2016). There are two ways in intelligent packaging systems, supporting data systems (bars, labels or radiofrequency identifcation plates) used to store or transmit data and indicators of incidents or biosensors in packaging that allow control of the environment and product packaging (Bagchi 2012). There are different smart indicators for food packaging available in society, i.e., time and temperature indicators, integrity indicators and freshness indicators, based on the variation of chemicals in food (Kim et al. 2016; Shillingford et al. 2016).

16.5 Conclusions and Future Prospects

Climate change and exponential growth in global population have increased the application of nantechnology in agriculture and allied sciences, which can potentially contribute to address the sustainability issue and make our agriculture system more resilient. Nanotechnology is concluded to be an emerging tool to contribute to efficient production, protection and reuse of waste in agricultural farming systems. For the sake of efficient management of resources, the technology have also been successfully implemented into precision agriculture via smart nanosensors technologies.

Application of nanotechnology to different fields of food and agriculture face limitations such as possible toxicity of NPs, complexity (physicochemical characteristics of NPs), cost and duration of procedures (single or multiple point study), compatibility with different labels and degree of details (selection of tissues, organs or cell). Adoption of nanotechnology is of ethical concern regarding health (human and animal) and environmental pollution, which suggests the development of reliable methods for characterization and quantification of the nanomaterials in different matrices of environment and health risk impacts. The task can be achieved through proper legislation based on stakeholders, non-governmental and consumers' concerns.

References

Abusalem, M., Awwad, A., Ayad, J. et al. (2019). Green synthesis of α-Fe_2O_3 nanoparticles using pistachio leaf extract influenced seed germination and seedling growth of tomatos. *Joran Journal of Earth and Environmental Sciences* 10: 161–166.

Acharya, P., Jayaprakasha, G.K., Crosby, K.M. et al. (2019). Green-synthesized nanoparticles enhanced seedling growth, yield, and quality of onion (*Allium cepa* L.). *ACS Sustainable Chemistry & Engineering* 7: 14580–14590.

Acharya, P., Jayaprakasha, G.K., Crosby, K.M. et al. (2020). Nanoparticle-mediated seed priming improves germination, growth, yield, and quality of watermelons (*Citrullus lanatus*) at multi-locations in Texas. *Scientific Reports* 10: 5037.

Adhikari, P., Prabin, P., Poudel, M. et al. (2020). CRISPR-Cas9 in agriculture: Approaches, applications, future perspectives, and associated challenges. *Malaysian Journal of Halal Research* 3 (1): 6–16. https://doi.org/10.2478/mjhr-2020-0002

Adibah, F.S., Jahan, M.S., and Fatihah, H.N. (2020). Betaine-rich nano fertilizer improves growth parameters of *Zea mays* var. saccharata and *Arabidopsis thaliana* under salt stress. *Bulgarian Journal of Agricultural Science* 26: 177–185.

Aghdaei, M., Salehi, H., and Sarmast, M.K. (2012). Effects of silver nanoparticles on *Tecomella undulata* (Roxb.) Seem. *Advances in Horticultural Science* 26: 21–24.

Ahmad, M.A., Javed, R., Adeel, M. et al. (2020). Engineered ZnO and CuO nanoparticles ameliorate morphological and biochemical response in tissue culture regenerants of candyleaf (*Stevia rebaudiana*). *Molecules (Basel, Switzerland)* 25: 1356.

Ahmad, N., Sharma, S., Alam, M.K. et al. (2010). Rapid synthesis of silver nanoparticles using dried medicinal plant of basil. *Colloid Surface B* 81: 81–86.

Ahuja, R., Sidhu, A., and Sharma, A. (2020). Effect of iron sulfide nanopriming in reducing *Drechslera* seed rot and seedling mortality in rice. *Agricultural Research Journal* 57: 60.

Alghuthaymi, M.A., Almoammar, H., Rai, M. et al. (2015). Myconanoparticles: Synthesis and their role in phytopathogens management. *Biotechnology & Biotechnological Equipment* 29: 221–236.

Almaary, K.S., Sayed, S.R.M., Abd-Elkader, O.H. et al. (2020). Complete green synthesis of silver-nanoparticles applying seed-borne *Penicillium duclauxii*. *Saudi Journal of Biological Sciences* 27: 1333–1339.

Amani, A., Zare, N., Asadi, A. et al. (2018). Ultrasound-enhanced gene delivery to alfalfa cells by hPAMAM dendrimer nanoparticles. *Turkish Journal of Biology* 42: 63–75.

Amist, N., Singh, N.B., Yadav, K. et al. (2017). Comparative studies of Al_3^+ ions and Al_2O_3 nanoparticles on growth and metabolism of cabbage seedlings. *Journal of Biotechnology* 254: 1–8.

Anastas, P.T. and Warner, J.C. (1998). *Green Chemistry: Theory and Practice*, Oxford University Press.

Aqaei, P., Weisany, W., Diyanat, M. et al. (2020a). Response of maize (*Zea mays* L.) to potassium nano-silica application under drought stress. *Journal of Plant Nutrition* 43: 1205–1216.

Arciniegas-Grijalba, P.A., Patiño-Portela, M.C., Mosquera-Sánchez, L.P. et al. (2019). ZnO-based nanofungicides: Synthesis, characterization and their effect on the coffee fungi *Mycena citricolor* and *Colletotrichum* sp. *Materials Science and Engineering* 98: 808–825.

Aref, M.S. and Salem, S.S. (2020). Bio-callus synthesis of silver nanoparticles, characterization, and antibacterial activities via *Cinnamomum camphora callus* culture. *Biocatalysis and Agricultural Biotechnology* 27: 101689.

Arulmani, S., Anandan, S., and Ashokkumar, M. (2018). Introduction to advanced nanomaterials. In: *Nanomaterials for Green Energy* (ed. B.A. Bhanvase, V.B. Pawade, S.J. Dhoble, et al.), Elsevier.

Azimi, R., Farzam, M., Feizi, H.et al. (2014), Interaction of SiO_2 nanoparticles with seed prechilling on germination and early seedling growth of tall wheatgrass (*Agropyron elongatum* L.). *Polish Journal of Chemical Technology* 16: 25–29.

Bagal-Kestwal, D., Kestwal, R.M., and Chiang, B.-H. (2015). Invertase-nanogold clusters decorated plant membranes for fluorescence-based sucrose sensor. *Journal of Nanobiotechnology* 13: 30.

Bagchi, A. (2012). Intelligent sensing and packaging of foods for enhancement of shelf life: Concepts and applications. *International Journal of Scientific & Engineering Research* 3: 1372–1384.

Beckers, F. and Rinklebe, J. (2017). Cycling of mercury in the environment: Sources, fate, and human health implications: A review. *Critical Reviews in Environmental Science and Technology* 47: 693–794.

Bharali, B., Chetia, H., Kalita, J.J. et al. (2019). Engineered nanomaterials in plants: Sensors, carriers, and bio-imaging. In: *Comprehensive Analytical Chemistry, Book 87* (ed. S.K. Verma and A.K. Das). Elsevier.

Bhaskaran, S., Sharma, N., Tiwari, P. et al. (2019). Fabrication of innocuous gold nanoparticles using plant cells in culture. *Scientific Reports* 9: 12040.

Bhattacharya, R. and Mukherjee, P. (2008). Biological properties of "naked" metal nanoparticles. *Advanced Drug Delivery Reviews* 60: 1289–1306.

Biji, K.B., Ravishankar, C.N., Mohan, C.O. et al. (2015). Smart packaging systems for food applications: A review. *Journal of Food Science and Technology* 52: 6125–6135.

Brockgreitens, J. and Abbas, A. (2016). Responsive food packaging: Recent progress and technological prospects. *Comprehensive Reviews in Food Science and Food Safety* 15: 3–15.

Cai, C., Zhao, M., Yu, Z. et al. (2019). Utilization of nanomaterials for *in-situ* remediation of heavy metal(loid) contaminated sediments: A review. *The Science of the Total Environment* 662: 205–217.

CdO, T., Paiva, R., Dos Reis, M.V. et al. (2019). Silver nanoparticles in the micropropagation of *Campomanesia rufa* (O. Berg) Nied. *Plant Cell, Tissue and Organ Culture* 137: 359–368.

Chang, C.Y., Yu, H.Y., Chen, J.J. et al. (2014). Accumulation of heavy metals in leaf vegetables from agricultural soils and associated potential health risks in the Pearl River Delta, South China. *Environmental Monitoring and Assessment* 186: 1547–1560.

Choupanian, M.M., Omar, D.D., Basri, M.M. et al. (2017). Preparation and characterization of neem oil nanoemulsion formulations against *Sitophilus oryzae* and *Tribolium castaneum* adults. *Pesticide Science* 42: 158–165.

Clarkson, T.W. and Magos, L. (2006). The toxicology of mercury and its chemical compounds. *Critical Reviews in Toxicology* 36: 609–662.

Cui, J., Liu, T., Li, F. et al. (2017). Silica nanoparticles alleviate cadmium toxicity in rice cells: Mechanisms and size effects. *Environmental Pollution* 228: 363–369.

Cunningham, F.J., Goh, N.S., Demirer, G.S. et al. (2018). Nanoparticle-mediated delivery towards advancing plant genetic engineering. *Trends in Biotechnology* 36: 882–897.

Dang, V.M., Joseph, S., Van, H.T. et al. (2019). Immobilization of heavy metals in contaminated soil after mining activity by using biochar and other industrial by-products: The significant role of minerals on the biochar surfaces. *Environmental Technology* 40: 3200–3215.

Das, C.K., Srivastava, G., Dubey, A. et al. (2016). Nano-iron pyrite seed dressing: A sustainable intervention to reduce fertilizer consumption in vegetable (beetroot, carrot), spice (fenugreek), fodder (alfalfa), and oilseed (mustard, sesamum) crops. *Nanotechnology for Environmental Engineering* 1: 2.

De Gregorio, P.R., Michavila, G., Ricciardi Muller, L. et al. (2017) Beneficial rhizobacteria immobilized in nanofibers for potential application as soybean seed bioinoculants. *PLOS ONE* 12: e0176930.

Demirer, G.S., Zhang, H., Matos, J.L. et al. (2019). High aspect ratio nanomaterials enable delivery of functional genetic material without DNA integration in mature plants. *Nature Nanotechnology* 14: 456–464.

Deng, H., Huang, W., and Zhang, Z. (2019). Nanotechnology based CRISPR/Cas9 system delivery for genome editing: Progress and prospect. *Nano Research* 12: 2437–2450.

Din, M.I., Nabi, A.G., Rani, A. et al. (2018). Single step green synthesis of stable nickel and nickel oxide nanoparticles from *Calotropis gigantea*: Catalytic and antimicrobial potentials. *Environmental Nanotechnology, Monitoring & Management* 9: 29–36.

Ditta, A. and Arshad, M. (2016). Applications and perspectives of using nanomaterials for sustainable plant nutrition. *Nanotechnology Reviews* 5: 209.

Djanaguiraman, M., Belliraj, N., Bossmann, S.H. et al. (2018). High-temperature stress alleviation by selenium nanoparticle treatment in grain sorghum. *ACS Omega* 3: 2479–2491.

Emeka, E., Uzoh, I., Onwudiwe, D. et al. (2019). The role of nanotechnology in the fortification of plant nutrients and improvement of crop production. *Applied Sciences* 9: 499.

Faizan M., Hayat S., and Pichtel J. (2020) Effects of zinc oxide nanoparticles on crop plants: A perspective analysis. In: *Sustainable Agriculture Reviews* (ed. Hayat, S., Pichtel, J., Faizan, M. et al.), vol. 41. Cham: Springer International Publishing. https://doi.org/10.1007/978-3-030-33996-8_4

Farooq, M., Gogoi, N., Hussain, M. et al. (2017). Effects, tolerance mechanisms and management of salt stress in grain legumes. *Plant Physiology and Biochemistry* 118: 199–217.

Fischer, R.A., Byerlee, D., and Edmeades, G. (2014). *Crop Yields and Global Food Security: Will Yield Increase Continue to Feed the World? ACIAR Monograph No. 158*. Canberra: Australian Centre for International Agricultural Research.

Gericke, M. and Pinches, A. (2006). Biological synthesis of metal nanoparticles. *Hydrometallurgy* 83: 132–140.

Gil-Díaz, M., Alonso, J., Rodríguez-Valdés, E. et al. (2017). Comparing different commercial zero valent iron nanoparticles to immobilize As and Hg in brownfield soil. *The Science of the Total Environment* 584–585: 1324–1332.

Gil-Díaz, M., Ortiz, L.T., Costa, G. et al. (2014). Immobilization and leaching of Pb and Zn in an acidic soil treated with zerovalent iron nanoparticles (nZVI): Physicochemical and toxicological analysis of leachates. *Water, Air, and Soil Pollution* 225: 1990.

Giraldo, J.P., Landry, M.P., Faltermeier, S.M. et al. (2014). Plant nanobionics approach to augment photosynthesis and biochemical sensing. *Nature Materials* 13: 400–408.

Giraldo, J.P., Landry, M.P., Kwak, S.Y. et al. (2015). A ratiometric sensor using single chirality near-infrared fluorescent carbon nanotubes: Application to *in vivo* monitoring. *Small (Weinheim an Der Bergstrasse, Germany)* 11: 3973–3984.

Giraldo, J.P., Wu, H., Newkirk, G.M. et al. (2019). Nanobiotechnology approaches for engineering smart plant sensors. *Nature Nanotechnology* 14: 541–553.

Gohari, G., Mohammadi, A., Akbari, A. et al. (2020). Titanium dioxide nanoparticles (TiO_2 NPs) promote growth and ameliorate salinity stress effects on essential oil profile and biochemical attributes of *Dracocephalum moldavica*. *Scientific Reports* 10: 912.

Gong, Y., Liu, Y., Xiong, Z. et al. (2014). Immobilization of mercury by carboxymethyl cellulose stabilized iron sulfide nanoparticles: Reaction mechanisms and effects of stabilizer and water chemistry. *Environmental Sciences* 48: 3986–3994.

Guilger, M., Pasquoto-Stigliani, T., Bilesky-Jose, N. et al. (2017). Biogenic silver nanoparticles based on *Tichoderma harzianum*: Synthesis, characterization, toxicity evaluation and biological activity. *Scientific Reports* 7: 44421.

Guilger-Casagrande, M. and Lima, R. (2019). Synthesis of silver nanoparticles mediated by fungi: A review. *Frontiers in Bioengineering and Biotechnology* 7: 287.

Haghighi, M. Abolghasemi, R., and Teixeira da Silva, J.A. (2014) Low and high temperature stress affect the growth characteristics of tomato in hydroponic culture with Se and nano-Se amendment. *Scientia Horticulturae* 178: 231–240.

Hao, Y., Yang, X., Shi, Y. et al. (2013). Magnetic gold nanoparticles as a vehicle for fluorescein isothiocyanate and DNA delivery into plant cells. *Botany* 91: 457–466.

Hasanpour, H., Maali-Amir, R., and Zeinali, H. (2015). Effect of TiO_2 nanoparticles on metabolic limitations to photosynthesis under cold in chickpea. *Russian Journal of Plant Physiology* 62: 779–787.

Hasanuzzaman, M., Nahar, K., and Fujita, M. (2013). Extreme temperature responses, oxidative stress and antioxidant defense in plants. In: *Abiotic Stress-Plant Responses and Applications in Agriculture* (ed. V. Kourosh and L. Charles), 169–205. IntechOpen.

Hofacker, A.F., Voegelin, A., Kaegi, R. et al. (2013). Mercury mobilization in a flooded soil by incorporation into metallic copper and metal sulfide nanoparticles. *Environmental Science & Technology* 47: 7739–7746.

Hossain, A., Hong, X., Ibrahim, E. et al. (2019). Green synthesis of silver nanoparticles with culture supernatant of a bacterium *Pseudomonas rhodesiae* and their antibacterial activity against soft rot pathogen *Dickeya dadantii*. *Molecules (Basel, Switzerland)* 24: 2303.

Iqbal, M., Raja, N.I., Mashwani, Z.-U.-R. et al. (2019). Effect of silver nanoparticles on growth of wheat under heat stress. *Iranian Journal of Science and Technology, Transactions A* 43: 387–395.

Iravani, S. (2014). Bacteria in nanoparticle synthesis: Current status and future prospects. *International Scholarly Research Notices* (2014): 359316.

Ivani, R., Sanaei Nejad, S.H., Ghahraman, B. et al. (2018). Role of bulk and nanosized SiO_2 to overcome salt stress during Fenugreek germination (*Trigonella foenum-graceum* L.). *Plant Signaling & Behavior* 13: e1044190.

Jaberzadeh, A., Moaveni, P., Moghadam, H.R.T. et al. (2013). Influence of bulk and nanoparticles titanium foliar application on some agronomic traits, seed gluten and starch contents of wheat subjected to water deficit stress. *Notulae Botanicae Horti Agrobotanici Cluj-Napoca* 41: 201–207.

JLd, O., Campos, E.V.R., Pereira, A.E.S., Pasquoto, T. et al. (2018). Zein nanoparticles as eco-friendly carrier systems for botanical repellents aiming sustainable agriculture. *Journal of Agricultural and Food Chemistry* 66: 1330–1340.

Joshi, A., Kaur, S., Dharamvir, K. et al. (2018). Multi-walled carbon nanotubes applied through seed-priming influence early germination, root hair, growth and yield of bread wheat (*Triticum aestivum* L.). *Journal of the Science of Food and Agriculture* 98: 3148–3160.

Kang, J., Tahir, A., Wang, H. et al. (2021). Applications of nanotechnology in virus detection, tracking, and infection mechanisms. In: *Wiley Interdisciplinary Reviews: Nanomedicine and Nanobiotechnology* 13 (4): e1700.

Kasana, R.C., Panwar, N.R., Kaul, R.K. et al. (2017). Biosynthesis and effects of copper nanoparticles on plants. *Environmental Chemistry Letters* 15: 233–240.

Khalifa, G.S., Abdelrassoul, M., Hegazi, A.M. et al. (2016). Attenuation of negative effects of saline stress in two lettuce cultivars by salicylic acid and glycine betaine. *Gesunde Pflanzen* 68: 177–189.

Khan, I., Saeed, K., and Khan, I. (2019). Nanoparticles: Properties, applications and toxicities. *Arabian Journal of Chemistry* 12: 908–931.

Khan, N. and Bano, A. (2016). Role of plant growth promoting rhizobacteria and Ag-nano particle in the bioremediation of heavy metals and maize growth under municipal wastewater irrigation. *International Journal of Phytoremediation* 18: 211–221.

Khodakovskaya, M.V., De Silva, K., Nedosekin, D.A. et al. (2011) Complex genetic, photothermal, and photoacoustic analysis of nanoparticle-plant interactions. *Proceedings of the National Academy of Sciences of the United States of America* 108:1028–1033

Kim, J.-H., Lee, Y., Kim, E.-J. et al. (2014). Exposure of iron nanoparticles to arabidopsis thaliana enhances root elongation by triggering cell wall loosening. *Environmental Science & Technology* 48: 3477–3485.

Kim, J.U., Ghafoor, K., Ahn, J. et al. (2016). Kinetic modeling and characterization of a diffusion-based time-temperature indicator (TTI) for monitoring microbial quality of non-pasteurized angelica juice. *LWT – Food Science and Technology* 67: 143–150.

Kohan-Baghkheirati, E. and Geisler-Lee, J. (2015). Gene expression, protein function and pathways of *Arabidopsis thaliana* responding to silver nanoparticles in comparison to silver ions, cold, salt, drought, and heat. *Nanomaterials (Basel)* 5: 436–467.

Kokina, I., Mickeviča, I., Jermaļonoka, M. et al. (2017). Case study of somaclonal variation in resistance genes Mlo and Pme3 in flaxseed (*Linum usitatissimum* L.) induced by nanoparticles. *International Journal of Genomics* 2017: 1–5.

Kokina, I., Plaksenkova, I., Jermaļonoka, M. et al. (2020). Impact of iron oxide nanoparticles on yellow medick (*Medicago falcata* L.) plants. *Journal of Plant Interactions* 15: 1–7.

Kumar, V. and Arora, K. (2020). Trends in nano-inspired biosensors for plants. *Materials Science for Energy Technologies* 3: 255–273.

Kwak, S.-Y., Wong, M.H., Lew, T.T.S. et al. (2017). Nanosensor technology applied to living plant systems. *Annual Review of Analytical Chemistry* 10: 113–140.

Li, S., Shen, Y., Xie, A. et al. (2007). Green synthesis of silver nanoparticles using *Capsicum annuum* L. extract. *Green Chemistry* 9: 852–858.

Lin, D. and Xing, B. (2007). Phytotoxicity of nanoparticles: Inhibition of seed germination and root growth. *Environmental Pollution* 150: 243–250.

Liu, J., Wang, J., Ning, Y. et al. (2019). Methylmercury production in a paddy soil and its uptake by rice plants as affected by different geochemical mercury pools. *Environment International* 129: 461–469.

Lobell, D.B., Cassman, K.G., and Field, C.B. (2009). Crop yield gaps: Their importance, magnitudes, and causes. *Annual Review of Environment and Resources* 34: 179–204.

Lu, K., Yang, X., Shen, J. et al. (2014). Effect of bamboo and rice straw biochars on the bioavailability of Cd, Cu, Pb and Zn to *Sedum plumbizincicola*. *Agriculture, Ecosystems & Environment* 191: 124–132.

Mahakham, W., Sarmah, A.K., Maensiri, S. et al. (2017). Nanopriming technology for enhancing germination and starch metabolism of aged rice seeds using phytosynthesized silver nanoparticles. *Scientific Reports* 7: 8263.

Mahendran, D., Geetha, N., and Venkatachalam, P. (2019). Role of silver nitrate and silver nanoparticles on tissue culture medium and enhanced the plant growth and development. In: *Vitro Plant Breeding Towards Novel Agronomic Traits: Biotic and Abiotic Stress Tolerance* (ed. M. Kumar, A. Muthusamy, V. Kumar, et al.). Singapore: Springer.

Mahil, T., Baburai, N., and Kumar, A. (2019). Foliar application of nanofertilizers in agricultural crops: A review. *Journal of Farm Sciences* 32: 239–249.

Marchiol, L. (2012). Synthesis of metal nanoparticles in living plants. *Italian Journal of Agronomy* 7: e37.

Mathur, P., Jha, S., Ramteke, S. et al. (2018). Pharmaceutical aspects of silver nanoparticles. *Artificial Cells, Nanomedicine, and Biotechnology* 46: 115–126.

Milewska-Hendel, A., Gawecki, R., Zubko, M. et al. (2016). Diverse influence of nanoparticles on plant growth with a particular emphasis on crop plants. *Acta Agrobotanica* 69: 1694.

Mitter, N., Worrall, E.A., Robinson, K.E. et al. (2017). Clay nanosheets for topical delivery of RNAi for sustained protection against plant viruses. *Nature Plants* 3: 16207.

Moghanloo, M., Iranbakhsh, A., Ebadi, M. et al. (2019). Seed priming with cold plasma and supplementation of culture medium with silicon nanoparticle modified growth, physiology, and anatomy in *Astragalus fridae* as an endangered species. *Acta Physiologiae Plantarum* 41: 54.

Mohammadi, M., As'habi, M.A., Salehi, P. et al. (2018). Immobilization of laccase on epoxy-functionalized silica and its application in biodegradation of phenolic compounds. *International Journal of Biological Macromolecules* 109: 443–447.

Mohammadi, R., Maali-Amiri, R., and Abbasi, A. (2013). Effect of TiO_2 nanoparticles on chickpea response to cold stress. *Biological Trace Element Research* 152: 403–410.

Mohammadi, R., Maali-Amiri, R., and Mantri, N.L. (2014). Effect of TiO_2 nanoparticles on oxidative damage and antioxidant defense systems in chickpea seedlings during cold stress. *Russian Journal of Plant Physiology* 61: 768–775.

Monteith, J.L., Moss, C.J., Cooke, G.W. et al. (1977). Climate and the efficiency of crop production in Britain. *Philosophical Transactions of the Royal Society of London. B, Biological Sciences* 281: 277–294.

Mustafa, G., Sakata, K., and Komatsu, S. (2015). Proteomic analysis of flooded soybean root exposed to aluminum oxide nanoparticles. *Journal of Proteomics* 128: 280–297.

Nanjwade, B.K., Derkar, G.K., Bechra, H.M. et al. (2011). Design and characterization of nanocrystals of lovastatin for solubility and dissolution enhancement. *Journal of Nanomedicine Nanotechnology* 2: 107.

Naqvi, S., Maitra, A.N., Abdin, M.Z. et al. (2012). Calcium phosphate nanoparticle mediated genetic transformation in plants. *Journal of Materials Chemistry* 22: 3500–3507.

Niknejad, F., Nabili, M., Daie Ghazvini, R. et al. (2015). Green synthesis of silver nanoparticles: Advantages of the yeast *Saccharomyces cerevisiae* model. *Current Medical Mycology* 1: 17–24.

Panigrahi, S., Kundu, S., Ghosh, S. et al. (2004). General method of synthesis for metal nanoparticles. *Journal of Nanoparticle Research* 6: 411–414.

Pariona, N., Martinez, A., Sánchez-Rangel, D. et al. (2019). Green-synthesized copper nanoparticles as a potential antifungal against plant pathogens. *RSC Advances* 9: 18835–18843.

Park, E.-J., Yi, J., Chung, K.-H. et al. (2008). Oxidative stress and apoptosis induced by titanium dioxide nanoparticles in cultured BEAS-2B cells. *Toxicology Letters* 180: 222–229.

Parkash, V., Gaur, A., and Agnihotri, R. (2020). Nanoparticles from fungal resources: Importance and applications. In: *Nanotechnology for Food, Agriculture, and Environment* (ed. D. Thangadurai, J. Sangeetha and R. Prasad). Cham: Springer International Publishing.

Pasupathy, K., Lin, S., Hu, Q. et al. (2008). Direct plant gene delivery with a poly(amidoamine) dendrimer. *Biotechnology Journal* 3: 1078–1082.

Pei, Z., Ming, D., Liu, D. et al. (2009). Silicon improves the tolerance to water-deficit stress induced by polyethylene glycol in wheat (*Triticum aestivum* L.) seedlings. *Journal of Plant Growth Regulation* 29: 106–115.

Pillai, H.P.S. and Kottekottil, J. (2016). Nano-phytotechnological remediation of endosulfan using zero valent iron nanoparticles. *Journal of Environmental Protection* 07: 734–744.

Piterková, J., Luhová, L., Mieslerová, B. et al. (2013). Nitric oxide and reactive oxygen species regulate the accumulation of heat shock proteins in tomato leaves in response to heat shock and pathogen infection. *Plant Science : An International Journal of Experimental Plant Biology* 207: 57–65.

Ponmurugan, P., Manjukarunambika, K., Elango, V. et al. (2016). Antifungal activity of biosynthesised copper nanoparticles evaluated against red root-rot disease in tea plants. *Journal of Experimental Nanoscience* 11: 1019–1031.

Pour, M.M., Saberi-Riseh, R., Mohammadinejad, R. et al. (2019). Nano-encapsulation of plant growth-promoting rhizobacteria and their metabolites using alginate-silica nanoparticles and carbon nanotube improves ucb1 pistachio micropropagation. *Journal of Microbiology and Biotechnology* 29: 1096–1103.

Pradhan, S., Patra, P., Mitra, S. et al. (2014). Manganese nanoparticles: Impact on non-nodulated plant as a potent enhancer in nitrogen metabolism and toxicity study both *in vivo* and *in vitro*. *Journal of Agricultural and Food Chemistry* 62: 8777–8785.

Prasad, R., Bhattacharyya, A., and Nguyen, Q.D. (2017). Nanotechnology in sustainable agriculture: Recent developments, challenges, and perspectives. *Frontiers in Microbiology* 8: 1014.

Prasad, R., Kumar, V., and Prasad, K.S. (2014). Nanotechnology in sustainable agriculture: Present concerns and future aspects. *African Journal of Biotechnology* 13: 705–713.

Priyanka, P., Kumar, D., Yadav, A. et al. (2020). Nanobiotechnology and its application in agriculture and food production. In: *Nanotechnology for Food, Agriculture, and Environment* (ed. D. Thangadurai, J. Sangeetha and R. Prasad). Cham: Springer International Publishing.

Qi, M., Liu, Y., and Li, T. (2013). Nano-TiO_2 improve the photosynthesis of tomato leaves under mild heat stress. *Biological Trace Element Research* 156: 323–328.

Rafique, M., Jahangir, J., Amin, B.A.Z. et al. (2019). Investigation of photocatalytic and seed germination effects of TiO_2 nanoparticles synthesized by *Melia azedarach* L. leaf extract. *Journal of Inorganic and Organometallic Polymers and Materials* 29: 2133–2144.

Rahimi, D., Kartoolinejad, D., Nourmohammadi, K. et al. (2016). Increasing drought resistance of *Alnus subcordata* C.A. Mey. seeds using a nano priming technique with multi-walled carbon nanotubes. *Journal of Forest Science* 62: 269–278.

Rajaee Behbahani, S., Iranbakhsh, A., Ebadi, M. et al. (2020). Red elemental selenium nanoparticles mediated substantial variations in growth, tissue differentiation, metabolism, gene transcription, epigenetic cytosine DNA methylation, and callogenesis in bittermelon (*Momordica charantia*); an *in vitro* experiment. *PLOS ONE* 15: e0235556.

Rajput, V.D., Minkina, T.M., Behal, A. et al. (2018). Effects of zinc-oxide nanoparticles on soil, plants, animals and soil organisms: A review. *Environmental Nanotechnology, Monitoring & Management* 9: 76–84.

Raliya, R., Nair, R., Chavalmane, S. et al. (2015). Mechanistic evaluation of translocation and physiological impact of titanium dioxide and zinc oxide nanoparticles on the tomato (*Solanum lycopersicum* L.) plant. *Metallomics* 7: 1584–1594.

Ray, D.K., Mueller, N.D., West, P.C. et al. (2013). Yield trends are insufficient to double global crop production by 2050. *PLOS ONE* 8: e66428.

Reddy Pullagurala, V.L., Adisa, I.O., Rawat, S. et al. (2018). ZnO nanoparticles increase photosynthetic pigments and decrease lipid peroxidation in soil grown cilantro (*Coriandrum sativum*). *Plant Physiology Biochemistry* 132: 120–127. doi: 10.1016/j.plaphy.2018.08.037. Epub 2018 Aug 29. PMID: 30189415.

Rezvani, N.S., Sorooshzadeh, A., and Farhadi, N. (2012). Effect of nano-silver on growth of saffron in flooding stress. *World Academy of Science, Engineering and Technology* 61: 517–522.

Romeh, A.A.A. (2018). Green silver nanoparticles for enhancing the phytoremediation of soil and water contaminated by fipronil and degradation products. *Water, Air, and Soil Pollution* 229: 147.

Roy Chowdhury, N., Cowin, A.J., Zilm, P. et al. (2018). "Chocolate" gold nanoparticles: One pot synthesis and biocompatibility. *Nanomaterials* 8: 496.

Roychoudhury, A., Datta, K., and Datta, S. (2008) C4 plants and abiotic stress. *Proceedings of the Humboldt College on Global Warming in Context to the Indian Sub-continent* Humboldt Club, Calcutta: 86–105.

Sarmast, M.K. and Salehi, H. (2016). Silver nanoparticles: An influential element in plant nanobiotechnology. *Molecular Biotechnology* 58: 441–449.

Saxena, S. and D'Souza, S. (2006). Heavy metal pollution abatement using rock phosphate mineral. *Environment International* 32: 199–202.

Shafiq, F., Iqbal, M., Ali, M. et al. (2019). Seed pre-treatment with polyhydroxy fullerene nanoparticles confer salt tolerance in wheat through upregulation of H_2O_2 neutralizing enzymes and phosphorus uptake. *Journal of Soil Science and Plant Nutrition* 19: 734–742.

Shah, K., Singh, M., and Rai, A.C. (2013). Effect of heat-shock induced oxidative stress is suppressed in BcZAT12 expressing drought tolerant tomato. *Phytochemistry* 95: 109–117.

Shaik, A.M., David Raju, M., and Rama Sekhara Reddy, D. (2020). Green synthesis of zinc oxide nanoparticles using aqueous root extract of *Sphagneticola trilobata* Lin and investigate its role in toxic metal removal, sowing germination and fostering of plant growth. *Inorganic and Nano-Metal Chemistry* 50: 569–579.

Shang, Y., Hasan, M.K., Ahammed, G.J. et al. (2019). Applications of nanotechnology in plant growth and crop protection: A review. *Molecules (Basel, Switzerland)* 24: 2558.

Sharifi, M., Ghorbanli, M., and Ebrahimzadeh, H. (2007). Improved growth of salinity-stressed soybean after inoculation with salt pre-treated mycorrhizal fungi. *Journal of Plant Physiology* 164: 1144–1151.

Sherry Davis, A., Prakash, P., and Thamaraiselvi, K. (2017). Nanobioremediation technologies for sustainable environment. In: *Bioremediation and Sustainable Technologies for Cleaner Environment* (ed. M. Prashanthi, R. Sundaram, A. Jeyaseelan, et al.). Cham: Springer International Publishing.

Shillingford, C., Russell, C.W., Burgess, I.B. et al. (2016). Bioinspired artificial melanosomes as colorimetric indicators of oxygen exposure. *ACS Applied Materials & Interfaces* 8: 4314–4317.

Siddiqui, M.H. and Al-Whaibi, M.H. (2014). Role of nano-SiO_2 in germination of tomato (*Lycopersicum esculentum* seeds Mill.). *Saudi Journal of Biological Sciences* 21: 13–17.

Soares, C., Spormann, S., and Fidalgo, F. (2018). P-372 – Salicylic acid improves the performance of the enzymatic antioxidant system of barley exposed to glyphosate. *Free Radical Biology & Medicine* 120: S157.

Sotoodehnia-Korani, S., Iranbakhsh, A., Ebadi, M. et al. (2020). Selenium nanoparticles induced variations in growth, morphology, anatomy, biochemistry, gene expression, and epigenetic dna methylation in capsicum annuum; an *in vitro* study. *Environmental Pollution* 265: 114727.

Srivastava, G., Das, A., Kusurkar, T. et al. (2014). Iron pyrite, a potential photovoltaic material, increases plant biomass upon seed pretreatment. *Materials Express* 4: 23–31.

Stefaniuk, M., Oleszczuk, P., and Ok, Y.S. (2016). Review on nano zerovalent iron (nZVI): From synthesis to environmental applications. *Chemical Engineering Journal* 287: 618–632.

Subramanian, K.S., Manikandan, A., Thirunavukkarasu, M. et al. (2015). Nano-fertilizers for balanced crop nutrition. In: *Nanotechnologies in Food and Agriculture* (ed. M. Rai, C. Ribeiro, L. Mattoso, et al.). Cham: Springer International Publishing.

Suzuki, K., Nagasuga, K., and Okada, M. (2008). The chilling injury induced by high root temperature in the leaves of rice seedlings. *Plant & Cell Physiology* 49: 433–442.

Tan, Y.N., Lee, J.Y., and Wang, D.I.C. (2010). Uncovering the design rules for peptide synthesis of metal nanoparticles. *Journal of the American Chemical Society* 132: 5677–5686.

Thakkar, K.N., Mhatre, S.S., and Parikh, R.Y. (2010). Biological synthesis of metallic nanoparticles. *Nanomed-Nanotechnol* 6: 257–262.

Thomas, R., Basheer, J., and Nil, J.K.R. (2012). Extracellular synthesis of silver nanoparticles by endophytic *Bordetella* sp. isolated from *Piper nigrum* and its antibacterial activity analysis. *Nano Biomedicine and Engineering* 4: 183–197.

Tippannanavar, M., Verma, A., Kumar, R. et al. (2020). Preparation of nanofungicides based on imidazole drugs and their antifungal evaluation. *Journal of Agricultural and Food Chemistry* 68: 4566–4578.

Torabian, S., Zahedi, M., and Khoshgoftarmanesh, A. (2016). Effect of foliar spray of zinc oxide on some antioxidant enzymes activity of sunflower under salt stress. *Journal of Agriculture, Science and Technology-Iran* 18: 1013–1025.

Ullrich, S.M., Tanton, T.W., and Abdrashitova, S.A. (2001). Mercury in the aquatic environment: A review of factors affecting methylation. *Critical Reviews in Environmental Science and Technology* 31: 241–293.

Usman, M., Farooq, M., Wakeel, A. et al. (2020). Nanotechnology in agriculture: Current status, challenges and future opportunities. *The Science of the Total Environment* 721: 137778.

Varma, A. and Uma, K.M. (2017). Role of nanoparticles on plant growth with special emphasis on piriformospora indica: A review. In: *Nanoscience and Plant–soil Systems* (ed. M. Ghorbanpour, K. Manika and A. Varma). Cham: Springer International Publishing.

Venkatachalam, P., Jayaraj, M., Manikandan, R. et al. (2017). Zinc oxide nanoparticles (ZnONPs) alleviate heavy metal-induced toxicity in *Leucaena leucocephala* seedlings: A physiochemical analysis. *Plant Physiology and Biochemistry* 110: 59–69.

Wan, T., Niu, D., Wu, C., Xu, F.-J. et al. (2019). Material solutions for delivery of CRISPR/Cas-based genome editing tools: Current status and future outlook. *Materials Today* 26: 40–66.

Wang, P., Rahman, M.A., Zhao, Z. et al. (2018). Visualization of the cellular uptake and trafficking of DNA origami nanostructures in cancer cells. *Journal of the American Chemical Society* 140: 2478–2484.

Wang, S., Wang, F., and Gao, S. (2015). Foliar application with nano-silicon alleviates Cd toxicity in rice seedlings. *Environmental Science and Pollution Research International* 22: 2837–2845.

Wang, Y., Liu, Y., Zhan, W. et al. (2020). Stabilization of heavy metal-contaminated soils by biochar: Challenges and recommendations. *The Science of the Total Environment* 729: 139060.

Welti, R., Li, W., Li, M et al. (2002). Profiling membrane lipids in plant stress responses. Role of phospholipase D alpha in freezing-induced lipid changes in *Arabidopsis*. *The Journal of Biological Chemistry* 277: 31994–32002.

Willett, W., Rockström, J., Loken, B. et al. (2019). Food in the anthropocene: The EAT-*Lancet* Commission on healthy diets from sustainable food systems. *Lancet* 393: 447–492.

Wong, M.H., Giraldo, J.P., Kwak, S.-Y. et al. (2017). Nitroaromatic detection and infrared communication from wild-type plants using plant nanobionics. *Nature Materials* 16: 264–272.

Worrall, E.A., Hamid, A., Mody, K.T. et al. (2018). Nanotechnology for plant disease management. *Agronomy* 8: 285.

Xiong, J.-L., Li, J., Wang, H.-C., Zhang, C.-L. et al. (2018). Fullerol improves seed germination, biomass accumulation, photosynthesis and antioxidant system in *Brassica napus* L. under water stress. *Plant Physiology and Biochemistry* PPB 129: 130–140.

Xu, J., Yang, J., Duan, X., Jiang, Y. et al. (2014). Increased expression of native cytosolic Cu/Zn superoxide dismutase and ascorbate peroxidase improves tolerance to oxidative and chilling stresses in cassava (*Manihot esculenta* Crantz). *BMC Plant Biology* 14: 208.

Yadav, K.K., Singh, J., Gupta, N. et al. (2017). A review of nanobioremediation technologies for environmental cleanup: A novel biological approach. *Journal of Materials and Environmental Science* 8: 740–757.

Yang, J., Cao, W., and Rui, Y. (2017). Interactions between nanoparticles and plants: Phytotoxicity and defense mechanisms. *Journal of Plant Interactions* 12: 158–169.

Yuan, J., Chen, Y., Li, H. et al. (2018). New insights into the cellular responses to iron nanoparticles in *Capsicum annuum*. *Scientific Reports* 8: 3228.

Zaytseva, O. and Neumann, G. (2016). Carbon nanomaterials: Production, impact on plant development, agricultural and environmental applications. *Chemical and Biological Technologies in Agriculture* 3: 17.

Zhang, H., Demirer, G.S., Zhang, H. et al. (2019). DNA nanostructures coordinate gene silencing in mature plants. *Proceedings of the National Academy of Sciences* 116: 7543–7548.

Zhang, X., Yan, S., Tyagi, R.D. et al. (2011). Synthesis of nanoparticles by microorganisms and their application in enhancing microbiological reaction rates. *Chemosphere* 82: 489–494.

Zhang, X.-F., Liu, Z.-G., Shen, W. et al. (2016). Silver nanoparticles: Synthesis, characterization, properties, applications, and therapeutic approaches. *International Journal of Molecular Sciences* 17: 1534.

Zhao, L., Peng, B., Hernandez-Viezcas, J.A. et al. (2012). Stress response and tolerance of zea mays to CeO_2 nanoparticles: Cross talk among H_2O_2, heat shock protein, and lipid peroxidation. *ACS Nano* 6: 9615–9622.

Zhao, X., Meng, Z., Wang, Y. et al. (2017). Pollen magnetofection for genetic modification with magnetic nanoparticles as gene carriers. *Nature Plants* 3: 956–964.

Zhou, B. and Chen, X. (2017). Effect of nano-carbon on water holding capacity in a sandy soil of the loess plateau. *Earth Sciences Research Journal* 21: 189–195.

Zörb, C., Senbayram, M., and Peiter, E. (2014). Potassium in agriculture: Status and perspectives. *Journal of Plant Physiology*' 171: 656–669.

17

Nanoscience

A Boon for Reviving Agriculture

Afrozah Hassan, Shabana Gulzar, Hanan Javid and Irshad Ahmad Nawchoo

Plant Reproductive Biology, Genetic Diversity and Phytochemistry Research Laboratory, Department of Botany, University of Kashmir, Srinagar, Jammu and Kashmir, India

CHAPTER MENU

17.1 Introduction

Nanoparticles are contrived materials. The field of nanoscience functions at a range of 100 nanometers (nm) or less (Moraru et al. 2003), but has an enormous impact (Bhattacharyya et al 2011). The term "nanomaterial" comes from the Greek word "nano" meaning "dwarf." Specifically, the word nano signifies 10–9 or 1 billionth of a meter; therefore, nanomaterial is commonly applied to materials whose size range is between 1 and 100 nm (Rai et al. 2012). Nanotechnology is commonly applied to the observation, measurement, manipulation and manufacturing of objects at the nanometer scale (Mongillo 2007). Nanotechnologies hold enormous capacity for generating novel materials with strengthened properties. An assortment of products grounded on nanotechnology have relevance in trade and commerce, which include medical instruments, sports, imaging, biosensing, drugs, electronics, makeup (NCPI 2011), farming, fabrics and food (Sharon et al. 2010; Sastry et al. 2010).

Nanotechnology provides novel materials suitable for the textile, packaging, food and transportation businesses (Roco and Bainbridge 2002). Nanotechnology, the cutting edge technological innovation, has a variety of uses such as in food transformation, farming and the environment. In nanotechnology, different materials such as food, seeds and pesticides are broken down and manipulated, and further food packing is done at microscale, and in doing so a nano-food and a revolution in farming is driven (Friends of the Earth 2008). The Royal Society has defined nanotechnologies as the plan, representation, creation and usage of structures, devices and systems by controlling shape and size at the

Bioresource Technology: Concept, Tools and Experiences, First Edition. Tanveer Bilal Pirzadah, Bisma Malik, Rouf Ahmad Bhat, and Khalid Rehman Hakeem.
© 2022 John Wiley & Sons Ltd. Published 2022 by John Wiley & Sons Ltd.

nanometer scale (Rsrae 2004). In 2011, the European Commission defined nanomaterial as an untreated, related or artificial material containing particles, in an emancipated state or in assortment, where most of the particles (50% or more) are in the range of 1 nm–100 nm.

17.2 Agriculture: A Growing Need

In a world with huge technological advancement in both the food and health sector, about 800 million people are still suffering from starvation and undernourishment. A goal was set by the United Nations, which aimed to cut this number to half by 2015. Today, the most ultimate challenge faced by society is the eradication of hunger and malnutrition (Lomborg 2004). Food is the basic requirement for population survival contrasting with the simulated value of meager substances such as gold. Thus, agriculture is considered as a first-level priority. With the advancement in science and technology, human civilization evolved from "Gatherer" to "Food producer." The concept of "scientific agriculture" dates back to journals by Liebig in 1840 and Johnston in 1842, who presumed that chemistry plays a significant role in agriculture (Pesek 1993).

Agriculture is one of the chief sources of income, as it provides employment for 60% of the world's total labor force (FAO 2013). About 22% of India's population is undernourished as estimated by The Food and Agriculture Organization (FAO 2011). Additionally, malnutrition in India is a serious issue. In 2008, UNICEF projected that out of all the developing countries, in India, 42% of children were underweight (low weight-for-age) and 32% among them showed stunted growth (low height-for-age) (UNICEF 2009). Agriculture is the principal source of all micronutrients to human beings. Agricultural systems must be delivered to the unstable food systems that are dwindling to meet the nutritional needs of everyone (Welch et al. 1997). Food availability designates the amount of food at the international, national or local level. It does not depict the proficiency of human beings to purchase it. There are different bases of food supply, such as household production for feeding and marketable food production (Diskin 1994). Improvement and growth in agricultural productivity are the central focus of rural development strategy. Agriculture plays a significant role in rural financial growth in many low-income countries, therefore has an important role in eradicating poverty in rural areas (Timmer 1997; World Bank 2001).

17.2.1 Advanced Agriculture System Through Nanoscience

Agriculture is the main source of food in nature. In India, agriculture has legalized extensive implementation of genetically-modified seeds, pesticides, chemical fertilizers, rigorous irrigation, ideal agronomic settings and up-to-date machinery, to limited crops. Food security is also an important factor to endure food quality with respect to its huge production. It is equally essential to maintain the production, quality and security of food which will be achieved by familiarizing nanoscience in the present era. Thus nanoscience has a huge influence and is identified as nanotechnology (Bhattacharyya et al. 2011). Nanotechnology allows customers to alter the food based on their nutritional requirements. This is achieved by using nano encapsulated substances that are being established as a portion of interactive foods (Lyons 2010). Nanotechnology finds use in changing the properties and attributes of food. Applications of nanotechnology are being reformed so that it can influence international markets for mineral, agricultural and other items (Figure 17.1). Many authors have reported that applications of nanotechnology could influence universal demand, which would lead to the opening up of innovative markets for merchandize. Such developments could have extensive socio-economic and many other effects (Dunford 2010). Nano biotechnologies allow us to generate directed nanoparticles, nanofibers and nanocapsules, which can transport foreign DNA and nano chemicals. This foreign DNA and nanochemical would ultimately alter the genes of insects and their hosts (Torney et al. 2007). In more recent times, a novel branch of technoscience grounded on the practices of nanotechnology, genetic engineering and informatics has emerged, which is known as synthetic biology. By this innovation, researchers can now entirely replace the hereditary content of one bacterium with another. This would eventually transform one species to another (Lyons 2010).

17.2.2 Nanofertilizers for Agriculture

Emerging fertilizers that are nano-based would solve the revolutionized agricultural productivity in which fertilizers are the main concern. These fertilizers are provided so that they can deliver three main constituents in equal ratios, which include nitrogen, phosphorous and potassium (Corradini et al. 2010). Nanofertilizers comprise nutrients and growth promoters compressed in nanoscale polymers. In this form they have measured and directed competent release. Nanotechnology is collecting data of atoms in the range

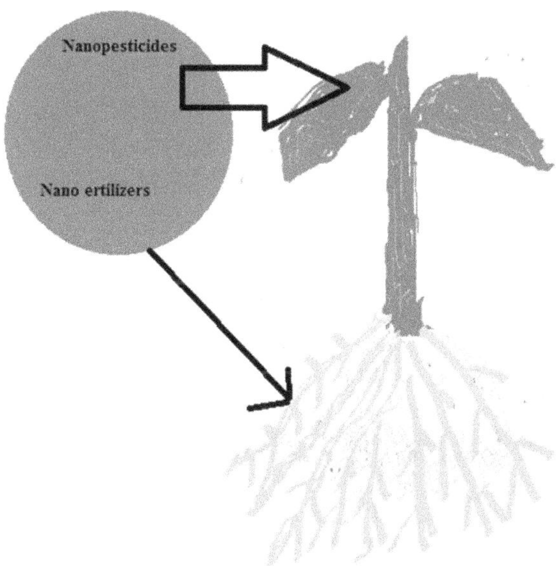

Figure 17.1 Application of nanofertilizers and nano pesticides helps in growth of plants.

of nanoscale, containing their catalytic, physical, optical and magnetic properties (Sadik et al. 2009). Nanofertilizers are less costly and are needed in small quantities as compared to chemical fertilizers which are costly and required in large quantities (Ibtisam 2001). Nanoporous zeolite, when used with urea, shows significant intensification in the uptake of nitrogen (Manik and Subramanian 2014). It is also recounted that ordinary fertilizers are less proficient than nanofertilizers (Liu et al. 2006).

In 2013, Subramanian and Rahale studied the nutrient release pattern of nanofertilizer formulations carrying nitrogen as a fertilizer. The results have revealed that the nano-clay-based fertilizer formulations (montmorillonite and zeolite having dimensions 30–40 nm) are more efficient than conventional fertilizers. They have the ability to release the nutrients over a longer duration of time (>1,000 h), particularly nitrogen (Subramanian and Sharmila Rahale 2013).

17.3 Nano Herbicides and Agriculture

Herbicides are used to kill weeds, but the problem associated with conventional weed-killers is that they affect both unwanted and useful plants. This limitation of conventional herbicides causes a huge loss in crop yield. With the advancement in nanotechnology, this problem can be tackled by using nano herbicides. Nano herbicides are minute particles whose size ranges between 1 and 100 nm. These minute particles intersperse with the soil particles and terminate the whole weed by destroying their roots without harming other useful crops. As the nanoparticles are explicit in destroying only the target plants they are thus useful in targeting the weeds and destroying them (Prasad et al. 2014). Nanotechnology has other applications, such as preventing surfaces from deteriorating by weather and bacteria. Surfaces that are painted can last for a longer duration, of up to 21 years or more, by using eco-friendly nano solutions. Nanomaterials can also control the discharge of the fertilizers, and this technique prevents the loss of nutrients in water or soil. Thus, nutrients are only taken up by the plants (Pradeep 2007).

A significant pest, *Spodopteralitoralis*, disturbs all chief plant kingdoms and is resistant to most of the pesticides. This pest was competently controlled by using nano silica-coated tomato, which was hydrophobic. It showed excellent results by devastating the pests at the concentration of 300–350 ppm (El-bendary and El-Helaly 2013). Multi-walled carbon nanotubes and citric acid molecules were combined, including some pesticides such as Zineb and Mancozeb which were confined in aqueous solution. This led to the establishment of encapsulated pesticides and posed a severe threat to the fungi, *Alternaria alternate* (Sarlak et al. 2014). In agriculture, the aim of nanomaterials is to diminish the spraying of chemical products. This can be achieved by efficient transport of vigorous ingredients and minimizing nutrient loss in the process of fertilization (Gogos et al. 2012). In genetic transformation and plant breeding, nanotechnology-derived devices are being surveyed for further improvements (Torney et al. 2007).

Research on the applications of agricultural nanotechnology has been ongoing for decades. Humankind is constantly probing for elucidations to many agricultural and environmental trials, which include competent varieties, sustainability and improved productivity. Some researchers have highlighted the budding inclination toward both patents and scientific publications in the field of agricultural nanotechnology (Sastry et al. 2010). Currently, the most critical obsession is created by poverty, agriculture and the process of attaining nutritious food. Hence, novel technology must be developed that decidedly emphasizes attaining better agricultural production (Yunlong and Smit 1994). This cannot be attained in isolation with agrochemicals which include fertilizers, pesticides, etc. Irrespective of this, agrochemical has some limitations, such as impurity of water or harmful deposits in food products. This contamination threatens the health of all living beings and the environment. Therefore, effective administration and monitoring of inputs can decrease these risks (Kah 2015).

The expansion of progressive agricultural systems along with the practice of engineered nanotools can prove beneficial in causing a revolution in agricultural practices. This would in turn reduce or eradicate the harmful effects on the environment by modern agriculture. It would enhance not only the quality but also the yields (Sekhon 2014; Liu and Lai 2015). Pathogen detection and diagnosis can be done by nanoscale biosensors. Additionally, this nanotechnology can stream bioactive constituents in food products to their hosts. This would lead to the enhancement of awareness of food products at the nanoscale (Martirosyan and Schneider 2014). Nickel and other metallic kinds can be allied to usual short-ordered surfaces of TiO_2, aluminosilicates, aromatic compounds and humic acids, by MWCNTs. These links are recognized as an operational method of bioremediation in the field of nano agriculture (Raliya et al. 2013; Hajirostamlo et al. 2015). The degree of development in technology and the rate of improvement in the coming years will positively affect the stability, and ultimately the throughput of agriculture (Hutchins and Gehring 1993).

17.4 Nanotechnology Leading to Sustainable Agriculture

Sustainable development means meeting the needs of the present without posing a threat to the needs of future generations (Corradini et al. 2010). Sustainability should involve various aspects, the people, the environment and the economy. To attain sustainability, it is important to take into consideration the complex linkages between the "social systems." A social system includes the institutions that support human existence on Earth, the global system and all those factors which impact the health and welfare of humans (Atalah et al 2013). Nanoscale manufacturing will provide the means for sustainable development: fewer materials, less water, less energy, less manufacturing waste, and new methods to convert energy and filter water (Gogos et al. 2012). In 2008, the United States spent $1,165 billion on food consumption (US Department of Agriculture 2009). Another food-related cost is that of food-borne illness, which was estimated to be around $152 billion a year in the United States (Scharff 2010).

Godfray and colleagues (FAO 2011) suggested, in order to meet the global demand for adequate food, the world will face a huge challenge over the next 40 years. This is because the population is continuously increasing and will reach approximately 9 billion by 2050. Advances in nanotechnology can result in major developments in the technologies which are used to grow, store, process and distribute food (Sarlak et al. 2014. Agricultural nanotechnology includes nanomaterials whose main function is to reduce the amount of chemical spraying products. This can be achieved by the smart delivery of active ingredients, minimizing nutrient losses in fertilization. Also, increasing the yields by optimizing water and nutrient management can reduce the need for chemical spraying.

New devices derived from nanotechnology are also being investigated in the field of plant breeding and genetic transformation. Furthermore, in agriculture, bio nanocomposites can be used for bio-industrial purposes. These particles have enhanced physical and mechanical properties based on traditionally harvested materials such as wheat straw and soy hulls. Subramanian et al. (2008) reported that nano-fertilizers and nanocomposites can be used to control the release of nutrients from the fertilizer granules. This would improve the nutrient use efficiency by preventing the nutrient ions from either becoming fixed or lost in the environment. Nano-fertilizers are highly efficient and can be delivered to the rhizospheric target on time. Nano-silica particles which are absorbed by the roots can form films at the cell walls. This will improve the plant's resistance to stress and lead to superior yields (DeRosa et al. 2010). As we know, agriculture is also under constant developmental transformation to meet the ever-increasing demands. Opara (2004) and Robinson and Zadrazilova (2010) reported that among several advances in science, nanotechnology is a rapidly evolving field, as it can revolutionize food production systems.

In 2011, Chena and Yada pointed out that nanoscale science and nanotechnology have the capability to provide novel and enhanced solutions to some chief challenges in agriculture (Chena and Yada 2011). Harmful effects of agrochemicals on the agroecosystems (i.e., soil, flora, fauna and water bodies), the well-being of the farmers using them, and the people consuming the chemically grown food, have become prominent concerns worldwide (Joseph and Morrison 2006). As a result, it has become important to apply green technology to mitigate environmental damage. The largest part of green technology is the use of bio-based technologies which benefit human needs, such as food, medicine, energy, commodities, pollution control, etc. This can be done through bioconversion and bio refinery processes highlighting pollution-free strategies (Feehan et al. 2009; Jones et al. 2009; Siegmann et al. 2008). The use of nanotechnology in agriculture has created much interest. This will increase the potential for significantly enhanced agricultural productivity and efficiency with reduced cost (Kah 2015; Scott and Chen 2013).

In 2009, Khodakovskaya et al. reported that the germination of tomato seeds can be improved by using carbon nanotubes, as they cause efficient infusion of moisture (Khodakovskaya et al. 2009). Their study showed that carbon nanotubes function as new pores. These pores assist in water permeation by penetrating the seed coat. Therefore, nanotubes function as a path to move water into the seeds from the substrate. These processes assist germination which can be integrated into a rain-fed agronomic structure. Nanotechnology, when directed as a tool in relation to further measures, can resolve some of the world's extreme sustainable development issues related to health and management of the ecosystem, including water, energy, environment, agriculture and biodiversity. Together, these five spheres are known as WEHAB. They were first recognized in the year 2002, by the United Nations Johannesburg summit on Sustainable Development. Preliminary results suggest that nanotechnology can help in achieving balanced fertilization (Tarafdar and Raliya 2012). The polymer-based CPA nanocapsule sprays can preclude this infection. This is achieved as the CPA response is triggered by the plant cell stress enzymes by acting as stimuli (Uzu et al. 2010). Some wetland plants are involved in copper biomineralization that leads to the transformation of copper into metallic nanoparticles. This process occurs at the root–soil interface, in the presence of some endomycorrhizal fungi. These endomycorrhizal fungi have the capability to diminish copper toxicity in the unclean soils (Manceau et al. 2008). Nanocomposites provide much of reactivity to soil biological and abiotic processes because of their physicochemical properties (Navrotsky 2004).

In 2006, Zhang et al. revealed that the outcome of nutrients, when wheat plants were treated with the five kinds of bulk-blended slow release fertilizers, absorbed NPK content by wheat in the shoots was greater than that of chemical fertilizer containing equal amounts of NPK (Zhang et al. 2006). This was coherent with the rule of nutrients required by wheat. Liu et al. (2005) reported that the addition of nano-sub nanocomposites proved advantageous to the soil and advances the utilization efficiency of fertilizer because of its exceptional characteristics.

17.5 Conclusion

The improvement of the innovative agronomic system with the use of concocted, efficient nanotools could prove advantageous in transforming the field of agricultural. This would in turn decrease or exclude the impact of prevailing agriculture on the surroundings. It would also augment the quality and quantity in terms of yield. New devices derived from nanotechnology are being scrutinized in the subjects of plant breeding and chromosomal (genetic) transformation. Additionally, for bio-industrial purposes, bio nano-composites can be used. These particles have improved physical and motorized properties that are grounded on conventionally harvested materials. The nanoscale science and nanotechnology have the capability to provide innovative and enhanced elucidations to some chief tasks in agriculture. To mention a few are the nanotubes that direct the water from the substrate into the seeds by acting as a passage for the water. Investigation on the applications of agricultural nanotechnology has been continuing for several decades. Mankind is constantly trying to find solutions to several agronomic and environmental problems, which include sustainability, competent varieties and improved productivity. Many researchers have shown the increasing trends in the field of agricultural nanotechnology, most importantly in the management of diseases and protection of crops. Nanotechnology can address some of the most perilous sustainable development issues in the areas of health, water energy, environment, biodiversity, agriculture and ecosystem management.

References

Atalah, J., Bennett, H., Hopkins, G.A. et al. (2013). Evaluation of the sea anemone *Anthothoe lbocincta* as an augmentative biocontrol agent for biofouling on artificial structures. *Biofouling* 29 (5): 559–571.

Bhattacharyya, A., Datta, P.S., Chaudhuri, P. et al. (2011). Nanotechnology: A new frontier for food security in socio economic development. In: *Proceedings of the Disaster, Risk And Vulnerability Conference*, pp. 559–571.

Chena, H. and Yada, R. (2011). Nanotechnologies in agriculture: New tools for sustainable development. *Trends in Food Science and Technology* 22 (11): 585–594.

Corradini, E., De Moura, M.R., and Mattoso, L.H.C. (2010). A Preliminary study of the incorporation of NPK fertilizer into chitosan nanoparticles. *Express Polymer Letters* 4 (8): 509–515.

DeRosa, M.C., Monreal, C., Schnitzer, M. et al. (2010). Nanotechnology in fertilizers. *Nature Nanotechnology* 5 (2): 91–91.

Diskin, P.K. (1994). Understanding linkages among food availability, access, consumption and nutrition in Africa: Empirical findings and issues from the literature. MSU International Development Working Paper 46. East Lansing, MI: Michigan State University. (No. 1096-2016-88334).

Dunford, N. (2010). *Nanotechnology and Opportunities for Agriculture and Food Systems*, Oklahoma Cooperative Extension Service. Division of Agricultural Sciences and Natural Resources. 139–140.

El-bendary, H.M. and El-Helaly, A.A. (2013). First record nanotechnology in agricultural: Silica nanoparticles a potential new insecticide for pest control. *Applied Scientific Reports* 4 (3): 241–246.

FAO (2011). Food security statistics. Available at: http://www.fao.org/economic/ess/ess-fs/fs-data/ess-fadata/en

FAO (2013). Synthesis of Guiding Principles on Agriculture Programming for Nutrition.

Feehan, J., Harley, M., and Minnen, J. (2009). Climate change in Europe. 1: Impact on terrestrial ecosystems and biodiversity. A review. *Agronomy for Sustainable Development* 29 (3): 409–421.

Friends of the Earth (2008). *Out of the Laboratory and on to Our Plates*. Australia: Nanotechnology in Food and Agriculture.

Gogos, A., Knauer, K., and Bucheli, T.D. (2012). Nanomaterials in plant protection and fertilization: Current state, foreseen applications, and research priorities. *Journal of Agricultural and Food Chemistry* 60 (39): 9781–9792.

Hajirostamlo, B., Mirsaeedghazi, N., Arefnia, M. et al. (2015). The role of research and development in agriculture and its dependent concepts in agriculture. *Asian Journal of Applied Science and Engineering* 4 (3): 78–80.

Hutchins, S.H. and Gehring, P.J. (1993). Perspective on the value, regulation, and objective utilization of pest control technology. *American Entomologist* 39 (1): 12–15.

Ibtisam, E.T. (2001). Biosensors developments and potential applications in the agricultural diagnosis sector. *Elsiver Science, Computers and Electronics in Agriculture* 30: 205–218.

Jones, A., Stolbovoy, V., Rusco, E. et al. (2009). Climate change in Europe. 2: Impact on soil. A review. *Agronomy for Sustainable Development* 29 (3): 423–432.

Joseph, T. and Morrison, M., 2006. Nanoforum report: nanotechnology in agriculture and food. *European Nanotechnology Gateway*.

Kah, M. (2015). Nanopesticides and nanofertilizers: Emerging contaminants or opportunities for risk mitigation? *Frontier Chemistry* 3: 64.

Khodakovskaya, M., Dervishi, E., Mahmood, M. et al. (2009). Carbon nanotubes are able to penetrate plant seed coat and dramatically affect seed germination and plant growth. *ACS Nano* 3 (10): 3221–3227.

Liu, R. and Lai, R. (2015). Potentials of engineered nanoparticles as fertilizers for increasing agronomic productions. *The Science of the Total Environment* 514: 131–139.

Liu, X.M, Zhang, F.D., Zhang, S.Q. S.Q.S.Q. et al. (2005). Study on adsorption and desorption properties of nano-kaoline to nitrogen, phosphorus, potash and organic carbon. *Scientia Agricultural Sinica* 1.

Liu, X.M., Feng, Z.B., Zhang, F.D. et al. (2006). Preparation and testing of cementing and coating nano-subnanocomposites of slow/controlled-release fertilizer. *Agricultural Sciences in China* 5 (9): 700–706.

Lomborg, B. (ed.) (2004). Global Crises, Global Solutions. Cambridge University Press.

Lyons, K. (2010). Nanotechnology: Transforming food and the environment. *Food First Backgrounder* 16 (1): 1–4.

Manceau, A., Nagy, K.L., Marcus, M.A. et al. (2008). Formation of metallic copper nanoparticles at the soil–root interface. *Environmental Science & Technology* 42 (5): 1766–1772.

Manik, A. and Subramanian, K.S. (2014). Fabrication and characterisation of nanoporous zeolite based N fertilizer. *African Journal of Agricultural Research* 9 (2): 276–284.

Martirosyan, A. and Schneider, Y.J. (2014). Engineered nanomaterials in food: Implications for food safety and consumer health. *International Journal of Environmental Research and Public Health* 11 (6): 5720–5750.

Mongillo, J.F. (2007). *Nanotechnology 101*. ABC-CLIO.

Moraru, C.I., Panchapakesan, C.P., Huang, Q. et al. (2003). Nanotechnology: A new frontier in food science understanding the special properties of materials of nanometer size will allow food scientists to design new, healthier, tastier, and safer foods. *Nanotechnology* 57 (12): 24–29.

Navrotsky, A. (2004). Environmental nanoparticles. *Dekker Encyclopedia of Nanoscience and Nanotechnology*, vol. 2 (ed. J.A. Schwarz and C.I. Contescu), pp. 1147–1155. New York.

NCPI (2011). Nanotechnology consumer products inventory. Available at: http://www.nanotechproject.org Observatory NANO, Briefing No. 20. Statistical Patent Analysis. Patents: an indicator of nanotechnology innovation.

Opara, L.U. (2004). Emerging technological innovation triad for smart agriculture in the 21st century. Prospects and impacts of nanotechnology in agriculture. *Agricultural-Engineering International* 33 (2): 101–111. 33 (20).

Pesek, J. (1993). Historical perspective. In: *Sustainable Agriculture Systems* (ed. J.L. Hatfield and D.L. Karlen). Boca Raton, FL: CRC Press.

Pradeep, T. (2007). *NANO: The Essential: Understanding Nanoscience and Nanotechnology*. McGraw Hill.

Prasad, R., Kumar, V., and Suranjit, K.P. (2014). Nanotechnology in sustainable agriculture: Present concerns and future aspects. *Academic Journal of Biotechnology* 13 (6): 705–713.

Rai, M., Deshmukh, S., and Gade, A. (2012). Strategic nanoparticle-mediated gene transfer in plants and animals: A novel approach. *Current Nanoscience* 8 (1): 170–179.

Raliya, R., Tarafdar, J.C., Gulecha, K. et al. (2013). Review article: Scope of nanoscience and nanotechnology in agriculture: A Review. *Journal of Applied Biology and Biotechnology* 1 (3): 41–44.

Robinson, D.K.R. and Zadrazilova, G.S. (2010). Nanotechnologies for nutrient and biocide delivery in agricultural production. *Working Paper Version* 285–297.

Roco, M.C. and Bainbridge, W.S. (2002). Converging technologies for improving human performance: Integrating from the nanoscale. *Journal of Nanoparticle Research* 4 (4): 281–295.

Rsrae, N. (2004). *Nanotechnologies: Opportunities and Uncertainties RS Policy Document,* 19/04. London: The Royal Society and Royal Academy of Engineering.

Sadik, O.A., Zhou, A.L., Kikandi, S. et al. (2009). Sensors as tools for quantitation, nanotoxicity and nanomonitoring assessment of engineered nanomaterials. *Journal of Environmental Monitoring* 11 (10): 1782–1800.

Sarlak, N., Taherifar, A., and Salehi, F. (2014). Synthesis of nanopesticides by encapsulating pesticide nanoparticles using functionalized carbon nanotubes and application of new nanocomposite for plant disease treatment. *Journal of Agricultural and Food Chemistry* 62 (21): 4833–4838.

Sastry, R.K., Rashmi, H.B., and Rao, N.H. (2010). Integrating nanotechnology into agri-food systems research in India: A conceptual framework. *Technological Forecasting and Social Change* 77 (4): 639–648.

Scharff, R.L. (2010). *Health-Related Costs from Foodborne Illness in the United States.* Washington DC: Produce Safety Project.

Scott, N. and Chen, H. (2013). Nanoscale science and engineering for agriculture and food systems. *Industrial Biotechnology* 9 (1): 17–18.

Sekhon, B.S. (2014). Nanotechnology in agri-food production: An overview. *Nanotechnology, Science and Applications* 7: 31–53.

Sharon, M., Choudhary, A.K., and Kumar, R. (2010). Nanotechnology in agricultural diseases and food safety. *Journal of Phytology* 2 (4).

Siegmann, P., Acevedo, F.J., Siegmann, K. et al. (2008). A probabilistic source attribution model for nanoparticles in air suspension applied on the main roads of Madrid and Mexico City. *Atmospheric Environment* 42 (17): 3937–3948.

Subramanian, K.S. and Sharmila Rahale, C. (2013). Nano-fertilizers: Synthesis, characterization and application. In: *Nanotechnology in Soil Science and Plantnutrition* (ed. T. Adhikari and S. Rao), pp. 69–80. New Delhi, India: New India Publishing Agency.

Subramanian, K.S., Paulraj, C., and Natarajan, S. (2008). Nanotechnological approaches in nutrient management. In: *Nanotechnology Applications in Agriculture, TNAU Technical Bulletin* (ed. C.R. Chinnamuthu, B. Chandrasekaran, and C. Ramasamy), pp. 37–42. Coimbatore: TNAU.

Tarafdar, J.C., Raliya, R., and Rathore, I. (2012). Microbial synthesis of phosphorous nanoparticle from tri-calcium phosphate using *Aspergillus tubingensis* TFR-5. *Journal of Bionanoscience* 6 (2): 84–89.

Timmer, C.P. (1997). How well do the poor connect to the growth process? Consulting Assistance on Economic Reform Discussion Paper 178. Cambridge, MA: Harvard Institute for International Development.

Torney, F., Trewyn, B.G., Lin, V.S.Y. et al. (2007). Mesoporous silica nanoparticles deliver DNA and chemicals into plants. *Nature Nanotechnology* 2 (5): 295–300.

UNICEF (2009). *State of the World's Children.* New York.

US Department of Agriculture (2009). Economic Research Service. (USDA-ERS).

Uzu, G., Sobanska, S., Sarret, G. et al. (2010). Foliar lead uptake by lettuce exposed to atmospheric pollution. *Environmental Science & Technology* 44 (3): 1036–1042.

Welch, R.M., Combs, G.F., and Duxbury, J.M. (1997). Toward a "Greener" revolution. *Issues in Science and Technology* 14 (1): 59050–58.

World Bank (2001). *Making Sustainable Commitments: An Environment Strategy for the World Bank*. Washington DC: World Bank Publications.

Yunlong, C. and Smit, B. (1994). Sustainability in agriculture: A general review. *Agriculture, Ecosystems & Environment* 49 (3): 299–307.

Zhang, F., Wang, R., Xiao, Q. et al. (2006). Effects of slow/controlled-release fertilizer cemented and coated by nano-materials on plants. *Nanoscience* 11: 18–26.

18

Profitability and Economics Analysis of Bioresource Management

Ghulam Mustafa

Department of Economics and Business Administration, Division of Arts and Social Sciences, University of Education, Lahore, Pakistan

CHAPTER MENU

18.1 Introduction

Bioresources are usually associated with production, processing, consumption and use of materials. However, Krozer and Lordkipanidze (2019) reported that non-material activities should also be included in bioresources and they were of the opinion that cultural ecosystem services is emerging in terms of economics development and scientific interest. Bioresources can be produced endlessly if nature is used with care. Bioresources contribute to the regulation of environmental qualities and sustain life on Earth. These provide food and energy, and other materials such as quality of life, leisure, relaxation and such-like amenities that foster tranquillity. Bio-based resources also add cultural values that include basic resources for artistic, educational and scientific inspiration, and thus support ethical, spiritual and esthetical behavior.

18.2 Bioeconomy

A bioeconomy can be defined as an economic systems based on renewable biological resources including energy, chemical and the basic building blocks for materials

Bioresource Technology: Concept, Tools and Experiences, First Edition. Tanveer Bilal Pirzadah, Bisma Malik, Rouf Ahmad Bhat, and Khalid Rehman Hakeem.
© 2022 John Wiley & Sons Ltd. Published 2022 by John Wiley & Sons Ltd.

(McCormick and Kautto 2013). In more simple words, an economy based on biomass rather than fossil fuels. Different studies used terms such as knowledge-based bioeconomy (KBBE) or bio-based economy for bioeconomy. However, these concepts can be understood as an economy founded on renewable bioresources, such as plant and animal sources (EuropaBio 2011; European Commission 2012b).

Europe is considered a pioneer and the global leader in a number of fields of bioresources and its related products (ENDS Europe 2012; European Commission 2012b). Bioeconomy is essential for smart and green growth worldwide, particularly in European countries. According to the European Commission (2012a), bioeconomy has a huge market of 2 trillion Euros in Europe and furnishes 22 million jobs in the multiple sector, including bioenergy, chemical, agriculture, food and forestry (Bio-Economy Technology Platforms 2011; Clever Consult 2010). It benefits from almost 9% of total EU labor (European Commission 2012a). These statistics describe not only the magnitude of bioeconomy to the European economy, but also narrate opportunities to amalgamate different sectors and flourish of bio-based products. However, the USA and some countries in Asia, like China, are investing woodenly into the bioeconomy. This is highlighted by a National bioeconomy Blueprint published in 2012 to rehabilitate bioeconomy and bio-based products. The EC fears that long-term competitiveness of Europe is in jeopardy, as Europe holds back these countries in market deployment (European Commission 2012b).

Depletion of greenhouse gas (GHG) emissions, decline in fossil fuels dependence, sagacious management of natural resources and enhanced food security are advantages of transition to a bio-based economy. It also gives rise to employment opportunities, both in rural and urban settings. Moreover, the establishment of non-food markets (bioenergy) in collaboration with existing food markets and alternative income sources for farmers can give a major boost to rural areas (Langeveld and Sanders 2010). The technical prospective of bioeconomy is spectacular, as portrayed by Bünger (2013). He noted that over 90% of bioresource-based alternatives could be replaced with oil-based products. However, the issue is to enhance the scale of activities (e.g., in terms of biomass production) in parallel with meeting key sustainability goals.

18.3 Profitability Analysis of Bioresource-based Business

Building business models and their profitability analysis are useful tools for assessing opportunities in the markets of bioresources. The bioresources markets are founded on the consumption and production of wood products, meat and crops fibers, food, fodder and residues.

The world bioresource production is about 14,768 million tonnes per year, where half of it comprises of food crops. The EU is the largest producer of bio-based products, with 1965 million tonnes per year. Its regional production is 13% worldwide, which constitutes approximately 7% of the world population. The EU earns €2357 billion annually from the production, processing and sale of bioresources-based products, while these activities generate about 22 million jobs (Scarlat et al. 2015).

Bioenergy is the lowest price category among the few categories of total market of bioresources. There are a few classifications of the bioresource market, in which bioenergy rank last. The EU has provided 390 million tons of biomass to the market, equivalent to about 123 million tonnes of oil. Through which consumption has been increased from 18% in 2015 to 63% of today, all this information has been collected from European statistical data (Eurostat).

Industrial products are higher in price and more diverse than bioenergy, although smaller in scale. Forestry and wood products are also included in industrial products where wooden products alone account for almost US$ 500 billion a year. Moreover, these also include agro-industrial products based on surfactants, solvents, lubricants, bio-based chemicals, sugar, starch, bioplastic and other biochemical (~US$ 50 billion per year). In the EU, about US$ 230 billion is generated by agro-culture and where around US$ 1000 per tonne biomass is the usual price (Biddy et al. 2016; Haveren et al. 2008).

Nutrition of people and cattle comprise the largest and most diversified bioresourses market. In the EU, it uses almost 1 billion tons biomass a year and accounts for more than US$ 1000 billion a year. Although the markets of high value products for personal care and medicine are large and diversified, the share of bioresourses in these sectors is still unknown. Per ton prices are many times greater than food prices, sometimes up to US$ 25,000 per tonne (Bio-biased Industries 2016).

18.3.1 Short Rotation Cultivation (SRC)

When fast-growing species are cultivated in short rotation, the output can escalate. Short rotation cultivations mainly use Acacia (Robinia), Eucalyptus (Eucaliptus), Poplar (Populus) and willows (Salix). These SRC species produce a high yield of bioresources within a short span of time. These can be harvested a few years after sowing and hence are ready to sell and generate cash flows for the farming business (Bartha et al. 2017). This is important for impoverished regions where people with low incomes can generate a livelihood and cannot afford a delay in the payments due to them.

The bioresources extracted from woodland are used in the production of conventional wood products, such as lumber, laminated panels, round wood and structural timber. Additionally, by-products from forests are not wasted but modified into pellets and briquettes for biofuel utilized in heating and electricity generation that further the sale in the national electricity market. Those uses can be secured due to afforestation, and thanks to the SRC on plantations that focus on biofuels. The producers of bioresources are small farmers who own less than 50 ha land. They produce the bioresources for biofuel in addition to various other uses, particularly for foods which support regional and local farming.

SRS is a substitute of agricultural monocultures and opportunity for smallholdings and disadvantaged lands with poor soil quality close to villages, particularly when SRC plantations are often amalgamated with agroforestry. These marginal lands can be inappropriate for housing and suchlike uses, or constitute areas prone to flooding (Hart 2001; Heimlich 1989). Therefore a country, such as Romanian, authorizes SRC species only on poor and less fertile lands after treating with fertilizer, pest control and good preparation. The employment of poor lands for SRC is a choice that helps sustainable regional development because it sustains ecosystem services if good soil management techniques are applied, stimulates water and soil quality, increases carbon sequestration, and boosts economics for the local people (Lal 2004).

Bioresources from SRC are sold to the local businessmen who are usually large owners of cogeneration installations for heat and power (CHP). They can also be responsible for district heating on a subsistence level. Since bioresources are less costly and easy to handle in contrast to CHP for heating and electricity, they use woodchips for heating. District heating boilers are less energy-efficient than co-generators so they lose more heat as compared to CHP. However, district heating can still be used for village requirements with minimum costs. Agricultural wastes can be used for bioenergy production, but such technological advancements are rarely available in developing countries. For instance, approximately 12.6 million tons biomass per year is left unattended or burned in Romania, which could not be used more efficiently (Scarlat et al. 2011).

18.3.2 Ecotourism

Ecotourism is rapidly grown and constitutes a huge economic volume which, for example, covered about 24% of all tourism expenditure in 2015, which means that approximately US$ 900 billion is spent on the use of bioresources for leisure and recreational purposes (Krozer and Lordkipanidze 2019). Other researchers reported US$ 600 billion expenditure on ecotourism in the form of bioresources (Balmford et al. 2015). The number of tourists using bioresources are increasing with per capita consumption of ecotourism of US$ 117 per year (Balmford et al. 2009).

Aesthetic natural and environmental qualities enhance ecotourism and hence national income of any country that contributes to such an environmental-friendly environment. For instance, Krozer and Lordkipanidze (2019) found that there was substantial income from national parks when governments took environmental-friendly initiatives based on bioresources and were allowed to attract visitors into the framework of nature conservation. They further estimated that income generation from visitors usually exceeded that from governmental support.

Ecotourism is growing fast when compared to other bioresource products such as biofuels. For instance, in terms of US$ 100 price per barrel of oil equivalents, ecotourism proved itself a bigger industry than that of the value of biofuels. On the one hand, ecotourism generated more demand for bioresources. However, a large numbers of visitors are putting pressure on bioresources that degrade biodiversity, silence, beauty and other facilities. Therefore, it is necessary to overcome the challenges posed by the large numbers of visitors to nature and environment for sustainable development of ecotourism. This can be done if tourists receive environmental education about how to protect nature (Krozer and Christensen-Redzepovic 2006).

18.3.3 District Heating

The bioresources for energy production and hence district heating in many areas are available in many countries. For instance, Tihamer (2019) estimated energy demands and supplies based on local bioresources regarding district heating. He developed district heating based bioenergy on local bioresources in seven villages in Romania,

Slovenia, Macedonia, Croatia and Serbia. He found that such projections generated positive NPV (net present value), which means that any such project is financially good if developed.

18.3.4 Aquatic Biorefinery

The aquatic biorefinery business ecosystem is based on aquaculture and includes algae industry, dryland, and freshwater fisheries and marine resources. With the expansion of assembling merchandize for the food business, the hydroponics can work with crude materials for various enterprises. In the Nordic countries such as Iceland and Norway, the attainable quality of aquatic raw material is their greatest strength. It is necessary for a successful aquatic biorefinery business and its value chain (from production to end consumer) that raw materials should be situated nearby and should be fresh when processed. The aquatic biomass in such biorefineries is purified and turned efficiently into multiple products for many end users and numerous industries. For instance, fish waste can be utilized for algae and omega oils in dietary supplements. It can be used in biodiesel production, despite its traditional use as goods for food and feed industries.

A full industrial hub can be developed near aquaculture ecosystems, where all raw materials can be utilized, energy can be produced from wastes, and nutrients can be recycled to their full potential. For instance, organic wastes and residue flow from aquaculture can act as inputs for biogas production and this further refines the rejected waters of the biogas unit into a biofertitilzer for farming. In this way, nutrients can be recycled. Additionally, energy can be produced from biogas plants which can be used for electricity and heat generation. Transportation fuel can also be generated if the biogas plant is suitably upgraded. For any successful business model, there should be close connections to many industries. The aquatic biorefinery business is the best example of this. There are numerous possibilities for synergies between nutrient production, chemical generation, energy and biofuel production, food production, fish farming and algae production. The fish farming business can benefit from the algae industry by utilizing the oxygen provided by growing algae and from close proximity of greenhouses to utilize their excess heat.

One critical angle to consider in aquatic biorefineries treatment facilities is the linkage with reasonable organization of the oceanic assets. Sustainable management aims at keeping the strength of the aquatic ecosystems business in the long run and minimization of any antagonistic impacts are necessary for sustainable development of oceanic bioresources. Aquatic ecosystem-based ecotourism and other synergistic uses such as nature excursions are necessary for successful future aquatic biorefineries.

The organizations within the fisheries and aquaculture sector tend to be small. Financing is often a bottleneck and co-activity is obligatory in order to have the option to fund improvements and new ventures. The organizations chosen to exemplify the possibilities in the aquatic biorefineries business ecosystem were Icelandic Ocean Cluster (Iceland) and the Sybimar Oy (Finland). Sybimar was included as an innovative SME in a business which combines many different concepts within the bioeconomy, having the aquatic biorefinery as a core of the concept (Rönnlund et al. 2014).

18.4 Food Waste to Bioresource Businesses and Their Efficacies

18.4.1 Biofertilizer and Biogas Production

There are four ways to handle the food wastes from mass catering: i) donation to charity; ii) animal feed; iii) incineration; and iv) landfill. The small portion of the total surplus is going to charity at a cost involving its distribution. However, most of the food distributed to charity is unhygienic and rotten, so cannot be used and is thus wasted. Usually, caterers invite feed companies to collect food waste for animal feed at nominal or zero cost. Unfortunately, most feed companies rarely take advantage of collecting from caterers. Moreover, studies have found that food waste used for animal feed can result in severe health issues for the animals that will receive the feed. For instance, Salemdeeb et al. (2017) reported that food waste for animal feed caused diseases such as mad cow disease and scrapie. Thus, they were of the opinion that it can a be large threat, not only to animal health but to the public also. Perhaps, for this reason, untreated food waste for animal feed is now banned in the USA and Canada. Moreover, China recently passed the Chinese Animal Industry Act which articulated that food waste before high-temperature treatment is not allowed to be used as animal feed (Ma and Liu 2019). Another two methods of handling food waste in Pakistan are incineration and landfilling. Incineration and landfill are the main management practices adopted historically to handle food waste; however, these tactics have serious impacts on economic, social and environmental systems.

All four management techniques are used to handle food waste in Pakistan. However, other countries, particularly developing countries, are using food waste as potential fertilizers (Gutierrez et al. 2018; Lam et al. 2018). In this regard, anaerobic digestion, composting and fertilizer production are the promising techniques to produce energy that further can be used for heating, electricity and fertilizing agricultural land. Furthermore, these techniques relieve burdens on climate change that may be caused due to incineration and landfilling. Although bio-fertilizer demand through composting is contributing to agricultural and horticultural industries, it takes a long processing time (15–30 days). Compost also creates other environmental issues such as its carbon footprint and secondary pollutants. Therefore, anaerobic digestion that is also eco-friendly can be used for large commercial purposes. Anaerobic digestion has a large treatment capacity of more than 150 tons of food waste per day. It has more energy conversion efficiency and can be used for small events to large multi-storey buildings having a capacity for thousands of guests. Moreover, food waste can be used for bioenergy recovery that can be good business, as can be seen in Box 18.1.

18.4.2 Biomethane

Biogas cleaned from contaminants and upgraded through removal of CO_2 is known as biomethane. Biomethane can be produced by using a biogas purification system using water scrubbers, membranes and bio-filters. It can be used as an alternative to natural gas in natural gas vehicles (NGVs) and its usage has grown rapidly worldwide. Over 18 million vehicles were operating on natural gas by year 2015 in many countries such as Brazil, Argentina, China, Pakistan and Iran (Khan et al. 2015). In Mexico, "the cleanest and securest fuel in the country" was adopted and the number of NGVs are supposed to increase as SEMARNAT invests in projects to encourage the use of fuels with a low carbon footprint (SEMARNAT 2017). This bio-based product craze has been developed in Europe and natural gas is replacing traditional fuels (petrol and diesel) in buses (Bord Gais 2010). The NGV fleet in Sweden is estimated at 44,000 vehicles (Le Fevre 2014). Biomethane has been used in all the metropolitan city transports of Linkoping since 2002. Linkoping used an anaerobic digestion plant that treats a combination of industrial organic waste, abattoir and household wastes, and animal manure (IEA 2005).

Biomethane from food waste potentially produced 42.32 PJ per year, equivalent to 6.5% of the energy content of diesel used in transport in 2015 in Mexico (Gutiérrez et al. 2018). By replacing diesel with biomethane from food waste, a reduction of 17.91 $MtCO_2e$ can be effected, 6.06% of the 2050 GHG emissions target (Gutiérrez et al. 2018). Furthermore, Gutiérrez et al. (2018) estimated the economic feasibility of a biomethane plant for a Mexican city using two scenarios. In the first scenario, economic performance based on net present value (NPV) gave a positive outcome. In the second scenario, the NPV value was negative and the authors recommended that this situation can be economically viable if a subsidy of $US 1.38/GJ is applied, equivalent to 5% of the cost of diesel.

18.4.3 Bioethanol Fermentation

Commercial gasoline prices are on a rising trend despite its environmental implications. For instance, only in China prices rose to CNY 6200/ton in 2016 from CNY3750/ton in 2005. Therefore, demand for bioethanol and its price both increased steadily. Fossil fuels can partially be replaced when gasoline is blended with bioethanol to combat environmental issues because of its density and combustibility (Manzetti and Andersen 2015). Due to its efficacy and environmentally-friendly behavior, more than 40 countries have adopted the uses of biofuel-ethanol and automotive ethanol gasoline. It accounts for 60% world consumption of gasoline with the annual rate of about 600 million tons of bioethanol (Jiao et al. 2019).

Sugar and corn are the main cops that act as feedstocks for bioethanol production in many countries. However, these too have environmental complications due to many obvious reasons, and thus this approach is not environmentally sustainable and economically viable. Therefore, it is needed to explore alternative, sustainable and green feedstocks for bioethanol production toward sustainable economical and environmental development. Food waste management is one of the green approaches that can act as feedstock for bioethanol production through anaerobic or aerobic fermentation, as food waste is rich in carbohydrate. For instance, Kiran et al. (2015) produced bioethanol with a concentration of 58 $g\ L^{-1}$ from food waste fermentation, although 30–40% of solid residue still required to be managed further. Similarly, Ma et al. (2017) found an innovative approach, fungal mash (pre-treatment with *in-situ* produced hydrolytic enzymes), for bioethanol production through food waste. This approach is more cost-effective as compared to the approach of Kiran et al. (2015), with about 90% total residue reduction of food waste, while it has produced a higher bioethanol concentration of 71.8 $g\ L^{-1}$. Thus,

Box 18.1 Potential of food waste to bioenergy recovery: An emerging business

Currently, 1.3 billion tonnes of food per year is being wasted worldwide (FAO 2014) and about one-third (33.3%) of total food production (from food processing to consumption) is wasting globally (FAO 2018). As per the FAO, food wastage is higher at the processing level in developing countries, while it is higher at the retail and consumer levels in developed countries. Studies have shown that the per capita food loss varies from region to region and even within the same region. For instance, it is 280–300 kg per year in Europe and North America, while it is 120–170 kg per year in Sub-Saharan Africa, South Asia and Southeast Asia. The United Nations has targeted to halve the per capita global food waste at the retail and consumer levels by 2030, defined under the Sustainable Development Goal 12 (UN 2015). However, food wastage is still a real conundrum as the study reported that food waste in Asian countries might rise to 416 million tonnes in 2025 (Melikoglu et al. 2013).

In Pakistan, conditions are worste, for instance, the Agha Khan University (2011) reported that 44% of children under 5 years of age were stunted and about 60% of the population is food insecure in that country. Similarly, Dawn (2016) reported that about 40% of the cooked food in Pakistan is wasted. Therefore, such inefficiencies should be curbed as the country is already suffering from malnutrition and food insecurity.

It has been estimated that 3.3 billion tonnes of CO_2 is accumulating in the atmosphere every year due to food waste and thus making this one of the major sources of environmental pollution. There are a number of reasons for CO_2 emission, such as dumping of food waste in open areas or into landfill. Hence, food waste becomes a major part of the municipal waste. Conventionally, CO_2 emissions increase in bulk quantity when this municipal waste is incinerated (Agarwal et al. 2005; Kumar and Goel 2009; Kumar et al. 2009; Pattnaik and Reddy 2010). Food waste has a high moisture contents, so when it is incinerated it releases many dioxins which may further lead to health and environmental issues (Katami et al. 2004). Therefore, proper management of food waste is urgent (Tayyab et al. 2019).

Food waste is higher at the stage of consumption (56%) as compared to the manufacturing phase (38%) throughout the food supply chain (Monier et al. 2010). At the consumption stage, restaurants, cafeterias, hotels, food processing plants, household and commercial kitchens (catering) are the main contributors of food waste. Among these, catering services contribute about 30% to the total food served (FAO 2014). Catering services are heavily used for large events attended by thousands of people. These catering services are mostly provided internally or sometimes externally. Thus, the demand for catering services are increasing day-by-day. Eventually, food waste increases with these growing catering services (Tayyab et al. 2019). It is estimated that catering services alone had a value of US$ 125.5 billion in 2016 and further have increased to US$ 153.2 billion in 2021 (Business Wire 2017). Particularly, catering services increased in South East Asia and Pakistan due to cultural practices that significantly increased the number of large catering events (SMEDA 2016). Therefore, food waste is one of the greatest opportunities for renewable energy production, and so proper management is required to handle these wastes.

In developing countries, biogas plants from animal dung have been successfully established in rural areas. These have been used for tube-wells (for pumping waters), electricity generation for agricultural operations and domestic use of biogas in rural areas. Food waste that is enriched in biogas as compared to animal dung has more potential and is used for commercial purposes, particularly for catering services. Food waste is rich in minerals, starch, lipid, protein, nutrients, cellulose and trace elements (Ayomoh et al. 2008; Hoornweg and Bhada-Tata 2012). Food waste can be used to produce multi-functional protein feeds known as dehydrated feed and biochemical feed (San Martin et al. 2016). Therefore, it has more potential to produce biogas that further can be used for other auxiliary purposes. Generation of biogas from food waste has already been adopted in developed and other developing countries, but in Pakistan there is still a need to utilize it on a commercial basis.

it seems obvious that the food waste fermentation for bioethanol production could not be an environmentally-friendly solution in the near future if the post-concentration and purification approaches cannot be adapted to meet various commercial requirements.

18.5 Bioresources for Risk Prevention and Poverty Alleviation

Metropolitan regions need a huge volume of bioresources in playgrounds, orchards, gardens and plantations, and such vegetation is regularly abundant in biodiversity. Their main role is for the well-being and prosperity of the population; however, they additionally generate other commodities, for example, biofuels. Bioresources encourage healthy living through the natural habitat. Trees give evaporation and shade, which have a cooling impact. Such a natural cooling impact is especially significant in dense metropolitan territories where "heat islands" may create health risks. Plants additionally add to flood avoidance as they slow down the action of waves, temper water flow and create spongy soil. Moreover, bioresources provide other cultural ecosystem services such as combinations of the cultural and regulatory ecosystem services usually known as "building with nature." Bioresources are also useful in wastewater gardens, constructed wetlands and rooftop vegetation. Such vegetation is useful to control water and air pollution in combination with education, scientific and aesthetic services. Moreover, in education, healthcare and tourism, there are more direct and indirect uses of bio-based resources.

Bioresources help to reduce poverty, particularly in developing countries. For instance, Badola and Aitken (2010) found that bioresources had helped with poverty alleviation in the Indian Himalaya. They studied the use of wild biological bioresources such as rhododendrons, orchids, edibles, medicinal plants and ethnobiological knowledge packages to reduce poverty in the Himalaya, Sikkim and Himachal Pradesh states. Among others, herbal medicinal plants have a global market value of US$ 43 billion per year (Christie 2001). Many bioresources-based plant varieties create a livelihood for local communities but in recent years over utilization of fodder, medicinal plants, food sources and firewood has reduced their sustainable development. For centuries, the sustainable use of biodiversity has been the mainstay for both poverty reduction and biological conservation. For any project to be successful, particularly in relation to bioresources and biodiversity, involvement of society as a whole is necessary. The main role of civil society is to create harmony and support bioresource-based entrepreneurships that help poverty reduction and sustainable development of bioresources.

18.6 Conclusion

The bioresources markets have a pivotal role in both developing and developed economies. The current chapter highlights the profitability of bioresource-based products and their efficacies. The chapter discusses how many countries have efficiently produced and utilized bio-based products and earned multi billions in the national income. Hence, many economies are going bioeconomic based on bioresources. Such economies are not only helpful in poverty reduction but also business and environmental risks are reduced to a great extent that further ensures food availability for ever-increase populations. In most businesses, NPV values are positive, which represents the soundness of bio-based business. Although in some businesses NPV was negative, for which subsidy is recommended, due to high positive externality for the environment where social benefits are more than that of social costs.

Food waste converted to bioenergy recovery is spreading around the globe. There is still much potential of food waste for biofertilizer, biogas and biofuel production. Managing food waste in such a way will ensure a safer and cleaner environment and reduce the burden on municipal waste. In developed countries, biogas production from food wastes has been developed but still there is a need to build such units in developing countries to fulfil their demand for energy.

References

Agarwal, A., Singhmar, A., Kulshrestha, M. et al. (2005). Municipal solid waste recycling and associated markets in Delhi, India. *Resources, Conservation and Recycling* 44 (1): 73–90.

Ayomoh, M.K.O., Oke, S.A., Adedeji, W.O. et al. (2008). An approach to tackling the environmental and health impacts of municipal solid waste disposal in developing countries. *Journal of Environmental Management* 88 (1): 108–114.

Badola, H.K. and Aitken, S. (2010). Biological resources and poverty alleviation in the Indian Himalayas. *Biodiversity* 11 (3–4): 8–18.

Balmford, A., Beresford, J., Green, J. et al. (2009). A global perspective on trends in nature-based tourism. *PLoS Biology* 7 (6): e1000144.

Balmford, A., Green, J.M., Anderson, M. et al. (2015). Walk on the wild side: Estimating the global magnitude of visits to protected areas. *PLoS Biology* 13 (2): e1002074. https://doi.org/10.1371/journal.pbio.1002074.

Bartha, S., Vajda, B., Duarte, L.C. et al. (2017). Environmental and Economical Assessment Analysis of the Energy Willow. A New Biomass Source for the Biorefinery Industry.

Resilient Society, Multidisciplinary Contributions from Economic, Law, Policy, Engineering, Agriculture and Life Sciences. Retrieved from http://www.editura.bioflux.com.ro/docs/Binder1.pdf#page=10 (accessed 7 April 2021).

Biddy, M.J., Scarlata, C., and Kinchin, C. (2016). Chemicals from Biomass: A Market Assessment Of Bioproducts With Near-Term Potential. (No. NREL/TP-5100-65509). Golden, CO: National Renewable Energy Lab. (NREL).

Bio-based Industries (2016). *A Public–Private Partnership on Bio-based Industries, Bio-based for Growth*. p. 12. http://www.biobasedeconomy.nl/wp-content/uploads/2012/07/Bio-Based-Industries-PPP-Vision-doc.pdf (accessed 5 April 2021).

Bio-Economy Technology Platforms-BECOTEPS (2011). *The European Bioeconomy in 2030: Delivering Sustainable Growth by Addressing the Grand Societal Challenges*. Brussels, Belgium: Bio-Economy Technology Platforms.

Bord Gais (2010). *Compressed Natural Gas Transport Fuel for Your Vehicles*. http://www.gasnetworks.ie/Global/Gas.Industry/NGV/CNG_Info_Sheet.pdf.

Bünger, M. (2013). *Biofuels: Putting Pressure on Petrol*. Available online: http://www.renewableenergyworld.com/rea/news/article/2010/06/biofuels-putting-pressure-on-petrol (accessed 30 March 2021).

Business Wire (2017). *Technavio Reveals Cost Saving Opportunities for the Global Catering Services Market 2017–2021*. https://www.businesswire.com/news/home/20170131005506/en/Technavio-Reveals-Cost-avingOpportunities-Global-Catering (accessed 4 February 2019).

Christie, A. (2001). Herbs for health, but how safe are they. *Bulletin of the World Health Organization* 79 (7): 691–692.

Clever Consult (2010). The Knowledge Based Bio-Economy in Europe: Achievements and Challenges. Brussels, Belgium: Clever Consult.

Dawn (2012). *Food Wastage at Weddings*. Retrieved from https://www.dawn.com/news/746508 (accessed on 28 October, 2021).

ENDS Europe. (2012). EU biotech research plan. Available online: http://www.endseurope.com (accessed 29 March 2021).

EPA (2014). Landfill Methane Outreach Program. http://epaglobalmethanestage.icfwebservices.com/lmop/international/mexicano.html. (accessed 4 April 2021).

EuropaBio-European Association for Bioindustries (2011). *Building a Bio-based Economy for Europe in 2020*. Brussels, Belgium: European Association for Bioindustries.

European Commission (2012b). *Staff Working Document on Innovating for Sustainable Growth: A Bioeconomy for Europe*. SWD (2012) final. Brussels, Belgium: European Commission.

European Commission-EC (2012a). *Innovating for Sustainable Growth: A Bioeconomy for Europe*. COM (2012) final. Brussels, Belgium: European Commission.

FAO (2014). *Food Losses and Waste in the Latin America and the Caribbean*.

FAO (2018). *Save Food: Global Initiative on Food Loss and Waste Reduction*. http://www.fao.org/save-food/resources/keyfindings/en. (accessed 22 March 2018).

Gutiérrez, E.C., Wall, D.M., O'Shea, R. et al. (2018). An economic and carbon analysis of biomethane production from food waste to be used as a transport fuel in Mexico. *Journal of Cleaner Production* 196: 852–862.

Hart, J.F. (2001). Half a century of cropland change. *Geographical Review* 91 (3): 525–543. https://doi.org/10.1111/j.1931-0846.2001.tb00239.x.

Haveren, J.V., Scott, E.L., and Sanders, J. (2008). Bulk chemicals from biomass. *Biofuels, Bioproducts and Biorefining: Innovation for a Sustainable Economy* 2 (1): 41–57.

Heimlich, R.E. (1989). *Productivity and Erodibility of US Cropland* (No. 604). US Department of Agriculture, Economic Research Service.

Hoornweg, D. and Bhada Tata, P. (2012). *What a Waste: A Global Review of Solid Waste Management* Vol. 15. 116. Washington, DC: World Bank.

IEA (2005). 100 % Biogas for Urban Transport in Link€oping, Sweden Biogas in Buses, Cars and Trains, IEA BIOENERGY TASK 37 Energy from Biogas and Landfill Gas. Retrieved from http://www.iea-biogas.net/files/daten-redaktion/download/linkoping_final.pdf (accessed 28 October, 2021).

Jiao, J., Li, J., and Bai, Y. (2019). Uncertainty analysis in the life cycle assessment of cassava ethanol in China. *Journal of Cleaner Production* 206: 438–451.

Katami, T., Yasuhara, A., and Shibamoto, T. (2004). Formation of dioxins from incineration of foods found in domestic garbage. *Environmental Science & Technology* 38 (4): 1062–1065.

Khan, M.I., Yasmin, T., and Shakoor, A. (2015). Technical overview of compressed natural gas (CNG) as a transportation fuel. *Renewable and Sustainable Energy Reviews* 51: 785–797.

Kiran, E.U., Trzcinski, A.P., and Liu, Y. (2015). Enhancing the hydrolysis and methane production potential of mixed food waste by an effective enzymatic pretreatment. *Bioresource Technology* 183: 47–52.

Krozer, Y. and Christensen-Redzepovic, E. (2006). Sustainable innovations at tourist destinations. *Tourism Review International* 10 (1–2): 113–124.

Krozer, Y. and Lordkipanidze, M. (2019). Bioresources for cultural services. In: *Economics of Bioresources*, 195–205. Cham: Springer.

Kumar, K.N. and Goel, S. (2009). Characterization of municipal solid waste (MSW) and a proposed management plan for Kharagpur, West Bengal, India. *Resources, Conservation and Recycling* 53 (3): 166–174.

Kumar, S., Bhattacharyya, J.K., Vaidya, A.N. et al. (2009). Assessment of the status of municipal solid waste management in metro cities, state capitals, class I cities, and class II towns in India: An insight. *Waste Management* 29 (2): 883–895.

Lal, R. (2004). Soil carbon sequestration to mitigate climate change. *Geoderma* 123 (1–2): 1–22.

Lam, C.M., Iris, K.M., Medel, F. et al. (2018). Life-cycle cost-benefit analysis on sustainable food waste management: The case of Hong Kong International Airport. *Journal of Cleaner Production* 187: 751–762. doi: https://doi.org/10.1016/j.jclepro.2018.03.160

Langeveld, J.W. and Sanders, J.P.M. (2010). General Introduction. In: *The Biobased Economy: Biofuels, Materials and Chemicals in the Post-oil Era*, vol. 2010 (ed. J.W. Langeveld, J.P.M. Sanders, and M. Meeusen), 3–17. London, UK: Earthscan.

Le Fevre, C.N. (2014). *The Prospects for Natural Gas as a Transportation Fuel in Europe.* Oxford Institute for Energy Studies.

Ma, Y. and Liu, Y. (2019). Turning food waste to energy and resources towards a great environmental and economic sustainability: An innovative integrated biological approach. *Biotechnology Advances* 37 (7): 107414.

Ma, Y., Cai, W., and Liu, Y. (2017). An integrated engineering system for maximizing bioenergy production from food waste. *Applied Energy* 206: 83–89.

Manzetti, S. and Andersen, O. (2015). A review of emission products from bioethanol and its blends with gasoline. Background for new guidelines for emission control. *Fuel* 140: 293–301.

McCormick, K. and Kautto, N. (2013). The bioeconomy in Europe: An overview. *Sustainability* 5 (6): 2589–2608.

Melikoglu, M., Lin, C., and Webb, C. (2013). Analysing global food waste problem: Pinpointing the facts and estimating the energy content. *Open Engineering* 3 (2): 157–164. https://doi.org/10.2478/s13531-012-0058-5.

Monier, V., Mudgal, S., Escalon, V. et al. (2010). Preparatory study on food waste across EU 27. Report for the European Commission [DG ENV – Directorate C].

Pattnaik, S. and Reddy, M.V. (2010). Assessment of municipal solid waste management in Puducherry (Pondicherry), India. *Resources, Conservation and Recycling* 54 (8): 512–520.

Rönnlund, I., Pursula, T., Bröckl, M. et al. (2014). Creating Value from Bioresources: Innovation in Nordic Bioeconomy. Nordic Innovation.

Salemdeeb, R., Vivanco, D.F., Al-Tabbaa, A. et al. (2017). A holistic approach to the environmental evaluation of food waste prevention. *Waste Management* 59: 442–450.

San Martin, D., Ramos, S., and Zufía, J. (2016). Valorisation of food waste to produce new raw materials for animal feed. *Food Chemistry* 198: 68–74.

Scarlat, N., Blujdea, V., and Dallemand, J.F. (2011). Assessment of the availability of agricultural and forest residues for bioenergy production in Romania. *Biomass & Bioenergy* 35 (5): 1995–2005.

Scarlat, N., Dallemand, J.F., Monforti-Ferrario, F. et al. (2015). The role of biomass and bioenergy in a future bioeconomy: Policies and facts. *Environmental Development* 15: 3–34.

SEMARNAT (2017). Anuncia SEMARNAT 160 MDP para impulsar transporte público que use combustibles de baja huella de carbon. https://www.gob.mx/semarnat/prensa/anuncia-semarnat-160-mdp-para-impulsartransporte-publico-que-use-combustibles-de-baja-huella-de-carbono. (accessed 5 March 2021).

SMEDA. (2016). Small and Medium Enterprises Development Authority, Ministry of Industries & Production, Government of Pakistan. http://www.commerce.gov.pk/wpcontent/uploads/pdf/Banquet-Hall-500-Hosts.pdf.

Tayyab, A., Ahmad, Z., Mahmood, T. et al. (2019). Anaerobic co-digestion of catering food waste utilizing *Parthenium hysterophorus* as co-substrate for biogas production. *Biomass & Bioenergy* 124: 74–82.

Tihamer, S. (2019). District Heating in Villages. In: *Economics of Bioresources* 125–134. Cham: Springer.

UN (2015). *Transforming Our World: The 2030 Agenda for Sustainable Development.* Resolution adopted by the General Assembly. Available at https://sustainabledevelopment.un.org/content/documents/21252030%20Agenda%20for%20Sustainable%20Development%20web.pdf.

Index

Bioresource Technology: Concept, Tools and Experiences, First Edition. Tanveer Bilal Pirzadah, Bisma Malik, Rouf Ahmad Bhat, and Khalid Rehman Hakeem.
© 2022 John Wiley & Sons Ltd. Published 2022 by John Wiley & Sons Ltd.